BioMEMS and Biomedical Nanotechnology

Volume II
Micro/Nano Technology for Genomics
and Proteomics

BioMEMS and Biomedical Nanotechnology

Mauro Ferrari, Ph.D., Editor-in-Chief
Professor, Brown Institute of Molecular Medicine Chairman
Department of Biomedical Engineering
University of Texas Health Science Center, Houston, TX

Professor of Experimental Therapeutics
University of Texas M.D. Anderson Cancer Center, Houston, TX

Professor of Bioengineering
Rice University, Houston, TX

Professor of Biochemistry and Molecular Biology
University of Texas Medical Branch, Galveston, TX

President, the Texas Alliance for NanoHealth
Houston, TX

Volume II
Micro/Nano Technology for Genomics and Proteomics

Edited by

Mihrimah Ozkan
Dept. of Electrical Engineering
University of California, Riverside
Riverside, California USA

Michael J. Heller
Dept. of Bioengineering
University of California, San Diego
La Jolla, California USA

Mihrimah Ozkan
University of California, Riverside
Riverside, California

Michael Heller
University of California, Riverside
Riverside, California

Mauro Ferrari
Ohio State University
Columbus, Ohio

Library of Congress Cataloging-in-Publication Data

Volume II
ISBN-10: 0-387-25564-8   e-ISBN 10: 0-387-25843-4   Printed on acid-free paper.
ISBN-13: 978-0387-25564-4   e-ISBN-13: 978-0387-25843-0
Set
ISBN-10: 0-387-25661-3   e-ISBN:10: 0-387-25749-7
ISBN-13: 978-0387-25561-3   e-ISBN:13: 978-0387-25749-5

© 2006 Springer Science+Business Media, LLC
All rights reserved. This work may not be translated or copied in whole or in part without the written permission of the publisher (Springer Science+Business Media LLC, 233 Spring Street, New York, NY 10013, USA), except for brief excerpts in connection with reviews or scholarly analysis. Use in connection with any form of information storage and retrieval, electronic adaptation, computer software, or by similar or dissimilar methodology now known or hereafter developed is forbidden.
The use in this publication of trade names, trademarks, service marks and similar terms, even if they are not identified as such, is not to be taken as an expression of opinion as to whether or not they are subject to proprietary rights.

9 8 7 6 5 4 3 2 1   SPIN 11407157

springer.com

Dedicated to Richard Smalley (1943–2005), in Memoriam

To Rick,

father founder of nanotechnology
prime inspiration for its applications to medicine
gracious mentor to its researchers
our light—forever in the trenches with us

(Rick Smalley received the 1996 Chemistry Nobel Prize
for the co-discovery of carbon-60 buckeyballs)

Contents

List of Contributors xvii
Foreword xxi
Preface xxiii

I. Application of Microarray Technologies 1

1. Electronic Microarray Technology and Applications in Genomics and Proteomics 3
Ying Huang, Dalibor Hodko, Daniel Smolko, and Graham Lidgard
1.1 Introduction 3
1.2 Overview of Electronic Microarray Technology 4
    1.2.1 NanoChip® Array and NanoChip® Workstation 5
    1.2.2 Capabilities of the NanoChip® Electronic Microarrays 7
1.3 Applications 10
    1.3.1 Single Nucleotide Polymorphisms (SNPs)—Based Diagnostics 10
    1.3.2 Forensic Detection 10
    1.3.3 Gene Expression Profiling 12
    1.3.4 Cell Separation 12
    1.3.5 Electronic Immunoassays 14
    1.3.6 Miniaturization of Electronic Microarray Technology and Applications 15
    1.3.7 Applications in Proteomics 18
1.4 Summary and Outlook 19
    References 19

2. Gene Expression Profiling Utilizing Microarray Technology and RT-PCR 23
Dominick Sinicropi, Maureen Cronin, and Mei-Lan Liu
2.1 Introduction 23
2.2 Real-Time PCR 25
    2.2.1 Detection Systems 25
    2.2.2 Real-Time RT-PCR Data Analysis 31
    2.2.3 Qualification of Gene Panels Using Real-Time RT-PCR 32
    2.2.4 Real-Time RT-PCR Summary 34

2.3 Microarrays.. 35
    2.3.1 Technology Platforms 35
    2.3.2 Target Amplification and Labeling............................ 37
    2.3.3 Applications... 40
2.4 Comparison of Gene Expression Profiling Methods..................... 41
    2.4.1 Comparison of cDNA Arrays with Other Gene Expression Profiling Methods 41
    2.4.2 Comparison of Oligonucleotide Arrays with Other Gene Expression Profiling Methods......................... 42
    2.4.3 Comparison of cDNA and Oligonucleotide Microarray Expression Profiles... 44
2.5 Summary.. 44
Acknowledgements 45
References... 45

3. Microarray and Fluidic Chip for Extracellular Sensing 47
Mihrimah Ozkan, Cengiz S. Ozkan, Shalini Prasad, Mo Yang, and Xuan Zhang

3.1 Introduction 47
3.2 Antibody Based Biosensors 50
3.3 Nucleic Acid Based Biosensors... 51
3.4 Ion Channel Biosensors.. 51
3.5 Enzyme Based Biosensors 51
3.6 Cell Based Biosensors... 52
3.7 Cellular Microorganism Based Sensors................................. 52
3.8 Fluorescence Based Cell Biosensors..................................... 53
3.9 Cellular Metabolism Based Biosensors 55
3.10 Impedance Based Cellular Sensors 56
3.11 Intracellular Potential Based Biosensors 57
3.12 Extracllular Potential Based Biosensors 58
3.13 Cell Patterning Techniques.. 60
3.14 Dielectrophoresis for Cell Patterning 61
3.15 Basis of Dielectrophoresis.. 62
3.16 Microelectrodes and Dielectrophoresis 63
3.17 Dielectric Properties of Cells.. 64
3.18 Effect of Electric Fields on Cells.. 64
3.19 Cell Types and the Parameters for Dielectrophoretic Patterning 65
3.20 Biosensing System 66
3.21 Chip Assembly 66
3.22 Environmental Chamber 67
3.23 Experimental Measurement System 67
3.24 Cell Culture 67
    3.24.1 Neuron Culture... 67
    3.24.2 Primary Osteoblast Culture 68
3.25 Signal Processing 68
3.26 Selection of Chemical Agents 69

		3.26.1 Ethanol	69
		3.26.2 Hydrogen Peroxide	69
		3.26.3 Pyrethroid	70
		3.26.4 Ethylene Diamene Tetra Acetic Acid (EDTA)	70
	3.27	Chemical Agent Sensing	70
		3.27.1 Signature Pattern for Control Experiments	70
	3.28	Electrical Sensing Cycle	70
	3.29	Ethanol Sensing	71
		3.29.1 Single Neuron Sensing	71
		3.29.2 Single Osteoblast Sensing	71
	3.30	Hydrogen Peroxide Sensing	72
		3.30.1 Single Neuron Sensing	72
		3.30.2 Single Osteoblast Sensing	73
	3.31	Pyrethroid Sensing	74
		3.31.1 Single Neuron Sensing	74
		3.31.2 Single Osteoblast Sensing	75
	3.32	EDTA Sensing	76
		3.32.1 Single Neuron Sensing	76
		3.32.2 Single Osteoblast Sensing	76
	3.33	Immunohistochemistry	77
	3.34	Visualization of Physiological Changes Due to the Effect of the Chemical Analytes	80
		3.34.1 Effect of Ethanol on Neurons	80
		3.34.2 Effect of Ethanol on Osteoblasts	80
		3.34.3 Effect of Hydrogen Peroxide on Neurons	83
		3.34.4 Effect of Hydrogen Peroxide on Osteoblasts	84
		3.34.5 Effect of Pyrethroid on Neurons	86
		3.34.6 Effect of Pyrethroid on Osteoblasts	88
		3.34.7 Effect of EDTA on Neurons	89
		3.34.8 Effect of EDTA on Osteoblasts	91
	3.35	Discussion and Conclusions	93
		References	98
4.	**Cell Physiometry Tools based on Dielectrophoresis**		**103**
Ronald Pethig			
	4.1	Introduction	103
	4.2	Dielectrophoresis	104
	4.3	Dielectric Polarizability of Bioparticles	107
	4.4	Dynamics of Interfacial Polarization	107
	4.5	Surface Charge Effects	113
	4.6	Other Physiometric Effects	116
	4.7	Traveling Wave Dielectrophoresis	118
	4.8	Controlling Possible DEP-Induced Damage to Cells	120
		Concluding Comments	123
		References	124

5. Hitting the Spot: The Promise of Protein Microarrays ... 127
Joanna S. Albala
    5.1 Introduction ... 127
    5.2 Generation of Protein Microarrays ... 128
        5.2.1 Content ... 128
        5.2.2 Surface Chemistry ... 129
        5.2.3 Microarray Production ... 129
        5.2.4 Detection ... 130
    5.3 Protein Arrays for Analysis of Proteins Involved in Recombination & DNA Repair ... 130
        5.3.1 Protein Expression Microarrays ... 130
        5.3.2 Protein Interaction Arrays ... 132
    5.4 Summary: Protein arrays-Hope or hype? ... 133
    Acknowledgements ... 133
    References ... 133

6. Use of Electric Field Array Devices for Assisted Assembly of DNA Nanocomponents and Other Nanofabrication Applications ... 137
Michael J. Heller, Cengiz S. Ozkan, and Mihrimah Ozkan
    6.1 Introduction ... 138
    6.2 Active Microelectronic Array Hybridization Technology ... 141
    6.3 Electric Field Assisted Nanofabrication Process ... 146
    6.4 Integration of Optical Tweezers for Manupilation of Live Cells ... 153
    Conclusions ... 156
    Abbreviations ... 156
    Acknowledgements ... 157
    References ... 157

7. Peptide Arrays in Proteomics and Drug Discovery ... 161
Ulrich Reineke, Jens Schneider-Mergener, and Mike Schutkowski
    7.1 Introduction ... 161
    7.2 Generation of Peptide Arrays ... 162
        7.2.1 Coherent Surfaces and Surface Modification ... 163
        7.2.2 Generation of Micro-Structured Surfaces ... 173
        7.2.3 Peptide Array Preparation ... 182
        7.2.4 Techniques for Array Production with Pre-Synthesized Peptides ... 200
    7.3 Library Types ... 203
        7.3.1 Protein Sequence-Derived Libraries ... 204
        7.3.2 De Novo Approaches ... 210
    7.4 Assays for Peptide Arrays ... 214
        7.4.1 Screening ... 215
        7.4.2 Read-Out ... 219
    7.5 Applications of Peptide Arrays ... 221
        7.5.1 Antibodies ... 222
        7.5.2 Protein-Protein Interactions ... 224
        7.5.3 Enzyme-Substrate and Enzyme-Inhibitor Interactions ... 226

      7.5.4 Application of Peptide Arrays: Miscellaneous 228
      7.5.5 Peptidomimetics... 231
  7.6 Bibliography... 231
      References... 265

8. From One-Bead One-Compound Combinatorial Libraries to Chemical Microarrays 283
Kit S. Lam, Ruiwu Liu, Jan Marik, and Pappanaicken R. Kumaresan

  8.1 Introduction 283
  8.2 OBOC Peptide Libraries 284
  8.3 Encoded OBOC Small Molecule Combinatorial Libraries 287
  8.4 Peptide and Chemical Microarrays 289
      8.4.1 Immobilization Methods for Pre-Synthesized Libraries 289
      8.4.2 In Situ Synthesis of Microarrays 292
      8.4.3 CD, Microfluidics, Fiber Optic Microarray, Multiplex Beads 295
  8.5 Detection Methods in Chemical Microarrays 296
      8.5.1 Identification and Characterization of Bound Proteins 296
      8.5.2 Detection Methods to Identify Post-Translational Modification of Proteins or to Quantitate Enzyme Activity in Analytes 297
  8.6 Application of Chemical Microarray 297
      8.6.1 Protein Binding Studies 298
      8.6.2 Post-Translational Modification, Enzyme-Substrate and Inhibitor Studies 299
      8.6.3 Cell-Binding Studies 300
      8.6.4 Drug Discovery and Cell Signaling 300
      8.6.5 Diagnostic Studies 301
      8.6.6 Non-Biological Applications 301
  8.7 Future Directions 302
      Acknowledgements 303
      Abbreviations 303
      References 304

II. Advanced Microfluidic Devices and Human Genome Project 309

9. Plastic Microfluidic Devices for DNA and Protein Analyses 311
Z. Hugh Fan and Antonio J. Ricco

  9.1 Introduction 311
      9.1.1 Detection 311
      9.1.2 Materials 312
  9.2 Electrokinetic Pumping 312
  9.3 Plastic Devices 314
      9.3.1 Pumping and Detection 315
      9.3.2 Device Fabrication 316
  9.4 DNA Analyses 318
      9.4.1 Integrating PCR and DNA Fragment Separations 318

                9.4.2 DNA Sequencing 320
                9.4.3 DNA Sample Purification 321
        9.5 Protein Analyses 322
                9.5.1 Isoelectric Focusing for Studying Protein Interactions 323
                9.5.2 Enzymatic Digestion for Protein Mapping 324
                Concluding Remarks 326
                Acknowledgements 326
                References 326

10. Centrifuge Based Fluidic Platforms 329
    Jim V. Zoval and M.J. Madou
        10.1 Introduction 329
        10.2 Why Centrifuge as Fluid Propulsion Force? 330
        10.3 Compact Disc or Micro-Centrifuge Fluidics 333
                10.3.1 How it Works 333
        10.4 Some Simple Fluidic Function Demonstrated on a CD 334
                10.4.1 Mixing of Fluid 334
                10.4.2 Valving 335
                10.4.3 Volume Definition (Metering) and Common
                         Distribution Channels 338
                10.4.4 Packed Columns 339
        10.5 CD Applications 339
                10.5.1 Two-Point Calibration of an Optode-Based Detection System 339
                10.5.2 CD Platform for Enzyme-Linked Immunosorbant
                         Assays (ELISA) 340
                10.5.3 Multiple Parallel Assays 341
                10.5.4 Cellular Based Assays on CD Platform 342
                10.5.5 Automated Cell Lysis on a CD 344
                10.5.6 Integrated Nucleic Acid Sample Preparation and
                         PCR Amplification 356
                10.5.7 Sample Preparation for MALDI MS Analysis 358
                10.5.8 Modified Commercial CD/DVD Drives in
                         Analytical Measurements 359
                Conclusion 361
                Acknowledgements 362
                References 362

**11. Sequencing the Human Genome: A Historical Perspective
      On Challenges For Systems Integration** 365
    Lee Rowen
        11.1 Overview 365
        11.2 Approaches Used to Sequence the Human Genome 366
                11.2.1 Overview 366
                11.2.2 Strategy Used for Sequencing Source Clones 368
                11.2.3 Construction of the Chromosome Tiling Paths 379
                11.2.4 Data Sharing 379

	11.3 Challenges for Systems Integration	380
	11.3.1 Methodological Challenges for Sequencing Source Clones: 1990–1997	381
	11.3.2 Challenges for Sequencing the Entire Human Genome: 1998–2003	386
	11.4 Are there Lessons to be Learned from the Human Genome Project?	395
	Acknowledgements	397
	References	398

III. Nanoprobes for Imaging, Sensing and Therapy ... 401

12. Hairpin Nanoprobes for Gene Detection ... 403
Philip Santangelo, Nitin Nitin, Leslie LaConte, and Gang Bao

	12.1 Introduction	403
	12.2 Nanoprobe Design Issues for Homogeneous Assays	405
	12.3 In Vitro Gene Detection	408
	12.3.1 Pathogen Detection	409
	12.3.2 Mutation Detection and Allele Discrimination	409
	12.4 Intracellular RNA Targets	411
	12.4.1 Cytoplasmic and Nuclear RNA	411
	12.4.2 RNA Secondary Structure	418
	12.5 Living Cell RNA Detection	418
	12.5.1 Cellular Delivery of Probes	419
	12.5.2 Intracellular Probe Stability	424
	12.5.3 Intracellular mRNA Detection	428
	12.6 Opportunities and Challenges	431
	Acknowledgements	433
	References	433

13. Fluorescent Lanthanide Labels with Time-Resolved Fluorometry In DNA Analysis ... 437
Takuya Nishioka, Jingli Yuan, and Kazuko Matsumoto

	13.1 Introduction	437
	13.2 Lanthanide Fluorescent Complexes and Labels	438
	13.3 Time-Resolved Fluorometry of Lanthanide Complexes	441
	13.4 DNA Hybridization Assay	442
	Conclusion	445
	References	445

14. Role of SNPs and Haplotypes in Human Disease and Drug Development ... 447
Barkur S. Shastry

	14.1 Introduction	447
	14.2 SNP Discovery	448
	14.3 Detection of Genetic Variation	449
	14.4 Disease Gene Mapping	450

14.5 Evolution	450
14.6 Haplotypes	452
14.7 Drug Development	452
Concluding Remarks	454
References	454

15. Control of Biomolecular Activity by Nanoparticle Antennas ... 459
Kimberly Hamad-Schifferli

15.1 Background and Motivation	459
15.1.1 ATP Synthase as a Molecular Motor	459
15.1.2 Biological Self Assembly of Complex Hybrid Structures	461
15.1.3 DNA as a Medium for Computation	463
15.1.4 Light Powered Nanomechanical Devices	463
15.2 Nanoparticles as Antennas for Controlling Biomolecules	465
15.2.1 Technical Approach	468
15.2.2 Dehybridization of a DNA Oligonucleotide Reversibly by RFMF Heating of Nanoparticles	469
15.2.3 Determination of Effective Temperature by RFMF Heating of Nanoparticles	469
15.2.4 Selective Dehybridization of DNA Oligos by RFMF Heating of Nanoparticles	471
Conclusions and Future Work	473
References	474

16. Sequence Matters: The Influence of Basepair Sequence on DNA-protein Interactions ... 477
Yan Mei Wang, Shirley S. Chan, and Robert H. Austin

16.1 Introduction	477
16.2 Generalized Deformations of Objects	481
16.3 Sequence Dependent Aspects to the Double Helix Elastic Constants	484
16.4 Sequence Dependent Bending of the Double Helix and the Structure Atlas of DNA	485
16.5 Some Experimental Consequences of Sequence Dependent Elasticity	486
16.5.1 Phage 434 Binding Specificity and DNase I Cutting Rates	486
16.5.2 Nucleosome Formation: Sequence and Temperature Dependence	491
Conclusions	494
References	494

17. Engineered Ribozymes: Efficient Tools for Molecular Gene Therapy and Gene Discovery ... 497
Maki Shiota, Makoto Miyagishi, and Kazunari Taira

17.1 Introduction	497
17.2 Methods for the Introduction of Ribozymes into Cells	498
17.3 Ribozyme Expression Systems	499
17.3.1 The Pol III System	499

    17.3.2 Relationship Between the Higher-Order Structure of
           Ribozymes and their Activity 500
    17.3.3 Subcellular Localization and Efficacy of Ribozymes.................. 501
    17.3.4 Mechanism of the Export of tRNA-Ribozymes from the
           Nucleus to the Cytoplasm 504
17.4 RNA-Protein Hybrid Ribozymes.. 505
    17.4.1 Accessibility to Ribozymes of their Target mRNAs.................... 505
    17.4.2 Hybrid Ribozymes that Efficiently Cleave their Target mRNAs,
           Regardless of Secondary Structure 505
17.5 Maxizymes: Allosterically Controllable Ribozymes.................... 508
    17.5.1 Shortened Hammerhead Ribozymes that Function as Dimers........ 508
    17.5.2 Design of an Allosterically Controllable Maxizyme................... 509
    17.5.3 Inactivation of an Oncogene in a Mouse Model........................ 512
    17.5.4 Generality of the Maxizyme Technology 512
17.6 Identification of Genes Using Hybrid Ribozymes...................... 513
17.7 Summary and Prospects 515
    References.. 516

About the Editors.. 519
Index.. 521

List of Contributors

VOLUME II

Joanna S. Albala, Biology & Biotechnology Research Program, Lawerence Livermore National Laboratory, Livermore, California USA

Robert H. Austin, Dept. of Physics, Princeton University, Princeton, New Jersey USA

Gang Bao, Dept. of Biomedical Engineering, Georgia Institute of Technology and Emory University, Atlanta, Georgia USA

Shirley S. Chan, Dept. of Physics, Princeton University, Princeton, New Jersey USA

Maureen Cronin, Genomic Health, Inc., Redwood City, California USA

Z. Hugh Fan, Dept. of Mechanical and Aerospace Engineering, University of Florida, Gainesville, Florida USA

Mauro Ferrari, Ph.D., Professor, Brown Institute of Molecular Medicine Chairman, Department of Biomedical Engineering, University of Texas Health Science Center, Houston, TX; Professor of Experimental Therapeutics, University of Texas M.D. Anderson Cancer Center, Houston, TX; Professor of Bioengineering, Rice University, Houston, TX; Professor of Biochemistry and Molecular Biology, University of Texas Medical Branch, Galveston, TX; President, the Texas Alliance for NanoHealth, Houston, TX

Kimberly Hamad-Schifferli, Dept. of Mechanical Engineering, Massachusetts Institute of Technology, Cambridge, Massachusetts USA

Michael J. Heller, Dept. of Bioengineering, University of California, San Diego, La Jolla, California USA

Dalibor Hodko, Nanogen Inc., San Diego, California USA

Dr. Ying Huang, Nanogen Inc., San Diego, California USA

Pappanaicken R. Kumaresan, Division of Hematology/Oncology & Internal Medicine, University of California Davis, Sacramento, California USA

Leslie LaConte, Dept. of Biomedical Engineering, Georgia Institute of Technology and Emory University, Atlanta, Georgia USA

Kit S. Lam, Division of Hematology/Oncology & Internal Medicine, University of California Davis, Sacramento, California USA

Graham Lidgard, Nanogen Inc., San Diego, California USA

Mei-Lan Liu, Genomic Health, Inc., Redwood City, California USA

Ruiwu Liu, Division of Hematology/Oncology & Internal Medicine, University of California Davis, Sacramento, California USA

M.J. Madou, Dept. of Mechanical and Aerospace Engineering, University of California, Irvine, Irvine, California USA

Jan Marik, Division of Hematology/Oncology & Internal Medicine, University of California Davis, Sacramento, California USA

Kazuko Matsumoto, Dept. of Chemistry and Advanced Research Institute for Science & Engineering, Waseda University, Tokyo, Japan

Makoto Miyagishi, Dept. of Chemistry and Biotechnology, The University of Tokyo, Tokyo, Japan

Takuya Nishioka, Dept. of Chemistry and Advanced Research Institute for Science & Engineering, Waseda University, Tokyo, Japan

Nitin Nitin, Dept. of Biomedical Engineering, Georgia Institute of Technology and Emory University, Atlanta, Georgia USA

Cengiz S. Ozkan, Dept. of Mechanical Engineering, University of California, Riverside, Riverside, California USA

Prof. Mihrimah Ozkan, Dept. of Electrical Engineering, University of California, Riverside, Riverside, California USA

Prof. Ronald Pethig, School of Informatics, University of Wales, Bangor, Gwynedd, United Kingdom

Shalini Prasad, Dept. of Electrical Engineering, University of California, Riverside, Riverside, California USA

Ulrich Reineke, Jerini AG, Berlin, Germany

Antonio J. Ricco, NASA Ames Research Center, Mountain View, California USA

Lee Rowen, Institute for Systems Biology, Seattle, Washington USA

LIST OF CONTRIBUTORS

Philip Santangelo, Dept. of Biomedical Engineering, Georgia Institute of Technology and Emory University, Atlanta, Georgia USA

Jens Schneider-Mergener, Institut für Medizinische Immunologie, Universitätsklinikum Charité, Berlin, Germany

Mike Schutkowski, Jerini AG, Berlin, Germany

Barkur S. Shastry, Dept. of Biological Sciences, Oakland University, Rochester, Michigan USA

Maki Shiota, Dept. of Chemistry and Biotechnology, The University of Tokyo, Tokyo, Japan

Dominick Sinicropi, Genomic Health, Inc., Redwood City, California USA

Daniel Smolko, Nanogen Inc., San Diego, California USA

Kazunari Taira, Dept. of Chemistry and Biotechnology, The University of Tokyo, Tokyo, Japan

Yan Mei Wang, Dept. of Physics, Princeton University, Princeton, New Jersey USA

Mo Yang, Dept. of Mechanical Engineering, University of California, Riverside, Riverside, California USA

Jingli Yuan, Dept. of Chemistry and Advanced Research Institute for Science & Engineering, Waseda University, Tokyo, Japan

Xuan Zhang, Dept of Mechanical Engineering, University of California, Riverside, Riverside, California USA

Jim V. Zoval, Dept. of Mechanical and Aerospace Engineering, University of California, Irvine, Irvine, California USA

Foreword

Less than twenty years ago photolithography and medicine were total strangers to one another. They had not yet met, and not even looking each other up in the classifieds. And then, nucleic acid chips, microfluidics and microarrays entered the scene, and rapidly these strangers became indispensable partners in biomedicine.

As recently as ten years ago the notion of applying nanotechnology to the fight against disease was dominantly the province of the fiction writers. Thoughts of nanoparticle-vehicled delivery of therapeuticals to diseased sites were an exercise in scientific solitude, and grounds for questioning one's ability to think "like an established scientist". And today we have nanoparticulate paclitaxel as the prime option against metastatic breast cancer, proteomic profiling diagnostic tools based on target surface nanotexturing, nanoparticle contrast agents for all radiological modalities, nanotechnologies embedded in high-distribution laboratory equipment, and no less than 152 novel nanomedical entities in the regulatory pipeline in the US alone.

This is a transforming impact, by any measure, with clear evidence of further acceleration, supported by very vigorous investments by the public and private sectors throughout the world. Even joining the dots in a most conservative, linear fashion, it is easy to envision scenarios of personalized medicine such as the following:

- patient-specific prevention supplanting gross, faceless intervention strategies;
- early detection protocols identifying signs of developing disease at the time when the disease is most easily subdued;
- personally tailored intervention strategies that are so routinely and inexpensively realized, that access to them can be secured by everyone;
- technologies allowing for long lives in the company of disease, as good neighbors, without impairment of the quality of life itself.

These visions will become reality. The contributions from the worlds of small-scale technologies are required to realize them. Invaluable progress towards them was recorded by the very scientists that have joined forces to accomplish the effort presented in this 4-volume collection. It has been a great privilege for me to be at their service, and at the service of the readership, in aiding with its assembly. May I take this opportunity to express my gratitude to all of the contributing Chapter Authors, for their inspired and thorough work. For many of them, writing about the history of their specialty fields of *BioMEMS and Biomedical Nanotechnology* has really been reporting about their personal, individual adventures through scientific discovery and innovation—a sort of family album, with equations, diagrams, bibliographies and charts replacing Holiday pictures....

It has been a particular privilege to work with our Volume Editors: Sangeeta Bhatia, Rashid Bashir, Tejal Desai, Michael Heller, Abraham Lee, Jim Lee, Mihri Ozkan, and Steve Werely. They have been nothing short of outstanding in their dedication, scientific vision, and generosity. My gratitude goes to our Publisher, and in particular to Greg Franklin for his constant support and leadership, and to Angela De Pina for her assistance.

Most importantly, I wish to express my public gratitude in these pages to Paola, for her leadership, professional assistance throughout this effort, her support and her patience. To her, and our children Giacomo, Chiara, Kim, Ilaria and Federica, I dedicate my contribution to BioMEMS and Biomedical Nanotechnology.

With my very best wishes

Mauro Ferrari, Ph.D.
Professor, Brown Institute of Molecular Medicine Chairman
Department of Biomedical Engineering
University of Texas Health Science Center, Houston, TX

Professor of Experimental Therapeutics
University of Texas M.D. Anderson Cancer Center, Houston, TX

Professor of Bioengineering
Rice University, Houston, TX

Professor of Biochemistry and Molecular Biology
University of Texas Medical Branch, Galveston, TX

President, the Texas Alliance for NanoHealth
Houston, TX

Preface

Numerous miniaturized DNA microarray, DNA chip, Lab on a Chip and biosensor devices have been developed and commercialized. Such devices are improving the way many important genomic and proteomic analyses are performed in both research and clinical diagnostic laboratories. The development of these technologies was enabled by a synergistic combination of disciplines that include microfabrication, microfluidics, MEMS, organic chemistry and molecular biology. Some of these new devices and technologies utilize sophisticated microfabrication processes developed by the semiconductor industry. Microarrays with large numbers of test sites have been developed which employ photolithography combinatorial synthesis techniques or ink jet type printing deposition methods to produce high-density DNA microarrays. Other microarray technologies have incorporated microelectrodes to produce electric fields which are able to affect the transport and hybridization of DNA molecules on the surface of the device. As remarkable as this generation of devices and technological appears, the advent of new nanoscience and nanofabrication techniques will lead to even further miniaturization, higher integration and another generation of devices with higher performance properties. Thus, in some sense these devices and systems will follow a similar evolution as did microelectronics in going from 8 bit, to 16 bit to 32 bit technology. Where feature sizes for integrated components of microelectronic devices is now well into the submicron scale, nanoscale biodevices will soon follow. Likewise, the potential applications for this new generation of micro/nanoarray, lab on a chip and nanosensor devices is also broadening into areas of whole genome sequencing, biowarfare agent detection, and remote environmental sensing and monitoring. Today the possibility of making highly sophisticated smart micro/nano scale in-vivo diagnostic and therapeutic delivery devices is being seriously considered.

Nevertheless, considerable problems do exist. Unfortunately, many applications for these bioresearch or biomedically related devices do not have the large consumer markets that will drive and fund their development. The economic forces which drive the development of high volume retail consumer microelectronic and optoelectonic devices (such as computers, cell phones, digital cameras, and fiber optic communications), are not there for most bioresearch or biomedical devices. Thus, it is very common to see so-called "good" technologies in the bioresearch and biomedical device area fail somewhere along the arduous path to commercialization. This is particularly true for any biomedical device or system which has to go through the regulatory process. Frequently, the problem relates to the inability to economically manufacture a viable device for commercialization as opposed to a working prototype device. Thus, a key aspect for achieving final success for our new generation of bioresearch and biomedical micro/nano biodevices will be the corresponding development of both viable and efficient nanofabrication and micro/nano integration processes.

The Volume II: Micro/Nano Technologies for Genomics and Proteomics presents a wide range of exciting new science and technology, and includes key sections on DNA micro/nanoarrays which additional chapters on peptide arrays for proteomics and drug discovery, new dielectrophoretic cell separation systems and new nanofabrication and integration processes; advanced microfluidic devices for the human genome project (whole genome sequencing); and final section on nanoprobes for imaging and sensing. Overall this volume should be of considerable value for a wide range of multidisciplinary scientists and engineers who are either working in or interested in bionanotechnology and the next generation of micro/nano biomedical and clinical diagnostic devices.

Mihrimah Ozkan
Dept. of Electrical Engineering,
University of California, Riverside,
Riverside, California USA

Michael J. Heller
Dept. of Bioengineering,
University of California, San Diego,
La Jolla, California USA

Mauro Ferrari
Professor, Brown Institute of Molecular Medicine Chairman
Department of Biomedical Engineering
University of Texas Health Science Center, Houston, TX

Professor of Experimental Therapeutics
University of Texas M.D. Anderson Cancer Center, Houston, TX

Professor of Bioengineering, Rice University, Houston, TX

Professor of Biochemistry and Molecular Biology
University of Texas Medical Branch, Galveston, TX

President, the Texas Alliance for NanoHealth, Houston, TX

I

Application of Microarray Technologies

1

Electronic Microarray Technology and Applications in Genomics and Proteomics

Ying Huang, Dalibor Hodko, Daniel Smolko, and Graham Lidgard
Nanogen Inc., 10398 Pacific Center Court, San Diego, CA 92121, USA.

Keywords: Electronic microarray/ Miniaturization/ Single nucleotide polymorphism/ Gene expression profiling/ Cell separation/ Protein kinase/ Forensic detection/ Biological warfare

Electronic microarrays that contain planar arrays of microelectrodes have been developed to provide unique features of speed, accuracy and multiplexing for genomic and proteomic applications through utilizing electric field control to facilitate analytes concentration, DNA hybridization, stringency and multiplexing. An overview of electronic microarray technology is presented followed by its variety applications in genomic research and DNA diagnostics, forensic detection, biologic warfare, and proteomics.

1.1. INTRODUCTION

DNA microarrays have provided a new and powerful tool to perform important molecular biology and clinical diagnostic assays. The basic idea behind DNA microarray technology has been to immobilize known DNA sequences referred to as probes in micrometer-sized spots on a solid surface (microarray) and specifically hybridize a complementary sequence of the analyte DNA or a target. A fluorescently labeled reporter facilitates fluorescent detection of the presence or absence of a particular target or gene in the sample. By using laser-scanning and fluorescence detection devices such as CCD cameras, different target hybridization patterns can be read on the microarray and the results quantitatively analyzed. This chapter describes a specific microarray technology where an electric field and phenomena induced by the application of the electric field are used to direct and concentrate the DNA molecules through permeation layer [25] on the array.

Whereas, in the past, different technologies have been used to immobilize DNA probes including physical deposition [13, 40], photolithographic synthesis [Fodor et al., 1993, 7], and utilization of electric field [25]. Accordingly, several substrates have been used to generate different DNA microarrays ranging from glass slides, membrane to silicon. High density microarrays have been used to identify disease outcomes through relevant RNA expression patterns on thousands of genes [1] and for gene sequencing [33]. However, focused arrays, which often consists of 100–1,000 test sites, are better suited to detect a panel of genes for applications in point of care diagnostics, detection of infectious diseases, as well as identification of biological warfare agents. In these particular applications, speed, accuracy and multiplexing are basic requirements. Electronic microarrays, one type of the focused arrays, can meet all these requirements through utilizing electric field control to facilitate analytes concentration, DNA hybridization, stringency and multiplexing [17, 23–25, 45]. In this chapter, an overview of electronic microarray technology is presented followed by its applications in genomics and proteomics.

1.2. OVERVIEW OF ELECTRONIC MICROARRAY TECHNOLOGY

Nanogen, Inc. has developed an electronic micro-array based technology (NanoChip® Electronic Microarray) for manipulation, concentration and hybridization of biomolecules on the chip array (Figure 1.1). This approach extends the power of microarrays through the use of electronics by connecting each test site on the NanoChip® array to an electrode.

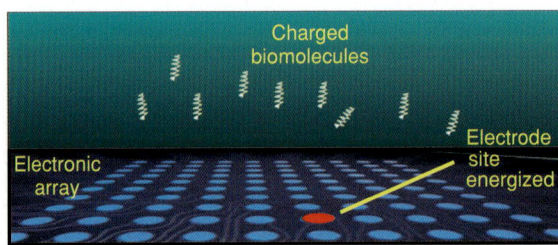

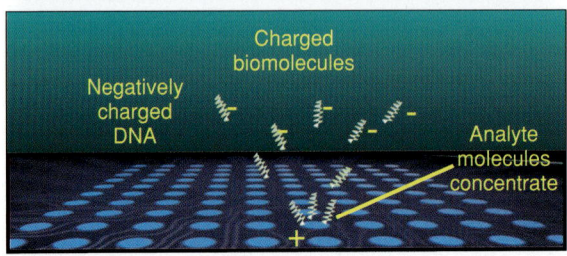

FIGURE 1.1. Nanochip® micro-array technology uses electronic addressing of charged biomolecules on the electrode array to separate and concentrate analyte targets. Negatively charged DNA targets and molecular probes (top) are moved to a particular site by energizing the electrodes at a reverse potential (bottom). Targeted molecules concentrate at the array site where they can be bound chemically or hybridized to a DNA probe. Fluorescent signal is obtained from the reporter probes hybridized to the target DNA and signal proportional to the concentration of analyte DNA is measured.

Most biological molecules have a natural positive or negative charge. When biological molecules are exposed to an electric field (Figure 1.1), molecules with a positive charge move to electrodes with a negative potential, and molecules with a negative charge move to electrodes with a positive potential. Current and voltages are applied to the test sites via individual electrode activation to facilitate the rapid and controlled transport of charged molecules to any test sites. Additional advantages of electrically facilitated transport include (1) the ability to produce reconfigurable electric fields on the microarray surface that allows the rapid and controlled transport of charged molecules to any test sites [22, 25]; (2) the ability to carry out site selective DNA or oligonucleotide addressing and hybridization [11]; (3) the ability to significantly increase DNA hybridization rate by concentration of target at the test sites (Kassegne et al., 2003); and (4) the ability to use electronic stringency to improve hybridization specificity (Sosnowski, et al., 1997).

1.2.1. NanoChip® Array and NanoChip® Workstation

1.2.1.1. Fabrication Electronic microarrays consist of an array of electrodes that have been fabricated on silicon with array sizes ranging from 5 to 10,000 individual electrodes or test sites [24, 25, 42]. Figure 1.2 shows a number of electronic microarrays with arrays ranging from 4 to 100 electrodes or test sites. These arrays have been designed

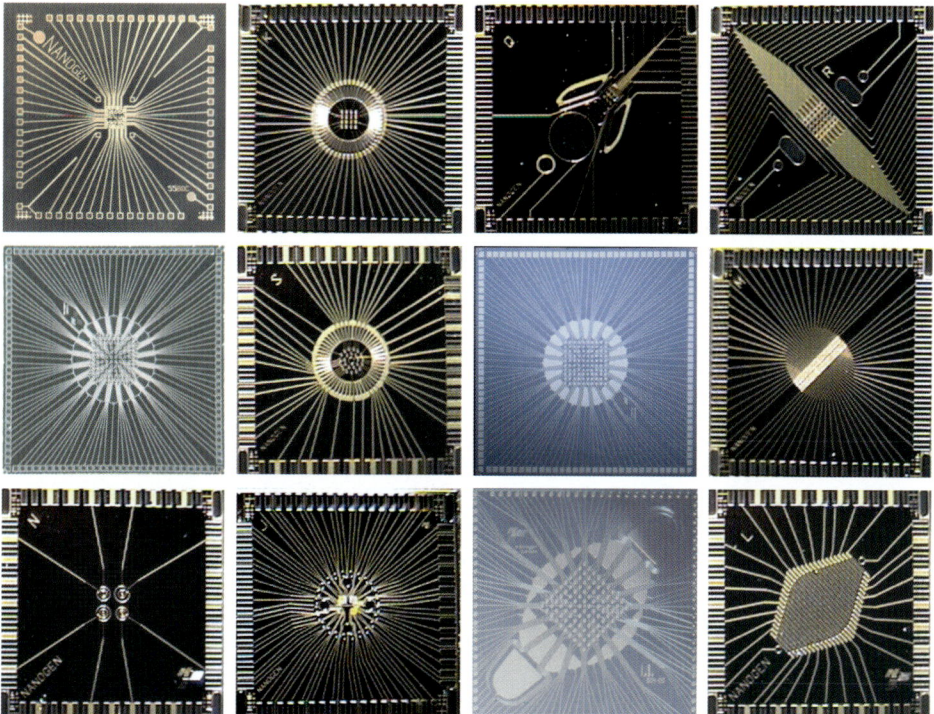

FIGURE 1.2. A series of designs of Nanogen's silicon microarray chips. Chip sizes shown range from 4–100 sites. These include different electrode geometries as well as chips designed for particles and or biomolecular microseparations (last column of chips).

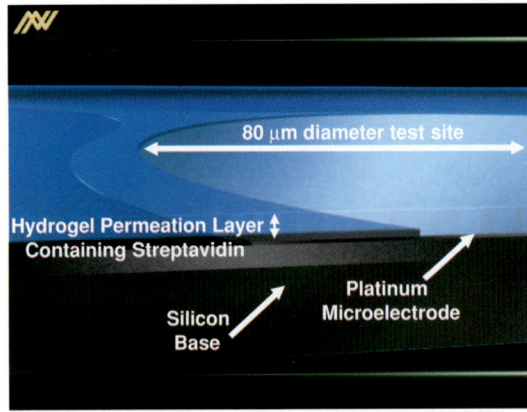

FIGURE 1.3. Cross-section of a single platinum micro-electrode pad on the NanoChip® microarray. A hydrogel permeation layer loaded with streptavidin covers the electrode array and serves for capturing electronically addressed molecules.

for both microassays and microseparations. The current commercialized NanoChip® array comprising 100 platinum microelectrodes with additional 20 outer microelectrodes acting as counter-electrodes.

The electrodes on the NanoChip® array are fabricated on a silicon substrate using standard photolithography and deposition processes. Each electrode is 80 μm in diameter with 200 μm center-to-center space between and is connected to the outside edge of the chip by a platinum wire trace. Figure 1.3 shows a cross section of a single electrode pad on the NanoChip® array. The base structure of the array consists of silicon over which an insulating layer of silicon dioxide is applied. Platinum is deposited and selectively held in place to form electrodes and accompanying electrical traces. These wire traces terminate at the edges of the chip forming electrical contact pads. Additional layers of silicon dioxide and silicon nitride are deposited to electrically insulate the platinum electrical traces, leaving the central array of 80 μm diameter microelectrodes, outer microelectrodes and contacts pads exposed. The chips are flip-chip bonded to a ceramic substrate which contains contacts to pogo-pins.

1.2.1.2. Permeation Layer Typically, on the surface of the array, a 10 μm thick hydrogel permeation layer (Figure 1.3). containing co-polymerized streptavidin is deposited by microreaction molding. This permeation layer serves two main functions [24]. First it protects the sensitive analytes from the adverse electrochemical effects at the platinum electrode surface during active operation. These electrochemical products include the generation of hydrogen ions (H^+) and oxygen at the positively biased (anode) microelectrodes and hydroxyl ions (OH^-) and hydrogen at the negatively biased electrodes (cathode), as well as various free radical entities. Secondly, the permeation layer also serves as a matrix for the attachment of biotinylated molecules e.g., analytes capture oligos, antibodies, and primers through biotin and streptavidin binding [24, 25]. Figure 1.4 is a close-up photograph of the 100-site array covered with the permeation layer.

ELECTRONIC MICROARRAY TECHNOLOGY AND APPLICATIONS

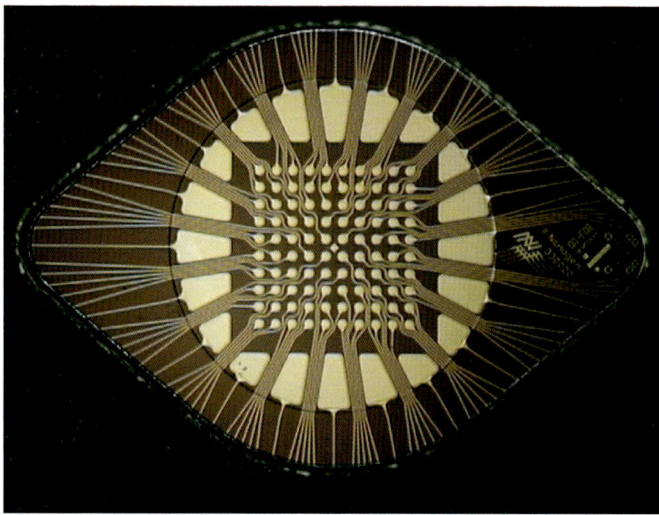

FIGURE 1.4. Photograph of the 100-site NanoChip® microarray. A hydrogel permeation layer covers the electrode array including working and counter electrodes. The hydrogel layer is visible as a circle surrounding counter electrodes (yellow pads).

1.2.1.3. NanoChip® Cartridge The 100-site array is assembled into a complete NanoChip® cartridge (Figure 1.5a) by ultrasonically welding two molded polymethyl methacrylate (PMMA) cartridge bodies that contain fluidic channels and inlet and outlet ports. The cartridge eliminates sample evaporation, prevents sample contamination and provides a fluidic interface to the NanoChip® Workstation.

1.2.1.4. NanoChip® Workstation The NanoChip® electronic microarrays are operated through a fully integrated and automated NanoChip® Workstation (Figure 1.5b). The system consists of three major subsystems: (1) the NanoChip® Loader for loading patient samples on one to four NanoChip® Cartridges, (2) the NanoChip® Reader, a highly sensitive, laser-based fluorescence scanner for detection of assay results and (3) computer hardware and software which automates import, analysis and export of sample information making data analysis simple.

1.2.2. Capabilities of the NanoChip® Electronic Microarrays

Using electric field, Nanochip® electronic microarrays have provided many unique features over other passive microarrays (Table 1.1). Electronic addressing allows users to quickly customize arrays in their own laboratory. Electronic hybridization provides an extremely accurate and specific hybridization process by creating optimal electric and pH conditions at the hybridization sites [11, 25]. Electronic stringency, in combination with thermal control, enables researchers to remove unbound and nonspecifically-bound DNA quickly and easily after hybridization at the microarrays, achieving rapid determination of single base mismatch mutations in DNA hybrids [44].

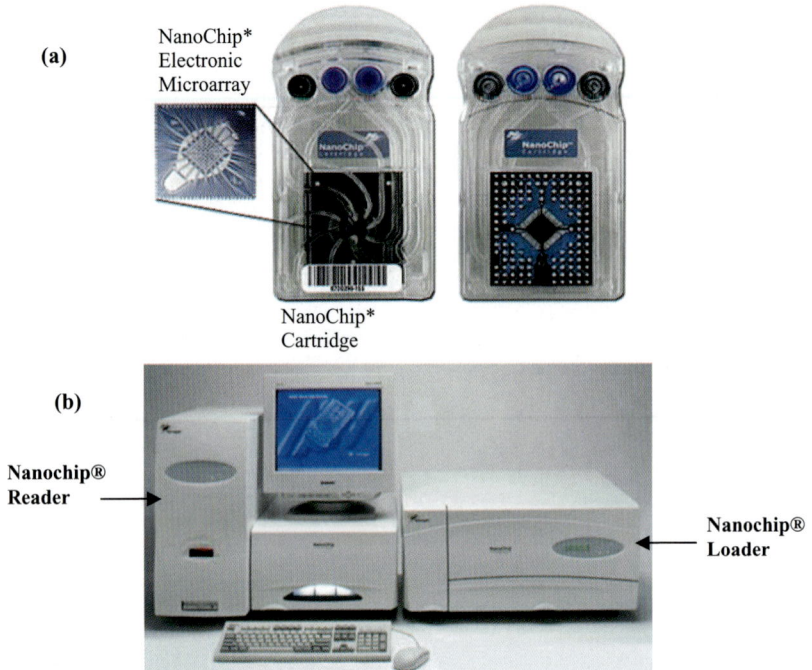

FIGURE 1.5. Photograph of the Nanochip® cartridge containing the electronic microarray, and b) Nanogen's Nanochip® Workstation which allows fully automated processing of 4 cartridges simultaneously in the loader and fluorescent detection in the reader.

1.2.2.1. Assay Formats on NanoChip® Electronic Microarrays The open architecture of electronic microarray enables flexibility in assay design. Since each individual electrode can be selectively activated, different assays can be generated depending on different analytes to be addressed (Figure 1.6). For example, a dot blot assay [19] is conducted when biotinylated PCR amplicon is addressed to the selected electrodes and remained embedded through interaction with strepavidin in the permeation layer. The DNA at each electrode is then hybridized to mixtures of allelespecific oligonucleotides (discriminators) and fluorescently labeled oligonucleotides. The thermal or electronic stringency is used to discriminate

TABLE 1.1. Comparison between NanoChip® microarray active hybridization technology and passive hybridization technologies.

	Hybridization time	Concentration of targets	Concentration factor at a site	Stringency control
NanoChip® active hybridization	10–100 seconds	Directed and localized at the array sites; Ability to control individual sites	>1000 times	Electronic, Thermal Chemical
Passive hybridization	1–2 hours	Non-directed; Sites cannot be controlled independently	Low, diffusion dependent	Thermal, Chemical

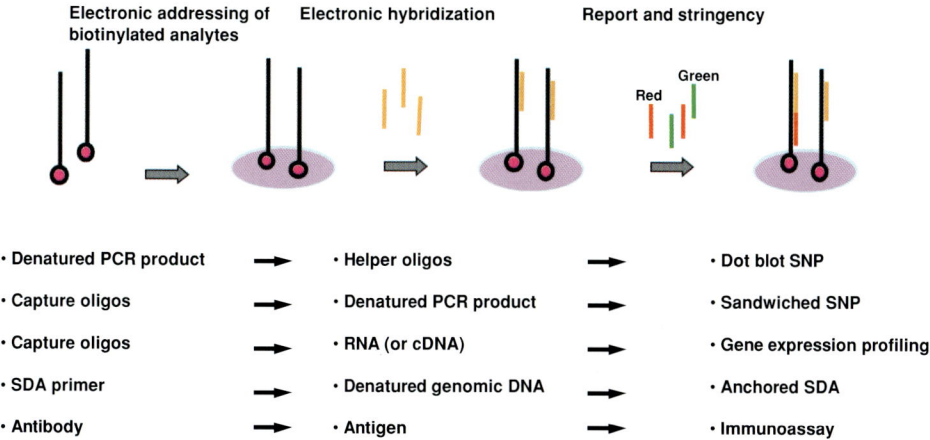

FIGURE 1.6. The open platform of the electronic microarray enables flexibility in assay design. Depending on the nature of the biotinylated analytes (denatured PCR products, capture oligos, SDA primers or antibody) in electronic addressing and on the targets (helper oligos, denatured PCR products, RNA, denatured genomic DNA or antigen), different assays (dot blot SNP, sandwiched SNP, gene expression profiling, anchored SDA or immunoassay) can be performed on the electronic microarrays.

the SNP. In this assay format, multiple samples can be analyzed at different electrodes on a single array.

A "sandwich" assay is created when biotinylated capture probes are addressed to selective electrodes. The capture probes can be sequence specific oligos or antibodies. The embedded capture probes are then electronically hybridized to the targets. Depending on the specific targets (PCR amplicons, or RNA, or antigens), the assays can be SNP analysis, gene expression profiling, or immunoassay.

Most recently, a special assay, anchored strand displacement amplification (aSDA), which integrates the amplification and discrimination, is demonstrated on electronic microarrays [12, 29, 50, 51]. In this assay format, biotinylated SDA primers were addressed and anchored on selective electrodes. These anchored primers were then electronically hybridized to the denatured genomic DNA. Target DNA was amplified over the electrodes when an enzyme mix containing restriction endonuclease and DNA polymerase and dNTPs was introduced to the array. After amplification the final discrimination was determined using same base-stacking principle [37]. Using aSDA, multiple genes from one sample or one gene from multiple samples can be simultaneously amplified and detected on a single electronic microarray.

1.2.2.2. Electronic Multiplexing By electronically controlling each test site, the electronic microarrays also provide a platform to perform different types of multiplexed assays:

(1) single array multiplexing where multiple genes from one sample can be analyzed;
(2) single array multiplexing where multiple samples with one gene of interest can be analyzed;
(3) single array multiplexing where multiple samples with multiple genes of interest can be analyzed;

(4) single site multiplexing where several targets are discriminated on the same site using different fluorescent probes;
(5) single site multiplexing where several targets are addressed, different discriminator oligonucleotides hybridized and reporter addressed—this method allows the use of a universal reporter for different targets.

The ability to electronically control individual test sites permits biochemically unrelated molecules to be used simultaneously on the same microchip. In contrast, sites on a conventional DNA array cannot be controlled separately, and all process steps must be performed on an entire array. Nanogen's electronic microarray technology delivers increased versatility over such conventional methods. This is particularly important in applications such as biological warfare and infectious disease detection since the accurate identification of biological agents requires determination of two or more (often five) characteristic genes of a particular agent.

1.3. APPLICATIONS

1.3.1. Single Nucleotide Polymorphisms (SNPs)—Based Diagnostics

Given the advances in genomic studies, more and more single nucleotide polymophisms (SNPs) are found to be contributory factors for human disease and can be used as genetic markers for molecular diagnostics. The speed, accuracy and flexibility provided by electronic microarrays have received great interest from clinical diagnostic researchers for a variety of genotyping applications (Table 1.2). Using the Nanochip® system, researchers at American Medical Laboratories analyzed 635 clinical samples for the G1691A mutation on the factor V Leiden, associated with thrombosis with 100% accurate in characterizing wild-type, heterozygous, and homozygous samples [15]. Researchers at ARUP Laboratories have evaluated 3 thrombosis associated SNPs, factor V (Leiden), factor II (prothrombin), and methylenetetrhydrofolate reductase (MTHFR), on 225 samples with 100% accuracy [14]. Schrijver et al at Stanford University Medical Center found that the SNP analysis based on Nanogen electronic microarray for factor V (Ledein) and factor II (prothrombin) on 800 samples were comparable with other SNP analysis methods such as restriction enzyme digestion (RFLP) and the Roche LightCycler [41]. Researchers at Children's Nantional Medical Center, Washington DC, genotyped 8 common MeCP2 mutations associated to Rett syndrome on 362 samples with 100% specificity [47]. Using electronic microarray, researchers at Mayo Clinic Cancer Center have developed genotyping assays for 5 different cytokine polymorphisms [43]. Moreover, 8 SNPs distributed within a highly variable region of the polC gene from six isolates of *Staphylococcus aureus* were analyzed on electronic microarrays [9].

1.3.2. Forensic Detection

Short tandem repeats (STRs) represent another type of polymorphism with important applications in forensic DNA identification. In 1990, the FBI created a combined DNA index system (CODIS), which consist of 13 polymorphic STR loci, to provide a database of forensic DNA profile for nearly all forensic laboratories in the United States [5, 6]. The

TABLE 1.2. Examples of Nanochip® technology application in genomics.

Relevance	Test	Ref
Cystic Fibrosis	CFTR	http://www.nanogen.com/products/cystic_fibrosis.htm
Thrombosis	Factor V Leiden	[41]
	Factor II (prothrombin)	[15]
	MTHFR	[14]
	Factor V/Prothrombin	http://www.nanogen.com/products/Factor_vII.htm
Hereditary Hemochromatosis	HFE	http://www.nanogen.com/products/HH.htm
Alzheimer's Disease	ApoE	http://www.nanogen.com/products/apoe.htm
		http://biz.yahoo.com/prnews/040106/latu005_1.html
β-thalassemia	Factor VII	[39]
Asthma and chronic obstructive pulmonary disease	B(2)-adrenergic receptor	[53]
Ulcerative colitis	N-acetyltransferase 1 (NAT 1)	[38]
	N-acetyltransferase 2 (NAT 2)	
Cancer	p53	[3]
Rett syndrome	MeCP2	[47]
Immunologic defect	Mannose binding protein (MBP)	[19]
Parkinson's discease	CYP1A2/CYP2E1	[10]
Cytokine	Tumor necrosis factor-α (TNF-α)	[43]
	IL-4	
	Interferon-γ (CA)n repeats	
	IL-1 RN VNTR	
	CCR5	
Bacterial ID	*Staphylococcus aureus* pol C	[9]
	Escherichia coli gyrA	[50]
	Salmonella gyrA	
	Campylobacter gyrA	
	E. coli parC	
	Staphylococcus mecA	

typical STR loci are selected groups of four nucleotide repeats that are represented in the human population by 4-15 alleles distinguished by a different number of repeat units [8]. Unlike SNPs, STRs are more difficult to analyze by conversional passive hybridization techniques [20, 34]. However, electronic hybridization techniques have been proven to overcome these problems and allow STR analysis to be performed on electronic microarray in a rapid and high fidelity fashion [37]. Multiplex hybridization analysis of three STR loci (CSF1PO, TH01 and TPOX) was achieved for 12 individuals, 100% concordant with genotyping results of an accredited forensic laboratory.

Given the recent discovery of abundance SNPs and the ease of automation and miniaturization of detection techniques, SNP assays have started to be implemented in DNA forensic analysis. According to Chakraboty *et al.* [6], somewhere in the range of 30-60 SNP loci would be needed to equal the power of the 13 STR loci with regard to genotypic match probability and/or paternity exclusion. The flexibility of electronic microarray will permit accelerated development of SNPs for DNA forensic analysis by allowing an easy

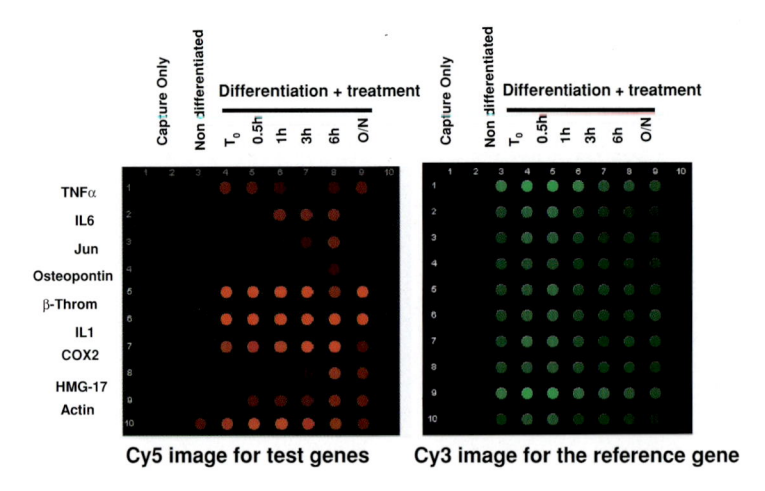

FIGURE 1.7. Multiplexed gene expression profiling on the electronic microarray. Ten target genes plus the β-la reference gene from the U937 cell lines at various time points after differentiation and LPS treatment. The genes were electronically hybridized to target-specific capture probes and to the β-la capture probe. The left panel represents the Cy5 fluorescent signals from the 10 target genes, whereas the right panel shows the Cy3 signal corresponding to the β-la reference gene. Each target gene exhibits a distinct expression pattern over the time course of LPS treatment, while the β-la reference gene remains relatively unchanged. Reprinted with permission from Weidenhammer et al. Copyright [49] American Association for Clinical Chemistry.

method of adapting new loci. Currently, supported by government funding, assays are under development to detect 35 SNPs and 6 STRs on a single electronic microarray.

1.3.3. Gene Expression Profiling

Another array-based method in genomic studies is to simultaneously monitor global gene expression profiling of cells under a condition of interest and to identify a set of gene markers for specific disease [1]. Towards this trend, a multiplex, targeted gene expression profiling method has been developed using electronic field-facilitated hybridization on Nanochip® electronic microarray [49]. In this method, target mRNA generated from T7 RNA polymerase-mediated amplification were detected by hybridization to sequence-specific capture oligonucleotides on electronic microarray. The expression of a model set of 10 target genes in the U937 cell line was analyzed during lipopolysaccharide-mediated differentiation with 2-fold changes in concentration and 64-fold range of concentration (Figure 1.7). This electronic array based expression analyzing method allows simultaneous assessment of target concentrations from multiple sample sources.

1.3.4. Cell Separation

In addition to dc current, ac voltage can also be applied to the electronic microarray in certain ways to create a dielectrophoretic (DEP) force applicable to analytes such as cells. Depending on the frequency of the ac voltage and the dielectric properties of the cells, the DEP forces can be either positive (moving cells towards electrodes) or negative (moving

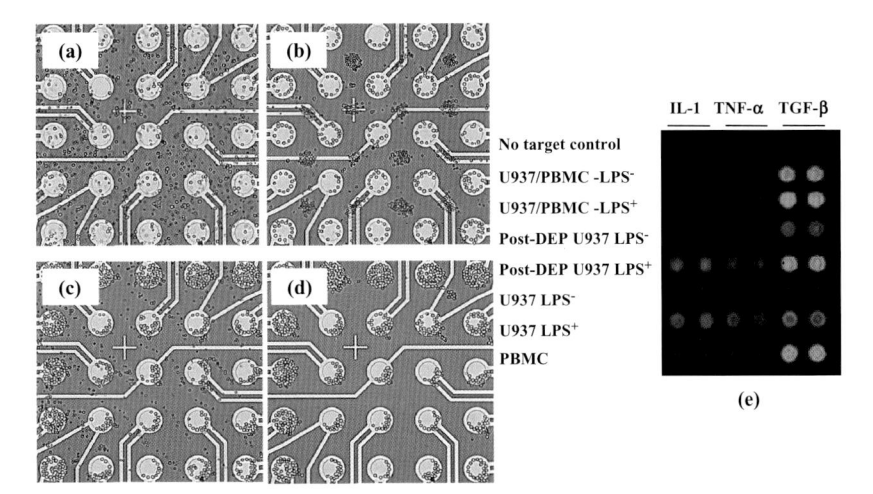

FIGURE 1.8. The procedure of DEP separation for U937 and PBMC mixture and gene expression profiling of three genes for U937 cells before and after DEP separation on electronic microarrays. (a) Mixture is introduced to the microarray. (b) U937 cells are separated from PBMC on array by dielectrophoresis 5 min after an ac voltage is applied. U937 cells are collected on the electrodes and PBMC are accumulated at the space between the electrodes. (c) Buffer is introduced from the reservoir to the array by fluid flow while the voltage is kept on. PBMC are carried away with the fluid stream. (d) PBMC are washed off from the array and U937 cells are retained on the electrodes after 10 min of washing. (e) Fluorescent image of three genes expression profiling before and after DEP separation. The specific signals for IL-1, TNF-α, and TGF-β from different samples after electronic hybridization are indicated. Reprinted with permission from Huang et al. Copyright (2002) American Chemical Society.

cells towards spaces) resulting in special separation of different types of cells on electronic microarrays [8, 26]. Using DEP, U937 and PBMC were separated on the array [22]. The separation procedure is illustrated in Figure 1.8. After the mixture was introduced to the array, the flow was stopped (Figure 1.8a). Five minutes after an ac voltage of 500 kHz, 7 Vpp (volts, peak to peak) was applied to the array, U937 cells were separated from PBMC on chip (Figure 1.8b). Then by introducing the fluidic flow of 40 µl/min and keeping on the voltage, only PBMC were washed away with the buffer (Figure 1.8c). Ten minutes after washing, only U937 cells were remained on the electrodes by positive dielectrophoresis (Figure 1.8d). These U937 cells could then be released from the electrodes by fluidic flow and subsequently collected for gene expression analysis if the applied voltage was turned off.

RNA from DEP separated U937 cells was extracted, and the expression levels of IL-1, TNF-α, and TGF-β were quantitatively monitored on a 10×10 microelectronic chip array using a targeted gene expression profiling assay. The gene expression levels of IL-1, TNF-α, and TGF-β for LPS treated or untreated U937 cells were compared with those of U937 and PBMC mixture on a 10×10 array before and after DEP separation (Figure 1.8e). As demonstrated in Figure 1.8e, upon LPS treatment, U937 cells exhibited a significant increase in expression levels of IL-1, TNF-α, and TGF-β. After mixing U937 cells with PBMC at 1:5 ratio (U937 cells to PBMC), the expression patterns of the three genes had changed. In the mixed sample, the LPS induction of IL-1 and TNF-α expression could not be detected, and the induction of TGF-β was decreased. This reduction of the TGF-β expression level in the mixed samples was most likely due to the high expression level of TGF-β gene in PBMC. Apparently, in heterogeneous samples, the gene expression levels in a cell population of interest are not simply diluted by unrelated cells but are masked by the expression patterns of the contaminating cell populations. Notably, in the DEP separated U937 cells, the LPS induction of IL-1, TNF-α, and TGF-β expression was observed. This result indicates that DEP separation can improve the accuracy of gene expression profiling by purifying out cells of interest.

1.3.5. Electronic Immunoassays

The flexibility of the microarray platform in terms of the ability to electrophoretically transport charged molecules to any site on the planar surface of the array has been exploited for developing electronic immunoassays. The following is an example of an electric field-driven immunoassay developed for two biological toxins—Staphylococcal enterotoxin B (SEB) and Cholera toxin B (CTB) for biological warefare applications [16]. A 25-site electronic microarray (Figure 1.9) was transformed into an immunoassay array by electronically biasing electrodes at user defined microlocation to direct the transport, concentration, and binding of biotinylated monoclonal capture antibodies specific for SEB and CTB to streptavidin in the hydrogel layer. The detection of fl-SEB and fl-CTB were accomplished in only 6 minutes, including 1-minute electronic addressing step to bind fl-SEB and fl-CTB, followed by a 5-minute washing step to reduce nonspecific binding. More noticeably, electronic addressing permitted the detection of CTB down to concentrations of 18 nM. No fluorescence

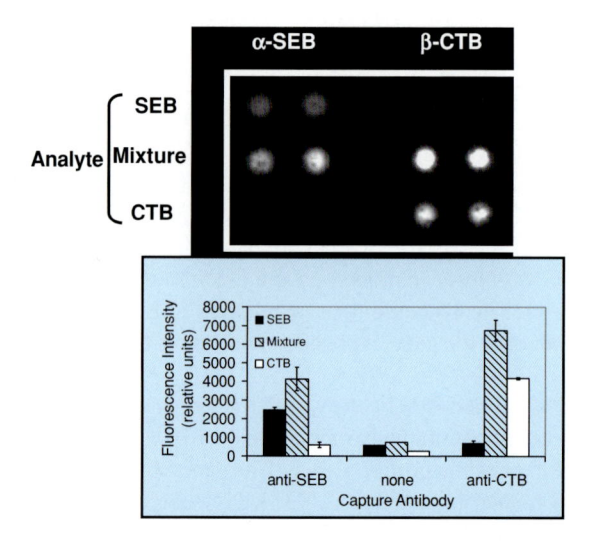

FIGURE 1.9. Electronic assay for fl-SEB and fl-CTB on the 10,000 site CMOS array. Biotinylated anti-SEB capture antibody was electronically addressed to the microlocations in the two columns on the left side of the image. Biotinylated anti-CTB capture antibody was electronically addressed to each of the microlocations in the two columns on the right side of the image. The column of microlocations in the center was left free. A mixture of fl-SEB and fl-CTB (20 nM in 50 mM histidine, pH 7.5) was applied to the chip and electronically addressed to all 15 microlocations of the chip. After washing, the fluorescence intensity at each microlocation was measured.

ELECTRONIC MICROARRAY TECHNOLOGY AND APPLICATIONS

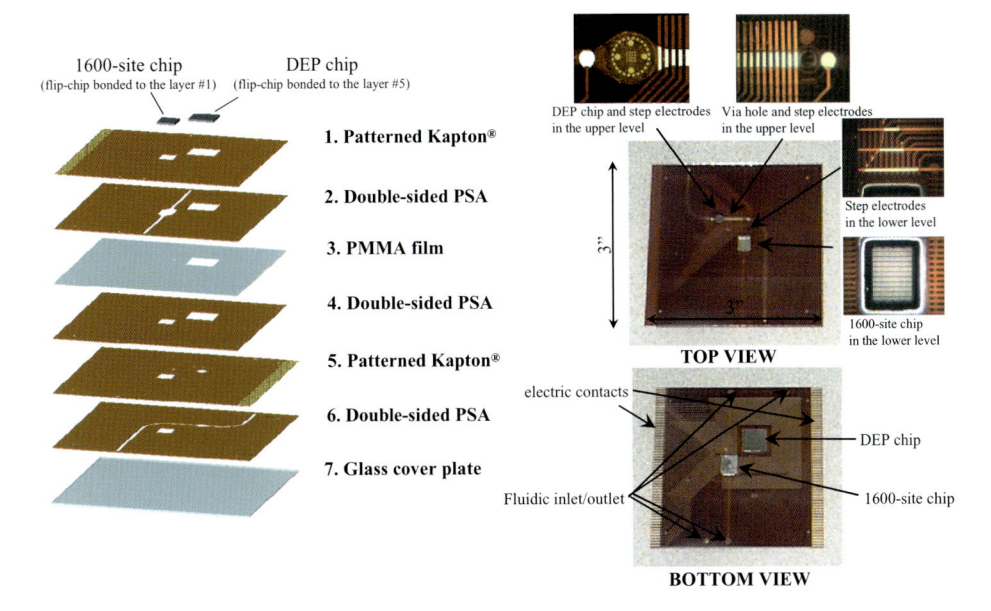

FIGURE 1.10. The stacked microlaboratory : (left) Fabrication of the stacked structure; (right) Photographs of the completed stacked structure showing the top and the bottom views.

attributable to antigen binding was observed at any other location on the chip other than the sites addressed with the appropriate capture antibody. It was possible to detect both toxins from a mixture in a single electronic addressing step (Figure 1.10). The ability to perform a rapid, electric field-mediated immunoassay for multiple analytes provides an advantage over existing approaches in terms of sample volume, speed and system complexity.

1.3.6. Miniaturization of Electronic Microarray Technology and Applications

DNA microarray technology provides a highly specific analytical response that enables the identification of a particular sequence within the DNA target molecules. Using these characteristic sequences the identification of biological agents, including biological warfare agents and infectious disease pathogens can be performed with the highest specificity and accuracy. It is of emergent interest to make portable instruments capable of performing rapid DNA analysis in the field or in hospitals. Such trends call for the development of the so-called point-of-care or field-portable instruments which could be used immediately at the patient bed or by first responders in the case of biological attacks. The requirements for such instrumentation are very stringent including high sensitivity and specificity, simultaneous detection of multiple targets, automated operation, and ease of use. Encountering various types of samples in the field, there is an urgent need to integrate the sample preparation processes with the detection and create simple to operate field portable sample-to-answer instruments [2]. Using the basic principles of the electronic separation and addressing of the targets we have developed a number of different approaches to miniaturize detection systems as well as the integration of sample preparation with detection on the electronic microarray.

1.3.6.1. Stacked Microlaboratory A sample-to-answer prototype instrument was developed to perform pathogen isolation [28], DNA hybridization and protein immunoassays from mixed samples [52]. At the core of this instrument is a small two-level-stacked microlaboratory that is 76×76 mm^2 and 2.77 mm thick. The stacked microfluidic structure was constructed from a set of laminated flexible substrates with fluidic cutouts, pressure sensitive adhesive layers, electrode arrays, and two Si chips. The completed stacked structure has the DEP chip in the upper level chamber for dielectrophoretic collection of bioparticles and the 1600-site chip in the lower level for performing automated electric-field-driven assays (Figure 1.10). Various automated assays have been demonstrated on this instrument. *E. coli* bacteria and Alexa-labeled protein toxin SEB were detected by electronic immunoassay. The identification of SLT1 gene from *E. coli* was accomplished in 2.5 hours starting from a dielectrophoretic concentration of intact *E. coli* bacteria and finishing with an electronic DNA hybridization assay [52].

1.3.6.2. Miniaturized Electronic Microarray System Rapid identification of pathogens or biological agents, including biological warfare agents, would ideally incorporate small automated portable systems. Supported through Department of Defense funding, a simple miniaturized system for the detection of DNA targets from pathogenic microorganisms based on the electronic microarray technology was developed. The first generation of the instrument is an automated portable DNA analysis system designed to operate with the Nanogen's 400-site NanoChip® array and cartridge. The 400-site array (Figure 1.11) is a

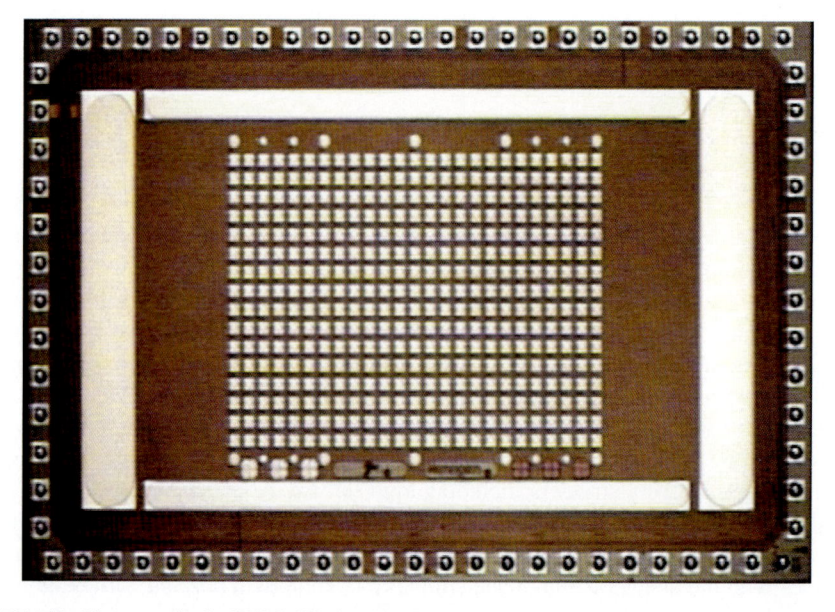

FIGURE 1.11. Photogoraph of a CMOS 400-site chip. Counter electrodes surround the central 400-site working electrode array.

Cartridge loading Port   **400-site ACV-chip and cartridge**

FIGURE 1.12. Photograph of the electronic microarray system with the laptop used to operate the instrument and perform data storage and processing. The arrow indicates a port for the 400-site cartridge introduction.

second generation CMOS chip developed at Nanogen [46]. The advantages of the CMOS chip over the first generation design include simple control circuitry, improved electrode current and voltage control, four times as many assay sites and on-chip temperature sensing. The CMOS chip has an array of 16×25 (400-sites); each electrode being 50 μm in diameter with a 150-μm center-to-center distance. Voltage on each electrode can be individually controlled and measured. Furthermore, a temperature sensor is built into the chip's CMOS circuitry.

A photograph of the Nanogen's portable electronic microarray detection instrument is shown in Figure 1.12. It is a compact, fully enclosed instrument featuring automated DNA sample and reagent delivery to the chip, temperature control of the chip and the fluidic cell, electronic control of the chip array, and optical detection system. Any standard laptop or desktop computer can be used to operate and control the system. The overall dimensions of the system are (not including laptop) ca. 14 inch $\times$ 12 inch by 6 inch. The total weight including the laptop which provide automated control of the system and data acquisition, ca. 30 lbs.

The portable instrument and electronic microarray platform allows development of a number of assays which can support PCR amplification of the target DNA from the sample, as well as solution strand displacement amplification (SDA) and anchored SDA. Anchored SDA provides an extremely convenient way to integrate the sample preparation steps. In this arrangement, the DNA amplification step and fluorescent detection steps are performed on the microarray chip. Highly sensitive assays were developed for detection of typical biological warfare agents such as anthrax and vaccinia. A number of other assays for viral and bacterial DNA identification, e.g., including *Staphylococcus* enterotoxin a (SEA) and b (SEB), *Yersinia pestis* (plague), *E. Coli, Salmonella typhimurium, Streptococcus pneumoniae*, and others are being optimized for sensitivity using the same technology. Table 1.3 shows results of a sensitivity study performed for determination of *Bacillus anthracis*

TABLE 1.3. Results of a sensitivity study for Bacillus anthracis, Vollum, CapB gene*.

Sample	Number positives/number replicates				
	Chip 1	Chip 2	Chip 3	Chip 4	Total
1 fg/µl	1/10	1/10	1/10	1/10	4/40
10 fg/µl	8/10	7/10	7/10	8/10	30/40
100 fg/µl	10/10	10/10	10/10	10/10	40/40

* Study performed as an independent evaluation of the method by, Midwest Research Institute, Florida.

Vollum DNA. A PCR amplification protocol was developed and CapB gene was used as a confirmation gene. Table 1.3 demonstrates that 100 % correct calls were made for 100 fg/µl DNA and 75% for 10 fg/µl DNA. These values correspond to ca. 170 and 17 copies of anthrax DNA. The achieved sensitivity is well within the requirements for field detection of biological warfare agents. Similar sensitivity has been achieved for other biowarfare agents.

1.3.7. Applications in Proteomics

After the completion of the human genome project, the attention of the scientific community has turned toward the gene products within the cell and tissue matrix, namely proteins. The field of proteomics is an evolving area, which may shed light on the proteins associated with diseases and tumors. Protein kinases are of particular interest because they have been shown to be key regulators of many cell functions and have been one of the main targets in drug industry. Several high-throughput screening (HTS) kinase assays have been developed based on either antibodies or radioactivity for detection. Recently, an electronic, fluorescent assay for kinases, phosphatases and proteases has been developed for the serine/threonine kinase PKA [30]. This ElectroCapture™ PKA assay combines electric field separation of a substrate and a reaction product of different net charge and subsequent capture of the reaction product on a capture matrix (Figure 1.13). The Lissamine-rhodamine labeled Kemptide peptide substrate contains a (+1) charge. Upon phosphorylation, the

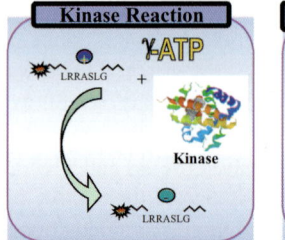

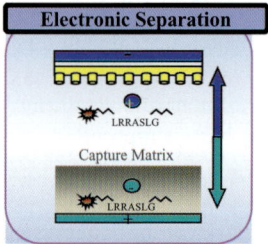

FIGURE 1.13. The ElectroCapture™ Assay. The fluorescently labelled substrate contains a (+1) charge. Upon phosphorylation, the peptide substrate undergoes a charge inversion from a (+1) charge to a (−1) charge on the product of the reaction. When an electric field is applied, the unphosphorylated peptide substrate (+1) migrates towards the negative electrode and the product (−1) migrates towards the positive electrode. The (−1) charged phosphorylated peptides migrate towards the (+) electrode from the solution, through the diffusion barrier and bind to the membrane.

peptide substrate undergoes a charge inversion from a (+1) to a (−1) charge on the product of the reaction. When an electric field is applied, the positively charged unphosphorylated peptide substrate migrates towards the negative electrode and the neagtively charged phosphorylated peptides product migrates towards the positive electrode from the solution, through the diffusion barrier and bind to the membrane. After the electrophoretic separation, the amount of Lissamine-rhodamine labeled phosphorylated peptides product can be quantitated, e.g., using a Tecan Ultra 384 fluorescence reader. The ElectroCapture™ PKA assay was validated with both known PKA inhibitors and with library compounds. The pK_i results obtained in the ElectroCapture™ assay were comparable to those generated in our current radioactive Filter Binding assay and antibody-based competitive Fluorescence Polarization (FP) PKA assay formats [30].

1.4. SUMMARY AND OUTLOOK

Future applications of focused arrays will involve improvements in speed, sample preparation and systems integration. This evolution in technology will allow the user to process a variety of samples in the disposable cartridge or directly on the embedded array. As we have demonstrated in this review, complete sample to answer systems based on site-specific electrophoresis, dielectrophoresis, DNA amplification and detection can be optimized, miniaturized and integrated into a complete sample-to-answer system. The usage of integrated and portable biological detection systems is expected to increase several fold in the next few years including point of care diagnostic applications for genotyping, pharmacogenomics, proteomics, detection of infectious agents as well as identification of biological warfare agents. The electronic microarray technology offering high speed of hybridization, target concentration and multiplexing has all the advantages to be one of highly competitive technologies in future miniaturized instruments for molecular and clinical diagnostics.

REFERENCES

[1] A.A. Alzadeh, M.B. Elsen, R.E. Davis, C. Ma, I.S. Lossos, and A. Rosenwald et al. *Nature*, 403:503, 2000.
[2] R. Anderson, X. Su, G. Bogdan, and J. Fenton. *Nucleic Acids Res.*, 28:12, 2000.
[3] H.A. Behrensdorf, M. Pignot, N. Windhab, and A. Kappel. *Nucleic Acids Res.*, 30:e64, 2002.
[4] J. Boguslavsky. Lab-on-a-Chip: Easier, Faster, Smaller, Drug Discovery & Development, July/August, p. 32, 2001.
[5] L. Carey and L. Mitnik. *Electrophoresis*, 23:1386, 2002.
[6] R. Chakraboty, D.N. Stivers, B. Su, Y. Zhong, and B. Budowle. *Electrophoresis*, 20:1682, 1999.
[7] M. Chee, R. Yang, E. Hubbell, A. Berno, X.C. Huang et al. *Science*, 274:610, 1996.
[8] J. Cheng, E.L. Sheldon, L. Wu, A. Uribe, L.O. Gerrue, J. Carrino, M.J. Heller, and J.P. O'Connell. *Nat. Biotech.*, 16:541, 1998.
[9] K.L. Cooper and R.V. Goering. *J. Mol. Diagn.*, 5:28, 2003.
[10] B. Dukek and D.J. O'Kane. IVD *Technol.*, 47:Jan/Feb, 2004.
[11] C.F. Edman, D.E. Raymond, D.J. Wu, E. Tu, R.G. Sosnowski, W.F. Butler, M. Nerenberg, and M.J. Heller. *Nucleic Acids Res.*, 25:4907, 1997.
[12] C.F. Edman, P. Mehta, R. Press, C.A. Spargo, G.T. Walker, and M. Nerenberg. *J. Investig. Med.*, 48:93, 2000.

[13] M. Eggers, M. Hogan, R.K. Reich, J.B. Lamture, D. Ehrlich et al. *Biotechniques*, 17:516, 1994.
[14] M. Erali, B. Schmidt, E. Lyon, and C. Wittwer. *Clin. Chem.*, 49:732, 2003.
[15] J.G. Evans and C. Lee-Tataseo. *Clin. Chem.*, 48:1406, 2002.
[16] K.L. Ewalt, R.W. Haigis, R. Rooney, D. Ackley, and M. Krihak. *Anal. Biochem.*, 289:162, 2001.
[17] L. Feng and M. Nerenberg. *Gene. Ther. Mol. Biol.*, 4:183, 1999.
[18] S.P. Fodor, R.P. Rava, X.C. Huang, A.C. Pease, C.P. Holmes, and C.L. Adams. *Nature*, 251:767, 1991.
[19] P.N. Gilles, D.J. Wu, C.B. Foster, P.J. Dillon, and S.J. Chanock. *Nat. Biotechnol.*, 17:365, 1999.
[20] J.G. Hacia. *Nat. Biotechnol.*, 21(suppl):42, 1999.
[21] M.J. Heller. *IEEE Eng. Med. Biol.*, 100: March/April, 1996.
[22] M.J. Heller. *Annu. Rev. Biomed. Eng.*, 4:129, 2002.
[23] M.J. Heller, E. Tu, R. Martinsons, R.R. Anderson, C. Gurtner, A.H. Forster, and R. Sosnowski. In M.J. Heller and A. Guttman (ed.), *Integrated Microfabricated Biodevices*. Marcel Dekker, New York, p. 223, 2002.
[24] M. Heller, A.H. Forster, and E. Tu. *Electrophoresis*, 21:157, 2000.
[25] M.J. Heller, E. Tu. U.S. Patent # 5,605,662 Nanogen, Inc., San Diego, CA, 1997.
[26] Y. Huang, K.L. Ewalt, M. Tirado, R. Haigis, A. Forster, D. Ackley, M.J. Heller, J.P. O'Connell, and M. Krihak. *Anal. Chem.*, 73:1549, 2001.
[27] Y. Huang, S. Joo, M. Duhon, M. Heller, B. Wallace, and X. Xu. *Anal. Chem.*, 74:3362, 2002.
[28] Y. Huang, J.M. Yang, P.J. Hopkins, S. Kassegne, M. Tirado, A.H. Forster, and H. Reese. *Biomed. Microdev.*, 3:217, 2003.
[29] Y. Huang, J. Shirajian, A. Schroder, Z. Yao, T. Summers, D. Hodko, and R. Sosnowski. *Electrophoresis*, 25:3106, 2004.
[30] K. Huss, R.M. Campbell, S. Miick, S. Jalali, D. Thomas, and M. Jimenez. *9th Annual SBC Conference*, Poster 1025, Portland, OR, 2003.
[31] S.K. Kassegne, H. Reese, D. Hodko, J.M. Yang, K. Sarkar, D. Smolko, P. Swanson, D.E. Raymond, M.J. Heller, and M.J. Madou. *Sens. Actu. B*, 94:81, 2003.
[32] K. Liszewski. Broader Uses for Microfluidics Technologies, *Genet. Eng. News*, vol. 23, no 9, p. 40, 2003.
[33] M. McCormick. *Genet. Eng. News*, vol. 23, no. 15, p. 34, 2003.
[34] C.E. Pearson, Y.H. Wang, J.D. Griffith, and R.R. Sinden. *Nucleic Acids Res.*, vol. 26, 816, 1988.
[35] E.S. Pollak, L. Feng, H. Ahadian, and P. Fortina. *Ital. Heart J.*, 2:568, 2001.
[36] S. Raddatz, J. Mueller-Ibeler, J. Kluge, L. Wass, G. Burdinski, J.R. Haven, T.J. Onofrey, D. Wang, and M. Schweitzer. *Nucleic Acids Res.*, 30:4793, 2002.
[37] R. Radtkey, L. Feng, M. Muralhidar, M. Duhon, D. Canter, D. DiPierro, S. Fallon, E. Tu, K. McElfresh, M. Nerenberg, and R. Sosnowski. *Nucleic Acids Res.*, 28:e17, 2000.
[38] E. Ricart, W.R. Taylor, E.V. Loftus, D. O'Kane, R.M., Weinshilboum, W.J. Tremaine, W.S. Harmsen, A.R. Zinsmeister, and W.J. Sandborn. *Am. J. Gastroenterol.*, 97:1763, 2002.
[39] R. Santacroce, A. ratti, F. Caroli, B. Foglieni, A. Ferraris et al. *Clin. Chem.*, 48:2124, 2002.
[40] M. Schena, D. Shalon, R.W. Davis, and P.O. Brown. *Science*, 270:467, 1995.
[41] I. Schrijver, M.J. Lay, and J.L. Zehnder. *Am. J. Clin. Pathol.* 119:490, 2003.
[42] E. Sheldon, et al. In electronic sample handling. In S.A. Minden and L.M. Savage (ed.), *Diagnostic Gene Detection & Quantification Technologies for Infectious Agents and Human Genetic Diseases*, IBC Library Series, pp. 225–238, 1997.
[43] Y.R. Sohni, J.R. Cerhan, and D. O'Kane. *Hum. Immunol.* 64:2003.
[44] R. Sosnowski, E. Tu, W.F. Butler, J. O'Connell, and M. J. Heller. *Proc. Natl. Acad. Sci.* USA, vol. 94, p. 1119, 1997.
[45] R. Sosnowski, M.J. Heller, E. Tu, A.H. Forster, and R. Radtkey. *Psychiatr. Genet.*, 12:181, 2002.
[46] P. Swanson, R. Gelbart, A.E. Atlas, L. Yang, T. Grogan, W.F. Butler, D.E. Ackley, and E. Sheldon. *Sens. Actur. B.*, 64:22, 2000.
[47] W.A. Thistlethwaite, L.M. Moses, K.C. Hoffbuhr, J.M. Devaney, and E.P. Hoffman. *J. Mol. Diagn.*, 5:121, 2003.
[48] V.W. Weedn and J.W. Hicks. *Natl. Inst. Justics J.*, 234:16, 1997.
[49] E.M. Weidenhammer, B.F. Kahl, L. Wang, L. Wang, M. Duhon, J.A. Jackson, M. Slater, and X. Xu. *Clin. Chem.*, 48:1873, 2002.
[50] L. Westin, X. Xu, C. Miller, L. Wang, C.F. Edman, and M. Nerenberg. *Nat. Biotechnol.*, 18:199, 2000.
[51] L. Westin, C. Miller, D. Vollmer, D. Canter, R. Radtkey, M. Nerenberg, and J.P. O'Connell. *J. Clin. Microbiol.*, 39:1097, 2001.

[52] J.M. Yang, J. Bell, Y. Huang, M. Tirado, T. Thomas, A.H. Forster, R.W. Haigis, P.D. Swanson, R.B. Wallace, B. Martinsons, and M. Krihak. *Biosen. Bioelect.*, 17:605, 2002.
[53] N. Yoshida, Y. Nishimaki, M. Sugiyama, T. Abe, T. Tatsumi, A. Tanoue, A. Hirasawa, and G. Tsujimoto. *J. Hum. Genet.*, 47:500, 2002.
[54] K. Zimmermann, T. Eiter, and F. Scheiflinger. *J. Microbio. Methods*, 55:471, 2003.

2

Gene Expression Profiling Utilizing Microarray Technology and RT-PCR

Dominick Sinicropi, Maureen Cronin, and Mei-Lan Liu
Genomic Health, Inc., Redwood City, California USA

2.1. INTRODUCTION

Over the last decade microscale technologies for molecular analysis have become the springboard for a new era in biological investigation. In parallel with nucleic acid sequencing technology improvements that enabled completion of the human genome project years ahead of schedule [1, 2], methods were developed for high throughput analysis of genetic variation and gene expression. These new molecular analytical tools have stimulated a resurgence of non-hypothesis driven biological research and promise to play a key role in the emerging field of personalized medicine [3]. Knowledge gained from the most common type of genetic variation in DNA, single nucleotide polymorphisms (SNPs), has had an enormous impact on the identification of genes involved in disease and is beginning to be of value in tailoring therapeutic regimens for an individual's genetic composition [4]. Application of technologies for gene expression analysis, the subject of this review, has lagged behind analysis of genetic variation, primarily due to the intrinsic complexity of gene expression measurement. However, the number of studies employing gene expression analysis has expanded in the last few years as the available analytical methods mature and become more reliable and affordable.

A variety of methods have been used for gene expression quantification. The first methods to be widely adopted include northern blotting, RNA protection, differential display, serial analysis of gene expression (SAGE), and quantitative competitive reverse transcription-polymerase chain reaction (RT-PCR). Although all of these methods are still used today, real-time RT-PCR and DNA microarrays have achieved prominence in recent years and are the focus of this review. Both of these methods are also useful for analyzing genetic variation; however, for this review we will concentrate on the use of these technologies for gene expression analysis.

Real-time RT-PCR, though not the first PCR-based method for gene expression analysis, has emerged as the "gold standard" against which other methods are compared. "Real-time", or "kinetic", PCR methods are those that measure the accumulation of PCR product with each PCR cycle. The original real-time PCR method measured the fluorescence of ethidium bromide intercalated in the double stranded DNA products of PCR amplification [5]. Subsequently, a variety of alternative methods for real-time quantitation of PCR products (reviewed below) have appeared as commercial products. The primary advantage of real-time quantitation is the relative simplicity of experiments as compared with other PCR-based methods. The ability to obtain a quantitative result in a single reaction is responsible, in large part, for the popularity of real-time PCR as compared with other PCR-based methods that require multiple step reactions for each sample that is analyzed. Another benefit is that commercial systems for real-time RT-PCR are configured for simultaneous analysis of hundreds of samples (or genes) simultaneously. These features, taken together with the high degree of analytical precision that is possible, have made real-time RT-PCR the method of choice for quantitative expression profiling.

Interest in DNA microarray technologies for measuring gene expression has exploded in recent years. Without question, the biggest advantage microarray technology has to offer is the large number of transcripts that can be quantified in a single experiment. DNA microarrays are capable of making tens of thousands of gene expression measurements simultaneously. Major commercial suppliers of DNA microarrays have recently released products in which the entire complement of known expressed human genes (the "transcriptome"; approximately 40,000 expressed sequences) can be measured on a single microarray. The unprecedented ability to monitor the expression of entire genomes has led to biological discoveries that would not have been possible by other methods. Nevertheless, microarray technology has limitations including its relatively high cost and inability to analyze more than one sample per array experiment.

In biological research it is often desirable to explore the expression profiles of two or more conditions (disease vs. normal, treated vs. untreated, etc.) with no hypothesis about which genes may be differentially expressed. Microarray technology is ideally suited for this early "discovery" phase of biological research due to the large number of genes that can be analyzed simultaneously. Microarray studies can reveal changes in expression of a smaller number of genes that are used for subsequent hypothesis generation and testing. Once identified, the smaller gene set can be analyzed by real-time RT-PCR, which is better suited to analyzing multiple samples.

Increasingly, the capabilities of real-time RT-PCR and microarray technologies are beginning to overlap. Several companies that offer high-density microarrays have recently introduced microarray products that are designed for the analysis of relatively small numbers of genes. In addition, several microarray manufacturers are designing products intended for high throughput analysis of multiple samples, such as "arrays of microarrays" in standard microtiter plate format. Likewise, advances in real-time PCR technology have been introduced to enable simultaneous analysis of larger numbers of genes or samples.

In the remainder of this chapter, we will review the most common methods used for real-time RT-PCR followed by a review of alternative microarray technologies. We will

2.2. REAL-TIME PCR

As PCR methodology matures, new technical developments refining its use continue to emerge. One of these is "real-time" PCR product quantification, made possible by the development of fluorescence-kinetic detection methods. In this technique, PCR product is measured as it accumulates during sequential amplification cycles. The strength of this method is that no post-PCR manipulation or product analysis is required since the quantitative measurement of the reaction product is known at the end of the cycling process. The term "reverse transcriptase-PCR" (RT-PCR) is reserved for quantification of gene expression at the level of mRNA. In real-time RT-PCR, a sample containing the mRNA target of interest is first reverse transcribed to cDNA that is subsequently amplified by real-time PCR.

More recently, a number of detection technologies, allowing both non-specific and specific target detection, have been developed for real-time PCR applications. Non-specific detection systems detect all double-stranded DNA generated during the PCR reaction. Specific detection systems distinguish specific target sequences from primer dimers or non-specific amplification products.

2.2.1. Detection Systems

2.2.1.1. Non-Specific Detection Systems DNA intercalating dyes, such as ethidium bromide [5] and SYBR Green®, have been used to detect PCR products non-specifically [6]. During the polymerization phase of a PCR reaction, these dyes bind to the newly synthesized double-stranded DNA, resulting in increased fluorescence emission that can be detected in real-time as PCR cycling progresses. The level of specificity is limited to that of the reaction primers since dye intercalation detects all double-stranded DNA and is not amplicon sequence specific. When intercalating dyes are used for allelic discrimination or to test PCR specificity, PCR products are assessed by running an end-point analysis of the DNA melting curve after thermal cycling concludes. PCR products of different lengths or different sequence will give distinct dissociation curves as they melt at different temperatures (Tm). The temperature at which the double-stranded DNA becomes single-stranded is measured as fluorescence reduction when the intercalator dissociates from the melting double-stranded DNA. A single melting peak is expected in an optimized PCR reaction since it implies a single specific product and the absence of self-priming artifacts. Therefore, a SYBR Green assay is frequently employed to evaluate PCR performance of difference primer pairs, varying cycling conditions or to optimize primer concentrations in PCR reactions. Melting curve analysis on the Applied Biosystems 7900HT Sequence Detection System can typically resolve amplicon Tm values that differ by one or two degrees. The SYBR Green assay is also commonly used for SNP detection since short distinct sequences, differing by even a single nucleotide, melt at slightly different temperatures. Peak SYBR Green fluorescence measured over a series of PCR cycles can also be plotted to derive a quantitative PCR result similar to the original real-time PCR applications developed with ethidium bromide detection of dsDNA PCR products. However, sensitivity, specificity, accuracy, and precision of this method do not compare to results that can be achieved using specific detection probes labeled with fluorescent dyes as described in the next section. Nonspecific detection systems work best for short amplicons since Tm resolution decreases as amplicon length increases.

2.2.1.2. Specific Detection Systems Specific PCR detection systems generally rely on using detection probes that are complementary to the target sequence amplicon generated by the forward and reverse PCR primers. The following sections describe some common probe chemistries including hydrolysis probes, Molecular Beacons, Scorpion™ probes and FRET hybridization probes.

Hydrolysis Probes Hydrolysis probes, often called TaqMan® probes, are probably the most widely used detection method for real-time PCR. Typical probe structure is shown in Figure 2.1A. These probes are oligonucleotides with a fluorescent reporter dye conjugated to the 5' end and a quencher dye conjugated to the 3' end that absorbs the light emission of the reporter dye conjugated to the 3' end. The reporter may be any one of a number of fluorescent molecules but most often is FAM™ (6-carboxyfluorescein), TET™ (tetrachloro-6-carboxyfluorescein), JOE™ (2,7-dimethoxy-4,5-dichloro-6-carboxyfluorescein), HEX™ (hexacholoro-6-carboxyfluuo-rescein), or VIC®, a proprietary dye from Applied Biosystems (Foster City,CA). The 3' quencher may be either a fluorescent dye with an emission

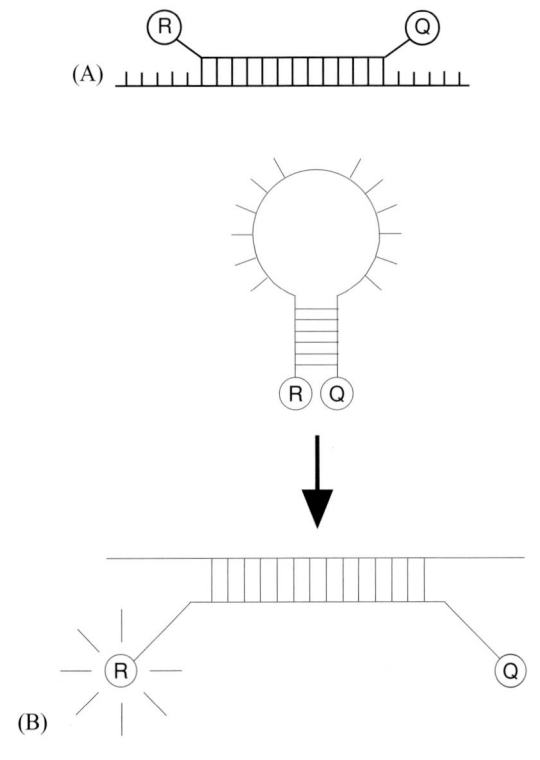

FIGURE 2.1. A. Hydrolysis probe. R, fluorescent reporter dye; Q, quencher dye. B. Molecular beacon shown in nonfluorescent stem-loop structure and as fluorescent hybrid with DNA target.

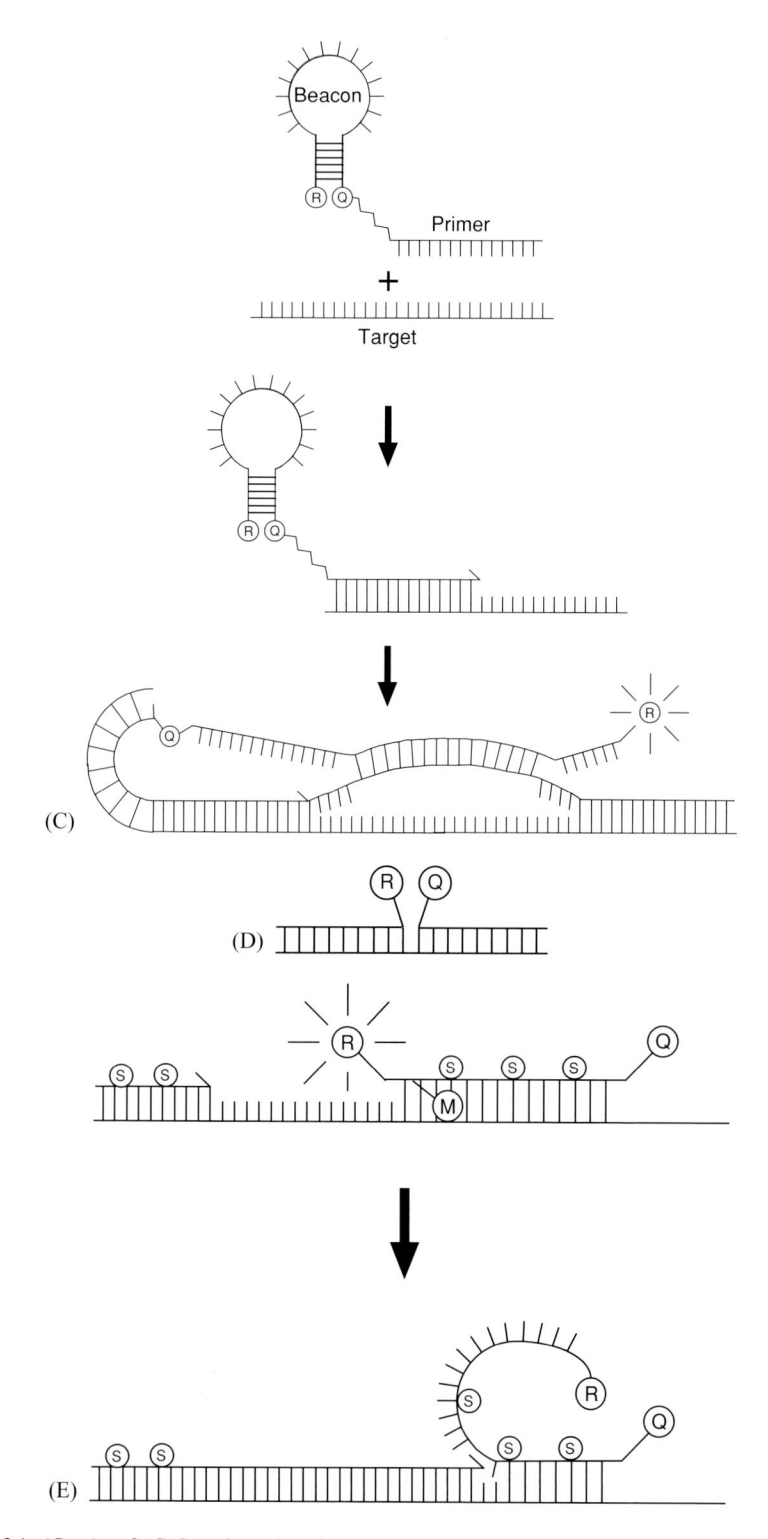

FIGURE 2.1. (*Continued*) C. Scorpion Primer/Probe; D. Hybridization probes; E. Qiagen QuantiProbe™. M, minor groove binder; S, Superbases™.

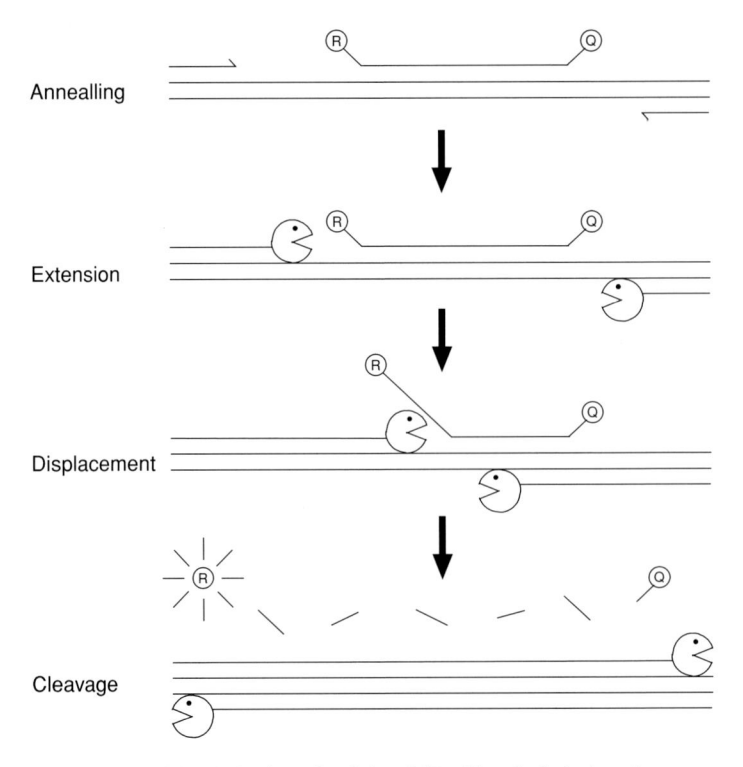

FIGURE 2.2. Mechanism of real-time PCR with an hydrolysis probe.

wavelength well separated from that of FAM and whose excitation wavelength overlaps the FAM emission wavelength, such as TAMRA™ (6-carboxytetramethylrhodamine), or a non-fluorescent ("dark") quencher, such as Black Hole Quencher™ (Biosearch Technologies, Inc.) or DABCYL [7]. Dark quenchers absorb the light emission from the reporter dye and release it as energy other than fluorescence. Consequently, the dark quenchers tend to provide improved signal-to-noise ratios since they do not contribute to background signals as the fluorescent quenchers do. These probe constructs rely on a form of fluorescence resonance energy transfer (FRET) [8]. When hydrolysis probes are intact, the report dye and quencher dye are in close proximity. The quencher dye absorbs the fluorescence emitted by the reporter dye, which results in a non-fluorescent probe. During the PCR reaction, the probe anneals specifically to the target template when it is present as shown in Figure 2.2. The 5'-exonuclease activity of Taq DNA polymerase displaces and hydrolyzes the probe while it polymerizes a replicate of the template on which the probe is bound. The cleaved reporter dye becomes separated from its quencher resulting in fluorescence emission. Accumulation of the PCR product is detected by monitoring the increase in fluorescence emission from the cleaved reporter dye at each cycle and is directly proportional to the amount of target present in the sample as illustrated in Figure 2.3. A threshold fluorescence value is established within the exponential amplification stage of the reaction. The PCR cycle number at which the fluorescence emission is equal to the threshold value is termed the threshold cycle (C_T). As shown in Figure 2.4, the C_T value is linearly correlated with the logarithm of the mass of RNA added to the initial reaction mixture. Therefore, TaqMan real-time RT-PCR can be

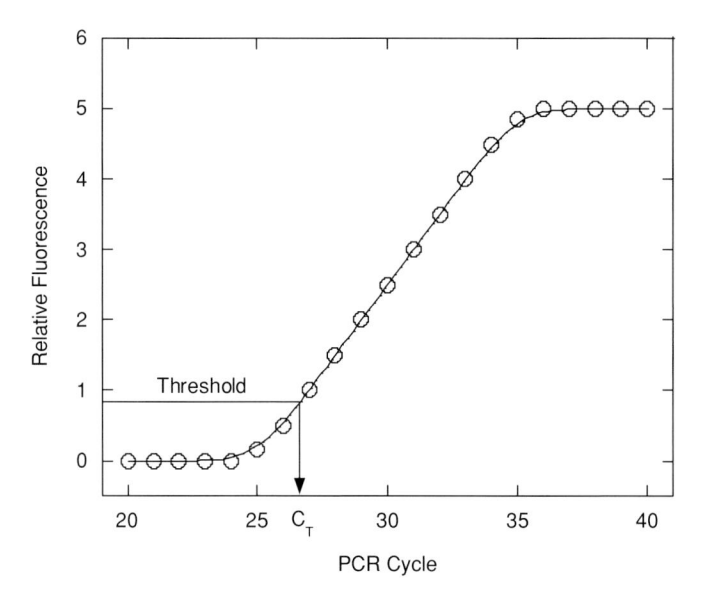

FIGURE 2.3. Idealized amplification plot. Symbols are defined in the text.

used as a quantitative assay for a specific RNA target in a complex background of other sequences. TaqMan hydrolysis probes have been widely used in both research applications, particularly genotyping and gene expression analysis and diagnostic applications, such as HIV, HCV and other RNA viral load teststesting.

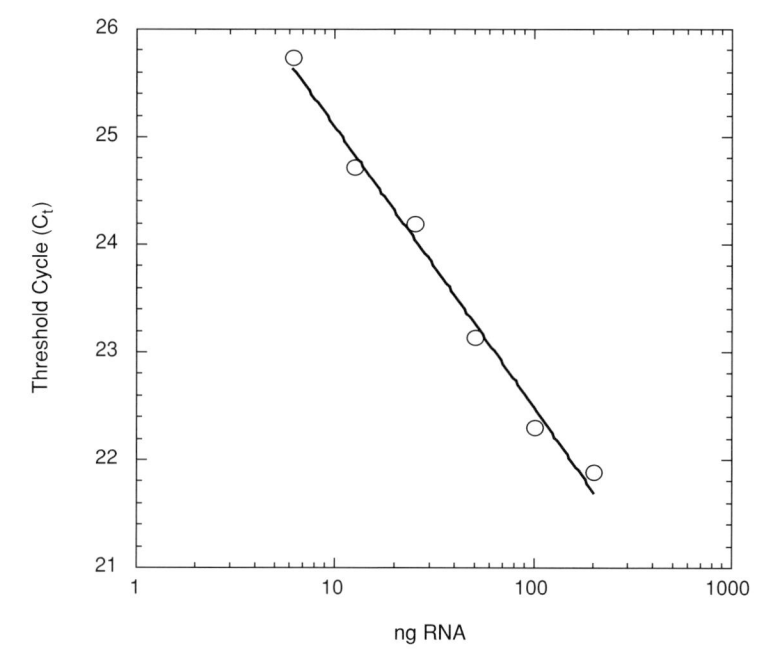

FIGURE 2.4. Calibration curve.

Molecular Beacons Molecular beacon probes [9] differ from the hydrolysis probes described above in both their structure and their mode of action. These oligonucleotide probes are designed to form stem-loop structures when they are free in solution as shown in the example in Figure 2.1B. The loop structure contains sequences complementary to the target template while the stem portion consists of a self-complementary stretch of approximately six bases (mainly C's and G's), which holds the probe in a hairpin like configuration. A fluorescent dye is linked to one end of the molecule and a quencher dye to the other end. In solution while the molecular beacon is in a hairpin structure, the stem holds the fluorophore and quencher in close proximity allowing efficient quenching of the fluorophore by FRET. During the PCR reaction, the molecular beacon encounters its complementary sequence in the amplicon and the probe sequence in the loop anneals to the complementary target sequence. A conformational change in the molecule results as the hybridization linearizes the probe, FRET no longer occurs and an increase in fluorescence emission is observed. The probe-target duplex is designed to be more thermodynamically stable than the hairpin structure. Unlike the hydrolysis probes, the increase in fluorescence with molecular beacons is reversible as the probe will dissociate from the target sequence and close back to the hairpin structure if the temperature is increased (for example during the denaturing cycle of PCR). Molecular beacon probes are particularly suitable for detecting point mutations or polymorphisms in target sequences because the probe-target duplex will not form at annealing temperature when a mismatch is present in the duplex.

Scorpion^{TM} Primer/Probes Scorpions are fluorescent PCR primers with a sequence in the hairpin loop structure at the 5' end of the oligonucleotide that acts as a detection probe [10, 11]. In this real-time PCR detection system, the PCR reaction is carried out using one standard primer and a second Scorpion primer, which serves the dual functions of both a primer and reporter probe. An example of a Scorpion probe structure and function can be found in Figure 2.1C. The fundamental elements of Scorpion probes include (i) a PCR primer in series with, (ii) a PCR "stopper" which prevents PCR read-through of the hairpin loop probe structure, (iii) a specific detection probe sequence complementary to the target held in a hairpin configuration by complementary stem sequences and (iv) a fluorescence detection system consisting of a fluorophore and a quencher held in close proximity by the stem structure. After extension of the Scorpion primer during PCR amplification, the probe sequence located in the loop structure binds to the complementary sequence in the amplicon within the same strand of DNA. This results in increased fluorescence emission because the quencher is no longer in the vicinity of the fluorophore. Scorpions perform better under fast cycling conditions [11] than hydrolysis (TaqMan) probes and molecular beacons. This may result from the unimolecular hybridization mechanism, which is kinetically highly favorable and not dependent on enzyme hydrolysis activity. Scorpion detection technology has been used successfully in allelic discrimination [11] and in SNP genotyping [12].

2.2.1.3. *Hybridization Probes* This FRET-based real-time PCR reaction system requires two primers and two probes. The two hybridization probes are designed to bind to the target template adjacent to each other as shown in Figure 2.1D. One probe is labeled with a donor fluorophore at its 5' end and the second probe with an acceptor fluorophore at its 3' end. In the presence of target sequence, hybridization brings the two probes into close proximity allowing energy transfer via FRET from the donor fluorophore emission to the acceptor fluorophore, which then emits the detected fluorescent signal. Fluorescence emission increases during sequential PCR cycles in proportion to the quantity of amplicon synthesized during the PCR reaction. This technology has been optimized and validated using the LightCycler™ instrument [13]. This detection strategy is particularly suitable for multiplexing since a single donor molecule can be used to excite multiple acceptor fluorophores.

More recently, another type of hybridization probe has been developed based on the MGB Eclipse™ (Epoch Biosciences) probe chemistry, namely the QuantiProbe™ assays (Qiagen). In this detection system, only one probe is required. QuantiProbe detection reagents are dually labeled oligonucleotides with a fluorophore at the 3' end of the probe and a non-fluorescent quencher at the 5' end (Dark Quencher™, Epoch Biosciences). When these probes are not bound to a target sequence, they form a random structure in solution allowing the fluorophore and quencher to come into close enough proximity to prevent fluorescence. During the annealing step of the PCR reaction, these probes hybridize to the target sequence separating the fluorophore from the quencher and allowing fluorescence emission. An example of this type of probe construct is shown in Figure 2.1E. The amount of fluorescence measured in real time is proportional to the amount of target sequence. During the extension step in PCR, the bound probe is displaced from the target sequence allowing the fluorophore to be quenched during that phase of the reaction. QuantiProbes have both minor groove binder (MGB) and Superbase™ nucleotide modifications designed to stabilize their hybridization to target amplicons. The MGB moiety is a protein that associates with DNA by either hydrogen bonding or hydrophobic interactions along the duplex minor groove [14–17]. When the MGB moiety is attached to the 3' or 5' end of a DNA probe, it folds back into the minor groove formed by DNA hybridization and effectively increases the Tm of the probe. This chemistry allows shorter probes to be used which increases specificity, especially for polymorphism discrimination assays. Modified nucleotides known as Superbases are analogs of the corresponding natural bases [18]. Super A™ and Super T™ form stronger bonds with their natural complements in the target sequence providing further DNA duplex stabilization and allowing short, discriminating probes to be successfully used in these assay formats.

2.2.2. Real-Time RT-PCR Data Analysis

One of two methods is commonly used to quantitatively analyze data obtained by real-time PCR, either the standard curve method or the comparative threshold method.

2.2.2.1. Standard Curve Method   This method requires using a standard template of known concentration to generate a standard curve of threshold cycle (C_T) values relative to input target quantity as was illustrated in Figure 2.4. The quantity of RNA target in test samples is calculated by comparing the threshold cycle value for the sample run under the same conditions as the known standard and assigning the corresponding input copy number from the standard curve. There are several different approaches to making standards including synthesizing a single-stranded oligonucleotide of the amplicon sequence, making *in vitro* transcribed RNA, or purifying plasmid DNA with the target sequence inserted. When absolute quantitation of mRNA expression is required, for example when quantifying viral load or analyzing gene expression, an *in vitro* transcribed RNA standard is preferred to control for variations that may be introduced by the reverse transcription reaction. This involves constructing a cDNA clone with an RNA polymerase promoter sequence that can serve as a template for *in vitro* transcribing copy RNA. One advantage of including a standard curve in each RT- PCR run is having an internal process control measurement of the PCR efficiency during that individual run.

2.2.2.2. Comparative Threshold Method The comparative threshold method for relative RT- PCR quantitation relates the fluorescence signal generated from a test sample to that of an internal control template. Examples where this is particularly useful would include study designs similar to those used in competitive hybridization on microarrays, such as treated and untreated control samples or the time zero control sample in a time-course study that must be compared to post treatment time point samples. Derivation and validation of the arithmetic formula ($2^{-\Delta\Delta C_T}$) used for quantifying the relative change in gene expression using real-time PCR have been described [19]. This method is valid under the assumption that amplification efficiencies of the target gene and reference gene are approximately equal. The $\Delta C_T = (C_T$ of the target gene $-C_T$ of the reference gene), and the $\Delta\Delta C_T = (\Delta C_T$ (sample) $-\Delta C_T$ (calibrator)). This equation represents the normalized expression of the target gene in the test sample, relative to the normalized expression of the calibrator sample.

The reference gene is used as an internal control for normalization of the amount of RNA input for the RT-PCR reaction and the efficiency of each individual run. The reference can be single or multiple genes; common examples include β-actin, glyceraldehydes-3-phosphate dehydrogenase (GAPDH), or a ribosomal RNA sequence. Selecting a suitable housekeeping gene is crucial to ensure the reliability of the experimental results. It is important to verify that the expression level of the chosen reference gene(s) is not affected by the treatment or tissue type of the experimental design.

2.2.3. Qualification of Gene Panels Using Real-Time RT-PCR

Here we describe a study performed in our laboratory to evaluate the analytical performance of the RT-PCR assay for quantifying gene expression in RNA samples isolated from human tumors. In this study, TaqMan technology was employed for quantitative RT-PCR measured on the ABI Prism 7900HT SDS instrument (Applied Biosystems, Inc.). To obtain sufficient RNA for the study, a pooled sample was prepared by combining equal amounts of RNA extracts from 52 breast cancer tumors. The expression levels for sixteen target genes and five reference genes in the pooled RNA sample were assessed. The real-time RT-PCR assay was carried out for fifteen different RNA input levels, ranging from 0 to 8 ng of total RNA per reaction. Assay performance characteristics were measured as described in the next sections.

2.2.3.1. Amplification Efficiency As described previously, for reference normalization to be valid and for panels of genes to be comparable with one another for relative expression, the amplification efficiencies of the target gene(s) and reference gene(s) must be very similar. Estimates of amplification efficiencies were obtained for each of sixteen

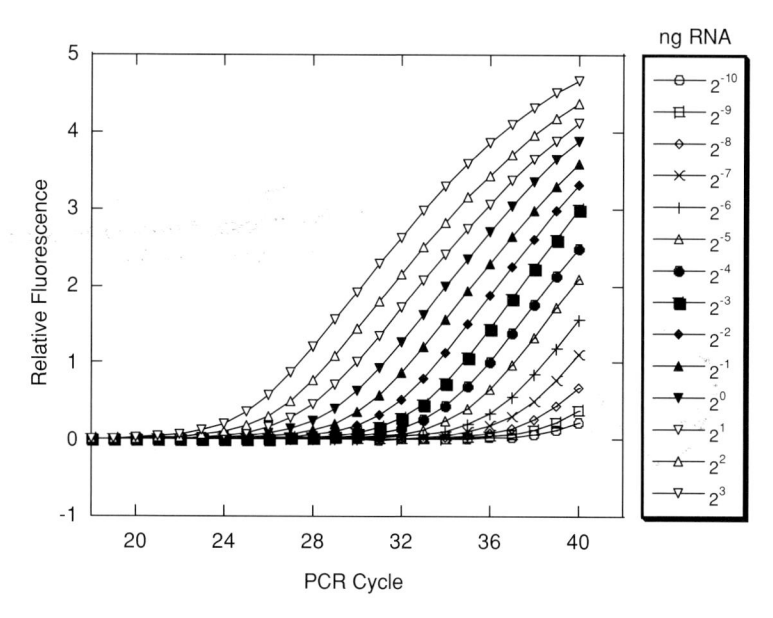

FIGURE 2.5. Amplification plots for serial dilutions of GAPDH target.

target genes and five reference genes according to the formula:

$$\text{Efficiency} = 2^{-1/\text{slope}} - 1$$

where slope is estimated from the regression of C_T measurements and $\log_2$ RNA concentration.

The results showed that amplification efficiencies for the sixteen target genes was 96% while the average efficiency among the reference gene set was 88%. This indicates consistent amplification and strong agreement among the target and reference genes.

2.2.3.2. Analytical Sensitivity and Dynamic Range The linearity of the C_T value relative to RNA concentration was evaluated for each individual gene. Data indicated that for both target and reference genes C_T measurements were proportional to input RNA amounts over at least a 10 $\log_2$ range and generally were consistent over a range of nearly 14 $\log_2$. A typical set of amplification curves from an input target dilution series is shown in Figure 2.5.

The accuracy and precision of the assay were estimated from calibration curves. Regression of calibration data (Figure 2.4 is an example) was used to calculate predictions of input RNA concentration based upon observed C_T measurements. Bias for each gene and RNA concentration level was then calculated as the percent difference between the predicted and input RNA concentrations.

$$\%Bias = \frac{\text{Predicted RNA Concentration} - \text{Input RNA Concentration}}{\text{Input RNA Concentration}} \times 100\%$$

Similarly, for each gene and input RNA concentration level, precision was estimated as the coefficient of variation (CV) in predicted RNA concentrations versus input RNA concentration level.

$$\text{CV} = \frac{s \text{ (Predicted RNA Concentrations)}}{\text{Mean (Predicted RNA Concentrations)}} \times 100\%$$

Analytical sensitivity of the assay expressed in terms of the assay limit of detection (LOD) and limit of quantitation (LOQ), were calculated for each individual gene in the panel under standard assay conditions. Results were quite consistent across the full assay panel with the sixteen test genes averaging a LOD C_T of 40 and a LOQ C_T of 38.4 while the five reference genes gave an average LOD C_T of 39.6 and a LOQ C_T of 38.3.

The dynamic range of expression for each gene was defined by the maximum range of C_T measurements extending from the LOQ for the specific gene that maintained acceptable amplification efficiency, accuracy and precision. If acceptance criteria are set to have a mean bias within ± 20% and coefficient of variation less than 20%, the test genes in the panel could be reliably measured over a dynamic range of about 0.1 to 8 ng of input RNA while the reference genes had a range of 0.03 to 8 ng of input RNA. The dynamic range would increased for most of these genes by another 2–3 $\log_2$ if acceptance criteria for accuracy and precision were relaxed to mean bias within ± 30% and coefficient of variation less than 40%; a range often considered typical for microarray data.

2.2.3.3. Reproducibility and Precision Reproducibility for this RT-PCR assay was evaluated at 2 ng of input RNA to measure how variable the RT and PCR reactions were as different operators assembled and ran them on different instruments over time. Precision was assessed by estimating between-day, between-run (within day), and within-run variability as well as total variability over the test period. Reproducibility was assessed by estimating differences in mean C_T's separately for each gene over the duration of the test period. Analyses were performed on both non-normalized and reference normalized C_T measurements. The combined largest variation ranged from 0.1 to 0.4 C_T (average = 0.28 C_T) and 0.1 to 0.4 C_T (average = 0.26 C_T) for the sixteen target genes and five reference genes, respectively. Overall, the differences in non-normalized C_T between instruments and between operators across the entire gene panel were small. The variability in the reference normalized C_T measurements were even smaller than the variability in the non-normalized C_T measurements. These results indicate that real-time RT-PCR analysis can be highly precise and reproducible.

2.2.4. Real-Time RT-PCR Summary

Based on these analyses, it can be concluded that quantitative RT-PCR is a fast, sensitive and accurate assay with a broad quantitative range in total RNA samples. Although RT-PCR is a very powerful and precise technique, achieving performance levels such as those described here requires careful optimization. Factors that have to be taken into consideration for RNA analysis include preventing co-amplification from genomic DNA, evaluating and choosing the optimal primer and probe designs, and identifying appropriate reference gene(s) for normalization. Co-amplifying genomic DNA in RT-PCR assays can be avoided in two ways. The first is to design primers and probes that generate amplicons spanning more than one exon. In this case, the additional length of an intervening intron separates PCR primers when they prime genomic DNA, which generally results in signal being detected only from the RNA target. Alternatively, a reliable "no RT" control assay may be developed for evaluating the contribution of residual genomic DNA in the RNA sample on RT-PCR performance and indicate when DNase treatment is required.

It is generally useful to evaluate more than one primer and probe set for each intended target to obtain one with optimal performance. Desirable characteristics include high amplification efficiency, absence of primer dimer or non-specific amplification products, high target sensitivity. Finally, if reference normalization is used for data analysis, it is crucial to confirm that the chosen reference gene(s) are stable in the experimental sample set. Once the factors discussed here are verified and all parameters are optimized, real-time RT-PCR is able to provide accurate, reproducible results while being capable of considerable sample throughput. The value of being able to fully characterize and optimize RT-PCR assay performance should not be underestimated in clinical applications.

2.3. MICROARRAYS

2.3.1. Technology Platforms

Microarray technology is based on the principles of nucleic acid hybridization. Since base pairing permits identification of complementary sequence within complex mixtures, single stranded labeled "probes" of known sequence were used to detect the presence of their complements in unknown samples. Adaptation of solution phase nucleic acid hybridization to solid supports was the direct precursor to present microarray technologies [20]. Early macroscopic hybridization methods used flexible, porous filters as the solid support to immobilize detection probes. The advent of solid supports such as glass and the miniaturization of immobilized nucleic acid features marked the arrival of the first microarrays. We will use the terminology established with the publication of a special supplement on microarrays [21] in which the immobilized strand of nucleic acid is referred to as the "probe" and the complementary solution phase strand is termed the "target".

Microarrays composed of three different types of probes are in common use: cDNA (purified PCR products of cDNA clones), pre-synthesized oligonucleotides, and in situ synthesized oligonucleotides. In the sections that follow, the properties of each of these microarray types are presented separately.

2.3.1.1. cDNA Arrays   cDNA arrays are typically made by "printing" PCR products generated from cloned cDNA libraries,, on specially prepared glass microscope slides [22]. This process involves PCR amplification using primers modified to enable covalent coupling of the amplicons to the derivatized substrate surface. These PCR products are generally hundreds to thousands of bases long and are immobilized on individual array features that are typically in the range of 80–200 μm in diameter. The number of features on an array can vary from under one hundred to more than 20,000. Usually, there is a one-to-one correspondence between features on the array and target molecules, i.e., each target sequence is complementary to only a single probe sequence on the array. The immobilized probes are double-stranded when printed onto the array and are subsequently denatured; therefore, they can be used for hybridization with either sense or antisense targets. The most common strategy for preparing a sample for cDNA array analysis is to reverse transcribe mRNA in the presence of one or more fluorescently labeled deoxynucleotide triphosphates to produce a fluorescently labeled cDNA target.

Detailed methods for microscale printing of nucleic acid probes onto array substrates have been published elsewhere [22–24]. Alternatively, cDNA arrays can be purchased from a few commercial suppliers. cDNA arrays have proven difficult to quality control since PCR products are often heterogeneous, are subject to contamination by other probes and often are only inefficiently coupled to array substrates. Many, but not all, investigators have discontinued use of cDNA arrays in favor of oligonucleotide arrays, which generally have a more carefully and accurately characterized probe composition that perform more uniformly. Among the reasons cDNA arrays continue to be popular, especially with academic investigators is that the technology and materials for making them is easily accessible and affordable. In addition, since the investigator can easily customize cDNA arrays, they offer a degree of flexibility that is not easily achieved with commercial arrays, particularly for investigators studying organisms whose expressed genome is not represented on commercial arrays.

2.3.1.2. Printed Oligonucleotide Microarrays Oligonucleotide microarrays are available with probes either "printed" or synthesized "in situ" on the array substrate surface. The latter are synthesized directly on a glass surface using proprietary chemical methods discussed in the next section. Printed oligonucleotide microarrays are made using techniques similar to those used for cDNA microarrays. An important property of printed oligonucleotide arrays is that the nucleic acid probes are pre-synthesized using standard phosphoramidite chemistry including the addition of a reactive group to permit covalent coupling to the array surface. An advantage that synthetic oligonucleotide probes have over PCR products is they can be synthesized with great homogeneity and chemical purity. Consequently, printed oligonucleotide probes are generally more uniform in length and sequence composition as compared even with in-situ synthesized oligonucleotide probes. The size and density of individual features is similar to that of cDNA arrays: up to 40,000 probe features per array that are 80–200 μm diameter. Probe sequences for oligonucleotide arrays are typically designed for optimal target specificity and uniform hybridization performance. Factors contributing to specificity and uniformity include the length of the probe (see below), selection of a sequence that does not have significant homology with other transcripts in the genome, similar probe length, and similar base composition. Therefore it should be sufficient to print a single feature on the array for hybridization to each target.

The length and sequence orientation of printed oligonucleotide arrays are determined by the manufacturer. Practical limits for oligonucleotide length are 25–80 bases. In theory, probes 30–80 nucleotides in length provide greater hybridization specificity than either cDNA [25]. or shorter oligonucleotide probes without compromising sensitivity. The probes deposited on commercially available microarrays from Agilent and MWG Biotech are single-stranded oligonucleotides, 50–70 bases in length, in the sense orientation and are suitable for hybridization with antisense cDNA or aRNA (amplified RNA, defined below) targets. Recently, Affymetrix, MWG Biotech, Agilent, and Applied Biosystems have announced the release of whole "human genome" microarrays for simultaneous monitoring of 40,000–61,000 transcripts.

2.3.1.3. In Situ Synthesized Oligonucleotide Microarrays   Affymetrix, Inc. pioneered the development of microarrays that contain oligonucleotide probes synthesized in situ on the array surface. Alternative methods for in situ synthesis of oligonucleotides were developed subsequently [26–29]; however, Affymetrix microarrays are the predominant platform in this category. Affymetrix GeneChip® microarrays are manufactured by a proprietary light-directed method [30]. The process is similar to the photolithographic method used for mass production of semiconductor chips. Oligonucleotide probes are synthesized in situ by a modification of the phosphoramidite method. The most recent generation of GeneChip microarrays have square features, 11 μm on each side, and the oligonucleotide probes are uniformly 25 bases in length. Twenty-two nucleic acid features are designated as a "probe set" for analysis of each target transcript to be detected. Eleven features in each probe set, called perfect match (or PM) features, are exact complements to the target sequence that hybridize to distinct, albeit sometimes overlapping, sequences of the target nucleic acid. Each PM feature has a paired mismatch (or MM) feature that is identical in sequence except for the middle nucleotide base which is substituted to cause a homonucleotide mismatch (A:A, C:C, G:G or T:T). Any signal generated from the MM feature is used in the computational algorithms as a measure of nonspecific hybridization to its paired PM feature. The recently released "Human Genome U133 Plus 2.0" array contains approximately 1,300,000 features in more than 54,000 probe sets representing 47,400 human transcripts.

In contrast to many cDNA and printed oligonucleotide arrays, specialized Affymetrix scanners are needed for analysis of GeneChip microarrays. The procedures for amplification and labeling RNA are similar to the methods used for printed oligonucleotide arrays but have been optimized for GeneChip arrays and scanners. The biggest difference in sample processing for Affymetrix arrays as compared with most printed oligonucleotide arrays is that only one sample is hybridized to each array; therefore, only one type of fluorophore (ie., one color) is needed. Dual sample hybridization, with differentially labeled samples, one serving as a reference and one as a test sample is a common approach used with printed arrays to compensate for variability in feature probe content and quality. The target amplification and labeling procedures for both types of experimental design are discussed below.

2.3.2. Target Amplification and Labeling

There are a variety of different methods available for amplifying and labeling nucleic acids that are to be hybridized to different types of microarrays. In addition, different microarray platforms recommend a variety of different approaches to experimental design and data analysis. In the sections that follow, we discuss the two general experimental designs that use either one or two fluorophores for labeling nucleic acid targets.

2.3.2.1. Single Fluorophore Experimental Designs   Affymetrix GeneChip microarrays were designed to be hybridized with an amplified mRNA (aRNA) target labeled with a single fluorophore. Labeled RNA offers two advantages as a hybridization sample. First, single stranded targets are more available for hybridization than denatured, double stranded targets such as PCR products, which tend to self anneal rather than hybridize with array probes. Secondly, RNA: DNA duplexes are more stable than DNA:DNA duplexes. Test samples to be compared with one another are labeled separately and hybridized to different

```
Sample A                                          Sample B
   ↓         Reverse Transcription,                  ↓
             Oligo dT-T7 primer

   ↓         In Vitro Transcription                  ↓
             (Amplification)
 aRNA_A                                           aRNA_B

   ↓         Hybridize on Arrays                     ↓

 Signal       Single Color                        Signal
(gene X_A)    Fluorescence                       (gene X_B)
      ↘                                          ↙
             Relative Expression = Fold-Change
```

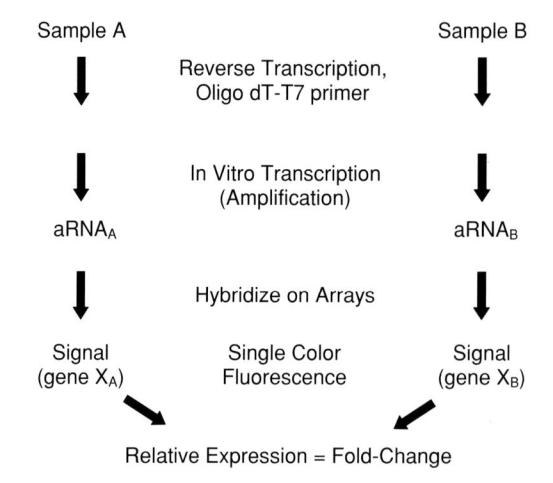

FIGURE 2.6. Single fluorophore experimental design.

microarrays as illustrated in Figure 2.6. The amplification procedure involves reverse transcription of purified RNA followed by in vitro transcription of cDNA using a strategy originally developed by Eberwine and coworkers [31,32]. Briefly, reverse transcription is primed using an oligo-dT primer tailed with a T7 RNA polymerase recognition sequence on the 3' end. A second cDNA strand is then synthesized by the method of Gubler and Hoffman [33]. In vitro transcription of the double stranded cDNA product with T7 RNA polymerase in the presence of biotinylated UTP results in the production of 100–1000 copies of amplified, biotin labeled RNA (aRNA) from each original mRNA target molecule. The biotinylated targets are hybridized to the array followed by detection of RNA: DNA duplexes with the fluorophore phycoerythrin conjugated to streptavidin and biotinylated goat anti-streptavidin antibody.

Affymetrix GeneChip microarrays require the use of a custom scanner that cannot scan other types of microarrays. Several types of controls are used in the GeneChip design to validate the various steps of a microarray experiment. Verification of microarray quality and image orientation is are accomplished by hybridization of a biotinylated control oligonucleotide to complementary probes on the microarray. Other biotinylated oligonucleotide controls are added to the hybridization mixture to qualify the performance and sensitivity of each microarray. Synthetic mRNA controls constructed from bacterial genes are added to RNA samples to qualify the amplification and fluorescent labeling steps of the procedure.

The computational algorithms used for processing the raw fluorescence data is beyond the scope of this review but is described in detail elsewhere (www.affymetrix.com). In addition to the software provided by Affymetrix, alternative computational algorithms for analyzing data from these arrays have been developed by independent investigators [34–36]. The most recent version of the software provided by Affymetrix computes both qualitative and quantitative metrics of gene expression. The qualitative results classify each transcript as either present, marginal, or absent in an individual sample and will not be considered further here. The quantitative measure of gene expression, termed the signal value, is computed for each transcript by combining the fluorescent intensity data obtained for the 22 probes in each probe set. Comparative analysis of the probe set data in different samples can also be done to calculate the relative expression or fold-change in expression of that gene. A variety of software packages are available from Affymetrix and other vendors for comparative analysis of more than two samples.

Comparison of samples hybridized on different microarrays requires consistency in the process for manufacturing the arrays as well as the use of the controls mentioned above for GeneChip microarrays to normalize for variation in the immobilized probes, target labeling reactions, and sample hybridization conditions. Similar controls are used with printed cDNA and oligonucleotide microarrays. Many of these sources of variation in printed microarray experiments are controlled by competitive hybridization of two samples that have been labeled with different fluorescent dyes (see below). However, as the methods for production of printed cDNA and oligonucleotide microarrays have improved it is also possible to use them with a single fluorophore experimental design.

2.3.2.2. Two Fluorophore Experimental Designs Most experimenters employ dual fluorophore experimental designs for use with cDNA and printed oligonucleotide microarrays. These designs involve competitive hybridization of two samples labeled separately with different fluorophores as illustrated in Figure 2.7. In this example, two different samples were amplified and labeled for co-hybridization to a cDNA microarray although the same strategy applies to printed oligonucleotide microarrays. Each sample is reverse transcribed and amplified independently following the "Eberwine" strategy discussed above for single fluorophore experiments. A variety of methods have been published for fluorescent labeling of nucleic acid targets [37]. In our laboratory, we have made labeled cDNA by reverse

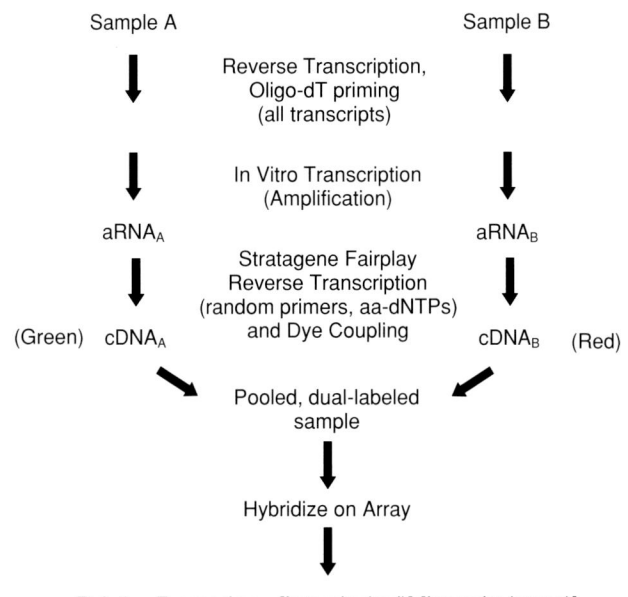

FIGURE 2.7. Two fluorophore experimental design.

transcribing amplified RNA in the presence of aminoallyl-nucleotide triphosphates using a commercially available kit (Stratagene, Inc.; FairPlay™ Microarray Labeling Kit) followed by reaction with N-hydroxysuccinimide esters of fluorescent dyes. Use of aminoallyl NTPs in the labeling reaction eliminates bias that can arise if samples are labeled using different fluorophore conjugated NTP's with different efficiencies of incorporation. In the example shown in Figure 2.2, the fluorophores are designated as green and red, the two colors that are commonly used for labeling microarray targets. The samples are mixed and hybridized to the microarray then the red and green fluorescence intensities are measured separately for each feature on the array. The primary measure of relative gene expression between the two samples is the ratio of fluorescence intensities at each array feature, sometimes referred to as the fold-change or fold-difference. A more useful data transformation is the base 2 logarithm of the intensity ratio, called the signal log ratio, which is symmetrical for increases and decreases of gene expression in one sample versus another [38].

As mentioned above, competitive hybridization of two samples on the same microarray controls for many potential sources of experimental variation. One drawback of dual fluorophore experimental designs; however, is that the results from multiple samples cannot be compared with one another unless they are all hybridized competitively against the same reference sample. To address this need, "universal standards" have been proposed and evaluated [39]. A commercially available mixture of RNA isolated from 10 human cell lines has been widely adopted for this purpose (Stratagene, Inc.).

2.3.3. Applications

Numerous studies have been published over the past five years in which microarrays have been used for gene expression profiling. The underlying theme of these studies is that expression data from multiple genes provides much more informative power than can be obtained from a single gene. A formidable challenge is the development of computational and statistical methods to analyze the large datasets generated by such studies. Consequently, the development of algorithms for "mining" data from microarray experiments has become a field of specialization [40]. Despite the continuing evolution of data mining algorithms, gene expression profiling has already revealed biological insights that have not been achieved by other methods. Two recent examples of such studies are discussed below.

In a landmark study, Golub and coworkers [41] used Affymetrix microarrays to profile the expression of 6817 genes in bone marrow samples from 38 patients with acute leukemia. The expression of approximately 1100 genes was found to correlate with the leukemia classification (based on a combination of morphological, histochemical, immunohistochemical, and cytogenetic analyses) as either acute lymphoblastic leukemia (ALL) or acute myeloid leukemia (AML). From this set of 1100 genes, a "class predictor" set of 50 informative genes was defined. The 50 gene class predictor was then tested on an independent group of 34 leukemia samples and was found to be 100% accurate in distinguishing between ALL and AML. Further "class discovery" analysis was able to further subclassify ALL into B-lineage ALL and T-lineage ALL. Although AML and ALL can be distinguished based on a combination of histological and cytogenetic criteria (see above), this was the first study to demonstrate that clinically relevant classifications were possible using only gene expression profiles.

Another study demonstrated that gene expression profiles could be used to predict future clinical outcome. Alizadeh and coworkers [42] designed a specialized cDNA microarray, named the "lymphochip", containing 17,856 clones derived from a B-cell library as well as other genes believed to be important in immunology or cancer. Lymphocyte samples were obtained at biopsy from patients with diffuse large B-cell lymphoma (DLBCL) prior to a regimen of standard multi-agent chemotherapy. Hierarchical clustering of the microarray data identified two new subtypes of DLBCL that correlated with long-term (8–10 year) patient survival, similar to the International Prognostic Indicator (an index that takes into account patient age, disease severity, disease location as well as other clinical parameters). Some patients defined as low risk based on the International Prognostic Indicator had gene expression subtypes that predicted substantially worse survival. Thus, gene expression profiling was able to predict poor prognosis in a subgroup that was not predicted by clinical criteria alone. Identification of these high-risk patients by gene expression profiling may, in the future, influence treatment decisions as well as patient selection in clinical trials of new therapeutics.

2.4. COMPARISON OF GENE EXPRESSION PROFILING METHODS

Validation of gene expression profiles generated using microarrays presents a challenge due to the high multiplicity of genes that are represented on a single array. It is not feasible to validate the results obtained using microarrays for thousands of genes by comparison with an independent method that measures expression only one gene at a time. Consequently, most investigators have compared their microarray results with data obtained using other technologies for only small numbers of genes.

Generalizations about the comparability of data obtained using different methods for gene expression profiling are complicated due to the use of different technology platforms and experimental design. It is generally accepted that RT-PCR is more sensitive, precise, and able to resolve smaller differences in gene expression than DNA microarrays. However, qualitative and quantitative comparisons of the same samples by different technology platforms have yielded varying results. A recent review concludes that the results of microarray experiments are only partially consistent with RT-PCR or Northern blot analyses in several laboratories [43]. The discussion below cites additional examples of comparisons between technology platforms as well as data generated in our laboratory.

2.4.1. Comparison of cDNA Arrays with Other Gene Expression Profiling Methods

In a study of the temporal response of fibroblasts to serum growth factors, Iyer and coworkers [44] were among the first to verify expression differences determined using microarrays by an independent method. Expression changes in five genes, determined using custom cDNA microarrays, were qualitatively very similar to expression profiles measured by real time RT-PCR. They found that fold-change values determined by real time RT-PCR were quantitatively larger than fold-change values determined by microarrays for four of the five genes; the fifth was a housekeeping gene whose expression was unchanged throughout the study period. The larger dynamic range of expression profiles generated by RT-PCR as compared with microarrays has been noted in subsequent studies. [45, 46] and may reflect the exponential amplification inherent in the former method as compared with the relatively compressed range of hybridization. In one study, expression changes measured by microarrays were biased towards underestimation of the changes measured by real time

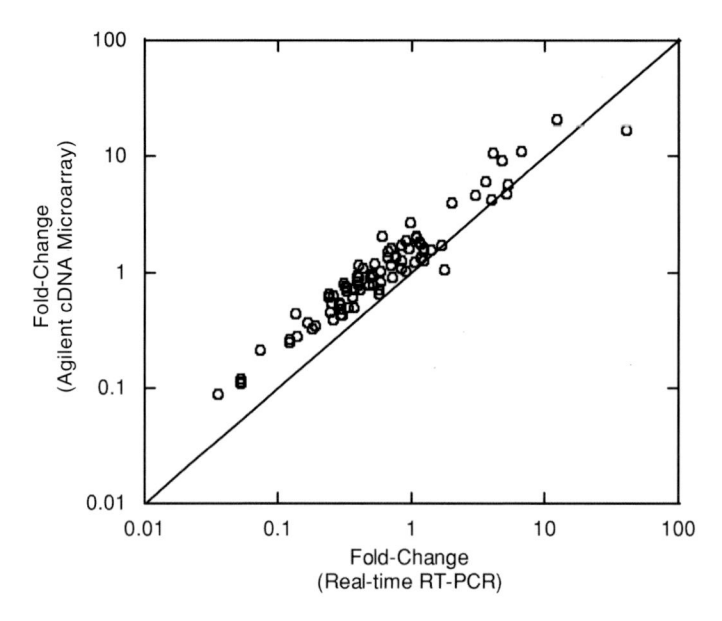

FIGURE 2.8. Comparison of gene expression profiles determined by cDNA microarrays and TaqMan RT-PCR.

RT-PCR, although the degree of bias for 47 individual transcripts was not predictable [46]. The Spearman rank correlation coefficient (our calculation from the supplemental data supplied by Yuen and coworkers) between expression fold-changes for all 47 genes measured by cDNA arrays and real time RT-PCR was 0.725.

Results obtained in our laboratory comparing cDNA microarray and TaqMan expression profiles were in general agreement with the studies cited above. The two samples we analyzed were different mixtures of RNA from various human tissues that we refer to as Human Reference RNA A and B. The relative expression ratio of 91 genes in the samples was determined by TaqMan RT-PCR and was compared with the expression ratio determined by Agilent cDNA microarrays as shown in Figure 2.8. Although the agreement for individual transcripts can vary quantitatively, an excellent correlation was observed between the rank orders of expression profiles determined by these two technology platforms (Spearman's $R = 0.943$). Similar to the results of Yuen and coworkers, we observed a systematic bias in the magnitude of expression ratios measured by microarrays as compared with RT-PCR. Overall, our results provide further support for the qualitative validity of expression differences measured by cDNA microarrays.

2.4.2. Comparison of Oligonucleotide Arrays with Other Gene Expression Profiling Methods

Published studies report varying degrees of correlation between expression profiles determined by oligonucleotide arrays and RT-PCR. Yuen and coworkers reported qualitatively similar expression profiles using Affymetrix GeneChip arrays and real time RT-PCR. However, as with cDNA microarrays, the magnitude of expression differences measured by oligonucleotide microarrays tended to underestimate the expression differences measured by real time RT-PCR [46]. The Spearman rank correlation coefficient (our calculation from the supplemental data supplied by Yuen and coworkers) between expression fold-changes for all 47 genes measured by Affymetrix arrays and real time RT-PCR was 0.683. As noted above for cDNA arrays, the degree of bias varied for individual genes. In another study, Van den Boom and coworkers [47] studied the expression profiles of different grades of astrocytomas using Affymetrix microarrays and compared their results for 12 genes determined by real-time RT-PCR. Eight pairs of gliomas were profiled by both methods. Correlation coefficients calculated for each of the 12 genes were in the range of 0.48–0.98. In contrast, Baum and coworkers measured the expression profile of 56 genes induced by osmotic shock in yeast [29] using a different in situ synthesized oligonucleotide microarray. They observed a Spearman rank correlation coefficient of 0.972 when their microarray results were compared with the expression profile for the same samples determined by real-time RT-PCR.

In our laboratory, we compared the expression profiles of the Human Reference RNA samples mentioned above as determined using Affymetrix microarrays and TaqMan real-time RT-PCR. Prior to hybridization, the RNA samples were carried through two cycles of amplification before labeling the final aRNA product. The original RNA samples were evaluated by RT-PCR for 180 genes. Comparison of the expression differences determined by the two methods produced a Spearman rank correlation coefficient of 0.911 although disagreement between the methods was observed for several individual genes as seen in Figure 2.9. In addition, the fold-change measured by microarrays was biased towards underestimation of the fold-change measured by real time RT-PCR.

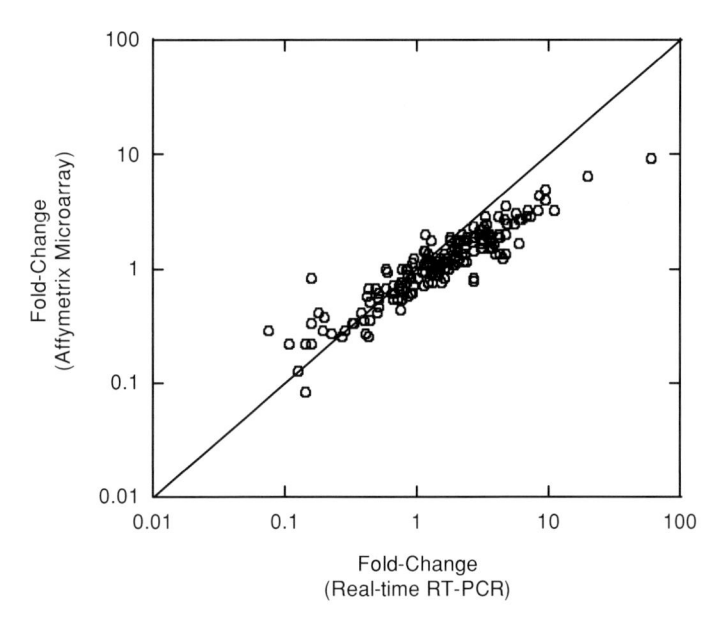

FIGURE 2.9. Comparison of gene expression profiles determined by Affymetrix oligonucleotide microarrays and TaqMan RT-PCR.

2.4.3. Comparison of cDNA and Oligonucleotide Microarray Expression Profiles

Recently, Lee and coworkers [48] proposed a method for validation of thousands of gene expression levels at a time by comparing results obtained with cDNA and oligonucleotide microarrays that are subject to different artifacts. In their study, several thousand transcripts were profiled in 60 human cancer cell lines (the NCI-60 panel) using both cDNA and Affymetrix GeneChip microarrays. The investigators posit that agreement of results obtained for a single transcript across many samples provides support for the validity of data obtained using both technology platforms. On the other hand, disagreement for an individual transcript does not indicate which (or if either) of the technology platforms generated a valid result. The correlation coefficients calculated for 2,344 Unigene-matched transcripts on the two microarray platforms were broadly distributed between -0.5 and 1.0. A consensus set of transcripts was identified that produced similar expression profiles on both cDNA and Affymetrix GeneChip microarrays. The observation by Lee and coworkers that expression profiles generated by cDNA and oligonucleotide microarrays are discordant for many transcripts supports the view that inaccuracies can arise from cross-hybridization, sequence variability of hybridization efficiency, as well as variability in the design, synthesis, manufacture of probes and target labeling. Other investigators have reached similar conclusions based on comparisons of expression profiles for the same samples generated on different microarray platforms [46, 49]. These studies demonstrate that one cannot compare expression differences identified using different microarray platforms without first cross-validating the methods for the specific genes of interest.

2.5. SUMMARY

The advent of technologies for expression profiling of multiple genes has launched a new era of biological research. Real time RT-PCR and DNA microarrays are among the most widely adopted methods employed in this new era. As originally developed, real-time RT-PCR and DNA microarrays were considered complementary technologies. Real-time RT-PCR is ideal for studies involving moderate numbers of genes (up to several hundred) in many biological specimens whereas DNA microarrays are better suited to analysis of many genes (tens of thousands) in fewer biological specimens. Given these characteristics DNA microarrays have more often been applied in the discovery phase of biological research with the aim of identifying the most informative genes. A relatively small number of genes, typically less than 50, can be identified whose differential expression is sufficient for the biological inquiry [50]. Once identified, the expression profile of this smaller set of informative genes can be screened with greater precision, better resolution, and more economically in a larger number of specimens by real-time RT-PCR. Concordance of results obtained from DNA microarray and real-time RT-PCR is critical if the former is used to identify smaller gene sets that will be screened subsequently by the latter technology. Existing data indicate that, although good overall correlations between technology platforms are possible, substantially different results can occur for individual genes. Thus, expression differences identified by microarrays must be verified if they are to be analyzed subsequently by another technology platform.

Microarray and real-time RT-PCR technology continues to evolve and improve. Increasingly, so-called "low density" microarrays targeting a small number of selected genes are being adopted in formats suitable for analyzing large numbers of specimens. Conversely, improvements in RT-PCR technology and instrumentation have enabled simultaneous analysis of larger gene sets. Thus, we expect that both of these technologies will be used in the future for gene "discovery" as well as for quantitative analysis of gene expression profiles.

ACKNOWLEDGEMENTS

John Morlan, Ken Hoyt, Debjani Dutta, Jennie Jeong, Anhthu Nguyen and Mylan Pho are thanked for providing microarray and TaqMan data. Chithra Sangli is thanked for assistance with statistical analysis and Joffre Baker is thanked for helpful discussions.

REFERENCES

[1] E.S. Lander et al. *Nature*, 409:860–921, 2001.
[2] J.C. Venter et al. *Science*, 291:1304–1351, 2001.
[3] L. Mancinelli, M. Cronin, W. Sadee. *AAPS. Pharm Sci.*, 2:E4, 2000.
[4] A.D. Roses. *Nature*, 405:857–865, 2000.
[5] R. Higuchi, C. Fockler, G. Dollinger, and R. Watson. *Biotechnology (N.Y.)*, 11:1026–1030, 1993.
[6] T.B. Morrison, J.J. Weis, and C.T. Wittwer. *Biotechniques*, 24:954–8, 960, 962, 1998.
[7] S. Nasarabadi, F. Milanovich, J. Richards, and P. Belgrader, *Biotechniques*, 27:1116–1118, 1999.
[8] J. Ju, C. Ruan, C.W. Fuller, A.N. Glazer, and R.A. Mathies. *Proc. Natl. Acad. Sci. U.S.A*, 92:4347–4351, 1995.
[9] S. Tyagi and F.R. Kramer. *Nat. Biotechnol.*, 14:303–308, 1996.
[10] D. Whitcombe, J. Theaker, S.P. Guy, T. Brown, and S. Little. *Nat. Biotechnol.*, 17:804–807, 1999.
[11] N. Thelwell, S. Millington, A. Solinas, J. Booth, and T. Brown. *Nucleic Acids Res.*, 28:3752–3761, 2000.
[12] A. Solinas et al. *Nucleic Acids Res.*, 29:E96, 2001.
[13] B.E. Caplin, R.P. Rasmussen, P.S. Bernard, and C.T. Wittwer. *Biochemica*, 1:5–8, 1999.
[14] E.A. Lukhtanov, I.V. Kutyavin, H.B. Gamper, R.B. Meyer, Jr. *Bioconjug. Chem.*, 6:418–426, 1995.
[15] I. Afonina, I. Kutyavin, E. Lukhtanov, R. B. Meyer, and H. Gamper. *Proc. Natl. Acad. Sci. U.S.A*, 93:3199–3204, 1996.
[16] I.V. Kutyavin et al. *Nucleic Acids Res.*, 28:655–661, 2000.
[17] I.A. Afonina, M.W. Reed, E. Lusby, I.G. Shishkina, and Y.S. Belousov. *Biotechniques*, 32:940–949, 2002.
[18] E.A. Lukhtanov, M.A. Podyminogin, I.V. Kutyavin, R.B. Meyer, and H.B. Gamper. *Nucleic Acids Res.*, 24:683–687, 1996.
[19] K.J. Livak and T.D. Schmittgen. *Methods*, 25:402–408, 2001.
[20] E. Southern, K. Mir, and M. Shchepinov. *Nat. Genet.*, 21:5–9, 1999.
[21] B. Phimister. *Nat. Genet.*, 21:1, 1999.
[22] M. Schena, D. Shalon, R.W. Davis, and P.O. Brown. *Science*, 270:467–470, 1995.
[23] M.B. Eisen and P.O. Brown. *Methods Enzymol.*, 303:179–205, 1999.
[24] V.G. Cheung et al. *Nat. Genet.*, 21:15–19, 1999.
[25] M.D. Kane et al. *Nucleic Acids Res.*, 28:4552–4557, 2000.
[26] M. Beier and J.D. Hoheisel. *Nucleic Acids Res.*, 28:E11, 2000.
[27] S. Singh-Gasson et al. *Nat. Biotechnol.*, 17:974–978, 1999.
[28] T.R. Hughes et al. *Nat. Biotechnol.*, 19:342–347, 2001.
[29] M. Baum et al. *Nucleic Acids Res.*, 31:e151, 2003.
[30] S.P. Fodor et al. *Science*, 251:767–773, 1991.
[31] R.N. Van Gelder et al. *Proc. Natl. Acad. Sci. U.S.A*, 87:1663–1667, 1990.
[32] J. Eberwine et al. *Proc. Natl. Acad. Sci. U.S.A*, 89:3010–3014, 1992.
[33] U. Gubler and B.J. Hoffman, *Gene*, 25:263–269, 1983.
[34] S. Dudoit, R.C. Gentleman, and J. Quackenbush. *Biotechniques*, Suppl: 45–51, 2003.

[35] R.A. Irizarry et al. *Nucleic Acids Res.*, 31:e15, 2003.
[36] M.A. Zapala et al. *Genome Biol.*, 3:SOFTWARE0001, 2002.
[37] A. Richter et al. *Biotechniques*, 33:620–8, 630, 2002.
[38] J. Quackenbush, *Nat. Genet.*, 32 Suppl:496–501, 2002.
[39] M.R. Weil, T. Macatee, and H.R. Garner. *Biotechniques*, 32:1310–1314, 2002.
[40] G. Stolovitzky. *Curr. Opin. Struct. Biol.*, 13:370–376, 2003.
[41] T.R. Golub et al. *Science*, 286:531–537, 1999.
[42] A.A. Alizadeh et al. *Nature*, 403:503–511, 2000.
[43] P.J. van der Spek, A. Kremer, L. Murry, and M.G. Walker, *Geno. Prot. and Bioinfo.*, 1:9–14, 2003.
[44] V.R. Iyer et al. *Science*, 283:83–87, 1999.
[45] M.S. Rajeevan, S.D. Vernon, N. Taysavang, and E.R. Unger, *J. Mol. Diagn.*, 3:26–31, 2001.
[46] T. Yuen, E. Wurmbach, R.L. Pfeffer, B.J. Ebersole, and S.C. Sealfon, *Nucleic Acids Res.*, 30:e48, 2002.
[47] B.J. van den et al. *Am. J. Pathol.*, 163:1033–1043, 2003.
[48] J.K. Lee et al. *Genome Biol.*, 4: R82, 2003.
[49] W.P. Kuo, T.K. Jenssen, A.J. Butte, L. Ohno-Machado, and I.S. Kohane, *Bioinformatics.*, 18:405–412, 2002.
[50] C.H. Chung, P.S. Bernard, and C.M. Perou. *Nat. Genet.*, 32 Suppl:533–540, 2002.

3

Microarray and Fluidic Chip for Extracellular Sensing

Mihrimah Ozkan[1], Cengiz S. Ozkan[2], Shalini Prasad[1],
Mo Yang[2], and Xuan Zhang[2]
[1]*Department of Electrical Engineering, University of California Riverside, Riverside CA 92521*
[2]*Department of Mechanical Engineering, University of California Riverside, Riverside CA 92521*

In the past two decades, the biological and medical fields have seen great advances in the development of biosensors and biochips capable of characterizing and quantifying biomolecules. Biosensors incorporate a biological sensing element that converts a change in an immediate environment to signals conducive for processing. Biosensors have been implemented for a number of applications ranging from environmental pollutant detection to defense monitoring. This chapter first provides an overview of the various types of biosensors and biochips that have been developed for biological applications, along with significant advances over the last several years in these technologies. It also describes the various classification schemes that can be used for categorizing the different biosensors and provides relevant examples of these classification schemes from recent literature. Finally it elucidates a sensing scheme based on cell based sensors. This technique is based on the development of single cell arrays that are used as biosensors that show parts per billion sensitivity and have the capability of identifying specific chemical analytes based on unique electrical identification tags also known as "Signature Patterns". The reliability of this technique is verified using conventional fluorescence based techniques.

3.1. INTRODUCTION

According to a recently proposed IUPAC definition [136], "A biosensor is a self-contained integrated device which is capable of providing specific quantitative or semi-quantitative analytical information using a biological recognition element (biochemical receptor) which is in direct spatial contact with a transducer element. A biosensor should be clearly distinguished from a bioanalytical system, which requires additional processing steps, such as reagent addition. Furthermore, a biosensor should be distinguished from a bioprobe which is either disposable after one measurement, i.e., single use, or unable to continuously monitor the analyte concentration". Biosensors that include transducers based on integrated circuit microchips are known as biochips [139].

Specificity and sensitivity are the main properties of any proposed biosensor. The first depends entirely on the inherent binding capabilities of the bioreceptor molecule whereas sensitivity will depend on both the nature of the biological element and the type of transducer used to detect this reaction [12]. In general, depending on the recognition properties of most biological components, two biosensor categories are recognized [48, 83, 119].

The first class of biosensors is the catalytic biosensors. These are also known as metabolism sensors and are kinetic devices based on the achievement of a steady-state concentration of a transducer-detectable species. The progress of the biocatalyzed reaction is related to the concentration of the analyte, which can be measured by monitoring the rate of formation of a product, the disappearance of a reactant, or the inhibition of the reaction. The biocatalyst can be an isolated enzyme, a microorganism, a sub cellular organelle, or a tissue slice. The second class of biosensors is the affinity biosensors. In these the receptor molecule binds the analyte "irreversibly" and non-catalytically. The binding event between the target molecule and the bioreceptor, for instance an antibody, a nucleic acid, or a hormone receptor, is the origin of a physicochemical change that will be measured by the transducer. Biosensor development is driven by the continuous need for simple, rapid, and continuous in-situ monitoring techniques in a broad range of areas. Biosensors can be classified according to either the nature of the bioreceptor element or the principle of operation of the transducer. The main types of transducer used in the development of biosensors can be divided into four groups [58] (1) optical, (2) electrochemical, (3) mass-sensitive, and (4) thermometric. Each group can be further subdivided into different categories, because of the broad spectrum of methods used to monitor analyte–receptor interactions.

The bioreceptor component can be classified into five groups [114].

(1) Enzymes, proteins that catalyze specific chemical reactions. These can be used in a purified form or be present in a microorganism or in a slice of intact tissue. The mechanisms of operation of these bioreceptors can involve: (a) conversion of the analyte into a sensor-detectable product, (b) detection of an analyte that acts as enzyme inhibitor or activator, or (c) evaluation of the modification of enzyme properties upon interaction with the analyte.
(2) Antibodies and antigens. An antigen is a molecule that triggers the immune response of an organism to produce an antibody, a glycoprotein produced by lymphocyte B cells which will specifically recognize the antigen that stimulated its production (Aga 1997).
(3) Nucleic acids. The recognition process is based on the complementary nature of the base pairs (adenine and thymine or cytosine and guanine) of adjacent strands in the double helix of DNA. These sensors are usually known as genosensors.

Alternatively, interaction of small pollutants with DNA can generate the recognition signal [140].
(4) Cellular structures or whole cells. The whole microorganism or a specific cellular component, for example a non-catalytic receptor protein, is used as the biorecognition element.
(5) Biomimetic receptors. Recognition is achieved by use of receptors, for instance, genetically engineered molecules [5, 10], artificial membranes [17], or molecularly imprinted polymers (MIP), that mimic a bioreceptor. The most recent investigations in artificial receptors include application of a combined approach of computer (molecular) modeling and MIP and the application of combinatorial synthesis for the development of new sensing layers.

Conventional methods for detecting environmental threats are primarily based on chemical, antibody- or nucleic acid-based assays. Biosensors incorporate a biological sensing element that converts a change in an immediate environment to signals conducive for processing. They generally rely on chemical properties or molecular recognition to identify a particular agent [98]. Modern approaches to biosensors can provide detection to a wide variety of analytes over a broad range of concentrations. However, as yet the current techniques involved in assessing risks to humans in areas contaminated with pollutants, pathogens or other agents are lacking technology. This is evidenced by our inability to: (1) simultaneously detect large numbers of possible threats, especially unknown or unanticipated ones; (2) characterize the functionality of known agents or analytes that have been identified using conventional techniques; and (3) predict human performance decrements caused by low levels of agents or synergistic effects of environmental toxicants.

In the past ten years, substantial progress has been made in the functional characterization of drugs, pathogens and toxicants using cultured biological cells. This has resulted due to a growing interest in the use of biosensors in environmental [115], medical [145] toxicological [2, 7] and defense applications [98]. For researchers involved in the development of next generation sensors, biosensors have two intriguing characteristics. First, biosensor recognition elements need to make use of sensing elements that have a naturally evolved selectivity to biological or biologically active analytes. Second, biosensors need to have the capacity to respond to analytes in a physiologically relevant manner. Biosensors have the potential of providing rapid, sensitive, low-cost measurement technology for monitoring bioavailable analyte concentrations. Major issues impeding the widespread acceptance of biosensor technology have been previously identified as stability and reproducibility [84]. A desirable characteristic for any biosensor implementation is the capacity for continuous monitoring or, at a minimum, multiple use ability with little sensor regeneration or renewal. Associated, but not exclusive, features of ideal biosensors include rapid response times, automation, and portability [98].

In all the classes of biosensors discussed so far with the exception of cellular and tissue based biosensors there are many shortcomings the major one being that they are highly specific to the recognition of a specific analyte. A certain degree of success in terms of high selectivity and rapid response times have been achieved in the other biosensor categories chiefly due to a series of revolutionary advances in cell and molecular biology and technologies that have allowed for in-depth molecular interrogation. However, there is still a major gap between our ability to perform functional assays using cells in the laboratory compared with in the field. This occurs due to the long response times associated with the molecular recognition elements, non-reusability of the detector performing functional assays and high specificity and excessive emphasis on ambient conditions for the accurate functioning of the sensors.

Cellular and tissue-based biosensors on the other hand incorporate isolated cells or tissue derived from a wide range of plant and animal sources. Also these biosensors use the cell as both the sensing as well as the transduction element. Thus elaborate signal transduction and coupling schemes can be eliminated and such sensors can be used in a diverse range of applications. Another added advantage of this category of sensors is that it is possible to monitor in-situ the physiological reaction to a particular chemical analayte along with simultaneous sensing. We elaborate on the sensing schemes using the other previously mentioned techniques of bioreceptor detection. We postulate the inherent drawbacks in each scheme and finally we elaborate upon the sensing technique based on detection and analysis of extra cellular potentials also known as "electrical sensing".

3.2. ANTIBODY BASED BIOSENSORS

Antibody-based technology takes advantage of specific interactions with antigenic regions of an analyte to achieve high selectivity. Antibody-based approaches that require additional reagents for each measurement fall into the category of traditional immunoassays such as enzyme linked immunosorbant assay (ELISA), colorimetric test strips, etc. There are several approaches for detecting antigen: antibody binding ranging from conventional optical and piezoelectric [129] to more complex methods involving antibody-modified ion channel switches [21]. Fiber optic immunosensors detect binding of an analyte via modulation of evanescent wave properties yielding rapid and specific detection [131, 132]. Fiber optic sensors are particularly suited for applications where a low number of positive reactions are expected (drug screening or environmental sampling) since they can be reused multiple times as long as antibody sites remain unoccupied. Once significant positive signals are measured, the consumable elements of the immunosensor must be replaced. Challenges limiting the application of immunosensors include: antibody manufacturability, inherent antibody instability, and limited reversibility of binding. Technologies exploiting genetically engineered antibodies, e.g., phage display, [56] have demonstrated the most promise in addressing limitations in traditional antibody production. In addition, specific protein-binding nucleic acid sequences (aptamers) have been developed which may be well suited for sensor applications that previously relied on antibodies [109]. Whereas antibodies are raised in the circulation under in vivo conditions, aptamers can be evolved in the actual test media of interest. With regard to reusability, successful regeneration of immunosensors by chemical elution has been demonstrated as feasible, but not particularly practical in high throughput applications. 109. Commercially available systems such as a resonant mirror-based biosensor (Lab Systems) and a surface plasmon resonsance biosensor (BIAcore) utilize an automated fluidic regeneration system to recycle binding surfaces. There have been reports of exploiting the natural reversibility of antibodies for "recycling" sensors; (Hanbury et al., 1997) however, most immunosensors cannot operate continuously.

3.3. NUCLEIC ACID BASED BIOSENSORS

Nucleic acid technology is based on the hybridization of known molecular DNA probes or sequences with complementary strands ina sample under test. Developments in sensors that exploit nucleic acid binding events (DNA sensors) have been generally limited to the antibody-based technology. Nucleic acid analysis in general requires extensive sample preparation, amplification, hybridization, and detection. In theory, nucleic acid analysis provides a higher degree of certainty than traditional antibody technologies because antibodies occasionally exhibit cross reactivity with antigens other than the analyte of interest. Near real time detection of hybridization events has been demonstrated in numerous optical [Pollard-Knight et al., 1997, 142] or electrochemical systems [24, 49, 85]. Further, the feasibility of sensor regeneration and reuse has been demonstrated in several optical-based systems, including both fiber optic [41, Piunno et al., 1992] and resonant mirror applications [142]. In practice, however, development of nucleic acid sensor systems has been hampered by the challenges presented in sample preparation. Nucleic acid isolation remains the limiting step for all of the state-of-the-art molecular analyses. Typically, cells must be mechanically or chemically disrupted and treated with enzymes to remove associated proteins before nucleic acid can be isolated for hybridization to specific probes. As a result, rapid and automatable isolation of nucleic acid is an area of intense development at present.

3.4. ION CHANNEL BIOSENSORS

Membrane ion channels are targets of a range of transmitters, toxins, and potential pharmaceutical agentsThe fact that many ion channels and receptors can be purified and reconstituted in black lipid membranes (BLMs) for studies of function and pharmacology [86] has spurred initial interest in the development of channel/receptor-based biosensors. [74, 96, 126] However, ion channels especially those pertaining to mammalian physiology, cannot be considered robust in BLMs or isolated membrane patches due to the well-known property of ion channel "rundown" or "washout" [116]. In the absence of integral intracellular machinery provided by cells needed to maintain function, ion channels typical of mammalian physiology presently do not constitute practical biosensors.

3.5. ENZYME BASED BIOSENSORS

Enzyme-based technology relies upon a natural specificity of given enzymatic protein to react biochemically with a target substrate or substrates. Like ion channels, there are many enzymes that participate in cellular signaling and, in some cases, are targeted by compounds associated with environmental toxicity. In the medical diagnostic field, several manufacturers have marketed biosensors for measurement of common blood chemistry components including glucose, urea, lactate, and creatinine. ([64], Foch-Anderson et al., 1997). In general, enzyme-based biosensors employ semipermeable membranes through which target analytes diffuse toward a solid-phase immobilized enzyme compartment. The major drawback with this type of sensors is that many enzymes are inherently unstable, thus necessitating packaging approaches to limit degradation of biosensor performance.

3.6. CELL BASED BIOSENSORS

Cell based biosensors on the other hand offer a broad spectrum detection capability. Moreover by using cells as the sensing elements provides the advantage of in-situ physiological monitoring along with analyte sensing and detection. A cell by itself encapsulates an array of molecular sensors. Receptors, channels, and enzymes that may be sensitive to an analyte are maintained in a physiologically stable manner by native cellular machinery. In contrast with antibody-based approaches, cell-based sensors are expected to respond optimally to functional, biologically active analytes. Cell-based biosensors have been implemented using microorganisms, particularly for environmental monitoring of pollutants. Sensors incorporating mammalian cells have a distinct advantage of responding in a manner that can offer insight into the physiological effect of an analyte. Several approaches for transduction of cell sensor signals are available these approaches include measures of cell fluorescence, metabolism, impedance, intracellular potentials, and extracellular potentials. Finally the technique of using extracellular sensing on single mammalian cells to detect specific chemical analytes is discussed in detail. The associated signal analyses that results in a unique identification tag associated with each cell type for a specific chemical agent also called "Signature Patterns" is described. The advantages of this technique in comparison to other cell based techniques namely stems from speed of response as well as accuracy and reliability in analyte identification. This is also verified with the conventional fluorescent techniques. The technique of chemical detection based on individually patterned cell's extracellular potential variations is also known as electrical sensing. The viability and the reliability of this technique are discussed.

3.7. CELLULAR MICROORGANISM BASED SENSORS

Microorganism pathways are activated by some analytes, such as pollutants. These pathways are involved in metabolism or nonspecific cell stress that result in the expression of one or more genes (Belkin et al., 1993) Immobilized yeast his one of the most commonly used sensor. It has been used in the detection of formaldehyde [69] and toxicity measurements of cholanic acids [13]. The changes in metabolism indicative of the analyte were detected via O_2 electrode measurements or extracellular acidification rates. One of the areas where microbial biosensors are widely used is in environmental treatment processes. This is done by detecting the biochemical oxygen on demand (BOD) [65, 66, 73, 133]. Most of the BOD sensors consist of a synthetic membrane with immobilised microorganisms as the biological recognition element. The bio-oxidation process is registered in most cases by means of a dissolved O_2 electrode. A wide variety of microorganisms have been screened during the construction of BOD sensors. The microbial strains selected are chosen for their ability to assimilate a suitable spectrum of substrates. BOD sensors based on a pure culture have the advantages of relatively good

Principle of cellular micro-organism based biosensor

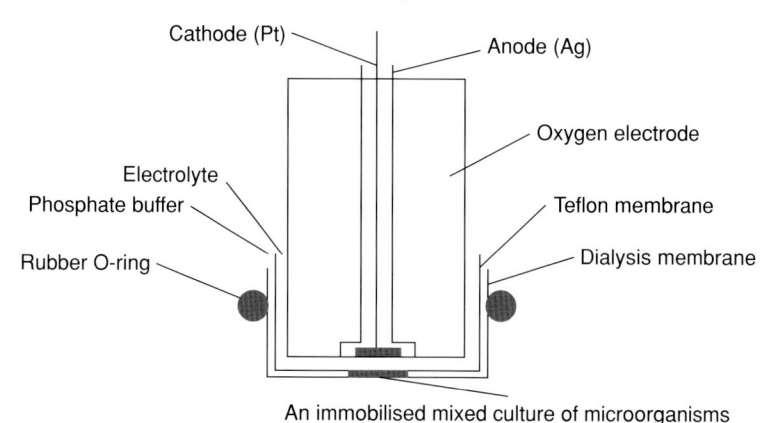

FIGURE 3.1. Schematic of the most commonly used cellular micro organism based sensor-Biochemical Oxygen Demand (BOD) sensor. An immobilized mixed culture of microorganisms in combination with a Clark-type oxygen electrode is used for analyzing the components of waste water. The BOD sensor is aimed at being highly capable of analyzing a sample of complex constituents with relatively low selectivity. A Clark-type probe for dissolved oxygen is used as the physical transducer, which consists of a platinum cathode as the working electrode, a silver anode as the reference electrode, and a 0.1 M potassium chloride (KCl) electrolyte. The Teflon side of the synthetic microbial membrane is attached to the cathode of the oxygen probe. The electrolyte is filled in the space between the synthetic biomembrane and those two electrodes [75].

stability and longer sensor lifetime, but are restricted by their limited detection capacity for a wide spectrum of substrates. The general schematic of a microbial biosensor is shown in figure 3.1 [75]. Modified bacteria have served as whole cell sensor elements for the detection of napthalene and its metabolic product salicylate [68] benzene [2] toluene [10] mercury [121] and middle chain alkanes such as octane.139. The touted advantage of the microorganism based sensors is that generic detection is possible as any alteration of a microorganism-based biosensor response is important and that insufficient selectivity actually offers the identification advantage [29]. The crux being that if generic detection is the ultimate goal then cell-based sensors that can give physiologically relevant information along with detection would offer a better advantage and capability.

3.8. FLUORESCENCE BASED CELL BIOSENSORS

Optical assays rely on absorbance, fluorescence or luminescence as read-outs. Instruments are available that can conveniently and rapidly measure light from standard 56 and 384 well micro titer plates. More customized systems have been developed to detect signals from very high density plates containing over a thousand individual wells. The migration to miniaturized assays (10 μL volume or less) and higher density formats favor non-invasive assay methods with the largest and brightest signals [72, 59]. Furthermore, genetically encoded probes offer the possibility of custom engineered biosensors for intracellular biochemistry, specifically localized targets, and protein-protein interactions. Fluorescence imaging has proven to be an invaluable tool for monitoring changes in the concentrations of ions and protein expression related to cellular signaling [23, 138].

Recently, fluorescence based technologies have been implemented in high-throughput screening [30]. New fluorescent reagents based on the combination of molecular biology, fluorescent probe chemistry, and protein chemistry is being developed for cell-based assays. Reporter gene constructs, such as green fluorescent protein, have been implemented in genetically engineered mammalian and nonmammalian cell types [150] to achieve measures of cell function rather than radioligand binding. (Giuliano et al., 1994). The fluorescent detection is generally based on fluorescence resonance energy transfer (FRET). The most basic use of fluorescent biosensors is the collection of photons from a cell or tissue to detect the occurrence of a process with temporal resolution.

Spatial information expands the usability of biosensors by adding subcellular and supracellular information. On the subcellular level, the read-out of the biosensor can be sampled with spatio-temporal resolution, enabling the morphological dissection of the studied process. An advantage of spatial resolution is the possibility to integrate data from different biosensors or other cell-state parameters to gain additional information—for example, on causal connections. The biological machinery inside cells can be investigated by various microscopic techniques and biosensors using fluorescent techniques [143]. The idea behind fluorescence detection is largely related to FRET. Figure 3.2 gives the concept of using FRET in a simple biosensor design consisting of a minimal protein

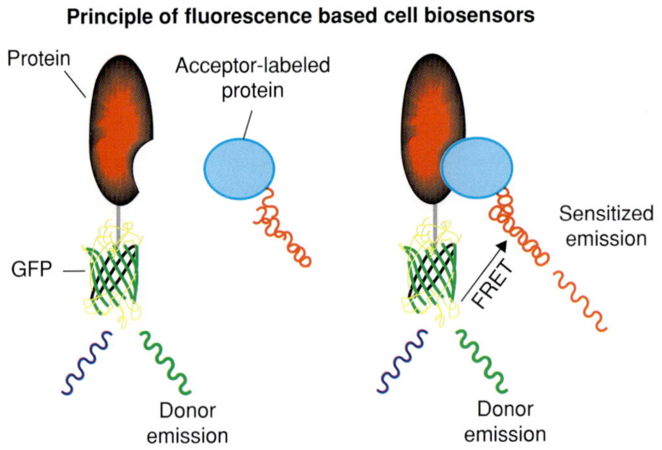

FIGURE 3.2. Fluorescence spectroscopy approaches, are progressively making their way into the field of cell biology. This novel development adds an aspect other than spatial resolution to microscopy—detection of protein activity in the cell. Due to the availability of an ever-increasing range of intrinsically fluorescent proteins that can be genetically fused to virtually any protein of interest, the area of their application as fluorescent biosensors has reached the inner workings of the living cell. The most basic use of fluorescent biosensors is the collection of photons from a cell or tissue to detect the occurrence of a process with temporal resolution. This can be achieved using fluorescence energy transfer techniques (FRET). The principle of FRET is shown in figure 3.2. FRET is a photophysical phenomenon where energy is transferred non-radiatively from a donor fluorophore to an acceptor fluorophore [147].

domain fused to green-fluorescent protein (GFP) that interacts specifically with molecules that are transiently generated at specific sites in cells. This allows monitoring of second messenger generation by imaging translocation of the fluorescent protein molecule [147]. In spite of the obvious utility of fluorescent techniques, there are three important considerations:

First, in mammalian cell systems, the cells that are readily transfected are those belonging to the tumor derived type as a result the cell types that can be detected using this technique is limited. Second, cell loading with fluorescent dyes is generally considered to be a potentially invasive technique. Third, analytes of interest must be examined for autofluorescence to determine the feasibility of cellular fluorescent assays for the resolution of small effects.

3.9. CELLULAR METABOLISM BASED BIOSENSORS

Another category of cellular biosensors relies on the measurement of energy metabolism, a common feature of all living cells. This is especially useful in testing drugs as in cancer research. The combined application of microfabrication technology to microfluidics have aided in the development of portable sensors. The changes in the cell metabolism due to the effect of a chemical reagent are transduced into electrical signals that are read out and analyzed. McConnel et al. 1992, developed and described a microsensor-based device, called the Cytosensor Microphysiometer [71] that was reported to be useful in the assessment of chemosensitivity of different human tumor cell lines [20, 26]. The instrument integrates up to eight channels, and detects sensitively and continuously the rate of extracellular acidification of cellular specimens. According to published results, it appears to be suited for clinical applications. Figure 3.3 shows the cross section of the

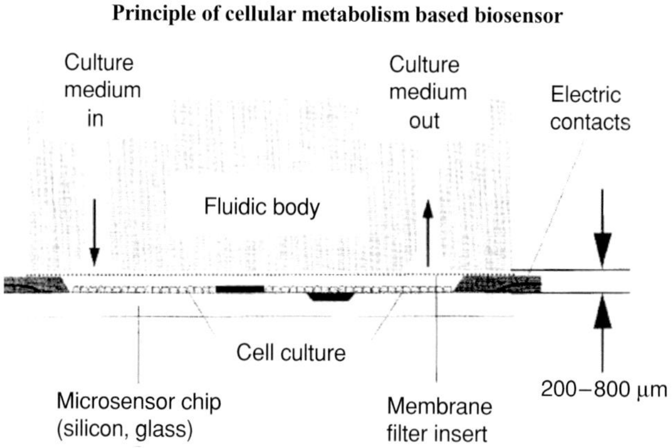

FIGURE 3.3. Two microsensor-based test systems for the dynamic analysis of cellular responses have been developed by the Henning group based on the principle of Cytosensor Microphysiometer®. One of them is equipped with transparent glass chips (GC), the other one with silicon chips (SC). The systems accommodate both adherent cell types and cell suspensions/tissue explants. The above schematic shows a cross-section of the chip and culture area found in both prototype versions [53].

general schematic of a morphological sensor used in cancer drug testing applications [53]. The disadvantage of this technique is that proper interpretation of data derived using this approach requires parallel experiments in the presence of known receptor antagonists that eliminate specific receptor responses.

3.10. IMPEDANCE BASED CELLULAR SENSORS

The membranes of biological materials including cells exhibit dielectric properties. By culturing cells over one or more electrode contacts, changes in the effective electrode impedance permits a noninvasive assay of cultured cell adhesion, spreading, and motility 41,98. By combining microfludics with microelectronics it has been shown that the physiological and morphological changes in primary mammalian cell cultures can be monitored [87]. It has been shown that cultures prepared from neonatal rat cerebral cortex were placed in a confined channel containing a balanced salt solution, and the electrical resistance of the channel was measured using an applied alternating current. If the volume of the cells increases, then the volume of the solution within the channel available for current flow decreases by the same amount, resulting in an increase in the measured resistance through the channel. If the volume of the cells decreases, a decrease in resistance would be recorded. This method allows continuous measurements of volume changes in real time [94]. Figure 3.4 shows the schematic of an impedance sensor. Electric cell—substrate impedance sensing (ECIS) is the technique that is used to monitor attachment and spreading of mammalian cells quantitatively and in real time. The method is based on measuring changes in AC impedance

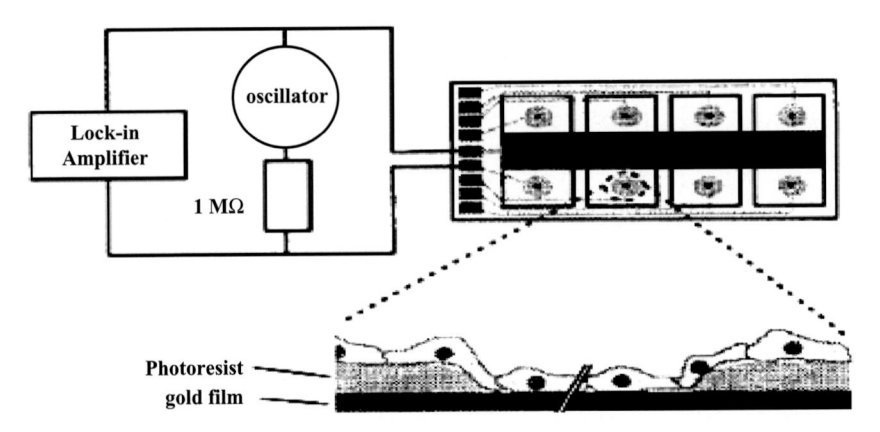

FIGURE 3.4. The schematic of an impedance sensor. Electric cell–substrate impedance sensing (ECIS) is the technique that is used to monitor attachment and spreading of mammalian cells quantitatively and in real time. The method is based on measuring changes in AC impedance of small gold-film electrodes deposited on a culture dish and used as growth substrate. The gold electrodes are immersed in the tissue culture medium. When cells attach and spread on the electrode, the measured electrical impedance changes because the cells constrain the current flow. This changing impedance is interpreted to reveal relevant information about cell behaviors, such as spreading, locomotion and motility. They involve the coordination of many biochemical events [38].

of small gold-film electrodes deposited on a culture dish and used as growth substrate. The gold electrodes are immersed in the tissue culture medium. When cells attach and spread on the electrode, the measured electrical impedance changes because the cells constrain the current flow. This changing impedance is interpreted to reveal relevant information about cell behaviors, such as spreading, locomotion and motility. They involve the coordination of many biochemical events. They are extremely sensitive to most external parameters such as temperature, pH, and a myriad of chemical compounds. The broad response to changes in the environment allows for this method to be used as a biosensor. The measurements are easily automated, and the general conditions of the cells can be monitored using a computer controlling the necessary automation (Keese and Giaever, 1986). Impedance techniques are theoretically capable of dynamic measurements of cellular movement at the nanometer level, a resolution above that of conventional time lapse microscopy [38]. Impedance measurements have been used to assess the effect of nitric oxide on endothelin-induced migration of endothelial cells (Noiri et al., 1994). From standpoint of biosensors, changes in cell migration or morphology tend to be somewhat slow; marked changes in impedance in the presence of cadmium emerged only after 2–3 h of exposure [25]. Thus for real time tracking and monitoring the effects of analytes a sensing technique based on impedance measurements would be slow and cumbersome.

3.11. INTRACELLULAR POTENTIAL BASED BIOSENSORS

An important aspect of the information that can be derived from cell-based biosensors relates to the functional or physiologic significance of the analyte to the organism. To this end, bioelectric phenomena, characteristic of excitable cells, have been used to relay functional information concerning cell status [43] Membrane excitability plays a key physiologic role in primary cells for the control of secretion and contraction, respectively. Thus, analytes that affect membrane excitability in excitable cells are expected to have profound effects on an organism. Furthermore, the nature of the changes in excitability can yield physiologic implications for the organism response to analytes. Direct monitoring of cell membrane potential can be achieved through the use of glass microelectrodes. Of particular interest was whether or not cell-based sensors could be used to rapidly detect chemical warfare agents such as VX and soman (GD). In bullfrog sympathetic ganglion neurons, both VX and soman have been shown to increase membrane excitability in a manner consistent with voltage-gated Ca^{2+} channel blockade [54, 55]. The basic principle behind intracellular measurements is that tissue slices are prepared and are exposed to chemical analytes under test and the electrical activity from excitable cells are measured using patch clamp technique. In this technique a giga ohm seal is created by inserting a microelectrode into the cell under study. Figure 3.5 shows electrical activity being measured from striatal cholinergic interneurons in young adult rats using recording pipettes. Figure 3.5A shows the control electrical activity, whereas figure 3.5B shows the electrical activity recorded from cells exposed to Ca^{2+} ions blockade agents [88]. This technique illustrates the utility of excitable cells as sensors with sensitivity to chemical warfare agents; however, the invasive nature of intracellular recording significantly limits the robustness of this approach for biosensor applications. Another drawback is that excitable cells assemble into coupled networks rather than acting as isolated elements; As a result, for certain sensing applications the ability to simultaneously

Principle of intracellular potential based biosensors

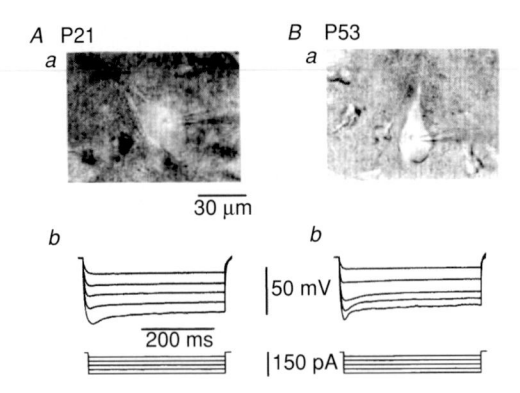

FIGURE 3.5. Measurement of extracellular potentials of striatial cholingeric internurons in young and adult rats. **5(A)** IR_DIC images of cholingeric interneurons in the dorsolateral striatium obtained from P....Aa. and P......rats. Ba. Recording pipettes were attached on the cell surface. Ab and Bb, membrane potential recorded fro neuronsin Aa and BA, respectively in response to applied chemical reagents namely magnesium chloroide (MgCl2). **5(B)** Voltage sags during hyperpolarization, characteristic of cholingeric neurons are indicated [88].

monitor two or more cells is essential as it permits measurements of membrane excitability and cell coupling. This is not possible using intracellular techniques. The advantage of the technique is that the physiological state can be assessed. Due to the invasiveness of the technique it is not possible to apply it for long term measurements. This can be rectified by using the exrtracellular potential as the sensing indicator. Extracellular sensing from excitable cells relies on microelectrode technology. This makes the technique non-invasive and in cases where the sensing from a network of cells is required the microelectrodes function as an array of sensing elements. Also long term sensing is possible using this technique.

3.12. EXTRACELLULAR POTENTIAL BASED BIOSENSORS

In recent years, the use of microfabricated extracellular electrodes to monitor electrical activity in cells has been used more frequently. Extracellular microelectrode arrays offer a noninvasive and long-term approach to the measurement of biopotentials [19]. Multielectrode arrays, typically consisting of 16–64 recording sites, present a tremendous conduit for data acquisition from networks of electrically active cells. The invasive nature of intracellular recording, as well as voltage-sensitive dyes, limits the utility of standard electrophysiological measurements and optical approaches. As a result, planar microelectrode arrays have emerged as a powerful tool for long term recording of network dynamics. Extracellular recordings have been achieved from dissociated cells as well; that is more useful in specific chemical agent sensing applications. The current microelectrode technology comprises of 96 microelectrodes fabricated using standard lithography techniques as shown in figure 3.6A

Principle of extra cellular potential based biosensors

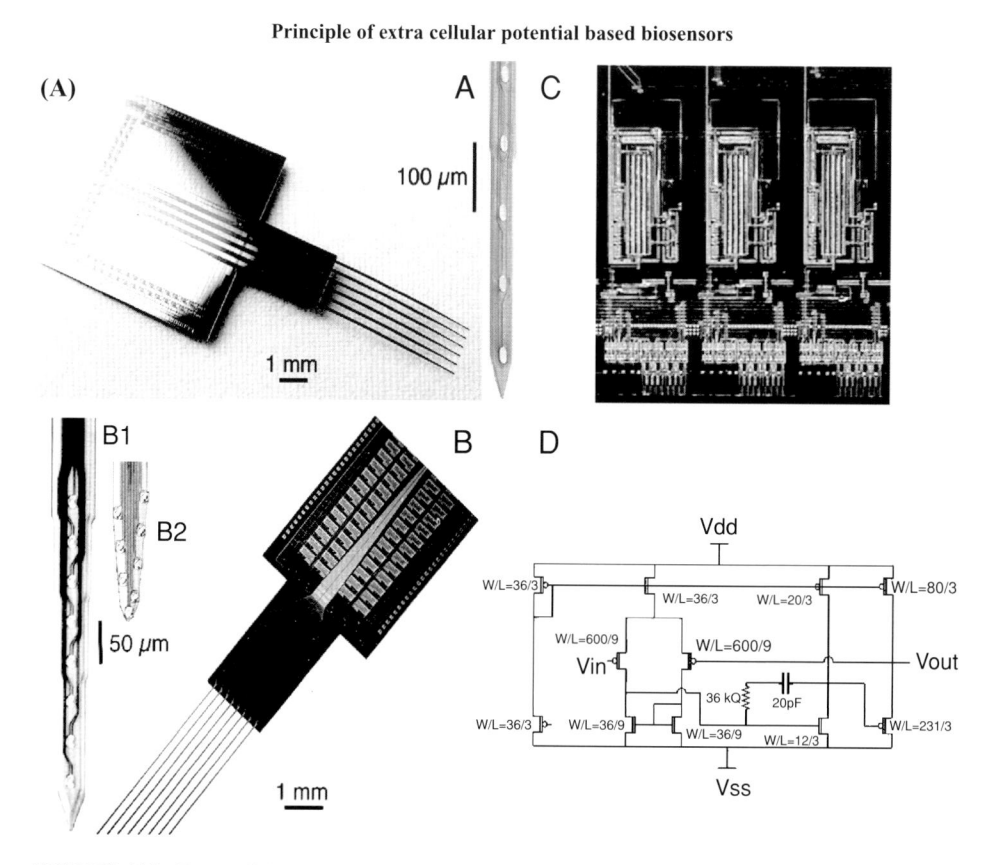

FIGURE 3.6(A). Extra cellular multiple-site recording probes. A: 6-shank, 96-site passive probe for 2-dimensional imaging of field activity. Recording sites (16 each; 100 μm vertical spacing) are shown at higher magnification. B: 8-shank, 64-site active probe. Two different recording site configurations (linear, B1 and staggered sites, B2) are shown as insets. C: close-up of on-chip buffering circuitry. Three of the 64 amplifiers and associated circuits are shown. D: circuit schematic of operational amplifier for buffering neural signals) [22].

[22]. More detailed work by Gross and colleagues at the University of North Texas over the past 20 yr have demonstrated the feasibility of neuronal networks for biosensor applications [43, 46]. In this work, transparent patterns of indium–tin–oxide conductors, 10 μm wide, were photoetched and passivated with a polysiloxane resin [42, 44]. Laser de-insulation of the resin resulted in 64 recording "craters" over an area of 1 mm^2, suitable for sampling the neuronal ensembles achieved in culture. Indeed, neurons cultured over microelectrode arrays have shown regular electrophysiological behavior and stable pharmacological sensitivity for over 9 months [45]. Figure 3.6B shows neuronal cultures on a 64 microelectrode array [99]. In fact, their precise methodological approach generates a co culture of glial support cells and randomly seeded neurons, resulting in spontaneous bioelectrical activity ranging from stochastic neuronal spiking to organized bursting and long-term oscillatory activity (Gross et al., 1994). Microelectrode arrays coupled with "turnkey" systems for

Principle of extra cellular potential based biosensors

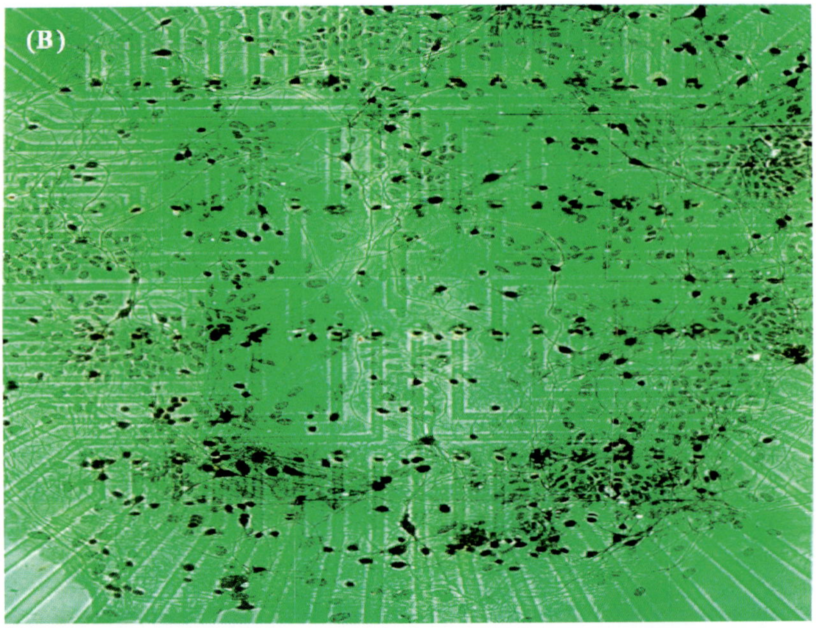

FIGURE 3.6(B). Neuronal cultures on a 64 microelectrode array. Laser de-insulation of the resin resulted in 64 recording "craters" over an area of 1 mm2, suitable for sampling the neuronal ensembles achieved in culture. neurons cultured over microelectrode arrays have shown regular electrophysiological behavior and stable pharmacological sensitivity for over 9 months [99].

signal processing and data acquisition are now commercially available. In spite of the obvious advantages of the microelectrode array technology for biosensing for determining the effect of chemical analytes at the single cell level it becomes essential to pattern the dissociated cells accurately over microelectrodes. Single cell sensing forms the basis for determining cellular sensitivity to wide range of chemical analytes and determining the cellular physiological changes. Also analyses of the extracellular electrical activity results in unique identification tags associated with cellular response to each specific chemical agent also known as "Signature Patterns".

3.13. CELL PATTERNING TECHNIQUES

There are three cell patterning methods that are currently in use. The first is a topographical method, which is based on the various microfabrication schemes involved in developing microstructures that enable the isolation and long-term containment of cells on the substrates [15, 76]. Other fabrication techniques used for cell patterning and the formation of ordered networks involves the development of the bio-microelectronic circuits, where the cell positioning sites function as field effect transistors (FET). This provides a non invasive interface between the cell and the microelectronic circuit [60, 95, 150]. These multi-electrode designs incorporating the topographical method have become increasingly complex, as the efficiency of cell patterning, has improved and hence fabrication has become more challenging and the devices are unsuitable for large-scale production. The other drawback is the need for an additional measurement electrode for determining the electrical activity from the electrically excitable cells. The second method is based on micro-contact printing (μCP) where simple photolithography techniques are coupled with the use of some growth permissive molecules (e.g., an aminosilane, laminin-derived synthetic peptide, Methacrylate and acrylamide polymers or poly-L-Lysine) that favor cell adhesion and growth and anti permissive molecules like fluorosilanes to form ordered cell networks [117, 118, Scholl et al., 2002, Wyrat et al., 2002, 146]. The disadvantage of this technique is the presence of multiple cells on a single patterned site that results in formation of a dense network of cell processes along the patterned areas. This in turn results in difficulties in measurement as well as determination of the electrical activity associated with a specific cell. The third method is based on using biocompatible silane elastomers like polydimethylsiloxane (PDMS). Cell arrays are formed using microfluidic patterning and cell growth is achieved through confinement within the PDMS structure. This technique is hybrid in the sense that it also incorporates μCP for promoting cell adhesion [47, 137]. The drawback of this technique is its complexity. As of today no single technique has been developed that (1) efficiently isolates and patterns individual cells onto single electrodes (2) provides simultaneous electrical and optical monitoring (3) achieves reliable on-site and non-invasive recordings using the same electrode array for both positioning as well as recording.

3.14. DIELECTROPHORESIS FOR CELL PATTERNING

The method that satisfies the above mentioned requirements for trapping cells is the use of dielectrophoretic forces [62, 80], it was determined that cells under the influence of low AC fields can be manipulated based on the variance in their dielctric properties and this process is termed as dielectrophoresis (DEP) [107]. Dielectrophoretic cell separation works on the principle of dielectrophoretic forces that are created on cells when a non-uniform electrical field interacts with the field induced electrical polarization on the cells. Depending on the dielectric properties of the cells relative to the suspending medium, these forces can be either positive or negative and can direct the cells toward strong or weak electrical fields respectively. Next to the (frequency-dependent) electrical properties of particle and medium the DEP force is determined by the particle dimensions and the gradient of the electric field. Field strengths between two and several hundred kV/m are required for trapping particles [35] shown that several types of living cells are capable to survive the rather high electric fields over longer periods of time up to two days. This was shown for red blood cells, mouse fibroblasts (3T3, L929), suspensor protoplast, bacteria, and yeast [3, 33, 80]. However, high temperatures, which can be caused by an electric field in a medium of high conductivity, can be disastrous [3, 36]. Previous research in the field of DEP has already shown that small particles and living cells can be manipulated by DEP [34, 80, 89, 104].

3.15. BASIS OF DIELECTROPHORESIS

The basic dielectrophoretic effect is demonstrated in figure 3.7, in which electrodes of spherical geometry are used to generate an inhomogeneous (non uniform) electric field. Many different electrode geometries have been used in DEP applications, but the spherical type is used generally as a model system because the non uniform field it generates can be described by a simple formula. The dipole moment m induced in a particle can be represented by the generation of equal and opposite charges $+q$ and $-q$ at the particle boundary. The magnitude of the induced charge q is small and is equivalent to 0.1% of the net surface charge normally carried by the cells and can be generated within a micro second [104]. If the applied electric field is non-uniform as shown in figure 3.7, the local electric field E_1 and the resulting force ($E_1 \delta q$) on each side of the particle will be different. Thus, depending on the relative polarizability of the particle with respect to the surrounding medium, it will be induced to move towards the inner electrode (from figure 3.7) or the region of high electric field (Positive DEP), or towards the outer electrode (from figure 3.7) where the field is weaker (negative DEP).

Following established theory [101] the DEP force FDEP acting on a spherical particle of radius r suspended in a fluid of absolute dielectric permittivity εm is given by equation 3.1

$$F_{DEP} = 2\pi r^3 \varepsilon_m \alpha \nabla E^2 \qquad (3.1)$$

where α is a parameter defining the effective polarizability of the particle and the factor

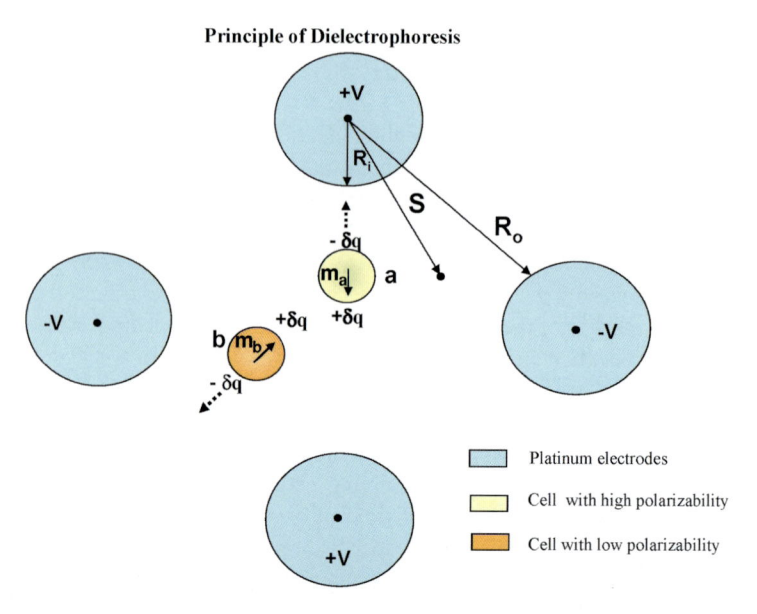

FIGURE 3.7. Principle of generation of positive and negative dielectrophoretic traps. The dielectrophoretic force is dependent on the polarizability of the particles and the surrounding medium and the applied root mean square voltage. Particle a experiences positive dielectrophoretic force and is trapped over the electrode. Particle b experiences negative dielectrophoretic force and is trapped in regions between the electrodes [102].

In experiments to demonstrate the DEP separation of viable and unviable yeast, the subsequent viability of the cells was checked by staining with methylene blue and plate counts [79]. The viabilities of erythrocytes dielectrophoretically separated from leukemia cells have also been confirmed with trypan blue dye (Becker et al., 1997), and CD34+ cells have been successfully cultured following DEP separation [127]. Fuhr et al. have also demonstrated that fibroblasts can be successfully cultivated without any significant changes to their viability, motility, anchorage, cell-cycle time, when exposed continuously over a period of 3 days to fields generate by types of microelectrodes used in DEP [34]. Even bacteria have shown no significant changes o their physiological characteristics after long exposures to DEP forces [79].

Considering the effect of DEP and the use of microfabrication technology, it is possible to integrate the two techniques to develop electrical sensors using mammalian cells as both the sensing as well as the transduction elements. To develop the technique of electrical sensing it first becomes essential to determine the parameters of DEP that are essential to isolate and position the individual cell types and then expose them to the chemical agents under test. The detection limit for a specific chemical associated with a specific cell type can be determined using this technique. The variation in the extracellular electrical signal is analyzed using fast fourier transformation (FFT) techniques that yield identification tags also known as "Signature Patterns". Finally the veracity of the technique is verified using conventional fluorescent chemistry methods.

3.19. CELL TYPES AND THE PARAMETERS FOR DIELECTROPHORETIC PATTERNING

In our experiments we have used two types of mammalian cells that have electrically excitable cell membranes. The first are the rat hippocampal cells obtained from the H19-7 cell line from ATCC. The second is the primary rat osteoblast culture. The parameters for DEP isolation and positioning are determined in each case and are summed up in table 3.1. A gradient AC field is set up among the electrodes by adjusting the parameters of the applied frequency, peak-to-peak voltage and conductivity of the separation buffer. Cells under the absence of an electric field have a uniformly distributed negative charge along the membrane surface. On applying the gradient AC field a dipole is induced based on the cell's

TABLE 3.1. Parameters for positive and negative DEP for neurons and osteoblasts

Cell Type	Separation buffer for DEP	Conductivity of buffer solution (m S/cm)	Positive DEP frequency	Negative DEP frequency	Cross over frequency	V_{pp} (Volts)
Neurons	250 m M Sucrose/ 1640 RPMI	1.2	4.6 MHz	300 kHz	500 kHz	8
Osteoblasts	250m M Sucrose/ Dubecco's modified Eagele Medium (DMEM)	6.07	1.2 MHz	75 kHz	120 kHz	2

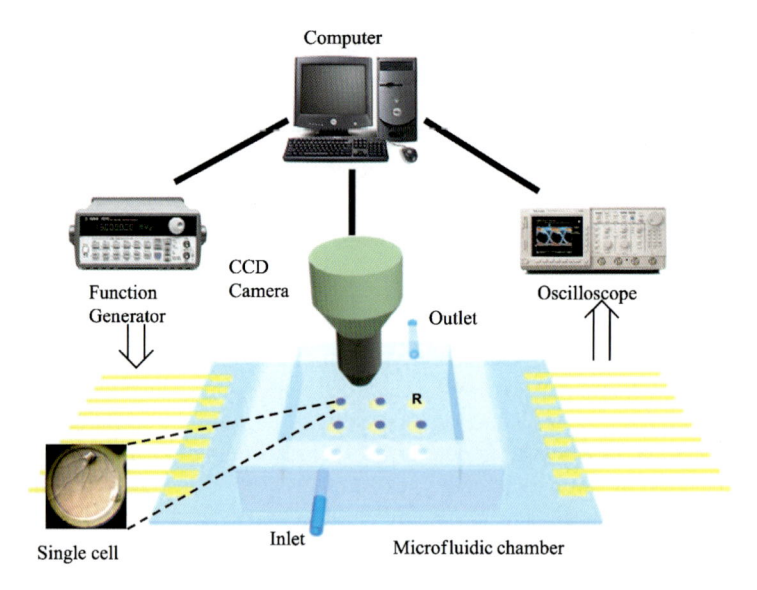

FIGURE 3.8. Schematic diagram of the integrated recording/measurement system. Single cell arrays are formed over the platinum electrodes/ sensing sites by setting up positive dielectrophoretic traps over the sensing sites. Gradient electric fields are set up over the sensing platform using a function generator. The chemical analyte mixture is circulated over the sensing platform via the microfluidic chamber using a pumping system. Electrical monitoring of the sensing sites is achieved by connecting micromanipulator probes to the electrode leads emerging from each sensing site. The extracellular electrical activity is measured with respect to a reference electrode denoted by R. The electrical activity is monitored using an oscilloscope. The activity is recorded using LabVIEW. Simultaneous optical monitoring is achieved through a CCD camera [111].

dielectric properties and due to the non uniform electric field distribution the electrically excitable cells experience a positive dielectrophoretic force that causes their migration to the electrodes, which are the regions of high electric fields [110]. In this manner neurons and osteoblasts are isolated and positioned over electrodes.

3.20. BIOSENSING SYSTEM

The biosensing system comprises of a chip assembly and an environmental chamber to maintain a stable local environment for accurate data acquisition. The biosensing system is schematically represented in figure 3.8.

3.21. CHIP ASSEMBLY

The microelectrode array fabrication and cell patterning has been achieved using a previous procedure [110] and is now briefly described. A 5 × 5 microelectrode array comprising of platinum electrodes (diameter: 80μm, center-to-center spacing: 200μm) spanning a surface area of $0.88 \times 0.88mm^2$ on a silicon/silicon nitride substrate with electrode leads (6 μm thick) terminating at electrode pads (100 μm × 120 μm) is fabricated using standard lithography techniques. To achieve a stable local microenvironment the microelectrode array is integrated with a silicone chamber (16mm × 16mm × 2.5mm) with a microfluidic channel (50 μm, wide), to pump in the testing agent and pump out the test buffer once the sensing process has been completed. The flow rate of the buffer is 40μL/min. The silicone chamber is provided with an opening (8mm × 8mm × 2.5mm) and is covered by a glass cover slip for in-situ monitoring. Simultaneous electrical and optical monitoring is achieved by using a Microzoom™(Nyoptics Inc, Danville, CA) optical probe station under 8× and 25× magnification. The Electrical stimulation and measurements are achieved utilizing micromanipulators (Signatone, Gilroy, CA).

3.22. ENVIRONMENTAL CHAMBER

The optical probe station along with the chip assembly is enclosed by an acrylic chamber (S&W Plastics, Riverside, CA). The environment in the chamber is controlled so as to maintain a constant temperature of 37°C. A heat gun (McMaster, Santa Fe Springs, CA) inside the chamber heats the air in the chamber and this is linked to a temperature controller (Cole Parmer, Vernon Hills, Illinois) that stops the heat gun from functioning above the desired temperature. A 6" fan (McMaster, Santa Fe Springs, CA) inside the chamber circulates the hot air to maintain temperature uniformity throughout the chamber and is monitored by a J-type thermocouple probe attached to the temperature controller. The carbon dioxide concentration inside the chamber is maintained at 5% and is humidified to prevent excessive evaporation of the medium. This chamber with all of its components will ensure cell viability over long periods of time and stable cell physiology in the absence of the chemical agents.

3.23. EXPERIMENTAL MEASUREMENT SYSTEM

The measurement system comprises of extracellular positioning, stimulating and recording units. The cells were isolated and positioned over single electrodes by setting up a gradient AC field using an extracellular positioning system comprising of a pulse generator (HP 33120A) and micromanipulators (Signatone, Gilroy, CA). The signal from the pulse generator was fed to the electrode pads of the selected electrodes using the micromanipulators. The extracellular recordings from the individual osteoblasts obtained from the electrode pads were amplified and recorded on an oscilloscope (HP 54600B, 100 MHz). The supply and measurement systems are integrated using general purpose interface bus (GPIB) control and controlled through LabVIEW (National Instruments, Austin, TX).

3.24. CELL CULTURE

3.24.1. Neuron Culture

The H19-7 cell line is derived from hippocampi dissected from embryonic day 17 (E17) Holtzman rat embryos and immortalized by retroviral transduction of temperature sensitive tsA58 SV40 large T antigen. H19-7 cells grow at the permissive temperature

(34°C) in epidermal growth factor or serum. They differentiate to a neuronal phenotype at the nonpermissive temperature (39C) when induced by basic fibroblast growth factor (bFGF) in N2 medium (DMEM-high glucose medium with supplements). H19-7/IGF-IR cells are established by infecting H19-7 cells with a retroviral vector expressing the human type I insulin-like growth factor receptor (IGF-IR). The cells are selected in medium containing puromycin.H19-7/IGF-IR cells express the IGF-IR protein. IGF-IR is known to send two seemingly contradictory signals inducing either cell proliferation or cell differentiation, depending on cell type and/or conditions. At 39°C, expression of the human IGF-IR in H19-7 cells induces an insulin-like growth factor (IGF) I dependent differentiation. The cells extend neuritis and show increased expression of NF68. This cell line does not express detectable levels of the SV40 T antigen.Following spin at $100 \times g$ for 10 minutes at room temperature; cells were re-suspended in a separation buffer (see Table 3.1). The density of the re-suspended cells (2500 cell/mL) ensured single cell positioning over individual electrodes. Separation buffer used for neurons contained 250 mM sucrose/1640 RPMI (Roswell Park Memorial Institute), with a conductivity of 1.2 mS/cm and a pH of 7.48. The separation buffer was replaced by a buffer comprising of minimum essential medium/10% Fetal Bovine Serum (FBS)/5% Phosphate buffer saline (PBS) of conductivity 2.48 mS/cm and pH of 7.4 suitable for cell viability.

3.24.2. Primary Osteoblast Culture

Primary rat osteoblast cells are cultured to a concentration of 2,500 cells in 1mL for sensing experiments. To achieve the patterning of a single cell over a single electrode, a 10 μL of cell culture solution was mixed with 500 μL Dulbeco modified eagle medium (DMEM; Gibco, Grand Island NY) supplemented with 10% fetal bovine serum (FBS; Gibco, Grand Island NY), 100μg/mL penicillin, and 100gμg/mL streptomycin (P/S; Gibco, Grand Island NY). The cells were centrifuged and re-suspended in 1mL of separation buffer consisting of 1:9 dilutions of Phosphate Buffer Saline/250mM Sucrose (Sigma, St Louis) and de-ionized water (w/v). The conductivity of the separation buffer was 4.09mS/cm and with a pH of 7.5. The separation buffer is replaced with a test buffer ((DMEM)/ Fetal Bovine Serum (FBS)/Phosphate Buffer Saline (PBS)) with conductivity of 2.5 mS/cm and a pH of 7.4.

3.25. SIGNAL PROCESSING

Changes in the extracellular potential shape have been used to monitor the cellular response to the action of environmental agents and toxins. The extracellular electrical activities of a single osteoblast cell are recorded both in the presence and absence of chemical agents and the modulation in the electrical activity is determined. However, the complexity of this signal makes interpretation of the cellular response to a specific chemical agent difficult to interpret. It is essential to characterize the signal both in time domain and frequency domain for extracting the relevant functional information.

The use of power spectral density analysis as a tool for classifying the action of a chemically active agent was investigated and found to offer a suitable technique for data analysis. The power of the extracellular potential is a better indicator of general shape than the peak-to-peak amplitude monitored in this work.

Spatial and temporal modifications to the extracellular potential have been used to monitor the cellular response to the action of chemical and biological analytes. The extracellular electrical activity from single neurons and osteoblasts are recorded from the sensing sites both in the presence and absence of the chemical analytes under study. The modulation in the electrical activity is then determined in each case. However, the complexity of this signal makes interpretation of the cellular physiological response to a specific chemical analyte difficult to interpret. Hence there arises a requirement for signal characterization in both the time domain and frequency domain for extracting the pertinent functional information. Power spectral density analysis of the acquired data was found to be a suitable and reliable tool for data analysis. The power of the extracellular potential is a better indicator of general shape and variations due to the effect of the chemical analyte than the monitored peak-to-peak amplitude of the signal. This is achieved by examining the RMS power in different frequency bands.

Using Fast Fourier Transformation (FFT) analysis, the shifts in the signal's power spectrum are analyzed. The FFT analysis indicates the modulation in the frequency of the extracellular potential burst rate and hence is termed as "frequency modulation" and generates the signature pattern vector (SP) (Yang et al., 2003). This SP is unique to the chemical analyte and the cell type.

3.26. SELECTION OF CHEMICAL AGENTS

To obtain the effect of a broad spectrum of chemical agents ranging from highly toxic and physiologically damaging to relatively less toxic to determine and evaluate the time window of response of a particular cell type for a specific known agent based on varying concentrations and finally determines the limit of detection for a specific chemical agent. All the experiments were conducted based on the hypothesis that a unique SP would be generated for each cell type for a specific chemical. This was hypothesized as it has been scientifically proven that different chemicals bind to different ion channel receptors thus, modifying the electrical response of the cell in a unique manner. We present here the responses of single excitable cells to the effect of the following chemical agents: ethanol, hydrogen peroxide, ethylene diamene tetra acetic acid (EDTA), and pyrethroid.

3.26.1. Ethanol

Ethanol produces anesthetic effects but in a milder form as compared to pentobarbitone and ketamine, though the mechanism of action is essentially assumed to be the same [123]. We hypothesized that determination of single cell ethanol sensitivity would help us identify the lowest threshold concentration, for the family of chemicals whose physiological response mechanism would mimic that of ethanol.

3.26.2. Hydrogen Peroxide

It is one of the major metabolically active oxidants present in the body and leads to apoptosis. Hydrogen peroxide also leads to the degradation of cells. As the behavior of hydrogen peroxide *in-vivo* is similar to the behavioral responses obtained from exposure to carcinogenic chemicals like rotenone, we estimated that hydrogen peroxide would make an ideal candidate for sensing studies [39].

3.26.3. Pyrethroid

They are active ingredients in most of the commercially used pesticides. The pyrethroid share similar modes of action, resembling that of DDT. Pyrethroid is expected to produce a "knock down" effect *in-vivo*; the exact *in-vitro* response at a cellular level has not yet been understood. Hence they are ideal candidates for the analysis of this genre of chemicals.

3.26.4. Ethylene Diamene Tetra Acetic Acid (EDTA)

EDTA belongs to a class of synthetic, phosphate-alternative compounds that are not readily biodegradable and once introduced into the general environment can re-dissolve toxic heavy metals. Target specificity of EDTA in a single electrically excitable cell has not been electrically analyzed to date.

3.27. CHEMICAL AGENT SENSING

3.27.1. Signature Pattern for Control Experiments

In order to determine the signature pattern vector corresponding to a specific chemical, the initial activity pattern vector for each cell type was determined. Using the process of Dielectrophoresis, a single cell was positioned over a single electrode and its initial electrical activity was recorded. As the first stage, control experiments were performed in which single osteoblasts and neurons were exposed to the sensing buffer in the absence of chemical agents, and the extracellular signal was recorded and analyzed to generate the initial (background) SP pertaining to osteoblast's and neuron's characteristic burst rate depending on its physiological condition. FFT analysis extracted the characteristic burst frequency from the firing pattern. The characteristic burst frequency was determined to be at 668 Hz for a single osteoblast and 626 Hz for a single neuron. This corresponds with neuronal electrical activity determined from other topographical methods [63, 76].

3.28. ELECTRICAL SENSING CYCLE

Chemical agents are first premixed individually with the sensing buffer and introduced into the sensor system. The modified electrical activity due to presence of chemical agents was recorded. Testing of a specific chemical agent was performed in a cyclic manner with each cycle comprising of three phases. The time duration of each phase was on an average of 60 seconds. The data presented here is averaged over fifteen cycles (n = 15). The action of each chemical agent at decrementing concentration ranges (step size in the higher concentration range: 500 ppm, lower concentration range (<1000 ppm): 50 ppm) was determined by monitoring the electrical activity at 5 seconds intervals for the first 30 seconds and then at 30 seconds intervals over a period of 180 seconds. This constitutes a single sensing cycle. In the presence of each specific chemical agent (ethanol concentrations ranging from 5000 ppm to 5 ppm for neurons and 5000 ppm to 15 ppm for osteoblasts, hydrogen peroxide: 5000 ppm to 10 ppm for neurons and 5000 ppm to 20 ppm for osteoblasts, pyrethroid: 5000 ppm to 250 ppb for neurons and 5000 ppm to 850 ppb for osteoblasts, EDTA: 5000 ppm to 150 ppm for neurons and 5000 ppm to 250 ppm for osteoblasts), pronounced modifications in the extracellular action potentials were observed. The detection limits for a single neuron was—ethanol: 9 ppm, hydrogen peroxide: 19 ppm, pyrethroid: 280 ppb, and EDTA: 180 ppm. Similarly the detection limits for a single osteoblast was- ethanol: 19 ppm, hydrogen peroxide: 25 ppm, pyrethroid: 890 ppb and EDTA: 280 ppm. The lowest single neuron sensitivity as estimated theoretically by the existing methods of averaging and iteration indicate the lowest concentrations determined that are—ethanol (MW = 46.07): 25 ppm (Maldve et al., 2002) as compared to the experimentally obtained detection limit of 9ppm hydrogen peroxide (MW = 34.01): 45 ppm as compared to 19 ppm [9], pyrethroid (MW = 38.3): 550 ppb as compared to 280 ppb [144], EDTA (MW = 292.2): 300 ppm to 180 M [130].

3.29. ETHANOL SENSING

3.29.1. Single Neuron Sensing

The initial concentration of ethanol used was 5000ppm and the modified electrical activity was recorded. The concentration of ethanol was decremented in a stepwise manner and in each case the modified electrical activity was recorded. The lowest concentration of ethanol sensed by a single neuron was 9ppm. The SP corresponding to the response of a single neuron to a specific agent in the frequency domain is represented in figure 3.9(A). The obtained SP is unique to a specific chemical agent and remains unchanged for varying concentrations of the specific agent. The SP obtained from a single neuron in the absence of a chemical agent indicates the initial control characteristic burst rate of 626 Hz. Addition of ethanol leads to its binding to M1 and M2 regions on the outside face of the $GABA_A$ and Glycine receptor gated Cl^- ion channels [77]. This increases the duration of the channel openings causing a strong inhibitory ionic current associated with Cl^- influx and decreased the frequency of firing to 314 Hz (Fig 3.9(A)).

3.29.2. Single Osteoblast Sensing

Single osteoblast cells were positioned over individual electrodes. The sensing agent was then introduced onto the microelectrode array using the microfluidic inlet channel. The initial concentration of ethanol used was 5000ppm and the modified electrical activity was recorded. The concentration of ethanol was decremented in a stepwise manner and in each case the modified electrical activity was recorded. The lowest concentration of ethanol sensed by a single osteoblast was 19 ppm. FFT analysis was performed on the acquired data pertaining to the modified extracellular potential to yield the SP. The instant at which the chemical is added to the chip system is denoted by t = 0sec. Figure 3.9(B) represents the SP for a single osteoblast due to the action of ethanol at 19ppm. Osteoblasts have an unmodulated firing rate of 668Hz. This corresponds to the frequency of firing of osteoblast in the absence of a chemical agent. There are two eigen vectors (514Hz, and 722Hz) in the

Neuron-Ethanol Frequency Spectrum

Concentration: 9ppm

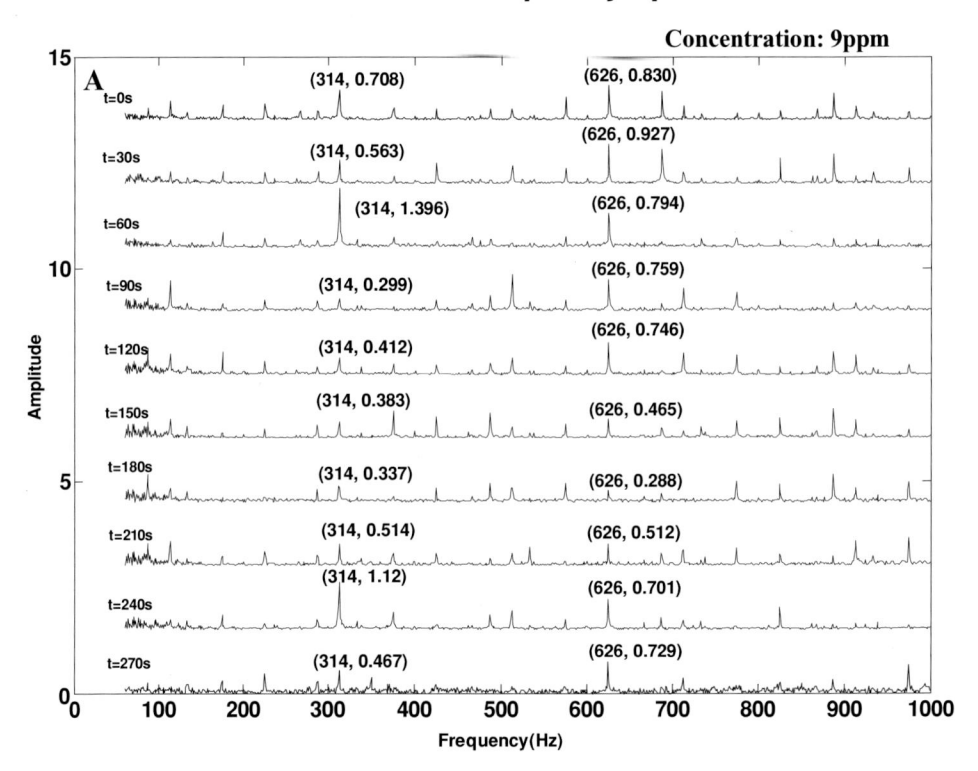

Peak values by time:

- t=0s: (314, 0.708), (626, 0.830)
- t=30s: (314, 0.563), (626, 0.927)
- t=60s: (314, 1.396), (626, 0.794)
- t=90s: (314, 0.299), (626, 0.759)
- t=120s: (314, 0.412), (626, 0.746)
- t=150s: (314, 0.383), (626, 0.465)
- t=180s: (314, 0.337), (626, 0.288)
- t=210s: (314, 0.514), (626, 0.512)
- t=240s: (314, 1.12), (626, 0.701)
- t=270s: (314, 0.467), (626, 0.729)

FIGURE 3.9(A). Signature pattern represented is obtained from a single neuron due to the action of ethanol at its detection limit of 9 ppm. SP due to the action of ethanol produces an Eigen Vector at 314 Hz [111].

SP corresponding to the modulated firing rate of the osteoblast. During the first phase of the sensing cycle (t = (0, 60) sec) the modulated firing rate is focused at 722Hz. During the second phase of the sensing cycle (t = (60,120)sec) the modulated firing rate shifts towards the lower frequency value (514Hz). During the third phase of the sensing cycle (t = (120, 180)sec), the modulated frequency shifts back to the original higher frequency bursting (668Hz and 722Hz) as observed in the first phase. As the concentration of ethanol is very low the cell quickly recovers and on re-introducing the chemical at t = 180sec the SP starts to repeat itself (t = (180,240) sec).

3.30. HYDROGEN PEROXIDE SENSING

3.30.1. Single Neuron Sensing

Addition of hydrogen peroxide causes it's binding to the α subunit of the APMA gated Na^+ ion channels which produce a rapid ionic depolarization current. It simultaneously acts

Osteoblast-Ethanol Frequency Spectrum

Concentration: 19 ppm

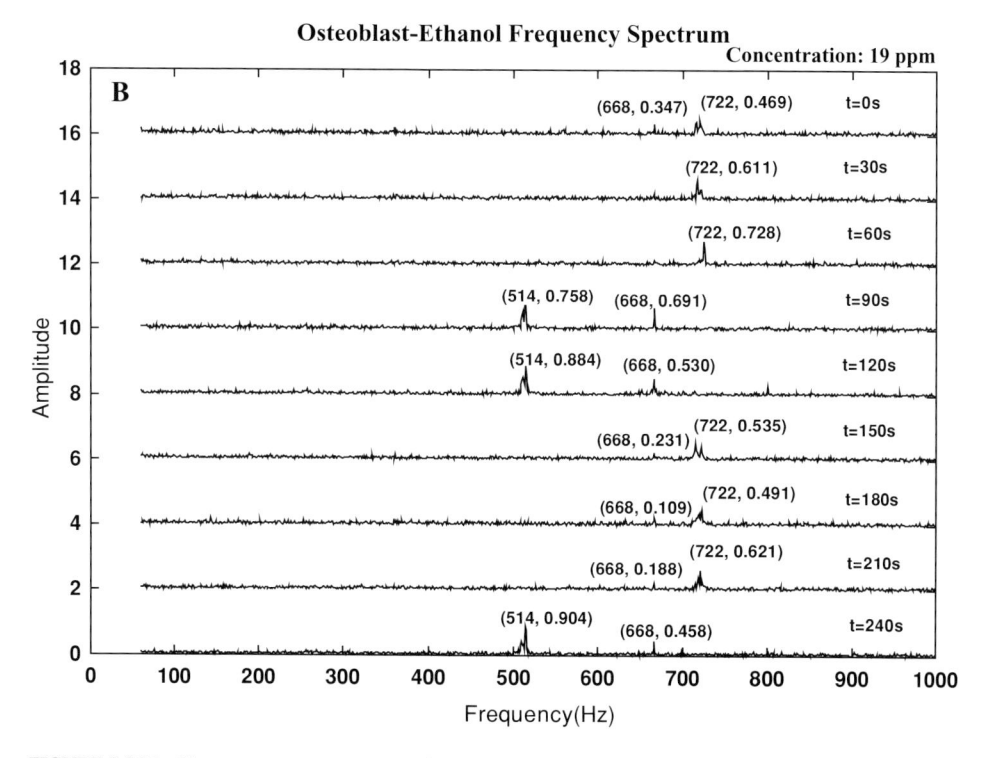

FIGURE 3.9(B). Signature pattern represented is obtained from a single osteoblast due to the action of ethanol at its detection limit of 19 ppm. SP due to the action of ethanol produces an Eigen Vectors at 514 Hz and 722 Hz [149].

upon the NMDA gated channels which triggers the entry of Ca^{++} ions into the cell, which causes the transmembrane release of glutamate and a steep increase of intracellular levels of Ca^{++} [4] The low frequency Eigen Vectors (175 Hz, 227 Hz, and 349 Hz) indicates the initial activation of APMA gated channels due to initial short binding transients of hydrogen peroxide. The mid frequency Eigen Vector (453 Hz) corresponds to the activation of NMDA gated channels and the longer duration of binding of hydrogen peroxide to NMDA receptors. The high frequency Eigen Vectors (749 Hz and 975 Hz) correspond to the induced excitotoxicity (Fig 3.10(a)) (Stout et al., 1998). The lowest concentration of hydrogen peroxide detected by the neuronal cell membrane is 19ppm.

3.30.2. Single Osteoblast Sensing

The single osteoblast cells were isolated and positioned over individual electrodes in a manner previously described. The initial concentration of hydrogen peroxide used was 5000ppm and the modified electrical activity was recorded. The concentration of hydrogen peroxide was decremented in a stepwise manner and in each case the modified electrical activity was recorded. The lowest concentration of hydrogen peroxide sensed by a single

Neuron-Hydrogen Peroxide Frequency Spectrum

Concentration: 19ppm

A
- t=0s: (175, 0.387), (227, 0.388), (349, 0.614), (453, 0.293), (626, 0.285), (749, 0.620), (853, 1.048), (975, 0.318)
- t=30s: (175, 0.141), (227, 0.489), (349, 0.427), (453, 0.612), (626, 0.352), (749, 0.651), (853, 1.454), (975, 0.264)
- t=60s: (175, 0.319), (227, 0.240), (349, 0.666), (453, 1.054), (626, 0.462), (749, 1.191), (853, 0.433), (975, 0.279)
- t=90s: (175, 0.253), (227, 0.207), (349, 0.546), (453, 0.295), (626, 0.436), (749, 0.435), (853, 0.464), (975, 0.177)
- t=120s: (175, 0.232), (227, 0.278), (349, 0.762), (453, 0.682), (626, 0.064), (749, 0.544), (853, 1.148), (975, 0.222)
- t=150s: (175, 0.205), (227, 0.503), (349, 0.554), (453, 0.129), (626, 0.228), (749, 1.589), (853, 0.811), (975, 0.364)

Frequency(Hz): 0–1000
Amplitude: 0–12

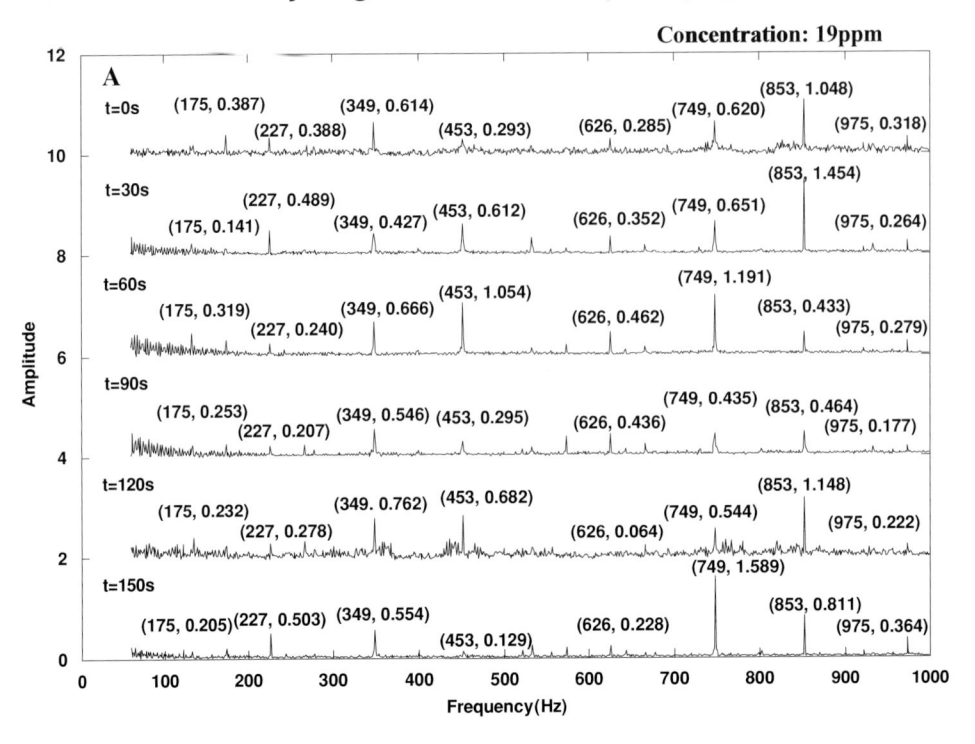

FIGURE 3.10(A). Signature pattern represented is obtained from a single neuron due to the action of hydrogen peroxide at its detection limit of 19 ppm. SP due to the action of hydrogen peroxide produces low frequency Eigen Vectors at 175 Hz, 227 Hz, and 349 Hz, mid frequency Eigen Vector at 453 Hz and high frequency Eigen Vectors at 749 Hz and 975 Hz [111].

osteoblast was 25ppm. FFT analysis was performed on the acquired data pertaining to the modified extracellular potential to yield the SP. The instant at which the chemical is added to the chip system is denoted by t = 0sec. Figure 3.10(b) represents the signature pattern. There are two eigen vectors (257Hz, and 873Hz) in the SP corresponding to the modulated firing rate of the osteoblast. The frequency of 668Hz corresponds to the osteoblast firing rate in the absence of a chemical agent. During the first half of the cycle, the low frequency subsidiary peaks (129Hz, 334Hz, and 437Hz) are expressed. During the second half of the cycle the high frequency subsidiary peak (565Hz) is expressed.

3.31. PYRETHROID SENSING

3.31.1. Single Neuron Sensing

Addition of pyrethroid results in the activation of the NMDA gated channels. The negative charge along the membrane surface induces the binding of Mg^{++} ions causing the clogging of the channels thus preventing the flow of Na^+ and K^+ ions [4]. These results in the

Osteoblast-Hydrogen Peroxide Frequency Spectrum

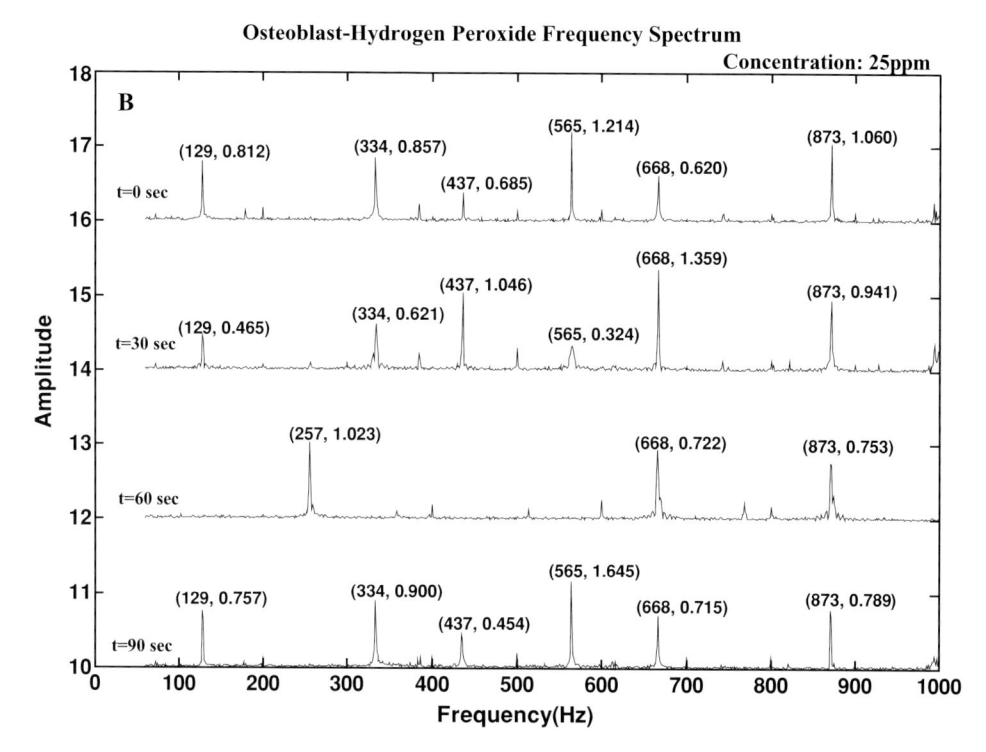

FIGURE 3.10(B). Signature pattern represented is obtained from a single neuron due to the action of hydrogen peroxide at its detection limit of 25 ppm. SP due to the action of hydrogen peroxide produces two Eigen Vectors at 257Hz, and 873Hz [149].

reduction of the depolarizing ionic current reducing the firing rate to 514 Hz (Fig 3.11(A)). SP due to the effect of pyrethroid produces an Eigen Vectors at 514 Hz and 576 Hz. The initial concentration of pyrethroid sensed was 5000ppm and the detection limit for the single cell was determined to be 280 ppb.

3.31.2. Single Osteoblast Sensing

The initial concentration of pyrethroid used was 5000ppm and the modified electrical activity was recorded. The concentration of pyrethroid was decremented in a stepwise manner and in each case the modified electrical activity was recorded. The lowest concentration of pyrethroid sensed by a single osteoblast was 890ppm. FFT analysis was performed on the acquired data pertaining to the modified extracellular potential to yield the SP. The instant at which the chemical is added to the chip system is denoted by t = 0sec. Figure 3.11(b) represents the signature pattern. There are two eigen vectors (129Hz, and 873Hz) in the SP corresponding to the modulated firing rate of the osteoblast The frequency of 668Hz corresponds to the osteoblast firing rate in the absence of a chemical agent. During the first half of the cycle there is a high frequency subsidiary peak corresponding to 565Hz. During the second half of the cycle there is a low frequency subsidiary peak of 257Hz.

Neuron-Pyrethroid Frequency Spectrum

Concentration: 280 ppb

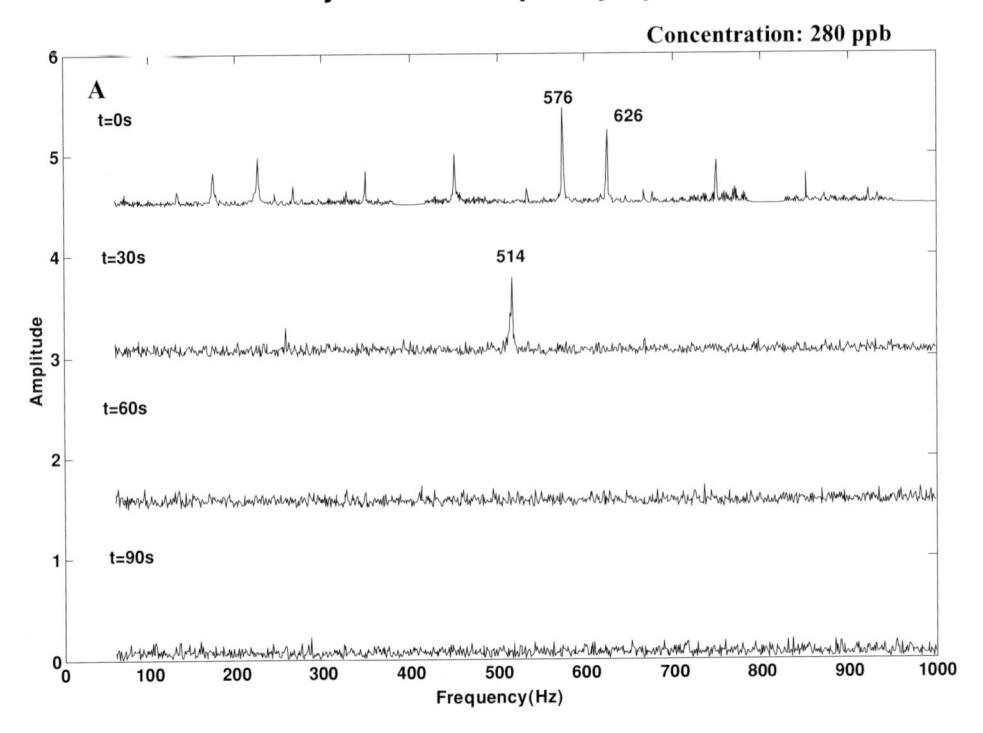

FIGURE 3.11(A). Signature pattern represented is obtained from a single neuron due to the action of pyrethroid at its detection limit of 280 ppb. SP due to the action of pyrethroid produces Eigen Vectors at 514 Hz and 576 Hz [111].

3.32. EDTA SENSING

3.32.1. Single Neuron Sensing

Addition of EDTA causes it's binding to the $GABA_A$ as well as the NMDA receptor gated Cl^- and Na^+ ion channels, respectively [130]. The $GABA_A$ activation produces a hyperpolarizing current resulting in low frequency bursting at 227 Hz and NMDA activation produces a high frequency burst of 873 Hz (Fig 3.12(a)). The initial concentration of EDTA used is 5000ppm and the detection limit obtained for the single neuron is 180ppm.

3.32.2. Single Osteoblast Sensing

The initial concentration of EDTA used was 5000ppm and the modified electrical activity was recorded. The concentration of EDTA was decremented in a stepwise manner and in each case the modified electrical activity was recorded. The lowest concentration of EDTA sensed by a single osteoblast was 280ppm. FFT analysis was performed on the acquired data pertaining to the modified extracellular potential to yield the SPV. The instant at which EDTA is added to the chip system is denoted by t = 0sec. Figure 3.12(b) represents

Osteoblast-Pyrethroid Frequency Spectrum

Concentration: 890 ppb

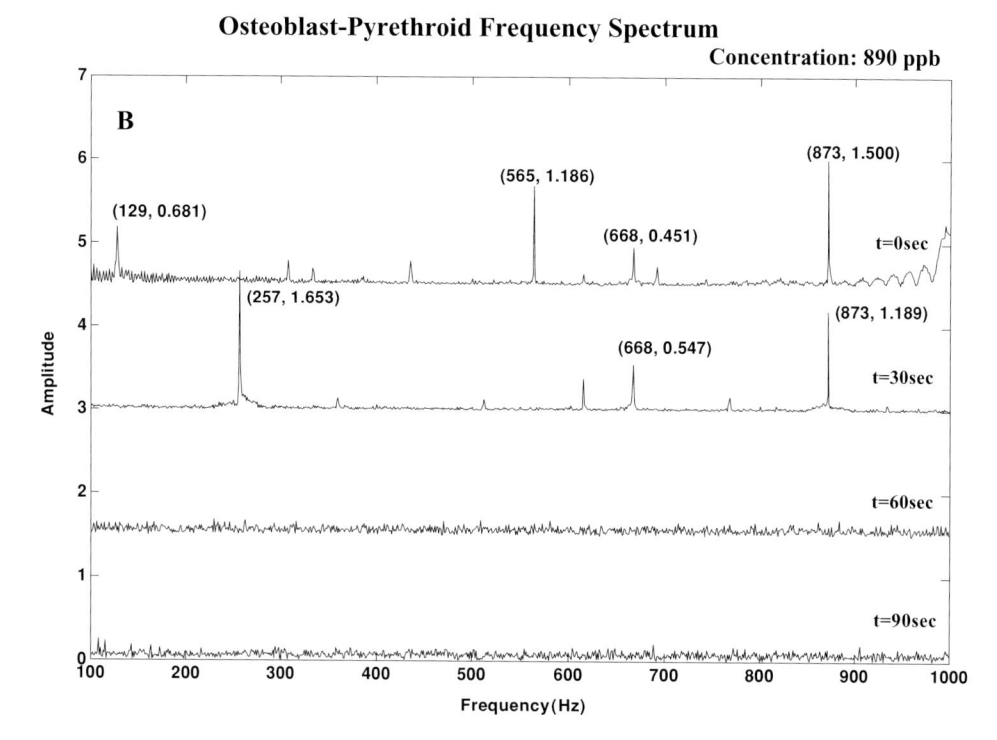

FIGURE 3.11(B). Signature pattern represented is obtained from a single osteoblast due to the action of pyrethroid at its detection limit of 890 ppb. SP due to the action of pyrethroid produces Eigen Vectors at 129 Hz and 873 Hz [149].

the signature pattern. The initial peak in the frequency spectrum is observed at 514Hz corresponding to the first eigen vector. This is obtained at t = 0, after the immediate application of EDTA. Osteoblast cells then regain their control of the firing rate, corresponding to 667Hz. The next two eigen vectors of 258Hz and 872Hz are obtained in the time interval (t = (60, 90) sec). Subsidiary low frequency peaks are observed at 129Hz, 334H, 437Hz and high frequency peaks are observed at 514Hz and 565Hz.

3.33. IMMUNOHISTOCHEMISTRY

To verify the associated changes physiological changes related to the exposure to varying dosages of diesel and gasoline at the cellular level. The cells are first exposed to the specific chemical agents for the sensing cycle period as described in detail elsewhere [110]. The cells are then fixed using standard immunohistochemistry procedures. Briefly the patterned cells on the sensing platform were exposed to the specific chemical analytes namely either diesel or gasoline for a period of 270 sec (or 4.5 minutes). This time period constitutes to one sensing cycle which was determined previously from other sensing experiments performed on single chemical analytes [Yang et al., 2003, 110]. The cell array is then washed twice with Phosphate Buffered Saline (1X). The wash cycle duration was 3 minutes each.

Neuron-EDTA Frequency Spectrum

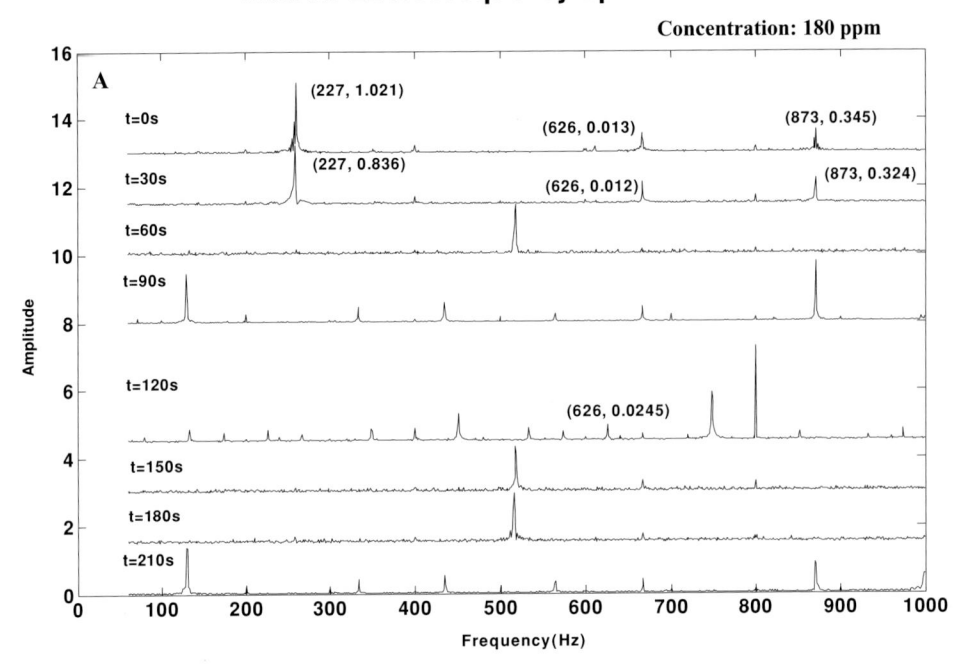

FIGURE 3.12(A). Signature pattern represented is obtained from a single neuron due to the action of EDTA at its detection limit of 180 ppm. SP due to the action of EDTA produces a Eigen Vectors at 227Hz and 873 Hz [111].

After aspiration the cells are exposed to 1–3% formalin for a period of 20 minutes. After aspiration the cells were exposed to 1% triton –X in the case of neurons and 10% triton –X in case of osteoblasts for a period of 25 minutes. This functions as a cell permeating agent that allows the easier penetration of the dye by increasing the permeability of the cell membrane. The stain then is provided with an easier access to the locations of interest namely the presence of caged calcium emissions which are the expected biological changes at the cellular level. The cells were then aspirated with 1X PBS three times for a period of 3 minutes each. Cells were then stained with Fluo-3 (Molecular Probes, Eugene, OR) which bind to the intracellular calcium channels that produce calcium emissions due to excitation by the chemical analytes under analysis (Coward et al., 1999). Cells are incubated with Fluo-3(1 mg/0.5 ml) at a 1:1000 dilution for a period of 30 min. The cells are then aspirated with 1X PBS 3 times for a period of 2 minutes each. The cells are then imaged using Leica SP2 UV confocal microscope (Leica Microsystems, Inc., Germany).

Similarly for staining the voltage sensitive sodium and potassium channels, the cells are fixed with 1–3% formalin using the same procedure as described above. The stock solution of the dye for cell loading is prepared by dissolving the material in anhydrous DMSO to a concentration of 10 mM. The same procedure is followed for the sodium ion stain which is the AM ester of SBFI (Molecular Probes, Eugene, OR) and the potassium ion stain which is the AM ester of PBFI (Molecular Probes, Eugene, OR). The molecular weights of SBFI AM and PBFI AM are 1127 and 1171, respectively. The use of Pluronic® F-127 is essential

Osteoblast-EDTA Frequency Spectrum

Concentration: 280 ppm

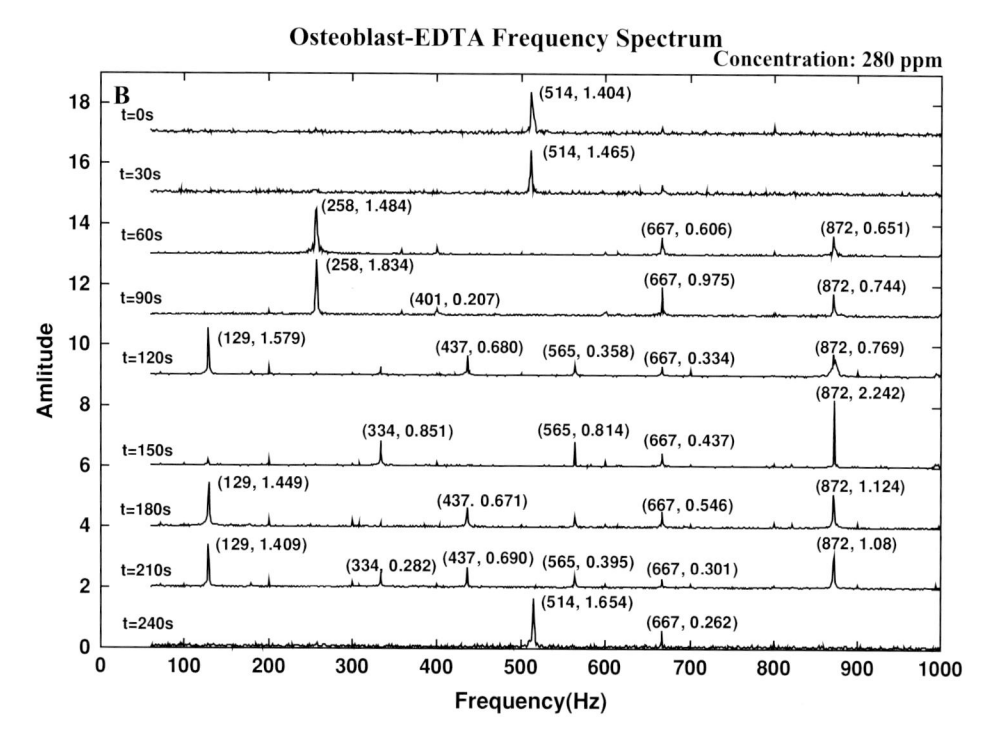

FIGURE 3.12(B). Signature pattern represented is obtained from a single osteoblast due to the action of EDTA at its detection limit of 280 ppm. SP due to the action of EDTA produces Eigen Vectors at 258Hz, 514Hz, and 872Hz. [149].

for optimal cell loading of both SBFI AM and PBFI AM due to the poor aqueous solubility of the dyes. 1 mL of a 20% (w/v) solution of Pluronic F-127 in DMSO (P-3000) is used in this case. The indicator stock solution in DMSO is mixed with an equal volume of 25% w/v Pluronic F-127 solution immediately prior to its addition to the cell loading buffer (Tsubokawa et al., 1999). The loading concentration used for both the indicators is 5 µM. The cells are incubated with the indicators for a period of 40 minutes before washing them twice with 1X PBS for a period of 2 minutes each. In the case of dual staining the cells are first loaded with the sodium and potassium indicators and then loaded with the calcium indicator.

Finally to verify the apoptotic state due to the effect of the chemical agents. The cells after exposure to the chemical agent at the detection limit are stained with Hoechst dye. The effect of the chemical induces apoptosis which can me imaged using the spectral shift that is noted with the Hoechst dye. As mentioned in literature [27, 141] a simple interpretation rests on the well-documented increased permeability of apoptotic cells to Hoescht, and if more dye permeates then some concentration-dependent dye-dye interaction may result in a red shift. This is consistent with the observation that the cells with redder emission appear only as dye loading saturates. The dye concentration is the critical parameter controlling the shift. Hence taking into account this effect 0.12 µg/ml of the stock solution is prepared for ideal imaging of the DNA alteration which is the expected effect of the chemical agent on the nucleic material of the cell. The protocol for Hoechst is similar to the protocols for calcium, sodium and potassium staining. The patterned cells after being exposed to the chemical analyte for a period corresponding to the sensing cycle are fixed using 1.5% formaldehyde for a period of 20 minutes. The cells are then aspirated twice using 1X PBS for a period of 3 minutes each. The cells are then incubated with 100% methanol for a period of 20 minutes. Methanol is a cell permeant and it allows the easy penetration of the Hoechst dye into the cell nucleus. The function of the Hoechst dye is to bind to the diffused chromatin that is formed due to the apoptosis of the cell due to the effect of the chemical analyte. Approximately 5ml of 0.12 µg/ml Hoechst 33382 dye (Santa Cruz Biotechnology, Santa Cruz, California) is prepared from a stock solution of 0.5 µg/ml dye. The cells are incubated with the dye for a period of 30 minutes at 37°C. The cells are then aspirated thrice with 1X PBS for a period of 2 minutes each. Hoechst dye is visualized by exciting the dye with UV (352 nm). The emission is in blue. The cells are imaged using a confocal microscope (Leica, Germany).

3.34. VISUALIZATION OF PHYSIOLOGICAL CHANGES DUE TO THE EFFECT OF THE CHEMICAL ANALYTES

3.34.1. Effect of Ethanol on Neurons

Addition of ethanol leads to its binding to M1 and M2 regions on the outside face of the $GABA_A$ and Glycine receptor gated Cl^- ion channels [77]. This increases the duration of the channel openings causing a strong inhibitory ionic current associated with Cl^- influx, which in turn causes the intracellular release of calcium as it has been established that intoxicating concentrations of ethanol inhibit N-methyl-d-aspartate (NMDA) receptor-dependent long-term potentiation, an interaction thought to underlie a major component of the central nervous system actions of ethanol [52]. At the cellular level this can be by fluo-3 that binds to the intracellular calcium ions released into the cytoplasm close to the nucleus. The patterned cells are exposed to 9 ppm concentration of ethanol before the immunohistochemistry procedures. Figure 3.13(a) is a confocal image of the neuron stained with fluo-3 indicating the intracellular calcium transients. Also shown in figure 3.13(b) is the three dimensional contour vector representation of the confocal image. The z axis represents the intensity profile. It shows that maximum intensity is obtained from the cytoplasm correlating to the position of the intracellular calcium transients. The Hoecsht stain indicates the dissociation of DNA associated with apoptotic state of the cells [28]. The cell death associated due to the effect of ethanol results in clumping of chromatin. Hoechst binds to the dissociated nucleic material and is visualized using confocal microscopy and is shown in figure 3.13(c).

3.34.2. Effect of Ethanol on Osteoblasts

In-vitro and in-vivo studies have shown the changes to the bone calcium density due to long term exposure to ethanol [18]. At a cellular level this translates to the presence of apoptotic bundles in the cytoplasm of the cells. The patterned cells are treated to 19ppm ethanol and then after cell fixation is incubated with fluo-3 calcium indicator that binds to free calcium ions. Figure 3.13(d) shows the location of free calcium imaged due to

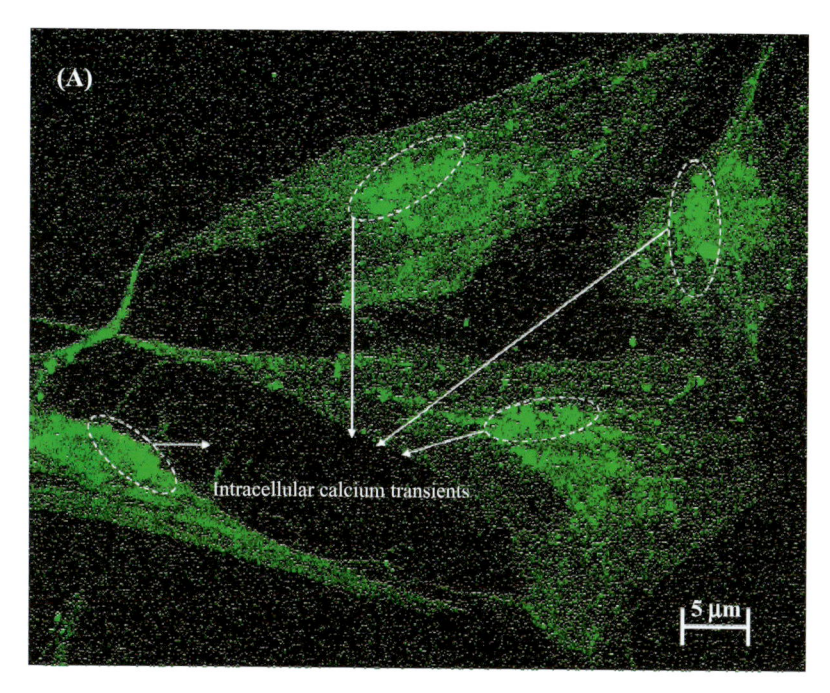

FIGURE 3.13(A). Physiological effect of ethanol at 9 ppm on individual neurons. The cells are stained with fluo-3 and visualized using FITC mode on Leica TCS SP2. Intracellular calcium transients are observed at the cytoplasm near the nucleus.

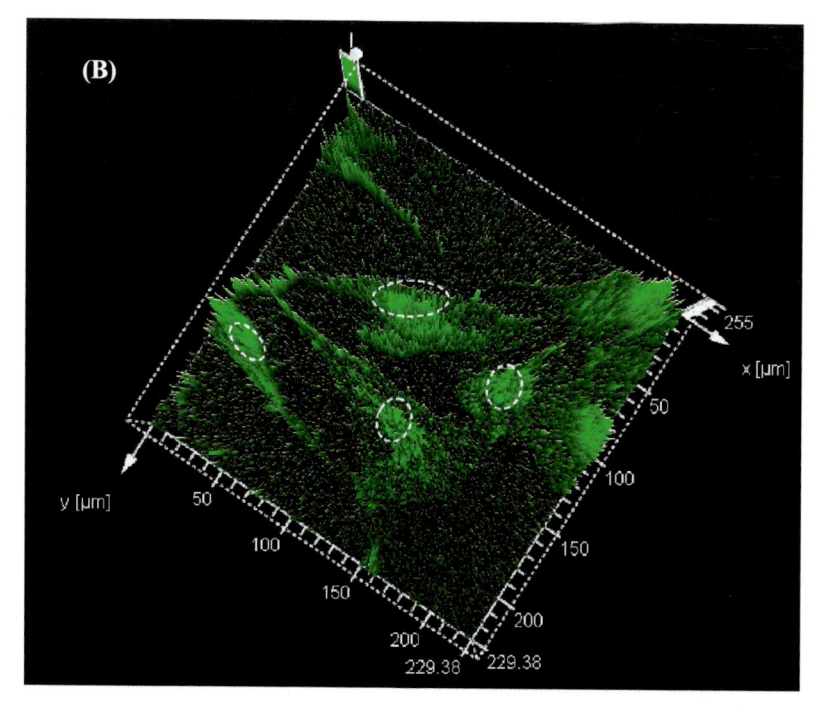

FIGURE 3.13(B). Three dimensional vector representtion of the fluorescence intensity. The circled areas represent the hight intensity peaks that correspond to the areas of localization of intracellular calcium transients.

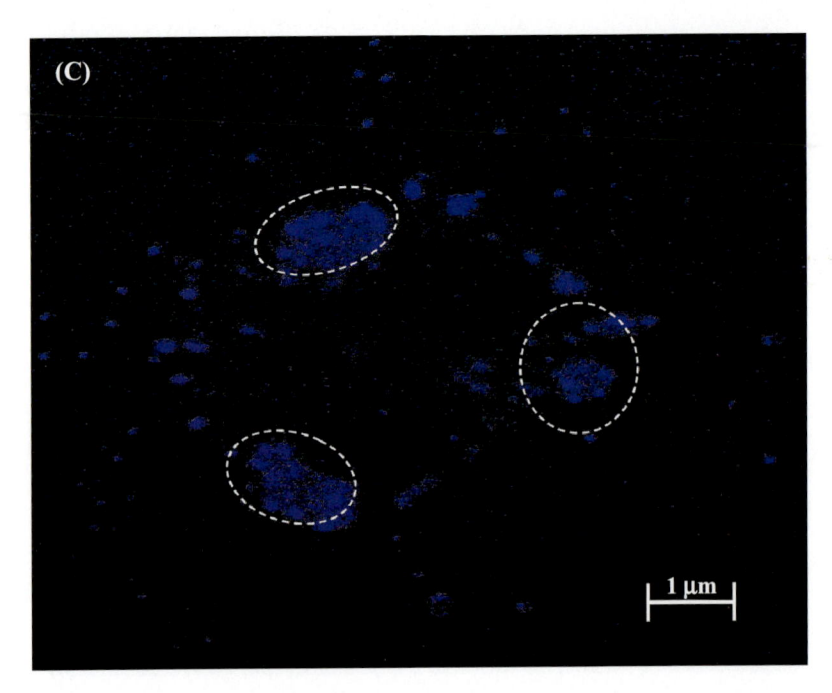

FIGURE 3.13(C). Hoechst staining reveals the dissociation of the nucleic matter due to apoptosis induced by the exposure of the neurons to 9 ppm ethanol. The circled areas represented the dissociated nucleic matter due to apoptosis.

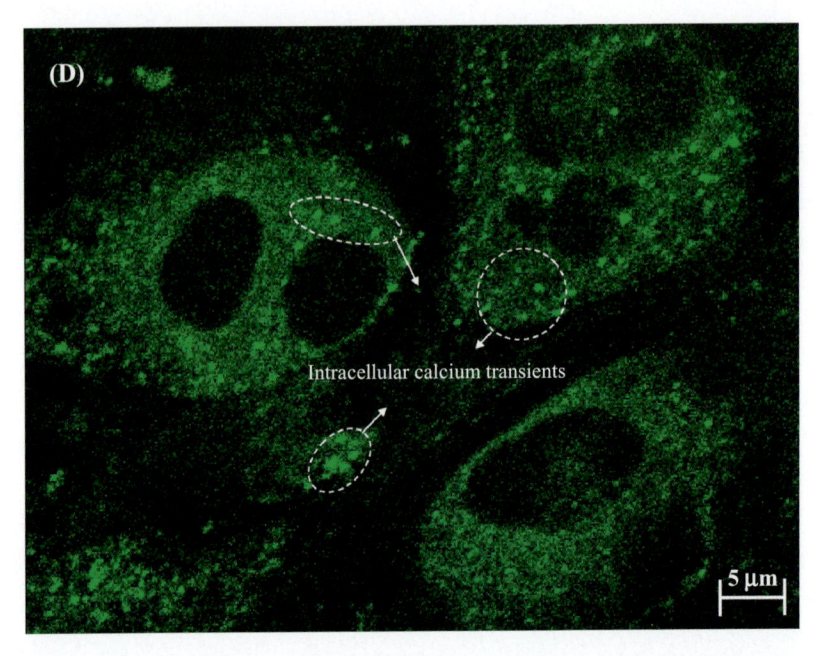

FIGURE 3.13(D). Physiological effect of ethanol at 19 ppm on individual osteoblasts. The cells are stained with fluo-3 and visualized using FITC mode on Leica TCS SP2. Apoptotic bundles are seen distributed throughout the cytoplasm.

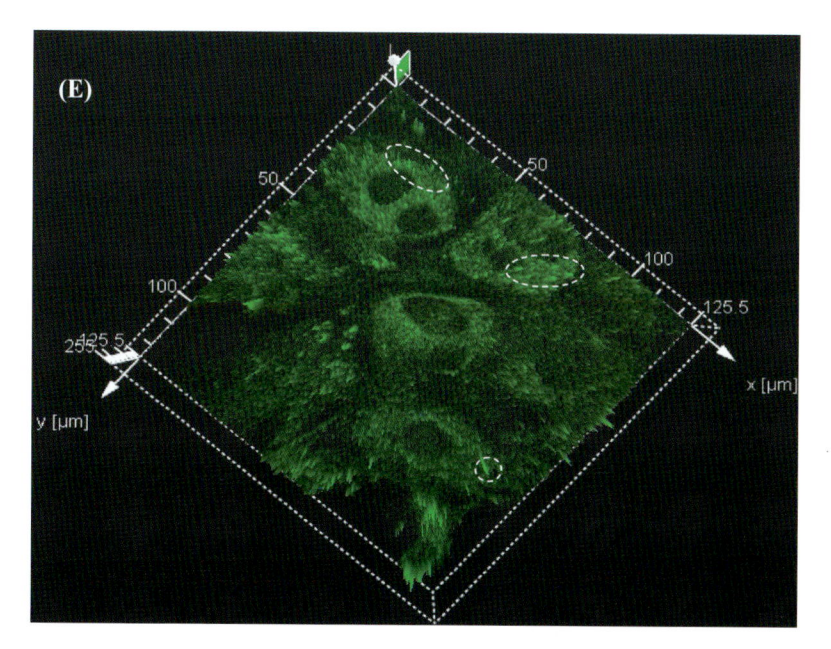

FIGURE 3.13(E). Three dimensional vector representation of the fluorescence intensity. The circled areas represent the high intensity peaks that correspond to the areas of localization of the apoptotic bundles.

the binding of fluo-3 to the ions. Figure 3.13(e) shows the three dimensional vector representation of the fluorescence intensity. The intensity levels are indicated by the z axis. The higher intensity peaks are associated with the apoptotic bundles. Figure 3.13(f) is the confocal micrograph of the dissociated chromosomes due to the effect of ethanol. Ethanol causes the differentiation of the bone matrix in large cell populations [134] but in single cell level it can be visualized using Hoechst stain as dissociated nucleic matter.

3.34.3. Effect of Hydrogen Peroxide on Neurons

Addition of hydrogen peroxide causes it's binding to the α subunit of the APMA gated Na^+ ion channels which produce a rapid ionic depolarization current. It simultaneously acts upon the NMDA gated channels which triggers the entry of Ca^{++} ions into the cell, which causes the transmembrane release of glutamate and a steep increase of intracellular levels of Ca^{++} [4]. Previous research has shown that large populations of hippocampal neurons are particularly vulnerable to hypoxia/ischaemia-induced damage, and free radicals are thought to be prime mediators of this neuronal destruction. It has been shown that hydrogen peroxide), through the production of free radicals, induces cell death by activation of a non-selective cation channel, which leads to irreversible cell depolarization and unregulated Ca^{2+} entry into the cell [124]. In single cells groups of free sodium potassium and calcium ions are expected localized at different regions of the cell. The sodium and potassium ions are expected to be localized near the cell membrane. The calcium ions are distributed throughout the cytoplasm with higher concentrations near the nucleus and the pre-synaptic

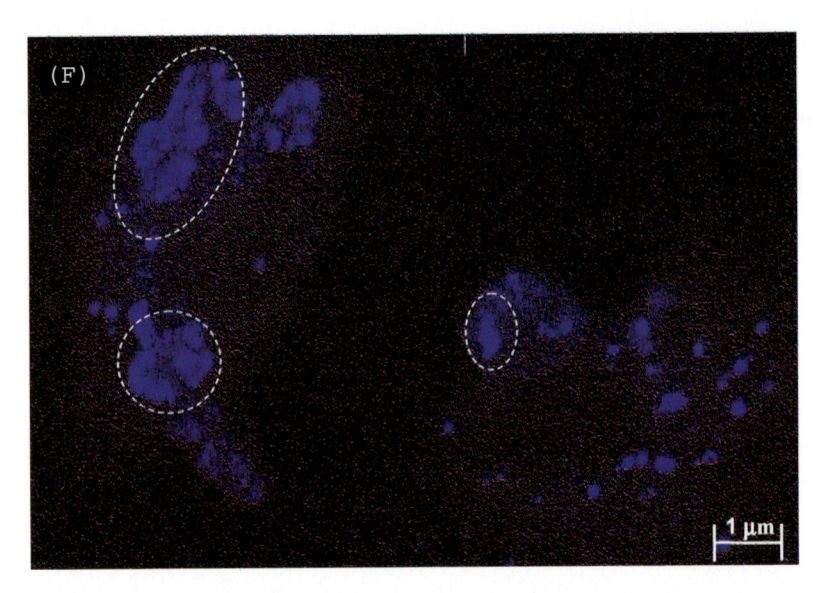

FIGURE 3.13(F). Hoechst staining reveals the dissociation of the nucleic matter due to apoptosis induced by the exposure of the osteoblasts to 19 ppm ethanol. The circled areas represent the dissociated nucleic matter due to apoptosis.

terminals. The cells are doubly stained for the three ion types. Based on the location of the apoptotic bundles the ion type is determined. Figure 3.14(a) is a confocal optical micrograph of neurons stained for the three ions after being exposed to hydrogen peroxide at 19ppm for the period of one sensing cycle. Figure 3.14(b) is a three dimensional vector representation of the fluorescence intensity of the apoptotic bundles. The z axis gives the various intensity levels of fluorescence. Previous in-vivo studies have established the presence and attributes of apoptosis and necrosis in the anatomically well-defined cortical region after exposure to hydrogen peroxide at physiologically relevant concentrations. These were verified under light microscopy with Hoechst 33342, staining [57]. Cells exhibiting apoptotic morphology with chromatin condensation and apoptotic bodies and necrotic ghost appearance were observed. In single cell in-vitro studies condensed chromatin is expected to be visualized using confocal microscopy. This is verified by figure 3.14(c).

3.34.4. Effect of Hydrogen Peroxide on Osteoblasts

Previous research has established that the physiological activity of osteoblasts is known to be closely related to increased intracellular Ca^{2+} activity ($[Ca^{2+}]i$) in osteoblasts. The cellular regulation of $[Ca^{2+}]i$ in osteoblasts is mediated by Ca^{2+} movements associated with Ca^{2+} release from intracellular Ca^{2+} stores, and transmembrane Ca^{2+} influx via Na^+-Ca^{2+} exchanger, and Ca^{2+} ATPase. Reactive oxygen species, such as hydrogen peroxide, play an important role in the regulation of cellular functions, and act as signaling molecules or toxins in cells. This behavior has been seen in-vivo and in-vitro in large cell populations [91]. Similarly after exposure to 25 ppm of hydrogen peroxide for the period corresponding

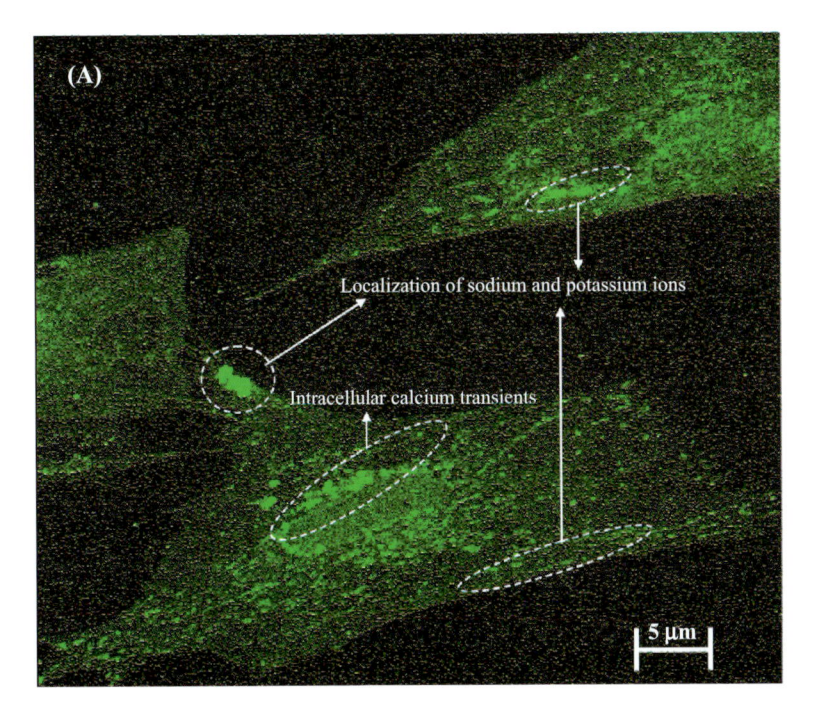

MICROARRAY AND FLUIDIC CHIP FOR EXTRACELLULAR SENSING 85

FIGURE 3.14(A). Physiological effect of hydrogen peroxide at 19 ppm on individual neurons. The cells are stained with SBFI and PBFI esters and fluo-3 for visualizing free sodium potassium and calcium ions respectively. Double immunohistochemistry is performed after cell fixation. The sodium and potassium ions are localized near the cell membrane and the pre-synaptic terminals where as the free calcium ions are localized near the nucleus.

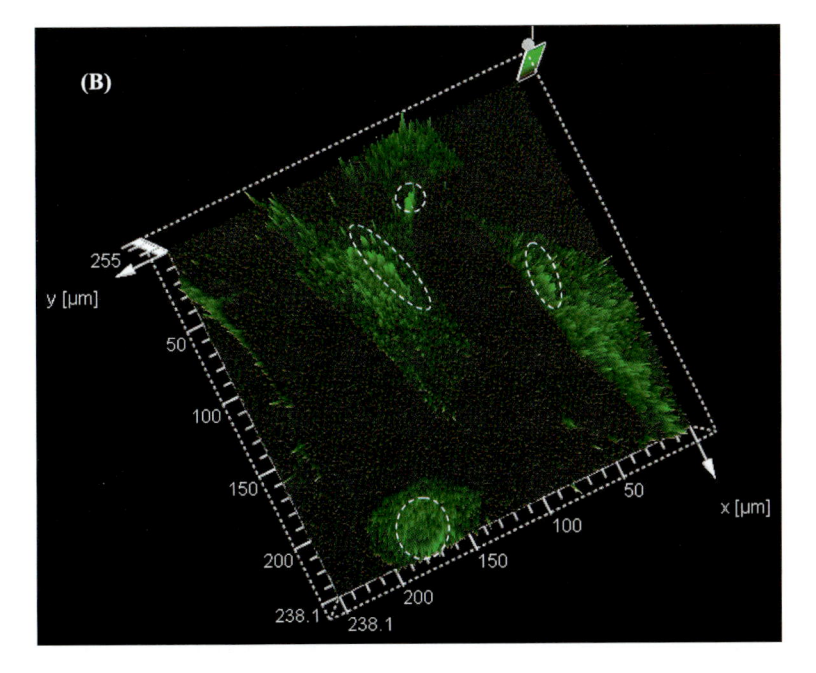

FIGURE 3.14(B). Three dimensional vector representation of the fluorescence intensity. The circled areas represent the high intensity peaks. The peak with the maximum intensity corresponds to the localization of the sodium and potassium ions at the pre-synaptic terminals. The intensity peaks around the nucleus correspond to the localization of the calcium ions.

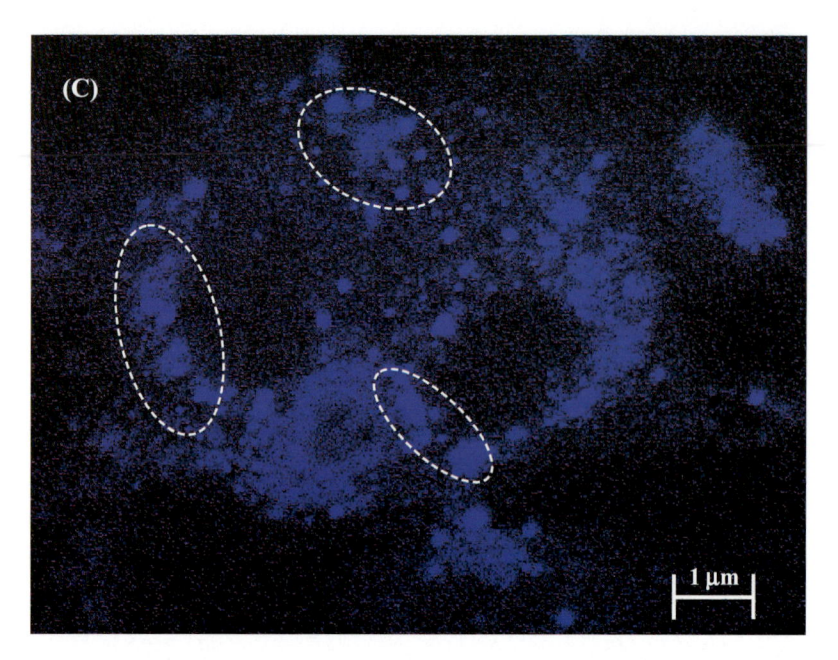

FIGURE 3.14(C). Hoechst staining reveals the dissociation of the nucleic matter due to apoptosis induced by the exposure of the individual neurons to 19 ppm hydrogen peroxide. The circleed areas represent the dissociated nucleic matter due to apoptosis.

to one sensing cycle the fixed cells stained for free sodium and calcium ions show apoptotic bundles throughout the cytoplasm. This corresponds to the expected physiological behavior due to the effect of reactive species like hydrogen peroxide on excitable mammalian cells like osteoblasts. This behavior is seen in figure 3.14(d) which is a confocal optical micrograph that identifies apoptotic bundles formed due to the localization sodium and calcium ions. Figure 3.14(e) is the three dimensional vector representation of the fluorescence intensity. The higher intensity levels correspond to the areas of localization of the apoptotic bundles. Previous research indicates marked nuclear condensation and fragmentation of chromatin due to exposure to reactive species like hydrogen peroxide at physiologically relevant concentrations were observed by Hoechst 33382 stain, for large cell populations [113]. Similarly condensed chromatin was observed in single osteoblasts after exposure to hydrogen peroxide at 25 ppm concentration after exposure for a period of one sensing cycle. Figure 3.14(f) is the confocal micrograph of osteoblasts after hydrogen peroxide exposure and stained with Hoechst 33382.

3.34.5. Effect of Pyrethroid on Neurons

Addition of pyrethroid results in the activation of the NMDA gated channels. The negative charge along the membrane surface induces the binding of Mg^{++} ions causing the clogging of the channels thus preventing the flow of Na^+ and K^+ ions. Previous research has shown that the treatment of mammalian cells in-vivo with pyrethroid results in

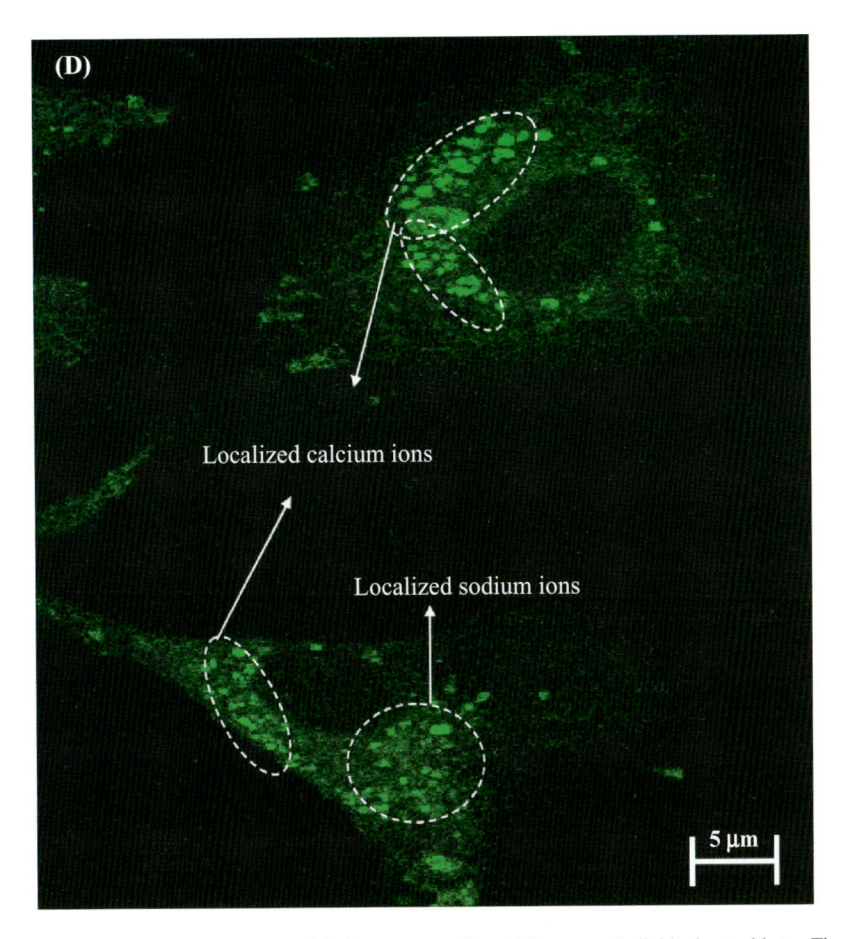

FIGURE 3.14(D). Physiological effect of hydrogen peroxide at 25 ppm on individual osteoblasts. The cells are stained with SBFI ester and fluo-3 for visualizing free sodium potassium and calcium ions respectively. Double immunohistochemistry is performed after cell fixation. The sodium ions are localized throughout the cytoplasm with a higher concentration away from the nucleus where as the free calcium ions are localized near the nucleus.

mitochondrial damage. The mitochondrial swelling was accompanied by the accumulation of electron dense granules. In addition, the neuropiles contained secondary lysosomes which increased in size and number with the progress of poisoning and showed signs of depletion of synaptic vesicles. This is imaged with free calcium binding stains. These stains show uniform permeability throughout the cell thus indicating the slow onset of apoptosis [122]. Similar behavior is observed at the single cell level. The patterned hippocampal neurons after exposure to pyrethroid at 280 ppb concentration for a period of one sensing cycle are stained with fluo-3 after fixation. The dye permeates throughout the cytoplasm indicating the slow onset of apoptosis associated with the effect of pyrethroid. Figure 3.15(a) is the confocal optical micrograph of patterned neurons undergoing fluorescence under FITC mode stained with fluo-3. Figure 3.15(b) is the three dimensional vector representation of

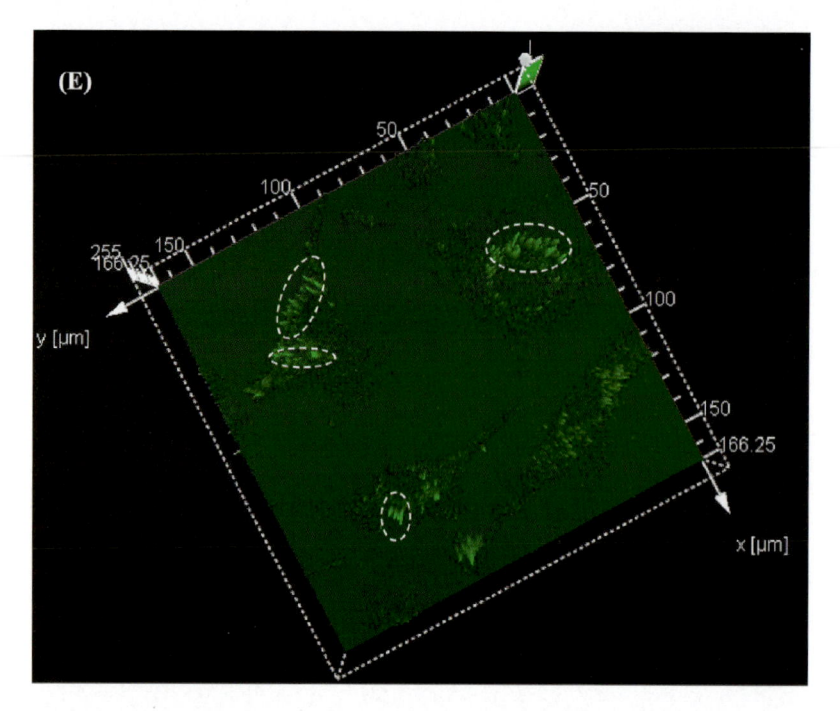

FIGURE 3.14(E). Three dimensional vector representation of the fluorescence intensity. The circled areas represent the high intensity peaks. The peak with the maximum intensity near the nucleus corresponds to the locallized sodium ions whereas the intensity peaks near the nucleus.

the fluorescence intensity. The fluorescence peaks are evenly distributed throughout the cell indicating the slow onsent of apoptosis due to the effect of pyrethroid. It has been shown that Organophosphorus (OP) compounds have been shown to be cytotoxic to neuroblastoma cell cultures. The mechanisms involved in OP compound-induced cell death (apoptosis versus necrosis) have been assessed morphologically by looking at nuclear fragmentation and budding using the fluorescent stain Hoechst 33342 Hoechst staining revealed induced time-dependent apoptosis. In many cells OP has also induced nuclear condensation with little fragmentation or budding [16]. Similarly in individual neurons treated with pyrethroid also show nuclear condensation in keeping with the previous results as shown by the confocal optical micrograph in figure 3.15(c).

3.34.6. Effect of Pyrethroid on Osteoblasts

Previous research has shown that in in-vivo studies specific features of osteogenesis under a toxic action of pesticide was observed in experiments with inbred albino rats. This resulted in accumulation of potassium clusters that was visualized using autoradiography with 3H-thymidine label [70]. Similarly exposure of the patterned osteoblast to 890 ppb concentration of pyrethroid for the period of one sensing cycle and staining with PBFI ester after cell fixation results in the identification of potassium ion bundles that are formed due

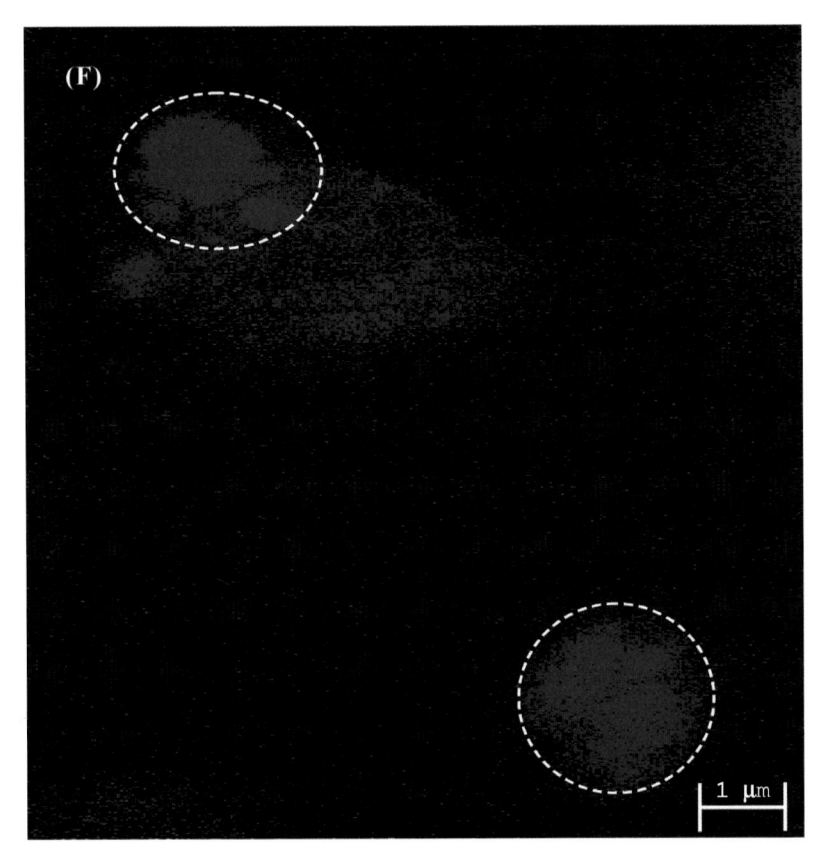

MICROARRAY AND FLUIDIC CHIP FOR EXTRACELLULAR SENSING 89

FIGURE 3.14(F). Single osteoblasts stained with Hoechst 33382 stain after being exposed to hydrogen peroxide at 19 ppm concentration for a period of one sensing cycle. The circled areas indicate the regions where the chromatin has condensed due to apoptosis.

to the effect of the pyrethroid. Figure 3.15(d) is the confocal optical micrograph that shows the presence of potassium bundles after exposure to PBFI stain. Figure 3.15(e) is the three dimensional vector representation of the fluorescence intensity. The high intensity peaks correspond to the localization of the potassium ions. Based on previous research, in-vivo administration of pyrethroid results in the development of bone lesions. This translates cellularly to the development of nucleic material condensation [125]. Similar behavior is seen in single osteoblasts after exposure to pyrethroid and stained with Hoechst 33382 and this is shown in figure 3.15(f).

3.34.7. Effect of EDTA on Neurons

EDTA, a chelator of Zn and calcium (a modulator of AbetaP-mediated toxicity) induced a reversible change in the Zn-mediated aggregation. Thus based on previous research the effect of physiologically relevant amounts of EDTA causes the depolarization of the

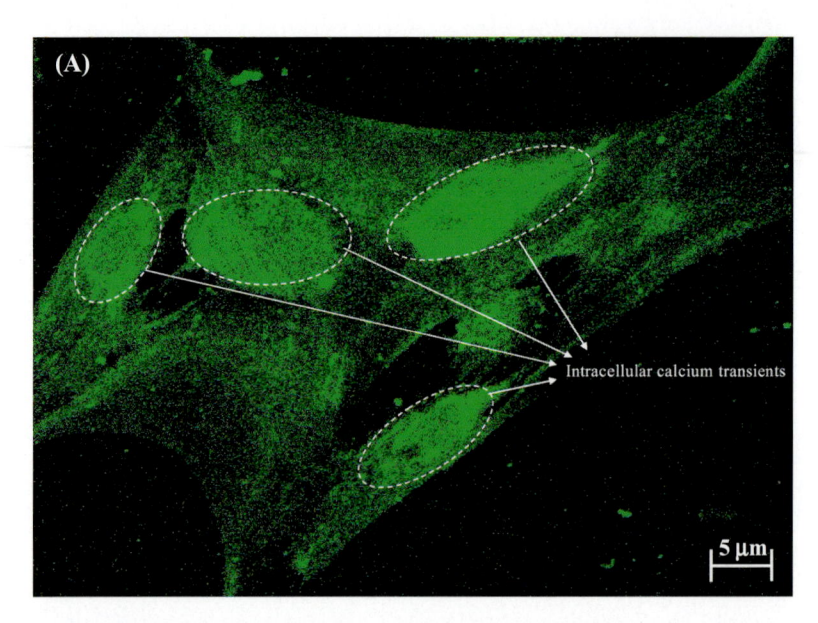

FIGURE 3.15(A). Physiological effect of pyrethroid at 280 ppb on individual neurons. The cells are stained with fluo-3 for visualizing free calcium ion transients. A uniform permeation of the fluo-3 stain is observed within the cytoplasm of individual cells this establishes the onset of apoptosis. This behavior correlates to the in-vitro behavior of large cell populations after exposure to organophosphates.

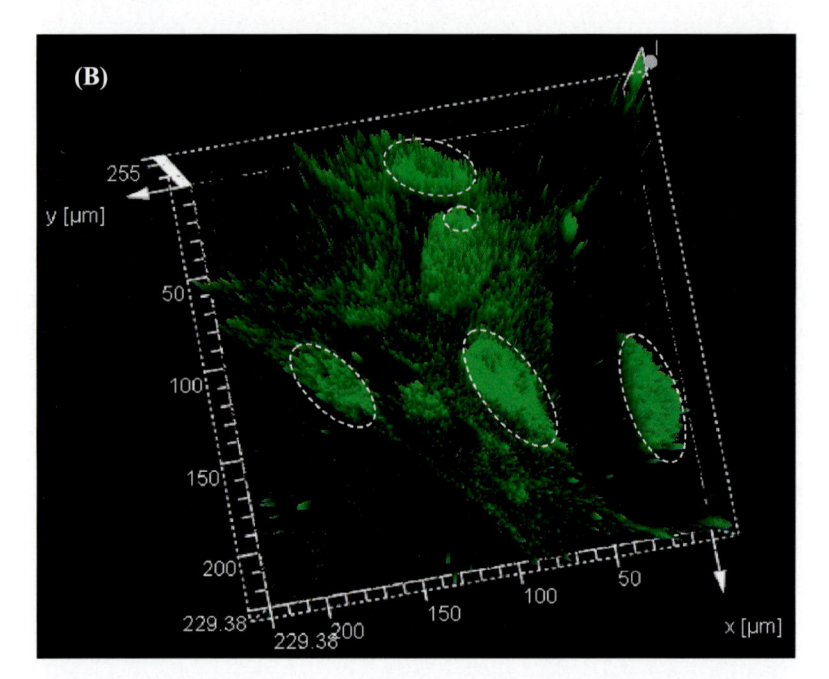

FIGURE 3.15(B). Three dimensional vector representation of the fluorescence intensity. The circled areas represent the high intensity peaks. The high intensity peaks are distributed throughout the cytoplasm of the cell indicating the uniform distribution of free calcium ions throughout the cytoplasm that is associated with the onset of apoptosis due to the effect of pyrethroid.

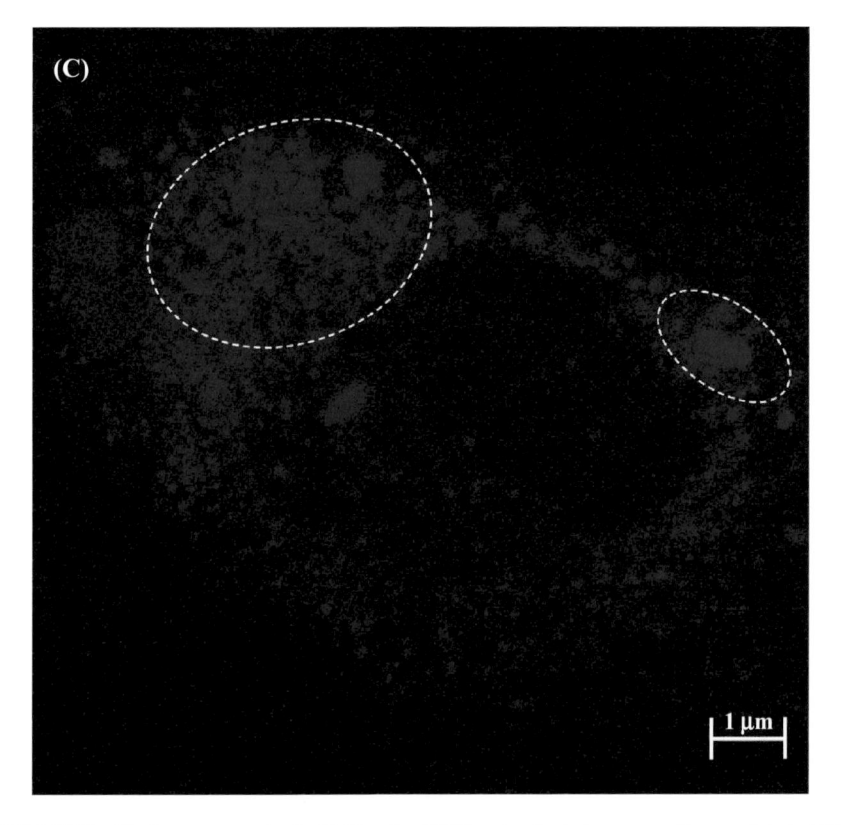

FIGURE 3.15(C). Single neurons stained with Hoechst 33382 stain after being exposed to pyrethriod at 280 ppb concentration for a period of one sensing cycle. The circled areas indicate the regions of nuclear condensation in keeping with the previous results obtained from in-vitro studies with exposure to physiologically relevant amounts of organophosphates.

membrane due to calcium emission. This results in excitotoxicity that eventually results in cell death [100]. The patterned neurons are exposed to 180 ppm concentration of EDTA over a period of one sensing cycle. The cells are then fixed and stained with fluo-3. This is a calcium indicator and binds to the free calcium ions generated throughout the cytoplasm due to the action of EDTA. Figure 3.16(a) is the confocal optical micrograph of patterned neurons stained with fluo-3. Figure 3.16(b) is the three dimensional vector representation of the fluorescence intensity. The z axis gives the intensity of fluorescence. The peaks indicate the location of the accumulation of free calcium ions. Previous research shows that nuclear spreads obtained from neurons of zebrafish indicate the formation of condensed chromatin due to apoptosis caused by the exposure to EDTA [14] Figure 3.16(c) is the confocal micrograph of neurons that show condensed chromation due to apoptosis caused by exposure to 180 ppm EDTA.

3.34.8. Effect of EDTA on Osteoblasts

Previous in-vivo research indicated that timed immersion in buffered EDTA selectively altered the mineral content at each level in the cortical bone structural hierarchy. This affected

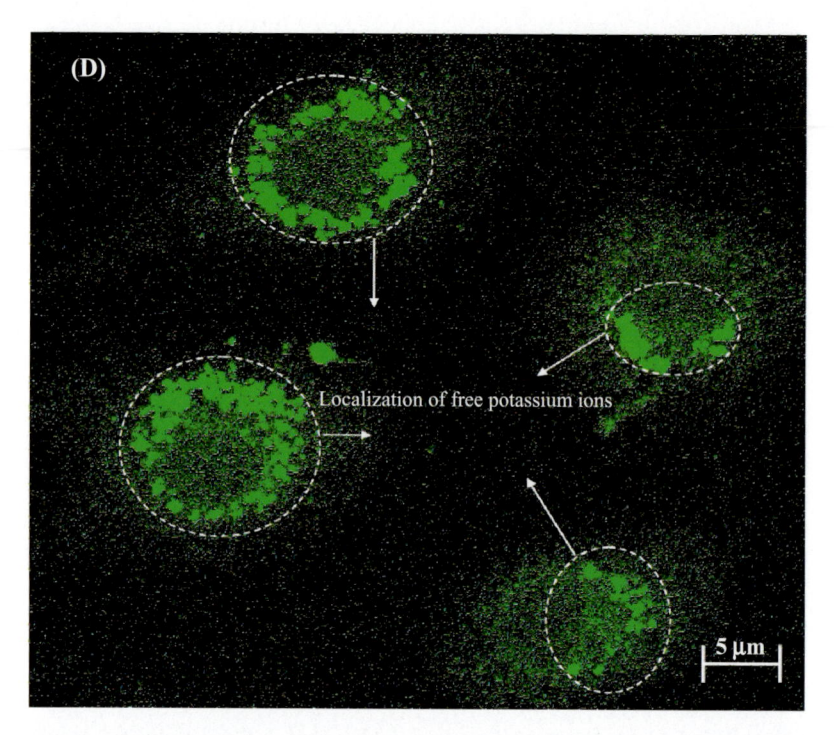

FIGURE 3.15(D). Physiological effect of pyrethroid at 890 ppb on osteoblasts. The cells are stained with PBFI stain for visualizing free potassium ion transients. The confocal optical micrograph that shows the presence of potassium bundles after exposure to PBFI stain. This behavior correlates to the in-vitro behavior of large cell populations after exposure to organophosphates.

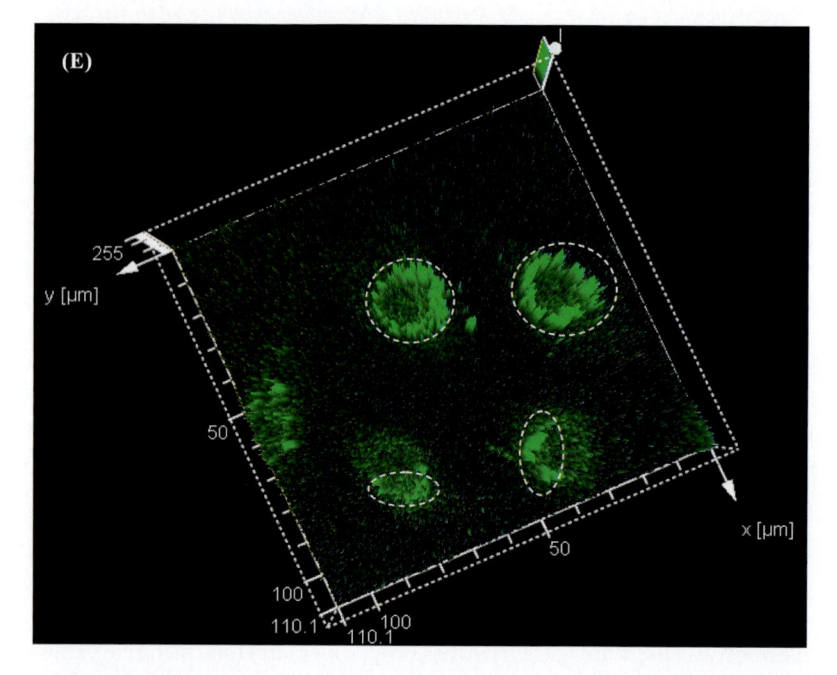

FIGURE 3.15(E). Three dimensional vector representation of the fluorescence intensity. The circled areas represent the high intensity peaks. The high intensity peaks are distributed uniformly near the cell membrane. These areas correspond to the areas of localization of the free potassium ions released due to the exposure to pyrethroid.

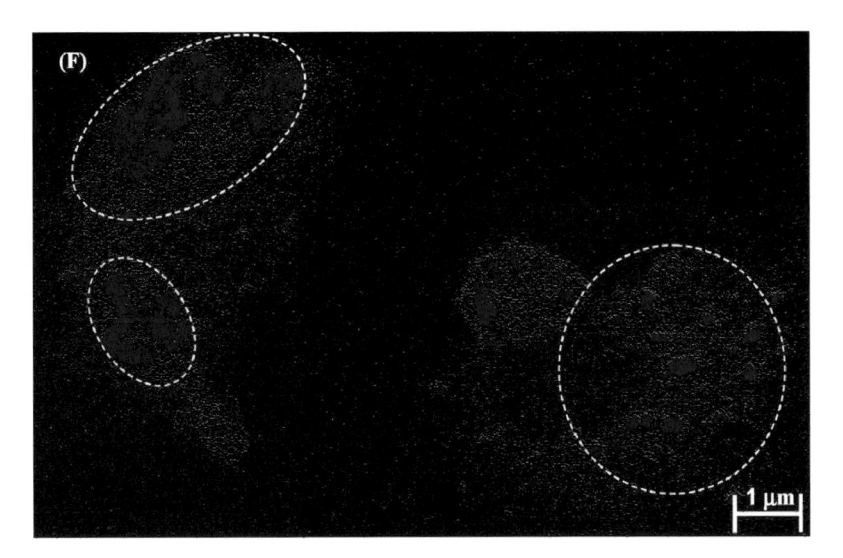

MICROARRAY AND FLUIDIC CHIP FOR EXTRACELLULAR SENSING 93

FIGURE 3.15(F). Single osteoblasts stained with Hoechst 33382 stain after being exposed to pyrethroid at 890 ppb concentration for a period of one sensing cycle. The circled areas indicate the regions where the nucleic matter has condensed due to apoptosis. This behavior correlates to the in-vivo apoptotic behavior in osteoblasts after exposure to organophosphates at physiologically relevant concentrations.

the mechanical behavior of the bone. The bone samples became less brittle as the time of exposure to EDTA increased [8]. This translates into intracellular calcium emission after exposure to EDTA. His can be imaged by using fluo-3. This is a calcium indicator that binds to the generated free calcium ions. Figure 3.16(d) is the confocal optical micrograph that shows the calcium transients due to cell exposure to EDTA at 280 ppb for a period of one sensing cycle. Figure 3.16(e) is the three dimensional vector transformation of the confocal optical micrograph. The z axis gives the intensity levels. The fluorescence intensity peaks correspond to the locations of accumulation of calcium ions. EDTA also results in the DNA ladder formation. This occurs due to the clumping of the nucleic matter that happens due to the on-set of apoptosis [135]. Nucleic acid fragmentation is observed in the patterned osteoblasts exposed to 280 ppm concentration EDTA and stained with Hoechst 33382. Figure 3.16(f) is the confocal optical micrograph indicating this behavior.

3.35. DISCUSSION AND CONCLUSIONS

The past decade has seen great advancements in the field of bioanalysis along many fronts. Among the most rapidly advancing of these fronts is the area of biosensing, whether it is single analyte detection methods or multiarray-based biochip technology. The 1990s have seen the development of biosensors for many different analyses, and even seen them begin to advance to clinical and in some cases commercially available technologies [78, 97].

This great interest in the field of biosensors and biochips has revealed a great deal of information about the biology of all living things and may eventually provide an easy method

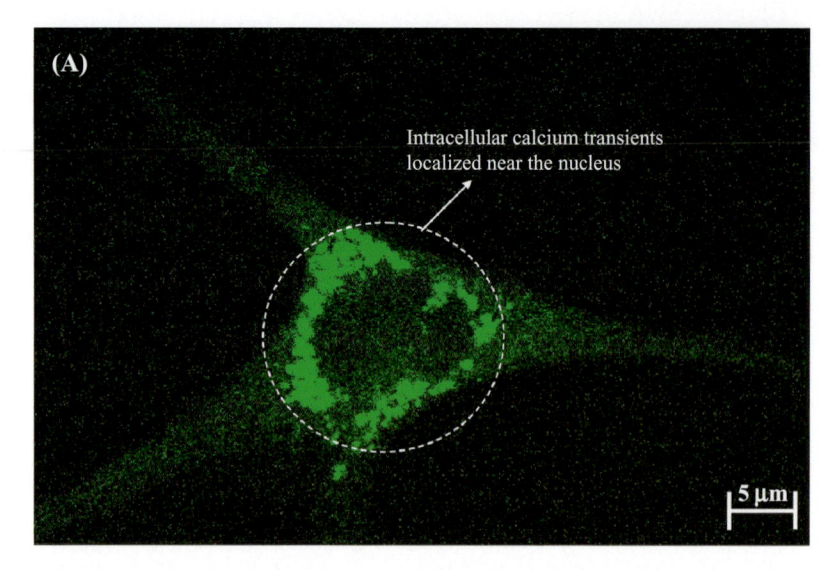

FIGURE 3.16(A). Physiological effect of EDTA at 180 ppm on individual neurons. The cells are stained with fluo-3 for visualizing free calcium ion transients. The calcium transients are localized in areas near the nucleus.

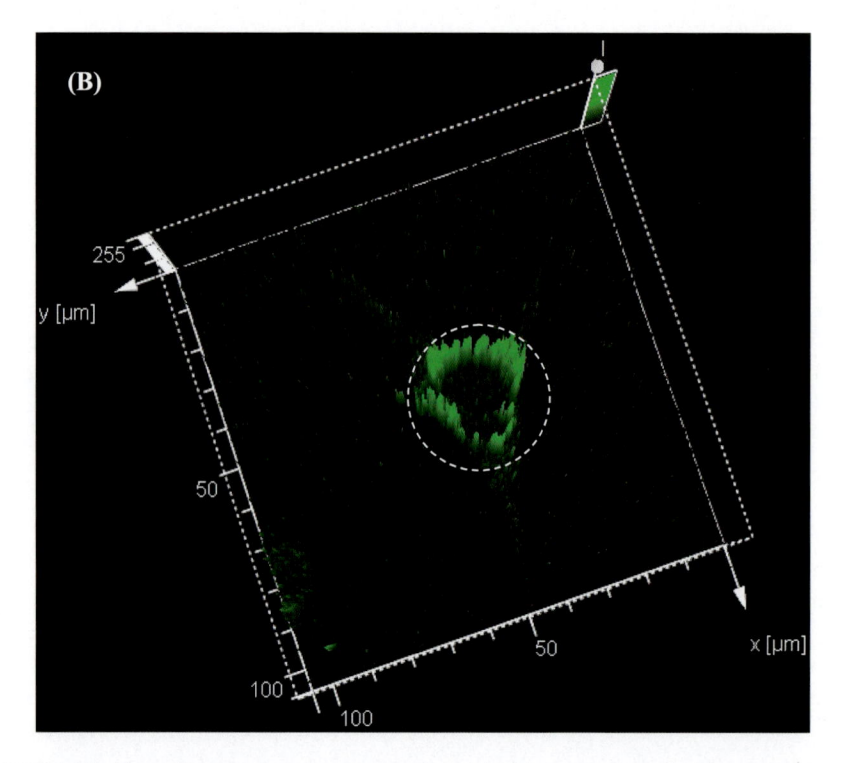

FIGURE 3.16(B). Three dimensional vector representation of the fluorescence intensity. The circled area represents the high intensity peaks. The high intensity peaks are distributed uniformly near the nucleus. These areas correspond to the areas of localization of the free calcium ions released due to the exposure to EDTA.

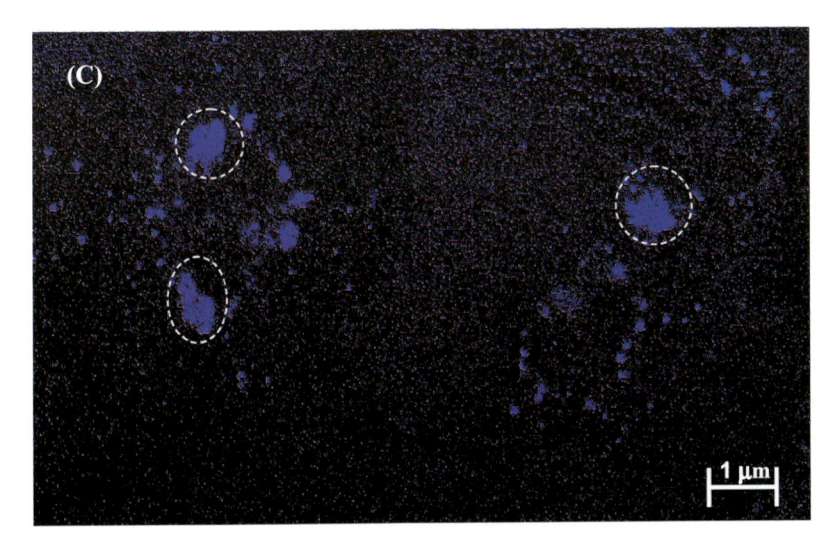

MICROARRAY AND FLUIDIC CHIP FOR EXTRACELLULAR SENSING 95

FIGURE 3.16(C). Single neurons stained with Hoechst 33382 stain after being exposed to EDTA at 180 ppm concentration for a period of one sensing cycle. The circled areas indicate the regions where the chromatin has condensed due to apoptosis.

for people to one day test themselves for certain illnesses at home or aid in the understanding of genetically transmitted illnesses. For practical medical diagnostic applications, there is currently a strong need for a truly integrated biochip system that comprises probes, samplers, detector as well as amplifier and logic circuitry. Such a system will be useful in physician's offices and could be used by relatively unskilled personnel.

Biochip technologies offer a unique combination of performance capabilities and analytical features of merit not available in any other bioanalytical system currently available. With its multichannel capability, biochip technology allows simultaneous detection of multiple biotargets. Biochip systems have great promise to offer several advantages in size, performance, fabrication, analysis and production cost due to their integrated optical sensing microchip. The small sizes of the probes (microliter to nanoliter) minimize sample requirement and reduce reagent and waste requirement. Highly integrated systems lead to a reduction in noise and an increase in signal due to the improved efficiency of sample collection and the reduction of interfaces. The capability of large-scale production using low-cost integrated circuit (IC) technology is an important advantage. The assembly process of various components is made simple by integration of several elements on a single chip. For medical applications, this cost advantage will allow the development of extremely low cost, disposable biochips that can be used for in-home medical diagnostics of diseases without the need of sending samples to a laboratory for analysis.

Bioreceptors are the key to specificity for biosensor technologies. They are responsible for binding the analyte of interest to the sensor for the measurement. These bioreceptors can take many forms and the different bioreceptors that have been used are as numerous as the different analytes that have been monitored using biosensors. Current receptor technology based on enzyme-substrate interactions, nucleic acid methods, antibody-antigen interactions are highly specific. Broad spectrum detection of analytes is not possible using these

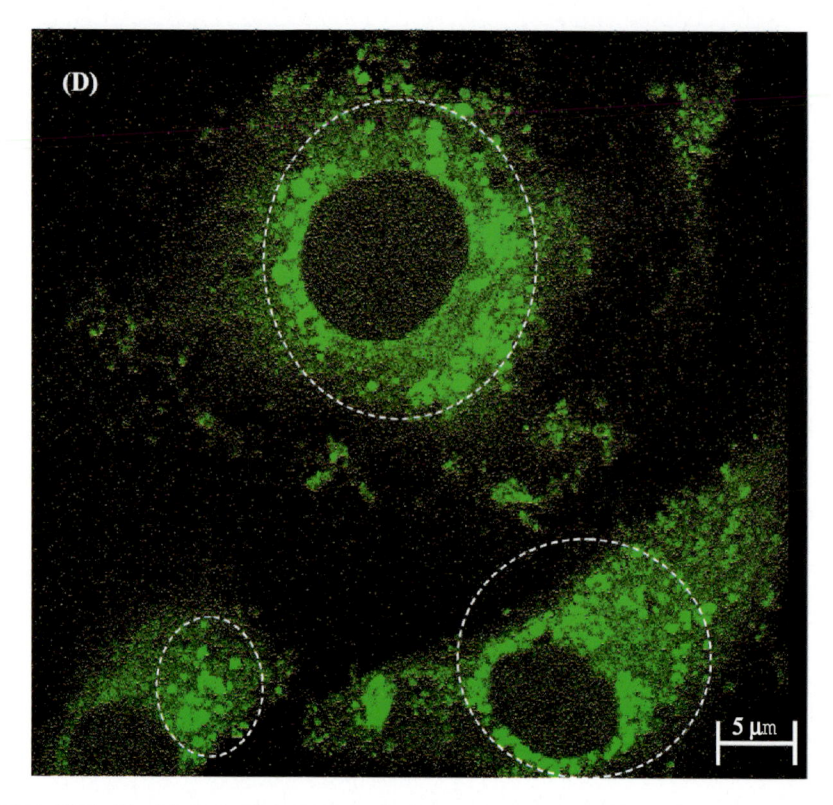

FIGURE 3.16(D). Physiological effect of EDTA at 280 ppm on osteoblasts. The cells are stained with fluo-3 for visualizing free calcium ion transients. The circled areas in the confocal optical micrograph show the presence of calcium bundles in the cytoplasm after exposure to fluo-3.

techniques. Moreover these methods are laboratory bound making them difficult to compartmentalize and commercialize other than for very specific applications as mentioned in previous sections. Thus there is a need for a different genre of sensors that have the broad based identification capability along with the associated advantages of biochip technology. Cell based biosensors offer this dual capability.

Cell-based biosensors constitute a promising field that has numerous applications ranging from pharmaceutical screening to environmental monitoring. Cells provide an array of naturally evolved receptors and pathways that can respond to an analyte in a physiologically relevant manner. Enzymes, receptors, channels, and other signaling proteins that may be targets of an analyte are maintained and, as necessary, regenerated by the molecular machinery present in cells. The array of signaling systems characteristic of cell-based sensors yields generic sensitivity that is a distinguishing feature in comparison to other molecular biosensor approaches. In addition, cell-based sensors offer an advantage of constituting a function-based assay that can yield insight into the physiologic action of an analyte of interest. Three important issues that constitute barriers for the use of cell-based sensors have been presented and discussed. There are certainly other areas that will require attention, as cell-based sensors move from the laboratory environment; namely, cell delivery and/or

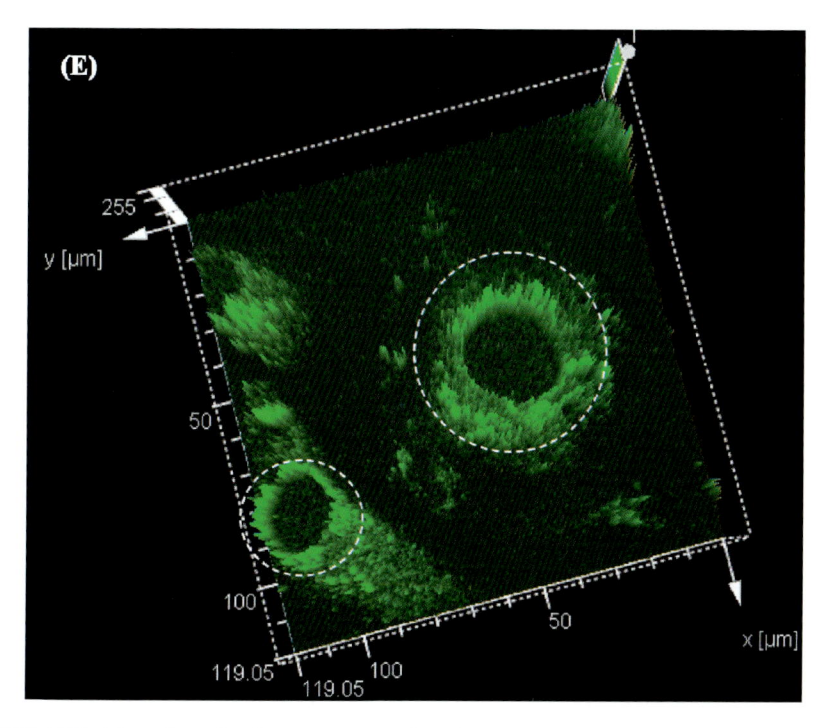

FIGURE 3.16(E). Three dimensional vector representation of the fluorescence intensity. The circled areas represent the high intensity peaks. The high intensity peaks are distributed uniformly throughout the cytoplasm. These areas correspond to the areas of localization of the free calcium ions released into the osteoblast's cytoplasm due to the exposure to EDTA.

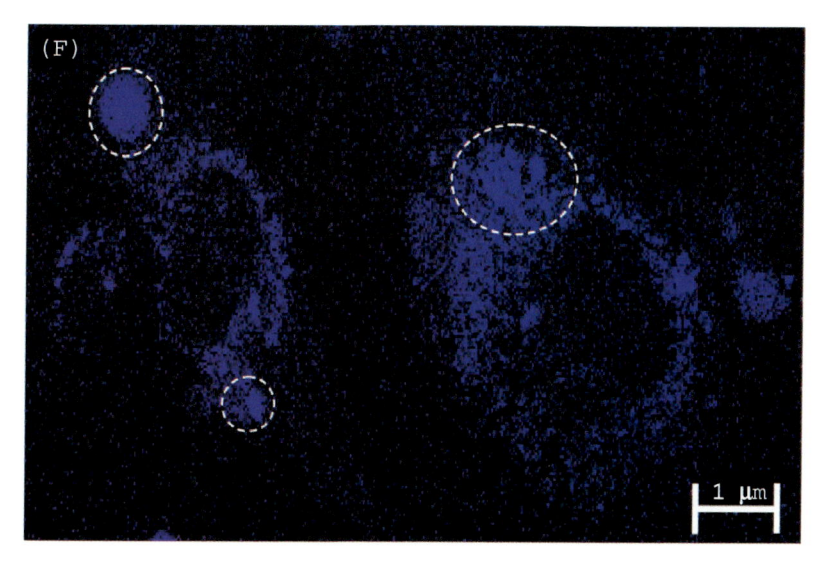

FIGURE 3.16(F). Single osteoblasts stained with Hoechst 33382 stain after being exposed to EDTA at 280 ppm concentration for a period of one sensing cycle. The circled areas indicate the regions of fragmentation of the nucleic acid. This behavior correlates with the results obtained from large cell populations after exposure to EDTA.

preservation technologies. As further progress is made to address fundamental challenges, cell-based biosensors and related cellular function based assays will undoubtedly become increasingly important and useful. It is of popular belief that such function-based assays will become an indispensable tool for monitoring in environmental, medical, and defense applications. Strategies relying on a single population of excitable cells appear most well suited for measurements of acute and direct effects of receptor agonist/antagonists. Compounds that fall within this category include ion channel modulators, metals, ligand–receptor blockers, and neurotransmitters. In fact, the detection of acute and direct effects of compounds may be sufficient and relevant for certain operational situations, such as a battlefield environment or the floor of an assembly plant, where cognitive function is absolutely critical. The prospect of detecting all physiologically active analytes using a single cell or tissue type is improbable. It is possible that particular analytes may undergo biotransformation, resulting in a secondary or tertiary compound of substantial physiologic effect. In spite of that drawback single cell based sensors are highly reliable. This has been shown by the newly developed single cell based sensing technique. This technique functions on the principle of integrating a fundamental biological tool like dielectrophoresis to biochip technology. Single cell arrays of the same biological state and differentiation can be developed using this method. Simultaneous sensing can be achieved, which reduces false alarms. Unique identification tags have been generated for identifying specific chemical analytes using this technique. These are known as Signature Patterns. The greatest advantage of this technique is its high sensitivity and speed of response. Chemical analytes of concentrations in the order of parts per billion have been detected. To determine the veracity and reliability of the sensor simultaneous fluorescence detection techniques have also been implemented at the detection limit obtained from the single cell based sensor. The physiological behavior corroborates the sensing. This establishes the viability of this technique for potential commercial implementation.

Finally, the threat of biological weapons has become a major concern to both the civilian and military populations [51]. All of the present biological warfare and environmental agent rapid detection systems, in field use or under prototype development, rely on structural recognition approaches to identify anticipated agents. Cell based sensor technology utilizing biochip capability can be thought to be one potentially reliable solution.

REFERENCES

[1] D. Aga. Environmental immunoassays: alternative techniques for soil and water analysis. In D. Aga, and E.M. Thurman (eds.), *Immunochemical Technology for Environmental Applications*, vol. 657, American Chemical Society, p. 212, 1997.
[2] B.M. Applegate, S.R. Kermeyer, and G.S. Sayler. *Appl. Environ. Microbiol.*, 64:2730, 1998.
[3] J.Z. Bao, C.C. Davis, and R.E. Schmukler. *Biophys. J.*, 61:1427, 1992.
[4] M.F. Bear, C.W. Barry, and M.A. Paradiso. *Neuroscience: Exploring the Brain*, 2nd ed., Lippincott, Williams and Wilkins, Baltimore, MD, p.147, 1999.
[5] F.F Becker, X.B. Wang, Y. Huang, R. Pethig, J. Vykoukal, and P.R.C. Gascoyne. *J. Phys. D. Appl. Phys.* 27:2659, 1994.
[6] S. Belkin, D.R. Smulski, S. Dadon, A.C. Vollmer, T.K. Van Dyk, and R.A. Larossa. *Water Res.*, 31:3009, 1997.
[7] L. Bousse. *Sens. Actu. B*, 34:270, 1996.
[8] J.J. Broz, S.J. Simske, and A.R. Greenberg. *J. Biomech.*, 28(11):1357, 1995.

[9] L.I. Bruijn, M.K. Houseweart, S. Kato, K.L. Anderson, S.D. Anderson, E. Ohama, A.G. Reaume, R.W. Scott, and D.W. Cleveland. *Science*, 281:1851, 1998.
[10] R.S. Burlage, A.V. Palumbo, A. Heitzer, and G. Sayler. *Appl. Microbiol. Biotechnol.*, 45:731, 1994.
[11] J.P.H. Burt, R. Pethig, M.S. Talary, and J.A. Tame. *International Progress in Engineered Precision*, Elseveir, Netherlands, p. 476, 1995.
[12] M.P. Byfield and R.A. Abuknesha. *Biosens. Bioelectron.*, 9:373, 1994.
[13] L. Campanella, G. Favero, D. Mastrofini, and M. Tomasetti. *J. Pharm. Biomed. Anal.*, 14:1007, 1996.
[14] D.W. Chan and T.D. Yager. *Chromosomal*, 107(1):39, 1998.
[15] J.C. Chang, G.J. Brewer, and B.C. Wheeler. *J. Biomed. Microdev.*, 2(4):245, 2000.
[16] K. Carlson, B.S. Jortner, and M. Ehrich. *Toxicol. Appl. Pharmacol.*, 168(2):102, 2000.
[17] D. Charych, Q. Cheng, A. Reichert, G. Kuziemko, M. Stroh, J.O. Nagy, W. Spevak, and R.C. Stevens. *Chem. Biol.*, 3:113, 1996.
[18] H. Chen, D. Hayakawa, S. Emura, Y. Ozawa, H. Taguchi, R. Yano, and S. Shoumura. *Histol. Histopathol.*, 6(3):763, 2001.
[19] P. Connolly, G.R. Moores, W. Monaghan, J. Shen, S. Britland, and P. Clark. *Sens. Actu.*, B6:113, 1992.
[20] D. Cooke and R. O'Kennedy. *Anal. Biochem.*, 274:188, 1999.
[21] B.A. Cornell, V.L. Braach Maksvytis, L.G. King, P.D. Osman, B. Raguse, L. Wieczorek, and R.J. Pace. *Nature*, 387:580, 1997.
[22] J. Csicsvari, D.A. Henze, B. Jamieson, K.D. Harris, A. Sirota, P. Bartho, K.D. Wise, and G. Buzsaki. *J. Neurophysiol.*, 90:1314, 2003.
[23] K.W. Dunn, S. Mayor, J.N. Myers, and F.R. Maxfield. *FASEB J.*, 8:573, 1994.
[24] C.F. Edman, D.E. Raymond, D.J. Wu, E.G. Tu, R.G.Sosnowski, W.F. Butler, M. Nerenberg, and M.J. Heller. *Nucleic Acids Res.*, 25:4907, 1997.
[25] R. Ehret, W. Baumann, M. Brischwein, A. Schwinde, K. Stegbauer, and B. Wolf. *Biosens. Bioelectron.*, 12:29, 1997.
[26] S. Ekelund, P. Nygren, and R. Larsson. *Anti-Cancer Drugs*, 9:531, 1998.
[27] J.W. Ellwart and P. Dormer. *Cytometry*, 11:239, 1990.
[28] M. Emgard, K. Blomgren, and P. Brundin. *Neuroscience*, 115(4):1177, 2002.
[29] G.A. Evtugyn, E.P. Rizaeva, E.E. Stoikova, V.Z. Latipova, and H.C. Budnikov. *Electroanalysis*, 9:1, 1997.
[30] P.B. Fernandes. *Curr. Opin. Chem. Biol.*, 2:597, 1998.
[31] F. Foch-Anderson and P. D'Orazio. *Clin. Chem.*, 44:655, 1998.
[32] G. Fuhr, R. Hagedorn, T. Müller, W. Benecke, B. Wagner, and J. Gisma. *Stud. Biophys.*, 120:79, 1991.
[33] G. Fuhr, H. Glasser, T. Muller, and T. Schnelle. *Biochim. Biophys. Acta*, 1201:353, 1994.
[34] G. Fuhr, T.Müller, T. Schnelle, R. Hagedorn, A. Voigt, S. Fiedler, W.M. Arnold, U. Zimmermann, B. Wagner, and A. Heuberger. *Naturwissenschaften*, 81:528, 1994.
[35] G. Fuhr and S.G. Shirley. *J. Micromech. Microeng.*, 5:77, 1995.
[36] G. Fuhr, A. Voigt, T. Müller, B. Wagner, K. Reimer, and T. Lisec. *Sens. Actu. B*, 26–27:468, 1995.
[37] I. Giaever and C.R. Keese. *IEEE Trans. Biomed. Eng.*, 33:242, 1986.
[38] I. Giaever and C.R. Keese. *Proc. Natl. Acad. Sci. USA*, 88:7896, 1991.
[39] W. Gopel, *A comprehensive survey, of sensors, Trends in sensor technology/sensor markets*, Elseiver, Netherlands, p. 295, 1995.
[40] K.A. Giuliano, R.L. DeBiasio, R.T. Dunlay, A. Gough, J.M. Volosky, J. Zock, G.N. Pavlakis, and D.L. Taylor. *J. Biomol. Screening*, 2:249, 1997.
[41] C.R. Graham, D. Leslie, and D.J. Squirrell. *Biosens. Bioelectron.*, 7:487, 1992.
[42] G.W. Gross, W. Wen, and J. Lin. *J. Neurosci. Meth.*, 15:243, 1985.
[43] G.W. Gross, B.K. Rhoades, and R. Jordan. *Sens. Actu. B*, 6:1, 1992.
[44] G.W. Gross, B.K. Rhoades, D.L. Reust, F.U. Schwalm. *J. Neurosci. Meth.*, 50:131, 1993.
[45] G.W. Gross. Internal dynamics of randomized mammalian neuronal networks in culture. In D.A. Stenger and T.M. McKenna (eds.), *Enabling Technologies for Cultured Neuronal Networks*, San Diego: Academic, San Diego, p. 277, 1994.
[46] G.W. Gross, B.K. Rhoades, H.M.E. Azzazy, and M.C. Wu. *Biosens. Bioelectron.*, 10:553, 1995.
[47] L. Griscom, P. Degenaar, B. LePioufle, E. Tamiya, and H. Fujita. *Sens. Actu. B*, 83(1-3):15, 2002.
[48] E.A.H. Hall. *Biosensors*, Prentice Hall, Englewood Cliffs, New Jersey, 1991.
[49] J.M. Hall, J. More Smith, J.V. Bannister, and I.J. Higgins. *Biochem. Mol. Biol. Int.*, 32:21, 1994.
[50] C.M. Hanbury, W.G. Miller, and R.B. Harris. *Clin. Chem.*, 3:2128, 1997.

[51] D.A. Henderson. *Science*, 283:1279, 1999.
[52] A.W. Hendricson, M.P. Thomas, M.J. Lippmann, and R.A. Morrisett. *J. Pharmacol. Exp. Ther.*, 307(2):550, 2003.
[53] T. Henning, M. Brischwein, W. Baumann, R. Ehret, I. Freund, R. Kammerer, M. Lehmann, A. Schwinde, and B. Wolf. *Anticancer Drugs.*, 12(1):21, 2001.
[54] T.J. Heppner and J.F. Fiekers. *Brain Res.*, 563:303, 1991.
[55] T.J. Heppner and J.F. Fiekers. *Comp. Biochem. Physiol.*, 102C:335, 1992.
[56] H.R. Hoogenboom, A.P. de Bruine, S.E. Hufton, R.M. Hoet, J.W. Arends, and R.C. Roovers. *Immunotechnology*, 4:1, 1998.
[57] X. Hu, I.M. Johansson, T. Brannstrom, T. Olsson, and P. Wester. *Acta Neuropathol.*, 104(5):462, 2002.
[58] A. Hulanicki, S. Glab, and F. Ingman. *Pure Appl. Chem.*, 63:1247, 1991.
[59] C.K. Javawickreme, S.P. Javawickreme, and M.R. Lerner. *Methods Mol. Biol.*, 87:119, 1998.
[60] M. Jenker, B. Muller, and P. Fromherz. *Biol. Cybernetics.*, 84:239, 2001.
[61] T.B. Jones. *Electromechanical Properties of Particles*, Cambridge University Press, UK, p. 5, 1995.
[62] T.B. Jones and M. Washizu. *J. Electrostat.*, 37:121, 1996.
[63] H. Kamioka, E. Maeda, Y. Jimbo, H.P.C. Robinson, and A. Kawana. *Neurosci. Lett.*, 206:109, 1996.
[64] I.S. Kampa and P. Keffer. *Clin. Chem.*, 44:884, 1998.
[65] I. Karube, K. Yokoyama, K. Sode, and E. Tamiya. *Anal. Lett.* 22(4):791, 1989.
[66] I. Karube, T. Matsunaga, S. Mitsuda, and S. Suzuki. *Biotechnol. Bioeng.*, 19(10):1535, 1977.
[67] C.R. Keese and I. Giaever. *IEEE Proc. Biomed. Eng.*, 12:500, 1990.
[68] J.M.H. King, P.M. DiGrazia, B. Applegate, R. Burlage, J. Sanseverino, P. Dunbar, F. Larimer, and G.S. Sayler. *Science*, 249:778, 1990.
[69] Y.I. Korpan, M.V. Gonchar, N.F. Starodub, A.A. Shul'ga, A.A. Sibirny, and A.V. El'skaya. *Anal. Biochem.*, 215:216, 1993.
[70] V.G. Koveshnikov and V.S. Pikaliuk. *Morfologiia*, 104(3-4):364, 1993.
[71] C.M. Kurbacher, I.A Cree, and H.W. Bruckne. *Anti-Cancer Drugs*, 9:51, 1998.
[72] M.R. Lerner. *Trends Neurosci.*, 17:142, 1994.
[73] Y.R. Li and J. Chu. *Appl. Biochem. Biotechnol.*, 28-29:855, 1991.
[74] F.S. Ligler, T.L. Fare, K.D. Seib, J.W. Smuda, A. Singh, P. Ahl, M.E. Ayers, A. Dalziel, and P. Yager. *Med. Instrum.*, 22:247, 1988.
[75] J. Liu, L. Björnsson and B. Mattiasson. *Biosens. Bioelectron.*, 14:883, 2000.
[76] M.P. Maher, J. Pine, J. Wright, and Y.C. Tai. *J. Neurosci. Meth.*, 87:45, 1999.
[77] R.E. Maldve, T.A. Zhang, K. Ferrani-Kile, S.S. Schreiber, M.J. Lippmann, G.L. Snyder, A.A. Fienberg, S.W. Leslie, R.A. Gonzales, and R.A. Morrisett. *Nat. Neurosci.*, 5:641, 2002.
[78] M. Malmquist. *M. Biochem. Soc. T.*, 27:335, 1999.
[79] G.H. Markx, H. Talary, and R. Pethig. *J. Biotechnology.*, 32:29, 1994.
[80] G.H. Markx and R. Pethig. *Biotech. Bioeng.*, 45:337, 1995.
[81] S. Masuda, M. Washizu, and T. Nanba. *IEEE Trans. Ind. Appl.*, 25:732, 1989.
[82] H.M. McConnell, J.C. Owicki, J.W. Parce, D.L. Miller, G.T. Baxter, H.G. Wada, and S. Pitchford. *Science*, 257:1906, 1992.
[83] M. Mehrvar, C. Bis, J.M. Scharer, M. Moo-Young, and J.H. Luong. *Anal. Sci.*, 16:678, 2000.
[84] M. Meusel and T. Vering. *Biosens. Bioelectron.*, 13:IX, 1998.
[85] K.M. Millan and S.R. Mikkelsen. *Anal. Chem.*, 65:2317, 1993.
[86] C. Miller, *Ion Channel Reconstitution*, Plenum Press, New York, 1986.
[87] P. Mitra, C.R. Keese, and I. Giaever. *Biotechniques*, 1(4):504, 1991.
[88] T. Momiyama, *Folia Pharmacol. Japon.*, 121(3):174, 2003.
[89] T. Müller, A. Gerardino, T. Schnelle, S.G. Shirley, F. Bordoni, G. De Gasperis, R. Leoni, and G. Fuhr. *J. Phys. D: Appl. Phys.*, 29:340, 1996.
[90] T. Müller, S. Fiedler, T. Schnelle, K. Ludwig, H. Jung, and G. Fuhr. *Biotechnol. Tech.*, 10:221, 1996.
[91] S.H. Nam, S.Y. Jung, C.M. Yoo, E.H. Ahn, and C.K. Suh. *Yonsei. Med. J.*, 43(2):229, 2002.
[92] E. Neuwmann and A.E. Sowers. *Electroporation and Electrofusion in Cell Biology*, C.A. Jordan (ed.), Plenum Press, New York, NY, 1989.
[93] E. Noiri, Y. Hu, W.F. Bahou, C.R. Keese, I. Giaever, and M.S. Goligorsky. *J. Biol. Chem.*, 272:1747, 1997.
[94] E.R. O'Connor, H.K. Kimelberg, C.R. Keese, and I. Giaever. *Am. J. Physiol.* 264(2 Pt 1):C471, 1993.
[95] A. Offenhausser, C. Sprossler, M. Matsuzawa, and W. Knoll. *Biosens. Bioelectron.*, 12(8):819, 1997.

[96] O. Orwar, K. Jardemark, I. Jacobson, A. Moscho, H.A. Fishman, R.H. Scheller, and R.N. Zare. *Science*, 272:1779, 1996.
[97] A. Ota and S. Ueda. *Hybridoma*, 17:471, 1998.
[98] B.M. Paddle. *Biosens. Bioelectron.*, 11:1079, 1996.
[99] J.J. Pancrazio, S.A Gray, Y.S. Shubin, N. Kulagina, D.S. Cuttino, K.M. Shaffer, K. Eisemann, A. Curran, B. Zim, G.W. Gross, and T.J. O'Shaughnessy. *Biosens. Bioelectron.*, 18(11):1339, 2003.
[100] A. Parbhu, H. Lin, J. Thimm, and R. Lal. *Peptides*, 23(7):1265, 2002.
[101] R. Pethig. *Dielectric and Electronic Properties of Biological Materials*, J. Wiley and sons, United Kingdom, p. 186, 1979.
[102] R. Pethig and D.B. Kell. *Phys. Med. Biol.*, 32:933, 1987.
[103] R. Pethig. Application of AC electrical fields to the manipulation and characterization of cells. In I. Karube (ed.), *Automation in Biotechnology—Proc. 4th Conf.*, Elsevier, Netherlands, p. 159, 1990.
[104] R. Pethig. Automation in Biotechnology, I. Kaurbe (ed.), Elsevier, Netherlands, p. 159, 1991.
[105] R.T. Piervincenzi, W.M. Reichert, and H.W. Hellinga. *Biosens. Bioelectron.*, 13:305, 1988.
[106] P.A. Piunno, U.J. Krull, R.H. Hudson, M.J. Damha, and H. Cohen. *Clin. Chem.*, 67:2635, 1995.
[107] H.A. Pohl, *Dielectrophoresis*, Cambridge University Press, United Kingdom, 1978.
[108] D. Pollard-Knight, E. Hawkins, D. Yeung, D.P. Pashby, M. Simpson, A. McDougall, P. Buckle, and S.A. Charles. *Ann. Biol. Clin.*, 8:642, 1990.
[109] R.A. Potyrailo, R.C. Conrad, A.D. Ellington, and G.M. Hieftje. *Anal. Chem.*, 70:3419–3425, 1998.
[110] S. Prasad, M. Yang, X. Zhang, C.S. Ozkan, and M. Ozkan. *J. Biomed. Microdevices.*, 5(2):125, 2003.
[111] S. Prasad, X. Zhang, M. Yang, Y. Ni, V. Parpura, C.S. Ozkan, and M. Ozkan. *J. Neurosci. Meth.* (accepted)
[112] J.A.R. Price, J.P.H Burt, and R. Pethig. *Biochim. Biophys. Acta*, 964:221, 1988.
[113] B. Pucci, L. Bellincampi, M. Tafani, V. Masciullo, G. Melino, and A. Giordano. *Exp. Cell Res.*, 252(1):134, 1999.
[114] K.R. Rogers and M. Mascini. *Field. Anal. Chem. Technol.*, 2:317, 1998.
[115] K.R. Rogers. *Biosens. Bioelectron.*, 10:533, 1995.
[116] C. Rosenmund and G.L. Westbrook. *J. Physiol.*, 470:705, 1993.
[117] T.G. Ruardij, M.H. Goedbloed, and W.L.C. Rutten. *Med. Biol. Eng. Comput.* 41(2):227, 2003.
[118] J. Ruhe, R. Yano, J.S. Lee, P. Koberle, W. Knoll, and A. Offenhausser. *J. Biomater. Sci. Polym. Ed.*, 10(8):859, 1999.
[119] F. Scheller and F. Schubert, *Biosensors*, Elsevier, Netherlands, 1992.
[120] M. Scholl, C. Sprossler, M. Denyer, M. Krause, K. Nakajima, A. Maelicke, W. Knoll, and A. Offenhausser. *J. Neurosci. Meth.*, 104:65, 2000.
[121] O. Selifonova, R.S. Burlage, and T. Barkay. *Appl. Environ. Microbiol.* 59:3083, 1993.
[122] G.J. Singh and B. Singh. *Neurobehav. Toxicol. Teratol.*, 6(3):201, 1984.
[123] J. Singh, P. Khosala, and R.K. Srivastava. *Ind. J. Pharmacology.*, 32:206, 2000.
[124] M.A. Smith, P.S. Herson, K. Lee, R.D. Pinnock, and M.L. Ashford. *J. Physiol.*, 547(Pt 2):417, 2003.
[125] A. Steinstrasser, A. Schwarz, C. Alexander, R. Berberich, and M. Zimmer. *Nuklearmedizin*, 31(5):164, 1992.
[126] D.A. Stenger, D.H. Cribbs, and T.L. Fare. *Biosens. Bioelectron.*, 6:425, 1991.
[127] M. Stephens, M.S. Talary, R. Pethig, A.K. Burnett, and K.I. Mills. *Bone marrow Transplant.*, 18:777, 1996.
[128] A.K. Stout, H.M. Raphael, B.I. Kanterewicz, E. Klann, and I.J. Reynolds. *Nature. Neurosci.*, 1(5):366, 1998.
[129] E. Sullivan, E.M. Tucker, and I.L. Dale. *Methods Mol. Biol.*, 114:125, 1999.
[130] J.R. Subramaniam, W.E. Lyons, J. Liu, T.B. Bartnikas, J. Rothstein, D.L. Price, D.W. Cleveland, J.D. Gitlin, and P.C. Wong. *Nature Neurosci.*, 5:301, 2002.
[131] R.M. Sutherland, C. Dähne, J.F. Place, and A.R. Ringrose. *Clin. Chem.*, 30:1533, 1984a.
[132] R.M. Sutherland, C. Dähne, J.F. Place, and A.R. Ringrose. *J. Immunol. Methods.*, 74:253, 1984b.
[133] T.C. Tan and Z.R. Qian. *Sens. Acuta B*, 40:65, 1997.
[134] T. Tanaka, Y. Taniguchi, K. Gotoh, R. Satoh, M. Inazu, and H. Ozawa. *Bone*, 14(2):117, 1993.
[135] Y. Taniguchi, T. Tanaka, K. Gotoh, R. Satoh, and M. Inazu. *Calcif. Tissue Int.*, 53(2):122, 1993.
[136] D.R. Thevenot, K. Toth, R.A. Durst, and G.S. Wilson. *Pure Appl. Chem.*, 71:2333, 1999.
[137] P. Thiebaud, L. Lauer, W. Knoll, and A. Offenhausser. *Biosens. Bioelectron.* 17(1-2):87, 2002.
[138] R.Y. Tsien. *Trends Neurosci.*, 11:419, 1988.
[139] T. Vo-Dihn, B. Cullum, and B. Fresenius. *J. Anal. Chem.*, 366:540, 2000.

[140] J. Wang, G. Rivas, X. Cai, E. Palecek, P. Nielsen, H. Shiraishi, N. Dontha, D. Luo, C. Parrado, M. Chicharro, P.A.M. Farias, F.S. Valera, D.H. Grant, M. Ozsoz, and M.N. Flair. *Anal. Chim. Acta*, 347:1, 1997.
[141] J.V.A. Watson, S.H. Nakeff, H. Chambers, and P. Smith. *Cytometry*, 6:310, 1985.
[142] H.J. Watts, D. Yeung, and H. Parkes. *Anal. Chem.*, 67:4283, 1995.
[143] C.J. Weijer. *Science*, 300:96, 2003.
[144] R. Wegeroff. *From Membrane to Mind*, 1st ed., Thieme, Stuttgart, p. 10, 1997.
[145] E. Wilkins and P. Atanasov. *Med. Eng. Phys.*, 18:273, 1996.
[146] B.C. Wheeler, J.M. Corey, G.J. Brewer, and D.W. Branch. *J. Biomech. Eng.*, 121:73, 1999.
[147] F.S. Wouters, P.J. Verveer, and P.I.H. Bastiaens, *Trends Cell Biol.*, 11(5):212, 2001.
[148] C. Wyart, C. Ybert, L. Bourdieu, C. Herr, C. Prinz, and D. Chattenay. *J. Neurosci. Meth.*, 117(2):23, 1997.
[149] M.Yang, S. Prasad, X. Zhang, A. Morgan, M. Ozkan, and C.S. Ozkan. *Sens. Mat.* (accepted)
[150] G. Zeck and P. Fromherz. *P. Nat'l. Acad. Sci. USA*, 98(18):10457, 2001.
[151] J.R. Zysk, and W.R. Baumbach. *Comb. Chem.*, 1:171, 1998.

4

Cell Physiometry Tools based on Dielectrophoresis

Ronald Pethig
School of Informatics, University of Wales, Dean Street, Bangor, Gwynedd LL57 1UT, UK

Keywords: Cell-based assays; cell separation; dielectrophoresis, microelectrodes.

Dielectrophoresis is a technique for moving cells and other particles using radio-frequency electric fields. The usefulness of this method depends on the ability to generate highly non-uniform electric fields using microelectrodes, and also on the intrinsic dielectric properties of the cells and their surrounding medium. Selective cell isolation or concentration can be achieved without the need for biochemical labels, dyes or other markers and tags, and the cells remain viable after this process. Changes in cell state, such as those associated with activation, apoptosis, differentiation, necrosis, as well as responses to chemical and physical agents for example, can be monitored by observing changes in dielectrophoretic behavior. The basic theories and experimental techniques of dielectrophoresis are described in this chapter, and a summary is given of our present understanding of how the dielectrophoretic behavior of cells relate to their physiological and physico-chemical properties.

4.1. INTRODUCTION

Although dielectrophoresis (DEP) can be considered to be a mature subject (with publications extending back more than 50 years), it has until recently mainly been a topic of interest for a relatively small number of engineers and biophysicists. There is now, however, a growing interest in exploiting the modes of interaction of electrokinetic forces with biological particles (such as cells, bacteria, viruses, proteins and DNA) with the objective of addressing key opportunities in medical diagnostics, drug discovery, cell therapeutics and biothreat defence. It has been demonstrated that DEP is capable of selectively isolating, concentrating, or purifying target bioparticles when present in complex mixtures. Examples include the isolation of stem cells, cancer cells and bacteria from blood for therapy or further analysis. DEP also lends itself readily to miniaturization and automation, either as stand-alone microdevices or as the means for rapid and efficient sample collection and preparation.

The purpose of this chapter is to provide, for interdisciplinary biomedical scientists, a *top level* description of how dielectrophoresis (DEP) can be used to *manipulate* and *characterize* cells and other bioparticles. The term *manipulate* signifies the selective separation, fractionation, enrichment, concentration, assembly or positioning of cells. Although the method may be used as part of a process to achieve *genetic* manipulation (e.g., positioning cells prior to electrofusion or electroporation) DEP acting alone cannot perform this function. The term *characterize* refers to those physico-chemical or physiological properties of bioparticles that have so far been found amenable to DEP investigation. An objective here is to outline our understanding of how the DEP properties of cells may relate to their cellular states of function, viability and differentiation, or serve as a means of phenotyping.

In line with the objective to give a *top level* description, a comprehensive review of dielectroophoresis is not attempted here. The relevant methodologies and theories are stripped down to bare essentials. Readers interested in further details can usefully refer to monographs [41, 47, 59, 66, 79] and recent reviews [25, 42, 68, 69]. The methods used in the microfabrication of devices are not described at all here, but many of the referenced papers contain such details. Finally, as the term *dielectrophoresis* implies, we are dealing with cells *being carried* as a result of their *dielectric* properties. There is an extensive literature on the dielectric properties of cells and other biological materials (e.g., tissue, axons, bacteria, DNA, proteins, amino acids) extending back to the early 1900's. The quickest way into this literature is through the monographs of Cole [14], Hasted [31], Grant *et al* [28], Pethig [66], Takashima [96] and Grimnes & Martinsen [29].

4.2. DIELECTROPHORESIS

Dielectrophoresis (DEP) is defined [78, 79] as the translational motion of neutral matter caused by polarization effects in a nonuniform electric field. This effect was observed by the ancient Greeks (e.g., Thales of Miletus, 600 B.C.) in the action of 'animated' amber attracting small particles. Polishing a piece of amber with cloth creates electric charges on its surface, and the resulting electric field can in turn polarize nearby dielectric particles into small electrets. These electrified particles (characterized by an induced electric dipole moment) can then be attracted to regions of high electric field strength on the amber surface. This is analogous to magnetophoresis, commonly observed as the attraction of iron filings to a magnet.

The dielectrophoretic collection of particles at electrode edges is shown in figure 4.1(a). An effect not possible by magnetophoretic manipulation of iron filings, namely their repulsion from a magnet followed by concentration in an aqueous medium, is shown in figure 4.1(b). The ability to attract or repel particles from electrodes is an important aspect of DEP. The translational forces producing these effects arise from the interaction of the

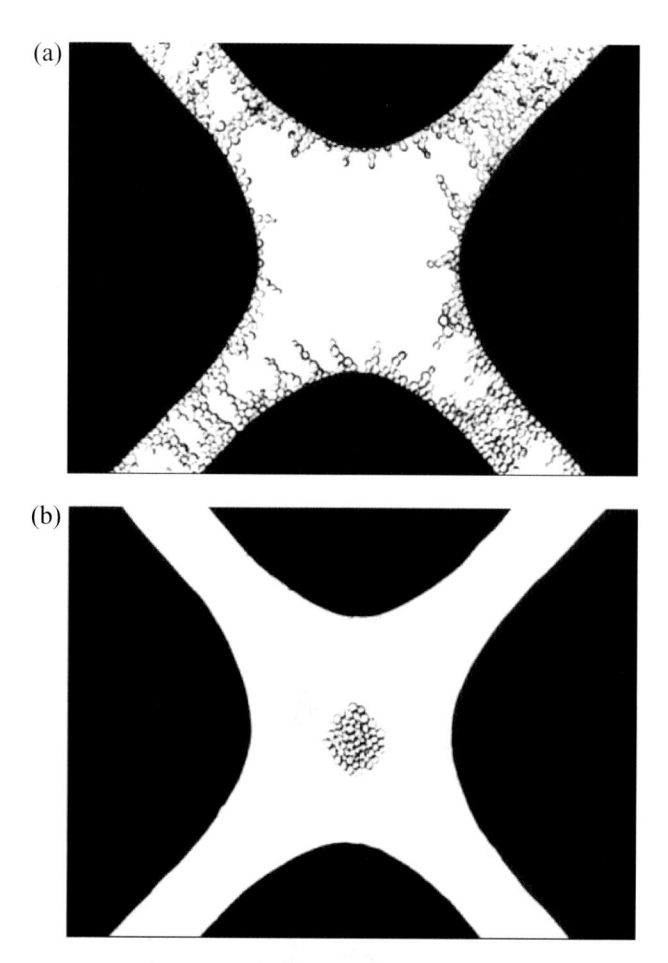

FIGURE 4.1. *Top (a)*: Yeast cells collecting at electrode edges under the influence of positive dielectrophoresis. *Bottom (b)*: Yeast cells directed away from electrode edges, into a field *cage*, under the action of negative dielectrophoresis. The *polynomial* microelectrodes used here [33] are based on isomotive electrodes used to characterize macroscopic objects [80].

particle's dipole moment m (permanent or induced) with the nonuniform electric field. This force is given by:

$$F = (m.\nabla)E \qquad (4.1)$$

where ∇ is the grad vector operator defining the gradient of the local electric field E (rms or D.C. value). The dipole moment m induced in a particle, of volume v, is given by:

$$m = pvE \qquad (4.2)$$

where p is the effective polarizability (per unit volume) of the particle, generally referred to as the Clausius-Mosotti factor (e.g., [17, 90]).

From (4.1) and (4.2) we can write the translational (DEP) force as:

$$F = pv(E.\nabla)E = \frac{1}{2}pv\nabla |E|^2 \quad (4.3)$$

This well known result in electrostatics (e.g., [1], p. 91) can be interpreted as the energy which must be expended to withdraw the particle from the local field E into a region where there is no field. Eqn. 4.3 provides several important facts, namely that the DEP force is zero if the field is uniform (i.e., for $\nabla E = 0$), and the force depends on:

- The polarizability of the particle
- The *effective* volume of the particle
- The *square* of the applied electric field magnitude
- The *geometry* of the electrodes producing the nonuniform field

The polarizability of a particle is a sensitive function of its physico-chemical properties and structure. This means that particles can be selectively manipulated as a result of their own intrinsic properties, without the need for biochemical labels, beads, dyes or other markers and tags. The dependence on particle volume indicates that we are dealing with a ponderomotive effect—with all other factors remaining constant the larger the object the greater will be the DEP force acting on it. Indeed, for objects the size of a pebble or pencil, the DEP force can be measured using a conventional weighing balance [60, 80]. The term *effective* volume reminds us that the dominant polarizabilities of small particles (e.g., proteins, DNA, viruses) are often dominated by relaxations of their electrical double layers whose total volume can be comparable to or greater than the particle. The dependence on the *square* of the field emphasizes that the DEP force is independent of the polarity of the applied field—dielectrophoresis can thus be observed using either A.C. or D.C. fields. The range of frequencies that can be employed is large, extending from around 100 Hz or less, up through radio wave frequencies and well into the microwave region (100 MHz and above).

The strong dependence on electrode geometry arises because the factor $(E.\nabla)E$ in eqn. 4.3 has dimensions of Volt^2/m^3. This instructs us that the same DEP force can be produced using a smaller applied field if the electrode dimensions are scaled down accordingly. For example, a hundred-times smaller field can produce the same effect if the electrodes are scaled down one thousand-fold (e.g., ten-fold reduction in size in each dimension). This has important implications for minimizing Joule heating effects when using DEP to characterize and manipulate cells and other biological particles in aqueous media. This was the reason, in our laboratory, to explore the use of microelectrode arrays fabricated by photolithography [81] and then excimer laser micromachining [72]. As discussed by Pethig and Markx [69] a value for the factor $(E.\nabla)E$ of around 10^{13} V^2/m^3 is required to produce a significant DEP force, and this can be achieved using microelectrodes with applied voltages of less than 10V, corresponding to an applied field of less than 10 kV/m.

Care has been taken above to emphasize the expression 'applied field' rather than 'applied voltage'. Electrode polarization effects commonly influence dielectric experiments, and are manifested as voltage drops across electrical double layers that form at electrode-solution interfaces. Electrode polarization therefore reduces the effective field in the bulk solution, and along with it a reduction of the DEP force. Electroosmotic fluid flow is also induced at the electrode-solution interface, and this can either impede or enhance the desired DEP manipulation of particles. The electrode polarization impedance has a frequency dependence of $\omega^{-\beta}$, where $\beta = 0.7 \sim 0.9$ (e.g., [4]) and so its effect increases with decreasing frequency. As a rough working rule, for solutions of conductivity 100 mS/m and lower, electrode polarization effects become apparent at frequencies below around 10 kHz. This can either be corrected by determining the magnitude of the electrode polarization (e.g., [9]) or the effect can be reduced by the well known method of applying platinum black to the electrodes (e.g., [37]). Electroosmotic fluid flow effects become increasingly important as the dimensional scale of a DEP device is reduced below the micron range (e.g., [8, 59, 82]).

4.3. DIELECTRIC POLARIZABILITY OF BIOPARTICLES

Some biological particles (e.g., proteins, peptides) possess permanent dipole moments, whereas others (e.g., DNA, RNA) do not, but can have large dipoles induced in them as a result of counter-ion displacement effects [69]. Other bioparticles (e.g., viruses, bacteria, cells, yeast, parasites) are assumed to possess no permanent dipole moment, and the accepted method of understanding their dielectric and dielectrophoretic behavior is to treat them as particles having a multilayered (multi-shell) structure. Each concentric layer represents a constituent component, such as a membrane, cell wall, cytoplasm, nucleus, vacuole, etc. [45, 47, 67, 86, 95].

When cells are subjected to an electric field, charges are induced to appear at the interfaces defining their gross structure. The induced charge density can be calculated [67] as typically some three decade orders of magnitude smaller than the uniformly distributed net surface charge that exists ($\sim -1\mu C \cdot cm^{-2}$) on the surface of cells and micro-organisms. The important fact is that the induced charge is not uniformly distributed about the cell surface, but is distributed so as to form an effective dipole moment. An alternative approach is to quantify this effect in terms of the electromagnetic momentum balance *via* the Maxwell stress tensor ([1], pp. 104–108; [87]). The effective dipole moment approach leads to the same phenomenological results and is more 'user friendly'. Jones [47] provides an excellent treatment of the effective dipole moment method for understanding DEP and related phenomena.

4.4. DYNAMICS OF INTERFACIAL POLARIZATION

One of the important factors controlling the frequency-dependent behavior of DEP is related to the dynamics of the interfacial charging effect. This effect is known as Maxwell-Wagner interfacial polarization, formulated by Maxwell [56] and Wagner [100]. The induced charge is generated with a characteristic relaxation time τ given by:

$$\tau = \frac{\varepsilon_p + 2\varepsilon_m}{\sigma_p + 2\sigma_m}\varepsilon_o \quad (4.4)$$

where ε_o is the permittivity of free space, and ε_m and σ_m are the relative permittivity and conductivity of the suspending medium, respectively. The parameters ε_p and σ_p are the *effective* relative permittivity and conductivity of the particle, respectively. The term *effective* is used to signify that a heterogeneous (multi-shell) particle may be replaced conceptually with one having homogeneous *smeared-out* bulk properties, such that substitution of one particle with the other would not alter the electric field in the surrounding medium. For red blood cells, suspended in an aqueous electrolyte, eqn. 4.4 yields values for τ of about 0.1 µs, signifying the presence of a dielectric dispersion centered at a frequency $(2\pi\tau)^{-1}$ of about 2 MHz.

Various formulations of the polarization dynamics are given in the literature, and one that leads to simple insights into frequency-dependent behavior gives the following expression for the interfacial (Maxwell-Wagner) polarizability $p(mw)$ of a spherical particle [44]:

$$p(mw) = \varepsilon_o \varepsilon_m \left[\frac{\omega^2 \tau^2}{1+\omega^2\tau^2} \left(\frac{\varepsilon_p - \varepsilon_m}{\varepsilon_p + 2\varepsilon_m} \right) + \frac{1}{1+\omega^2\tau^2} \left(\frac{\sigma_p - \sigma_m}{\sigma_p + 2\sigma_m} \right) \right] \quad (4.5)$$

where ω is the radian frequency ($2\pi f$) of the applied field. For low values of the frequency ($\omega\tau \ll 1$) eqn. 4.5 reduces to:

$$p(mw) \approx \varepsilon_o \varepsilon_m \left(\frac{\sigma_p - \sigma_m}{\sigma_p + 2\sigma_m} \right) \quad (4.6)$$

and at high frequencies ($\omega\tau \gg 1$) we have:

$$p(mw) \approx \varepsilon_o \varepsilon_m \left(\frac{\varepsilon_p - \varepsilon_m}{\varepsilon_p + 2\varepsilon_m} \right) \quad (4.7)$$

The low frequency DEP force arising from interfacial polarization thus depends on the conductive properties of the particle and suspending medium, whilst at high frequencies the permittivity values are important. At intermediate frequencies, both the conductive and dielectric properties of the medium and particle dictate the magnitude and polarity of the DEP force. The low frequency polarization stabilizes when continuity of electric current densities are established across the various interfaces that define the particle structure and its immediate environment. This involves the movement of free ions around and through the particle, and thus the particle size and its effective surface area are important factors. At a high frequency, which in the time-domain corresponds to the initial dielectric response of the particle to an imposed electric field, the system strives to attain continuity of so-called displacement flux densities. This involves perturbations of bound charges at the molecular scale, and so is not controlled by the physical size of the particle. It is important to remember that other polarizations (e.g., double layer relaxations, considered in the next section) can contribute significantly to the low frequency DEP response of particles. Eqn. 4.5 does not encompass such additional polarizations.

If the polarizability factor $p(mw)$ is positive (i.e., the particle is more polarizable than the surrounding medium) then the DEP force is positive and the particle is directed towards high field regions at electrode edges (For the range of frequencies and electrode dimensions commonly used in DEP applications, the field maxima are always located at electrode surfaces). If $p(mw)$ has a negative value, the DEP force is negative and the particle is directed towards field minima away from the electrode edges. These two effects are shown in

Figure 4.1. The concept shown in this figure, of directing and assembling colloidal particles into *field* cages under the action of negative DEP, has been significantly extended using 3-D electrode arrays [19, 21, 61, 88]. The continuous separation of particle mixtures can be achieved using synchronized pulses of fluid flow and electrode energization to selectively move particles into and out of positive and negative DEP traps [71] or using DEP combined with field flow fractionation [7, 37, 84] or electrokinesis [15]. Extruded quadrupolar traps [99], 'zipper electrodes' [32], high-density electrode arrays [16] and cell manipulation on a CMOS chip [51] have been described.

In DEP studies of mammalian cells (e.g., blood cells, cancer cells, stem cells) the suspending medium commonly takes the form of a low conductivity (10~180 mS/m) electrolyte containing sufficient concentrations of sugars (e.g., mannitol, sucrose, dextrose) to raise the osmolarity to the normal physiological level of around 280 mOs/kg. For viable cells the plasma membrane acts as an electrical insulator to passive ion conduction, and thus for frequencies below about 25 kHz the cell will appear as an insulating object suspended in a conducting medium. This corresponds to the situation $\sigma_p < \sigma_m$, so that the polarizability *p(mw)* given by eqn. 4.6 will be negative and viable cells will exhibit negative DEP at the lower frequencies. Physiological electrolytes typically have a conductivity and relative permittivity of the order 1.4 S/m and 79, respectively. For frequencies below around 1 MHz, the relative permittivity of the cytoplasm will be greater than 79 because of the presence of solvated polar molecules (proteins, peptides, amino-acids) and interfacial polarizations of membrane surfaces (e.g., endoplasmic reticulum, nucleus, mitochondria). We therefore commonly have the condition $\varepsilon_p > \varepsilon_m$, so that *p(mw)* given by eqn. 4.7 will be positive and the cells will exhibit positive DEP at frequencies above around 100 kHz. Positive DEP is enhanced by the fact that for frequencies above 100 kHz the electrical field penetrates into the cytoplasm. Electronic engineers will recognize this as capacitive coupling between the suspending medium and cytoplasm, where the effective capacitance of the plasma membrane shorts out the membrane resistance. The effective conductivity of the cytoplasm will be less than that (~1.4 S/m) of a pure physiological strength electrolyte because of the presence of insulating bodies and structures (e.g., protein cytoskeleton, lipid membranes). Values for σ_p in the range 0.1~0.5 S/m are commonly deduced for viable cells above 100 kHz, so that depending on the choice of suspending medium conductivity we can achieve the condition $\sigma_p > \sigma_m$. From this we deduce that a transition from negative to positive DEP may occur as the frequency is increased above 10 kHz. The frequency value, commonly referred to as the DEP cross-over frequency f_{xo}, of such a transition occurs when *p(mw)* is zero. From eqn. 4.5 this frequency is given by:

$$f_{xo} = \frac{1}{2\pi\varepsilon_o}\sqrt{\frac{(\sigma_m - \sigma_p)(\sigma_p + 2\sigma_m)}{(\varepsilon_p - \varepsilon_m)(\varepsilon_p + 2\varepsilon_m)}} \qquad (4.8)$$

The same result was derived by Jones & Kallio [46], using phasor notation to describe the induced dipole moment, in their investigation of the cut-off frequency for DEP levitation of particles. For viable cells with intact membranes of high electrical resistance the value for f_{xo} is given to a good approximation by:

$$f_{xo} = \frac{\sqrt{2}}{2\pi r C_{mem}}\sigma_m \qquad (4.9)$$

where r is the cell radius and C_{mem} is the membrane capacitance [74]. This dependence on the cell radius r and medium conductivity σ_m reflects the fact that the interfacial polarization at the lower frequencies is controlled by the movement of free ions around and through the particle. The concept of a membrane capacitance C_{mem} is introduced to account for the build-up of charge at the outer and inner membrane surfaces. We can think of the membrane as a very thin dielectric material separating two conducting mediums—namely the suspending electrolyte and the cytoplasm. This is somewhat similar to a conventional electrical capacitor formed by a dielectric held between two metal plates. The way in which the induced charges distribute themselves on the outer and inner membrane surfaces defines the magnitude and polarity of the resultant induced dipole moment m given in equations (4.1) and (4.2), and hence also the dielectrophoretic response of the cell. As for the case of a conventional electrical capacitor, where the magnitude of the induced charge is proportional to the electrode surface area, the effective surface area of the membrane will influence the value of C_{mem} in equation (4.9). The presence of microvilli or membrane folds will therefore influence the value of C_{mem}.

The situation for frequencies either side of f_{xo} is shown in figure 4.2. At low frequencies ($f < f_{xo}$) a mammalian cell with an intact, viable, plasma membrane will appear as an electrically insulating particle. The imposed electric field, and resulting ionic conduction flow, will skirt around the cell membrane and seek more conductive paths in the surrounding

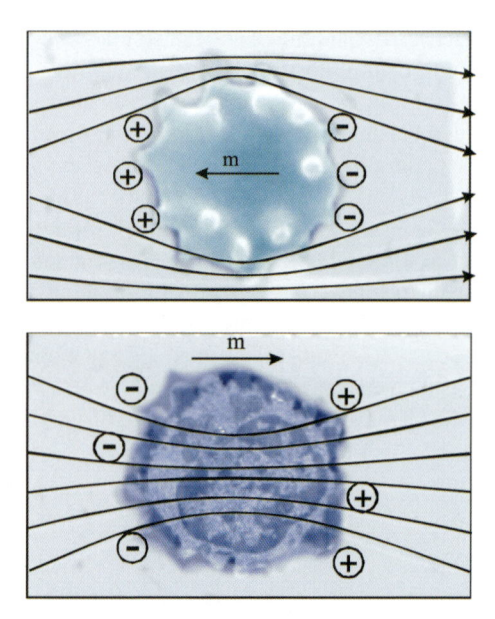

FIGURE 4.2. *Top:* At low frequencies a mammalian cell with an intact, viable, plasma membrane appears as an electrically insulating particle. The applied electric field, and resulting ionic currents, skirt around the cell membrane to seek more conductive paths in the surrounding electrolyte. Induced charges will appear at the membrane-electrolyte interface to produce a dipole moment opposing the applied field. The cell will exhibit negative DEP. *Bottom:* At high frequencies the cell appears as a more polarizable volume than the surrounding electrolyte, the field will penetrate the cell interior, and the induced dipole moment will be oriented in the same sense as the field. The cell will exhibit positive DEP.

electrolyte. Induced charges will appear at the cell surface to produce a dipole moment m opposing the applied field. The cell will exhibit negative DEP. At high frequencies ($f > f_{xo}$) the field will penetrate the cell interior and the cell will appear to be more polarizable than the surrounding electrolyte. The induced dipole moment will be oriented in the same sense as the field, and the cell will exhibit positive DEP.

At the cross-over frequency f_{xo} the induced dipole moment is zero—in other words the cell appears to be transparent to the applied field and no charges are induced on or within the cell. A very small change in either the cell state or physico-chemical properties of its membrane can result in a measurable change of f_{xo} (e.g., [24]). (A dramatic change will occur if the membrane becomes degraded and no longer acts as an insulting barrier to passive ion flow. Ions will leak out of the cell and the condition $\sigma_p > \sigma_m$ will no longer hold at high frequencies). A *PhysioNetics* instrument has been developed [74, 77] to measure such changes for individual cells in a cell suspension. This information can be used to refine protocols for separating different cell types from each other, or to monitor the effects of chemical agents added to a cell culture, for example. An early example of how knowledge of the f_{xo} values for different cell types can lead to their separation by DEP was demonstrated by Gascoyne *et al* [23] for mixtures of erythrocytes and erythroleukeia cells. More recently, Huang *et al* [40] have demonstrated that DEP separation significantly improves the accuracy of gene expression profiling by purifying out cells of interest in a complex cell population. Basically, by performing DEP at a frequency between the f_{xo} values of two cell types, one type will be repelled from the electrodes by negative DEP and the other type trapped at the electrodes by positive DEP. Examples of this are shown in figures 4.3–4.5. Particles

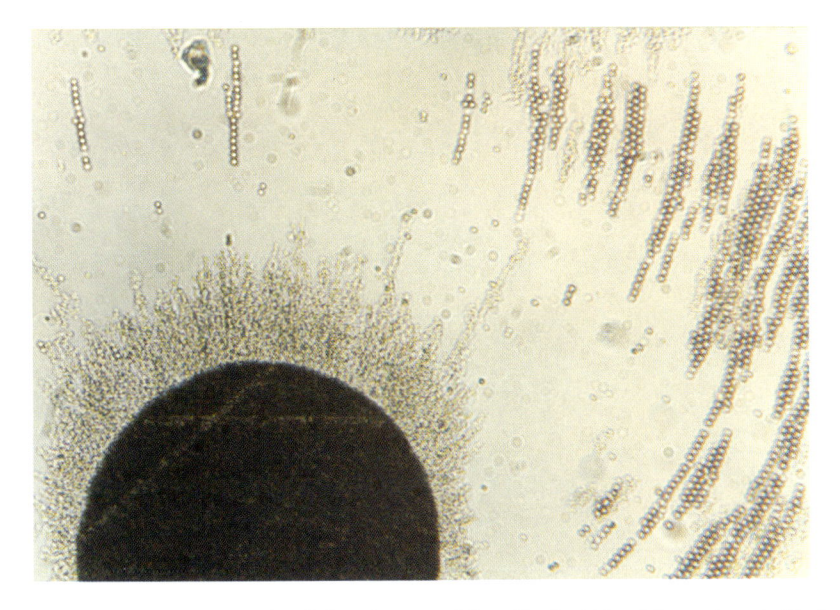

FIGURE 4.3. Bacteria (*Micrococcus lysodeikticus*) separated from blood cells by dielectrophoresis. The bacteria collect by positive DEP at the electrode, whilst erythrocytes are repelled under the action of negative DEP (unpublished work, related to Wang *et al* [104]).

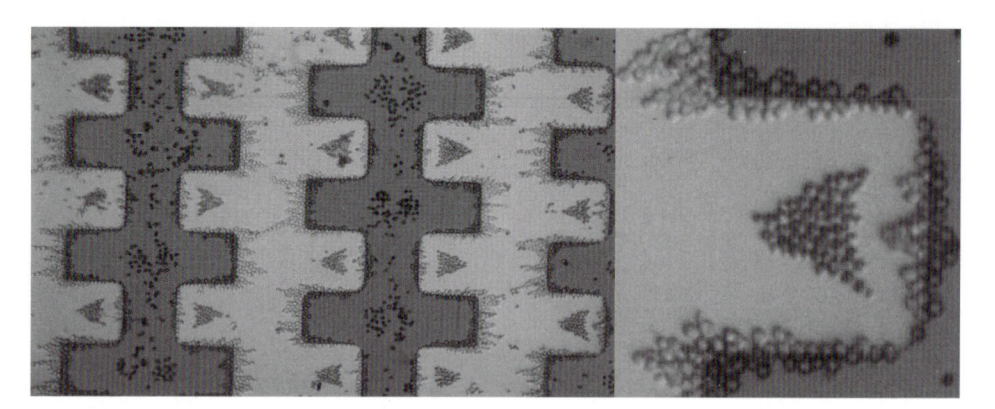

FIGURE 4.4. DEP separation of viable and non-viable yeast cells using interdigitated, castellated, electrodes. The viable cells collect at the electrode edges, whilst the non-viable cells collect in triangular aggregations between the electrodes and also along the central axes of the electrodes [53].

can be moved or assembled in electrode arrays by sequentially or concurrently applying one or more frequencies to vary the polarity and magnitude of the DEP force [70], and an example of how this can be used to manipulate a cell on a matrix array is shown in figure 4.6.

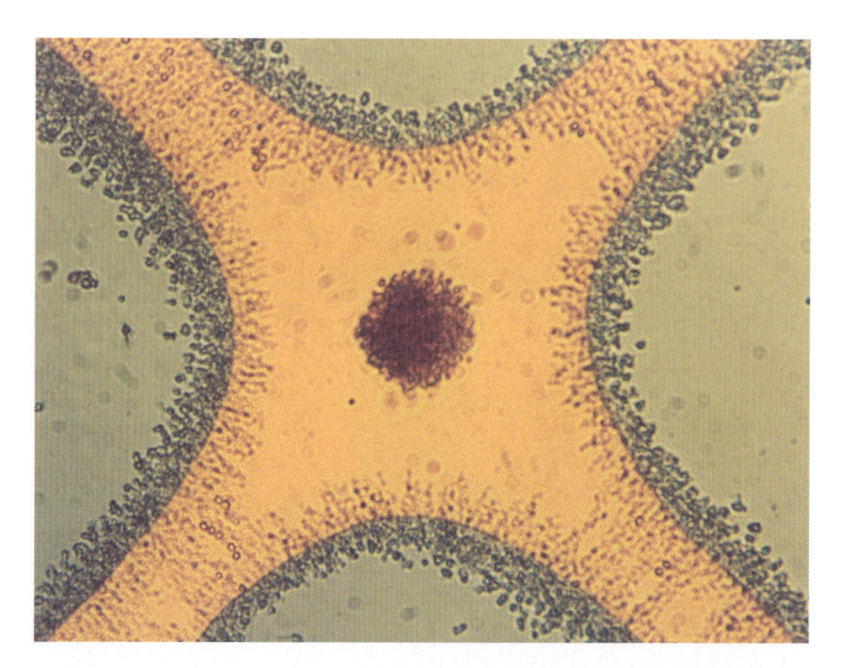

FIGURE 4.5. Separation of a mixture of Gram positive and Gram negative bacteria (*B. subtilis* and *E. coli*, respectively). Under the same experimental conditions, Gram positive bacteria usually exhibit the higher polarizability and can be trapped at the electrodes by positive DEP. Gram negative bacteria are forced away from the electrodes by negative DEP. (unpublished work, related to Markx *et al* [54]).

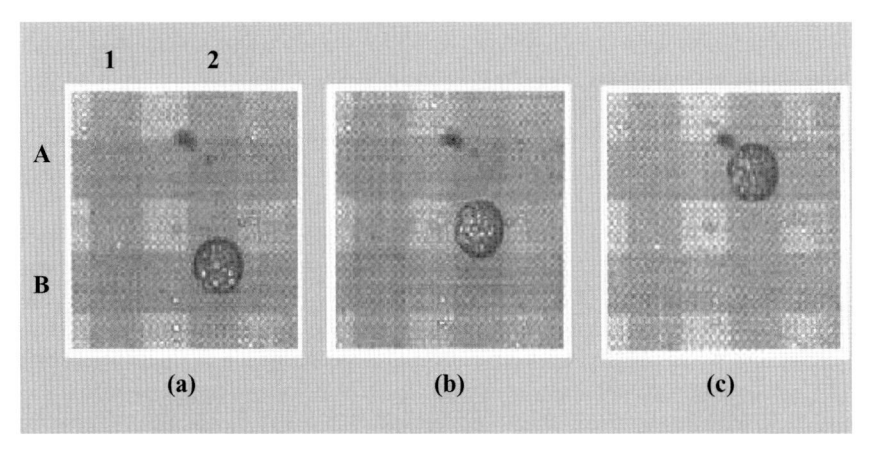

FIGURE 4.6. DEP manipulation of a single protoplast on a grid of transparent indium tin oxide (ITO) electrodes. The electrodes have been artificially darkened in these pictures [92].

4.5. SURFACE CHARGE EFFECTS

Bioparticles generally carry a net negative charge at physiological pH associated with the presence of charged polymers and acidic groups on their surfaces. This fixed charge can influence the DEP behavior through electrophoresis, and counter-ion relaxations in the double layer that forms around all charged particles when they are suspended in aqueous media. Counter-ion relaxation processes involve both ionic diffusion and ionic conduction around the particle surfaces. Valuable insights into the nature of dielectric relaxations associated with double layers and surface conductivity were obtained from electrorotation measurements of latex beads [2], and a comprehensive review of such effects is given by Lyklema [50]. (The electrode polarization effect described earlier represents an example of double layer polarization.) These processes should manifest themselves as additional polarizability terms to be added to eqn. 4.5, and thus contribute to the overall frequency-dependent DEP behavior of a particle. In fact, polarizations associated with surface charge effects can dominate over that given by eqn. 4.5, producing a so-called *anomalous* positive DEP effect [25, 64, 65]. The term *anomalous* is probably unfortunate, as it implies that the laws of physical chemistry are not being obeyed. Rather, it reflects more our incomplete knowledge of how electrical double-layer and surface conduction effects contribute to the *total* DEP polarizability of a particle.

A simple and rapid way to monitor the DEP response is shown in figure 4.7. In this method, small changes are measured of the intensity of the light transmitted through a suspension of particles held between two electrode arrays. The light output increases when particles collect at the electrodes by positive DEP, but decreases when negative DEP pushes the particles into a dense band in the centre of the chamber [9, 97]. Close inspection of the signal shown in figure 4.7, for the case of a 1 Hz signal applied to a suspension of bacteria (*M. lysodeikticus*), reveals an oscillatory pattern superimposed onto the absorbance decay. This oscillation represents electrophoretic motion of the bacteria, an effect which becomes insignificant (in this type of measurement at least) above 10 Hz. A simple analysis

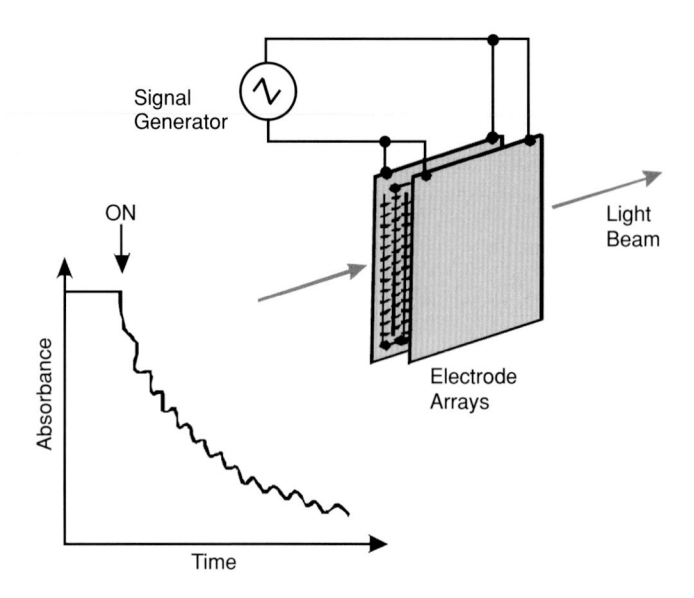

FIGURE 4.7. A simple method for monitoring the DEP behavior of cell suspensions involves measuring changes in the intensity of a light beam passing through the DEP chamber. The light input decreases if cells are attracted to the electrodes by positive DEP, and the rate of change of absorbance is proportional to the DEP force. The absorbance trace shown here was made after application of a 1 Hz voltage signal to the electrode array, and exhibits oscillations associated with electrophoresis effects [9].

demonstrated that this low-frequency electrophoretic response enhanced the DEP effect by driving the bacteria nearer to the electrodes [9].

The DEP frequency response for *Micrococcus lysodeikticus* is shown in figure 4.8 alongside the dielectric spectrum for these bacteria [16]. The regions marked α and β follow the conventional assignments of dielectric dispersions observed for biological materials [89]. The α-dispersion arises from ionic relaxations in electrical double-layers, and membrane ionic conduction effects. The β-dispersion is the Maxwell-Wagner interfacial polarization described by eqn. 4.5. It is of relevance to DEP studies to note the large magnitude of the α-dispersion, and that it is sensitive to metabolic state—often disappearing ahead of the β-dispersion on cell death [89]. It is also important to note, from figure 4.8, that the intact cell wall of a bacteria contributes significantly to its low frequency dielectric response. This strong dependence on the presence of an intact cell wall was also observed in DEP measurements [44]. Gram positive and Gram negative bacteria have differences in the structures of their outer cell walls and membranes, and this is reflected in their different DEP behavior [39, 53]. For example, Gram-positive bacteria synthesize a uniform peptidoglycan-structured cell wall, whilst Gram-negative cells have a more complicated wall structure which includes lipids and proteins forming an outer membrane covering a layer of peptidoglycan. The cell walls of Gram-positive bacteria are also characterized by the incorporation of covalently bound teichoic acid, teichuronic acid and proteins, forming open networks with high charge densities comparable with ion-exchange materials. The presence of such charged groups and the absence of an outer lipid membrane leads to the tendency of Gram positive bacteria to exhibit higher polarizabilities than Gram negative

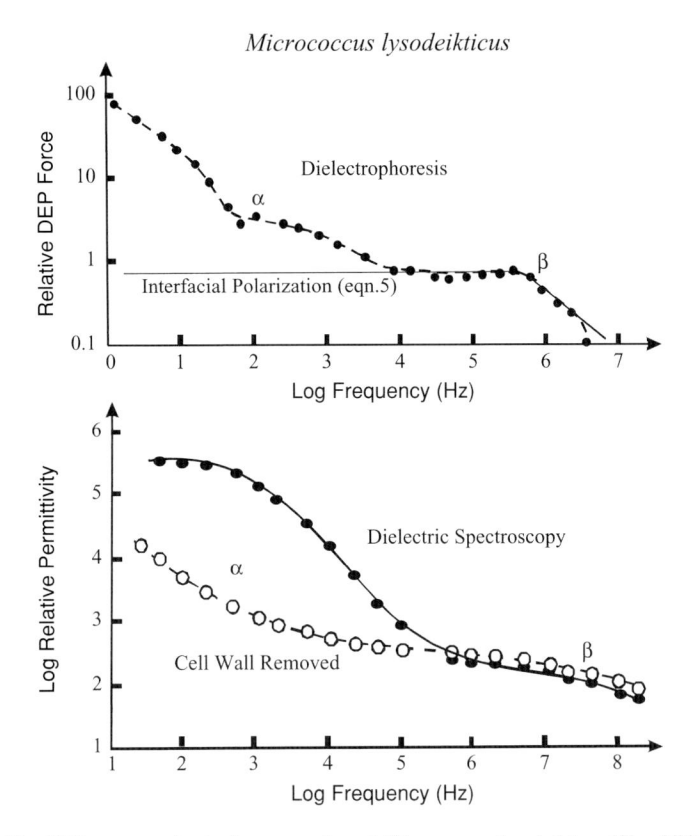

FIGURE 4.8. The DEP response (top) of a suspension of *Micrococcus lysodeikticus* [9] exhibits two regions of behavior reflected (bottom) in the dielectric properties [16]. The α-dispersion, arising from relaxation of the electrical double-layer and ionic conduction in the cell wall, builds onto the β-dispersion arising from interfacial polarizations. Removal of the cell wall significantly reduces the α-dispersion. The decade frequency shift between these two sets of results reflects differences in the ionic strength and osmolarity of the suspending media.

ones, and to their ready discrimination by DEP (e.g., figure 4.5). The cell walls of yeast cells consist mainly of uncharged polysaccharides, and this leads to their having a lower polarizability than bacteria [53].

We have noted above that the magnitude of the α-dispersion is sensitive to the metabolic state of the cell, and during the process of cell death disappears before the β-dispersion. We can expect similar behavior for bacteria. Apart from the degradation of charged groups, the loss of the membrane's ability to act as a barrier to passive ion flow will also influence the DEP behavior of a non-viable organism. DEP experiments are often performed in low conductivity media, and so impairment of the membrane can lead to a significant decrease in cytoplasmic conductivity as ions leak out of the cell. This process, rather than surface charge degradation, appears to be relevant to the fact that yeast and bacteria can be separated by DEP into viable and non-viable fractions (e.g., [48, 54, 93]). Selective separation and detection of bacteria by DEP can be enhanced using antigen-antibody reactions [94].

Mammalian cells do not possess a cell wall, so the α-dispersion should mainly arise from surface conduction effects and relaxations of the electrical double layer associated with fixed charges on their plasma membranes. This is supported by the finding that the low frequency DEP response for erythrocytes and erythroleukemia cells is reduced following neuraminidase-treatment to reduce cell membrane charge by 50~60% [10, 24]. Most mammalian cells have electrophoretic mobilities between 0.7 and 1.40×10^{-4} cm^2/Vs [4]. Following standard theory [27] these mobility values can be calculated to represent membrane surface charge densities ranging from -0.65 to -1.35 μC/cm^2 for a suspending medium of ionic strength 180 mM at 20°C. Cell surface charge appears to be regulated by cytokines which are predominantly produced during cell activation or differentiation [5]. From the studies of neuraminidase-treated cells, it would seem possible that such regulations of surface charge are amenable to DEP investigation. Complications may arise from the fact that, although the distribution of the negative charges associated with acidic groups on mammalian cell surfaces is fairly uniform, the positive charges arising from basic groups increases with increasing inward distance from the membrane-medium boundary [62]. How such a complicated surface charge arrangement influences double layer relaxations and surface conductance effects is as yet not fully understood.

DNA does not possess an intrinsic dipole moment, but it exhibits a large polarizability, assumed to be associated with field-induced fluctuations and mobility of counterions along the molecule's axis [66, 92]. Recent dielectric measurements have supported and refined this picture (e.g., [3, 85]). Washizu *et al* [108, 109] were the first to demonstrate that DNA (and proteins) can be manipulated by DEP, and in doing so opened up DEP studies of sub-micron particles. To date, DEP studies have been performed on enveloped and non-enveloped viruses, proteins, DNA and nano-beads. Electrical double-layer polarizations associated with surface charges dominate the DEP behavior of such particles. An excellent review of such work on nanoparticles, including their own important contributions, is given by Morgan and Green [59].

4.6. OTHER PHYSIOMETRIC EFFECTS

From considerations of the relevant dielectric and DEP studies, we can deduce that the most important cell parameters influencing the DEP behaviour of cells include:

- The viability of the cell (e.g., chemical and physical integrity of its plasma membrane)
- The size and shape of the cell
- The surface morphology of the cell plasma membrane
- The ratio of lipid to other molecular components in the cell membrane
- The internal composition of the cell (e.g., volume of nucleus, endoplasmic reticulum)

These various physiometric properties that contribute to the DEP spectrum for a typical mammalian cell (e.g., blood, cancer, stem cells) are summarized in figure 4.9. Different structural, physico-chemical and cell state-related properties are probed as the applied frequency is swept over a range from around 100 Hz to 500 MHz. Apart from the nature of its heterogeneous structure and surface charge properties, the gross size and shape of a cell, its fine membrane surface features, membrane composition and extent of cytoskeleton development, are important factors influencing DEP behavior at frequencies below around

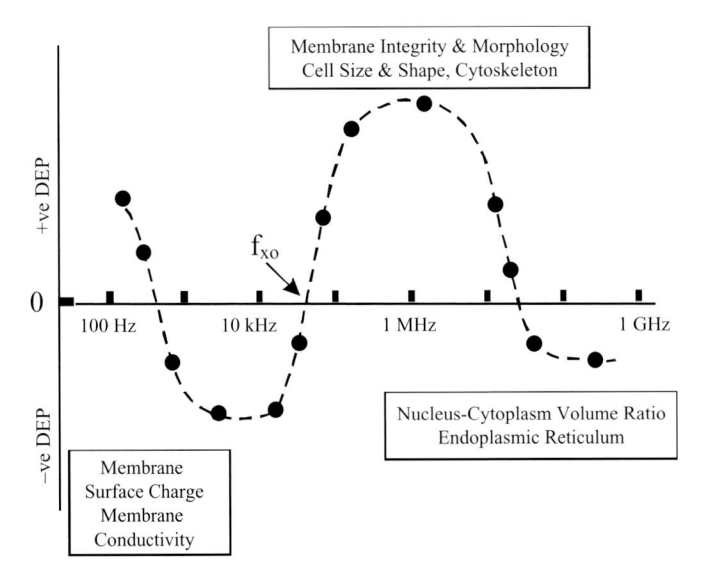

CELL PHYSIOMETRY TOOLS BASED ON DIELECTROPHORESIS 117

FIGURE 4.9. A typical DEP characteristic is shown for a mammalian cell, together with the main controlling physiometric parameters. The details of a DEP spectrum vary according to cell type and cell state, and with the dielectric properties of the suspending medium. The DEP cross-over frequency f_{xo} (given by eqn. 4.8) is an important parameter for phenotyping and formulating protocols for cell separations.

10 MHz. The relative permittivity of bulk water is around 79 for frequencies below 1 GHz. It is therefore interesting to note that values from 88 to 99 have been derived for the effective permittivity of the interior of erythroleukaemia cells [35] and from 104 to 154 for the interior of blood cells [110] using electrorotation measurements. Because electrorotation and DEP basically reflect the same dielectric properties of cells [47, 67] we can expect the DEP behavior of cells above 10 MHz to be influenced by their internal dielectric properties. Permittivity values of the order 150 can arise from the dielectric increment effect of solvated proteins, peptides and amino-acids [28, 66]. As the frequency is increased, so-called dielectric decrement effects will cause the permittivity to fall well below a value of 79. At the highest frequencies it is then possible to achieve the condition $\varepsilon_p < \varepsilon_m$ so that the cells exhibit negative DEP. The resulting transition from positive to negative DEP at a high frequency is depicted in figure 4.9. The nucleus-cytoplasm volume ratio and the extent of the endoplasmic reticulum will also influence matters. Due to experimental difficulties associated with generating the required voltages, very few DEP measurements for cells have been reported above 50 MHz. The usefulness of determining the DEP cross-over frequency in the upper MHz frequencies, as opposed to the lower frequency f_{xo} (given by eqns. 4.8 & 9) has therefore yet to be determined.

Cells at various stages of differentiation (e.g. stem cells, fetal cells) possess different forms of protein cytoskeleton. This will produce differences in the mechanical resilience of a cell, and the extent to which the membrane surface morphology (and hence DEP behavior) will change as a result of changes in medium osmolality or exposure to physical forces (e.g., [104]). Additional information can be obtained by changing the ionic strength, pH, and chemical composition of the suspending medium. Such DEP phenotyping can then be used to formulate protocols for selectively detecting and isolating cells or other bioparticles, for additional biochemical, physical or genetic analysis. Important examples include the separation and purification from blood or other biological fluids of cancer cells, fetal cells, stem cells, leukocyte subpopulations, cells in different stages of life cycle, bacteria and viruses.

The DEP cross-over frequency f_{xo} determined for mammalian cells, under normal experimental conditions, commonly falls in the frequency range from around 50 kHz to 500 kHz. At these frequencies effects associated with surface charge and surface conduction can usually be neglected. Also, polarizations associated with solvated polar molecules or organelles in the cytoplasm will contribute to the total effective value of the permittivity ε_p for the cell. In this way, eqn. 4.5 can be used for the total polarizability value, and for a fixed value of the medium conductivity, f_{xo} is determined by the effective capacitance of the cell. This in turn is influenced by the topography of the cell membrane surface, such as the extent of microvilli, bleb and membrane folds and ruffles [36, 38, 104]. Huang et al [38] observed that the f_{xo} values for resting T lymphocytes decreased following mitogenic stimulation from their normally resting G_o phase into the cell division cycle. The largest increase in f_{xo}, corresponding to a 77% increase in membrane capacitance (and a probable comparable increase in membrane surface 'roughness'), occurred as the cells progressed from the G_1 phase through the S phase. These results were mirrored in later studies by Pethig et al [74] where the membrane capacitance values of T lymphocytes were found to decrease following a decrease in S phase and an increase in G_1 cells. Changes in membrane capacitance and conductivity that occur with time after inducement of apoptosis have been determined by monitoring the f_{xo} values for HL-60 cells [107] and T lymphocytes [75, 77]. Similar studies have been reported for detecting cellular and intracellular responses to toxicants [83]. These various studies provide examples of how DEP can be used to monitor physiological changes, such as those that accompany transmembrane signaling events or gene-directed self-destruction, for example.

4.7. TRAVELING WAVE DIELECTROPHORESIS

The methods described so far for selectively trapping and fractionating bioparticles utilize the flow of a fluid through the DEP chamber. An excellent example of this is the DEP field-flow fractionation technique described by Gascoyne and co-workers [7, 37]. There are advantages for some applications of DEP to have a stationary fluid rather than a flowing fluid, and to selectively translocate particles using traveling electric fields rather than stationary fields.

Masuda et al [55] applied a traveling electric field, produced using poly-phase voltages and a parallel array of electrodes, to an aqueous suspension of blood cells. They found that they could control both translational and circular motions of the cells. The optimal frequency range for such motions was found to be 0.1–10 Hz. We can expect, from our discussion here and the effect shown in figure 4.7, that at such low frequencies these induced cell motions were largely electrophoretic in origin and dependent on cell size and electrical charge carried by the cells. Hagedorn et al [30] later demonstrated that quadrature-phase electric fields of frequency between 10 kHz and 30 MHz could produce linear motions of pollen and cellulose particles, and they coined the term traveling-wave dielectrophoresis (TWD) for

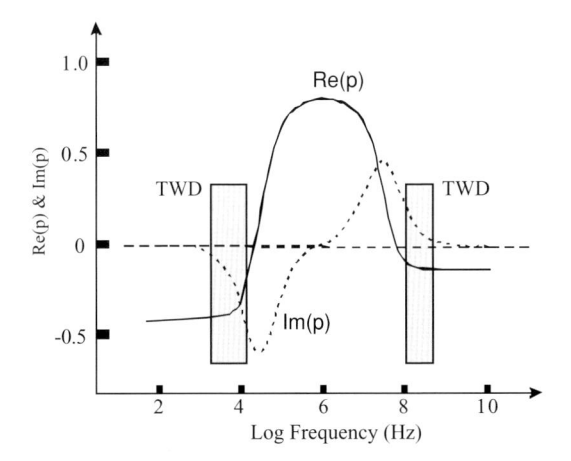

FIGURE 4.10. Frequency dependencies of the real, in-phase, component Re(p) and imaginary, out-of-phase, component Im(p) of the polarizability parameter of eqn. 4.5 for a simple model of a cell (based on [101]). Movement of a cell by conventional traveling wave dielectrophoresis (TWD) can occur only in those narrow frequency ranges where Re(p) is negative (so that the cell is levitated) and where Im(p) is of sufficient magnitude to produce a translational force on the cell [34].

this motion. TWD was found to be restricted to a frequency range where the particles were levitated above the electrodes under the influence of a negative DEP force. This finding was confirmed by Huang *et al* [34] who further deduced that the TWD velocity is proportional to the imaginary (out-of-phase) component of the Clausius-Mosotti polarizability factor p (eqn. 4.2), whilst the DEP force acting to either trap or levitate the particles above the electrode 'tracks' is proportional to the real (in-phase) component of p. This situation is shown in figure 4.10, from which it can be seen that TWD is restricted to narrow frequency ranges where the real component Re(p) of the DEP force is negative and where the imaginary component Im(p) is of sufficient magnitude to produce a translational force.

Theoretical models of TWD have been refined [43, 58, 105] and various practical devices have been developed for manipulating, separating and characterizing blood cells, cancer cells, yeast, and parasites [11, 12, 26, 57, 72, 98, 106]. By employing TWD it is possible to selectively move cells or other bioparticles along a channel in a stationary fluid. Selective movement arises from the intrinsic physico-chemical properties of the bioparticle, so that in many cases tagging the target particle with a biochemical label, bead, dye or other bioengineered marker is not required to achieve selective isolation of the target from other particle types. TWD electrodes can be fabricated in the form of spirals [26, 106], long tracks [57] and, as depicted in figure 4.11, junctions [71]. By applying different frequencies to each 'arm' of a TWD junction, particles can either be separated into separate paths, or brought together to engage in controlled particle-particle interactions [73]. Dense concentrations of cells, as well as single cells, can be manipulated and sorted by TWD [11]. Provided that sufficiently long TWD electrode tracks are used, a mixture of different particles can be fractionated into 'bands' with a high degree of separation [11, 57].

As shown in figure 4.10 'conventional' TWD motion is restricted to narrow frequency ranges. This restricts the ability to explore the full range of physiometric parameters shown in figure 4.9. This limitation can be overcome by employing the method of *Superposition-TWD*

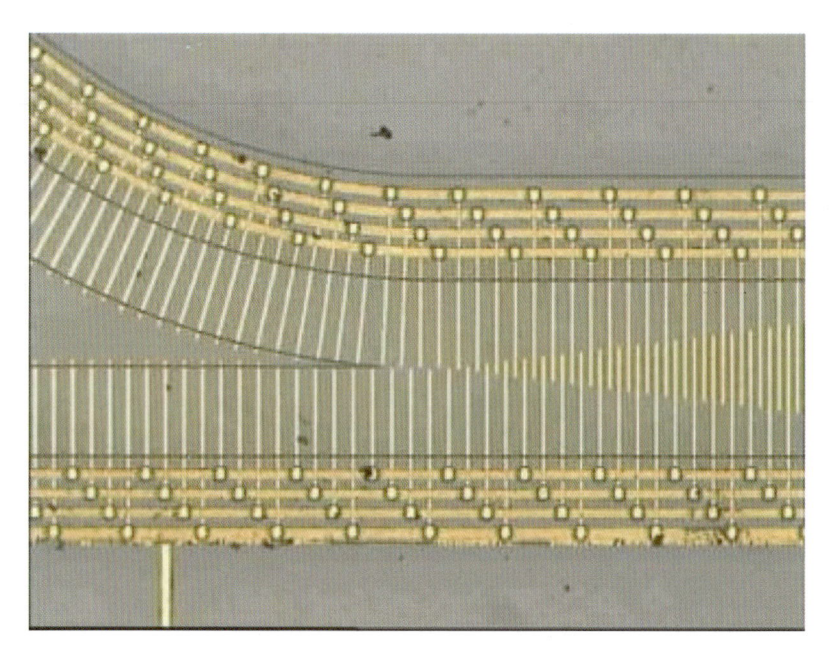

FIGURE 4.11. A TWD junction, fabricated by laser ablation, for separating or bringing together different particles. The electrode elements are 10 μm in width [70, 73].

[75]. If a DEP or second TWD signal is added to a primary TWD signal, the resulting electrostatic potential at the electrodes will be a superposition of the separate voltage functions. As shown in figure 4.12, the method of superposition of signals can be used to open up TWD to the full range of frequencies available to conventional DEP measurements. Basically, the Superposition-TWD method provides control of the levitation height of the cells above the electrode plane, and thus also of the range of frequencies over which TWD can occur. The direction of motion of cells of different phenotype or physiological state can also be controlled. The separation of T lymphocytes according to their cell cycle stage, as well as of the separation of mixtures of T lymphocytes and monocytes, has been demonstrated [75].

4.8. CONTROLLING POSSIBLE DEP-INDUCED DAMAGE TO CELLS

The possibility that, during the process of DEP separation or manipulation, cells may be irreversibly damaged is an important issue to address. The fact that mammalian cells can sometimes 'burst' when exposed to DEP forces is well known to workers in the field, and is sometimes reported (e.g., [24]). In figure 4.12, the frequency range is identified where T cells can, under certain conditions, either suffer irreversible damage or be physically destroyed when exposed to a TWD field. Selective cell bursting may be a desired objective as a means, for example, to release proteins or DNA from a target cell. In other applications, as for example in the purification of stem cells prior to cell therapy, damage is not acceptable.

DEP-induced cell damage can arise from at least three main sources, namely: effects associated with the cells being suspended in a non-physiological medium; stress induced

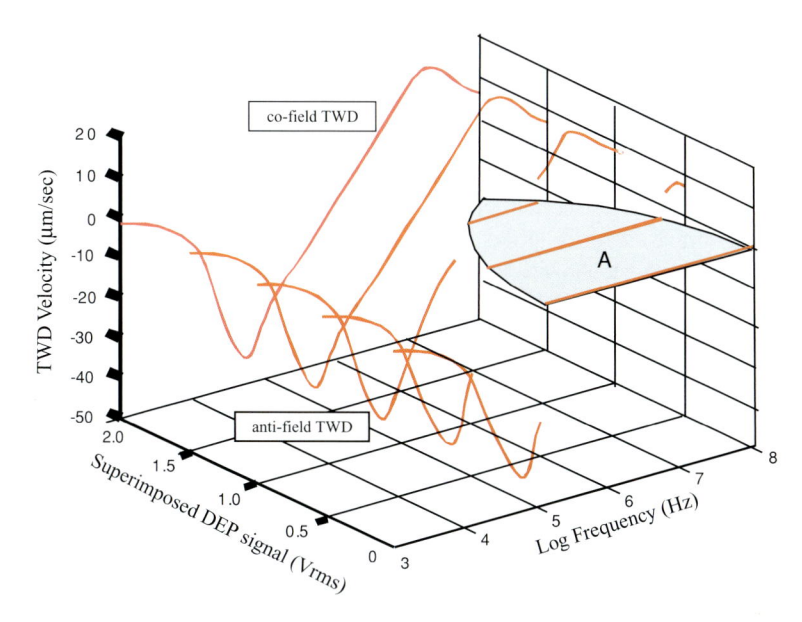

FIGURE 4.12. 3-D plot of the frequency variation of the TWD velocity for T cells, as a function of the magnitude of a superimposed 'stationary' DEP signal applied at a fixed frequency of 10 kHz. The superimposed DEP signal opens up the TWD 'window' to give access to the full frequency range available in normal DEP measurements. Cell velocity mirrors the Im(p) characteristic shown in figure 4.10, so that both co- and anti-field movement is available to facilitate cell separations. In frequency region 'A' cells experience a positive DEP force and can suffer damage from large shear forces when they are retained close to the electrodes. In the TWD regions, the cells are levitated off the electrodes and do not suffer from such shear forces [77].

by the applied electric field; shear stresses associated with fluid flow. Workers in the field are now competent in their choice of cell suspending media and levels of applied DEP forces. Under gentle DEP conditions, cell viability can be maintained. For example, the viability of erythrocytes separated from leukemia cells has been verified using trypan blue dye [6], and CD34+ cells have been successfully cultured following their DEP enrichment from bone marrow and peripheral stem cell harvests [91]. Fuhr *et al* [20] demonstrated that fibroblasts can be successfully cultivated, without significant change in their viability, motility, anchorage or cell-cycle time, when exposed continuously over a period of three days to DEP fields. Although very small increases were observed in the stress-related gene c-fos expression levels for glioma and neuroblastoma cells separated by DEP, subsequent culturing experiments demonstrated that there were no effects on cell growth [41]. Bacteria and yeast cells are protected by strong cell walls and can survive more severe DEP manipulations than mammalian cells. The viability of DEP-separated yeast cells has been checked by staining with methylene blue and plate counts [54], and DEP damage of bacteria has not been reported in the literature (or observed in our laboratories). The highest reported fluid flow rate for DEP cell separation appears to be 2.5 mL/min [52]. Because of the relatively small dimensions of a typical DEP separation chamber, this flow rate is well within the limits for laminar flow and corresponds to a shear stress exerted on the cells of around 0.3 N/m^2. This is well below the shear stress of 150 N/m^2 required to damage erythrocytes

[49] or of 20 N/m² for T cells [13]. Plant cells, because of their comparatively large size and the presence of large vacuoles, are more sensitive to shear stress (either from fluid flow or DEP forces) than mammalian cells [52].

It is well known that cell membranes can be disrupted by forced oscillation at frequencies greater than around 10 kHz, and this is the basis for using high-power sonication to disintegrate cells. Cell-destruction by DEP is primarily related to a field-induced breakdown of the physical integrity of the plasma membrane, as evidenced by the fact that the internal structure of the cell appears (under phase-contrast microscopy) to remain intact for some time and is even manipulable by DEP (e.g., [24]). Controlling the reversible electrical breakdown of membranes is an essential tool used in electroporation and electrofusion of cells (e.g., [63]). Electroporation (also called electropermeabilization) is typically achieved by subjecting the cells to a voltage pulse, where the field can range from around 5 MV/m for microsecond pulses, down to 0.1 MV/m for millisecond pulses, *depending on the cell type*. To obtain a significant DEP force, the factor $(E.\nabla)E$ in eqn. 4.3 should be of the order 10^{13} V²/m³, which corresponds to an applied field less than 10 kV/m [69]. The reason a relatively weak DEP field can cause an electrical breakdown of the plasma membrane is because the field is 'amplified' in the form of a potential induced across the membrane. For a model, spherical, cell the frequency dependence of the field E_m acting across the membrane is given ([67], and references cited therein) by:

$$E_m(\omega) = \frac{1.5(r/d)E \cos\theta}{(1 + \omega^2\tau^2)^{1/2}} \quad (4.10)$$

In eqn. 4.10 the factor r/d is the ratio of the cell radius r to the membrane thickness d, and τ is the relaxation time given by eqn. 4.4. Values for the membrane thickness are variously cited in the literature as being in the range 3–10 nm, so that the factor r/d can have a value of the order 10^3 and greater. Eqn. 4.10 therefore predicts that at low frequencies the field E_m acting across a mammalian cell membrane can exceed the applied field E by a factor of at least 10^3. Polarizations associated with the electrical double layer at the membrane surface are not included in eqn. 4.10, and so we can expect even larger values for E_m at the lowest frequencies. The greatest field stress is created across the membrane region that lies in a radial direction parallel with the field ($\theta = 0$). At high frequencies ($\omega\tau \gg 1$) the value for E_m falls to low values. The estimated frequency dependence of the membrane field E_m for T cells (Jurkat E6-1: ATCC, TIB-152) is shown in figure 4.13 for a range of values of $(E.\nabla)E$ commonly obtained in DEP experiments using microelectrodes and applied voltages up to 20 V r.m.s. Also shown in figure 4.13 is the DEP spectrum for the T cells and the f_{xo} value obtained with a medium conductivity of 40 mS/m [74], together with the frequency-field 'window' within which cell bursting has been observed (unpublished work). These results, together with observations for other cells (e.g., human blood cells; Bristol-8 B lymphoblastoid cells [ECACC 85011436]; HL-60 promyelocytic leukemia cells [ECACC 98070106]; and BeWo trophoblasts [ECACC 86082803]) suggest that the onset of irreversible damage (and cell bursting) can be induced using moderate to high values of a positive DEP force at frequencies where the field stress across the membrane approaches a value of 10^8 V/m or higher. Cancer cells appear to be more fragile than normal cells, and this is probably related to the fact they tend to have larger r/d values and therefore higher values of induced field stress E_m. Under standard experimental

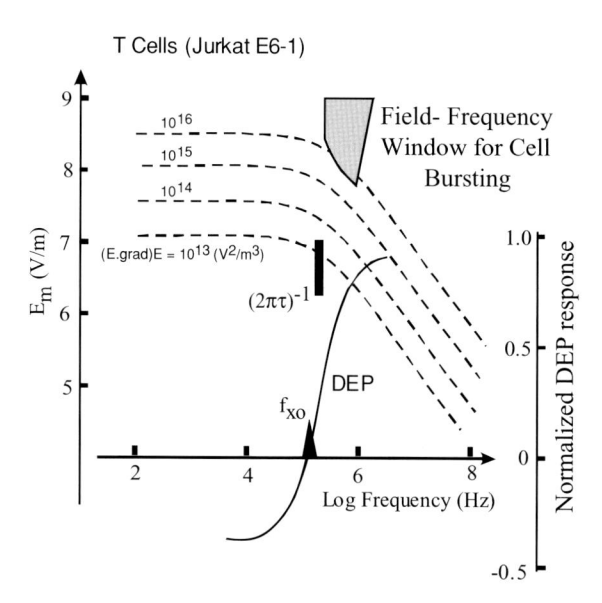

FIGURE 4.13. The frequency range where bursting (of more than 5% of total cell population) of T cells (Jurkat E6-1) occurs is plotted against the estimated induced membrane field stress E_m, derived using eqn. 4.10, for various values of the parameter $(E.\nabla)E$ of eqn. 4.3. Also shown is the DEP spectrum for the cells, the spread of f_{xo} values, and the characteristic frequency corresponding to $(2\pi\tau)^{-1}$. (Talery, Lee & Pethig, unpublished)

conditions, the E_m values experienced by cancer cells can also be larger because of their larger membrane capacitance values, associated with 'rougher' membrane surfaces than normal cells, which lowers the frequency where positive DEP occurs. The situation is more complicated for TWD experiments, because the bursting of a few cells can often be observed [77] at frequencies where the so-called FUN effect occurs [34]. This is characterized by unstable motions that can include cell rotation as well as rapid reversals of a cell's direction of TWD motion. Significant shear forces can be exerted on some types of cells during such behavior, and this will contribute to the onset of field-induced cell destruction.

CONCLUDING COMMENTS

DEP is worthy of being considered a competitive alternative to the more conventional methods of cell concentration and separation, such as centrifugation, filtration, fluorescence activated cell sorting, or optical tweezers. Because DEP can operate directly on native, unlabeled cells, it eliminates the expense, labor and time of labeling and tagging, as well as the development and validation of such labels and tags. The same basic DEP method has the (probably unique) capability of isolating and analyzing a wide range of particle types (cells, bacteria, viruses, DNA and proteins) using one basic procedure.

After their selective isolation and recovery by DEP, cells remain viable for further analysis, processing or cell therapy. Multiple parameters on individual live cells can be determined and, if desired, specific cell types can be collected. Because it utilizes electronic signals, the technology is capable of extensive automation, is inexpensive and portable. DEP can operate under sterile conditions. Although the ability to selectively isolate cells without harming them is important for many perceived applications, there are also advantages to be gained by being able to destroy selected target cells. DEP appears capable of achieving this objective. Just as there are limitations in the use of electrofusion and electroporation (mainly associated with the fact that not all cell types can be treated with the same ease, and with irreproducibility between different laboratories) we can expect that efficient and reproducible protocols for selective cell destruction by DEP will not be so easily achieved as the DEP isolation of viable target cells. But it is a worthwhile challenge.

REFERENCES

[1] M. Abraham. *The Classical Theory of Electricity and Magnetism*, English Translation of 8th German Ed., Hafner Publ. Co., New York, 1932.
[2] W.M. Arnold, H.P. Schwan, and U. Zimmermann. *Phys. Chem.*, 91:5093, 1987.
[3] D.J. Bakewell, I. Ermolina, H. Morgan, J.J. Milner, and Y. Feldman. *Biochim. Biophys. Acta.*, 1493:151, 2000.
[4] J.B. Bates, T.T. Chu, and W.T. Stribling. *Phys. Rev. Lett.*, 60:627, 1988.
[5] J. Bauer. In J. Bauer (ed.), *Cell Electrophoresis* CRC Press, Boca Raton, 1994.
[6] F.F. Becker, X.B. Wang, Y. Huang, R. Pethig, J. Vykoukal, and P.R.C. Gascoyne. *J. Phys. D: Appl. Phys.*, 27:2659, 1994.
[7] F.F. Becker, P.R.C. Gascoyne, Y. Huang, and X.B. Wang. US patent 5 993 630, 1999.
[8] D.J. Bennett, B. Khusid, C.D. James, P.C. Galambos, M. Okandan, D. Jacqmin, and A. Acrivos. *Appl. Phys. Lett.*, 83:4866, 2003.
[9] J.P.H. Burt, T.A.K. Al-Ameen, and R. Pethig. *J. Phys. E: Sci. Instrum.*, 22:952, 1989.
[10] J.P.H. Burt, R. Pethig, P.R.C. Gascoyne, and F.F. Becker. *Biochim. Biophys. Acta*, 1034:93, 1990.
[11] J.P.H. Burt, R. Pethig, and M.S. Talary. *Trans. Inst. MC*, 20:82, 1998.
[12] E.G. Cen, C. Dalton, Y. Li, S. Adamia, L. M. Pilarski, and K.V.I.S. Kaler. *J. Microbiol. Meth.*, 58:387, 2004.
[13] K.K Chittur, L.V. McIntire, and R.R. Rich. *Biotechnol. Prog.*, 4:89, 1988.
[14] K.S. Cole. *Membranes, Ions and Impulses*, Univ. of Calif. Press, 1972.
[15] E.B. Cummings. *IEEE Eng. Med. Biol. Magazine*, 22(6):75, 2003.
[16] C.W. Einolf and E.L. Carstensen. *Biophys. J.*, 9:634, 1969.
[17] M. Frénéa, S.P. Faure, B. Le Pioufle, Ph. Coquet, and H. Fujita. *Mat. Sci. Eng. C*, 23:597, 2003.
[18] H. Fröhlich. *Theory of Dielectrics*, Oxford University Press, Oxford, pp. 26–28, 1949.
[19] G. Fuhr, T. Müller, Th. Schnelle, R. Hagedorn, A. Voigt, S. Fiedler, W. M. Arnold, U. Zimmermann, B. Wagner, and A. Heuberger. *Naturwissenschaften*, 81:528, 1994a.
[20] G. Fuhr, H. Glasser, T. Müller, and T. Schnelle. *Biochim. Biophys. Acta*, 1201:353, 1994b.
[21] G. Fuhr, Th. Schnelle, R. Hagedorn, and S.G. Shirley. *Cell. Eng.*, 1:47, 1995.
[22] G. Fuhr, C. Reichle, T. Müller, K. Kahlke, K. Schutze, and M. Stuke, *Appl. Phys.*, A69:611, 1999.
[23] P.R.C. Gascoyne, Y. Huang, R. Pethig, J. Vykoukal, and F.F. Becker. *Meas. Sci. Technol.* 3:439, 1992.
[24] P.R.C. Gascoyne, R. Pethig, J.P.H. Burt, and F.F. Becker. *Biochim. Biophys. Acta*, 1149:119, 1993.
[25] P.R.C. Gascoyne and J. Vykoukal. *Electrophoresis* 23:1973, 2002.
[26] A.D. Goater, J.P.H. Burt, and R. Pethig. *J. Phys. D: Appl. Phys.*, 30:L65, 1997.
[27] D.C. Graham. *Chem. Rev.*, 41:441, 1947.
[28] E.H. Grant, R.J. Sheppard, and G.P. South. Clarendon Press, Oxford, 1978.
[29] S. Grimnes and O.G. Martinsen. *Bioimpedance and Bioelectricity*, Academic Press, San Diego, 2000.
[30] R. Hagedorn, G. Fuhr, T. Müller, and J. Gimsa. *Electrophoresis*, 13:49, 1992.
[31] J.B. Hasted. *Aqueous Dielectrics*, Chapman & Hall, London, 1973.
[32] K.F. Hoettges, M.P. Hughes, A. Cotton, N.A.E. Hopkins and M.B. McDonnell. *IEEE Eng. Med. Biol. Magazine*, 22(6):68, 2003.
[33] Y. Huang and R. Pethig. *Meas. Sci. Technol.*, 2:1142, 1991.
[34] Y. Huang, J.A. Tame, and R. Pethig. *J. Phys. D: Appl. Phys.*, 26:1528, 1993.

[35] Y. Huang, X.-B. Wang, R. Hölzel, F.F. Becker, and P.R.C. Gascoyne. *Phys. Med. Biol.*, 40:1789, 1995.
[36] Y. Huang, X.-B. Wang, F.F. Becker, and P.R.C. Gascoyne. *Biochim. Biophys. Acta*, 1282:76, 1996.
[37] Y. Huang, X.-B. Wang, F.F. Becker, and P.R.C. Gascoyne. *Biophys. J.*, 73:1118, 1997.
[38] Y. Huang, X.-B. Wang, P.R.C. Gascoyne, and F.F. Becker. *Biochim. Biophys. Acta*, 1417:51, 1999.
[39] Y. Huang, K.L. Ewaituhon, M. Tirado, R. Haigis, A. Foster, D. Ackley, M.J. Heller, J.P. O'Connell, and M. Krihak. *Anal. Chem.*, 73:1549, 2001.
[40] Y. Huang, S. Joo, M. Duhon, M. Heller, B. Wallace, and X. Xu. *Anal. Chem.*, 74:3362, 2002.
[41] M.P. Hughes. *Nanoelectromechanics in Engineering and Biology*, CRC Press, Boca Raton, 2002,a.
[42] M.P. Hughes. *Electrophoresis*, 23:2569, 2002,b.
[43] M.P. Hughes, R. Pethig, and X.-B. Wang. *J. Phys. D: Appl. Phys.*, 29:474, 1996.
[44] T. Inoue, R. Pethig, T.A.K. Al-Ameen, J.P.H. Burt, and J.A.R. Price. *J. Electrostat.*, 21:215, 1988.
[45] A. Irimajiri, T. Hanai, and A. Inouye. *J. Theor. Biol.*, 78:251, 1979.
[46] T.B. Jones and G.A. Kallio. *J. Electrostatics*, 6:207, 1979.
[47] T.B. Jones. *Electromechanics of Particles*, Cambridge University Press, Cambridge, 1995.
[48] H. Li and R. Bashir. *Sens. Actu. B: Chem.*, 86:215, 2002.
[49] L.B. Leverett, L.D. Hellums, C.P. Alfrey, and E.C. Lynch. *Biophys. J.*, 12:257, 1972.
[50] J. Lyklema. *Fundamentals of Interface and Colloid Science, Vol. 2: Solid-Liquid Interfaces*, Academic Press, San Diego, 1995.
[51] N. Manaresi, A. Romani, G. Medora, L. Altomare, A. Leonardi, and M. Tartagni. *IEEE J. Sol. St. Circ.*, 38:2297, 2003.
[52] G.H. Markx and R. Pethig. *Biotechnol. & Boieng.*, 45:337, 1995.
[53] G.H. Markx, Y. Huang, X.F. Zhou, and R. Pethig. *Microbiology*, 140: 1994a.
[54] G.H. Markx, M.S. Talary, and R. Pethig. *J. Biotechnology*, 32:29, 1994b.
[55] S. Masuda, M. Washizu, and I. Kawabata. *IEEE Trans. Ind. Appl.*, 24:214, 1988.
[56] J.C Maxwell. *A Treatise on Electricity and Magnetism*, 3rd Ed., Vol. 1, Ch. 9, Clarendon Press, Oxford, 1891.
[57] H. Morgan, N.G. Green, M.P. Hughes, W. Monaghan, W., and T.C. Tan. *J. Micromech. Microeng.*, 7:65, 1997.
[58] H. Morgan, A.G. Izquierdo, D. Bakewell, N.G. Green, and A. Ramos. *J. Phys. D: Appl. Phys.*, 34:1553, 2001.
[59] H. Morgan and N.G. Green. *AC Electrokinetics: Colloids and Nanoparticles*, Research Studies Press, Baldock, 2003.
[60] J.R. Morgan and R. Pethig. *J. Phys. E: Sci. Instrum.*, 12:1132, 1979.
[61] T. Müller, A. Pfennig, P. Klein, G. Gradl, M. Jäger, and T. Schnelle. *IEEE Eng. Med. Biol. Magazine*, 22(6):51, 2003.
[62] T. Nagahama, N. Muramatsu, H. Ohshima, and T. Kondo. *Colloid. Surf.*, 67:61, 1992.
[63] E. Neumann, A.E. Sowers, and C.A. Jordan. (Eds.) *Electroporation and Electrofusion in Cell Biology*, Plenum Press, New York, 1989.
[64] R. Paul, K.V.I.S. Kaler, and T.B. Jones. *J. Phys. Chem.*, 97:4745, 1993.
[65] S.J. Paddison, R. Paul, and K.V.I.S. Kaler. *J. Colloid Interface Sci.*, 183:78, 1996.
[66] R. Pethig. *Dielectric and Electronic Properties of Biological Materials*, J. Wiley, Chichester, 1979.
[67] R. Pethig. In I. Karube (ed.), Automation in Biotechnology, Elsevier Science Publishers B.V., 159–185, 1991.
[68] R Pethig. *Crit. Rev. Biotechnology*, 16:331, 1996.
[69] R. Pethig and G.H. Markx. *Trends in Biotechnology*, 15:426, 1997.
[70] R. Pethig and J.P.H. Burt. US Patent 5 795 457, 1998.
[71] R. Pethig and G.H. Markx. US Patent 5 814,200, 1998.
[72] R. Pethig, J.P.H. Burt, A. Parton, N. Rizvi, M.S. Talary, and J.A. Tame. *J. Micromech. Microeng.*, 8:57, 1998.
[73] R. Pethig and J.P.H. Burt. European Patent EP 0 898 493, 2000.
[74] R. Pethig, V. Bressler, C. Carswell-Crumpton, Y. Chen, L. Foster-Haje, M.E. Garicia-Ojeda, R.S. Lee, G.M. Lock, M.S. Talary, and K.M. Tate. *Electrophoresis*, 23:2057, 2002.
[75] R. Pethig, M.S. Talary, R.S. Lee, B. Kusler and C. Carswell-Crumpton. In M.A. Northrup, K.F. Jensen and D.J. Harrison (eds.), *Proc. µTAS 2003*, The Transducers Research Foundation, ISBN: 0-9743611-0-0) pp. 1065–1068, 2003.

[76] R. Pethig, M.S. Talary, and R.S. Lee. *IEEE Eng. Med. Biol. Magazine*, 22(6):43, 2003.
[77] R. Pethig, R.S. Lee, and M.S. Talary. *JALA*, 9:324, 2004.
[78] H.A. Pohl. *J. Appl. Phys.*, 22:869, 1951.
[79] H.A. Pohl. *Dielectrophoresis*, Cambridge University Press, Cambridge, 1978.
[80] H.A. Pohl and R. Pethig. *J. Phys. E: Sci. Instrum.*, 10:190, 1977.
[81] J.A.R. Price, J.P.H. Burt, and R. Pethig. *Biochim. Biophys. Acta*, 964:221, 1988.
[82] A. Ramos, H. Morgan, N.G. Green, and A. Castellanos. *J. Phys. D: Appl. Phys.*, 31:2338, 1998.
[83] K. Ratanachoo, P.R.C. Gascoyne, and M. Ruchirawat. *Biochim. Biophys. Acta*, 1564:449, 2002.
[84] J. Rousselet, G. H. Markx, and R. Pethig. *Colloids Surf. A: Physicochem. Eng. Aspects*, 140:209, 1998.
[85] B. Saif, R.K. Mohr, C.J. Montrose, and T.A. Litovitz. *Biopoymers*, 31:1171, 1991.
[86] M. Sancho, G. Martinez, and C. Martin. *J. Electrostat.*, 57:143, 2003.
[87] F.A. Sauer. In A. Chiabrera, C. Nicolini, and H. P. Schwan (eds.), *Interactions between Electromagnetic Fields and Cells*, Plenum Publ., New York, pp. 181–202, 1985.
[88] T. Schnelle, T. Müller, and G. Fuhr. *J. Electrostat.*, 50:17, 2000.
[89] H.P. Schwan. In A. Chiabrera, C. Nicolini, and H. P. Schwan (eds.), *Interactions between Electromagnetic Fields and Cells*, Plenum Publ., New York, pp. 75–97, 1985.
[90] J.W. Smith. *Electric Dipole Moments*, Butterworth Scientific Publ., London, 7–11, 1955.
[91] M. Stephens, M. Talary, R. Pethig, A.K. Burnett, and K.I. Mills. *Bone Marrow Transplant* 18:777, 1996.
[92] J. Suehiro and R. Pethig. *J. Phys. D: Appl. Phys.*, 31:3298, 1998.
[93] J. Suehiro, R. Hamada, D. Noutomi, M. Shutou, and M. Hara. *J. Electrostatics*, 57:157, 2003a.
[94] J. Suehiro, D. Noutomi, M. Shutou, and M. Hara. *J. Electrostatics*, 58:229, 2003b.
[95] V.L. Sukhorukov, G. Meedt, M. Kurschner, and U. Zimmermann. *J. Electrostat.*, 50:191, 2001.
[96] S. Takashima. *Electrical Properties of Biopolymers and Membranes*, Adam Hilger, Bristol, 1989.
[97] M.S. Talary and R. Pethig. *IEEE Proc. Sci. Meas. Technol.*, 141:395, 1994.
[98] M.S. Talary, J.P.H. Burt, J.A. Tame, and R. Pethig. *J. Phys. D: Appl. Phys.*, 29:2198, 1996.
[99] J. Voldman, M. Toner, M.L. Gray, and M.A. Schmidt. *J. Electrostat.*, 57:69, 2003.
[100] K.W. Wagner. *Archiv. Elektrotechnik*, 2:S371, 1914.
[101] X.B. Wang, R. Pethig, and T.B. Jones. *J. Phys. D: Appl. Phys.*, 25:905, 1992.
[102] X.B. Wang, Y. Huang, J.P.H. Burt, G.H. Markx, and R. Pethig. *J. Phys. D: Appl. Phys.*, 26:312, 1993a.
[103] X.B. Wang, Y. Huang, R. Hölzel, J.P.H. Burt, and R. Pethig. *J. Phys. D: Appl. Phys.*, 26:1278, 1993b.
[104] X.B. Wang, Y. Huang, P.R.C. Gascoyne, F.F. Becker, R. Hölzel, and R. Pethig. *Biochim. Biophys. Acta*, 1193:330, 1994a.
[105] X.B. Wang, Y. Huang, F.F. Becker, and P.R.C. Gascoyne. *J Phys. D: Appl. Phys.*, 27:1571, 1994b.
[106] X-B. Wang, Y. Huang, X. Wang, F.F. Becker, and P.R.C. Gascoyne. *Biophysical J.*, 72:1887, 1997.
[107] X.J. Wang, F.F. Becker, and P.R.C. Gascoyne. *Biochim. Biophys. Acta*, 1564:412, 2002.
[108] M. Washizu and O. Kurosawa. *IEEE Trans. Ind. Appl.*, 26:1165, 1990.
[109] M. Washizu, S. Suzuki, O. Kurosawa, T. Nishizaka, and T. Shinohara. *IEEE Trans. Ind. Appl.*, 30:835, 1994.
[110] J. Yang, Y. Huang, X. Wang, X. B. Wang, F.F. Becker, and P.R.C. Gascoyne. *Biophys. J.*, 76:3307, 1999.

5

Hitting the Spot: The Promise of Protein Microarrays

Joanna S. Albala

Biology & Biotechnology Research Program, Lawrence Livermore National Laboratory, 7000 East Avenue, L-448, Livermore, CA 94550

5.1. INTRODUCTION

With the thrust of scientific endeavor moving from genomics to proteomics, the protein array provides a powerful means by which to examine hundreds to thousands of proteins in parallel. A result of the many genome projects has been the advance of automation and robotic procedures to manipulate biomolecules using a high-throughput, systematic approach. The promise of the protein microarray is the ability to interrogate a large number of proteins simultaneously in a high-density format for disease diagnosis, prognosis or efficacy of therapeutic regime as well as for biochemical analysis. Similar to a DNA microarray, each spot on a protein array can be identified based on its addressability on the planar surface.

Protein microarrays can be generally classified into two categories: 1) protein profiling arrays and 2) functional protein arrays. Protein profiling arrays typically are arrays of antibodies used to compare two different cell types or disease states for novel biomarker identification, protein abundance or alteration [1] whereas functional protein arrays are usually comprised of recombinant proteins used to identify protein-protein, protein-DNA, or protein-small molecule interactions or other biochemical activities [2] and for review see [3–5]. Protein arrays have been designed to encompass the protein equivalent of whole organisms [6] and have been used for cell and tissue-based analyses as well [7, 8].

The concept of proteomic analysis by protein array has been emerging from both academic and industrial laboratories worldwide with increasing success. The generation of new technologies to manipulate the diversity of cellular proteins for protein array applications is advancing the ability to work with this diverse population of biomolecules as well as furthering a greater understanding of many different organisms and disease states. This review aims to provide a progress report of protein microarray advances to date and provides an example for the use of protein microarrays to identify protein-protein interactions in homologous recombination and DNA repair pathways.

5.2. GENERATION OF PROTEIN MICROARRAYS

There are many variables to consider in developing a protein array. These include content, surface chemistry, array production, and detection. Each of these will be discussed in turn below.

5.2.1. Content

Content has proven to be the greatest bottleneck for the generation of protein and antibody arrays. Although large collections of cDNAs are available from the I.M.A.G.E. cDNA Consortium at Lawrence Livermore National Laboratory [9], Mammalian Gene Collection from the National Cancer Institute [10] and the FLEXgene collection at the Harvard Institute of Proteomics [11], it is still a large undertaking to produce proteins in high-throughput.

5.2.1.1. Proteins   Most efforts have focused on generating recombinant protein using *E. coli* for its high productivity and ease of use [12]. High-throughput protein production using *E. coli* has been successful for proteins of several species from bacteria to man [13–16]. The majority of these studies have focused on the production of a 96-well plate of recombinant proteins and have applied both traditional as well as cell-free expression from bacteria [17–19]. Furthermore, several studies have successfully applied these techniques for the production of large numbers of proteins for high-density protein arrays [20]. Synder and co-workers have identified several novel kinases and lipid-binding proteins by generating a functional protein array encompassing nearly the entire proteome of *Saccharomeyces cerevisiae* [6]. Most recently, potential autoimmune disease-associated biomarkers were identified using a high-density array of 2400 human proteins devised by robotic spotting of the recombinant proteins onto glass microscope slides pretreated with a polyacrylamide coating [21].

5.2.1.2. Antibodies   The successful generation of protein arrays for protein profiling in a manner similar to that of a DNA microarray requires a different type of content for the generation of the array, the capture agent. Most protein arrays for this purpose have employed monoclonal antibodies due to their high specificity and affinity for the target from which they were generated; however, polyclonal antibodies have also been used successfully [22]. The development of monoclonal or polyclonal antibodies is a laborious and costly process. As a result, several other capture agents have been described for use in the development of protein microarrays. Fab/scFv fragments from phage display libraries have been used for protein profiling [23]. Phage-display libraries may allow for the production of a large and diverse population of capture agents on a proteomic scale more rapidly than is possible by immunization-based methods of antibody production [24]. In addition, *in vitro*-based methods of ribosome or mRNA display that physically links the capture agent to its cognate mRNA have been used successfully for protein microarray applications [25]. Aptamer arrays have also been demonstrated [26]. Aptamers are short stretches of DNA or RNA that are generated from an *in vitro* process termed "SELEX" for systematic evolution of ligands by exponential enrichment. It is possible to use this process to generate aptamers with high binding affinities to specific ligands including proteins [27]. It is noteworthy that the high-throughput production of capture agents relies on the successful generation of target proteins resulting in both the generation of the antibody arrays as well as for validation of the affinity, specificity, and cross-reactivity of each capture agent on the array.

5.2.2. Surface Chemistry

Protein arrays have been generated on PVDF [28], nitrocellulose [20], plastic [2] or more commonly on glass microscope slides [29] which are compatible with standard robotic arrayers and detection instrumentation used for printing and analyzing DNA microarrays. The surface to which proteins are bound is key in the development of protein microarrays and certain criteria must be considered. First, the surface needs to be advantageous for protein binding; second, it must be amenable to high-throughput production and application; third, low auto fluorescence of the surface is required as fluorescence is the most common means of detection using protein microarrays; and finally, the surface must strive to provide an environment suitable to maintain the native structure of the proteins or capture agents [30]. Several different surface coatings have been used for generating protein microarrays including glass slides activated with aldehyde [29], poly-lysine [31] or a homofunctional crosslinker [32], agarose [33], polyacrylamide gel pads [34], hydrogel [35]. Two recent reports have compared the use of standard surface chemistry with a hydrogel polyacrylamide support and found the hydrogel surface to have better sensitivity with lower background, thus proving to be a superior surface from which to produce protein arrays [30, 35]. Nitrocellulose-coated slides have also been employed for several protein array applications [20, 36].

Efforts have been made to orient proteins on the array to facilitate protein-protein interactions. For example, nickel-coated slides have been used for the generation of protein arrays of recombinant his-tagged proteins [6]. Protein G-coated and strepavidin-coated slides have been utilized to orient antibodies for the generation of antibody microarrays [37, 38], as have microfabricated pillars [39] and repeated application of the target and capture agent to the same address on the array to direct the application of the protein content onto the array surface [22].

5.2.3. Microarray Production

Protein microarray production requires careful preparation of samples and slides prior to printing, and attention to the humidity, temperature, and dust levels in the printing environment [5]. The implementation of automated protein array production includes deposition by pin-based or microdispensing liquid-handling robots [40], photolithography [41, 42], or ink-jet printing technology [43]. Piezoelectric dispensing of proteins from borosilicate glass capillaries is a common method of protein array production that deposits samples onto a planar surface by non-contact printing. It has been demonstrated that addition of bovine serum albumin (BSA) to a low ionic strength buffer can effectively minimize nonspecific protein adsorption to the glass capillaries, a common drawback of this approach [44].

Another new approach is affinity contact printing whereby different proteins are attached to three-dimensional features on an array that can then capture other proteins from a complex mixture. These 3D features can then be used to stamp the captured proteins onto another surface for further analysis by mass spectrometry [45].

5.2.4. Detection

There are several methods for detection of protein interactions using an array format. The most common rely on fluorescently-labeled antibodies or proteins applied to the array. A typical protein profiling experiment will employ an ELISA-based sandwich approach where antibodies are arrayed onto the slide, sample is applied, followed by addition of another capture antibody which is either directly labeled with a fluorochrome (direct detection) or a labeled secondary (indirect detection) is applied [46]. Alternatively, the sample material can be directly labeled and compared to a control sample population that has been labeled with a different fluorophore [1]. Other detection methods less commonly used include SPR (surface plasmon resonance) or RCA (rolling circle amplification). SPR relies on changes in the angle of reflection when a protein interaction occurs on the surface of a gold-coated chip or array and the light caused by the binding of the probe to the immobilized protein is measured. An advantage of SPR is the use of unlabeled probes; however, this method is less sensitive than methods using fluorescence, chemiluminescence or radioactivity for detection (for review see [5]). RCA employs a sandwich immunoassay in which the detection antibodies are linked to DNA sequences that can be amplified and hybridized with complimentary fluorescently-labeled oligonucleotides to detect the interaction between the antibody and its target [47]. Schweitzer et al. in a study of 51 cytokines from cultured dendritic cells demonstrated the differential regulation of several cytokines at biologically relevant concentrations [48].

5.3. PROTEIN ARRAYS FOR ANALYSIS OF PROTEINS INVOLVED IN RECOMBINATION & DNA REPAIR

5.3.1. Protein Expression Microarrays

We have developed protein expression microarrays for the production of recombinant human proteins using a baculovirus-based approach in a 96-well format. Heterologous protein production of mammalian proteins in *E. coli* often results in improperly modified, misfolded or insoluble protein [12]. For these reasons, we have focused our efforts on heterologous production of human protein using the baculovirus expression system [49]. Our process of protein production begins with the selection of cDNA clones from the I.M.A.G.E. collection. We produce recombinant baculoviruses with a PCR-based strategy into a custom transfer vector modified to contain an N-terminal histidine tag for purification of recombinant proteins. Automation is employed to set up the molecular biological reactions needed for PCR amplification, cleanup and cloning. A database has been developed to track the process of protein production. All reactions are performed in a 96-well plate format for miniaturization of the baculovirus paradigm. A 1.5 ml, 96-well insect cell culture system has been devised for the expression and purification of microgram amounts of protein that is ideal for protein microarray applications.

We have employed our miniaturized protein production strategy for the generation of protein fragments for mapping interactions among domains of critical DNA repair proteins in the homologous recombinational repair pathway [50] as well as for the development of protein interaction microarrays [51]. Rad51 is a central protein in the process of homologous recombinational repair (for review see [52, 53]). Several human proteins with homology to Rad51 have been identified. These proteins are known as the Rad51 paralogs and include Rad51B, Rad51C, Rad51D, Xrcc2 and Xrcc3 [54–57]. Recent studies have shown that the Rad51 paralogs exist in two complexes *in vivo* independent of Rad51, BCDX2 and CX3 [58–61]. We were interested in identifying the binding sites between the different Rad51 paralogs in the BCDX2 complex. To this end, we used computational modeling against the *Pyroccocus furious* crystal structure [62], to direct the generation of deletion mutants for the individual paralog proteins. Recombinant proteins for the deletion mutants of each of the Rad51 paralog proteins were produced using our miniaturized baculoviral expression system (i.e., protein expression microarrays). A western blot of the deletion mutant series is shown in Figure 5.1. These include the deletion mutants for Rad51B, 1–75 and 76–350, Rad51C, 1–79, 79–376, and Rad51D, 4–77, 77–379, as well as several point mutants of the Rad51B, Rad51C and Rad51D proteins. Using yeast two-hybrid and co-immunoprecipitation techniques, we have demonstrated that a deletion construct of Rad51B containing amino acids residues 1–75 interacts with the C-terminus of Rad51C amino acids

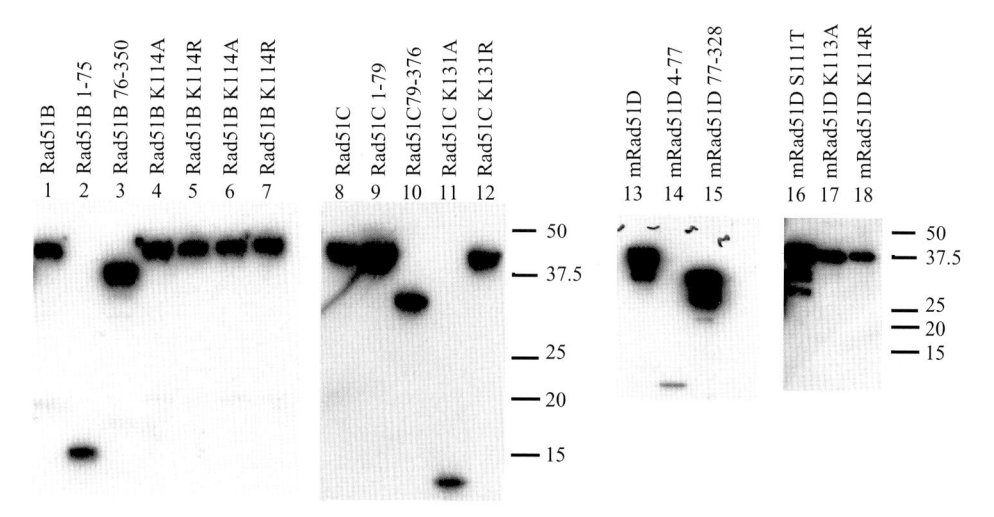

FIGURE 5.1. Rad51 paralog proteins produced using a miniaturized protein expression system in baculovirus (protein expression microarrays). Lane 1: Full-length Rad51B, Lanes 2–3: Deletion mutants of Rad51B, Lanes 4–7: Point mutants in the ATP binding site of Rad51B, Lane 8: Full-length Rad51C, Lanes 9–10: Deletion mutants of Rad51C; Lanes 11–12: Point mutants in the ATP binding site of Rad51C, Lane 13: Full-length Rad51D, Lanes 14–15: Deletion mutants of Rad51D, Lanes 16–18: Point mutants in the ATP binding site of Rad51D. Blots were incubated with a monoclonal antibody to an N-terminal his-tag (Sigma) and proteins were visualized using enhanced chemiluminescence (ECL, Amersham). Protein standards are shown representing molecular weight in kilodaltons (kDa).

residues 79–376, and this region of Rad51C also interacts with Rad51D and Xrcc3. We have also determined that the N-terminal domain of Rad51D, amino acid residues 4–77, binds to Xrcc2 while the C-terminal domain of Rad51D, amino acid residues 77–328, binds Rad51C [50].

5.3.2. Protein Interaction Arrays

We have generated protein interaction microarrays to examine protein-protein and protein-DNA interactions for the Rad51 paralogs. Protein microarrays were generated to contain various DNA repair proteins, histones, nucleosomes, various antibodies and DNA. We applied these various proteins, antibodies and DNA to functionalized glass slides and interrogated the slides with Rad51B and Rad51C to identify novel protein-protein interactions for proteins involved in DNA double-strand break repair.

Standard microarray spotting techniques were used to attach proteins to glass slides in a microarray format to analyze protein-protein interactions. Proteins were resuspended in spotting buffer and arrayed in duplicate on aminopropyl triethoxysilane and/or poly-lysine-coated glass slides using a robotic arrayer. Approximately two hundred proteins were spotted on the array. Scanning and analysis of the arrays were performed on a ScanArray 5000 (488 nm laser for FITC scans and 543 nm laser for rhodamine) using QuantArray software. All information regarding the construction and use of the protein arrays can be found at http://bbrp.llnl.gov/microarrays/external/index.html.

We have shown that the DNA repair protein RAD51B, and not its cognate partner RAD51C, interacts with histones and not nucleosomes (see Figure 5.2). Several proteins

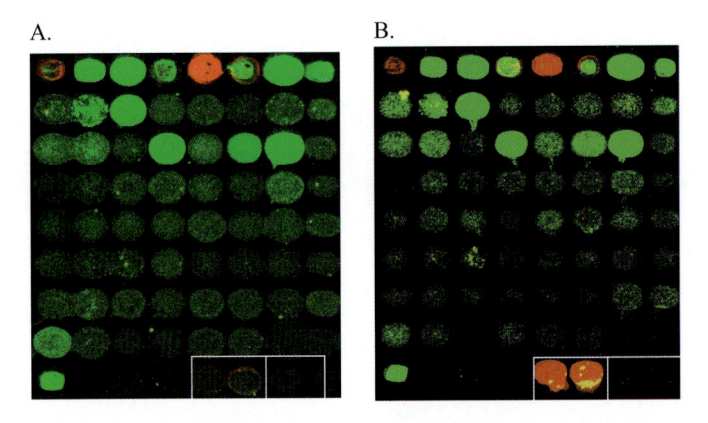

FIGURE 5.2. Protein interaction microarrays show that Rad51B interacts with histones and not nucleosomes. Protein arrays were generated to contain a variety of DNA repair proteins, histones, nucleosomes and antibodies. Proteins were generated in a miniaturized baculoviral expression system or as GFP fusion proteins by cell-free expression in *E. coli*. Panel A: Protein interaction microarray incubated without recombinant Rad51B. Panel B: Protein interaction microarray incubated with recombinant Rad51B protein. The two red spots indicate an interaction between Rad51B and histones on the array. No interaction is observed between Rad51B and the adjacent nucleosome proteins. Green spots are GFP fusion proteins. Red spots in the top row are rhodamine-labeled DNA markers for addressability of spots on the array. Two hundred and forty-two spots were robotically arrayed and only a fraction of these are shown.

on this array incorporated GFP into the final fusion protein [63]. Unique RAD51B-histone interactions were corroborated using Far Western analysis [51]. This is the first demonstration of an interaction between RAD51B and histone proteins that may be important for the successful repair of DNA double-strand breaks.

5.4. SUMMARY: PROTEIN ARRAYS-HOPE OR HYPE?

Protein microarrays hold great promise for accelerating basic biology, biomarker identification and drug discovery. Recent strides have been made in the use of protein arrays for diagnosis of autoimmune disorders [21, 31, 64, 65] as well as for cancer biomarker identification [66]. Of particular note is the use of reverse arrays pioneered by Petricion and Liotta where subpopulations of cells from a given tissue were obtained by use of laser capture microdissection, cell lysates were made, and the lysates interrogated for biomarker detection on arrays [67]. Another approach relies on the identification of panel of protein biomarkers or signatures for diagnosis or prognosis of disease [68].

There is still much to be learned in the protein array arena, particularly in standardization of array production and content for comparative analysis of results between labs. Future work will aim to utilize protein microarrays to understand complex subproteomes and the characterization of posttranslational modification of proteins. Although the work ahead is challenging, the payoff for miniaturized, high-throughput analysis for personalized medicine is enormous. Progress has been made, but the best may yet to come.

ACKNOWLEDGEMENTS

Thanks to Kristi Miller, Daniel Yoshikawa, Peter Beernick and Matthew Coleman for their contribution to the production of the proteins and protein interaction microarrays. The author would like to thank Marianne Kavanagh, Irene Jones and Christa Prange for critical review of this manuscript. This work was performed under the auspices of the U.S. Department of Energy managed by the University of California-Lawrence Livermore National Laboratory under Contract No. W-7405-Eng-48.

REFERENCES

[1] B.B. Haab, M.J. Dunham, and P.O. Brown. Protein microarrays for highly parallel detection and quantitation of specific proteins and antibodies in complex solutions. *Genome. Biol.*, 2:RESEARCH0004, 2001.
[2] H. Zhu, J.F. Klemic, S. Chang, P. Bertone, A. Casamayor, K.G. Klemic, D. Smith, M. Gerstein, M.A. Reed, and M. Snyder. Analysis of yeast protein kinases using protein chips. *Nat. Genet.*, 26:283–289, 2000.
[3] J.S. Albala. Array-based proteomics: the latest chip challenge. *Expert. Rev. Mol. Diagn.*, 1:145–152, 2001.
[4] D.J. Cahill. Protein and antibody arrays and their medical applications. *J. Immunol. Methods*, 250:81–91, 2001.
[5] B. Schweitzer, P. Predki, and M. Snyder. Microarrays to characterize protein interactions on a whole-proteome scale. *Proteomics*, 3:2190–2199, 2003.
[6] H. Zhu, M. Bilgin, R. Bangham, D. Hall, A. Casamayor, P. Bertone, N. Lan, R. Jansen, S. Bidlingmaier, and T. Houfek *et al.* Global analysis of protein activities using proteome chips. *Science*, 293:2101–2105, 2001.
[7] J. Ziauddin and D.M. Sabatini. Microarrays of cells expressing defined cDNAs. *Nature*, 411:107–110, 2001.

[8] J. Kononen, L. Bubendorf, A. Kallioniemi, M. Barlund, P. Schraml, S. Leighton, J. Torhorst, M.J. Mihatsch, G. Sauter, and O.P. Kallioniemi. Tissue microarrays for high-throughput molecular profiling of tumor specimens. *Nat. Med.*, 4:844–847, 1998.

[9] G. Lennon, C. Auffray, M. Polymeropoulos, and M.B. Soares. The I.M.A.G.E. Consortium: an integrated molecular analysis of genomes and their expression. *Genomics*, 33:151–152, 1996.

[10] R.L. Strausberg, E.A. Feingold, R.D. Klausner, and F.S. Collins. The mammalian gene collection. *Science*, 286:455–457, 1999.

[11] L. Brizuela, P. Braun, and J. LaBaer. FLEXGene repository: from sequenced genomes to gene repositories for high-throughput functional biology and proteomics. *Mol. Biochem. Parasitol.*, 118:155–165, 2001.

[12] M. Gilbert and J.S. Albala. Accelerating code to function: sizing up the protein production line. *Curr. Opin. Chem. Biol.*, 6:102–105, 2002.

[13] S.A. Lesley, P. Kuhn, A. Godzik, A.M. Deacon, I. Mathews, A. Kreusch, G. Spraggon, H.E. Klock, D. McMullan, and T. Shin *et al.* Structural genomics of the Thermotoga maritima proteome implemented in a high-throughput structure determination pipeline. *Proc. Natl. Acad. Sci. U.S.A.*, 99:11664–11669, 2002.

[14] R.Y. Huang, S.J. Boulton, M. Vidal, S.C. Almo, A.R. Bresnick, and M.R. Chance. High-throughput expression, purification, and characterization of recombinant Caenorhabditis elegans proteins. *Biochem. Biophys. Res. Commun.*, 307:928–934, 2003.

[15] P. Braun, Y. Hu, B. Shen, A. Halleck, M. Koundinya, E. Harlow, and J. LaBaer. Proteome-scale purification of human proteins from bacteria. *Proc. Natl. Acad. Sci. U.S.A.*, 99:2654–2659, 2002.

[16] P. Braun and J. LaBaer. High throughput protein production for functional proteomics. *Trends Biotechnol.*, 21:383–388, 2003.

[17] P. Sebastian, J. Wallwitz, and S. Schmidt. Semi automated production of a set of different recombinant GST-Streptag fusion proteins. *J. Chromatogr. B Analyt. Technol. Biomed. Life Sci.*, 786:343–355, 2003.

[18] Y.P. Shih, W.M. Kung, J.C. Chen, C.H. Yeh, A.H. Wang, and T.F. Wang. High-throughput screening of soluble recombinant proteins. *Protein Sci.*, 11:1714–1719, 2002.

[19] D. Busso, R. Kim, and S.H. Kim. Expression of soluble recombinant proteins in a cell-free system using a 96-well format. *J. Biochem. Biophys. Methods*, 55:233–240, 2003.

[20] L.J. Holt, K. Bussow, G. Walter, and I.M. Tomlinson. By-passing selection: direct screening for antibody-antigen interactions using protein arrays. *Nucleic Acids Res.*, 28:E72, 2000.

[21] A. Lueking, A. Possling, O. Huber, A. Beveridge, M. Horn, H. Eickhoff, J. Schuchardt, H. Lehrach, and D.J. Cahill. A Nonredundant Human Protein Chip for Antibody Screening and Serum Profiling. *Mol. Cell. Proteomics*, 2:1342–1349, 2003.

[22] P. Angenendt, J. Glokler, Z. Konthur, H. Lehrach, and D.J. Cahill. 3D protein microarrays: performing multiplex immunoassays on a single chip. *Anal. Chem.*, 75:4368–4372, 2003.

[23] N. Stich, G. van Steen, and T. Schalkhammer. Design and peptide-based validation of phage display antibodies for proteomic biochips. *Comb. Chem. High Throughput Screen*, 6:67–78, 2003.

[24] H.J. de Haard, N. van Neer, A. Reurs, S.E. Hufton, E.C. Roovers, P. Henderikx, A.P. de Bruine, J.W. Arends, and H.R. Hoogenboom. A large non-immunized human Fab fragment phage library that permits rapid isolation and kinetic analysis of high affinity antibodies. *J. Biol. Chem.*, 274:18218–18230, 1999.

[25] S. Weng, K. Gu, P.W. Hammond, P. Lohse, C. Rise, R.W. Wagner, M.C. Wright, and R.G. Kuimelis. Generating addressable protein microarrays with PROfusion covalent mRNA-protein fusion technology. *Proteomics*, 2:48–57, 2002.

[26] H. Petach and L. Gold. Dimensionality is the issue: use of photoaptamers in protein microarrays. *Curr. Opin. Biotechnol.*, 13:309–314, 2002.

[27] E.N. Brody and L. Gold. Aptamers as therapeutic and diagnostic agents. *J. Biotechnol.*, 74:5–13, 2000.

[28] Y. Lin, R. Huang, X. Cao, S.M. Wang, Q. Shi, and R.P. Huang. Detection of multiple cytokines by protein arrays from cell lysate and tissue lysate. *Clin. Chem. Lab. Med.*, 41:139–145, 2003.

[29] G. MacBeath and S.L. Schreiber. Printing proteins as microarrays for high-throughput function determination. *Science*, 289:1760–1763, 2000.

[30] P. Angenendt, J. Glokler, D. Murphy, H. Lehrach, and D.J. Cahill. Toward optimized antibody microarrays: a comparison of current microarray support materials. *Anal. Biochem.*, 309:253–260, 2002.

[31] T.O. Joos, M. Schrenk, P. Hopfl, K. Kroger, U. Chowdhury, D. Stoll, D. Schorner, M. Durr, K. Herick, and S. Rupp *et al.* A microarray enzyme-linked immunosorbent assay for autoimmune diagnostics. *Electrophoresis*, 21:2641–2650, 2000.

[32] L.G. Mendoza, P. McQuary, A. Mongan, R. Gangadharan, S. Brignac, and M. Eggers. High-throughput microarray-based enzyme-linked immunosorbent assay (ELISA). *Biotechniques*, 27:778–780, 782–776, 788, 1999.
[33] V. Afanassiev, V. Hanemann, and S. Wolfl. Preparation of DNA and protein micro arrays on glass slides coated with an agarose film. *Nucleic Acids Res.*, 28:E66, 2000.
[34] D. Guschin, G. Yershov, A. Zaslavsky, A. Gemmell, V. Shick, D. Proudnikov, P. Arenkov, and A. Mirzabekov. Manual manufacturing of oligonucleotide, DNA, and protein microchips. *Anal. Biochem.*, 250:203–211, 1997.
[35] J.C. Miller, H. Zhou, J. Kwekel, R. Cavallo, J. Burke, E.B. Butler, B.S. Teh, and B.B. Haab. Antibody microarray profiling of human prostate cancer sera: antibody screening and identification of potential biomarkers. *Proteomics*, 3:56–63, 2003.
[36] J. Madoz-Gurpide, H. Wang, D.E. Misek, F. Brichory, and S.M. Hanash. Protein based microarrays: a tool for probing the proteome of cancer cells and tissues. *Proteomics*, 1:1279–1287, 2001.
[37] J. Turkova. Oriented immobilization of biologically active proteins as a tool for revealing protein interactions and function. *J. Chromatogr. B Biomed. Sci. Appl.*, 722:11–31, 1999.
[38] P. Pavlickova, A. Knappik, D. Kambhampati, F. Ortigao, and H. Hug. Microarray of recombinant antibodies using a streptavidin sensor surface self-assembled onto a gold layer. *Biotechniques*, 34:124–130, 2003.
[39] P. Peluso, D.S. Wilson, D. Do, H. Tran, M. Venkatasubbaiah, D. Quincy, B. Heidecker, K. Poindexter, N. Tolani, M. Phelan et al. Optimizing antibody immobilization strategies for the construction of protein microarrays. *Anal. Biochem.*, 312:113–124, 2003.
[40] A. Lueking, M. Horn, H. Eickhoff, K. Bussow, H. Lehrach, and G. Walter. Protein microarrays for gene expression and antibody screening. *Anal. Biochem.*, 270:103–111, 1999.
[41] J.F. Mooney, A.J. Hunt, J.R. McIntosh, C.A. Liberko, D.M. Walba, and C.T. Rogers. Patterning of functional antibodies and other proteins by photolithography of silane monolayers. *Proc. Natl. Acad. Sci. U.S.A.*, 93:12287–12291, 1996.
[42] V.W. Jones, J.R. Kenseth, M.D. Porter, C.L. Mosher, and E. Henderson. Microminiaturized immunoassays using atomic force microscopy and compositionally patterned antigen arrays. *Anal. Chem.*, 70:1233–1241, 1998.
[43] A. Roda, M. Guardigli, C. Russo, P. Pasini, and M. Baraldini. Protein microdeposition using a conventional ink-jet printer. *Biotechniques*, 28:492–496, 2000.
[44] J.B. Delehanty and F.S. Ligler. Method for printing functional protein microarrays. *Biotechniques*, 34:380–385, 2003.
[45] D.S. Wilson and S. Nock. Recent developments in protein microarray technology. *Angew. Chem. Int. Ed. Engl.*, 42:494–500, 2003.
[46] T.O. Joos, D. Stoll, and M.F. Templin. Miniaturised multiplexed immunoassays. *Curr. Opin. Chem. Biol.*, 6:76–80, 2002.
[47] B. Schweitzer, S. Wiltshire, J. Lambert, S. O'Malley, K. Kukanskis, Z. Zhu, S.F. Kingsmore, P.M. Lizardi, and D.C. Ward. Inaugural article: immunoassays with rolling circle DNA amplification: a versatile platform for ultrasensitive antigen detection. *Proc. Natl. Acad. Sci. U.S.A.*, 97:10113–10119, 2000.
[48] B. Schweitzer, S. Roberts, B. Grimwade, W. Shao, M. Wang, Q. Fu, Q. Shu, I. Laroche, Z. Zhou, and V.T. Tchernev et al. Multiplexed protein profiling on microarrays by rolling-circle amplification. *Nat. Biotechnol.*, 20:359–365, 2002.
[49] J.S. Albala, K. Franke, I.R. McConnell, K.L. Pak, P.A. Folta, B. Rubinfeld, A.H. Davies, G.G. Lennon, and R. Clark. From genes to proteins: high-throughput expression and purification of the human proteome. *J. Cell. Biochem.*, 80:187–191, 2000.
[50] K.A. Miller, D. Sawicka, D. Barsky, and J.S. Albala. Domain mapping of the Rad51 paralog protein complexes. *Nucleic Acids Res.*, 32:169–178, 2004.
[51] M.A. Coleman, K.A. Miller, P.T. Beernink, D.M. Yoshikawa, and J.S. Albala. Identification of chromatin-related protein interactions using protein microarrays. *Proteomics*, 3:2101–2107, 2003.
[52] L.H. Thompson and D. Schild. Homologous recombinational repair of DNA ensures mammalian chromosome stability. *Mutat. Res.*, 477:131–153, 2001.
[53] S.C. West. Molecular views of recombination proteins and their control. *Nat. Rev. Mol. Cell. Biol.*, 4:435–445, 2003.
[54] D.L. Pittman, L.R. Weinberg, and J.C. Schimenti. Identification, characterization, and genetic mapping of Rad51d, a new mouse and human RAD51/RecA-related gene. *Genomics*, 49:103–111, 1998.

[55] N. Liu, J.E. Lamerdin, R.S. Tebbs, D. Schild, J.D. Tucker, M.R. Shen, K.W. Brookman, M.J. Siciliano, C.A. Walter, and W. Fan *et al*. XRCC2 and XRCC3, new human Rad51-family members, promote chromosome stability and protect against DNA cross-links and other damages. *Mol. Cell.*, 1:783–793, 1998.
[56] M.K. Dosanjh, D.W. Collins, W. Fan, G.G. Lennon, J.S. Albala, Z. Shen, and D. Schild. Isolation and characterization of RAD51C, a new human member of the RAD51 family of related genes. *Nucleic Acids Res.*, 26:1179–1184, 1998.
[57] J.S. Albala, M.P. Thelen, C. Prange, W. Fan, M. Christensen, L.H. Thompson, and G.G. Lennon. Identification of a novel human RAD51 homolog, RAD51B. *Genomics*, 46:476–479, 1997.
[58] N. Liu, D. Schild, M.P. Thelen, and L.H. Thompson. Involvement of Rad51C in two distinct protein complexes of Rad51 paralogs in human cells. *Nucleic Acids Res.*, 30:1009–1015, 2002.
[59] J.Y. Masson, M.C. Tarsounas, A.Z. Stasiak, A. Stasiak, R. Shah, M.J. McIlwraith, E.E. Benson, and S.C. West. Identification and purification of two distinct complexes containing the five RAD51 paralogs. *Genes Dev.*, 15:3296–3307, 2001.
[60] K.A. Miller, D.M. Yoshikawa, I.R. McConnell, R. Clark, D. Schild, and J.S. Albala. RAD51C Interacts with RAD51B and Is Central to a Larger Protein Complex in Vivo Exclusive of RAD51. *J. Biol. Chem.*, 277:8406–8411, 2002.
[61] C. Wiese, D.W. Collins, J.S. Albala, L.H. Thompson, A. Kronenberg, and D. Schild. Interactions involving the Rad51 paralogs Rad51C and XRCC3 in human cells. *Nucleic Acids Res.*, 30:1001–1008, 2002.
[62] D.S. Shin, L. Pellegrini, D.S. Daniels, B. Yelent, L. Craig, D. Bates, D.S. Yu, M.K. Shivji, C. Hitomi, A.S. Arvai, N. Volkmann, H. Tsuruta, T.L. Blundell, A.R. Venkitaraman, and J.A. Tainer. Full-length archaeal Rad51 structure and mutants: mechanisms for RAD51 assembly and control by BRCA2. *EMBO J.*, 22:4566–4576, 2003.
[63] P.T. Beernink, S.S. Krupka, V. Lao, G. Martin, and M.A. Coleman. Application of in vitro protein expression to human prote. *Sci. World J.*, 2:73–74, 2002.
[64] W.H. Robinson, L. Steinman, and P.J. Utz. Protein arrays for autoantibody profiling and fine-specificity mapping. *Proteomics*, 3:2077–2084, 2003.
[65] W.H. Robinson, C. DiGennaro, W. Hueber, B.B. Haab, M. Kamachi, E.J. Dean, S. Fournel, D. Fong, M.C. Genovese, and H.E. de Vegvar *et al*. Autoantigen microarrays for multiplex characterization of autoantibody responses. *Nat. Med.*, 8:295–301, 2002.
[66] J.C. Miller, E.B. Butler, B.S. Teh, and B.B. Haab. The application of protein microarrays to serum diagnostics: prostate cancer as a test case. *Dis. Markers*, 17:225–234, 2001.
[67] R.L. Grubb, V.S. Calvert, J.D. Wulkuhle, C.P. Paweletz, W.M. Linehan, J.L. Phillips, R. Chuaqui, A. Valasco, J. Gillespie, and M. Emmert-Buck *et al*. Signal pathway profiling of prostate cancer using reverse phase protein arrays. *Proteomics*, 3:2142–2146, 2003.
[68] E.F. Petricoin, A.M. Ardekani, B.A. Hitt, P.J. Levine, V.A. Fusaro, S.M. Steinberg, G.B. Mills, C. Simone, D.A. Fishman, and E.C. Kohn *et al*. Use of proteomic patterns in serum to identify ovarian cancer. *Lancet*, 359:572–577, 2002.

6

Use of Electric Field Array Devices for Assisted Assembly of DNA Nanocomponents and Other Nanofabrication Applications

Michael J. Heller[1], Cengiz S. Ozkan[2], and Mihrimah Ozkan[3]
[1]University of California San Diego, La Jolla, CA 92093, and Nanogen, San Diego, CA, 92121
[2]Department of Mechanical Engineering, University of California Riverside, Riverside CA 92521
[3]Department of Electrical Engineering, University of California Riverside, Riverside CA 92521

Microelectronic arrays utilizing electric field transport have been developed for DNA diagnostics (including infectious and genetic disease and cancer detection), for short tandem repeat (STR) forensics analysis, and for gene expression applications. In addition to these bioresearch and clinical diagnostic applications, such devices also have the potential to carry out the assisted assembly of a wide variety of molecular scale, nanoscale and microscale components into higher order structures. These microelectronic array devices are able to produce defined electric fields on their surfaces that allow molecules and other entities with high fidelity recognition properties to be transported to or from any site on the surface of the array. Such devices can utilize either DC electric fields which cause movement of entities by their relative charge, or AC electric fields which allow entities to be selectively positioned by their dielectric properties. An almost unlimited variety of molecules and nanocomponents can be utilized with these devices, including: DNA, DNA constructs with fluorescent, photonic or electronic transfer properties, RNA, RNA constructs, amino acids, peptides, proteins (antibodies, enzymes), nanoparticles (quantum dots, carbon nanotubes, nanowires), cells and even micron scale semiconductor components. Thus, electric field devices can be used for developing a unique highly parallel "Pick & Place" fabrication process by which a variety of heterogeneous molecules, nanocomponents and micron sized objects with intrinsic self-assembly properties can be organized into higher order 2D and 3D structures and devices. The process represents a unique synergy of combining the best aspects of a "top-down" process with a "bottom-up" process. Finally, integration of optical tweezers for manipulation of live cells and microspheres in a similar microarray setup is demonstrated for the applications of biological delivery and invasive manipulation of these species.

6.1. INTRODUCTION

Nanotechnology and nanoscience encompass a wide range of new concepts, ideas and potential applications for nanoelectronics, novel materials, more efficient energy conversion processes, and a new generation of sensors and biomedical devices [1]. Generally, molecular or nanoelectronic devices are envisioned as the more revolutionary outcome of this new field. Presently, there are a numerous examples of novel nanocomponents (organic electron transfer molecules, quantum dots, carbon nanotubes, nanowires, etc.), and some examples of a first level assembly of these nanocomponents into simple structures with some higher order properties [2–4]. Nevertheless, the larger issue of developing viable nanofabrication techniques that allow billions of molecular/nanoscale components to be assembled and interconnected into useful logic/memory devices still remains a considerable challenge. The one exception to this is the present silicon/CMOS fabrication technology which continues to shrink semiconductor feature sizes down to the nanoscale level, although it does appear that this photolithographic based "top-down" technique is being pressed to it's limit. In addition to the nanoelectronic applications, other new nanodevices and nanomaterials with higher order photonic, mechanical, mechanistic, sensory, chemical, catalytic, and therapeutic properties are also envisioned [1]. Again, a key problem in enabling such new devices and materials will most likely be in developing viable highly parallel fabrication technologies for organizing and integrating heterogeneous components of different sizes and compositions into higher level structures and devices.

Biology provides some of the best examples of nanotechnology in action, and biological self-assembly or self-organization might be considered the primer for developing so-called "bottom-up" processes for nanofabrication. Our present knowledge of the molecular biology of living systems demonstrates many unique molecular scale nanostructures and mechanisms which include efficient electron transfer and photonic transfer systems, energy conversion systems, high information content molecules (DNA and RNA), precision structural components (proteins, fatty acids, etc) and a wide variety of highly efficient chemomechanical (catalytic) molecules called enzymes. By way of example, in the photosynthetic energy conversion systems of plants and algae, the chloroplasts contain solid state antenna structures (50–100 nanometers) that are composed of hundreds of highly organized chromophore molecules and other nanostructures. These nanoscale antennas are able to collect photonic energy and transfer it to other structures with very high quantum efficiency. The process is quite different than the way a man-made device, such as charged coupled device (CCD) would capture photons and convert them into electrical energy (Figure 6.1). To date, it has not been possible to design a biomimetic model of these solid state antenna structures that retains any of the efficient properties of the biological systems. In some sense, the problem with enabling this type nanotechnology is like trying to implement

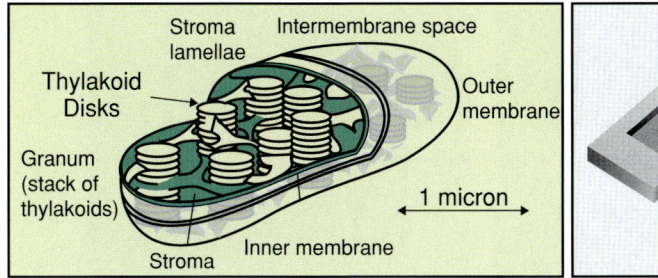

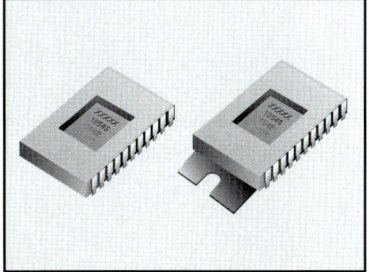

FIGURE 6.1. Comparison of one of nature's photonic conversion devices, a plant chloroplast, with a man-made charged coupled device (CCD). The chloroplast is a highly efficient structure which contains organized antenna structures that capture light energy and transfer it via the Forster resonant energy transfer process to one single molecule, which then converts the resonant energy into chemical energy. The CCD device also captures photons and converts them to an electrical signal. The overall efficiency of the chloroplast energy conversion process is much higher than for any man made device.

the microelectronics revolution with a basic understanding of semiconductor properties, but without the photolithography fabrication process being available to carry out the integration of components into the higher order structures. Living systems on the other hand have developed the ultimate "bottom-up" processes that allows component molecules and nanostructures with intrinsic self-assembly properties to be synthesized, and then further organized into more intricate three dimensional nanostructures, organelles, cells and final organisms.

Of all the biological based molecules available with high recognition and self-assembly properties, the nucleic acids represent one of the more promising materials which may be useful for creating molecular electronic/photonic circuits, organized nanostructures, and even integrated microelectronic and photonic devices [5–7]. The nucleic acids, deoxyribonucleic acid (DNA) and ribonucleic acid (RNA) are programmable molecules which, via their base sequence, have intrinsic molecular recognition and self-assembly properties. Short DNA sequences, called oligonucleotides, are easily synthesized and readily modified with a variety of functional groups. The ability to functionalize synthetic DNA molecules with fluorescent chromophore and charge transfer groups provides a means to incorporate electronic and photonic properties directly into these molecules. DNA can also be used to functionalize larger molecules, nanostructures (quantum dots, gold nanoparticles, carbon nanotubes, etc.) or even micron size components which then can self-assemble or be selectively attached to surfaces, including glass, silicon or other semiconductor materials. DNA molecules, in particular synthetic oligonucleotides represent an ideal type of "molecular legos" for self-assembly of nanocomponents into more complex two and three dimensional higher order structures. At a first level, DNA can be used for a kind of template directed assembly on solid surfaces. This technique involves taking complementary DNA strands and using them as a selective glue to bind other organic or inorganic structures together, or to surfaces. DNA molecules contain base moieties, cytosine, guanine, adenine, and thymine (C,G,A,T) which will only bind to each other in specific pairs: C with G, and A with T. This base pairing property allows single strands of DNA to recognize each other, and bind together to form double-stranded DNA structures. Consequently, a 5'-ATTTGC-3'

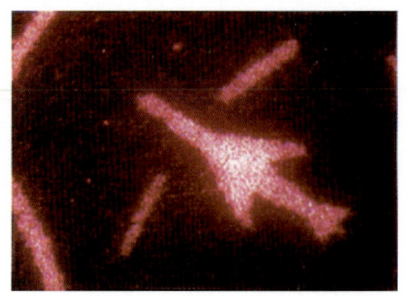

FIGURE 6.2. Shows two examples of DNA imprinting and patterning on a silicon substrate material. On the left, the logo was created by covering the silicon substrate with a DNA capture probe sequence, and then patterned using a mask and UV light to inactivate the exposed DNA sequences. The substrate was then hybridized with a fluorescent complementary DNA sequence, creating the fluorescent logo. The jet logo on the right was created by covering a silicon substrate with a DNA capture probe sequence and then patterned using a mask and UV light to inactivate the exposed DNA sequences. The substrate was then hybridized with fluorescent 200 nanometer nanospheres to which complementary DNA sequences were attached.

strand will only bind strongly with its complement 5'-GCAAAT-3" strand. DNA strands can be attached to surfaces (glass, silicon, gold, etc.) and then patterned by UV light exposure through a mask. Complementary DNA strands can be derivatized with fluorescent molecules or attached to nanostructures (quantum dots, fluorescent nanoparticles, metallic particles). When the specific fluorescent DNA or DNA nanostructures are hybridized to the complementary patterned DNA on a solid support, the patterned substrate material shows the fluorescent image. This DNA assembly method is best suited for the organization of small nanostructures on surfaces, but can be extended to structures in the micron size range. Figure 6.2 shows an example of UV patterned DNA silicon substrate hybridized with complementary fluorescent DNA sequences and with 200 nanometer fluorescent DNA nanospheres.

Active microelectronic arrays have been developed for a number of applications in bioresearch and DNA clinical diagnostics [8–17]. These active microelectronic devices have the ability to create almost any electric field transport geometry on the array surface, which then allows charged reagent and analyte molecules (DNA, RNA, proteins, enzymes), nanostructures, cells and micron-scale structures to be moved to or from any of the microscopic test sites on the device surface. When specific DNA hybridization reactions are carried out on the array, the device is leveraging the electric fields to direct the self-assembly of DNA molecules at the specified microlocation on the chip surface. Microelectronic arrays have been used to demonstrate the organization of complex fluorescent DNA molecular structures and mechanisms within selected microlocations on the array device. In principle these active microelectronic array devices are serving as "motherboards or hostboards" that can carry out the assisted self-assembly of DNA derivatized molecules, nanostructures or microscale structures into more complex two and three dimensional structures [18–25]. This electric field assisted assembly technique is a type of "Pick & Place" process that has potential applications from the nearer term heterogeneous integration of micron-scale photonic, microelectronic and MEMS devices; to the development of high density data storage devices; to the longer term nanofabrication of true molecular and nanoelectronic circuits and devices (Figure 6.3).

USE OF ELECTRIC FIELD ARRAY DEVICES FOR ASSISTED ASSEMBLY 141

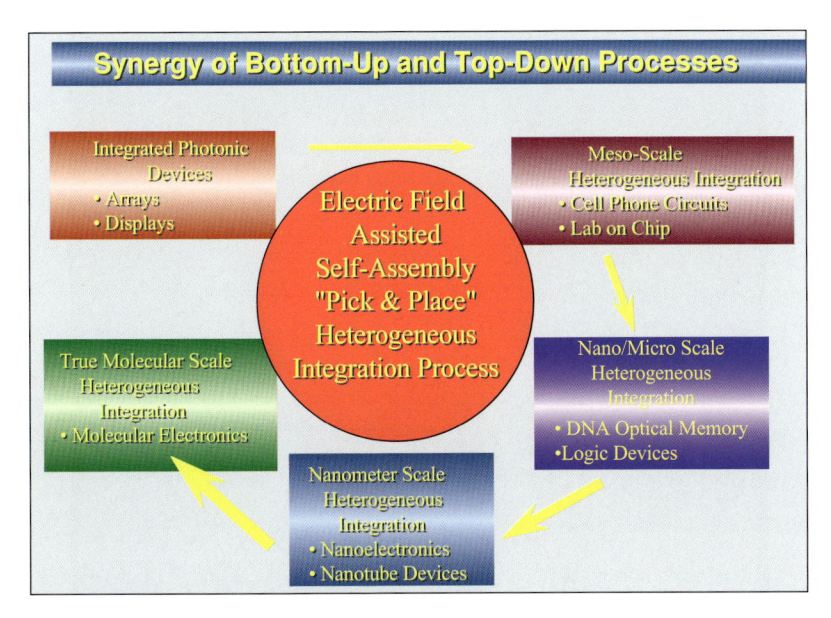

FIGURE 6.3. Chart shows some of the potential advantages and applications for using electric field assisted self-assembly to carry out the heterogeneous integration of molecular, nanoscale and microscale components into higher order materials and devices. Electric field assisted heterogeneous integration technology has the hierarchical logic of allowing one to control the organization and communication of structures and components from molecular level —> nanoscale level —> "Highly Integrated 3-D Structures and Devices" <— micron scale components and lift-off type devices.

6.2. ACTIVE MICROELECTRONIC ARRAY HYBRIDIZATION TECHNOLOGY

A variety of microelectronic arrays have been designed and fabricated by Nanogen, primarily for DNA genotyping diagnostic applications (8–17). These includes an early stage 25 test site microelectronic array device, and more advanced arrays with 100, 400 and 10,000 test sites or microlocations (Figure 6.4). The 100 test-site chip, which has

FIGURE 6.4. Shows Nanogen's 100 test site microelectronic array, a silicon wafer of 400 test site microarrays and a packaged 10,000 test site microarray. The 400 and 10,000 test site arrays have CMOS electronic control elements incorporated into the chips underlying structure.

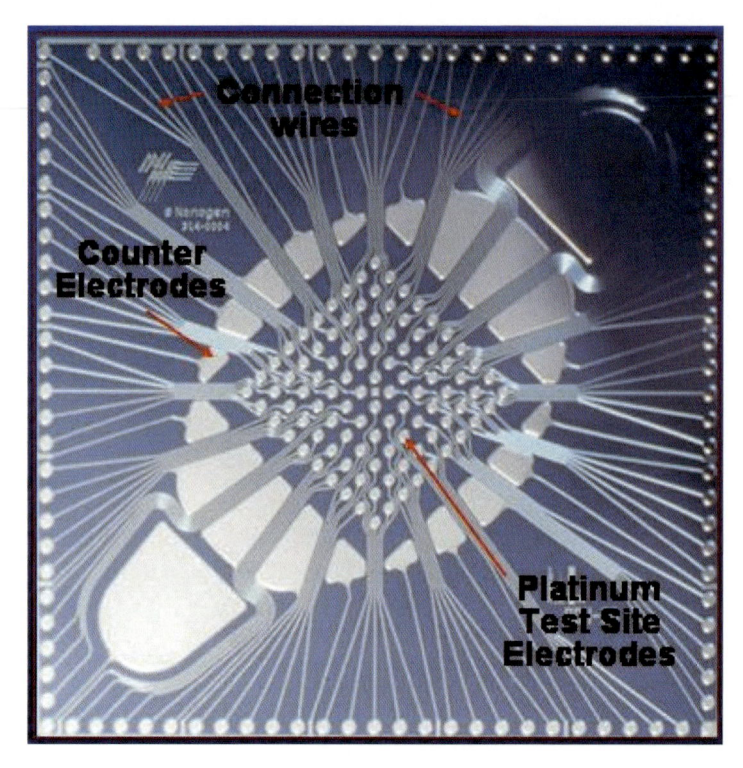

FIGURE 6.5. A more advanced version of Nanogen's microelectronic DNA chip device with 100 test-sites or microlocations. The test sites are approximately 80 microns in diameter, with underlying platinum microelectrodes. The outside ring of twenty micro-electrodes can be used as counter electrodes for the test-sites.

been commercialized, has an inner set of 80-micron diameter test sites with underlying platinum microelectrodes, and an outside set of auxiliary microelectrodes (Figure 6.5). The outer group of electrodes can be negatively biased, which allows DNA in the bulk sample solution to be concentrated at the specific internal test-sites (positively biased). Each microelectrode has an individual wire interconnect through which current and voltage are applied and regulated. The 100 test-site DNA chip is about 7 millimeters square in size, with an active test-site array area of about 2 millimeters. The chips are fabricated from silicon wafers, with insulating layers of silicon dioxide, platinum microelectrodes and gold connecting wires. Silicon dioxide/silicon nitride is used to cover and insulate the conducting wires, but not the surface of the platinum microelectrodes. The whole surface of the array is covered with several microns of hydrogel (agarose or polyacrylamide) which forms a permeation layer. The permeation layer is impregnated with a coupling agent (streptavidin) which allows attachment of biotinylated DNA probes or other entities [12–14, 17]. The ability to use silicon and microlithrography for fabrication of the DNA chips allows a wide variety of devices to be designed and tested (Figure 6.4). The higher density arrays (400 to 10,000) represent more sophisticated devices that have on-chip CMOS control elements for regulating the current and voltages to the microelectrode at each test site [26]. These control elements are located in the underlying silicon structure and are not exposed to the aqueous

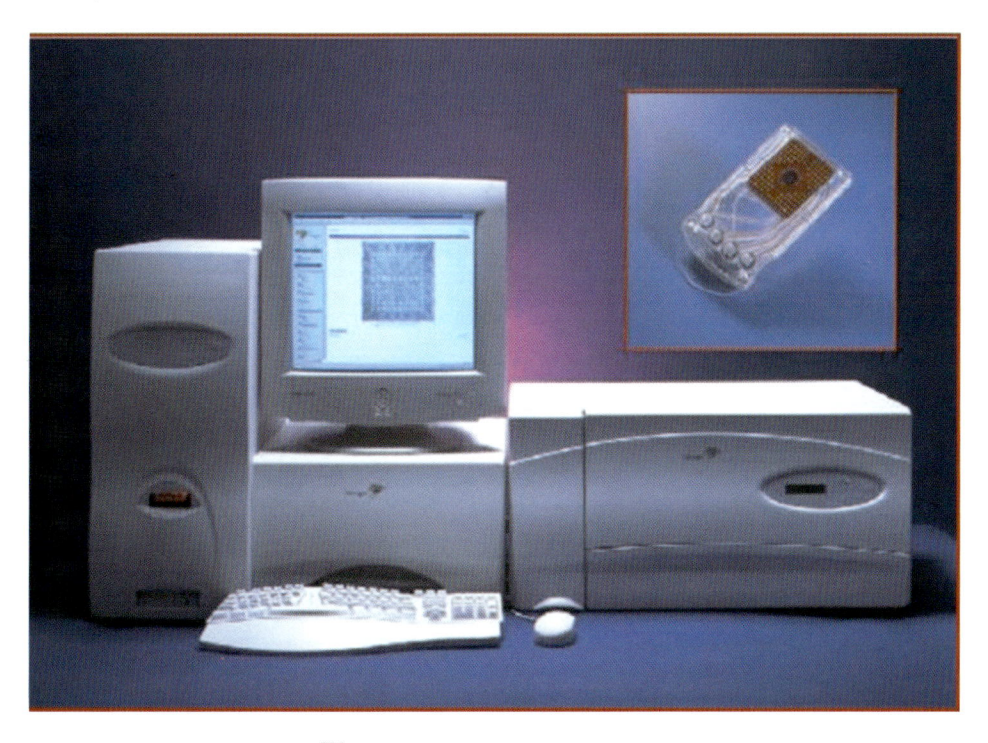

FIGURE 6.6. Shows the NanoChip™ Molecular Biology Workstation for bioresearch and DNA diagnostic applications. This system provides a chip loader component, fluorescent detection/reader component, computer control interface and data display component. The probe loader component is used for automated DNA probe addressing or spotting, and provides the end-user with "Make Your Own Chip" capabilities. Insert shows the NanoChip™ cartridge device. A 100 test site microelectronic array incorporated into a cartridge package that provides the electronic, optical, and fluidic interfacing.

samples that are applied to the chip surface when carrying out the DNA hybridization reactions.

The 100 test-site microelectronic array device has been incorporated into a cartridge package (NanoChip™ cartridge) which provides for the electronic, optical, and fluidic interfacing. The NanoChip™ cartridge assembly is shown as the insert in Figure 6.6. The chip itself is mounted (flip chip bonded) onto a ceramic plate and pinned out for electrical connections. The chip/ceramic plate component is mounted into a plastic cartridge that provides several fluidic input and output ports for addition and removal of DNA samples and reagents. The area over the active test-site portion of the array is an enclosed sample chamber covered with a quartz glass window. This window allows for fluorescent detection to be carried out on the hybridization reactions that occur at the test sites on the array surface. A complete instrument system (NanoChip™ Molecular Biology Workstation) provides a chip loader component, a fluorescent detection/reader component, a computer control interface and a data display screen component (Figure 6.6). The probe loading component allows DNA probes or target sequences (DNA, RNA, PCR amplicons) to be selectively addressed to the array test-sites, and provides a "make your own chip" capability. The automated probe loader system allows four 100 test-site NanoChip™ cartridges to be loaded

FIGURE 6.7. Demonstration of checker-boarding of fluorescent DNA molecules on a microelectronic array. Checker-boarding is achieved by DC biasing alternate test sites on the array positive and negative, and then switching the DC electric field bias every 6 seconds.

with DNA probes or samples from a ninety-six or three hundred eighty four well microtiter plates. Probes or target sequences are usually biotinylated, which allows them to become bound to streptavidin within the permeation layer of the specified test-site. In the electronic addressing procedure the probe loader component deliverers the desired biotinylated probe to specified test sites which are biased positive. The electric (electrophoretic) field causes the negatively charged DNA molecules to concentrate onto the positively activated test-sites, with subsequent binding via the biotin/streptavidin reaction. The ability of the electric field to concentrate DNA molecules on to specific test sites on the array is shown in Figure 6.7. The figure shows the controlled parallel movement (checker-boarding) of negatively charged fluorescent DNA molecules between alternating positively and negatively biased microlocations. The sample solution contains about 50 nM concentration of a 20-mer oligonucleotide sequence which is labeled in the 5'-terminal position with Bodipy Texas Red fluorophore (Ex 590 nm, Em 620 nm), the microelectrodes are biased positive and negative in a checkerboard fashion, and the field is reversed every six seconds. At about 3 volts DC the fluorescent DNA molecules (∼7 nm in length) are transported back and forth a distance of 200 microns during the six second switching time. In actuality, when DNA hybridization reactions are carried out on these active microelectronic arrays, controlled DC electric fields are first used to transport and address (spot) specific DNA molecules to the selected test-sites/microlocations on the array. Such oligonucleotide "capture" probes or target DNA sequences are usually functionalized with biotin molecules, and become strongly bound to streptavidin molecules which are cross-linked to the hydrogel layer covering the underlying microelectrode. In the next steps, electric fields are used to control and direct the hybridization of the other DNA molecules to the DNA sequences attached to the selected test sites. Electronic addressing also allows DNA probes to be spotted onto the array in a highly reproducible manner. Microelectronic arrays can be formatted in a variety of ways that include reverse dot blot format (capture/ identity sequences bound to test-sites), sandwich format (capture sequences bound to test-sites) and dot blot format (target sequences bound to test-sites). DNA hybridization assays involve the use of fluorescent reporter probes or target DNA sequences. The reporter groups are usually organic fluorophores that have been either attached to oligonucleotide probes or to the target/sample DNA/RNA sequences. After electronic addressing and hybridization are carried out, the chip is analyzed using the fluorescent detection system. The fluorescent detector has two different laser excitation sources (excitation 532 nm and excitation 635 nm). The laser beams are quickly scanned across the array using a confocal type optical system and the emissions from the fluorescent labeled probes or targets (550–600 nm "green" and 660–720 nm

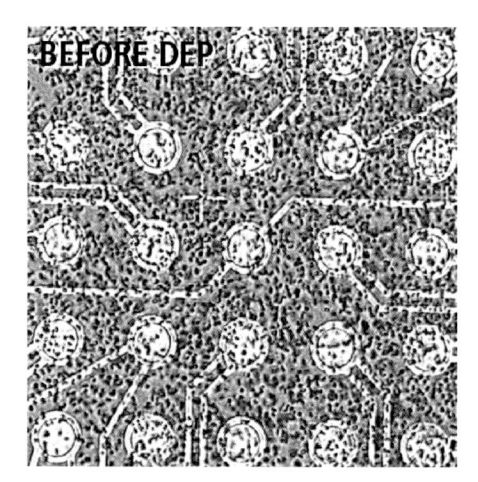

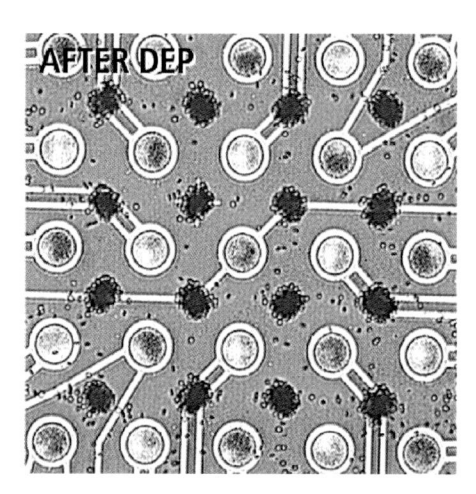

FIGURE 6.8. Shows the separation of Listeria from whole blood by DEP on a microelectronic array (Before and After DEP). Listeria cells become positioned on the high AC field (the microlocations), and the blood cells become aggregated in the low AC field regions between the microlocations.

"red") are detected. References [27–36] provide some examples of single nucleotide polymorphism (SNP) genotyping analysis that is carried out on these microelectronic arrays. For genotyping analysis, the selection rules are simple and no complex image processing is required for data analysis, match/mismatch calls are easy to make and results are highly reliable.

Microelectronic arrays can also be used for a variety of cell separation applications. Disease diagnostics frequently involve identifying a small number of specific bacteria or viruses in a blood sample (infectious disease), fetal cells in maternal blood (genetic diseases) or tumor cells in a background of normal cells (early cancer detection). While a number of techniques are available, electric field based methods have been developed and which can have considerable advantages. One basic electronic method for cell separation is called dielectrophoresis (DEP). This process involves the application of an asymmetric alternating current (AC) electric field to the cell population. Active microelectronic arrays have been used to achieve the separation of bacteria from whole blood [37], and for the separation of cervical carcinoma cells from blood [38] and for gene expression analysis [39]. Figure 6.8 shows a microelectronic array device utilizing high frequency AC fields to carry out the DEP separation of Listeria bacterial cells (~ 1 µm) from whole blood cells (~ 10 µm) in a highly parallel manner. At an AC frequency of about 10 kHz the Listeria bacterial cells are positioned on the microlocations (high field region), and the blood cells are positioned in the low field regions between the microelectrodes. The relative positioning of the cells between the high and low field regions is based on dielectric differences between the cell types. In Figure 6.8, the left panel shows microelectronic array with the mixed blood/bacteria sample, and the right panel shows the cell separation pattern which occurs after a high frequency AC electric field is applied. While maintaining the AC field, the microarray can be washed with a buffer solution that removes the blood cells (low field regions) from the more firmly bound bacteria (high field regions) near the microelectrodes. The bacteria can then be released and collected or electronically lysed to release the genomic DNA or RNA for further manipulation and analysis [37]. DEP represents a particularly useful process that allows difficult cell separation applications to be carried out rapidly and with high selectivity. The DEP process can also be extremely useful for nanofabrication purposes.

6.3. ELECTRIC FIELD ASSISTED NANOFABRICATION PROCESS

Enabling Nanotechnology—Generally, molecular or nanoelectronic devices and systems are envisioned as the more revolutionary application of this new technology. Presently, many examples of individual molecular components with appropriate basic properties exist and include entities such as carbon nanotubes, nanowires and various organic molecules with electronic switching capabilities. However, the larger issue with enabling molecular electronics is more likely to be the development of a viable technology which will allow billions of molecular or nanoelectronic components to be assembled and interconnected into useful logic/memory devices and systems. In addition to the electronic applications, nanodevices and nanosystems with higher order photonic, mechanical, mechanistic, sensory, chemical, catalytic, and therapeutic properties are also envisioned [1–4]. Again, key problems with enabling such devices and systems will most likely occur at the stage of organizing components for higher level functioning, rather than the availability of the molecular components. By way of example, Figure 6.1 shows a comparison between a biological photonic energy conversion system, "a plant chloroplast", and a man-made charged coupled device (CCD). The chloroplast is made by a self-assembly "bottom-up process, while the CCD device is made by microlithography, a top-down process. In the chloroplast, hundreds of relatively simple chromophore molecules are arranged into solid state antenna structures which collect photonic energy and transfer it through the structure with high quantum efficiency. The plant chloroplast (a few microns in size) represents a highly integrated light-capturing device composed of numerous self-organized molecular and nanoscale structures. In the man-made CCD device the feature size stops at about the micron-scale level. In addition to the chloroplast being composed of nanoscale structures, the energy transfer mechanism used within the structure is also very unique. The transfer of photonic energy through the antenna structures is carried out by what is called the Forster resonant energy transfer mechanism [40]. In some sense, this process might be thought of as the equivalent of molecular fiber optics, but it is quite different than any man-made process or device for photonic energy collection and transfer. Such a mechanism could potentially be very useful for communication between molecular and nanoscale components and devices. To date, it has been difficult to design a synthetic model of this type of solid state photonic transfer structure with any of the properties or efficiencies seen in the biological systems. Thus, a key problem with mimicking this type of bionanotechnology is the lack of a suitable "bottom-up" fabrication process to carry out the precision integration of diverse molecular and nanoscale components into viable higher order structures.

Electric Field Array Approach to Nanofabrication—As was stated earlier, microelectronic array devices have been developed for applications in genomic research and DNA diagnostics. These active microelectronic devices have the ability to create a variety of re-configurable electric field geometries on the surface of the array device. This capability allows almost any type of charged molecule or structure (DNA, RNA, protein, enzyme, nanostructure, cell, or micron-scale device) to be electrophoretically transported to or from

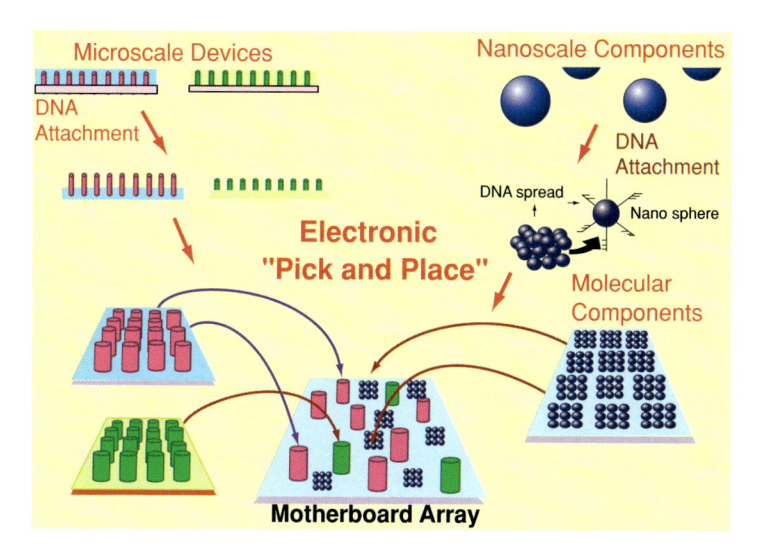

FIGURE 6.9. An electric field based "pick and place" process which uses a microelectronic "Mother Board" array to carry out the heterogeneous integration of molecular, nanoscale and microscale components into higher order devices and structures.

any of the microscopic sites (microlocations). When specific DNA hybridization reactions are carried out on this type of active microarray, the device is actually using electric fields to direct the self-assembly of DNA molecules at the specified test-site or microlocation on the chip surface. Thus, in principle these active devices serve as a "mother-board or host board" for the assisted assembly of DNA molecules into more complex structures. Since the DNA molecule with its intrinsic programmable self-assembly properties can be derivatized with electronic or photonic groups or attached to larger nanostructures (quantum dots, metallic nanoparticles, nanotubes), or microstructures, we have a unique process for nanofabrication. In essence, these active microelectronic mother-board arrays allow one to carry out a highly parallel electric field "pick & place" process for the heterogeneous integration of molecular, nanoscale and micron-scale components into complex three dimensional structures within the defined perimeters of larger silicon or semiconductor structures (Figure 6.9). The electric field assisted self-assembly technology is based on three key physical principles: (1) the use of functionalized DNA structures as "Lego" blocks for nanofabrication; (2) the use of DNA as a selective "glue" that provides intrinsic self-assembly properties to molecular and nanoscale electronic components and structures (carbon nanotubes, organic molecular electronic switches, etc.); and (3) the use of an active microelectronic array for electric field assisted self-assembly of any modified electronic/photonic components and structures into integrated structures [23–25]. Microelectronic arrays which are produced by the "top-down" process of photolithography, thus have an intrinsic capability to direct the assembly of molecular structures via a "bottom-up" process.

Microelectronic arrays have been used to direct the binding of derivatized nanospheres and microspheres on to selected locations on the microarray surface. In this case, fluorescent and non-fluorescent polystyrene nanospheres and microspheres derivatized with specific DNA oligonucleotides are transported and bound to selected test sites or microlocations

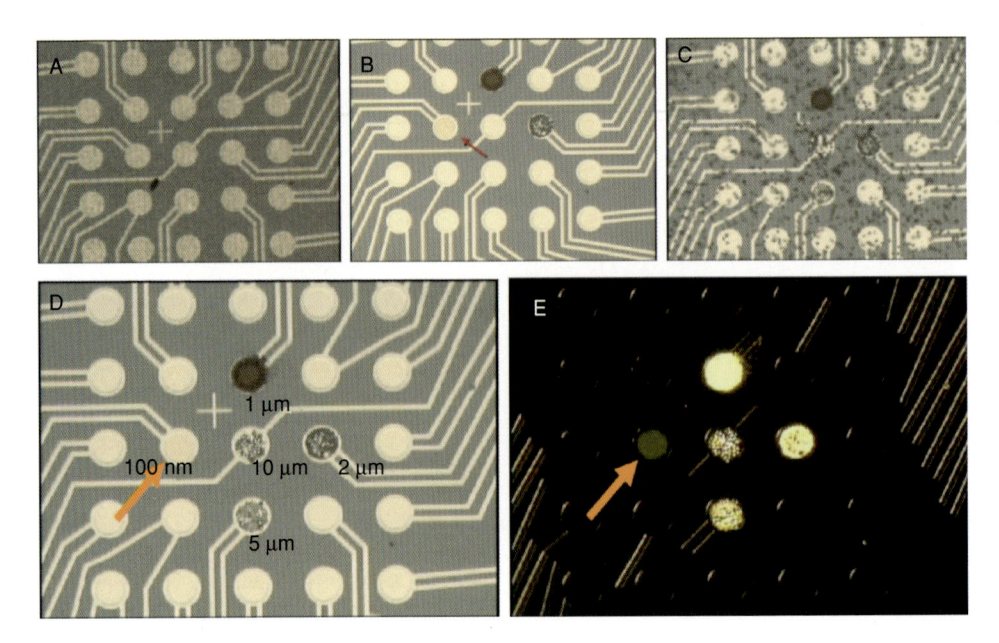

FIGURE 6.10. Demonstrates the selective addressing of 10, 5, 2 and 1 micron microspheres and 100 nanometer nanospheres (red arrows) using a 25 test site microelectronic array device. A. shows array before spheres are added. B, C, show stages in addition of the various spheres; D shows all spheres addressed, and E show all sphere spheres in dark field (nanospheres now visible).

derivatized with the specific complementary oligonucleotide sequences. Thus, microelectronic array devices are not just limited to selective transport and binding of small molecules such as fluorescent DNA, but also for selective transport and addressing of larger nanoparticles and microspheres, and even objects as large as 20 micron light emitting diode structures [21, 22, 24]. By way of example, Figure 6.10 shows a sequence of frames demonstrating the addressing of derivatized negatively charged polystyrene microspheres and nanospheres to selectively activated microlocations on a 25 test site microelectronic array. The addressing sequence is first, the one micron microspheres, second the two micron microspheres, third the 100 nanometer nanospheres, fourth the five micron microspheres and finally the ten micron microspheres. The addressing time for each group of spheres is about 10 to 15 seconds at about 3 volts DC. The final figure in the sequence shows a dark field contrast micrograph which allows the 100 nm nanospheres packed on the 80 μm diameter microlocation to be better visualized. Thus, microelectronic array devices can serve as "motherboards or hostboards" for the directed assembly of more complex nanostructures and microscale structures on their surface. A further advantage of the active microarray devices is that they can be operated in both DC and AC modes. In the DC electric field mode the electrophoretic transport of charged molecules and nanostructures is carried out between the positive and negative biased microlocations. The rate of transport is related to the strength of the electric field and the charge/mass ratio of the molecule or structure. When the microelectronic arrays are used to generate high frequency AC fields they can carry out the process of dielectrophoresis (DEP). With high AC fields, nanostructures, cells and other micron scale structures can be oriented and selectively positioned based on their

USE OF ELECTRIC FIELD ARRAY DEVICES FOR ASSISTED ASSEMBLY 149

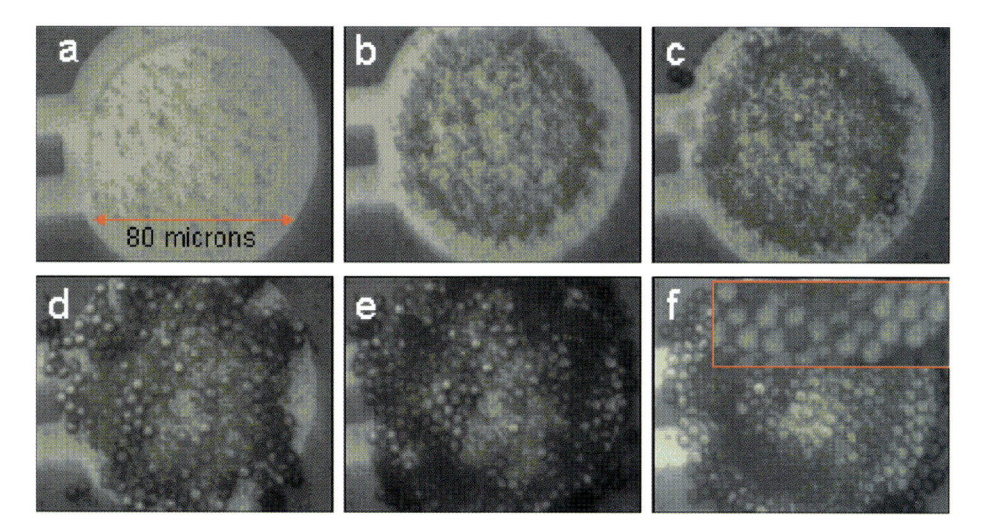

FIGURE 6.11. Shows the electric field addressing of derivatized negatively charged one micron size polystyrene microspheres to a selectively activated microlocation on a 25 test site microelectronic array, and then the addressing of the larger five micron microspheres over the one micron microspheres. Red box in (f) shows a magnified view of the packed five micron microsphere layer.

intrinsic dielectric properties. Figure 6.11 shows a sequence of frames demonstrating the addressing of derivatized negatively charged one micron size polystyrene microspheres to a selectively activated microlocation on a 25-site microelectronic array. After the microlocation is layered with the one micron polystyrene microspheres, a second addressing sequence is carried out in which larger five micron microspheres are now layered over the one micron sized polystyrene microspheres. Thus, it is possible to use electric field addressing to layer particles and other materials, allowing fabrication in the third dimension. In addition to DNA assisted assembly, other chemical and biological entities can be used advantageously for device forming, pick and place, as well as for micro and nanoactuation.

Present nanofabrication methods do not allow most nanostructures to be modified in a controlled or precise manner. For example, Figure 6.12 (upper section) shows a present scheme for modifying a nanoparticle with DNA or protein ligands using classical chemical functionalization procedures. While such a process allows a higher-order structure to be formed, there is little or no control as to how the DNA or protein ligands are arranged around the core nanostructure. Furthermore, it is not a viable process for arranging "different" DNA, protein or binding entities around the core nanostructure in a defined fashion. Microelectronic array devices may offer the opportunity to development processes that will allow core nanostructures to be selectively modified as show in Figures 6.12a & b. In this case, the nanostructures are derivatized at specific positions around the core with specific DNA sequences (oligonucleotides) proteins, or other ligands. Such modifications allow a given nanostructure to be joined with another nanostructure in a more precise manner, i.e., leaving other selected positions on the nanostructure available for binding different nanostructures (Figure 6.12c & d). The ability to produce precision nanocomponents with selective ligands as shown above is will be necessary in order to carry out further organization into more complex "heterogeneous" three dimensional structures. The need for precision placement

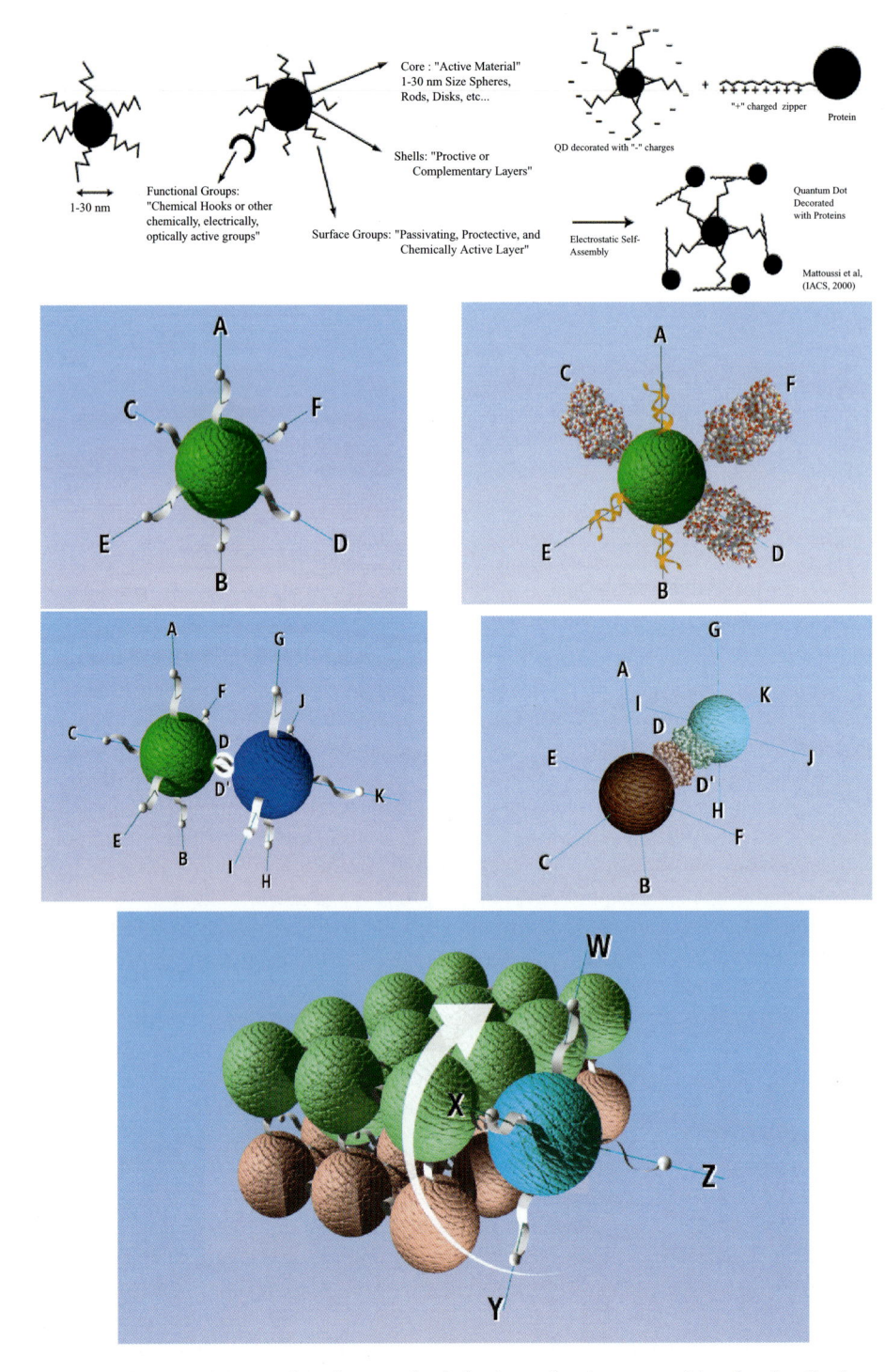

FIGURE 6.12. Upper figure section shows a classical scheme for the non-precision functionalization of nanospheres. Sections a, b, c & d shows a precision nanosphere functionalization scheme; with c&d showing the selective binding of two precision nanospheres. Final figure shows the type of heterogeneous 3D higher order structure that can only be obtained using precision nanostructures.

USE OF ELECTRIC FIELD ARRAY DEVICES FOR ASSISTED ASSEMBLY 151

of ligands on nanostructures becomes even more important when one observes that even a small amount of rotation around the axis of one ligand may disrupt the further formation of a higher order 3D structure (Figure 6.12d).

The proposed microelectronic array techniques have the capability to carry out the precision functionalization of nanostructures by processes which involve transporting and orienting the nanostructures onto surfaces containing the selected ligand molecules which are then reacted only with a selected portion of the nanostructure. By repeating the process and reorienting the nanostructures it will be possible to functionalize the core structure selectively with most biological and/or chemical groups. Such devices and processes allow one to design and create functionalized nanostructures with binding groups arranged in tetrahedral, hexagonal or other coordinate positions around the core nanostructure.

Microelectronic array devices may have the potential to carry out an electric field orientation synthesis process for producing precision nanostructures or microstructures (e.g., quantum dots, nanospheres, nanoparticles, nanocrystals) with multiple DNA identities [23–24]. These multiple identities can be in the form of specific DNA sequences which are located at different coordinates on the particle surface. These coordinates can be tetrahedral or hexagonal in their nature. Figure 6.13 shows the initial steps for producing a precision

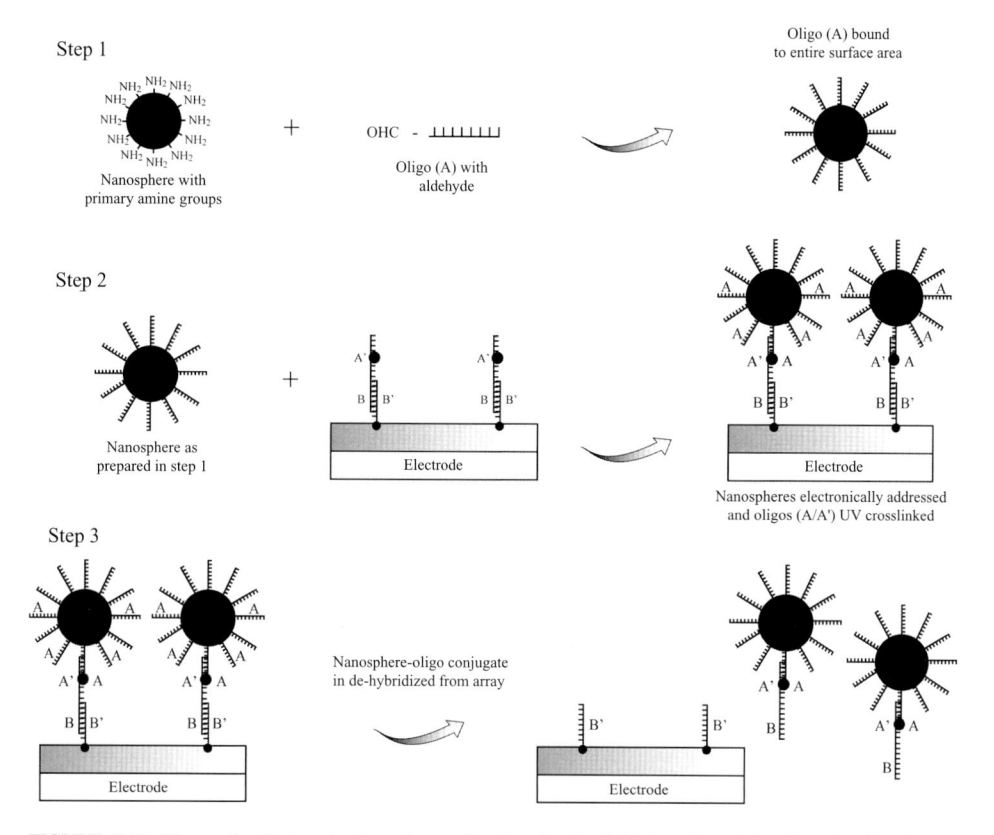

FIGURE 6.13. Shows the first part of a scheme for the electric field based precision functionalization of nanospheres with DNA oligonucleotide sequences [24].

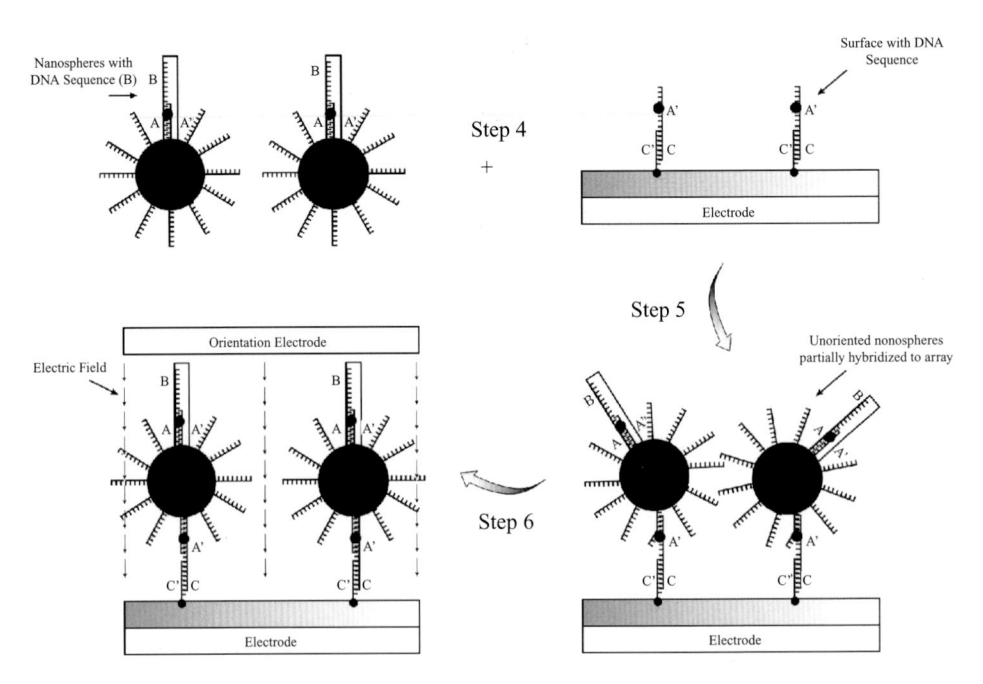

FIGURE 6.14. Shows the final part of a scheme for the electric field based precision functionalization of nanospheres with DNA oligonucleotide sequences [24].

functionalized DNA nanostructure. In step 1, suitably functionalized nanospheres (with primary amine groups) are reacted with aldehyde modified oligonucleotides with sequence identity (A). Identity (A) refers to a unique sequence of bases in the DNA; for example a 20-mer oligonucleotide with a 5'-GCACCGATTCGAT-3' sequence. In step 2, the oligonucleotide (A) modified nanospheres are now hybridized to a microlocation which has the complementary A' sequence (5'-CTACGGTATCGAATCGGTGC-3') attached. The (A') sequence contains a crosslinker agent (such as psoralen) and extends into a secondary sequence with the (B) identity (5'-TTCAGGCAATTGTCGTACA-3'), which was in turn hybridized to a (B') DNA sequence (5'-TGTACGATCAATTGCCTGAA-3') covalently linked to the surface. In step 3 the hybridized nanospheres are now given a short exposure of UV irradiation which causes the psoralen moiety within the (A/A') hybridized sequence to crosslink the structure. The nanospheres are released from the surface by thermal denaturation, and now have a (B) DNA sequence attached to a specific area of the nanostructure. Figure 6.14 now shows the continuation of the functionalization process. In steps 4 and 5, the (B) DNA sequence of the functionalized nanospheres are now "partially hybridized" to a new microlocation which has a (C-A') DNA sequence hybridized to a complementary C' DNA sequence which is covalently linked to the surface. The (C) sequence is different from the (A) and the (B) DNA sequences. The (B) DNA sequence nanospheres now become partially hybridized to the surface via the (A') DNA sequences, however they are not oriented in any particular fashion on the surface. Because the (B) DNA nanospheres have a non-uniform or asymmetric negative charge distribution on their surface (due to the extra charge from the (B) DNA), they can be oriented in an electric field. In step 6, a secondary electrode is

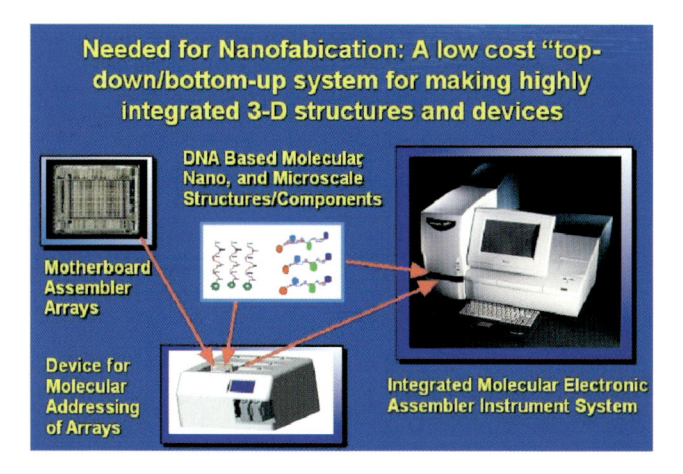

FIGURE 6.15. Future electronic pick and place system for the heterogeneous integration of nanostructures into higher order devices?

positioned above the lower electrode, and a DC electric field is applied which is just strong enough to orient the nanospheres, but does not de-hybridize them from the surface. While Figure 6.14 shows the functionalized nanospheres in a polar orientation, in terms of the (B) and (C) sequences; the relative positioning of the electrodes can produce electric fields which yield other angles for the relative position of the (B) and (C) DNA sequences. When the nanospheres are in their correct alignment, they can be completely hybridized (A'-C/C'), by lowering the temperature, and then exposed to UV irradiation to crosslink the (A/A') sequences. Upon de-hybridization, this process can produce nanospheres with (B) and (C) DNA sequences in relatively polar (north and south) positions. Repeating the process two more times should allow the production of nanospheres with specific (B), (C), (D), and (E) DNA sequences in polar/equatorial, tetrahedral or hexagonal coordinate positions. Thus, the electric field orientation process offers considerable promise for the manufacture of precision nanocomponents (quantum dots, photonic crystals, organic/metallic/semiconductor nanoparticles, and nanotubes), which can subsequently be used for further self-assembly into viable higher order heterogeneous structures and devices. Figure 6.15 shows a representation of a future electronic pick and place system for the heterogeneous integration of nanostructures into higher order devices.

6.4. INTEGRATION OF OPTICAL TWEEZERS FOR MANUPILATION OF LIVE CELLS

Custom designed miniature optical tweezers with the use of Vertical Cavity Surface Emitting Lasers (VCSEL) are integrated with the microelectrode array for the delivery or manipulation of live cells. For this individually addressable 2×2 arrays of VCSEL is focused through a series of lenses and directed onto the microarray chamber after a magnification objective. Figure 6.16 depicts the optical setup. Since the microarray chamber is

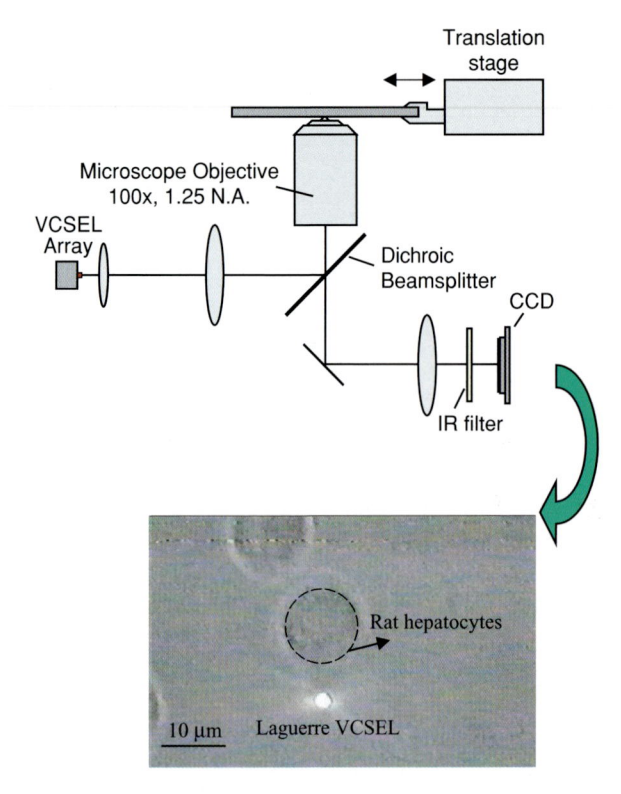

FIGURE 6.16. Schematic drawing of inverted VCSEL driven optical multi-beams setup. In-situ observations are made through a CCD camera.

fabricated with a transparent material (indium tin oxide), penetration of the optical beam is enabled. After optical trapping of the cells or other objects such as microspheres, precision controlled translation stage offers transport or manipulation of these objects without jeopardizing the sterility of the environment [41–43]. This remains to be the major issue with the mechanical manipulators.

For biological applications, mouse 10 μm diameter 3T3 fibroblast cells are manipulated with single Laguerre-Gaussian gradient VCSEL driven optical micro beam. The cells are continuously monitored for a week after they were exposed to the laser beam. There was no superficial evidence of cell damage from the laser beam, though we did not explicitly look at stress response genes. When cultured, cells attached, spread and underwent mitosis. As compared to microspheres made of polystyrene live fibroblast cell are held less strongly in the trap. We speculate that this is due to its lower dielectric constant and irregular shape. A 5 μm cell is transported with a speed of 2 μm/sec, which is about 4 times slower compared to the same size polystyrene sphere. The trapping force on the cell is estimated as ~ 0.1 pN. In this case since the refractive index of the surrounding medium, n_m, is close to the refractive index of cell, n, there is very limited force acting on the cell. Alternatively, when n is greater than n_m (polystyrene (1.58)-water (1.33) system), the sphere is always pushed out of the laser beam, which is also observed experimentally. During velocity measurements the glass

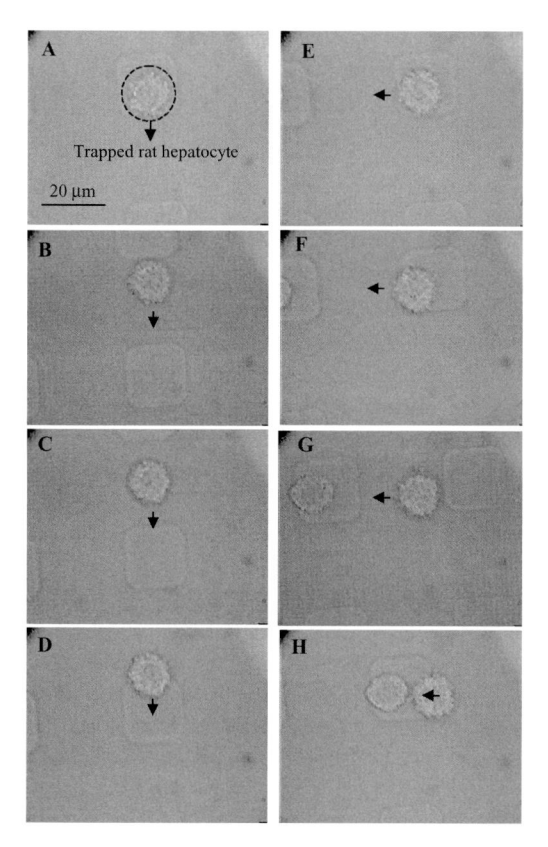

FIGURE 6.17. Primary rat hepatocyte in an electrically pre-patterned array is manipulated with 850 nm diode laser and transported to the next neighbor cell within the array.

substrate surface is pre-treated with a non-adhesion-promoting chemical to prevent possible measurement errors due to probable cell-substrate adhesion.

Since the index mismatch between cellular solution and the cover slip is more compared to the index mismatch between deionized water and the cover slip, the losses in the trapping force is more due to enhanced spherical aberration in the former case. This effect also contributes to the reason why trapping force for cells is predicted as less than the polystyrene model system with the same size.

Towards the practice of integrated system, electrically a pre-arrayed rat hepatocyte is manipulated with 850 nm single VCSEL beam and is transported next to a neighbor cell within the array. This experiment demonstrates ability of individual manipulation of cells within the array and also depicts the possibility of sample retrieval after a chemical or biological treatment in parallel. This experiment is summarized in Figure 6.17. In addition, mouse fibroblasts that are initially fluorescent tagged green and red are optically manipulated after patterning electrophoretically [44]. Red fluorescent labeled fibroblast is transported to the next die by using Laguerre mode VCSEL at 850 nm. This experiment demonstrates the first essential steps towards microscopic monitoring and micromanipulation of many live cells in real time in chip-based biosystems. This technology may find applications in cell-based functional genomics, high throughput phenotyping.

CONCLUSIONS

Active microelectronic array technology provides a number of distinct advantages for DNA hybridization diagnostics, DNA/protein/cell based affinity assays in molecular biology research and for potential nanofabrication applications. Microelectronic arrays have been designed and fabricated with 25 to 10,000 microscopic test sites. The higher density devices have CMOS elements incorporated into the underlying silicon structure that provide on-board control of current and voltage to each of the test sites on the device. Microelectronic chips are incorporated into a cartridge type device so as to be conveniently used with a probe loading station and fluorescent detection system. Active microelectronic arrays are differentiated from other DNA chip or array technologies by a number of important attributes. Active microelectronic arrays allow DNA molecules (genomic DNA, RNA, oligonucleotide probes, PCR amplicons), proteins, nanostructures, cells and microscale devices to be rapidly transported and selectively addressed (spotted) to any of the test sites on the microelectronic array surface. Microelectronic array devices have considerable potential for nanofabrication applications, including the directed self-assembly of molecular, nanoscale and microscale components into higher order mechanisms, structures, and devices. Electric field assisted self-assembly using active microelectronic arrays is a type of "Pick and Place" heterogeneous Integration process for the fabrication of 3D structures within defined perimeters of larger silicon or semiconductor structures. This technology provides the best aspects of a top-down and bottom-up process, and has the inherent hierarchical logic of allowing one to control the organization and assembly of components from molecular level —> to the nanoscale level —> to micro/macro scale 3D integrated structures and devices.

ABBREVIATIONS

AC	(Alternating Current)
CCD	(Charge Coupled Device)
CMOS	(Complementary metal-oxide-semiconductor field-effect transistors)
DEP	(Dielectophoresis)
DC	(Direct current)
DNA	(Deoxyribonucleic acid)
FRET	(Fluorescent resonant energy transfer)
MEMS	(Micro-Electromechanical Systems)
PCR	(Polymerase chain reaction)
SNP	(Single Nucleotide Polymorphism)
STR	(Short Tandem Repeats)
RNA	(ribonucleic acid)

ACKNOWLEDGEMENT

Authors would like to acknowledge the contributions of Dr. Esener, S and Dr. Bhatia, S.

REFERENCES

[1] Small Wonders, Endless Frontiers: Review of the Nanotional Nanotechnology Initiative, National Research Council, 2002.
[2] M.P. Hughes (ed.). *Nanoelectromechanics in Engineering and Biology*. CRC Press, Boca Raton, FL, 2003.
[3] Goddard, Brenner, Lyashevski, and Lafrate (ed.). *Handbook of Nanoscience, Engineering and Technology*. CRC Press, Boca Raton, FL, 2003.
[4] V. Balzani, M. Venturi, and A. Credi. *Molecular Devices and Mechanics—Journey into the Nanoworld*. Wiley-VCH, KGaA Weinheim, 2003.
[5] R. Bashir. Biological mediated assembly of artificial nanostructures and Microtructures. In Goddard, Brenner, Lyashevski and Lafrate (eds.), *Handbook of Nanoscience, Engineering and Technology*. CRC Press, Chapter 15, pp. 15–1 to 15–31, 2003.
[6] M.J. Heller and R.H. Tullis. *Nanotechnology*, 2:165–171, 1991.
[7] Daniel M. Hartmann, David Schwartz, Gene Tu, Mike Heller, Sadik C. Esener. Selective DNA attachment of particles to substrates. *J. Mat. Res.*, 17(2):473–478, 2002.
[8] M.J. Heller. An active microelectronics device for multiplex DNA analysis. *IEEE Eng. Med. Biol.*, 15:100–103, 1996.
[9] R.G. Sosnowski, E. Tu, W.F. Butler, J.P. O'Connell, and M.J. Heller. Rapid determination of single base mismatch in DNA hybrids by direct electric field control. *Proc. Nat. Acad. Sci. USA*, 94:1119–1123, 1997.
[10] C.F. Edman, D.E. Raymond, D.J. Wu, E. Tu, R.G. Sosnowski, W.F. Butler, M. Nerenberg, and M.J. Heller. Electric field directed nucleic acid hybridization on microchips. *Nucleic Acids Res.*, 25(24):4907–4914, 1997.
[11] M.J. Heller. An integrated microelectronic hybridization system for genomic research and diagnostic applications. In D.J. Harrison and A. van den Berg (eds.), *Micro Total Analysis Systems 98*, Kluwer Academic Publishers, pp. 221–224, 1998.
[12] M.J. Heller, E. Tu, A. Holmsen, R.G. Sosnowski, and J.P. O'Connell. Active Microelectronic Arrays for DNA Hybridization Analysis. In M. Schena (ed.), *DNA Microarrays: A Practical Approach*, Oxford University Press, pp. 167–185, 1999.
[13] M.J. Heller, A.H. Forster, E. Tu. Active microelectronic chip devices which utilize controlled electrophoretic fields for multiplex DNA hybridization and genomic applications. *Electrophoresis*, 21:157–64, 2000.
[14] C. Gurtner, E. Tu, N. Jamshidi, R. Haigis, T. Onofrey, C.F. Edman, R. Sosnowski, B. Wallace, and M.J. Heller. Microelectronic array devices and techniques for electric field enhanced DNA hybridization in low-conductance buffers, *Electrophoresis*, 23:1543–1550, 2002.
[15] M.J. Heller. DNA microarray technology: devices, systems and applications, *Ann. Rev. Biomed. Eng.*, 4:129–53, 2002.
[16] M.J. Heller, E. Tu, R. Martinsons, R.R. Anderson, C. Gurtner, A. Forster, and R. Sosnowski. Active microelectronic array systems for DNA hybridization, genotyping, pharmacogenomics and nanofabrication applications. In Heller and Guttman (eds.), *Integrated Microfabricated Devices*, Marcel Dekker, Chap. 10, pp. 223–270, 2002.
[17] S.K. Kassengne, H. Reese, D. Hodko, J.M. Yang, K. Sarkar, D.E. Swanson P. Raymond, M.J. Heller, and M.J. Madou. Numerical modeling of transport and accumulation of DNA on electronically active biochips. *Sens. Actu. B*, 94:81–98, 2003.
[18] S.C. Esener, D. Hartmann, M.J. Heller, and J.M. Cable. DNA Assisted Micro-Assembly: A Heterogeneous Integration Technology For Optoelectronics, *Proc. SPIE Critical Reviews of Optical Science and Technology, Heterogeneous Integration*, Ed. A. Hussain, CR70, Chapter 7, January 1998.
[19] C. Gurtner, C.F. Edman, R.E. Formosa, and M.J. Heller. Photoelectrophoretic Transport and Hybridization of DNA on Unpatterned Silicon Substrates. *J. Am. Chem. Soc.*, 122(36):8589–8594, 2000.

[20] Y. Huang, K.L. Ewalt, M. Tirado, R. Haigis. A. Forster, D. Ackley, M.J. Heller, J.P. O'Connell, and M. Krihak. Electric manipulation of bioparticles and macromolecules on microfabricated electrodes. *Anal. Chem.*, 73:1549–1559, 2001.
[21] C.F. Edman, C. Gurtner, R.E. Formosa, J.J. Coleman, and M.J. Heller. Electric-field-directed pick-and-place assembly. *HDI*, (3)10:30–35, 2000.
[22] C.F. Edman, R.B. Swint, C. Gurthner, R.E. Formosa, S.D. Roh, K.E. Lee, P.D. Swanson, D.E. Ackley, J.J. Colman, and M.J. Heller. Electric field directed assembly of an InGaAs LED onto silicon circuitry. *IEEE Photonics Tech. Lett.*, 12(9):1198–1200, 2000.
[23] US # 6,569,382 Methods and Apparatus for the Electronic Homogeneous Assembly and Fabrication of Devices, issued May 27, 2003.
[24] US # 6,652,808 Methods for the Electronic Assembly and Fabrication of Devices, issued Nov. 25, 2003.
[25] US #6,706,473 Systems and Devices for the Photoelectrophoretic Transport and Hybridization of Oligonucleotides, issued March 16, 2004.
[26] P. Swanson, R. Gelbart, E. Atlas, L. Yang, T. Grogan, W.F. Butler, D.E. Ackley, and E. Sheldon. A fully multiplexed CMOS biochip for DNA Analysis. *Sens. Actu. B*, 64:22–30, 2000.
[27] P.N. Gilles, D.J. Wu DJ, C.B. Foster, P.J. Dillion, and S.J. Channock. Single nucleotide polymorphic discrimination by an electronic dot blot assay on semiconductor microchips'. *Nat. Biotechnol.*, 17(4):365–370, 1999.
[28] N. Narasimhan and D. O'Kane. Validation of SNP Genotyping for Human Serum Paraoxonase Gene. *Clin. Chem.*, 34(7):589–592, 2001.
[29] R. Sosnowski, M.J. Heller, E. Tu, A. Forster, and R. Radtkey. Active Microelectronic Array System for DNA Hybridization, Genotyping and Pharmacogenomic Applications. *Psychiat. Genet.*, 12:181–192, 2002.
[30] Y.R. Sohni, J.R. Cerhan, and D.J. O'Kane. Microarray and Microfluidic Methodology for Genotyping Cytokine Gene Polymorphisms. *Hum. Immunol.*, 64:990–997, 2003.
[31] E.S. Pollak, L. Feng, H. Ahadian, and P. Fortina. Microarray-based genetic analysis for studying susceptibility to arterial and venous thrombotic disorders. *Ital. Heart J.*, 2:569–572, 2001.
[32] W.A. Thistlethwaite, L.M. Moses, K.C. Hoffbuhr, J.M. Devaney, and E.P. Hoffman. Rapid genotyping of common MeCP2 mutations with an electronic DNA microchip using serial differential hybridization. *J. Mol. Diag.*, 5(2):121–126, 2003.
[33] V.R. Mas, R.A. Fisher, D.G. Maluf, D.S. Wilkinson, T.G. Carleton, and A. Ferreira-Gonzalez. Hepatic artery thrombosis after liver transplantation and genetic factors: prothrombin G20210A polymorphism. *Transplanation*, 76(1):247–249, 2003.
[34] R. Santacroce, A. Ratti, F. Caroli, B. Foglieni, A. Ferraris, L. Cremonesi, M. Margaglione, M. Seri, R. Ravazzolo, G. Restagno, B. Dallapiccola, E. Rappaport, E.S. Pollak, S. Surrey, M. Ferrari, and P. Fortina. Analysis of clinically relevant single-nucleotide polymorphisms by use of microelectric array technology. *Clin. Chem.*, 48(12):2124–2130, 2002.
[35] A. Åsberg, K. Thorstensen, K. Hveem, and K. Bjerve. Hereditary hemochromatosis: the clinical significance of the S64C mutation. *Genet. Test.*, 6(1):59–62, 2002.
[36] J.G. Evans and C. Lee-Tataseo. Determination of the factor V leiden single-nucleotide polymorphism in a commercial clinical laboratory by use of NanoChip microelectric array technology. *Clin. Chem.*, 48(9):1406–1411, 2002.
[37] J. Cheng, E.L. Sheldon, L. Wu, A. Uribe, L.O. Gerrue, J. Carrino, M.J. Heller, and J.P. O'Connell. Electric field controlled preparation and hybridization analysis of DNA/RNA from E. coli on microfabricated bioelectronic chips. *Nat. Biotech.*, (16):541–546, 1998.
[38] J. Cheng, E.L. Sheldon, L. Wu, M.J. Heller, and J. O'Connell. Isolation of Cultured Cervical Carcinoma Cells Mixed with Peripheral Blood Cells on a Bioelectronic Chip. *Anal. Chem.*, (70):2321–2326, 1998.
[39] Y. Huang, J. Sunghae, M. Duhon, M.J. Heller, B. Wallace, and X. Xu. Dielectrophoretic separation and gene expression profiling on microelectronic chip arrays. *Anal. Chem.*, 74:3362–3371, 2002.
[40] T. Forster. *Dicuss. Faraday Soc.*, 27:7, 1959.
[41] R.A. Flynn, A.L. Birkbeck, M. Gross, M. Ozkan, M. Shao, M. Wang, and S.C. Esener. Parallel transport of biological cells using individually addressable VSCEL arrays as optical tweezers. *Sens. Actuat. B*, 6363:1–5, 2003.
[42] A.L. Birkbeck, R.A. Flynn, M. Ozkan, D. Song, M. Gross, and S.C. Esener. VCSEL arrays as micromanipulators in chip-based biosystem. *Biomed. Microdev.*, 5(1):61–67, 2003.

[43] M. Ozkan, M. Wang, C. Ozkan, R.A. Flynn, and S.C. Esener. Optical manipulation of objects and biological cells in microfluidic devices. *Biomed. Microdev.*, 5(1):47–54, 2003.
[44] M. Ozkan, T. Pisanic, J. Sheel, C. Barrow, S. Esener, and S. Bhatia. "Electro-Optical Platform for the Manipulation of Live Cells", *Special issue on the Biomolecular Interface, Langmuir*, 19(5):1532–1538, 2003.

7

Peptide Arrays in Proteomics and Drug Discovery

Ulrich Reineke[1], Jens Schneider-Mergener[1,2], and Mike Schutkowski[1]
[1]*Jerini AG, Invalidenstr. 130, 10115 Berlin, Germany*
[2]*Institut für Medizinische Immunologie, Universitätsklinikum Charité, Humboldt Universität zu Berlin, Schumannstr. 20–21, 10117 Berlin, Germany*

7.1. INTRODUCTION

Array technology has become a powerful tool for today's high throughput approaches in biology and chemistry. These large-scale technologies emerged mainly from the genomics field driven by the human genome project and other species sequencing efforts. The associated development of many new technologies led to the initiation of several additional "omics" fields such as proteomics, lipidomics, glycomics and others. Scientists in these fields subsequently demanded tools that allow rapid and reliable characterization of large numbers of molecules of very different natures including nucleic acids, proteins, carbohydrates, peptides, and small molecules with very small amounts of biological or synthetic material.

A paramount principle was the array technology emerging in the late 1980s with DNA arrays. Initial developments involving peptide arrays began almost simultaneously. Protein arrays arose later because several critical factors such as stability, native folding and activity of the immobilized proteins had to be addressed. The main array technology characteristics are spatially addressable immobilization of large numbers of different molecules (libraries), simultaneous analysis with one or more purified or crude biological samples used to probe the array, and a general tendency towards miniaturization, automated read out and integrated data analysis. The benefits are an unprecedented number of data points, rapid generation of data, and the extremely small biological sample volumes required per data point, providing opportunities for completely new insights within life sciences. This is reflected by an increasing market with an estimated market size for non-DNA biochips of about 200 million US$ and an annual growth rate of 36%.

In 1991 two different technologies for the preparation of peptide arrays were published (Section 7.2.3.1.). (1) Light-directed, spatially addressable parallel chemical synthesis [151] is a synthesis technology permitting extreme miniaturization of array formats but, conversely, involves sophisticated and rather tedious synthesis cycles. (2) The SPOT synthesis concept developed by Ronald Frank is the stepwise synthesis of peptides on planar supports (originally cellulose membranes) applying standard peptide chemistry [154, 155]. SPOT synthesis is technically very simple and flexible and does not require any expensive laboratory automation or synthesis hardware. In contrast, the degree of miniaturization is significantly lower. However, due to the simplicity of the technology more applications of the SPOT concept by far were published since 1991 compared to light-directed, spatially addressable parallel chemical synthesis (Section 7.6). In addition to these pioneering and meanwhile well established technologies, the last few years have seen the introduction of several other concepts for peptide array preparation accessing several new developments in microarray technologies inspired by DNA as well as protein array approaches.

In Section 7.2 the technologies for peptide array generation are described in detail, including coherent surfaces and surface modification, microstructured surfaces, peptide array preparation and printing techniques for peptide array production. Thereafter, the different types of peptide libraries such as protein sequence-derived and *de novo* approaches are described in Section 7.3. Section 7.4 summarizes assay principles for peptide arrays. This Section is divided into the subsections "screening" and "read-out" addressing either the molecular recognition event such as ligand binding or enzymatic conversion, or how one observes which peptide was bound and/or converted by an interaction partner or enzyme (chemoluminescence, fluorescence, radioactivity, chromogenic or label-free read-out). The different applications of peptide arrays such as antibody epitope and paratope mapping, protein-protein interaction mapping, identification of enzyme substrates and inhibitors, DNA and metal ion binding, chemical transformations, cell binding, and peptidomimetic alterations, are covered in Section 7.5. Finally, in Section 7.6 we present an extensive bibliography of publications describing peptide array applications (Table 7.4).

The scope of this review article is peptide arrays on planar supports. Thus, we do not consider the following topics: peptide synthesis on polymeric pins, surfaces modified with only one peptide rather than a peptide collection as often employed in material sciences, read-out methods such as surface-enhanced laser desorption/ionization time-of-flight mass spectrometry (SELDI-TOFMS) with peptides immobilized on the surface and surface plasmon resonance with immobilized peptides in standard throughput configurations (Biacore) as well as small molecule and protein arrays.

7.2. GENERATION OF PEPTIDE ARRAYS

Materials used for the preparation of peptides arrays are flexible porous planar supports such as cellulose [127, 153, 155], cotton [128, 492], polymeric films [34], disks [208, 337] and membranes [97, 578, 592], or rigid, non-porous materials, such as glass, gold-coated surfaces, titanium, aluminum oxide, silicon, and modified polymers such as polypropylene, polyethylene or polyurethane.

A planar support has to fulfill several requirements in order to be suitable for step by step solid phase peptide synthesis: these include chemical stability against a variety of solvents, reagents and reaction conditions as well as mechanical stability towards repetitive steps of solvent deposition, washing and deprotection. Moreover, surface functional groups must be accessible for chemical derivatization and the final surface properties should allow the biochemical interaction of the immobilized peptides with a binding partner during solid phase binding assays with limited non-specific interactions.

Cellulose-bound peptides paved the way in the field of peptide arrays. The SPOT technology, a very rapid, robust and easy-to-use method for preparing peptide arrays in a stepwise manner directly on cellulose or other membranes and its applications have been reviewed extensively (Wenschuh et al., 2000; [160–162, 443, 454]). Apart from several special applications these peptide arrays are mainly used for antibody epitope and paratope mapping, mapping of protein-protein interactions in general, and investigation of enzyme/substrate/inhibitor interactions (kinases, proteases, isomerases, and phosphatases). Obviously, for biochemical assays impermeable, rigid materials, such as glass, have a number of practical advantages over porous supports. Binding partners interacting with the immobilized peptides find immediate access to the ligands without diffusing into pores, thereby accelerating the binding process. Washing steps following incubation are not limited by diffusion, leading to improved reproducibility. Finally, the flatness and transparency of glass or plastic supports improve image acquisition due to the much better definition of probes compared to flexible membranes. High image definition is a critical factor for determining the final density that can be achieved on peptide microarrays. Additionally, rigidity permits incorporation into flow cells necessary for automated processing during high throughput analysis. Moreover, the interaction kinetics on peptide microarrays fixed on impermeable supports is not complicated by solvent and interacting partner diffusion into and/or out of pores (as described for DNA/DNA interactions [329]. The focus of this review will be peptide arrays composed of peptides preferably bound covalently to planar supports through one defined terminus.

7.2.1. Coherent Surfaces and Surface Modification

Polymer surface modification has been the subject of increased research in different areas such as the development of biocompatible implants [228] or biosensors, for affinity chromatography, and for immobilization of molecules in ELISA experiments.

Generally, the quality of peptide arrays is determined by surface properties such as homogeneity, roughness, hydrophobicity, density of functional groups, spacing between surface and biologically active compounds, and amenability to interaction with proteins or enzymes. Surface-sensitive analytical methods used to characterize nonporous materials such as silicon wafer, glass and titanium include goniometry (contact angle measurement), variable angle spectral ellipsometry, X-ray photoelectron spectroscopy, attenuated total reflection–Fourier transform infrared spectroscopy, secondary ion mass spectroscopy and atomic force microscopy.

In most cases amino acids (step by step synthesis) or peptides (immobilization of pre-synthesized peptides) cannot be coupled directly to the surface of the applied membranes or rigid materials (metal, glass or plastic). It is therefore necessary to transform such surfaces into amino functions or other appropriate reactive groups (hydroxy, mercapto, aldehyde, maleimide, epoxy) to allow either initiation of the solid phase peptide synthesis or immobilization of pre-synthesized peptides. A further important factor is that the penultimate amino acid linkage to the surface must have a chemical stability at least equal to that of the peptide backbone. Moreover, it is desirable that the surface is optically transparent since radioactivity has largely been replaced by spectroscopic, especially fluorescent detection techniques.

Polymer surfaces can be modified by silylation, corona treatment, γ-radiation or graft copolymerization (Figures 7.1, 7.2, 7.3, 7.19). Such treatments can enhance the polymer's reactivity and/or modify the hydrophobic/hydrophilic character of the surface. Polystyrene, for example, may be surface activated by glow discharge and graft co-polymerized with acrylic acid to provide a reactive surface for peptide immobilization. Polyurethane films can

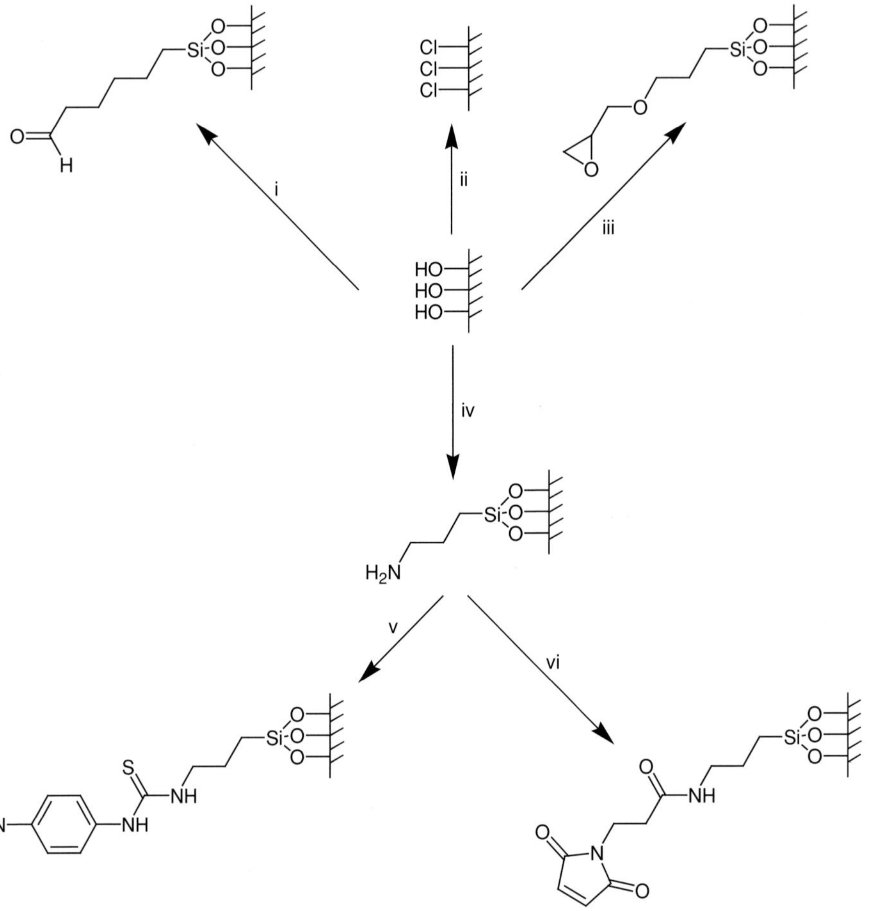

FIGURE 7.1. **Functionalization of glass surfaces**. i = 6-triethoxysilyl-hexanal; ii = H_2SO_4/H_2O_2, 16h, washing with water, thionyl cloride in THF, cat. DMF [204], iii = 3-glycidyloxypropyltrimethoxysilane/toluene [553]; iv = 3-aminopropyltrimethoxysilane/toluene; v = phenyldiisothiocyanate in pyridine/dimethylformamide 1:9, 2 h [31]; vi = N-succinimidyl-3-maleimidylpropionate in acetonitrile.

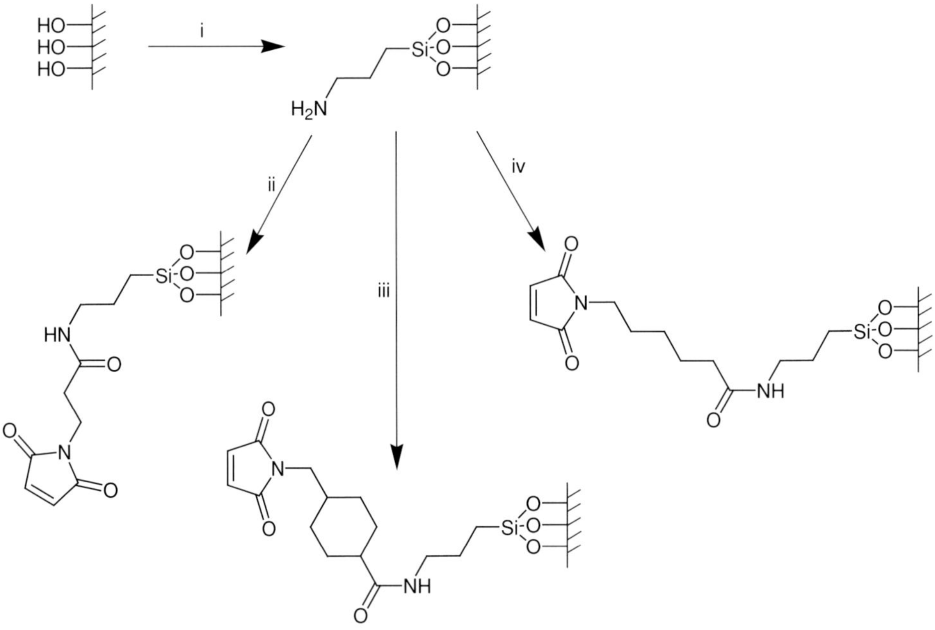

FIGURE 7.2. **Functionalization of titanium surfaces [603]**. i = silanization of water-vapor-plasma pretreated titanium using (3-aminopropyl)triethoxysilane in anhydrous toluene; ii = N-succinimidyl-3-maleimidylpropionate in acetonitrile, iii = N-succinimidyl *trans*-4-(maleimidylmethyl)cyclohexane-1-carboxylate in acetonitrile; iv = N-succinimidyl-6-maleimidylhexanoate in acetonitrile.

be modified by plasma deposition of 1,2-diaminocyclohexane followed by carbodiimide-mediated grafting of carboxylated dextran onto the amino-modified polyurethane yielding surface-bound carboxylic functions [513]. Polypropylene is readily aminated by plasma discharge in the presence of anhydrous ammonia [354]. Polypropylene and polyethylene can be functionalized by vapor phase photo-grafting of methyl methacrylate, acrylic acid or methacrylic acid in the presence of sensitizers such as benzoin ethyl ester or benzophenone [13, 383, 387]. The photochemical grafting of styrene onto polypropylene using photosensitizers is described [313]. Additionally, the photo-induced graft copolymerization of polypropylene with acrylates was reported to be effective for the generation of chemically defined ultrafiltration membranes with a hydrophilic surface modification [557]. Several different acrylates were used as monomers for the functionalization

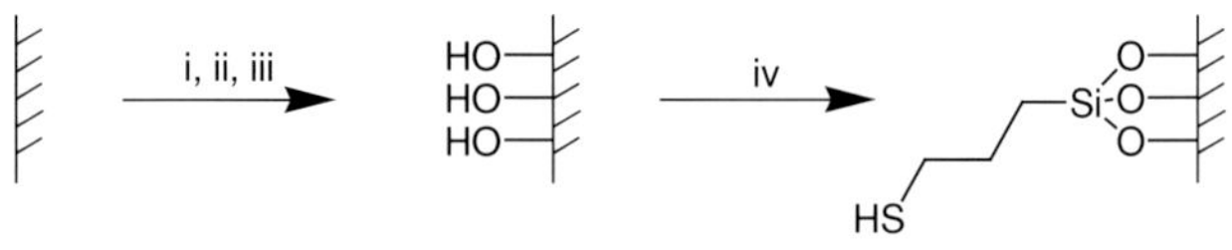

FIGURE 7.3. **Functionalization of silicon wafers [308]**. i = 90% H_2SO_4, 10% H_2O_2, 125°C, 10 min; ii = 2% hydrofluoric acid, 1 min; iii = potassium permanganate in concentrated H_2SO_4, 1.5 min; iv = 2-propanol/H_2O/(3-mercaptopropyl)trimethoxysilane 100:1:2.5 (v,v,v), reflux, 30 min, curing 115°C.

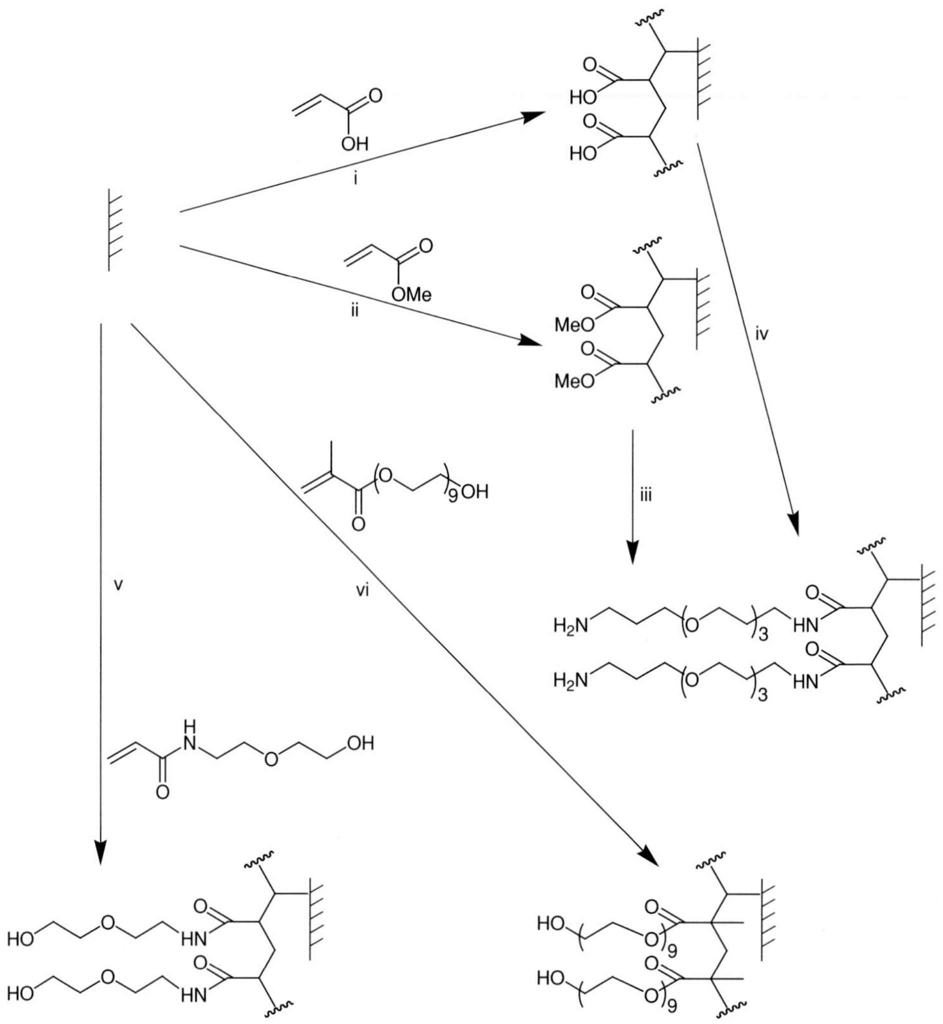

FIGURE 7.4. **Modification of polypropylene surfaces (Wenschuh et al., 2000)**. i = acrylic acid, ii = acrylic acid methylester, iii = 4,7,10-trioxa-1,13-tridecanediamine, 100°C, 72 h, iv = PCl$_5$ / methylene chloride / 4,7,10-trioxa-1,13-tridecanediamine, v = grafting using aryloyl-aminoethoxyethanol, vi = grafting using polyethylene glycol methacrylate.

of polypropylene membranes (Figure 7.4) or Teflon membranes [130]. Acrylic acid and acrylic acid methylester yielded carboxylic acid and methylester-modified polypropylene membranes, respectively (Figures 7.4i, 7.ii), with loadings between 7.5 and 12.5 nmol/mm^2 (Table 7.1) (Wenschuh et al., 2000). The acrylic acid modified membranes were activated and amidated with 4,7,10-trioxa-1,13-tridecanediamine yielding amino-functionalized polypropylene surfaces suitable for peptide synthesis (Figure 7.4iv). Alternatively, the methylester-modified membrane was treated directly with 4,7,10-trioxa-1,13-tridecanediamine resulting in an amino function linked to the membrane surface by a

TABLE 7.1. Reported loadings of different surfaces with functional groups

material	surface function	surface loading amount/mm^2	reference
paracyclophane coated stainless steel foil	isocyanate	7.6 pmol	[283]
trioxa cellulose	amino	1 nmol	[591]
glycinester cellulose	amino	12 nmol	[591]
epoxypropylamine cellulose	amino	5 nmol	[591]
trioxa polypropylene	amino	5 nmol	[591]
acylic acid modified polypropylene	carboxylic acid	12.5 nmol	[591]
acylate modified polypropylene	methylester	7 nmol	[591]
modified polypropylene	hydroxyl	3 nmol	[591]
poly(aminoethylmethacrylamide) coated polypropylene	amino	3–4 nmol	[97]
aminopropyldiethoxymethylsilane treated glass	amino	6 pmol	[388]
aminopropylmonomethoxydimethylsilane treated glass	amino	2 pmol	[388]
aziridine treated aminosilylated glass	amino	68 pmol	[388]
glycidyloxipropylsilylated glass	epoxide	5.7 pmol	[412]
hexadecanethiol and glycidyloxipropylsilylated glass	mercapto	5.5 pmol	[412]
diamino PEG and glycidyloxipropylsilylated glass	amino	1.1 pmol	[412]
aminosilylated titanium	amino	10 pmol	[603]
aminosilylated glass	amino	<1 pmol	[405]
RGD grafted polypyrrole on titanium	pyrrole	30 pmol	[100]
polypropylene	amino	10 pmol	[523]
aminosilylated glass	amino	100 fmol	[523]
self assembled monolayer on gold surfaces	benzoquinone	100 fmol	[218]
poly((D), L-lactide grafted with polyacylic acid	carboxyl	3–5 nmol	[525]
functionalized quartz surface	amino	40–60 fmol	[467]
3-(aminopropyl)diethoxymethylsilane treated Si(100) wavers	amino	5.5 pmol	[257]
3-(aminopropyl)dimethylethoxysilane treated Si(100) wavers	amino	2.5 pmol	[257]
aziridine treated aminosilylated Si(100) wavers	amino	80–110 pmol	[257]
glycidyloxipropylsilylated glass	epoxy	34 fmol	US-5474796
self-assembled monolayer of w-alkanethiol on gold surfaces	mercapto	250 fmol	[581]

hydrophilic spacer and a relatively stable amide bond (Figure 7.4iii). Moreover, acryloyl-aminoethoxyethanol or polyethylene glycol methacrylates gave hydroxy-functionalized surfaces by means of a one step grafting procedure (Figure 7.4v, 7.4vi; [591]). The quality of model peptides synthesized on these surfaces by SPOT synthesis was demonstrated to be similar or even better than that obtained with cellulose under comparable conditions [589]. In addition, it was shown that polypropylene membranes could be used to assemble peptoids, peptide nucleic acids, glycopeptides, and triazinyl peptides [590] using the SPOT technique. Further modifications (useful for solid phase peptide synthesis) of functionalized polypropylene (Figure 7.5; [97, 578]) and polypropylene coated with crosslinked polyhydroxypropylacrylate (Figure 7.6; [97]) are reported. A variety of plastic polymers can be functionalized directly by chemical treatment yielding carboxylic functions (Figure 7.7; [241]) that are easily transformed into pentafluorophenylesters. Recently, it was

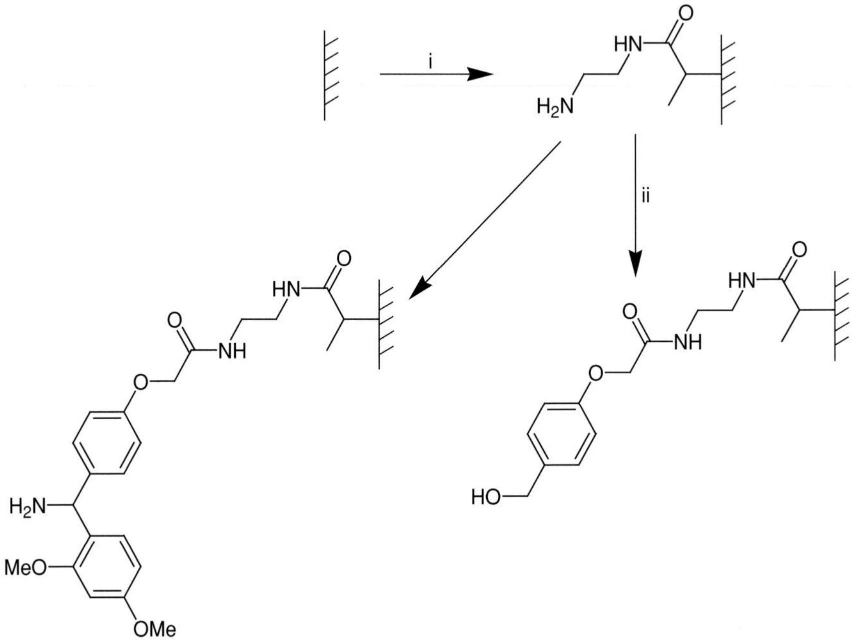

FIGURE 7.5. **Modification of polypropylene surfaces [97].** i = aminoethylmethacrylamide/ h*ν; ii = (4-hydroxymethyl)phenoxyacetic acid pentafluorophenyl ester.

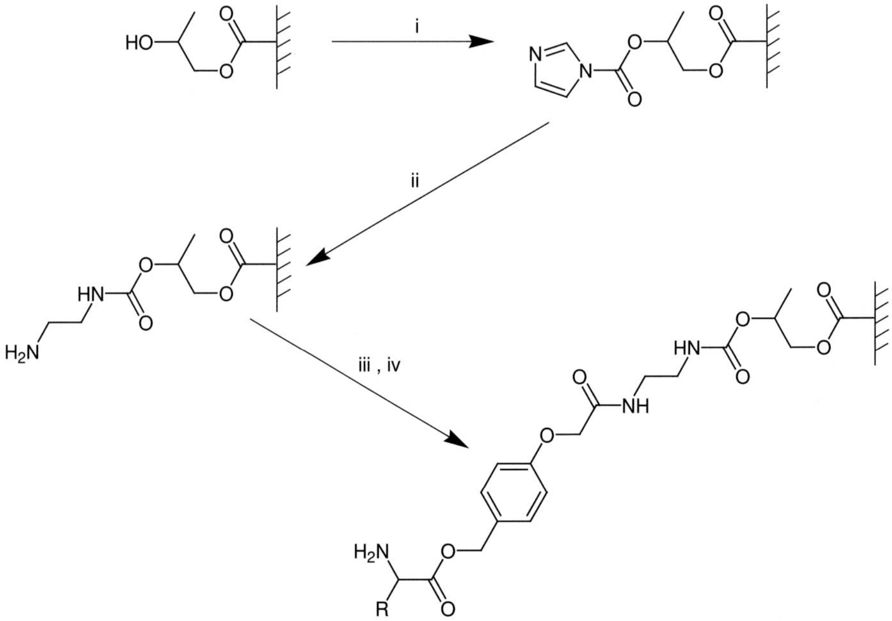

FIGURE 7.6. **Modification of polypropylene coated with crosslinked polyhydroxypropylacrylate [97, 578].** i = carbonyldiimidazole/DMF; ii = 1,3-diaminopropane/DMF; iii = Fmoc-amino acyl-oxymethylphenoxyacetic acid 2,4-dichlorophenylester/ 4 eq pyridine/DMF; iv = 20% piperidine/DMF.

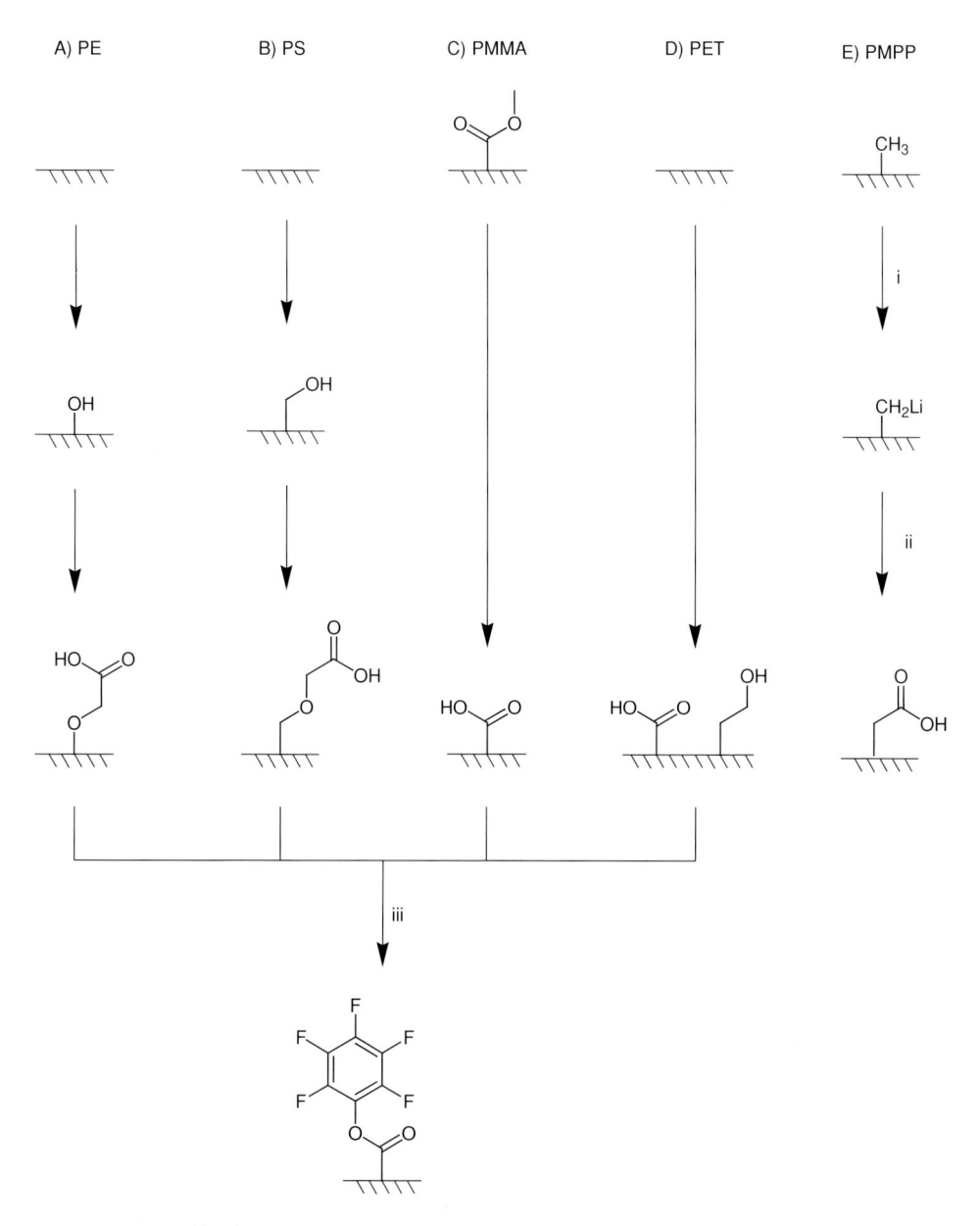

FIGURE 7.7. **Modification of different plastic surfaces**. A) PE = polyethylene; B) PS = polystyrene; C) PMMA = polymethylmethacrylate; D) PET = polyethyleneterephthalate; E) PMPP = polymethylphenylphosphazene; i = 2.5 M n butyl lithium, hexane, $-78°C$, 1 h, room temperature; ii = CO_2, 0.01 M HNO_3 [241]; iii = pentafluorophenol, DIC.

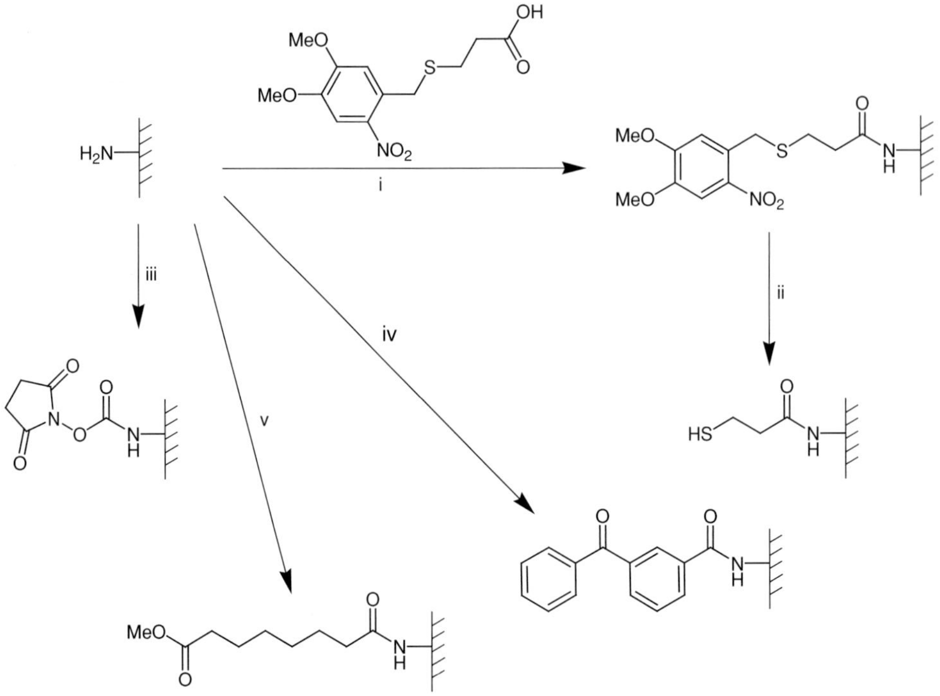

FIGURE 7.8. **Functionalization of aminosilylated glass surfaces.** i = (2-nitro-4,5-dimethoxy-benzyl)-3-mercaptopropionic acid/1-(3-dimethylaminopropyl)-3-ethylcarbodiimide hydrochloride; ii = h*ν; iii = disuccinimidylcarbonate, diisopropylamine, acetonitrile, 4 h [31]; iv = N-succinimidyl-3-benzoylbenzoate in DMF [480]; v = dimethylsuberimidate in saturated NaHCO$_3$, 1 h [31].

demonstrated that the polycarbonate surface of a normal CD-ROM could be converted into a chlorophosphate moiety that allows immobilization of biomolecules [282].

Glass, silicon dioxide, silica surfaces and transition metal oxides are readily functionalized by silanization with reactive silanes containing alkyl chains terminated with functional groups (Figure 7.1). Coatings with aminopropyltriethoxysilane yielding primary amino functions that can be easily modified are widely used (Figures 7.1v, 7.1vi, 7.8). Side reactions such as oligomerization and the formation of multiple layers normally prevent the formation of ordered alkyl silane monolayers [527] and result in hydrophobic silanized surfaces displaying strong unspecific adsorption [410]. This situation can be circumvented by either using different aminosilanes, such as aminopropyldiethoxymonomethylsilane or aminopropylmonoethoxydimethylsilane [388], or attaching ultra-thin layers of bifunctional polyethylene glycols to the surfaces [375, 411, 412, 608].

Additionally, multistep functionalization of pretreated titan surfaces has been described using aminopropyltriethoxysilane followed by reaction with N-succinimidyl-3-maleimidylpropionate, N-succinimidyl-6-maleimidylhexanoate or N-succinimidyl *trans*-4-(maleimidylmethyl)cyclohexane-1-carboxylate in acetonitrile (Figure 7.2; [603]). The resulting maleinimide-modified surface enables the directed immobilization of peptides containing cysteine by Michael addition. Moreover, the published modification of oxidized silicon wavers using (3-mercaptopropyl)trimethoxysilane in aqueous propanol [308] or in a vacuum chamber [333], yielded mercapto functions suitable for the covalent attachment of cysteinyl peptides via S-S-bond formation (Figure 7.3; [308]).

Functionalization of stainless steel foils or nickel/titanium alloys by chemical vapour deposition polymerization of substituted [2.2]paracyclophanes resulting in poly(amino-p-xylylene-co-p-xylylene) or poly(hydroxy methyl-p-xylylene-co-p-xylylene) coatings (Figure 7.9) was demonstrated by Lahann et al. [283, 284].

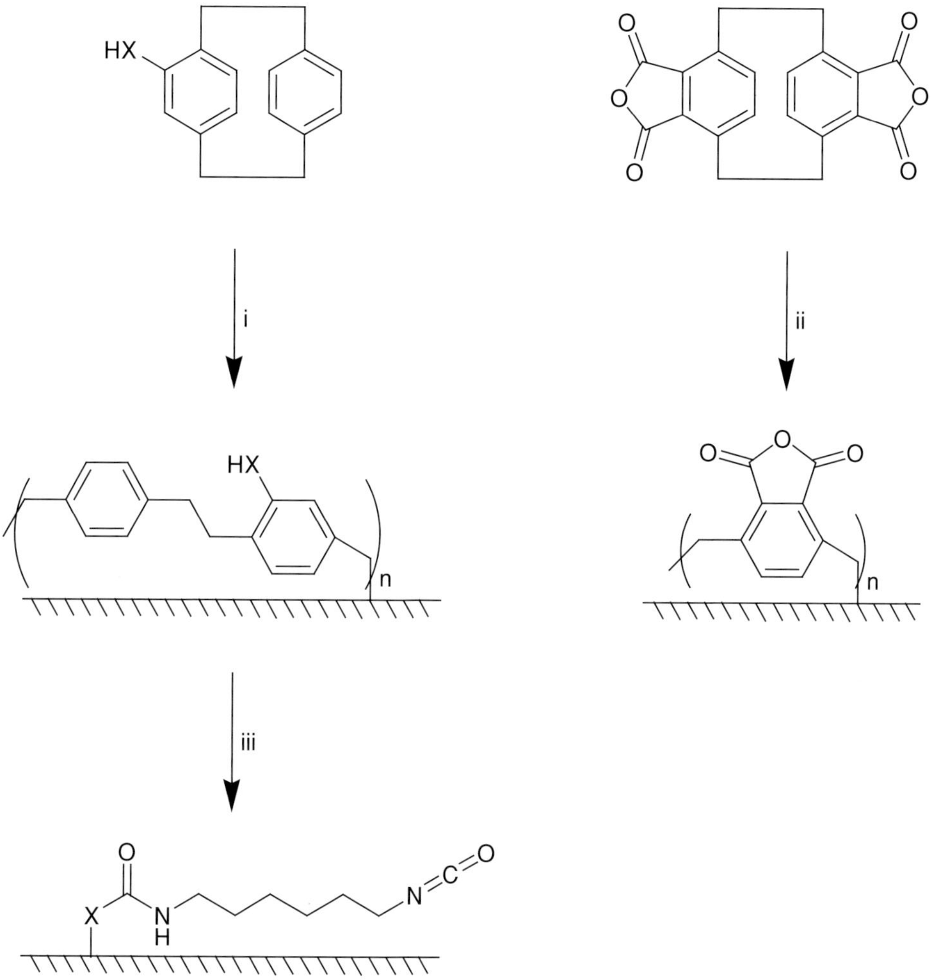

FIGURE 7.9. **Surface coatings on stainless steel foils or nickel/titanium alloy [283, 284].** i = chemical vapor deposition polymerization of 4-amino[2.2]paracyclophane (X = NH) or 4-hydroxymethyl[2.2]paracyclophane (X = CH_2-O) yielding poly(amino-p-xylylene-co-p-xylylene) or poly(hydroxy methyl-p-xylylene-co-p-xylylene) coatings, respectively; ii = chemical vapor deposition polymerization of [2.2]paracyclophane 4,5,12,13-tetracarboxylic acid anhydride yielding poly(p-xylylene-2,3-dicarboxylic acid anhydride coatings; iii = hexamethylene diisocyanate, diethylether.

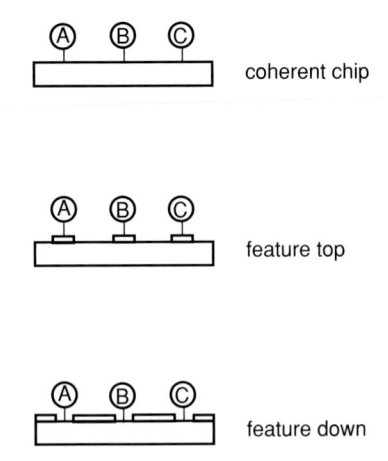

FIGURE 7.10. **General principles of array architecture.** The coherent chips represent the "standard" peptide array. Micropatterned arrays can be separated depending on the position of the immobilized peptides (A, B and C) into feature top or down arrays.

In principle, peptide arrays show different architectures: besides the classical coherent peptide array, structured arrays are also used (Figure 7.10). Two different types of structured arrays can be distinguished depending on the position of the features: feature top exemplified in gel pad arrays [136, 190, 367, 430, 565] or arrays of gold electrodes [345] and feature down [88, 225, 230, 392, 502, 604]. The direct attachment of peptides to the surface (Figure 7.11A) is only possible if the planar surface used is equipped with functional groups *per se*, such as hydroxyl groups of natural polymer cellulose, allowing

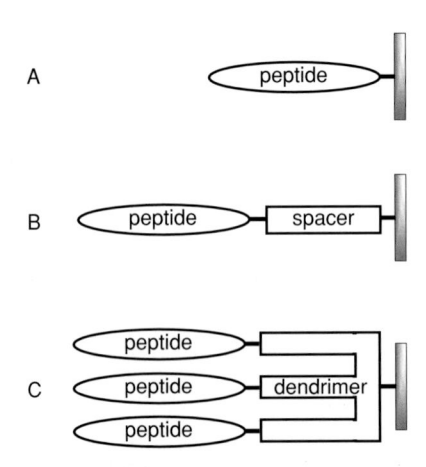

FIGURE 7.11. **Different possible architectures of immobilized peptides.** A = peptides directly attached to the surface, B = immobilized peptide is anchored apart from the surface by a chemical moiety introduced as spacer molecule, C = dendrimeric molecules inserted between the peptide and the surface enabling both appropriate spacing and increased density of immobilized peptide: combinations of C and B are used.

either the initiation of peptide synthesis or the immobilization of peptides. It was often demonstrated that the surface-bound peptides' accessibility to the proteins or enzymes used in screening is a critical factor. Insertion of a spacer between the peptide and the surface is an effective way to circumvent this drawback (Figure 7.11B). Generally, all linker molecules introduced to transform a given surface function into a functional group suitable for amino acid or peptide attachment can be considered as spacers. Such spacers can improve the efficiency of peptide/ligand interaction on surfaces as demonstrated with FLAG epitope peptides recognized by the monoclonal anti-FLAG M2 antibody [581]. The signal increased with the length of spacer introduced between the epitope and the surface. Additionally, for protein tyrosine kinase p60^{c-src} it was demonstrated that only incorporation of the long and hydrophilic 1-amino-4,7,10-trioxa-13-tridecanamine succinimic acid building block spacer allowed effective phosphorylation of the glass surface-bound peptides [143]. Moreover, insertion of hydrophilic dextran structures between the surface and the presented peptides (Figures 7.12 and 7.13) was described as necessary for efficient enzyme substrate interaction [337].

The loading of glass surfaces is too low for several biochemical reactions (Tables 7.1, 7.2). Several different approaches for generating dendrimeric spacers/linkers are used to increase the density of immobilized peptides (Figure 7.11C, 7.12–7.16). Alternatively, the effective surface per area can be increased by the use of porous silicon [306, 347, 462]. Examples for one-step generation of dendrimeric structures (Figure 7.11C) are represented by poly-lysine coated glass slides (Table 7.2) or aziridine polymerization onto aminopropylsilylated glass surfaces (Figure 7.16) [85, 257, 388]. An interesting approach for increasing the density of reactive functions is surface modification by adsorption of structured α-helical peptides [332] or proteins (bovine serum albumin, BSA; [338]) yielding amino modified polymers. The BSA amino functions were transformed into active esters by treatment with N,N-disuccinimidyl carbonate (Figure 7.15).

More sophisticated procedures for preparing multiple linker structures by multi-step functionalization of glass surfaces have been reported (Figures 7.14, 7.17) [31, 33, 197, 242, 258, 302, 410]). Alternatively, surface loading could be improved by coatings employing three-dimensional layers such as hydrogels (Rubina et al., 2003), semi-wet gels [258], agarose films [5], acrylamide gel pads [367, 430, 565] or gelatine pads [136].

7.2.2. Generation of Micro-Structured Surfaces

Micro-structured or micro-patterned surfaces represent an alternative way to achieve spatially addressed deposition of molecules. This could be useful for generating microreactors [604] or improving array regularity. Two major principles are used for micro-patterning: contact printing techniques (see Section 7.2.4.1.) and photolithografic technologies. The use of micro-contact printing for micro-patterning [36, 107, 225, 226, 240, 283, 284, 349, 507] has been reviewed extensively [60, 365, 366, 512, 601, 602] and is therefore not discussed in detail. Dip-Pen nanolithography/scanning probe lithography [51, 67, 249, 256, 302, 385, 519, 577, 580, 598] and microfluidic channel networks [83, 103, 105, 398] were used for nano- and micro-patterned immobilization of biomolecules such

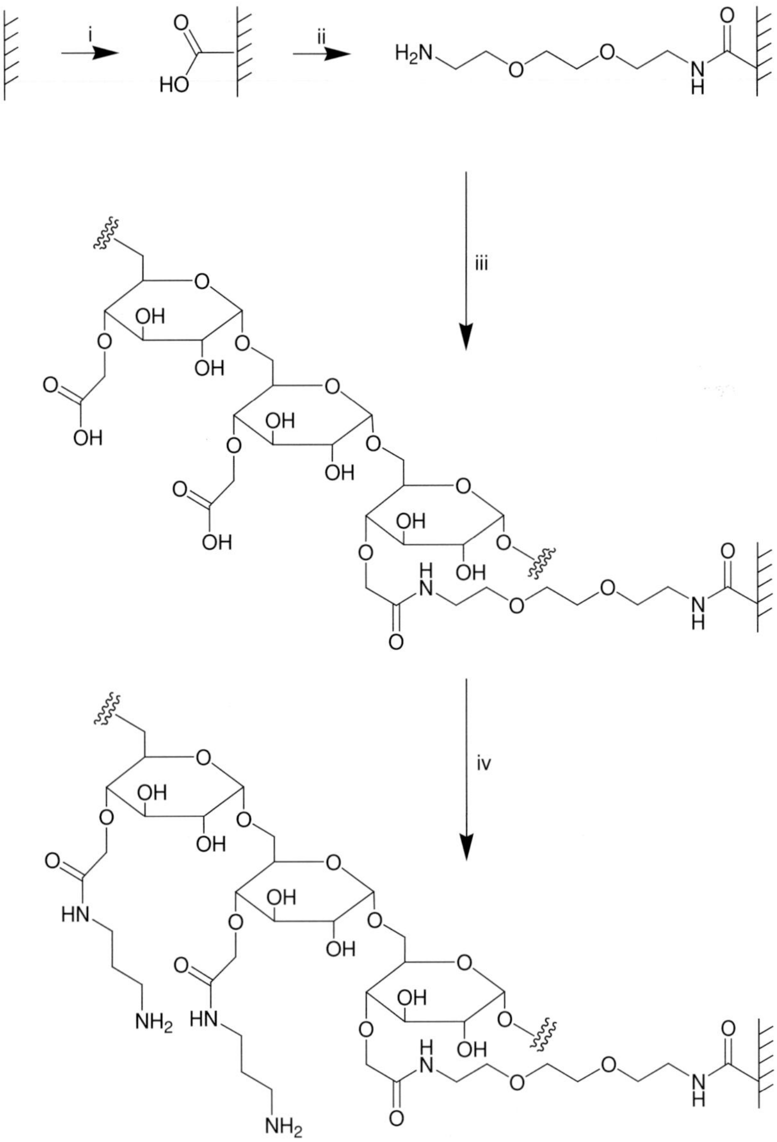

FIGURE 7.12. **Transformation of porous polypropylene into hydrophilic matrix suitable for enzyme profiling** [337]. i = oxidation using $Cr_2O_3/H_2SO_4/H_2O$, 54:54:80 (wt,wt,wt), ii = 5% oxalylchloride in CH_2Cl_2, 2 h, room temperature followed by treatment with 10% 2,2'-(ethylenedioxy)diethylamine in CH_2Cl_2, iii = carboxymethylated dextran, N-hydroxysuccinimide, 1-(3-dimethylaminopropyl)-3-ethylcarbodiimide hydrochloride in water, 12 h, room temperature, iv = carbonyl diimidazole, acetonitrile, 30 min, 0.3 M diaminopropane in acetonitrile, 1 h.

TABLE 7.2. Selected suppliers of modified surfaces

surface function	world wide web	dimension [mm]	material	loading
amino	www.aims-scientific-products.de	80 × 120	cellulose	400 nmol/ cm^2
amino		100 × 150	cellulose	400 nmol/ cm^2
amino		235 × 250	cellulose	400 nmol/ cm^2
streptavidin		100 × 150	cellulose	2 mmol/ cm^2
amino		240 × 240	polypropylene	80 nmol/ cm^2
amino		80 × 120	polypropylene	80 nmol/ cm^2
amino		240 × 240	polypropylene	600 nmol/ cm^2
amino		80 × 120	polypropylene	600 nmol/ cm^2
bromide		100 × 150	cellulose	800 nmol/ cm^2
carboxyl		100 × 150	cellulose	500 nmol/ cm^2
Fmoc-β-Alanine		100 × 150	cellulose	800 nmol/ cm^2
Fmoc-Proline		100 × 150	cellulose	600 nmol/ cm^2
mercapto		100 × 150	cellulose	600 nmol/ cm^2
hydroxyl		240 × 240	polypropylene	150 nmol/ cm^2
hydroxyl		80 × 120	polypropylene	150 nmol/ cm^2
hydroxyl		240 × 240	polypropylene	1500 nmol/ cm^2
hydroxyl		80 × 120	polypropylene	1500 nmol/ cm^2
epoxy	www.quantifoil.com	75 × 25	glass	n.a.
aldehyde		75 × 25	glass	n.a.
epoxy	www.noabdiagnostics.com	75 × 25	glass	n.a.
aldehyde		75 × 25	glass	n.a.
NHS ester		75 × 25	glass	n.a.
aldehyde	www.aat-array.com	75 × 25	glass	n.a.
amino	www.sigmaaldrich.com	75.5 × 25	glass	n.a.
poly-L-lysine		75.5 × 25.5	glass	n.a.
carboxyl	www.nuncbrand.com	76 × 25	heat stable polymer	n.a.
amino		76 × 25	glass	n.a.
epoxy		76 × 25	glass	n.a.
aldehyde		76 × 25	glass	n.a.
amino	www.eriesci.com	n.a.	glass	n.a.
poly-lysine		n.a.	glass	n.a.
epoxy		n.a.	glass	n.a.

(cont.)

TABLE 7.2. Continued

surface function	world wide web	dimension [mm]	material	loading
amino	www.arrayit.com	76 × 25	glass	8 pmol/cm²
amino		76 × 25	glass	8 pmol/cm²
aldehyde		76 × 25	glass	8 pmol/cm²
poly-L-Lysine		76 × 25	glass	n.a.
acrylic		76 × 25	glass	8 pmol/cm²
carboxyl		76 × 25	glass	8 pmol/cm²
cyanato		76 × 25	glass	8 pmol/cm²
epoxy		76 × 25	glass	8 pmol/cm²
mercapto		76 × 25	glass	8 pmol/cm²
amino	www.xenopore.com	75 × 25 or 22 × 40	glass	16 pmol/cm²
aldehyde		75 × 25 or 22 × 40	glass	16 pmol/cm²
epoxy		75 × 25 or 22 × 40	glass	16 pmol/cm²
maleimide		75 × 25 or 22 × 40	glass	16 pmol/cm²
nitrilotriacetic acid		75 × 25 or 22 × 40	glass	n.a.
streptavidin		75 × 25 or 22 × 40	glass	n.a.
biotin		75 × 25 or 22 × 40	glass	n.a.
mercapto		75 × 25 or 22 × 40	glass	16 pmol/cm²
amino	www.corning.com/lifesciences	75.5 × 25.3	injection molded polymer	n.a.
substituted antraquinones	www.exiqon.com	75 × 25	glass	n.a.
amino	www.greinerbioone.com	75 × 25	glass	n.a.
aldehyde			glass	n.a.
streptavidin			glass	n.a.
epoxy	www.asperbio.com	75 × 25 or 24 × 60	glass	n.a.
amino		75 × 25 or 24 × 60	glass	n.a.
isothiocyanate		75 × 25 or 24 × 60	glass	n.a.
epoxy	www.genescan.com	75.7 × 25.4	glass	16 pmol/mm²
amino	www.perkinelmer.com/ areas/proteomics/chem2.asp	75 × 25 active surface: 12 × 40 or 12 × 12 (two pad)	glass coated with modified acrylamide polymer	n.a.

epoxy	www.mwgbiotech.com	75 × 25	glass	n.a.
mercapto	www.ApogentDiscoveries.com	75 × 25	glass	n.a.
animo	www.csem.ch	75 × 25	glass	n.a.
carboxyl		22 × 20 available for print	metal oxides	4.2 pmol/mm^2
maleimide			metal sulfides	3.4 pmol/mm^2
mercapto			silicon	
			silicon nitride	
			plastics	
amino	www.us.schott.com	75.5 × 25	glass	n.a.
isothiocyanate	www.picorapid.de	76 × 26	glass	n.a.
amino-reactive	www.ucb-group.com	75 × 25	glass	n.a.
		variable size	cellophane	n.a.
		variable size	polypropylene	n.a.
amino	biolink@bellatlantic.net	120 × 80	highly crosslinked, polyaminated, polyurea coated polypropylene fleece	15 nmol/mm^2
amino	www.pall.com	75 × 25 (slide) 60 × 22 (membrane)	positive charged nylon membrane bound to glass	n.a.
aldehyde	www.pall.com	Ultrabind US450 membrane roll 3000 × 320	modified polyethersulfone	n.a.
amine reactive	www.accelr8.com	75 × 25	glass or silicon	n.a.
thiol-reactive				n.a.
biotin				n.a.
NHS ester	www1.amershambiosciences.com	75 × 25	glass	50 fmol/mm^2

n.a. not available

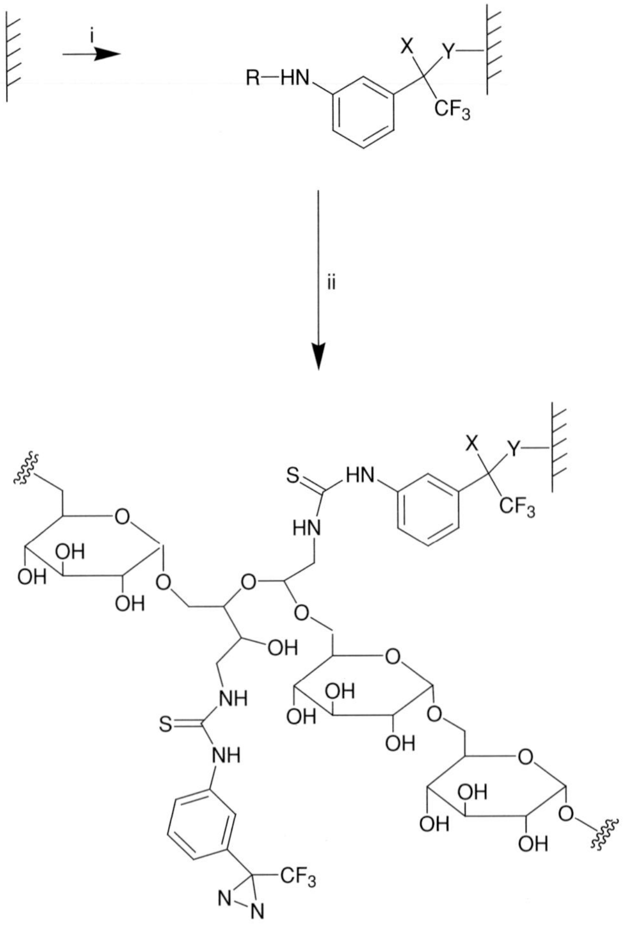

FIGURE 7.13. **Photobonding to and functionalization of surfaces such as glass, polystyrene, silicon nitride or polyurethane membranes**. i = photochemical coupling of N-substituted *m*-3-(trifluoromethyl)diaziridin-3-yl-anilines, R = 4-maleimidobutyryl residue [91, 442]; ii = Optodex treatment [71].

as proteins, but no applications for the creation of peptide arrays have been described so far.

Surfaces modified by amino groups protected by photosensitive protection groups can be used to generate microstructures simply by irradiation through a lithographic mask [480, 605]. A very similar approach [613] starts from the derivatization of glass surfaces with a photolabile self-assembled monolayer (Figure 7.18). Irradiation led to release of hydrophobic moieties yielding hydrophilic, carboxy modified areas (B) within a hydrophobic (A) environment.

Alternatively, photoresist coatings can be used in combination with photolithographic masks to create micro-patterns. Following irradiation, areas protected by photoresist coatings will not undergo modification upon treatment with tridecafluoro-1,1,2,2-tetrahydro)trichlorosilane in anhydrous toluene (Figure 7.19). After removal of the

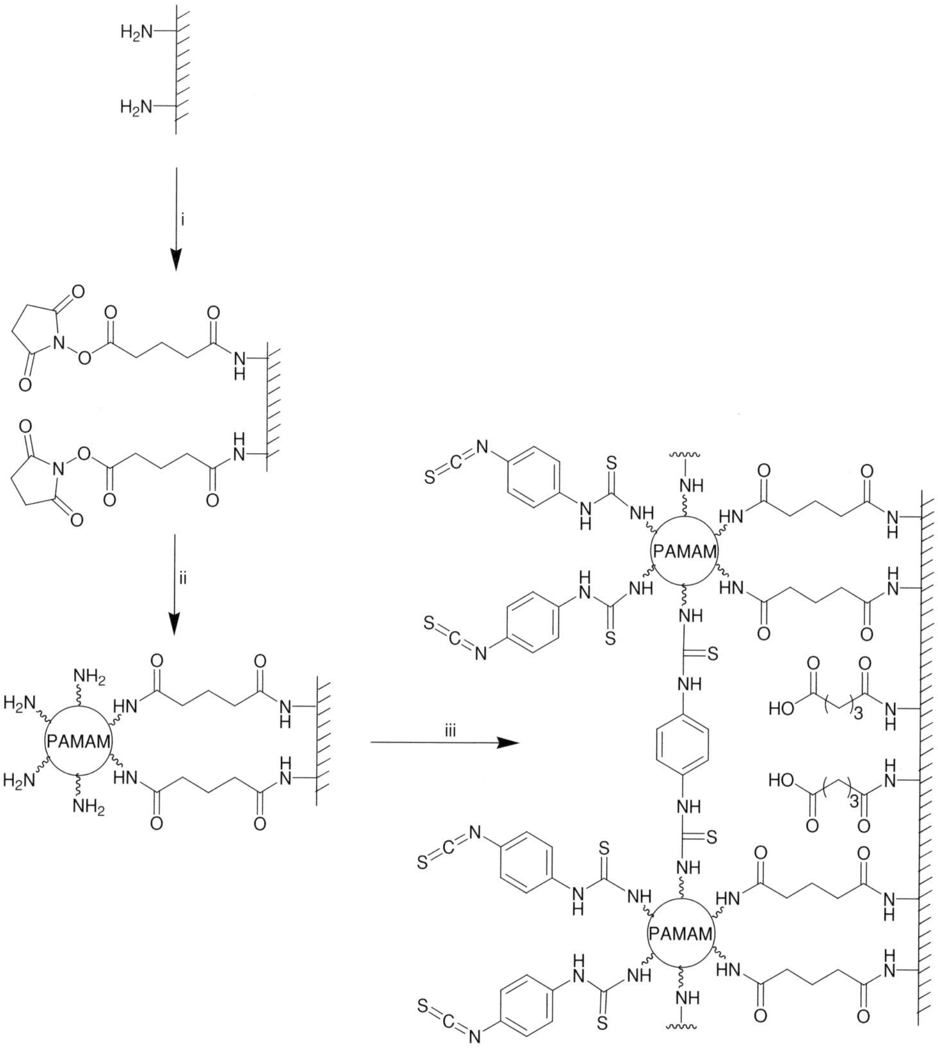

FIGURE 7.14. **Glass surface modification/activation via starbust dendrimer coating [33]**. i = treatment of aminosilylated glass with homobifunctional disuccinimidylglutarate in CH_2Cl_2 containing 1% diisopropylethylamine, 2 h; ii = 10% PANAM starbust monomer, 12 h; iii = activation and intermolecular crosslinking using homobifunctional 1,4-phenylenediisothiocyanate in CH_2Cl_2 containing 1% pyridine, 2 h.

protecting layer the newly exposed glass surfaces can be transformed into hydrophilic areas by aminopropylsilylation and acylation. Butler and coworkers produced so-called surface tension arrays [70] for the on-chip synthesis of oligonucleotides using a similar principle (Figure 7.20). The resulting hydrophilic areas (B), surrounded by hydrophobic areas (A) with a very low surface tension, represent micro-reactors holding the reaction solvent in area B due to differences in the wetting characteristics. The perfluorosilane modified areas A have wetting properties similar to Teflon. Alternatively, Brennan used a laser for

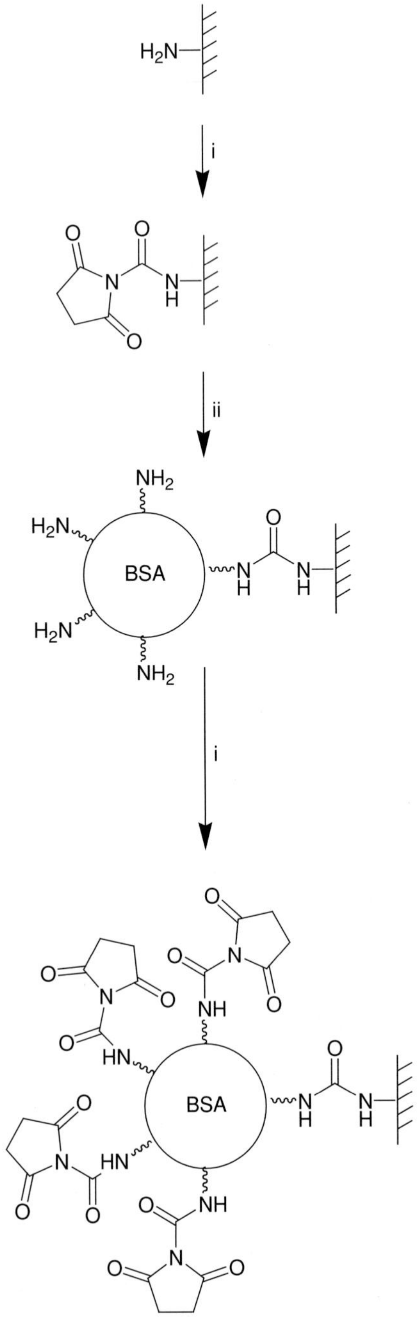

FIGURE 7.15. **Multistep functionalization of glass surfaces [338]**. i = N,N-disuccinimidyl carbonate, diisopropylamine, dimethylformamide, 3 h, room temperature; ii = PBS pH 7.5, 1% bovine serum albumin (BSA), 12 h, room temperature.

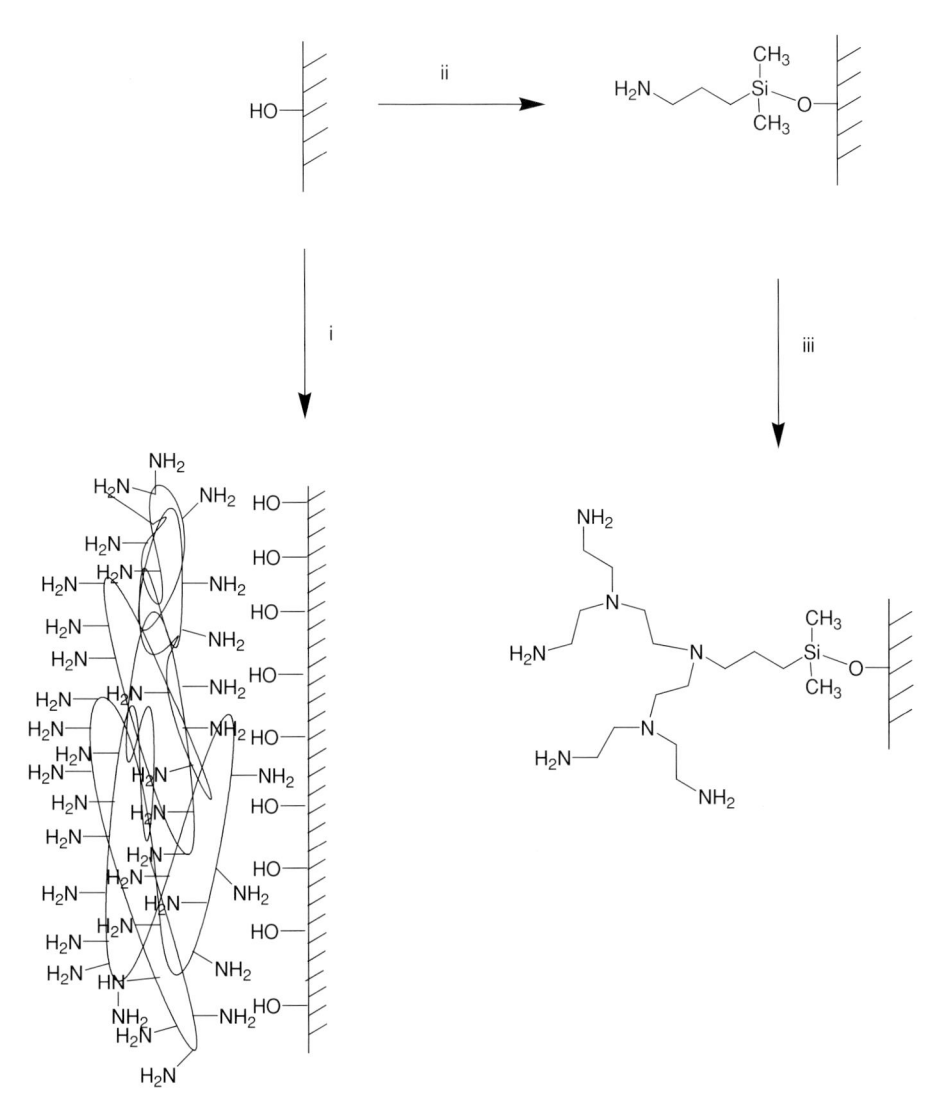

FIGURE 7.16. **Aziridine-polymerization [85, 257, 388] and poly-lysine coating**. i = the glass surface is coated by covering it with poly-L-lysine solution and drying the film; the polymer film is bound to the surface by electrostatic interactions between silyl-OH functions and positive charged amino groups in the polymer; ii = ethoxydimethylaminopropylsilane/toluene, iii = aziridine / methylene chloride / acetic acid.

ablation of fluorosiloxanes (Figure 7.21) to create micro-patterned surfaces starting from a uniformly hydrophobic modified glass slide [61]. Furthermore, photolithographic techniques have been reported for the micro-patterning of films ([188, 268, 381, 606]; reviewed in [186, 305]). Micro-mirror mediated patterning of biomolecules has been reported [303, 304, 510].

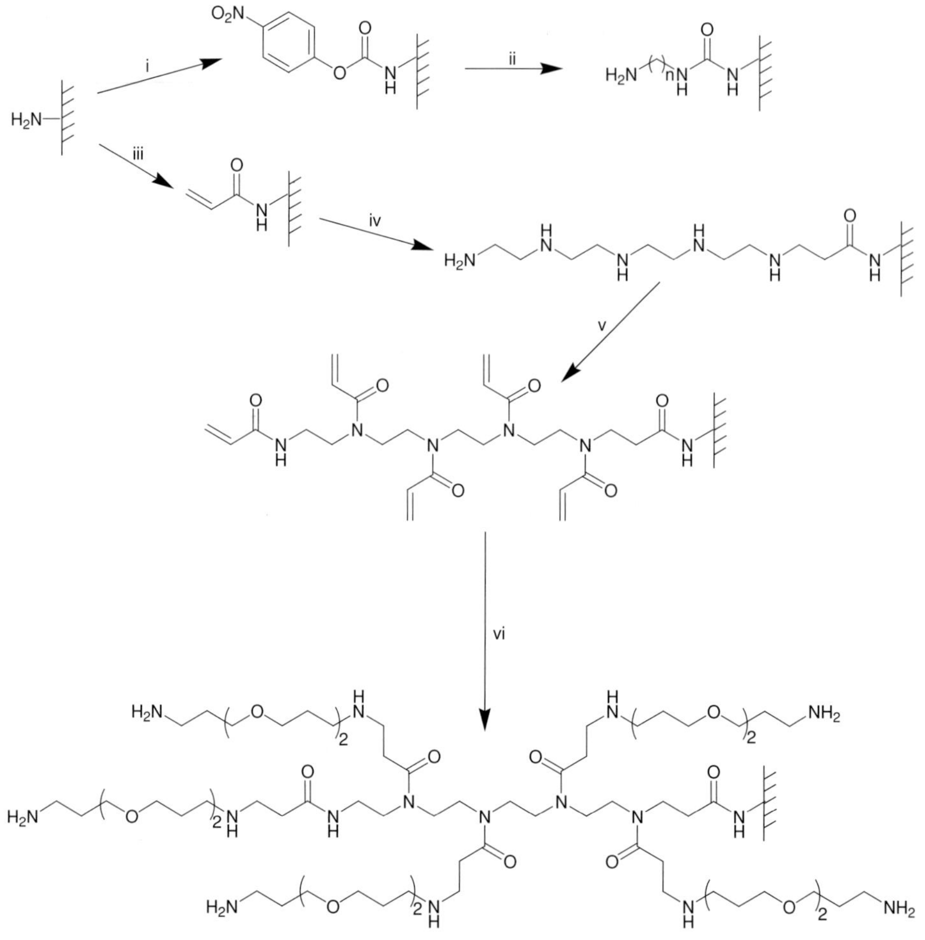

FIGURE 7.17. **Multistep functionalization of aminoalkylsilylated glass surfaces or plasma-aminated polypropylene surfaces [31].** i = 4-nitrophenylchloroformiate, diisopropylamine, CH$_2$Cl$_2$, 2 h; ii = diamine, dimethylformamide, 12 h; iii = acryloylchloride, diisopropylamine, dimethylformamide, 24 h, iv = tetraethylenepentamine, dimethylformamide, 12 h; v = acryloylchloride, diisopropylamine, dimethylformamide, 24 h; vi = 1,4-bis(3-aminopropoxy)-butane, dimethylformamide, 12 h.

7.2.3. Peptide Array Preparation

Peptides are usually synthesized step by step using appropriate side chain (permanent protecting group) and N-terminal protected (temporary protecting group) amino acid derivatives (Figure 7.22). Generally, fluorenyl-oxy-carbonyl- (Fmoc), *tert*-butyl-oxy-carbonyl- (Boc) and nitroveratryl-oxy-carbonyl (Nvoc) moieties are used as temporary protecting groups (PG) while preparing peptide arrays. Subsequent to amino acylation of surface-bound amino functions, the PG of the surface-bound amino acid derivative is removed by base (Fmoc), acid (Boc) or light (Nvoc) treatment. Repeated amino acylation and deprotection reactions yield a surface-bound target peptide in the side chain protected state. Final

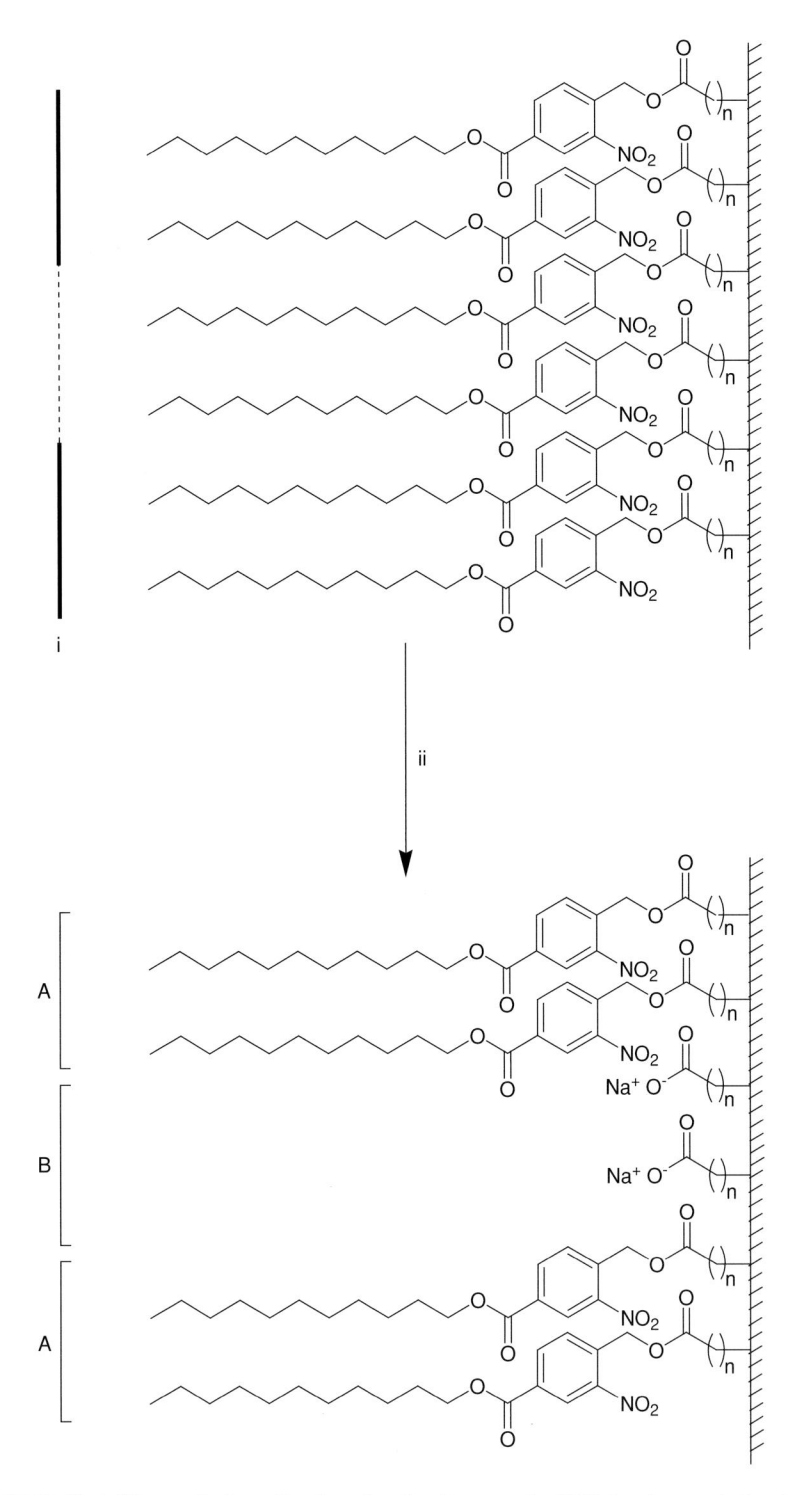

FIGURE 7.18. **Photolithography for patterning of surface free energies [613].** i = photomask placed on top of glass surface coated with photolabile self-assembled monolayer, ii = UV light (filter cube with band pass of 360 to 370 nm) in the presence of aqueous NaOH pH 11.7 yielding hydrophobic (A) and hydrophilic (B) functionalized regions.

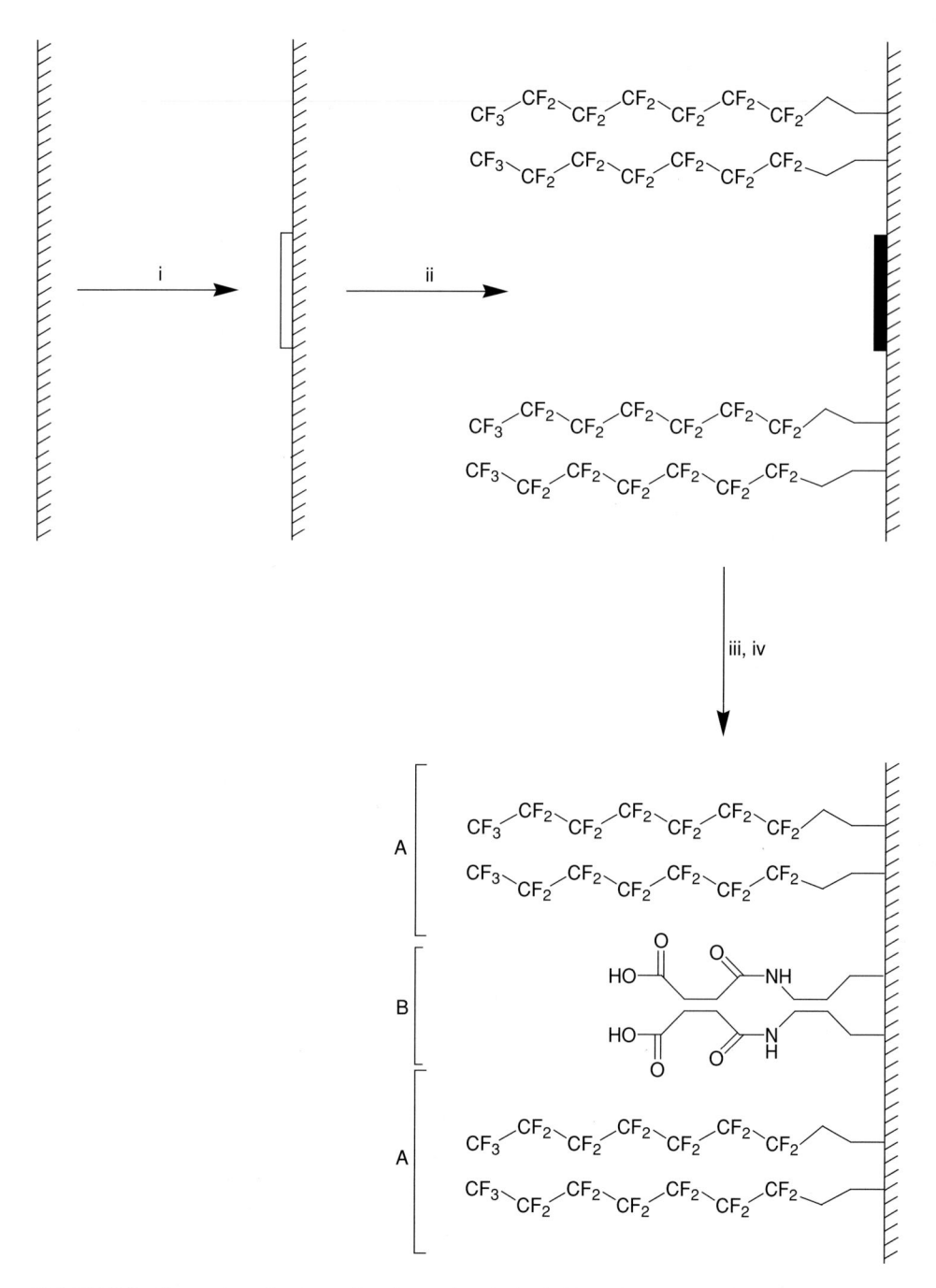

FIGURE 7.19. **Generation of patterned surfaces**. A = hydrophobic surface; B = hydrophilic surface; i = deposition of photoresist, irradiation and generation of exposed glass surface areas; ii = (tridecafluoro-1,1,2,2-tetrahydro)trichlorosilane in anhydrous toluene; iii = 3-aminopropyltrimethoxysilane in anhydrous toluene; iv = succinic anhydride.

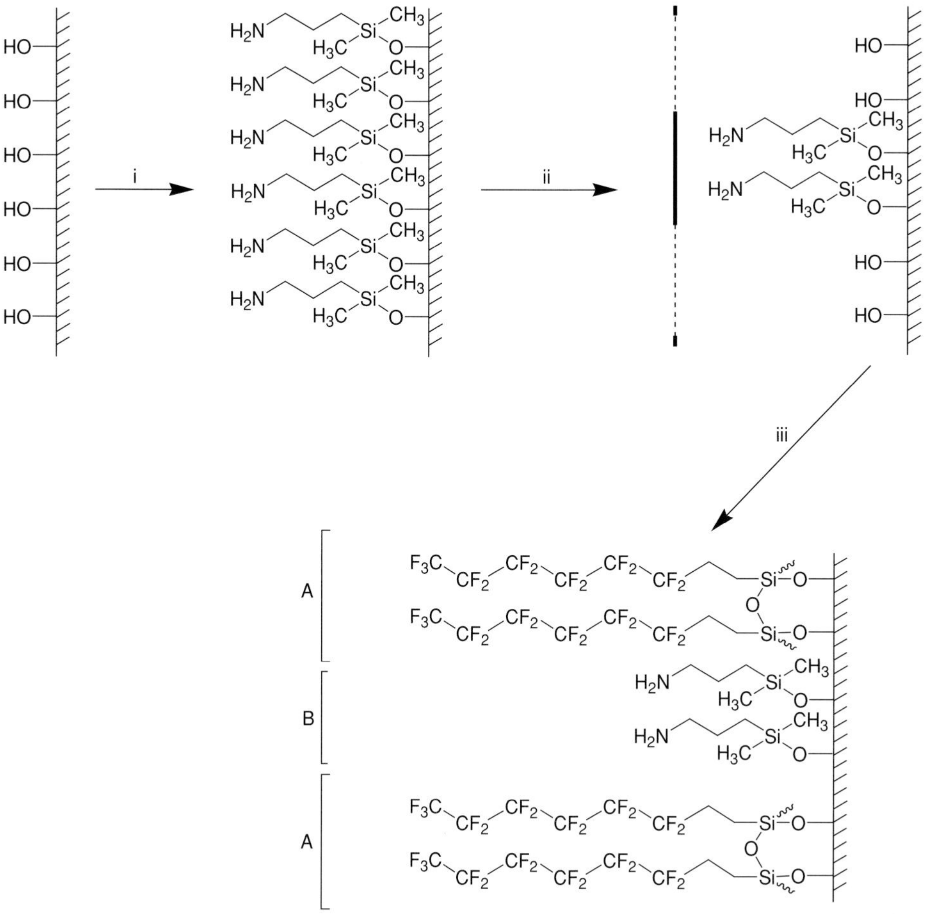

FIGURE 7.20. **Generation of surface tension arrays [70]**. A = hydrophobic surface; B = hydrophilic surface; i = 0.4% solution of 3-aminopropyldimethylethoxysilane in anhydrous toluene, 72 h; ii = coating with positive photoresist (2.5 μm) and exposure to near UV through a chromium-on-quartz mask followed by removal of the exposed photoresist and oxygen plasma treatment; iii = 0.25% solution of (tridecafluoro-1,1,2,2-tetrahydro)trichlorosilane in anhydrous toluene, 10 min, removal of residual photoresist.

removal of all the permanent protecting groups gives appropriate target peptides tethered to the surface via the C-terminal carboxylic acid function. Incomplete coupling/deprotection reactions during the peptide assembly result in a target peptide contaminated by a variety of deletion and truncation sequences.

7.2.3.1. Stepwise Synthesis on Coherent Surfaces In principle, *in situ* synthesis has a number of advantages compared to immobilization of pre-synthesized peptides. Normally, yields of peptide synthesis on surfaces are high and consistent over the entire support surface from one array region to another. It also permits combinatorial strategies for constructing large arrays of peptides in a few coupling steps. Several approaches have been used to

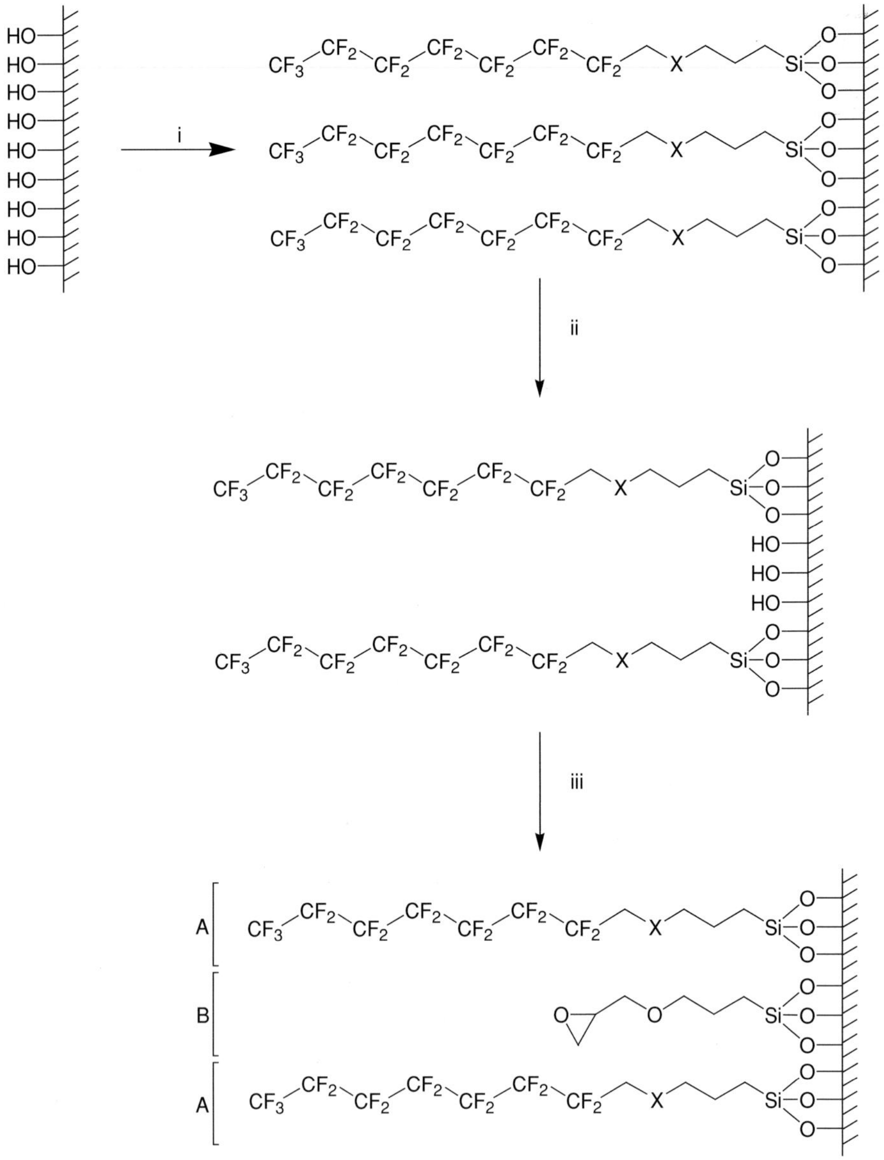

FIGURE 7.21. **Generation of surface tension arrays [61].** A = hydrophobic surface; B = hydrophilic surface, X = oxygen; i = 3-(1,1-dihydroperfluoroctyloxy)propyltriethoxysilane; ii = CO_2-laser for ablation of fluorosiloxane and exposure of glass surface; iii = glycidyloxypropyltriethoxysilane.

allow spatially addressed peptide synthesis on coherent surfaces and membranes for *in situ* fabrication of arrays.

Fodor and coworkers described a method for synthesizing large numbers of peptides bound to a planar, solid support by combining the techniques of solid phase peptide

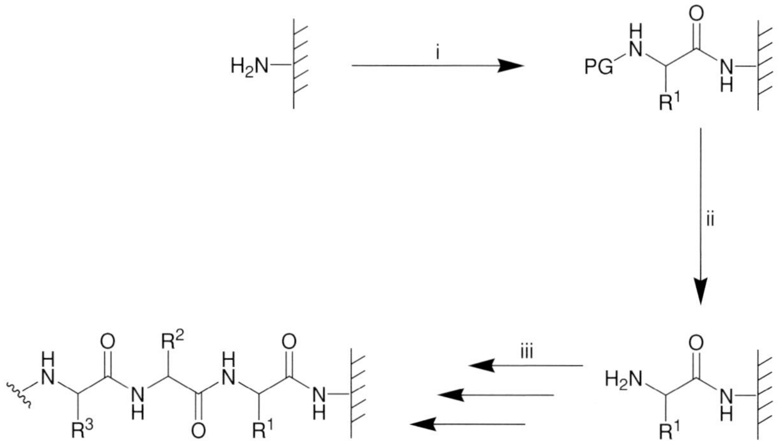

FIGURE 7.22. **Principle of peptide synthesis on coherent surfaces.** i = acylation of surface bound amino functions using appropriate side chain (permanent protecting group) and N-terminal protected (temporary protecting group [PG]), carboxyl-activated amino acid derivatives; ii = removal of PG; iii = repeated cycles of acylation and deprotection reactions followed by removal of all permanent protecting groups, yielding the surface bound target peptide sequence contaminated with impurities resulting from incomplete acylation or deprotection reactions (deletion and truncation sequences).

synthesis, photolabile protection, and photolithography, known as light-directed peptide synthesis (LDPS) or Very Large Scale Immobilized Polymer Synthesis [84, 151, 168, 324].

Here, a fraction of sites on a planar support carrying photo-detachable protecting groups, such as nitroveratryl-oxy-carbonyl (Figure 7.23), is exposed to light through a photolithographic mask. The fraction of sites thus deprotected is acylated with a specific

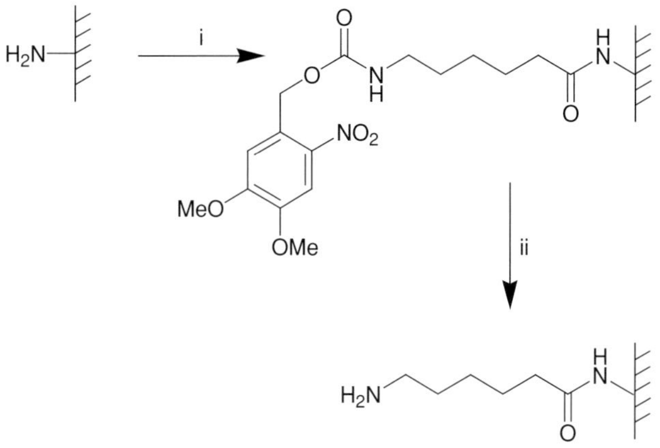

FIGURE 7.23. **Chemical principle enabeling photolithografic patterning of aminosilylated glass surfaces.** i = N-(nitroveratryl-oxy-carbonyl)-6-amino hexanoic acid N-hydroxysuccinimide ester; ii = UV-light (365 nm).

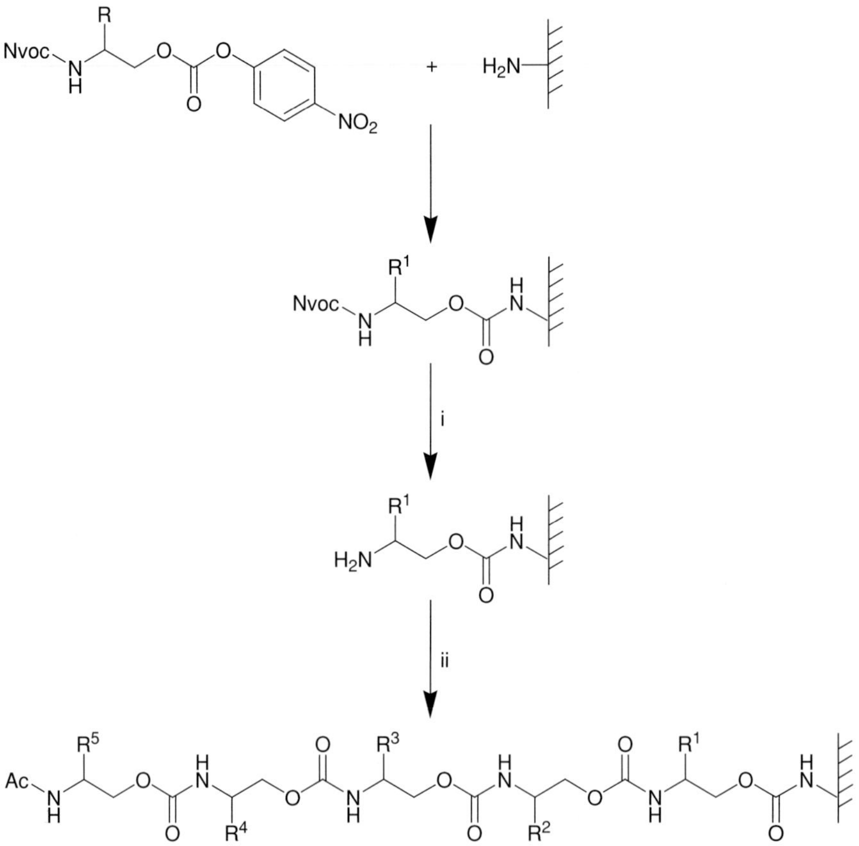

FIGURE 7.24. **Chemical principle of photolithografic oligocarbamate synthesis [84].** i = UV-light (365 nm), ii = repetitive cycles of nitrophenylcarbamate treatment and UV-light mediated deprotection of amino groups.

amino acid derivative or building block, itself carrying a photo-detachable protecting group (Figure 7.24). The photo-deprotection is repeated with the mask in a different position, or with a different mask, and a second amino acid derivative or building block is attached either to the functionalized support and/or to the first amino acid residue. After several cycles of acylation and deprotection, with careful attention to the pattern of masking, an array of peptides is built up on the planar support. The final peptide microarray is completely deprotected and exposed to the ligand of interest. Additionally, this technology could be successfully used for the generation of arrays of peptidomimetics such as oligocarbamates (Figure 7.24) [84]. However, this technology has some serious disadvantages. Light-directed peptide synthesis involves a novel set of chemistries, which have to be optimized otherwise the final quality of the surface-bound peptides will cause false positive (if an impurity is active) and/or false negative results (if the target peptide sequence was not synthesized). An interesting alternative to circumvent this limitation is the use of photo-generated acids in combination with Boc-chemistry [171, 265, 309,

406]. Nevertheless, the use of photolithografic masks combined with solid phase peptide synthesis is relatively labor intensive. Related techniques eliminating the mentioned disadvantages associated with photolithografic masks use LED-arrays [174], laser scanning by mirror-arrays [514] or a computer controlled micro-mirror projector [171, 309]. This latter method involves deprotecting part of a surface coated with an amino acid carrying a photo-detachable protecting group by a laser printer beam such as a HeCd laser. Spatial control of the laser beam is accomplished by reflecting the beam from a spinning mirror and employing a shutter to exclude the beam if desired. This programmable laser-activated parallel peptide synthesis permits automated synthesis of immobilized peptide libraries, avoiding the physical scale limitations of photolithography, but does not circumvent the disadvantage of using amino acids with photo-detachable protecting groups. Moreover, there are 20 separate deprotection steps necessary for each position to be varied within a peptide sequence. For example, the synthesis of all possible penta-peptides using the 20 natural amino acids (3.2 million peptides) requires 100 separate deprotection, washing and acylation steps.

A further approach involves physically locating the activated monomer by contacting a confined area of the functionalized surface with monomer solution using masks or physical barriers as demonstrated for the synthesis of oligonucleotides [351]. This allows the synthesis of complex microarrays comprising many different but related sequences within a few coupling steps by combinatorial methods. Flooding the activated monomers through intersecting micro-channels yielded arrays of all sequences of a chosen length [352, 521] but generating circular or diamond shaped reaction chambers by sealing an appropriate shaped mask against a functionalized surface such as glass or aminated polypropylene mounted on glass yielded so-called scanning or tiling-path arrays ([129]; reviewed in [522]).

A very elegant form of spatially addressed compound deposition makes use of modified color laser printers. The cartridges are filled with a solvent/amino acid derivative mixture (high melting point of the solvent yields toner-like powder) resulting in an activated amino acid solution during the laser induced melting process [428].

The use of ink-jet or bubble-jet technologies for the drop-on-demand liquid handling during stepwise oligonucleotide synthesis on glass microscope slides [46, 222, 541] or generation of combinatorial libraries on functionally graded ceramics [368, 369] is described. Analogous ink-jet delivery of activated amino acids to appropriate functionalized surfaces, such as membranes, microscope slides or spinning surfaces in a CD-format [4] for automated synthesis of peptides, have been developed by a number of companies, but is not yet commercially available.

A very simple but extremely robust method for the highly parallel synthesis of peptides on planar surfaces is the SPOT synthesis concept developed by R. Frank [153, 155] and commercialized by the company Jerini AG (www.jerini.com). This method is very flexible and economic relative to other techniques and was recently developed from a semi-automatic procedure [157] to a fully automated system (www.jerini.com). The basic principle involves the spatially addressed deposition of defined volumes of activated amino acid derivatives (or oligopeptides) directly onto a planar surface such as functionalized cellulose (Figures 7.25–7.27), aminated polypropylene or aminopropylsilylated glass slides. The areas contacted by the droplets represent individual micro-reactors allowing the formation of a covalent bond between the amino acid derivative and the surface function. The

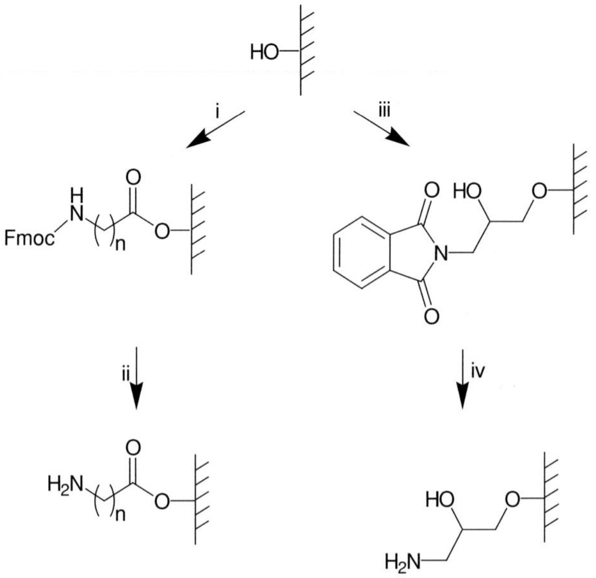

FIGURE 7.25. **Amino derivatization of cellulose.** i = Fmoc-glycine (n = 1) or Fmoc-6-aminohexane carboxylic acid (n = 5)/TBTU/DIPEA; ii = 20% piperidine in DMF; iii = N-phthaloyl-oxiranyl-methylamine, iv = hydrazine treatment.

resulting spot size is defined by the dispensed volume as well as the physical properties of the surface used. This SPOT synthesis has been reviewed extensively [158, 160–162, 273, 274, 443, 453, 454, 592]. Recent developments such as the introduction of novel polymeric surfaces [591], new linker and cleavage strategies [16, 320, 573] as well as automation (allowing the fully automated synthesis of up to 25,000 peptides within one run) have increased the value of this technique and led to the extension of SPOT synthesis to other molecule classes such as protein domains [546–548], peptide nucleic acids [356, 583], peptomers [16, 617], peptoids [16, 591, 202] and small heterocyclic compounds [488, 490]; see Section 7.5.5). Normally, peptides or peptidomimetics on arrays prepared by SPOT synthesis are assembled in a stepwise manner. For longer peptides this is a laborious and time-consuming procedure. Toepert and coworkers synthesized an oligopeptide array of several thousand triple-substituted variants of the human YAP-WW domain by a combination of classical stepwise SPOT synthesis and native chemical ligation [546–548]. This is one example of how to extend the concept of peptide arrays to arrays of chemically synthesized protein domains. Additionally, SPOT synthesis allows the use of defined amino acid mixtures for acylation reactions leading to the array of libraries concept. The problem of equal amounts of each library member could be circumvented by using either kinetically adjusted mixtures or sub-stoichiometric amounts of amino acid derivatives [269].

It is generally difficult to assess the quality of the peptides made on and bound to a planar surface. The amount of peptide is small for most materials used (Table 7.1). However, analysis of peptides synthesized step by step on cleavable linkers suggests a relatively

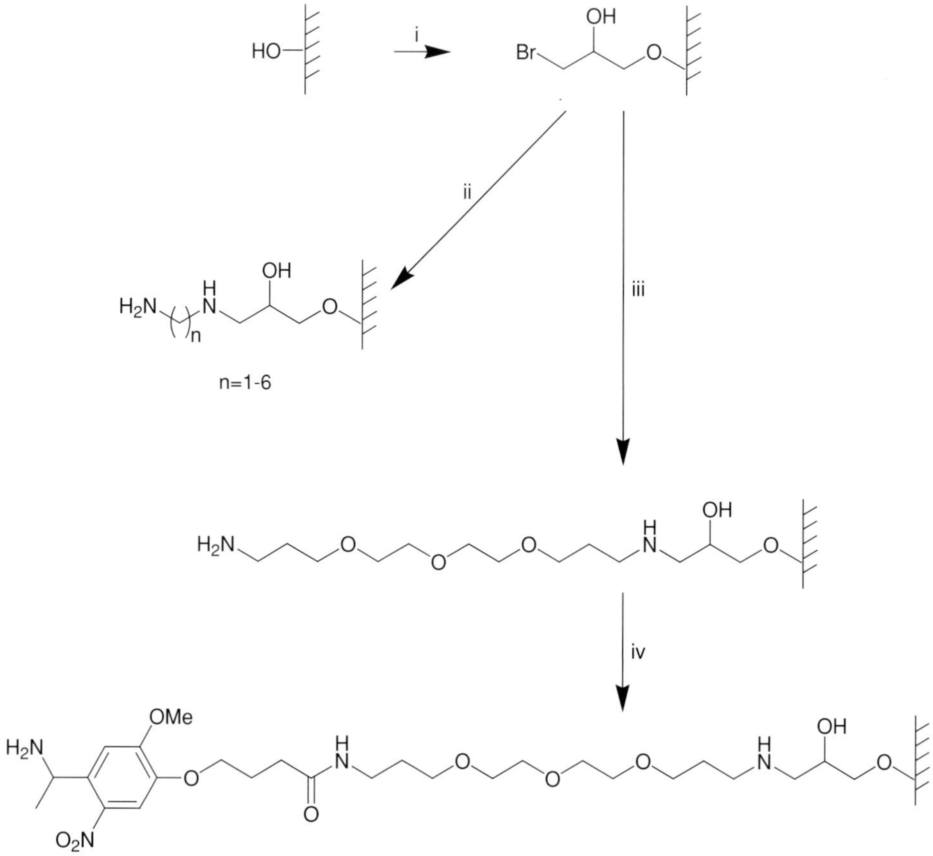

PEPTIDE ARRAYS IN PROTEOMICS AND DRUG DISCOVERY 191

FIGURE 7.26. **Amino derivatization of cellulose**. i = 2-bromomethyl-oxirane; ii = diamino-alkane treatment; iii = 4,7,10-trioxa-1,13-tridecanediamine, iv = 4-[4-(1-aminoethyl)-2-methoxy-5-nitro-phenoxy]-butyric acid /TBTU/DIPEA.

high quality (Wenschuh et al., 2000). Non-destructive measurements can be made by IR-spectroscopy, ellipsometry or interferometry [183].

In contrast, pre-synthesized peptides can be assessed before they are attached onto the surface, allowing for quality control. When large numbers of peptide arrays with the same sequences are needed, deposition of pre-synthesized peptides is more economical than *in situ* synthesis. Deposition is also the method of choice for long peptide sequences, which normally have to be purified to obtain high quality products (see Section 7.2.3.3.).

7.2.3.2. Non-Selective Immobilization of Peptides Non-specific immobilization has the advantage that no specific modification of the peptides is necessary. Additionally, the immobilized peptides within one spot are displayed as a mixture of differently attached molecules, reducing the probability of wrongly presented molecules. On the other hand, non-selective immobilization could prevent effective interaction with the screening probe

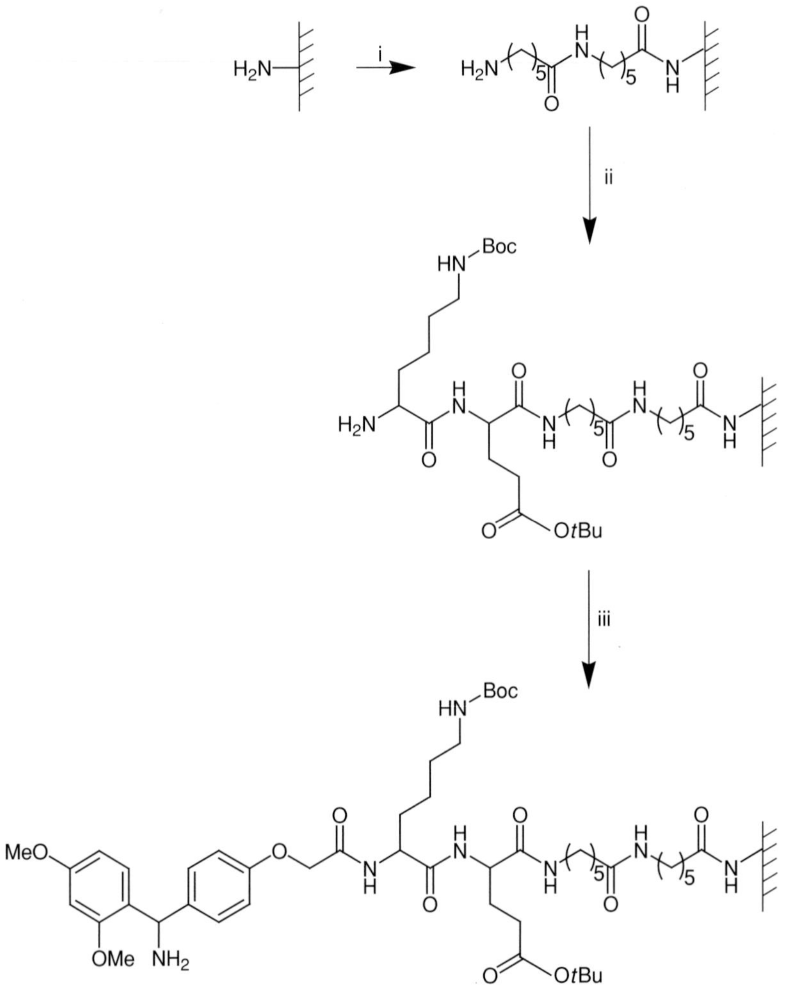

FIGURE 7.27. **Linker structure used for the automated SPOT-synthesis of PNA oligomers allowing the parallel synthesis and subsequent removal of individual sequences by treatment with either acid or protease trypsin [583].** i = Fmoc-6-aminohexanioc acid/DIC/HOAt/NMP, 20% piperidine/DMF, Fmoc-6-aminohexanioc acid/DIC/HOAt/NMP, 20% piperidine/DMF; ii = Fmoc-Glu(OtBu)-OH/ DIC/HOAt/NMP, 20% piperidine/DMF, Fmoc-Lys(Boc)-OH/DIC/HOAt/NMP, 20% piperidine/DMF; iii = Fmoc-{4-[amino-(2,4-dimethoxyphenyl)-methyl]-phenoxy}-acetic acid /DIC/HOAt/NMP, 20% piperidine/DMF.

if the functional group of the peptide used for the immobilization chemistry also represents a key residue.

Several non-selective chemistries have already been described for peptide array applications. In most cases the amino functions of the peptides (free N-terminus or lysine side chains) are targeted by surfaces carrying acid halides, active esters, isocyanates, isothiocyanates, activated double bonds, aldehydes or epoxides. Surface coatings with cyanuric chloride as described for immobilization of synthetic polynucleotides (Lee et al., 2002)

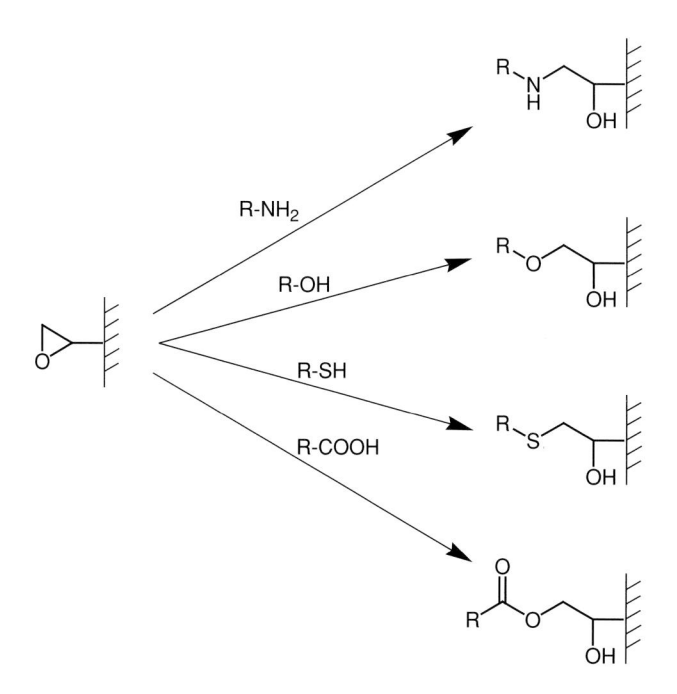

FIGURE 7.28. **Chemistry of epoxy-modified surfaces**. Versatility of epoxy functions demonstrated by reaction with primary amines, alcohols, mercaptanes and carboxylic acids yielding N-alkyl-aminoalcohols, monoalkylated diols, S-alkyl-mercaptoalcohols and monoacyl-diols, respectively.

should also be an attractive method for peptide immobilization. Compared with standard chemistry this method led to an approximately 35-fold higher biological signal subsequent to spotting a 10 µM sample solution.

The epoxy function represents the most versatile moiety for non-specific immobilization because it can react with nearly all possible side chain functionalities of the deposited peptide (Figure 7.28). Examples of non-specific immobilization chemistries are given in Figures 7.1i, 7.1ii, 7.1iii, 7.1v, 7.6i, 7.7v, 7.8iii, 7.9, 7.14, 7.15, 7.17i, 7.17iii, 7.17v, 7.29ii, 7.29iv, and 7.30i.

Another possibility for non-selective immobilization is the use of surface modifications that can be activated by light generating reactive surface functionalities. Treatment of silicone wavers with 2-nitro-5-[11-(trimethoxysilyl)undecyl]oxybenzyl methoxy poly(ethylene glycol) propanoate yielded surfaces covered with masked aldehyde functionalities [384]. Irradiation with UV-light (330 nm) generates the aldehyde that allows immobilization of amino-containing molecules. The light dependence of the reactive functions allows the generation of photolithografically defined immobilization sites. Alternatively, introduction of photoaffinity moieties such as benzoyl benzoic acid ([480]; Figure 7.8iv), perfluorophenyl azides [25, 26, 553] or 4-[3-(trifluoromethyl)-$3H$-diaziridin-3-yl]benzoic acid (Figure 7.13; [91, 82, 252, 442]) generates highly reactive species on the surfaces, allowing addition to peptides or other molecules in close proximity upon irradiation with UV-light.

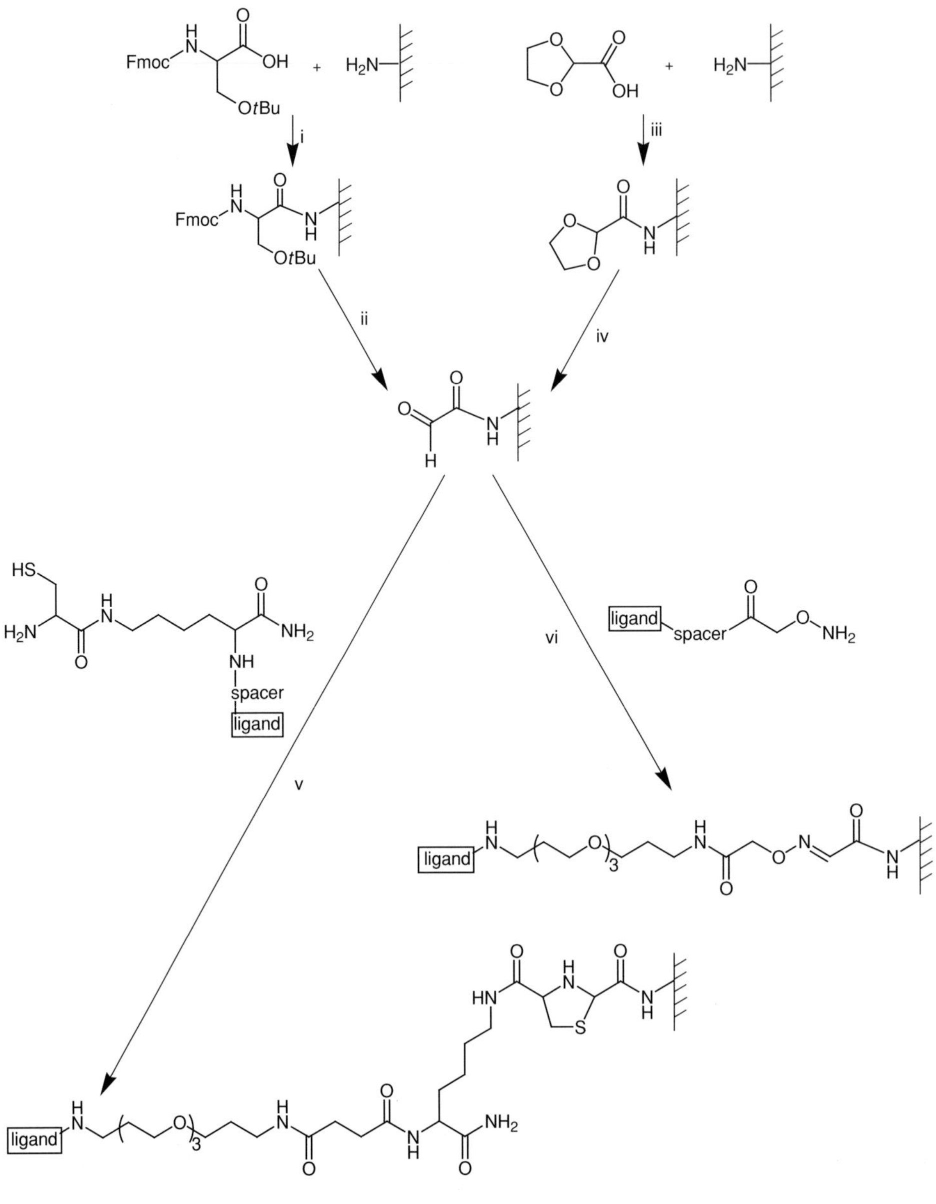

FIGURE 7.29. **Preparation of aldehyde functionalized glass surfaces and chemoselective immobilization of peptides (ligands) containing either amino-oxyacetyl function or an N-terminal free cysteine [143].** i = aminopropyl silylated glass and Fmoc-Ser(OtBu)-OH, DIC, HOBt, dimethylformamide, 1.5 h; ii = 95% trifluoroacetic acid, 2 h, 20% piperidine in dimethylformamide followed by oxidation using 100 mM $NaIO_4$ solution in PBS for 2 h, iii = protected glyoxylic acid, DIC, HOBt, dimethylformamide, 2 h, iv = 0.01 M HCl, room temperature, 2 h, v = chemoselective reaction with peptides containing an amino-oxyacetyl group, 250 mM NaOAc pH 5.2, 300 mM NaCl, vi = chemoselective reaction with ligands containing Lys(Cys)-amide at their carboxyl termini, 250 mM NaOAc pH 5.2, 300 mM NaCl.

FIGURE 7.30. **Preparation of semi-carbazide functionalized glass surfaces and chemoselective immobilization of glyoxylyl peptides [361].** i = aminopropyl silylated glass and triphosgen, DIEA, 1,2-dichloroethane; ii = Fmoc-hydrazide/dimethylformamide followed by treatment with piperidine/DBU in dimethylformamide; iii = chemoselective reaction with peptides containing an glyoxylyl group, 100 mM NaOAc pH 5.5.

7.2.3.3. Chemoselective Immobilization of Peptides   Chemoselective immobilization reactions (Figure 7.31) are of particular interest in the preparation of peptide arrays because they allow control over both the orientation of the attached peptide and the density of the immobilized biomolecule. The ideal reaction is absolutely selective, avoiding any impairment caused by amino, carboxyl, mercapto, guanidino or hydroxyl functionalities from the peptides' amino acid side chains. Additionally, the reaction has to be fast, minimizing the problems related to solvent evaporation during peptide microarray preparation. Recombinant tags represent a frequently used principle for chemoselective immobilization of proteins although they do not lead to the formation of a stable bond, for example via the oligohistidine tag or glutathione S-transferase conjugates. Alternatively, peptide immobilization using peptide tags (FLAG-tag, Myc-tag) or biotin moieties requires large and sensitive mediator proteins (antibodies, streptavidin).

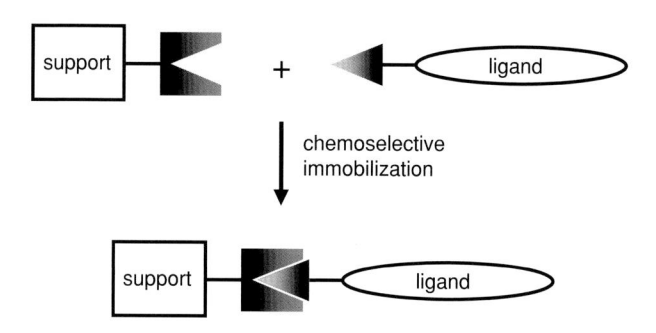

FIGURE 7.31. **Principle of chemoselective immobilization.**

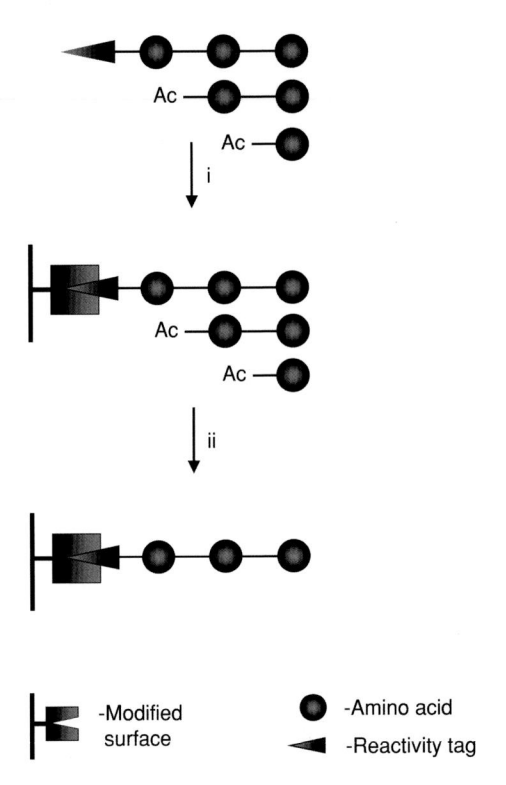

FIGURE 7.32. **Reactivity purification during chemoselective immobilization**. i = deposition of peptides mixtures and chemoselective immobilization: crude peptides normally contain truncated sequences resulting from incomplete coupling steps; introduced acetylations subsequent to every coupling reaction yield non-reactive truncated peptides together with the target peptide equipped with the reactivity tag; ii = washing steps subsequent to chemoselective immobilization yield purified covalently immobilized peptide.

One intrinsic advantage using chemoselective reactions is the formation of a covalent bond introduced as a result of reactivity purification (Figure 7.32). If the chemical moiety mediating the chemoselective reaction with the appropriately modified surface is attached to the N-terminus of the growing peptide, the peptide synthesis protocol can be modified to yield the target peptide equipped with the reactivity tag together with truncated, acetylated sequences resulting from incomplete coupling steps. Deposition of this mixture results in a covalent bond forming exclusively between the target peptide derivative and the surface. The chemically "inert" truncated sequences can be simply removed during subsequent washing steps. Thus, chemoselective reactions allow the generation of peptide arrays containing purified (free of truncated sequences) peptides.

Classical chemoselective immobilization reactions (Figure 7.33) are reviewed by Lemieux and Bertozzi [307]. Chemoselective reactions used for the preparation of peptide arrays or which are suited for the oriented immobilization of peptides are presented in Table 7.3. Examples of maleimide-surface modifications of glass slides and titan [603] are given in Figure 7.1vi and Figure 7.2, respectively. Figure 7.3 illustrates the transformation

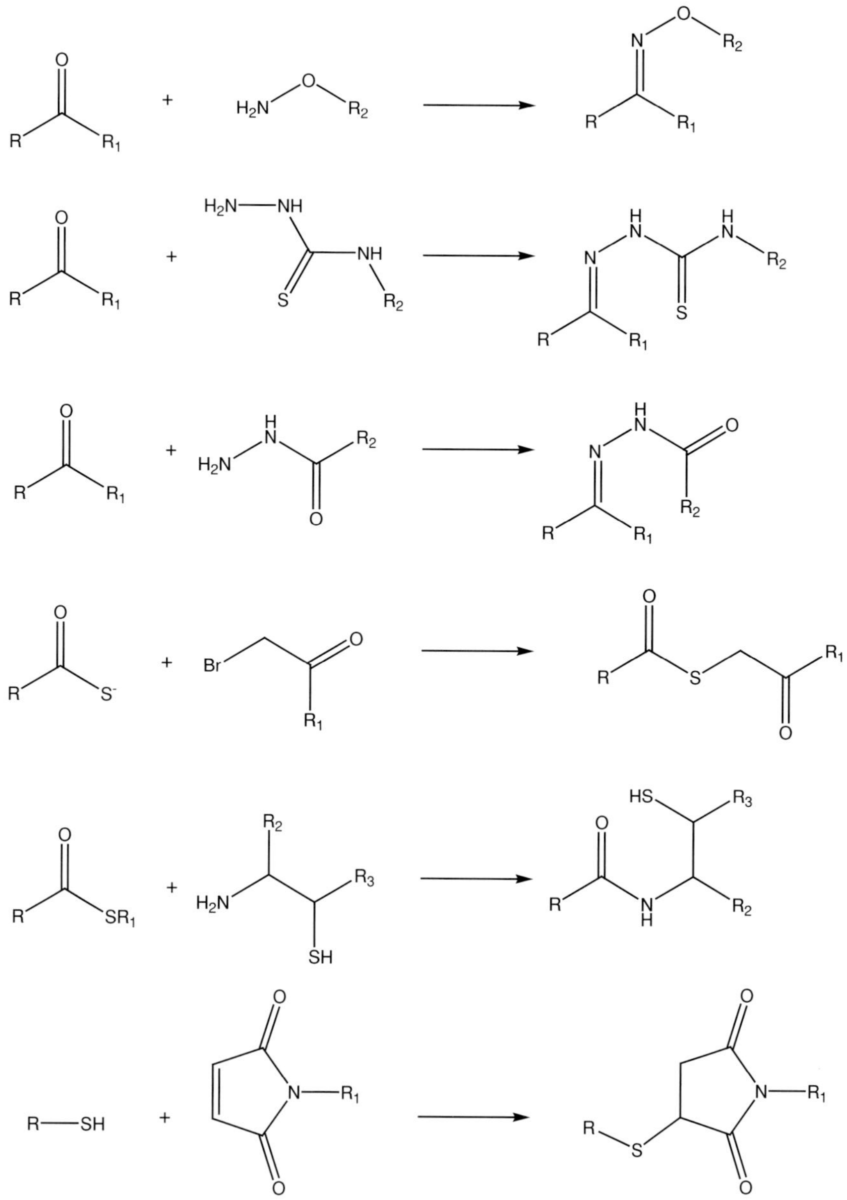

FIGURE 7.33. Chemical reactions useful for chemoselective immobilization of peptide derivatives.

of silicon wafers into mercapto-modified surfaces [308] allowing immobilization of maleimide-peptide conjugates. The formation of an oxime bond (Figure 7.29vi) using the chemoselective reaction between an amino-oxy-modified peptide derivative and an aldehyde surface has often been used in preparing peptide microarrays [143, 330, 395, 443, 484,

TABLE 7.3. Chemoselective reactions useful for the immobilization of peptides

educt surface function	educt (ligand) functional group of ligand allowing chemoselective reaction	product structure generated subsequent to ligation or immobilization reaction	reference
aldehyde/ketone	amino-oxy-alkyl	oxime	[1, 143, 330, 395, 443, 484, 486, 503]
aldehyde/ketone	β-amino-thiol	thiazolidine	[143, 325]
β-amino-thiol	aldehyde	thiazolidine	[189]
hydrazide	aldehyde	hydrazone	[167]
maleimide	mercapto	2-alkythio succinic imide	[603]
α-halocarbonyl	mercapto	thioether	[134]
mercapto	maleimide	2-alkythio succinic imide	[67]
thioester	β-amino-thiol	amide	[99, 310, 311, 335, 538, 558])
thiocarboxylate	α-halocarbonyl	thioester	[496]
benzoquinone	cyclopentadiene	Diels-Alder-Adduct	[607, Houseman, 2002]
Semicarbazide	aldehyde	semicarbazone	[361, 425]
salicylhydroxamic acid (SHA)	phenylboronic acid (PBA)	PBA*SHA complex	[530]

486, 503]. Alternatively, the reaction between aldehyde modified glass surfaces and peptides containing an N-terminal cysteine (Figure 7.29v) was also used successfully in peptide microarray preparation [143]. Melnyk and coworkers selected the reaction between semicarbazide glass slides and glyoxylyl peptide derivatives (Figure 7.30) to generate peptide microarrays [361]. It could be demonstrated that the native chemical ligation, introduced by Dawson et al. [99] is suited for effective attachment of peptides containing an N-terminal cysteine residue to thioester modified glass slides [310, 311, 558]. Finally, a special case of chemoselective immobilization of peptides is represented by the electrochemical copolymerization of pyrrole-modified peptides [200, 227, 328].

Recently, more sophisticated reactions have been introduced for oriented immobilization of peptide derivatives. A Diels-Alder reaction between benzoquinone groups on self-assembled monolayers and cyclopentadiene-peptide conjugates (Figure 7.34) led to

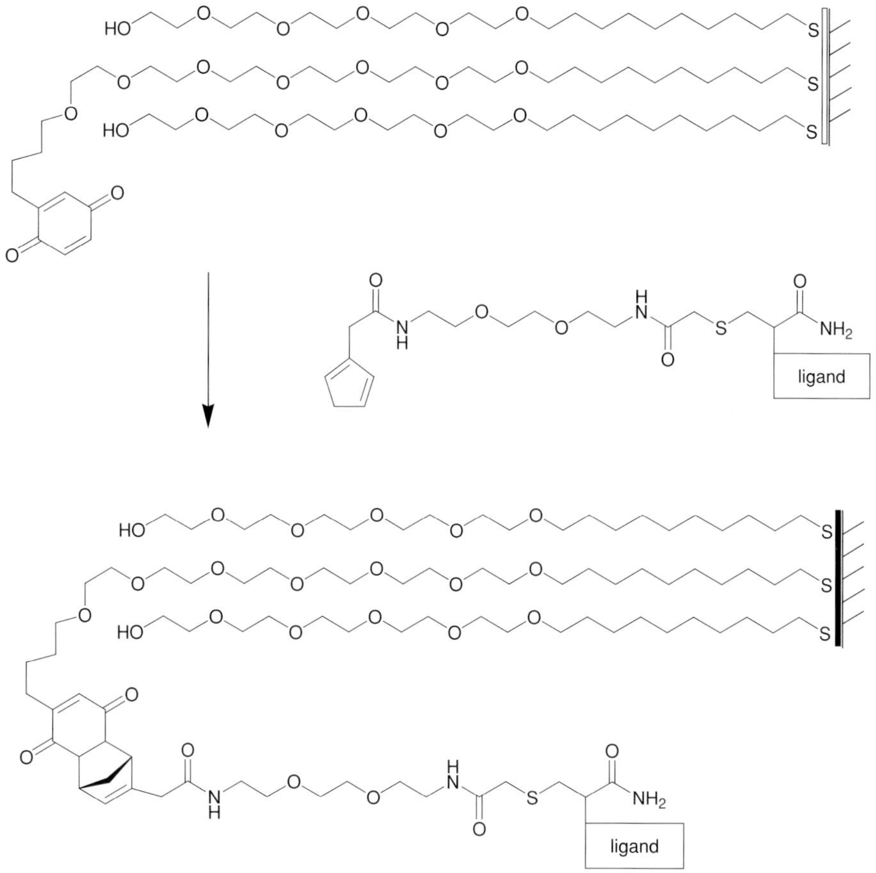

FIGURE 7.34. **Chemoselective immobilization of cyclopentadiene-peptide conjugates to surface bound benzoquinone functions [218]**. Diels-Alder reaction between self-assembled monolayers (prepared by reacting gold coated glass coverslips with a mixture of hydroquinone-oligo(ethylene glycol) and penta(ethylene glycol)-omega-mercaptoalkyl conjugates in methanol followed by oxidation to benzoquinones) and cyclopentadiene-peptide conjugates in aqueous solution yielding chemoselectively immobilized ligands (peptides).

efficient covalent attachment of peptides to the surface at a density of 100 fmol/mm^2 [218]. The formation of a 1:1 complex between surface-bound salicylhydroxamic acids and phenylboronic acid-conjugates was also used for chemoselective attachment of polypeptides to surfaces [530]. Moreover, selective cyclo-addition between biomolecules modified with an azido group and alkyne functions on the surface, resulting in the formation of a substituted 1H-[1,2,3]triazole [144], should be useful for peptide immobilization. A special case of immobilization is the selective interaction between surface-bound DNA molecules and peptide nucleic acid tags (PNA-tags). Schultz and coworkers used this interaction to assemble spatially addressed PNA-tagged peptidic protease inhibitors on an oligonucleotide microarray [596, 597].

7.2.4. Techniques for Array Production with Pre-Synthesized Peptides

In general, the different technical solutions available for the production of peptide arrays starting from pre-synthesized peptides can be divided into two classes: contact and non-contact printing. During contact printing processes the sample is loaded onto the tip of a pin (*Contact Tip Deposition Printing* and *Pin-and-Ring Printing*), a pen of an atomic force microscope (dip-pen nanolithography (DPN), a small stamp (*Micro Contact Printing*, µCP), or into the needle of a microsyringe (*Spotting*) and subsequently deposited to the functionalized coherent surface by direct physical contact of the pin/pen/stamp/needle with the surface (Figures 7.35, 7.36). A special case of contact printing techniques is the so-called *Micro-Wet-Printing* (µWP) of the company Clondiag [137] useful for either immobilization of pre-synthesized peptides or the stepwise *in situ* synthesis of peptide arrays (Figure 7.36C).

In contrast, during non-contact printing procedures small droplets generated by piezoelectric or bubble-jet devices or by a microsyringe pump are sprayed with high speed onto a functionalized coherent surface without direct contact with the dispenser (Figure 7.37). This

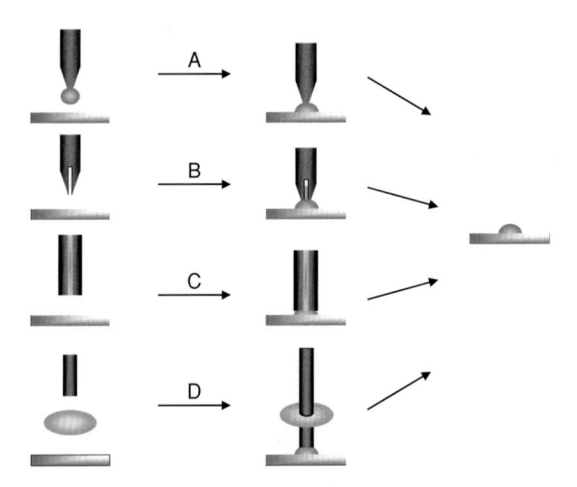

FIGURE 7.35. **Different principles of pin or capillary contact printing**. A = solid pin printing; B = split pin printing, C = capillary printing; D = pin-and-ring printing.

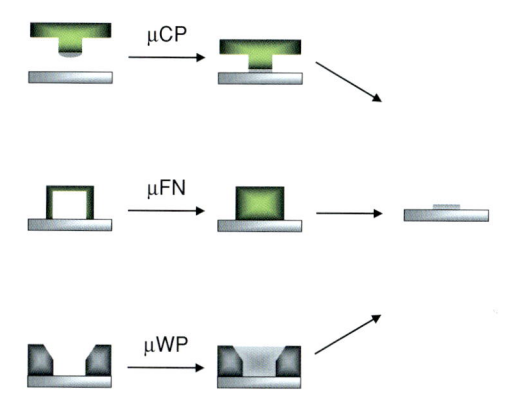

FIGURE 7.36. **Different principles of contact printing**. μCP = micro contact printing; μFN = micro fluidic networks; μWP = micro wet printing.

is a clear advantage because it avoids possible mechanical damage to either the dispenser or the sensitive surface.

7.2.4.1. Contact Printing The first step of *Contact Tip Deposition Printing* (Figure 7.35A) is dipping the tip of a solid pin into a sample solution. A defined amount of solution (depending on the material and on the shape and the dimensions of the pin) is adsorbed on the immersed tip. After moving to the appropriate printing position and establishing direct contact between the pin tip and the surface, a small spot of sample solution is generated on the surface (down to 50 μm spot size). If the same sample should be printed several times the pin has to be loaded again. However, so-called split pins were developed to circumvent

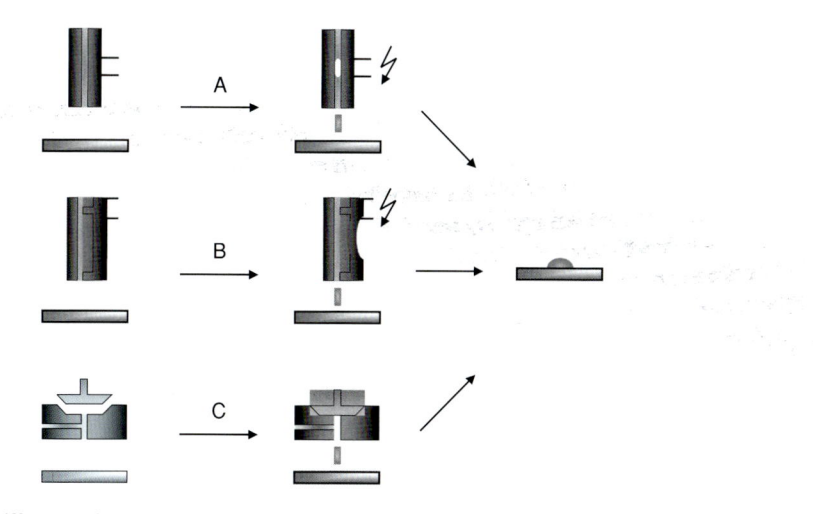

FIGURE 7.37. **Different principles of non-contact printing**. A = bubble jet: local heating generates a small bubble ejecting small droplets, B = piezoelectric device: compression of piezoelectric device generates a shock pulse in the fluidic chamber yielding a droplet, C = top-spot-printing: a microstamper ejects droplets simultaneously from array-like arranged micro-nozzles.

this situation (Figure 7.35B). The small slot of the split pin represents a reservoir that is loaded with sample solution by capillary forces. Contact of the split pin with the surface deposits a defined part of the sample solution. After moving to another position of the surface an additional defined amount of sample solution can be deposited without reloading the pin.

Different devices for the production of arrays are on the market using up to 256 pins in parallel (Biorobotics multiple printhead) for simultaneous loading with sample solutions provided in 1536-well microtitre plates and sample deposition on a number of slides. Alternatively, capillaries either alone or connected to microsyringe pumps [508] can be used for the deposition of small sample droplets onto functionalized surfaces (Figure 7.35C). A unique technology is so-called *Pin-and-Ring Printing* (Figure 7.35D; [479]) making use of a small ring filled with spotting solution and a solid pin going through the sample solution, thereby transporting a fraction of the sample to the surface in a multiple mode.

Microscale placement of samples on (even curved) surfaces with submicron precision is possible using inked stamps of structured elastomers such as polyurethans, polyimides and poly(dimethylsiloxan) [36, 105, 229, 240, 280, 285, 314, 315, 319, 437, 600, 613]. This μCP technology (Figure 7.36A) has been reviewed extensively [600, 601]. The μCP technique has widespread application for printing of alkanethiols onto metal surfaces to form micro-patterned self-assembling monolayers. Positive μCP using pentaerythritol-tetrakis(3-mercaptopropionate) as an ink for wetting PDMS stamps was described recently [106]. Closely related to μCP is the recently introduced nanotransfer printing (nTP) technique, enabling the generation of complex patterns on functional materials with nanometer resolution in one step [333]. Alternatively, the μFN method is based on conformal contact between the functionalized surface and a micro-machined PDMS substrate (Figure 7.36B) forming a network of micro-channels, which can be filled with sample solution by capillary forces. The geometry of the μFN and the limited amount of sample molecules flowing inside the channels lead to a gradient of immobilized compounds.

Micro-structured masks are used for the synthesis of oligonucleotides by the μWP method (Figure 7.36C; www.clondiag.com). This mask is positioned with extremely high precision on the functionalized surface. The mask is connected to a channel system allowing either the deposition of pre-synthesized biomolecules or stepwise synthesis directly on the surfaces by application of appropriate activated building blocks. Squares of 1 μm size are theoretically possible with this technology, allowing the immobilization (or synthesis) of up to one million biomolecules per cm^2.

A very special case of contact printing technologies is dip-pen nanolithography, making use of an inked atomic force microscope pen [6, 214, 302, 413]. Deposition of spots with diameters of around 50 nm onto gold or silicon oxide surfaces was demonstrated using this technology, resulting in a final density of more than 100,000 samples in an area of 100 μm by 100 μm [110].

7.2.4.2. Non-Contact Printing Several ink-jet modes of droplet generation have been used for the controlled delivery of small volumes of liquid to surfaces such as piezoelectric capillaries, piezoelectric cavities, nozzles with a thermal pressure transducer [14] and the nozzleless acoustic jet [541]. Of particular interest from the biochemical application standpoint is the acoustic jet method for droplet generation [132, 536] since the opportunity to eject extremely small droplets without possibly clogging a nozzle is unique. However, acoustic jets enabling to handle many different solutions in parallel have yet to be developed.

The prototype of a piezoelectric device is a glass capillary surrounded by a cylindrical piezoelectric device [3, 66, 509]. A control circuit sends a short pulse (typically 5 μs at 100 V) to the piezoelectric element, which causes compression of the piezoelectric device, generating shock pulses in the fluid chamber and forcing droplets out of the nozzle (Figure 7.37B). The size of the droplet depends upon the diameter of the nozzle, the magnitude of the pulse and the physical properties of the peptide solution. Micro-machined piezo or thermal jets [15, 251, 296, 302, 404] generate precisely sized droplets due to the highly defined orifice geometry, but unlike piezoelectric capillary jets these devices cannot be loaded directly from a microtitre plate.

One disadvantage of the piezoelectric ink-jets is the relatively large amount of peptide solution needed to generate reproducible droplets (typically 1–4 microliters). Therefore, these jets are used if one wants to deposit a large number of spots of the same peptide solution. Nevertheless, there are special piezoelectric micropipettes with extremely small dead volume available (GeSIMmbH, Großerkmannsdorf, Germany; [148]).

A major advantage of the ink-jet approach is the speed of printing. In the case of contact printing using pin tools the required close proximity of the reagent delivery tip and the surface makes a vertical motion necessary. During this vertical motion, touching the surface to deposit a droplet of the peptide solution, horizontal motion is stopped. Since motion must be stopped for each contact event, the time required to generate a microarray batch increases linearly with the number of peptide arrays in the batch. In contrast, with non-contact printing in the "on the fly" mode (no stop of horizontal motion necessary for delivering aliquots of peptide solution) total printing time increases much less rapidly with batch size.

Special cases of non-contact printing are the parallel direct displacement of liquids using an elastomer stamp [101] and the highly parallel top-spot-printing technology (Figure 7.37C; [124]) developed by IMTEK (University of Freiburg, Germany) and the HSG-IMIT Institute (Villingen-Schwenningen, Germany; www.hsg-imit.de). A pneumatic pulse to a specially designed mini-microtitre plate ejects up to 96 different sample droplets simultaneously onto a surface within one second. This high speed and the parallel mode makes top-spot-printing perfectly suited for applications that require an extremely high number of similar microarrays.

7.3. LIBRARY TYPES

A great variety of different library types has evolved, from the origin of multiple peptide synthesis through to the huge peptide collections generated and applied nowadays either in solution or immobilized as peptide arrays. In general, the design principles can be classified into protein sequence-based approaches dissecting or modifying the primary structure of a protein or peptide and *de novo* approaches exploring the entire or at least a significant and sometimes biased part of the potential sequence space. Only in exceptional cases are aspects of sequence-based and *de novo* strategies combined within one library (see 7.3.1.2. and 7.3.1.8.). So far, most investigations applying peptide arrays (Table 7.4) are based on the SPOT technology since it was the first generally amenable method suitable for even non-specialized laboratories. However, although many examples refer to publications using this technology, all library types described here can also be prepared by other newly developed peptide array production technologies.

7.3.1. Protein Sequence-Derived Libraries

Protein sequence-derived libraries provide the basic tools to elucidate interactions between a protein and a ligand, such as other proteins (enzymes), DNA, metal ions, cofactors and lipids. This provides detailed information about the ligand binding site. In addition, peptides derived from the binding site are often valuable starting compounds for peptide inhibitor or substrate development.

7.3.1.1. Scans of Overlapping Peptides The standard library to identify a protein's ligand binding site is a scan of overlapping peptides, also called a peptide scan or simply pepscan (Figure 7.38A; [176, 178]). The entire protein sequence, or a certain part of it corresponding to a particular domain perhaps, is synthesized as short, overlapping, linear peptides that are subsequently tested for ligand binding. Usually, this involves 6- to 15-mer peptides since most linear binding sites (see 7.5.1.1.) do not exceed this range ([113, 415, 458, 562]). Furthermore, longer peptides result in raw products with rather limited purities, which cannot be used for array preparation without expensive purification procedures. However, to identify discontinuous binding sites (see 7.5.1.2.) longer peptides are considered an advantage if the peptide covers a folding motif comparable to the native protein structure ("domain scan" http://www.pepscan.nl/html/outframeset.html). In addition to the peptide length, another important parameter of peptide scans is the number of overlapping amino acids between two consecutive peptides. Usually, the peptides are shifted by one to three positions along the protein sequence. With shorter overlaps important peptides may be overlooked. In peptide scans derived from proteins containing disulfide bonds or free cysteine residues, these residues are commonly exchanged by similar amino acids, such as serine, to avoid dimerization and oligomerization of the peptides or covalent linkage to thiols in the ligands [417].

7.3.1.2. Hybritope Scans The mapping of discontinuous (conformational) binding sites (see 7.5.1.2.) necessitates analyzing peptide-ligand interactions with very low affinities. This led to the introduction of the hybritope and duotope scan (see 7.3.1.3.). Discontinuous epitopes are composed of two or more binding regions separated in the primary structure. Upon folding they are brought together on the protein surface to form a composite epitope. In the hybritope scan (Figure 7.38B) the peptides of a peptide scan (see 7.3.1.1.) are N-, C- or N- and C-terminally flanked by randomized positions ([445]; for synthesis procedures of randomized positions refer to [269]). The rationale is that a juxtaposed binding region within the discontinuous binding site can be mimicked by the randomized positions or a subset of peptides with the appropriate amino acid composition in the randomized sequence. Deconvolution libraries (see 7.3.2.1.) must then be used to identify single peptides from these peptide mixtures.

7.3.1.3. Duotope Scans and Matrix Scans Flanking randomized positions of the hybritope scan may either contribute via sequence motifs homologous to binding regions

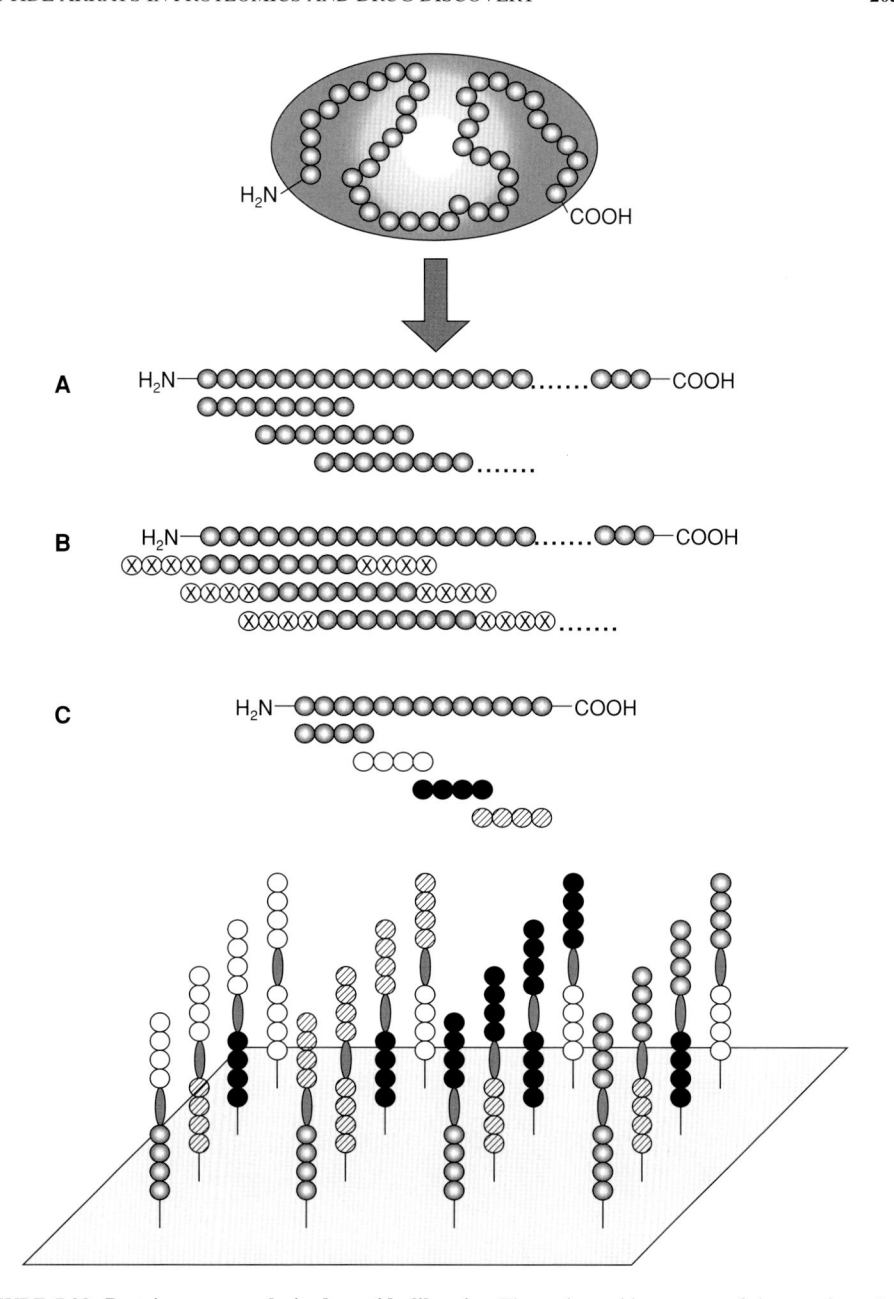

FIGURE 7.38. **Protein sequence-derived peptide libraries**. The amino acid sequence of the protein under investigation is used to generate short linear overlapping peptides. (A) Scan of overlapping peptides (peptide scan) (see 7.3.1.1.), (B) hybritope scan (see 7.3.1.2.), and (C) duotope scan or matrix scan (see 7.3.1.3.).

of the discontinuous binding site, mimicking sequences according to the "mimotope concept" [177, 363] or detract via adverse effects, e.g., by unfavorable charged amino acid side chains. The duotope scan (Figure 7.38C) was therefore introduced to provide a rational tool to identify peptide mimics for discontinuous binding sites. The concept is that these binding sites can only be mimicked adequately if two or more binding regions are connected in one molecule by a linker moiety resembling their spacing in the protein's three-dimensional structure. This means synthesizing all possible combinations of two overlapping peptides from a conventional peptide scan as one linear peptide for each combination, i.e., combinatorial chemistry with peptides as second level building blocks [451, 452]. This concept was validated by mapping the discontinuous epitope of the anti-hen-egg white lysozyme (HEL) antibody D1.3. A complete lysozyme duotope scan comprising all combinations of HEL-derived 10-mer overlapping peptides (offset by thee amino acids) combined via two β-alanine residues as a spacer moiety, synthesized by the SPOT method, resulted in an array of 41 by 41 (1681) 22-mer peptides, a number recommending array-based techniques for cost-effective experiments. Probing the array with mab D1.3 revealed a duotope peptide composed of two binding regions that exactly matched the structural epitope known from X-ray crystallography of the HEL-Fab D1.3 complex [40, 41]. The dissociation constant of the duotope scan peptide in a complex with mab D1.3 was determined by ELISA as 27 μM, a value significantly higher than that of the native HEL-D1.3 complex due to complete loss of the conformational stability conferred by the protein fold. Nevertheless, the duotope peptide clearly has a higher affinity compared to peptides spanning the single binding regions [451, 452].

If information about the binding region is available, e.g., from site-directed mutagenesis studies, one can carry out a partial duotope scan covering only the protein-derived sequences of interest. This strategy was applied to identify a mimic for a discontinuous epitope recognized by a neutralizing anti-interleukin-10 (IL-10) antibody [448, 451, 452]. Disadvantages of the duotope scan include having to assess a relatively large number of long peptides and possibly identifying false positives since hydrophobic residues that are often buried in the three-dimensional structure of the protein become exposed by dissecting the protein into duotope peptides and can cause unspecific interactions.

A concept analogous to the duotope approach called "matrix scan" has been developed (http://www.pepscan.nl/html/outframeset.html; [364]). However, experimental details were not available at the time of preparing this manuscript.

7.3.1.4. Amino Acid Substitution Scans The interaction of a peptide with a binding partner usually relies upon a limited number of amino acid residues that are effectively in contact with the binding partner. These amino acids contribute either to the binding free energy or to the specificity of the interaction and are referred to as key residues. If the peptide has to adopt a certain conformation upon or prior to binding, amino acids facilitating these conformations can also be regarded as critical. The concept of "alanine scanning" introduced to map protein-protein interactions by site-directed mutagenesis has been used to identify these residues (Figure 7.39A; [95]). Here, residues that cannot be exchanged by alanine without loss of binding and/or biological activity are regarded as key residues for the interaction. Scans with other amino acids are similarly used for alanine residues in the starting peptide itself or to explore the effect of charged residues, for example. It has to be considered that this only reveals effects that depend on the amino acid side

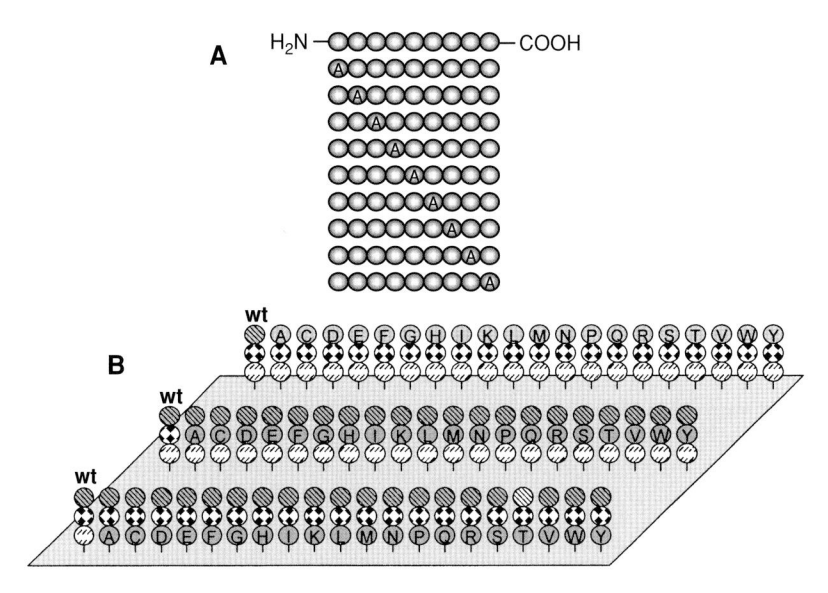

FIGURE 7.39. **Libraries of substitution analogs**. These peptide arrays of substitution analogs are used to identify the key residues required for the interaction with a binding partner or a certain biological activity. (A) Amino acid substitution scan (alanine scan) of a 9-mer peptide (see 7.3.1.4) and (B) complete substitutional analysis of a 3-mer peptide (see 7.3.1.5).

chains unless one incorporates building blocks that lead to a modified backbone. Among the naturally occurring amino acids proline plays a special role in amino acids substitution scans (synonym: replacement scan) and is therefore often used as a substitute. It can influence the pre-binding conformation by inducing a turn structure or preventing helical structures, providing indirect information about binding modes.

7.3.1.5. Substitutional Analyses If, for example, the amino acid substitution scanning approach employs all genetically encoded amino acids it is called (complete) substitutional analysis (synonyms: mutational analysis as referred to in some former publications, replacement analysis, analoguing). These experiments explore the effects of all possible single site substitutions of the starting peptide (Figure 7.39B; [415]; Frank and Overwin, 1996; [273, 447]). This identifies key residues that are those that cannot be substituted at all or only by physicochemically similar amino acids (e.g., leucine/isoleucine). Usually, other positions are not sensitive to substitutions and some may even lead to increased binding activity. Complete substitutional analyses are a rapid and effective way to delineate the structure-activity relationship of peptides and to simultaneously optimize the starting sequences with respect to the activity (binding, enzyme substrate properties, etc.) being screened for in the assay. In addition to substitutional analyses using the genetically encoded amino acids, D-amino acids [274], other unnatural amino acids [172, 265], or peptoidic building blocks [16, 456] are used to increase the diversity of side chain functionalities or backbone modifications. This often results in identifying substitution analogs that are stabilized against proteolytic degradation.

```
       A              B              C              D
  OOOOOOOOOO     OOOOOOOOOO     OOOOOOOOOO     OOOOOOOOOO
   OOOOOOOOO     -OOOOOOOOO     --OOOOOOOO     AOOOOOOOOO
    OOOOOOOO     O-OOOOOOOO     -O-OOOOOOO     AAOOOOOOOO
     OOOOOOO     OO-OOOOOOO     -OO-OOOOOO     AAAOOOOOOO
    . . . .      OOO-OOOOOO     -OOO-OOOOO     . . . .
                 OOOO-OOOOO     -OOOO-OOOO
   OOOOOOOOO     OOOOO-OOOO     -OOOOO-OOO      OOOOOOOOO
    OOOOOOOO     OOOOOO-OOO     . . . .         OOOOOOOOA
     OOOOOOO     OOOOOOO-OO                     OOOOOOOAA
      OOOOOO     OOOOOOOO-O     O--OOOOOOO      OOOOOOAAA
                 OOOOOOOOO-     O-O-OOOOOO
    . . . .                     O-OO-OOOOO      . . . .
                 --OOOOOOOO     O-OOO-OOOO
   OOOOOOOO      O--OOOOOOO     O-OOOO-OOO     AOOOOOOOOA
    OOOOOO       OO--OOOOOO     O-OOOOO-OO     AAOOOOOOAA
     OOOO        OOO--OOOOO     . . . .        AAAOOOOAAA
    . . . .      . . . .                       . . . .
```

FIGURE 7.40. **Analysis and optimization of peptide length (see 7.3.1.6.).** (A) Truncation library with N-terminal, C-terminal and bi-directional stepwise truncations; (B) deletion library (one or more consecutive amino acids deleted at all possible positions), (C) combinatorial deletion library comprising all peptides with two or more positions omitted independently all over the starting sequence, and (D) progressive alanine substitution library.

7.3.1.6. Truncation, Deletion, and Combinatorial Deletion Libraries Biologically active peptides identified, for example using a peptide scan or by other types of peptide libraries, including chemical and biological approaches, often contain a well-defined core of key residues. In addition, these peptides include other dispensable positions resulting from the predefined peptide length used in the library design. In order to narrow down the peptide to the "active principle" or to minimize the molecular weight to facilitate peptide-based drug design, three different types of libraries are useful: (1) Truncation libraries (synonyms: size scan, window scan) comprise peptides omitting one or more N-, C- or N- and C-terminal amino acids (Figure 7.40A; Frank and Overwin, 1996). (2) Peptides from libraries of deletion analogs (Figure 7.40B) have one or more consecutive amino acids deleted at all possible positions. (3) Compared to deletion libraries, combinatorial deletion libraries additionally cover peptides with two or more positions omitted independently all over the sequence (Figure 7.40C). It should be noted that the number of peptide analogs covered by a combinatorial deletion library rapidly increases depending on the number of deleted positions and the peptide length.

As an alternative to truncation analyses, a few authors used progressive substitutions by alanine while retaining the overall peptide length (Figure 7.40D). Such library types were called progressive alanine substitution or progressive alanine fill-up libraries [75, 139, 140].

7.3.1.7. Cyclization Scans A widespread strategy to optimize the binding free energy of a peptide interacting with a binding partner is to stabilize the binding conformation. This is often achieved by cyclization, for example via disulfide bonds [191, 448]. However, the binding conformation is usually unknown since structure determination experiments by X-ray crystallography or NMR are time-consuming and laborious. Furthermore, docking of peptides to binding partners *in silico* is one of the most complex modeling problems due to the tremendous intrinsic flexibility. Therefore, a large number of cyclic peptide analogs have to be synthesized and screened to seek out the proper conformation of a biologically active peptide. A systematic approach is the "cyclization scan" comprising all

```
OOOOOOOOOO
CCOOOOOOOO
COCOOOOOOO
COOCOOOOOO
COOOCOOOOO
COOOOCOOOO
COOOOOCOOO
 . . . .

OCCOOOOOOO
OCOCOOOOOO
OCOOCOOOOO
OCOOOCOOOO
OCOOOOCOOO
OCOOOOOCOO
 . . . .
```

FIGURE 7.41. **Disulfide cyclization scan**. The library covers all possible combinations of two cysteine residues within the starting sequence that are subsequently oxidized for cyclization (see 7.3.1.7.).

possible combinations of two cysteine residues within the starting peptide (Figure 7.41). An example of this approach is a library of 466 cyclic peptide analogs used to identify the optimal disulfide cyclic derivative of a linear 32-mer mimicking a discontinuous interleukin-10 (IL-10) epitope [448]. The affinity was increased by a factor of 10 although one disulfide bond within a 32-mer peptide had only a limited impact on the overall conformational freedom [587]. Cyclization of peptides by disulfide bonds via cysteine residues is the most amenable strategy and often applied for peptide arrays especially for stepwise *in situ* peptide array production. However, several other chemical cyclization strategies can be similarly applied as shown for amide bonds [191] and the entire chemical repertoire for peptide or peptidomimetic cyclization can be used for array production technologies with pre-synthesized compounds.

7.3.1.8. Library Types: Miscellaneous The focus of this review article is the field of peptide arrays. Protein arrays are extensively described in other Chapters of this volume. However, arrays of protein domains inhabit the borderland between peptide and protein arrays since their sequences lengths are still accessible to chemical synthesis, branding them "peptide-like" from the technological point of view. On the other hand these domains usually retain stable folding, a functional feature claimed by protein arrays. Only two protein domain array publications, described in more detail in Section 7.5.2.2., should be mentioned here: (1) a complete L-amino acid substitutional analysis of the human YAP WW domain (44-mer) resulting in an array of more than 800 single site substitution variants, which was used to identify the key residues for stable domain folding or WW domain ligand interaction [546, 547] and (2) an array of 11859 tri-substituted variants of the human YAP WW domain used to identify WW domains with novel binding specificities [548]. The substitutions were introduced in a combinatorial manner at three different positions within the WW domain sequence, in other words a combination of protein sequence-derived and combinatorial library techniques.

Another specialized library type was used to identify binding partners and characterize the binding specificities of PDZ domains, which mainly occur in proteins of the cytoskeleton and play a role in signal transduction. They predominantly interact via their binding partners' C-termini where the carboxyl group is essential for binding. Hoffmüller et al. described a SPOT synthesis peptide array of all known C-terminal peptides derived from the human proteins listed so far in the SWISSPROT databank [211]. Since peptides produced by standard SPOT synthesis protocols are attached to the solid support at their C-terminus, a novel synthetic pathway had to be established to generate free C-termini.

The last specialized peptide library type described here was applied to elucidate evolutionary transition pathways between three completely unrelated peptides recognized by the anti-p24 (HIV-1) mab CB4-1 [212]. These different peptide ligands were identified previously using combinatorial and deconvolution libraries [272]. The question of whether the different CB4-1 peptide ligands can be reciprocally converted into each other was answered by synthesizing and analyzing all possible (7,620,480) single step transition pathways (i.e. sequential conversion of one amino acid after the other) between the three ligands. The library comprising all 2560 possible transition peptides was designed with the software PepTrans. Complete L-amino acid substitutional analyses of all intermediates from the best transition pathways were performed in order to better understand the structural mechanisms involved in the sequence transformation. In this study the exceptional synthesis capacity of the SPOT method was exploited to analyze the sequence space between functionally related peptides with no sequence similarity.

7.3.2. De Novo Approaches

If a natural protein binding partner is not known, or if peptide ligands have to be identified without any previous knowledge, for example due to an intellectual property situation, one has to use combinatorial libraries with peptide mixtures or randomly generated libraries of single individual sequences. These strategies are summarized in this Section under "*de novo*" approaches. Although this review is restricted to approaches using peptide arrays and peptide chips it should be mentioned that most of the library types described in this Section were pioneered at the beginning of the combinatorial chemistry era and in the field of chemical libraries on beads [166, 215, 247, 286] or by using biological display techniques such as phage display [56, 118, 504]. Michal Lebl [299] has published a very lively historical review, with personal comments by the authors, of "classical" papers form the beginning of combinatorial chemistry.

The main problem for *de novo* identification of peptides is how to handle the immense number of potential peptide sequences, referred to as "combinatorial explosion". Even if only using the genetically encoded amino acids the number of possible sequences dramatically increases with the peptide length:

- dimers $20^2 = 400$
- trimers $20^3 = 8,000$
- tetramers $20^4 = 160,000$
- pentamers $20^5 = 3,200,000$
- hexamers $20^6 = 64,000,000$
- heptamers $20^7 = 1,280,000,000$
- octamers $20^8 = 25,600,000,000$

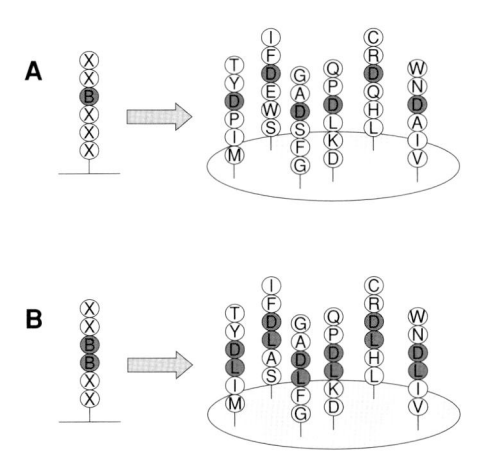

FIGURE 7.42. **Peptide mixtures**. Peptide mixtures with defined positions (B) and randomized position (X) (see 7.3.2.).

This Section describes two principles for the *de novo* identification of peptides with a predefined biological activity (mostly binding to another protein): (1) Using combinatorial libraries the aim is to completely cover the potential sequence space (see 7.3.2.1.). Although peptide arrays can be prepared with a high spot density (>40,000/cm^2; [151]) there is no technology yet available to synthesize and handle billions of different compounds individually. The solution is to synthesize peptide mixtures with degenerated or randomized positions by statistically incorporating amino acids of a certain set (Figure 7.42). Defined amino acids are only used at a limited number of positions. This results in a manageable number of peptide pools screened on the peptide arrays. The randomized positions of active pools must then be deconvoluted iteratively using deconvolution libraries individually designed for the project, ultimately selecting the active compounds. (2) Since deconvolution is a time-consuming process, arrays of randomly generated peptides (see 7.3.2.2.) have also been applied. Since such libraries only cover a small percentage of the potential sequence space, initially selected peptides often have low affinities to the binding partner and must subsequently be optimized, for example using substitutional analyses.

7.3.2.1. Combinatorial Libraries The five most critical parameters for identifying peptide ligands from combinatorial library arrays are: (1) the number of peptide mixtures tested, (2) the number of defined positions, (3) the ratio between defined and randomized positions, (4) the appropriate spacing of the defined positions within the entire sequence length, and (5) the overall length of the peptides. These parameters determine the ratio between active and inactive compounds in the peptide mixtures and consequently the signal to noise ratio and likelihood of identifying bioactive peptides.

The development of multiple peptide synthesis robots and devices for peptide array preparation have enabled a continuous increase in peptide library complexity, which can be extrapolated to the future based on novel developments in the field of high density microarray production. The first arrays to be developed were hexamer libraries with two defined positions (mostly abbreviated in the literature as "O" or "B") and four randomized, or mixed positions (X): e.g., XXO$_1$O$_2$XX [269]. Initially, these libraries were used to map

A

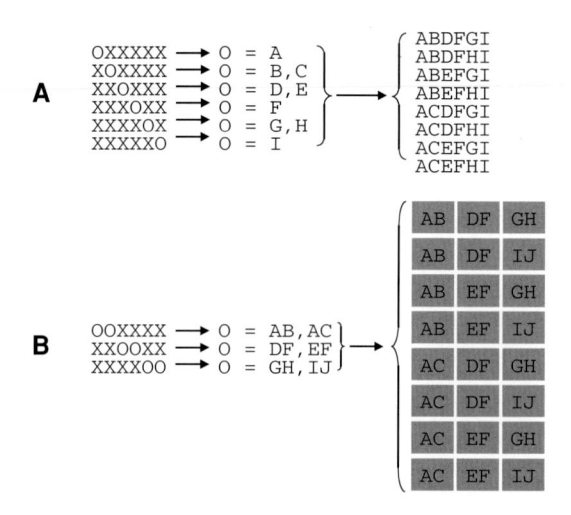

```
OXXXXX  →  O = A
XOXXXX  →  O = B,C
XXOXXX  →  O = D,E
XXXOXX  →  O = F
XXXXOX  →  O = G,H
XXXXXO  →  O = I
```
→
```
ABDFGI
ABDFHI
ABEFGI
ABEFHI
ACDFGI
ACDFHI
ACEFGI
ACEFHI
```

B

```
OOXXXX  →  O = AB,AC
XXOOXX  →  O = DF,EF
XXXXOO  →  O = GH,IJ
```
→
```
AB DF GH
AB DF IJ
AB EF GH
AB EF IJ
AC DF GH
AC DF IJ
AC EF GH
AC EF IJ
```

FIGURE 7.43. **Deconvolution of active peptide mixtures (see 7.3.2.1).** (A) In the positional scanning approach the most active amino acids at each position are identified from the initial library with peptide mixtures (X = randomized position; O = defined position with an individual amino acid). The deconvolution library consists of individual peptides representing all possible combinations of the most active amino acids. (B) In the dual positional scanning approach two positions are defined interdependently in the starting library.

linear antibody epitopes, but several other applications also emerged (see Section 7.5.; Table 7.4).

The randomized positions have to be deconvoluted to obtain single active peptides. Two general procedures have been described: (1) In the positional scanning approach (Figure 7.43A) the entire library is subdivided into a small number of peptide mixtures that have single amino acids at certain positions: O_1XXXXX, XO$_2$XXXX, XXO$_3$XXX, XXXO$_4$XX, XXXXO$_5$X and XXXXXO$_6$ (O and X as defined above). If the 20 naturally encoded amino acids are used for the defined positions (O) this library comprises $6 \times 20 = 120$ separate mixtures that are screened for binding, e.g., to an antibody [414]. Subsequently, individual peptides representing all possible combinations of the most active amino acids at each position are synthesized and screened. Alternatively, two (dual positional scanning approach) ([416]; Frank and Overwin, 1996; Figure 7.43B) or even more positions are defined in the first library. Although two defined positions involve greater synthesis efforts ($20^2 = 400$ peptide mixtures) the chance of successful primary screening is significantly better due to interactions with higher affinity and specificity. All randomized positions have to be deconvoluted in a second step based on the results with the starting library. Whereas the initial library is not predefined for a given screening molecule and can be applied universally, the follow-up libraries are tailor-made for specific purposes. The positional scanning approach assumes that the contributions of preferred amino acids at each position are additive or at least not interfering. However, this cannot be taken for granted in every system. (2) In order to circumvent this limitation, the randomized positions can be deconvoluted by an iterative process (Figure 7.44). Here, each deconvolution library is designed based on screening results from the starting or precursor library [269, 270]. Finally, a re-evaluation is recommended since there might be other amino acids at positions

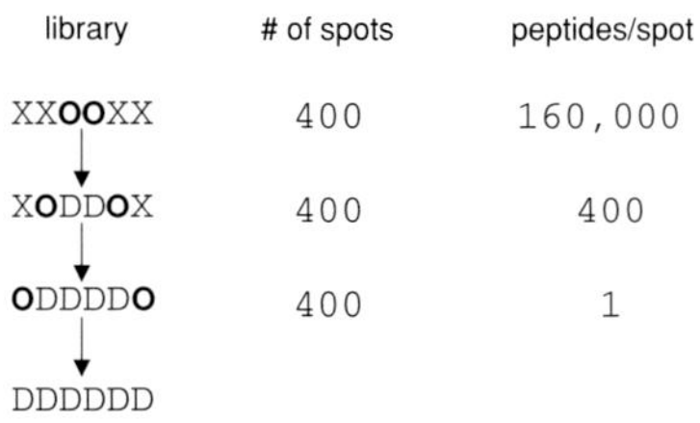

library	# of spots	peptides/spot
XXOOXX	400	160,000
↓		
XODDOX	400	400
↓		
ODDDDO	400	1
↓		
DDDDDD		

FIGURE 7.44. **Iterative deconvolution process of active peptide mixtures**. A starting hexamer library of the type XXOOXX (X = randomized position; O = defined position) is screened and the best dipeptide combination OO is selected for the first deconvolution library (XODDOX; D = defined position identified from the preceding library). Subsequently, the second deconvolution library ODDDDO is based on the best tetrapeptide motif ODDO from the preceding library and leads to a single peptide (see 7.3.2.1.).

defined early in the process that have a more positive effect on those defined later in the deconvolution.

A dramatic increase in the effectiveness of peptide and peptide mixture multiple automated syntheses paved the way for more complex libraries of the type XXXXO$_1$O$_2$O$_3$XXXX (8000 peptides mixtures) [496]. For example, the most complex library prepared by the SPOT technique described so far is of the type XXXX[3O3X]XXXX. The internal core [3O3X] is an abbreviation for three defined and three randomized positions arranged in all possible combinations XXXX[O$_1$O$_2$O$_3$XXX]XXXX; XXXX[O$_1$O$_2$XO$_3$XX]XXXX and so on [272]. This library comprised 68,000 spots and was used to identify not only antibody epitopes but also other peptides that bind to the antibody's paratope in a completely different way, referred to as mimotopes [177]. This was shown by structure determination of the peptide/antibody complexes by X-ray crystallography [255]. In many cases such complex libraries are essential for identifying peptide ligands that may require a certain number of key residues in a distinct pattern.

An alternative way to reduce the number of peptide mixtures that have to be prepared, yet match as many defined positions as practicable uses so-called combinatorial clustered amino acid peptide libraries [45, 271]. Each cluster contains physicochemically similar amino acids. The rationale of this approach is based on the assumption that physicochemically related amino acids contribute similarly to binding. For instance, grouping the amino acids into six clusters would lower the number of peptide mixtures in a combinatorial library containing four non-random positions from 20^4 (160,000, with four defined positions) to 6^4 (1,296, with four cluster positions). Kramer et al. described the epitope mapping of anti-transforming growth factor α (TGFα) mab Tab2 using a library of the type XC$_1$C$_2$C$_3$C$_4$X (C = one of six amino acid clusters [APG], [DE], [HKR], [NQST], [FYW], [ILVM]) in comparison to phage display techniques [271]. The peptide library array identified several motifs unrelated to the known TGFα-derived linear epitope sequence, whereas the phage display technique only revealed peptide ligands closely related to the wild-type epitope.

Several other combinatorial library techniques either as combinations or modifications of the principles described above or with unrelated design strategies were introduced for non-array technologies, but are similarly applicable for peptide arrays. A very interesting technique worth mentioning here is the so-called orthogonal library concept [111, 421]. The principle is that the same compound is represented in two different mixtures. Comparative activities of different mixtures observed after screening enables identification the compound responsible for activity.

In contrast to biological display techniques or *in vitro* translation systems, chemically prepared peptide arrays offer the opportunity to incorporate unnatural building blocks, e.g., D-amino acids, peptoidic building blocks, 1,3,5-trisubstituted hydantoins, 1,3,5-trisubstituted triazines, etc. A more detailed description of peptidomimetic classes applied so far is given in Section 7.5.5.

7.3.2.2. Random Libraries An alternative to protein sequence-derived or combinatorial peptide array libraries is to use sets of randomly generated peptide sequences. Recently, a peptide array approach was described using a library of 5520 randomly generated individual 15-mer peptides sequences prepared by SPOT synthesis and incorporating all genetically encoded amino acids except cysteine [458]. Of course, this only covers an extremely small fraction of the potential sequence repertoire. However, the peptide library array was successfully used to identify specifically binding peptide epitopes and mimotopes of three different antibodies (anti-IL-10 mab CB/RS/13, anti-TGFα mab Tab2, anti-p24 (HIV-1) mab CB4-1). Initially identified peptide ligands mostly had very low affinities for the antibodies with dissociation constants around 10^{-4} M. However, subsequent substitutional analyses revealed several analogs with dissociation constants in the low micromolar and high nanomolar range in a one step process. In two other studies 4450 randomly generated 12-mer peptides prepared on 10 "mini-Pepscan cards" (455 peptides per card) as well as a tripeptide library comprising the genetically encoded amino acids in all possible combinations were used to identify peptides binding to monoclonal antibodies against protein-S of transmissible gastroenteritis virus (TGEV) (mab 6A.A6 and 57.9), an EGF-like domain of the surface protein pfs25 of *P. falciparum* (mab 32F81), and the FLAG-tag (mab M2). Several peptides were identified as either homologous to the wild-type epitope sequence [516] or completely unrelated mimotopes [517]. Later, this approach was theoretically discussed and an algorithm to extract the amino acids required for binding was described [561].

In addition to randomly generated peptide library arrays, this approach was also used for peptidomimetics. Heine et al. described an 8000 membered hexapeptoid and hexapeptomer array used to identify peptidomimetics that bind the anti-TGFα mab Tab2 [202]. The best compound had a dissociation constant of 2.7 μM. The same antibody was used to probe an array of 8000 1,3,5-trisubstituted triazines, with the best hit having a dissociation constant of approximately 400 μM [488].

7.4. ASSAYS FOR PEPTIDE ARRAYS

This Section describes the general aspects of assay systems to probe peptide arrays. It is divided into the Sections "Screening" (4.1.) and "Read-out" (4.2.) addressing either the molecular recognition event, or how one observes which peptide was bound and/or converted by an interaction partner or enzyme. Of course, the quality of an assay system depends on the proper combination of screening and read-out methods.

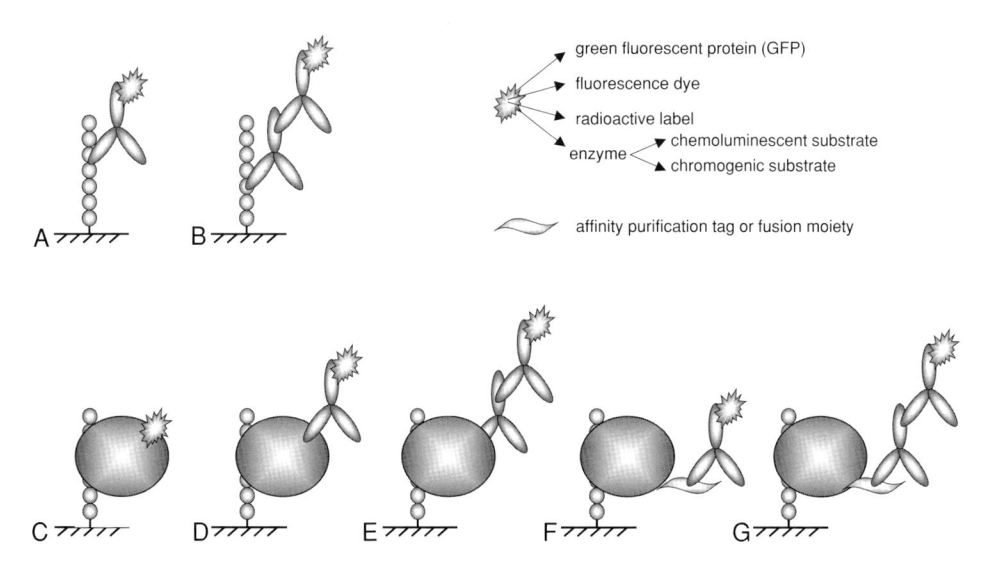

FIGURE 7.45. **Binding assays and detection methods to identify peptide-protein interactions (see 7.4.1.1.).** (A) Antibody epitope mapping with a directly labeled antibody; (B) antibody epitope mapping using a labeled secondary antibody; (C) detection with a directly labeled protein; (D) immunological detection of peptide-protein interactions using a directly labeled primary antibody; (E) immunological detection of peptide-protein interactions using a primary antibody in combination with a labeled secondary antibody; (F) immunological detection of peptide-protein interactions using a labeled antibody directed against an affinity purification tag or a fusion moiety; and (G) immunological detection of peptide-protein interactions using an antibody directed against an affinity purification tag or a fusion moiety in combination with a labeled secondary antibody.

7.4.1. Screening

Molecular recognition events on peptide arrays include either ligand binding (see 7.4.1.1.) or enzymatic conversion (see 7.4.1.2.). Chemical transformations of peptides on arrays (see 7.5.4.5.) or binding of intact cells (see 7.5.4.6.) are only described briefly.

7.4.1.1. Ligand Binding As shown in Table 7.4, peptide arrays are most frequently applied to study peptide binding by polyclonal or monoclonal antibodies. This can be achieved with directly labeled [593] primary antibodies (Figure 7.45A) or with labeled secondary antibodies (Figure 7.45B). Proteins in general can also be labeled directly (Figure 7.45C). Alternatively, they are detected using an antibody against the protein itself (Figure 7.45D) or recognition of a purification tag (poly-His-tag, Strep-tag, etc.) or a fused region, e.g., an Fc or GST fusion (Figure 7.45F). Similar to detection of peptide-antibody binding, proteins and fusion proteins can be recognized with labeled primary as well as secondary antibodies (Figures 7.45E, G). Generally, labeled protein A or G can be used as an alternative to labeled secondary antibodies [578]. For all immunological detection systems it is critical to rule out direct, nonspecific binding of the detection antibodies (or protein A/G) to the peptides. Incubation procedures as well as control experiments for peptide arrays prepared by SPOT synthesis have been described and are analogous in principal for all types of arrays

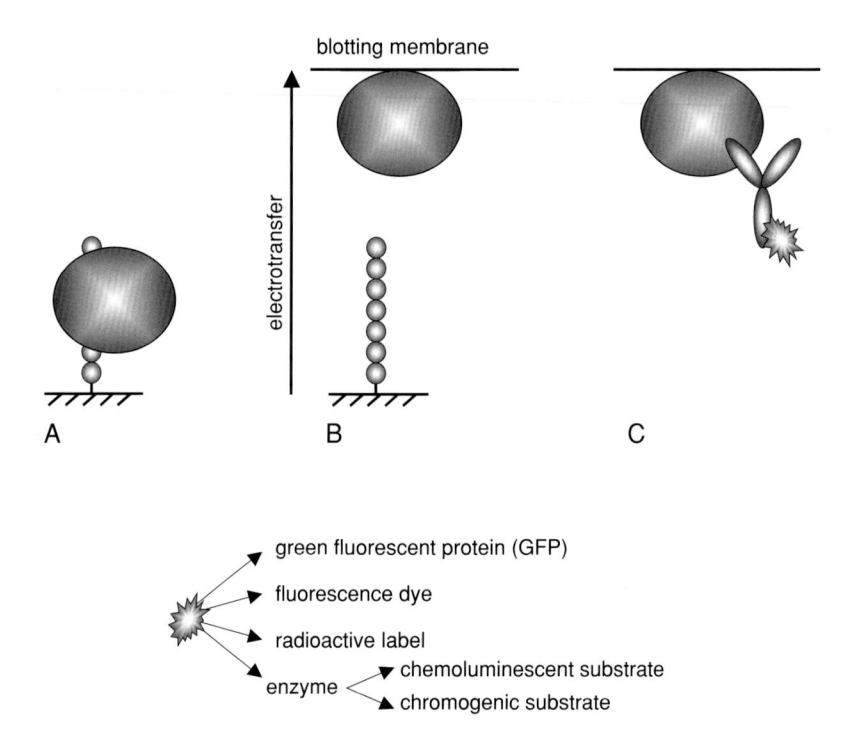

FIGURE 7.46. **Electrotransfer of peptide-bound proteins.** (A) Protein binding to the peptide array; (B) Protein electrotransfer onto a nitrocellulose or polyvinylene difluoride membrane; and (C) detection after immobilization on the blotting membrane (see 7.4.1.1.).

[453]. In addition, false-positive results can occur when using fusion proteins, and this must be checked by control incubations with the fusion moiety alone.

In most cases both screening and read-out are carried out directly on the peptide arrays. Alternatively, peptide-bound antibodies or proteins can be electrotransferred onto nitrocellulose or polyvinylene diflouride (PVDF) membranes (Figure 7.46; [453, 481]). This procedure, which is only applicable for peptide arrays on porous membranes, results in an exact mirror image of the ligands captured on the array. Subsequently, incubation with detection antibodies for example, and read-out by any of the methods described below are carried out on the blotting membrane. This procedure was extensively applied for probing peptide arrays with chaperones (see 7.5.2.3.; [481]) and is especially well suited for mapping protein homodimerization sites [205, 266]. Here, a peptide scan of the chosen protein is incubated with the protein itself. Detection with an antibody against this protein would usually lead to false-positives due to detection antibody interactions with protein sequence-derived peptides on the array (i.e., antibody epitope mapping), a problem avoided on the electrotransfer membrane. In addition, the electrotransfer procedure is extremely useful for detecting low-affinity peptide-protein interactions. This is due to the binding equilibrium shifting towards un-complexed detection antibody or protein during subsequent array processing. The advantage of the electrotransfer procedure is that all incubation and washing steps are carried out after transfer and immobilization of the primary incubation molecule. Detailed experimental protocols are described elsewhere [453].

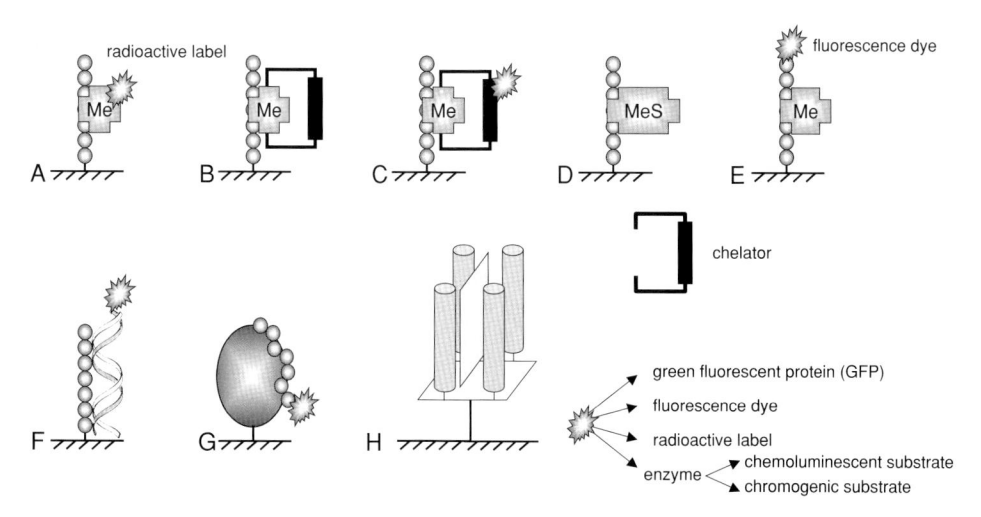

FIGURE 7.47. **Binding assays and detection of non-protein ligands binding to peptide arrays (see 7.4.1.1.).** (A) Binding of metal ions to peptide arrays is detected by radioactive isotopes; (B) detection via precipitating chromogenic chelators; (C) chelators that are labeled e.g., with peroxidase; (D) insoluble metal sulfides; (E) fluorescence dye labeled peptides. Other interactions identified on peptide arrays are (F) binding of labeled oligonucleotides; (G) binding of dye- or peroxidase-labeled peptides; and (H) binding of heme to arrays of synthetic four helix bundle mini-proteins.

Non-protein ligands used to probe peptide arrays include metal ions, DNA, peptides and protein cofactors. Screening with metal ions is carried out with radioactive isotopes detected by autoradiography (Figure 7.47A), employing precipitating chelators or insoluble metal sulfides that are quantified by densitometry (Figure 7.47B and D), with chelators that are coupled to enzymes (peroxidase) or other markers for detection (Figure 7.47C), or with peptide coupled fluorescence labels like dansyl that show an increased fluorescence intensity upon metal ion (Pb^{2+}, As^{3+}) binding (Figure 7.47E) [172, 265, 269, 270, 346, 494]. Oligonucleotide binding to peptides is detected using ^{32}P-labeled DNA or RNA (Figure 7.47F; [465, 466]). Peptide binding to protein domain arrays can be performed with peroxidase labeled [546, 547] or dye coupled peptides [548] (Figure 7.47G). Cofactor binding, for example heme binding to four-helix bundle protein arrays was carried out spectroscopically (Figure 7.47H; [441, 495]).

7.4.1.2. Enzymatic Conversion Measuring the activity of enzymes that modify peptides on arrays involves either enzymes adding something to the peptides (kinases, acetyl transferases, glycosyltransferases, etc.) or releasing a peptide part (proteases, phosphatases, etc.). In principle, all enzymes modifying peptides or proteins can be applied to screen peptide arrays. However, so far only assays for proteases, kinases, phosphatases, esterases and glycosyltransferases have been described, reflecting their respective importance in basic research and drug discovery. The different types of assays are described in detail below: kinases and phosphatases (7.5.3.1.; Figure 7.48) and proteases (7.5.3.2.; Figure 7.49). The binding specificity of proteins that convert or bind to peptides or proteins without covalent modifications, e.g., chaperones (see 7.5.2.3.) and isomerases (Table 7.4) is studied using binding assays as described above (see 7.4.1.1.).

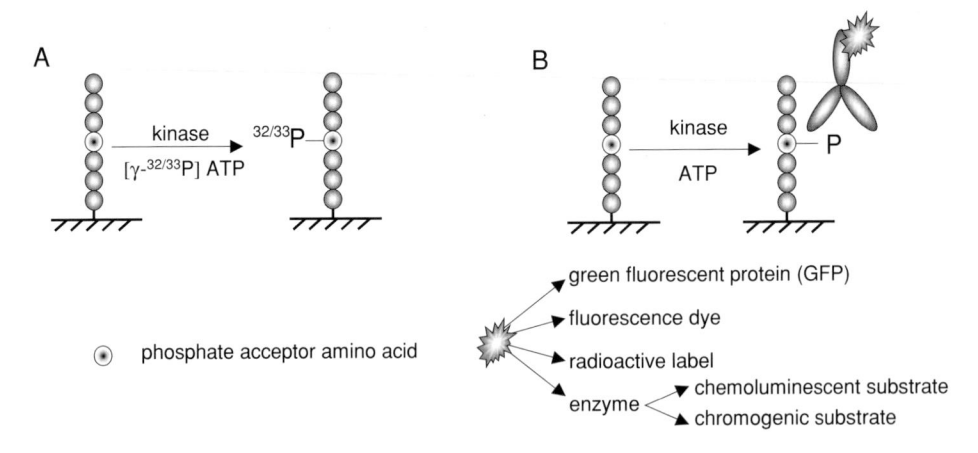

FIGURE 7.48. **Kinase assay on peptide arrays (see 7.5.3.1.).** (A) The array is incubated with the kinase of interest in the presence of [γ–^{32} or ^{33}P]ATP and detection is performed by autoradiography. (B) Alternatively, the phosphorylation is measured with a labeled anti-phospho-amino acid antibody.

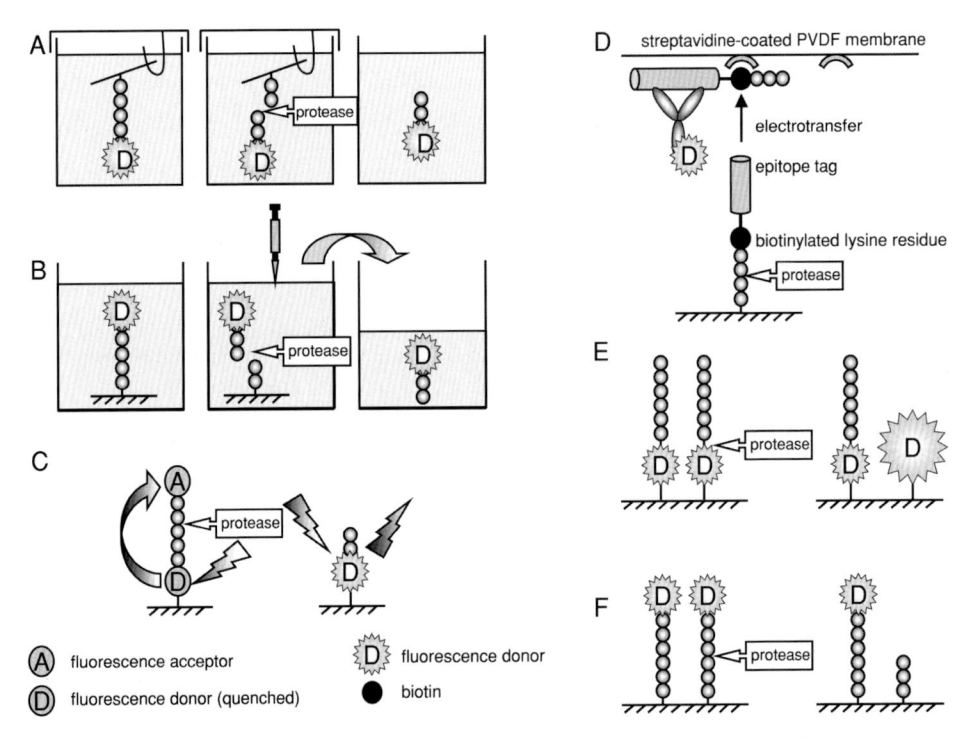

FIGURE 7.49. **Different types of protease assays (see 7.5.3.2.).** (A) Assay with N-terminally labeled immobilized peptides in the 96-well plate format [122]; (B) alternative assay with N-terminally labeled immobilized peptides in the 96-well plate format [254, 455]; (C) protease assay with internally quenched peptides on arrays [101, 254, 450]; (D) detection of peptide cleavage after electrotransfer of the released N-terminal peptide fragment [275]; (E) protease assay with peptides having a fluorogenic group C-terminal to the scissile bond [486, 616]; and (F) peptides with a fluorescence dye at the free terminus for array-based assays with decreasing signal upon cleavage.

7.4.2. Read-Out

Quantification of peptide-bound ligands or enzymatic peptide conversion employs chemoluminescent, fluorescent, radioactive, chromogenic, and label-free read-out methods such as surface plasmon resonance (SPR), mass spectrometry (MS), or atomic force microscopy (AFM). The following Sections describe these principles and discuss their advantages and drawbacks.

7.4.2.1. Chemoluminescence Detection of peptide-protein binding most often employs chemoluminescence read-out since a huge number of peroxidase-labeled monoclonal and polyclonal antibodies are commercially available. Very high sensitivity is achieved using a chemoluminescence substrate combined with either imaging systems or X-ray or photographic films. Luminol-based substrates can be mixed very cheaply or purchased from research reagent suppliers. In addition, ultra-high sensitivity substrates are commercially available. Fortunately, peroxidase itself shows almost no detectable binding to the peptides, and regeneration of peptide arrays, which is often done with SPOT synthesis-prepared arrays, is rather easy since no precipitates accumulate on the array support as occurs with chromogenic read-out methods (see 7.4.2.4.). Signal amplification and increased sensitivity is achieved by coupling more than one enzyme to the analyte or detection molecule, e.g., a second antibody or protein A or G.

7.4.2.2. Fluorescence The sensitivity of fluorescence read-out depends on the number of fluorescent moieties coupled per analyte molecule, the quantum yield of the fluorescent dye, the peptide loading, and the amplification achieved, for example by a sandwich assay with primary and secondary antibody. Fortunately, many secondary antibodies labeled with different fluorescence dyes are available. However, background fluorescence from the array support can be a severe drawback. Such background signals could be almost completely suppressed using planar waveguide technology in combination with fluorescence-based detection methods [400, 401, 584]. Glass chips usually have significantly lower intrinsic background compared to the cellulose or polypropylene membranes commonly used for arrays prepared by the SPOT technology. These porous membranes usually contain traces of fluorescent substances left over from the production process. Another source of background fluorescence can arise from the peptides, peptidomimetics or often side products from the synthesis process, e.g., side chain protecting groups. Generally, fluorescent dyes with longer emission wavelengths, i.e., in the red range of the visible spectrum, such as Texas Red® are preferable to avoid interference with background fluorescence from substances in complex mixtures such as cell lysates. If the peptides on the array are labeled themselves, as required for some protease assays (see 7.5.3.2.; Figure 7.49), bulky and often hydrophobic fluorescence labels with longer emission wavelengths can cause severe problems by affecting the analyte's interaction (e.g., a protease) with the peptides. In such cases aminobenzoic acid, with a molecular weight in the average range of natural amino acids, is often used in order to reduce assay and read-out artifacts [102, 254, 455].

7.4.2.3. Radioactivity Binding assays with radioactively labeled screening molecules are often used due to their very high detection sensitivity. Generally, proteins can be labeled with ^{125}I [126, 582], ^{35}S [44], ^{14}C [377], or phosphorylated with [γ-$^{32/33}$P]ATP

[75, 139, 539]. The major advantage compared to immunological detection methods is that no controls are needed to rule out false-positive results from detection antibodies binding directly to the peptides. The drawback is that some proteins are denatured or modified in their activity or binding properties by ^{125}I-labeling or ^{32}P/^{33}P phosphorylation, necessitating a control activity test with the radioactive protein.

Phosphorylation of peptides by kinases is the most important radioactivity-based assay (Figure 7.48A). Peptide arrays with a phosphate acceptor amino acid are incubated with the kinase of interest in the presence of $[\gamma$-$^{32/33}$P]ATP (see 7.5.3.1.; [50, 143, 330, 338, 395, 484, 503, 539]). Incorporated radioactivity is measured using a phosphoimager or by exposition to X-ray films.

7.4.2.4. Chromogenic Read-Out Several examples for chromogenic read-out and densitometric quantification on peptide arrays are described. These are (1) precipitating substrates, e.g., nitroblue tetrazolium (NBT)/bromochloroindolyl phosphate (BCIP) catalyzed by alkaline phosphatase [269] or bromochloroindolyl-β-D-galactopyranoside catalyzed by β-galactosidase [155]; (2) metal ion detection with chromogenic chelators such as nickel-dimethylglyoxime or as metal sulfides [269, 270, 494]; (3) heme [441] or metal ion [495] binding to four-helix bundle protein arrays; (4) binding of dye-coupled peptides to protein domain arrays [548]; and (5) chemical transformation such as crosslinking of advanced glycation end products (AGEs; [377]) or chemical ligation with dye-labeled aldehydes [538]. No expensive imager system is required for this type of read-out. Visual inspection is sufficient for quantification and documentation only requires a normal scanner. However, the dynamic range and sensitivity are far worse than, say, chemoluminescence or radioactivity. Another disadvantage is that regeneration of arrays on porous membranes treated with precipitating substrates, for example alkaline phosphatase or reagents for metal ion detection, is rather difficult since traces may be retained in the membrane structure.

7.4.2.5. Label-Free Read-Out Label-free read-out systems are the ultimate goal for screening peptide arrays. The screening molecule needs not to be modified, which is usually very tedious and may affect the biological activity of the analyte. Moreover, this excludes artifacts associated with the detection molecule, e.g. secondary antibodies or fluorescent markers, as discussed above. The most important label-free read-out systems are surface plasmon resonance (SPR), which can record kinetic data of the binding event [218, 581], mass spectrometry (MS), which can even identify a certain molecule out of a crude mixture, e.g., a cell lysate, with high sensitivity [278, 532], and atomic force microscopy (AFM) resulting in a three-dimensional image of the screening molecule bound to the array. So far, these technologies have mostly been used for DNA, protein, or small molecule arrays but it is just a question of time until they are used to a similar extend with peptide arrays.

Analysis of antibody-peptide interactions by SPR imaging using a peptide microarray was shown for interactions of the FLAG epitope and several of its variants with the anti-FLAG mab M2 [581]. First, a self-assembled monolayer (SAM) was prepared on a gold film and chemically modified to immobilize peptides. The FLAG peptide, three single substitution variants and an HA tag peptide as a negative control were tethered to this SAM as a "peptide-line" array using a microfluidics system. A second perpendicular microfluidics system was then used to deliver the analyte mab M2. In another study phosphorylation of a peptide substrate by the tyrosine kinase c-Src was measured by SPR [218].

The peptide was chemically immobilized on a SAM chip and phosphorylated by c-Src. Subsequently, phosphorylation was detected by SPR measurement of an anti-phospho-tyrosine antibody. Although in these studies only a few peptides were analyzed simultaneously, which does not really reach far beyond the capacity of standard SPR techniques as commercially supplied by Biacore for example, this read-out method has a huge potential for peptide microarrays.

Array-based read-out methods employing MS were first introduced as a robust method in the ProteinChip® Array Technology by the company Ciphergen [585]. Peptides or other molecules are immobilized on carriers with eight different positions. Subsequently, the chip is probed with either pure molecules or crude mixtures such as cell lysates, serum, or urine. Molecules that bind to the peptides or other molecules immobilized on the chip are retained during automated washing steps and are detected by **surface-enhanced laser desorption/ionization time-of-flight mass spectrometry (SELDI-TOFMS)**. This technology is not reviewed further here because it does not fit the scope of peptide arrays with respect to the number of peptides analyzed and the degree of potential miniaturization. The combination of mass spectrometry and array techniques in the strict sense has already been successfully applied in the field of protein arrays [54] and might similarly be used for peptide arrays in the future. In preliminary studies Su and Mrksich immobilized the peptide Ac-IYAAPKKKC-NH$_2$ on self-assembled monolayers presenting ethylene glycol chains with maleimide functions [532, 533]. However, only the success of immobilization was shown by MALDI-TOF MS. No further assays or detection were carried out by mass spectrometry. The application of mass spectrometry to identify binding partners from peptide arrays is described in a patent application [278]. A peptide scan derived from the intracellular domain of the erythropoietin (EPO) receptor was incubated with cell lysates from EPO stimulated and unstimulated cells as well as different purified signal transduction proteins. Subsequently, tryptic digestion, mass spectrometry, and databank analysis revealed the identity of the peptide-bound proteins.

Atomic force microscopy has not been used with peptide arrays so far, however proof-of concept has been confirmed with protein arrays (e.g., [64, 245, 373]). AFM was used as a read-out for peptide arrays in the broader sense in a study visualizing transmembrane peptides in planar phosphatidylcholine bilayers that were prepared via fusion of vesicles with a solid substrate [471]. However, there was no peptide sequence diversity and no clear-cut spatial addressability.

A very elegant label-free read-out method uses conventional compact disc technology [282]. Molecules, meaning also peptides, are immobilized on the polycarbonate surface of a CD. Read-out of the reflective metalized layer underneath by a standard polarized infrared laser as used in CD players is disrupted by binding of bulky ligands to the surface and leads to errors in the binary signal. Proof-of-concept was shown for streptavidin binding to immobilized biotin but the strategy should also be applicable for peptide arrays.

7.5. APPLICATIONS OF PEPTIDE ARRAYS

In this Section the main applications for peptide arrays are described in principal and illustrated by selected outstanding examples. In most of these examples the peptide arrays were prepared by the SPOT method. This is mainly because publications based on this technology dominate the literature, since the first full publication describing this synthesis concept dates back more than a decade now and such array preparation is not only simple but requires no specialized equipment. Most applications illustrated here can be applied in principle to all types of peptide arrays described in Section 7.2. For a comprehensive overview of most of the publications in the field please refer to Table 7.4 in Section 7.6.

7.5.1. Antibodies

The most frequent application of peptide arrays described so far is the mapping of antibody epitopes. This can be ascribed to the fact that antibody-antigen interactions are often used as model systems for the evaluation of novel peptide library techniques, for several reasons: (1) usually antibodies bind to their antigens with high affinity and specificity (disregarding cross-reactivity). (2) Many antibodies raised against proteins bind to linear epitopes, unlike other protein-protein interactions where this is only an exception (see 7.5.2.2.). (3) Antibodies can be easily detected using commercially available enzyme- or fluorescence dye-coupled secondary antibodies. (4) Antibodies are stable and easy to handle, and (5) state-of-the-art techniques allow rapid and cheap preparation of sufficient antibody amounts. Although antibody-antigen interactions are discussed separately in this Section many characteristics described here are also valid for protein-protein interactions in general (see 7.5.2.).

7.5.1.1. Monoclonal Antibody Epitope Mapping: Linear Epitopes In linear epitopes (also referred to as continuous or contiguous epitopes) [24] the key amino acids mediating antibody contacts are located within one part of the antigen's primary structure, usually a region not exceeding 15 amino acids in length. Peptides covering these sequences have affinities to the antibody within the range shown by the entire antigen.

Three-dimensional structures of antibody-antigen complexes obtained from X-ray crystallography reveal relatively large contact surfaces in a range between 500 and 1000 Å^2 with more than 15 amino acids in contact with the binding partner. This led to the definition of the "structural epitope" comprising all contact residues as observed in the complex structure without considering their energetic contribution. On the other hand, extensive site-directed mutagenesis studies have shown that only a few residues effectively contribute to the binding free energy. These residues are summarized as defining the "energetic epitope" building or "hot spot of binding" [108]. Here, it should be pointed out that linear peptide epitopes identified by protein sequence-derived peptide scans comprise the amino acids of the energetic epitope, as well as a few linking residues, rather than the structural epitope.

Table 7.4 summarizes the publications describing mapping and characterization of linear antibody epitopes. The peptide arrays used include peptide scans, amino acids scans, substitutional analyses, truncation libraries, deletion libraries, cyclization scans, all types of combinatorial libraries, and randomly generated libraries of single peptides. Today, experiments to identify and characterize linear antibody epitopes are standard techniques widely applied even in non-specialized laboratories.

As listed in Table 7.4 most of the peptide array applications for mapping antibody epitopes are based on the SPOT synthesis technique due to its simplicity and robustness. However, the earliest publication describing antibody binding to arrays of short linear peptides, in 1991, utilized light-directed spatially addressable peptide synthesis [151]. Furthermore, several papers have been published where microarrays were prepared by directed immobilization of pre-synthesized peptides (e.g., [143, 395, 503, 511]) or co-polymerization of pre-synthesized peptides [200, 328].

7.5.1.2. Monoclonal Antibody Epitope Mapping: Discontinuous Epitopes Compared to linear epitopes, discontinuous epitopes are much more difficult to map. In discontinuous (or conformational) binding sites the key residues are distributed over two or more binding regions separated in the primary structure [24]. Upon folding, these binding regions are brought together on the protein surface to form a composite epitope. Even if the complete epitope mediates a high affinity interaction, peptides covering only one binding region, as synthesized in a scan of overlapping peptides, have very low affinities, which often cannot be measured by normal ELISA or surface plasmon resonance (SPR) experiments. Therefore, very sensitive detection procedures have to be applied to map discontinuous epitopes using conventional peptides scans (Figure 7.38). Peptide arrays on cellulose membranes prepared by SPOT synthesis are especially suited for this purpose. This is mainly due to the extremely high peptide loading (see Table 7.1, 7.2; [276]), which correlates with a concentration in the millimolar range assuming an equal distribution of peptides over the membrane area. This extremely high peptide density facilitates detection of even weakly binding peptides due to avidity effects. These peptides often have dissociation constants in the high micromolar or even millimolar range. Such sensitive read-out can of course lead to the detection of unspecific peptide-protein interactions. This has to be ruled out by control incubations, visual inspection of the three-dimensional antigen structure if available, competitive ELISA experiments or SPR studies. These control experiments are described in detail elsewhere [444, 447].

In addition to the sensitive detection methods, advanced protein sequence-derived peptide scans for mapping or mimicking discontinuous antibody epitopes or protein-protein interaction sites in general have been introduced. These include the domain scan (see 7.3.1.1.), the hybritope scan (see 7.3.1.2.), and the duotope or matrix scan (see 7.3.1.3.) described in detail in these Sections.

There are considerably less publications describing the mapping or mimicking of discontinuous epitopes due to the obstacles mentioned above. Here, three outstanding publications are worth mentioning in detail: (1) Korth et al. identified a three-segmented binding site for mab 15B3, which specifically recognizes the disease form of bovine prion protein [267]. (2) An interleukin-10 (IL-10)-derived peptide scan was used to identify a discontinuous epitope on IL-10 composed of two segmented binding regions. Connection of these binding regions and further optimization by substitutional analyses followed by a disulfide cyclization scan resulted in an IL-10 mimicking peptide with a dissociation constant in the lower nanomolar range [448, 587]. (3) A special type of discontinuous epitope appears in the hinge region of antibodies where the analogous sequences of the two heavy chains have a parallel orientation with several disulfide bridges, depending on the origin species and the antibody class and subclass. Welschof et al. were able to identify and mimic epitopes of scFv antibodies that simultaneously bind to both heavy chains in the hinge region of human IgG antibodies [588]. These discontinuous binding sites were mimicked by branched peptides with two identical hinge region-derived sequences coupled to the α- and ε-amino group of a lysine residue immobilized on a cellulose membrane.

7.5.1.3. Antibody Paratope Mapping Alternatively, epitope mapping approaches can be applied to paratope mapping or to identify biologically active peptides from antibody complementarity determining regions (CDRs). For example, three peptides with CD4-binding capacity and HIV antiviral activity were identified using a scan of overlapping peptides covering the V_H and V_L domains of the murine IgG1κ anti-CD4 mab ST40 [374]. The same group also described the expedient application of this approach to anti-idiotype networks. The "paratope dissecting" approach was reviewed in detail recently [293].

7.5.1.4. Polyclonal Antibody Epitope Mapping Protein sequence-derived peptide scans are similarly applied to mapping polyclonal antibody epitopes. The only difference compared to monoclonal antibodies is that there is no way of distinguishing between linear epitopes and single binding regions of discontinuous antigenic determinants. In an outstanding publication Valle et al. described the epitope mapping of a polyclonal serum raised against the *Bacillus subtilis* bacteriophage Φ29 connector [559, 560]. Eleven immunodominant regions were identified using the SPOT approach. The membrane-bound spots were used to purify fractions of the serum specific for only one of the epitopes (for detailed experimental protocols of this approach see: [342]). Since only a low resolution structure of the Φ29 connector is available, these epitope-specific polyclonal antibodies proved extremely useful for topographical assignment of the epitope sequences using electron microscopy of Φ29 connector-antibody complexes. Several other studies using peptide arrays for epitope mapping of polyclonal antibodies from human patients as well as other species are listed in Table 7.4.

7.5.2. Protein-Protein Interactions

Most of the library types described in Section 7.3 have also been applied to protein-protein interaction mapping. Some parameters for the interaction, such as complementarity between the surfaces, size and shape of the binding site, and residue interface propensities or segmentation adopt different values for antibody-antigen and other protein-protein interactions [244]. However, the principles of the interaction are analogous. The overwhelming majority of protein-protein interactions are mediated via discontinuous binding sites. On the other hand, protein interaction domains (see 7.5.2.2.; [402, 403]) or chaperones (see 7.5.2.3.) usually interact with short linear sequences in their ligands. These interactions are therefore described separately.

7.5.2.1. Protein-Protein Interactions in General Among the first interactions to be mapped using peptide arrays were cytokine or growth factor interactions with their receptors: tumor necrosis factor (TNF)α/55-kDa TNF receptor [444], interleukin-6 (IL-6)/IL-6 receptor [582], IL-10/IL-10 receptor α [446], and vascular endothelial growth factor (VEGF)/VEGF receptor II [417]. Since then the interaction between many other protein classes and their respective binding partners have been studied (structural proteins, surface proteins of pathogens, etc., see Table 7.4). Interestingly, all these studies were based on protein sequence-derived peptide libraries and not on *de novo* approaches.

7.5.2.2. Protein Interaction Domain—Ligand Interactions Protein modules and other proteins involved in signal transduction often bind to short linear sequences [402, 403].

Therefore, these interactions are well suited to investigation using protein sequence-derived peptide scans and substitutional analyses. In fact, the binding specificity of many important protein modules has been characterized using peptide arrays. These are: (1) WW domains binding to proline-rich sequences with the consensus sequence Xaa-Pro-Pro-Xaa-Tyr (e.g., [323, 546–548]); (2) PDZ domains (named after the first letter of the proteins in which they were first discovered: **p**ost synaptic density, **d**isc large, **z**onula occludens) binding to the C-terminal 4–5 residues of their target proteins, frequently transmembrane receptors or ion channels [211, 501]; (3) SH2 (Src homology 2) domains of STATs (**s**ignal **t**ransducers and **a**ctivators of **t**ranscription) binding to phospho-tyrosine containing motifs [44]; (4) SH3 (Src-homology 3) domains binding to proline-rich peptides that form a left-handed polyproline type II helix with the minimal consensus motif Pro-Xaa-Pro [73]; (5) PTB domains (phospho-tyrosine binding) interacting with Asn-Pro-Xaa-phospho-Tyr motifs [220]; (6) EVH1 domains (**E**na-**V**ASP **h**omology domain 1) binding to Phe-Pro-Xaa-Pro-Pro motifs [19, 120, 121, 277, 382], and (7) TRAFs (**t**umor necrosis factor **r**eceptor **a**ssociated **f**actors) binding to different motifs [434]. Most of these studies used target protein-derived peptide scans as well as substitutional analyses.

Three publications are of particular interest here. The first is by Hoffmüller et al., who identified two novel PDZ domain interaction partners by screening a peptide array of all known C-termini derived from the human proteins listed so far in the SWISSPROT databank [211]. The second is by Toepert et al. describing the preparation of an array including more than 800 single substitution variants of the human YAP (Yes-associated protein) WW protein domain spanning 44 amino acids, which was successfully employed in a parallel ligand binding assay [546, 547]. This WW protein domain array was synthesized stepwise by the SPOT technique. In contrast to the other publications mentioned here, this was the first time that an array of natively folded protein modules was synthesized chemically. The stepwise synthesis of complete protein domains is, of course, a laborious and time-consuming procedure. Toepert et al. therefore synthesized an array of 11859 tri-substituted human YAP WW domains by a combination of stepwise synthesis (SPOT technology) and native chemical ligation [548]. The C-terminal part including the positions Leu30, His32, and Gln35 which were combinatorially substituted by other proteinogenic as well as unnatural amino acids was prepared by SPOT synthesis. The N-terminal portion was synthesized in advance and ligated to the peptides on the cellulose array. The array was incubated with 22 different dye-labeled peptide ligands (resulting in more than 250,000 binding experiments) to identify WW domains with novel binding specificities (Figure 7.47G). This is one example of how to extend the concept of peptide arrays to arrays of chemically synthesized proteins.

7.5.2.3. Chaperones In contrast to the specific recognition described for antibody-antigen, protein-protein, kinase and protease interactions, chaperones have only a degenerate recognition motif and usually bind to a huge number of substrates. However, a preference for certain motifs was identified for some chaperones such as DnaK, DnaJ, SecB and the chaperone activity of Triggerfactor (which also has an isomerase activity), all from *Escherichia coli*, using numerous peptide scans derived from substrate proteins as well as combinatorial peptide libraries prepared by the SPOT method [260, 359, 399, 481–483]. An example outlined here for illustration is the recognition motif of SecB (*Escherichia coli*) that was identified by screening 2688 peptides derived from 23 different substrate proteins.

The results were used to define a working recognition motif that can predict SecB binding peptides with up to 87% accuracy [260].

7.5.3. Enzyme-Substrate and Enzyme-Inhibitor Interactions

As well as being used for standard binding experiments as described so far in Section 7.5., peptide arrays are also amenable to identifying, characterizing and optimizing enzyme substrates and inhibitors. In principal, assays for all enzymes that modify peptides are possible but so far only studies with kinases, phosphatases, proteases [140, 141, 397, 486, 549, 616], isomerases [336], glycosyltransferases [144, 217, 219] ribosyltransferases [576], serine hydroxylases [80], epoxide hydrolases [616], and esterases [615] have been described using peptide arrays or peptide microarrays. Here, kinases and phosphatases (see 7.5.3.1.) and proteases (see 7.5.3.2.) are reviewed in detail.

7.5.3.1. Kinases and Phosphatases Kinase activity is detected by incubating arrays of potential substrates with the enzyme in the presence of $[\gamma$-$^{32/33}$P]ATP and subsequent autoradiography or phosphoimaging (Figure 7.48A). Experimental details have been described for peptide arrays prepared by SPOT synthesis [540] and for peptide microarrays [338, 395, 503]. This approach was used to determine the substrate specificity of cAMP- and cGMP-dependent kinases (PKA, PKG) [115, 539, 540, 550], calcium-dependent protein kinase [334], histidine kinase enzyme I of the phosphotransferase system [376], protein tyrosine kinase Lyn [534], protein kinase C [126], androgen receptor protein kinase [370], extracellular signal related kinase ERK-2 [140], and casein kinases I and II [550] using peptide arrays generated by SPOT synthesis. Combinatorial and deconvolution libraries led to substrates that are selectively phosphorylated by PKG [115]. Furthermore, a specific and similarly selective substrate-derived inhibitor was created by substituting the phosphate acceptor residue by alanine, resulting in a peptide with an inhibition constant of 7.5 μM for PKG and 750 μM for PKA [115]. Similar analyses identified the recognition motif of the kinase DYRK1A, which phosphorylates histone H3 [207]. The phosphate acceptor residue (threonine 45) was determined by mass spectrometry and a substitution analogue library of peptides containing this amino acid was synthesized and screened, revealing proline-directed substrate recognition. Recently, a very interesting approach introduced as the double peptide synthesis method was used to detect weak but synergistically acting moieties in kinase or phosphatase substrate proteins [141]. The authors used a modified SPOT synthesis protocol to prepare two defined peptide sequences per cellulose membrane-bound spot. While the substrate peptide was constant the second peptide was varied to detect weakly interacting but stimulating sequences. This approach allowed the efficient identification of so-called exo-sites in substrate proteins.

Proof-of-concept experiments using radioactively labeled ATP derivatives were published using peptide microarrays generated by immobilization of pre-synthesized peptides. MacBeath and Schreiber were able to demonstrate that p42 MAP kinase, protein kinase A, and casein kinase II recognize their substrates and transfer a phosphate moiety from the ATP to the peptide side chain hydroxyl function if these substrates are covalently immobilized on BSA coated surfaces (Figure 7.15; [338]). Falsey et al. used a chemoselective immobilization reaction (Figure 7.29) to demonstrate that protein tyrosine kinase p60 c-src is able to catalyze the phosphoryl-transfer to the microarray bound substrate EEIYGEFF, if a hydrophilic linker molecule is inserted between the surface and the peptide [143]. Alternatively, Houseman et al. were able to show that the same enzyme phosphorylates the substrate Ac-IYGEFKKKC selectively, if it is immobilized (Figure 7.34) and presented on a self-assembled monolayer [218]. Recently, more sophisticated peptide microarrays were used to profile different kinases using radioactive detection. Lizcano et al. investigated protein kinase NEK6 substrate specificity using a collection of 720 human phosphorylation site-derived peptides, which were chemoselectively immobilized in three identical subarrays (2160 peptides) on a functionalized standard microscope slide [330]. Protein tyrosine kinase Abl was analyzed using a similar peptide microarray [443] and a peptide microarray containing 1433 randomly generated peptides in three identical subarrays (4299 peptides per slide) [484]. Panse et al. analyzed casein kinase II substrate specificity using peptide microarrays prepared by chemoselective immobilization of peptide derivatives that were initially prepared by SPOT synthesis and cleaved off the original cellulose membrane. More than 13,000 immobilized peptides derived from human proteins were presented to the kinase in the presence of $[\gamma\text{-}^{32}P]$ATP, followed by phosphoimaging [395]. Finally, Schutkowski and coworkers were able to demonstrate that high-content peptide microarrays represent useful tools for detecting autophosphorylation sites (peptide scans through kinases), priming phosphorylation events (collections of kinase substrate peptides together with monophosphorylated derivatives, 2923 peptides) and potential downstream targets (activation loops of human kinases, 1394 peptides; [503]).

An alternative to detecting phosphorylation using radioactivity is the use of either fluorescence dye- or enzyme-labeled anti-phospho-amino acid antibodies [218, 310, 311, 350, 395, 443, 558] or phospho-specific fluorescence dyes [350] (Figure 7.48B). Unfortunately, antibody-based detection is limited to tyrosine phosphorylation because commercially available generic anti-phospho-serine or anti-phospho-threonine antibodies have well-defined sub-site specificities [253, 395, 526]. This limitation could be circumvented by the introduction of phospho-sensitive dyes [350]. Examples for successful profiling of tyrosine kinases using fluoresceine-labeled anti-phospho-tyrosine antibodies are listed in Table 7.4. Conversely, the recognition of cellulose membrane-bound phospho-peptides by anti-phospho-tyrosine antibodies was also used to detect protein tyrosine phosphatase activities [140, 141, 397].

7.5.3.2. Proteases Several different assay principles have been developed to measure protease activity on peptide arrays (Figure 7.49). Optimal assays lead to increased signal intensity upon substrate cleavage. Five assays of this type have been described: (1) the first assay was developed by Duan and Laursen and is based upon peptide arrays prepared on polyaminoethylmethacrylamide membranes by the SPOT method (Figure 7.49A) [122]. The array comprised all 400 possible dipeptides derived from genetically encoded amino acids with an N-terminally coupled fluoresceinyl thiocarbamyl moiety. These peptides were punched out and attached to pins in a microtiter plate lid. Subsequently, they were suspended in wells of a 96-well microtiter plate filled with protease solution. After specified reaction times, the spots were removed in order to quantify the fluorescence dye coupled to the cleaved-off N-terminal peptide fragment. This method was evaluated using chymotrypsin and papain [122] and for porcine pancreatic elastase [295]. (2) To avoid laborious pin attachment a modified assay involves immersing substrate spots (aminobenzoic acid as fluorescence dye) in wells filled with the protease solution (Figure 7.49B). At various times small aliquots are pipetted into new wells and cleavage is quantified using a fluorescence microtiter plate reader. This assay was employed to identify and characterize caspase-3 substrates using substitutional analyses of a known peptide substrate, a peptide scan, combinatorial libraries and randomly generated sets of peptides [254, 455, 457]. The major disadvantage of these two assay principles is that the peptide array has to be dissected, essentially abandoning the benefits of array technologies. (3) This led to the introduction of peptides arrays with internally quenched peptides (Figure 7.49C). Compartmentalization of the cleavage reaction is not necessary and increasing signal intensity is observed. This technique was evaluated using combinatorial peptide libraries and substitutional analyses of substrate peptides incubated with trypsin [102, 450] and subsequently employed to determine the substrate specificity of the integral membrane protease OmpT of *Escherichia coli* [102]. (4) A sophisticated but rather tedious procedure involves peptides coupled to cellulose membranes by their C-terminus and having an antibody epitope tag with a biotinylated lysine residue at the N-terminus (Figure 7.49D). Cleavage releases the N-terminal part of a substrate peptide including the epitope tag and the biotin moiety. This fragment is affinity-blotted onto a streptavidin-coated PVDF membrane and detected via an enzyme–conjugated antibody [275]. (5) Peptide derivatives containing a substituted fluorogenic group C-terminal to the scissile bond are immobilized on glass slides resulting in peptide microarrays (Figure 7.49E). In a proof-of-concept study it was demonstrated that the protease trypsin cleaved the amino acyl-fluorophore bond [615]. A very similar assay principle but using longer peptides successfully determined the substrate specificities of trypsin, granzyme B and thrombin employing peptide microarrays on glass slides generated by chemoselective peptide immobilization [486]. (6) Peptides with a fluorescence dye at the free terminus are applied for array-based protease assays with a decreasing signal upon cleavage (Figure 7.49F).

Recently, a novel principle was described for profiling proteolytic activities using semi-wet peptide microarrays and differences in the partition coefficients of peptide substrates and released fluorophores [258]. Lysyl endopeptidase treatment released an environmentally sensitive fluorophore resulting in a blue shift of the emission maximum from 540 nm to 508 nm, along with two-fold higher fluorescence intensity.

In a very elegant experiment, peptidic inhibitors tethered to fluorescence tagged peptide nucleic acids were used to profile inhibitor specificity against different cysteine proteases [596, 597]. The peptide nucleic acid tag encodes the structure of the attached peptide derivative and therefore allows spatially addressed deconvolution after hybridization to a GenFlex oligonucleotide microarray.

7.5.4. Application of Peptide Arrays: Miscellaneous

In addition to the broad fields of application described above, some very specific experiments using peptide arrays have also been described and these are summarized in this Section.

7.5.4.1. DNA Binding Peptides Reuter et al. identified the type II endonuclease *Eco*RII DNA-substrate binding site by incubating an *Eco*RII-derived peptide scan with ^{32}P-labeled oilgonucleotides ([465]; Figure 7.47F). An important part of this work addresses the problem of unspecific ionic interactions between oligonucleotides and peptides containing positively charged amino acids. Experimental protocols and procedures to rule out unspecific binding are described in detail [466].

7.5.4.2. Metal Ion Binding Peptides Combinatorial peptide libraries on cellulose membranes were utilized to study the interaction of peptides with different metal ions. This included chromogenic detection of Ni^{2+} (Figure 7.47B, C), Pb^{2+} (Figure 7.47D) and Ag^{1+} ions by reduction to the colloidal state as well as assays with radioactive isotopes, i.e., $^{45}Ca^{2+}$, $^{55}Fe^{3+}$ and technetium-99m ions (Figure 7.47A; [269, 270, 494]). Of particular interest is the identification of technetium-99m binding hexapeptides using an 8000-membered combinatorial peptide library array of the type $O_1XO_2XO_3X$ [346]. These peptides were intended for tumor imaging applications involving genetic fusion to an anti-CEA single chain Fv antibody (CEA = carcinoembryonic antigen). Binding of Pb^{2+} and As^{3+} to four different dansyl-labeled peptides prepared by light directed on-chip synthesis was recently shown [172, 265]. Binding was monitored by fluorescence measurement.

7.5.4.3. Arrays of 4 α-Helix Bundle Mini-Proteins Two publications describe the preparation of 4 α-helix bundle mini-protein arrays using a combination of the SPOT method and the TASP (template-assembled synthetic protein) concept (Figure 7.47H). Rau et al. prepared an array of 462 hemoproteins by attaching, in an anti-parallel manner, four peptides from a small pre-synthesized library to a cyclic decapeptide that was coupled to a cellulose membrane via a linker moiety [441]. This array was simply screened for heme binding properties by spectroscopic methods. As an extension of this concept the same group published the *de novo* identification of mono Cu^{2+} and Co^{2+} binding proteins using the same approach [495].

7.5.4.4. Identification, Characterization, and Optimization of Peptidic Ligands Peptide arrays have often been applied to identify peptide ligands for a predefined protein or binding partner in general. As the initial hits are usually sub-optimal with respect to affinity, specificity, or stability, substitutional analysis is an ideal tool to optimize these properties and simultaneously gather information about structure-activity relationships. This was published for the optimization of the *Strep*-tag II interacting with streptavidin [493] and for the identification and optimization of peptides binding blood coagulation factor VIII (FVIII) that could be used as affinity ligands in the purification process of FVIII, which is administered to treat hemophilia A patients [12].

Unnatural amino acids and especially D-amino acids can be incorporated into peptides in order to stabilize them against proteolytic degradation. However, the optimal amino acids required to retain the peptide's activity cannot be predicted. Therefore, Kramer et al., described a stepwise systematic transformation process starting from all-L-peptide epitopes through to all-D-peptides [274]. A complete array-based D-amino acid substitutional analysis was performed in the first transformation cycle. One peptide (containing one D-amino acid) with the same binding activity was selected and used for the next transformation cycle to convert a second position into a D-amino acid. Finally, all-D-peptides with completely different side chain functionalities but retained antibody binding specificity and affinity were identified. A similar procedure was described to optimize a vascular endothelial growth factor (VEGF) inhibiting peptide derived from the VEGF receptor II sequence [417] substituting four positions by D-amino acids [418].

7.5.4.5. Chemical Tranformations Reactions of reducing sugars on the N-termini or amino acid side chains of proteins are the most common non-enzymatic modifications of proteins. Combinatorial dipeptide libraries prepared by the SPOT method were used to elucidate how reactive the N-termini and amino acid side chains were to being modified by reducing sugars [377]. Following rearrangements, oxidations, and dehydration this leads to so-called advanced glycation end products (AGEs) or "Millard products" that are mostly colored and fluorescent. AGEs lead to changes in protein conformation, loss of function, and irreversible crosslinking and are mostly observed with long-lived proteins involved in the pathogenesis of various age-related diseases such as Alzheimer's. It was shown that the sugars preferentially react with cysteine or tryptophan residues when the peptide N-terminus was free, but cysteine, lysine and histidine residues were preferred with an acetylated N-terminus. Reaction activities were monitored using ^{14}C labeled sugars. In addition, the reaction activities of side chains from dipeptides with an acetylated N-terminus were assessed for crosslinking by bovine serum albumin (BSA)-AGE. Since the products are colored, reaction activities were monitored visually or using a standard scanner.

In a similar approach Tam et al. analyzed a SPOT library of 400 dipeptides to assess their suitability for the domain ligation strategy by incubating them with dye-labeled alanine-α-formylmethyl, -β-formylmethyl, and -β,β,β-dimethyl and formylethyl esters. The extent of the aldehyde reaction with the dipeptides was subsequently analyzed by image analysis [538].

7.5.4.6. Cell Binding to Peptide Arrays Falsey et al. described a proof-of-concept for an adhesion assay with intact cells on peptide microarrays [143]. This involved preparing an array by directed immobilization of pre-synthesized peptides, based on an all-D-amino acid 8-mer peptide previously selected to bind WEHI-231 murine lymphoma cells. Out of six different cell lines WEHI-231 were the only cells that bound to the array. Binding was observed by pre-staining the cells with Cell-Tracker Orange and measured using a microarray scanner. Furthermore, the method was extended to measure adhesion-dependent functional cell signaling. It was known that the above mentioned 8-mer peptide induces tyrosine phosphorylation in WEHI-231 cells. This activity was confirmed for the peptide immobilized on microarrays. Cells were fixed after the cell adhesion assay, and following permeation with nonionic detergent, a FITC-labeled anti-phospho-tyrosine antibody was added and the array was analyzed under a fluorescence microscope. In general, standard light microscopy can also be applied, as shown for an anti-CD (cluster of differentiation) antibody array used for immunophenotyping of different types of leukemia [32].

In another publication, Otvos et al. were able to show stimulated T-cell proliferation upon binding to peptides prepared by SPOT synthesis [394]. This study was not carried out using peptide arrays in the sense of this article since the membrane was cut into pieces before the assay and placed in segmented reaction vessels. However, this demonstrated the principle of cell binding to immobilized peptides and re-stimulated discussion about the paradigm that says such a T-cell assay requires free peptides in solution [362].

Finally, binding of 3T3 fibroblasts to the peptide Gly-Arg-Gly-Asp-Ser immobilized on a self assembled monolayer (SAM) was shown [609]. However, only one peptide was immobilized because the goal was to create spatially addressable cell arrays.

7.5.4.7. Identification of Protein-Protein Interaction Partners Peptide arrays have also been used to identify novel protein-protein interactions. Early in the development of peptide arrays Krönke and coworkers identified the novel WD-repeat protein FAN as a specific interaction partner of the tumor necrosis factor (TNF) 55 kDa receptor by a combination of the yeast interaction trap system and a peptide scan probed with ^{35}S-methionine-labeled protein extract from Jurkat cells [2]. Hoffmüller et al. synthesized an array of all known C-terminal peptides derived from the human proteins listed so far in the SWISSPROT databank ([211]; Section 7.3.1.8.). Krause et al. submitted a patent describing the identification of interaction partners for the intracellular domain of the erythropoietin (EPO) receptor ([278]; Section 7.4.2.5.). Very recently, a proof-of-concept for combining a phage display library with a peptide array library approach was shown [42]. Three different peptide ligands for protein domains and a negative control peptide were synthesized by SPOT synthesis. The respective protein domains were displayed on phages, which were used to probe the "array" in an optimized biopanning process. Peptide specific phage populations were eluted from the array elements, propagated, their DNA labeled with Cy3 or Cy5 dyes, and finally analyzed by hybridization to DNA microarrays. Although this was a relatively brief preliminary study it might be extended in the future to screen, for example, phage displayed cDNA libraries against peptide arrays representing a plethora of protein sequence-derived peptides. The basic principles for biopanning of phage libraries on peptide arrays prepared by SPOT synthesis were described previously [223]. In this study a pIII fusion library of calmodulin variants generated by error prone PCR was screened against mastoparan. In addition, a pIII fusion library of scFv fragments presenting the antibody repertoire of a mouse immunized with the recombinant extracellular domain of the human C3a receptor was screened against a partial peptide scan derived from the antigen used for immunization [199, 223].

7.5.5. Peptidomimetics

Examples of peptidomimetics arrays prepared by SPOT synthesis have been published for D-amino acids [272, 418], peptoids and peptomers [16, 202, 617], 1,3,5-trisubstituted hydantoins [201], amino-substituted 1,3,5-triazines [488, 490], branched peptides [490, 588], and non-disulfide bond cyclized peptides, e.g., N-terminus to side chain amide cyclizations [191]. The repertoire of peptidomimetic modifications and unnatural building blocks applicable for SPOT synthesis or stepwise *in situ* peptide array preparation in general is of course limited to synthesis procedures that can be carried out on planar surfaces, with the spot areas considered as open micro-reactors. On the other hand, all peptide/peptidomimetic microarray technologies that apply pre-synthesized compounds are open to the entire synthetically accessible diversity space.

7.6. BIBLIOGRAPHY

This Section gives an extensive list of publications where peptide arrays were used until finalization of this manuscript in January 2004. Patents, conference abstracts, or posters are only recorded when an important topic was not covered by an original publication or review. Websites are only cited in the text of the relevant Sections and in Table 7.2. We make no claims as to the comprehensiveness of Table 7.4 due to inherent difficulties in searching for publications that used peptide arrays. Unfortunately, the nomenclature in the field is not yet fully standardized and key words or abstracts sometimes do not cover the

TABLE 7.4.

Screening Molecule	Library	Technology	Reference
Antibodies			
Antibody Epitope Mapping: Linear Epitopes			
mab D32.39 and goat anti-mouse IgG-FITC conjugate or mab D32.39-FITC conjugate	deletions and truncations of fLRRQFKVVT (1024 peptides)	LDPS	[151]
mab 3E7 directed against ß-endorphin and goat anti-mouse IgG-FITC conjugate	alternating YGGFL and PGGFL		
First publication on the SPOT synthesis technique for peptide array generation and epitope mapping		SPOT	[154]
anti-CMV26 rabbit polyclonal serum	CMV26 peptide scan, truncation analysis of a selected peptide, alanine scan	SPOT	[155]
mAb 3E7 and goat anti-mouse IgG-FITC conjugate	deletions and truncations of YGPAFWGFMNLS (4096 peptides)	LDPS	[187]
rabbit antisera against bovine myelin proteolipid protein (PLP) (Ab45) and against PLP peptides (Ab32, Ab34, Ab46)	selected PLP peptide	SPOT	[578]
mAb 20D6.3 recognizing oligocarbamates and goat anti-mouse IgG-FITC conjugate	256 oligocarbamates derived from YKFL	LDPS	[84]
anti-HIV-1 MN envelope mab 58.2	substitutional analysis of a peptide identified by phage display library panning against mab 58.2	SPOT	[237]
anti-transforming growth factor (TGF) mab Tab2; silver(I); dsDNA	combinatorial libraries of the type XXB_1B_2XX and XB_1XB_2XX	SPOT	[269]
anti-V3 loop (gp120 from HIV-1) mab 50.1	substitutional analysis of a V3 loop-derived peptide	SPOT	[470]
anti-p60 (*Listeria monocytogenes*) polyclonal serum	partial peptide scans of 2 regions (PepA, PepD) of p60 of *Listeria monocytogenes* and of the PepD homologous region of *Listeria innocua*	SPOT	[65]
mabs against the coat protein of beet necrotic yellow vein virus (BNYVV)	peptide scans and alanine scans of the BNYVV coat protein	SPOT	[92]
mab 3E7 directed against ß-endorphin and goat anti-mouse IgG-FITC conjugate	alternating YGGFL and PGGFL	LDPS	[168]
technetium–99m; nickel(II); anti-transforming growth factor α (TGFα) mab Tab2	combinatorial with 2 defined positions (linear, cyclic, L- and D- amino acids) and deconvolution libraries	SPOT	[270]
anti- potato mop-top virus (PMTV) mabs SCR 76 and SCR 77	partial peptide scan of the PMTV coat protein (CP)	SPOT	[407]
anti-interleukin-4 (IL-4) mab	peptide scan of IL-4	SPOT	[464]
anti-p24 (HIV-1) mab CB4-1	substitutional analysis of the linear p24-derived mab CB4-1 epitope	SPOT	[572]
anti-porcine tubulin-tyrosine ligase (TTL) mab 1D3	dual positional scanning combinatorial libraries	SPOT	[157]

TABLE 7.4. Continued

Screening Molecule	Library	Technology	Reference
8 anti-sIL-6R mabs	partial peptide scans of the IL-6 receptor extracellular domain	SPOT	[192]
anti-dynorphin B mab D32.39 and goat anti-mouse IgG-FITC conjugate or mab D32.39-FITC conjugate	deletions and truncations of FLRRQFKVVT (1024 peptides)	LDPS	[213]
anti-transforming growth factor α (TGFα) mab Tab2	combinatorial clustered hexapeptide library and deconvolution libraries	SPOT	[271]
6 mabs against phosphoprotein (P-protein) of morbillivirus strains PDV and CDV	peptide scans of the P-proteins of morbillivirus strains PDV and CDV	SPOT	[348]
anti-protein-S of transmissible gastroenteritis virus (TGEV) mabs 6A.A6 and 57.9; anti-EGF-like domain of surface protein pfs25 of *Plasmodium falciparum* mab 32F81	4450 randomly generated 12-mer peptides; all 8000 tripeptides based on the genetically encoded amino acids	MPC	[516]
anti-gp41 (HIV-1) mab 3D6	complete L- and D amino acid substitutional analyses of the linear epitope	SPOT	[528]
6 anti-listeriolysin (*Listeria monocytogenes*) mabs	peptide scans of listriolysin	SPOT	[98]
anti-porcine tubulin-tyrosine ligase (TTL) mab 1D3	peptide scan of porcine tubulin-tyrosine ligase (TTL), substitutional and truncation analysis of a selected peptide, combinatorial library (XXB$_1$B$_2$XX)	SPOT	[158]
anti-porcine tubulin-tyrosine ligase (TTL) mab 1D3	substitutional and truncation analysis of a TTL-derived peptide; combinatorial library (XXB$_1$B$_2$XX) and deconvolution libraries	SPOT	[159]
streptavidin-alkaline phosphatase	combinatorial library (XXB$_1$B$_2$XX) and deconvolution libraries		
cAMP-dependent protein kinase	combinatorial library (XXB$_1$B$_2$XX) and deconvolution libraries		
5 polyclonal anti-U1 small nuclear ribonucleoprotein complex (U1 snRNP-C) antibodies (rabbit); 5 sera from patients with mixed connective tissue disease (MCTD) or systemic lupus erythematosus (SLE)	peptide scans of U1 snRNP-C	SPOT	[193]
rabbit serum against a peptide derived from the β-antigen of the c protein complex of group B streptococci	partial peptide scan of the β-antigen of the c protein complex of group B streptococci and truncation analysis of a selected peptide	SPOT	[239]
12 sera from patients with autoantibodies against intestinal alkaline phosphatase (IAP)	peptide scans of intestinal alkaline phosphatase (IAP)	SPOT	[264]
anti-gp41 (HIV-1) mab 2F5	complete L-amino acid substitutional analysis of the epitope peptide	SPOT	[436]

(cont.)

TABLE 7.4. Continued

Screening Molecule	Library	Technology	Reference
anti-transforming growth factor α (TGFα) mab Tab2; anti-p24 (HIV-1) mab CB4-1	substitutional analyses; combinatorial library of the type XXXXB$_1$B$_2$B$_3$XXXX; positional scanning combinatorial library	SPOT	[494]
lead(II); calcium(II); iron(III); silver(I); nickel(II); technetium-99m	combinatorial libraries of the type XB$_1$XB$_2$XX; deconvolution libraries; positional scanning library		
2 anti-presenilin 1 (PS1) polyclonal rabbit antisera	partial peptide scans of presenilin 1	SPOT	[542]
human antiserum against pertussis toxin	peptide scan of pertussis toxin	SPOT	[575]
3 anti-FcRIIb2 mabs	peptide scans of FcRIIb2	SPOT	[586]
pooled sera from 10 latex-sensitized health care workers; pooled sera from 10 patients with spina bifida	peptide scan of the latex allergy associated allergen prohevin	SPOT	[20]
pooled sera from latex-allergic patients; serum from a single latex allergic patient; high IgE control sera; pooled anti-latex rabbit sera	peptide scans of prohevein	SPOT	[29]
anti-RhopH3 (*Plasmodium falciparum*) mouse mabs 84, 85, 87, and 91	partial peptide scan (aa 816-836) of RhopH3	SPOT	[119]
serum from a juvenile rheumatoid arthritis (JRA) patient	partial peptide scan of the nucleosomal high mobility group protein 2 (HMG-2) and a glycine walk of a selected peptide	SPOT	[248]
anti-p24 (HIV-1) mab CB4-1	positional scanning combinatorial L- and D-amino acid libraries, deconvolution libraries, substitutional analyses	SPOT	[272]
2 mabs against the heavy chain of botulinum type E neurotoxin	peptide scans of two regions of botulinum type E neurotoxin	SPOT	[279]
9 sera from patients with Goodpasture's disease	peptide scan of the C-terminal noncollagenous domain of the α3 chain of type VI collagen	SPOT	[312]
anti-profilin mab 2H11	peptide scan of bovine profiling I	SPOT	[357]
anti-human thyroglobulin (hTg) mouse mabs Tg2 and Tg8	partial peptide scan of the central part of human thyroglobulin (hTg)	SPOT	[372]
mAb 3E7 and goat anti-mouse IgG-FITC conjugate	endorphin epitope derived peptides up to 16-mers (65536 peptides)	LDPS	[422]
anti-troponin I mabs	peptide scans of sceletal troponin I	SPOT	[440]
HIV-1 positive sera	peptide scans of the gp120 (HIV-1) V3 loop	SPOT	[497]
anti-protein-S of transmissible gastroenteritis virus (TGEV) mabs 6A.A6; anti-FLAG mab M2	4450 randomly generated 12-mer peptides; all 8000 tripeptides based on the genetically encoded amino acids	MPC	[517]
mab 755 against the α-chain of C3	peptide of the C-terminal region of the C3 α-chain	SPOT	[569]

TABLE 7.4. *Continued*

Screening Molecule	Library	Technology	Reference
10 antibodies against ras proteins	peptide scans of the human oncogene products Ha-ras, K-ras, and N-ras; truncation analyses and complete L-amino acid substitutional analyses of selected peptides	SPOT	[579]
anti-elongation factor TU (EF-TU) mab	peptide scan of EF-TU, truncation and substitutional analysis of a selected peptide	SPOT	[18]
2 anti-centrosome scleroderma autoimmune sera	peptide scans of the immunogenic regions of the centrosome autoantigen PCM-1	SPOT	[23]
anti-lipoprotein lipase (LPL) mab (*discontinuos epitope possible*)	peptide scan of LPL and substitutional analyses of selected peptides	SPOT	[74]
anti-glucoamylase (*thermoanaerobacterium thermosaccharolyticum*) polyclonal antibody	partial peptide scan of glucoamylase	SPOT	[123]
anti-M2 acetylcholine receptor mab	partial peptide scan of M2 acetylcholine receptor	SPOT	[131]
anti-cardiac troponin I monoclonal and polyclonal antibodies	peptide scans of cardiac troponin I	SPOT	[145]
28 anti-cardiac troponin I (cTnI) mabs	peptide scans of cTnI	SPOT	[150]
IgG fractions from 2 patients with erythema multiforme major, polyclonal serum from a rabbit immunized containing the epitope YSYSYS	partial peptide scan of the C-terminal part of desmoplakin	SPOT	[152]
3 anti-C3aR scFvs	peptide scans of the second extracellular loop of C3aR	SPOT	[198]
anti-alanine dehydrogenase *(AlaDH)* of *Mycobacterium tubercolosis* mab HBT-10	peptide scan of *AlaDH*	SPOT	[224]
anti-transforming growth factor α (TGFα) mab Tab2	peptide scan of TGFα, substitutional analysis of a selectd peptide, combinatorial library (XXB$_1$B$_2$XX)	SPOT	[273]
anti-cholera toxin mab TE33; anti-p24 (HIV-1) mab CB4-1	D-amino acid substitutional analyses for the transformation of linear wt-epitopes into all-D peptide analogs	SPOT	[274]
anti-cardiac troponin I monoclonal and polyclonal antibodies	peptide scans of cardiac troponin I	SPOT	[289]
anti-β2 adrenergic receptor mabs	peptide scans of the β2 adrenergic receptor and alanine scans of selected peptides	SPOT	[297]
5 anti-glycoprotein G of HSV-2 (gG-2) mabs; 8 anti-gG-2 antibodies purified from patients sera	peptide scans of gG-2	SPOT	[321]
biotinylated antibodies directed against ACTH derived peptides and strepavidin-phycoerythrin conjugate	pyrrole peptides (18–39) and (11–24) of ACTH	CPPP	[328]

(*cont.*)

TABLE 7.4. *Continued*

Screening Molecule	Library	Technology	Reference
8 anti-gliadin polyclonal rabbit sera; 6 human sera positive for gliadin antibodies	peptides identified by phage display (panning against anti-gliadin rabbit sera)	SPOT	[389]
anti-eukaryotic initiation factor 4E (eIF4E) mabs	peptide scans of eIF4E	SPOT	[431]
7 rabbit sera against coxsackievirus A9 and B3 and against echovirus 11	peptide scans of the capsid proteins VP1, VP2 and VP3 from coxsackievirus A9	SPOT	[435]
interleukin 10 (IL-10), interleukin 10 receptor a (IL-10Rα), 3 anti-IL-10 mabs, 1 anti-IL-10Rα mab	peptide scans of IL-10 and IL-10Rα	SPOT	[446]
sera from insulin-dependent diabetes mellitus (IDDM) patients	peptide scans of glutamic acid decarboxylase 65 (GAD65) and alanine scans of selected peptides	SPOT	[468]
15 sera from systemic lupus erythematosus (SLE) patients	peptide scans of the SLE autoantigen SmD1	SPOT	[469]
anti-interferon regulatory factor 1 (IRF-1) polyclonal antibody	peptide scan of interferon regulatory factor 1 (IRF-1)	SPOT	[487]
sera from patients with allergic bronchopulmonary aspergillosis (ABPA)	peptide scan of a major allergen (Asp f 2) of *Aspergillus fumigatus*; truncation scans of selected peptides	SPOT	[21]
sera from latex allergic patients and sera of rabbits immunized with latex proteins	peptide scans of the latex protein Hev b 5 from *Hevea brasiliensis*	SPOT	[30]
several sera from patients with primary biliary cirrhosis containing autoantibodies against nuclear protein sp100	partial peptide scans of the nuclear protein sp100	SPOT	[47]
68 murine IgG mabs recognizing linear epitopes	peptide scans, length analyses, L-substitutional analyses	SPOT	[113]
13 sera from multiple sclerosis patients positive for human transaldolase (TAL-H); 2 anti-TAL-H polyclonal rabbit antibodies	peptide scans of TAL-H	SPOT	[142]
anti-pneumolysin mab PLY-5	peptide scan of pneumolysin (PLY)	SPOT	[231]
sera from parvovirus B19 infected patients	peptide scans of the VP1 and VP2 parvovirus B19 capsid proteins; truncation analyses; alanine and glycine walks of selected peptides	SPOT	[250]
3 mouse mabs and a polyclonal rabbit serum against pVIII of filamenteous bacteriophage fd	peptide scans of pVIII; truncation analyses and glycine walks of selected peptides	SPOT	[259]
anti-pre-S1 and pre-S2 (hepatitis B virus) mabs 1F6 and MA 18/7	peptide scan of the pre-S domain, substitutional analyses of selected peptides	SPOT	[281]
anti-RNA polymerase II *(Drosophila melanogaster)* scFv215	substitutional analysis of the RNA polymerase II derived epitope	SPOT	[326]
2 anti-interleukin 10 (IL-10) mabs	peptide scans of IL-10 and substitutional analyses of selected peptides	SPOT	[449]

TABLE 7.4. *Continued*

Screening Molecule	Library	Technology	Reference
anti-midsize neurofilament NF-M mab 155	peptide scan of neurofilament NF-M	SPOT	[500]
anti-colera toxin mab TE33	substitutional analyses of mab TE33 binding peptides	SPOT	[529]
sera from patients receiving streptokinase (SK) therapy	peptide scans of streptokinase	SPOT	[552]
polyclonal serum against bacteriophage φ29 connector protein	peptide scans of the bacteriophage φ29 connector protein (p10)	SPOT	[559]
polyclonal serum against bacteriophage φ29 connector protein	peptide scans of the bacteriophage φ29 connector protein (p10)	SPOT	[560]
anti-β-amyloid (Aβ) polyclonal antibody	peptide scan of apolipoprotein E (apoE) each peptide being flanked by randomized positions (hybritope scan)	SPOT	[594]
sera from latex-sensitized health care workers; sera from patients with spina bifida	peptide scan of the latex allergy associated proteins Hev b 1 and Hev b 3	SPOT	[22]
CD4; anti-gp17 polyclonal antibody	peptide scan of gp17; alanine walks of selected peptides	SPOT	[28]
sera from patients with polymyositis-scleroderma overlap syndrome	partial peptide scan of the polymyositis-scleroderma overlap syndrome antigen PM/Scl-100; length analysis, glycine-walk and complete substitutional analysis of a selected positive peptide from the peptide scan	SPOT	[48]
mab 2E11 against a peptide 38 from the variable region of the Vb 6.2 T-cell receptor	truncation analysis of peptide 38	SPOT	[52]
mab 3E12 against Kx blood group antigen and spectrin β-chain	overlapping peptides derived from epitope regions in Kx blood group antigen and spectrin β-chain	SPOT	[72]
2 anti-tubulin-tyrosine ligase (TTL) mabs	peptide scan of pig TTL	SPOT	[135]
anti-cardiac troponin I mab 11E12	alanine walks of mab 11E12 binding peptides	SPOT	[147]
10 sera from patients with malaria	peptide scan of block 17 of merozoite surface glycoprotein-1 (MSP1) of *Plasmodium falciparum*	SPOT	[164]
anti-p24 (HIV-1) mab CB4-1	libraries with all possible single step transition pathways between 3 sequentially and structually unrelated mab CB4-1 binding peptides; substitutional analyses of peptides from selected transition pathways	SPOT	[212]
^{35}S-labeled calmodulin	mastoparan (Ac-INLKALAALAKKIL) and truncation and deletion analogs	SPOT	[223]

(*cont.*)

TABLE 7.4. *Continued*

Screening Molecule	Library	Technology	Reference
pIII phage display library of calmodulin variants	mastpoparan (Ac-INLKALAALAKKIL) and truncation and deletion analogs		
pIII phage display scFv library presenting the antibody repertoire of a mouse immunized with the recombinant extracellular domain of the C3a receptor	partial peptide scan of the extracellular domain of the C3a receptor		
anti-amyloid β (Aβ) mouse mabs WO-2, G2-10, and G2-11	peptide scans of residues 1-42 of amyloid β (Aβ)	SPOT	[238]
mouse and rabbit antisera against capsid proteins VP0, VP3, and VP1 of human parechovirus 1 (HPEV 1)	peptide scans of the capsid proteins VP0, VP3, and VP1 of HPEV 1	SPOT	[243]
anti-porcine circovirus (PCV) sera from pigs	partial peptide scans of ORF-1 and complete peptide scans of ORF-2 and ORF-3 from PCV1 and PCV2	SPOT	[339]
sera and affinity purified antibodies from patients with systemic sclerosis	peptide scans of centromeric protein A (CENP-A); length analyses of selected antigenic regions	SPOT	[340]
8 sera from anti-centromere autoantibody positive patients	partial peptide scans of the centromere protein CENP-A	SPOT	[378]
sera from coeliac disease patients containing anti-gliadin antibodies	peptide scans of α/β- and γ-gliadin; substitution analogs of a selected peptide	SPOT	[390]
polyclonal mouse antibodies against a protective epitope from measles virus fusion protein	peptide scan of the peptide used for immunization	SPOT	[396]
14 mabs against human or *Xenopus* p53	peptide scans of the N-terminal part of p53; truncation analyses and alanine walks of selected peptides	SPOT	[426]
11 sera from *Mycoplasma bovis* infected cows with symptoms of mastitis or pneumonia; 3 anti-variable surface lipoproteins (Vsps) mabs	peptides scans of Vsp; peptide truncation analyses of the Vsp repetitive regions	SPOT	[485]
anti-p193 (from major vault protein, MVP) mabs p193-4, p193-6, and p193-10	partial peptide scan of amino acids 408–611 of the p193 protein	MPC	[498]
pool of 10 sera from papilloma bearing hamsters; pool of 19 sera from hamster polyomavirus (HaPV)-infected papilloma-free hamsters; rabbit sera against against intact SV40 virions and JCV-VP1 virus like particles	peptide scan of the hamster polyomavirus (HaPV) major capsid protein VP1	SPOT	[515]
mab 26/9 against peptide HA175-110 of hemagglutinin of influenza (X47:HA1)	4450 randomly generated cyclic 12-mer peptides (XXCXXXXXXCXX), 4450 randomly generated linear 12-mer peptides and substitutional analyses of selected hits from the random libraries	MPC	[518]

TABLE 7.4. Continued

Screening Molecule	Library	Technology	Reference
phosphorylation specific anti-human tau protein mab AD2	peptide scan of phosphorylated and unphosphorylated tau protein and alanine walks of selected peptides	SPOT	[551]
anti-glycoprotein G-1 (gG-1) antibodies purified from herpes simplex virus 1 (HSV-1) patients	peptide scans of gG-1 (HSV-1)	SPOT	[554]
anti-p24 (HIV-1) scFv CB4-1 and 2 mutants thereof	substitutional analyses of the p24-derived wt epitope and a mimotope	SPOT	[595]
mouse sera against peptides from Der p2 (house dust mite *Dermatophogoides pteronyssinus*)	alanine walks of peptides derived from Der p2	SPOT	[599]
anti-plastidic phosphorylase (Pho 1a) from *Solanum tuberosum* L. mab	peptide scan of Pho 1a; complete substitutional analysis of the Pho 1a-derived epitope peptide	SPOT	[9]
anti-α-bungarotoxin scFv C12; nicotinic acetylcholine receptor nAchR	peptide scan of α-bungarotoxin	SPOT	[58]
sera from cow's milk allergy patients (IgG and IgE antibody detection)	peptide scans of α_{s1}-casein	SPOT	[77]
polyclonal sera from mice, rabbits and horses against a non-toxic protein from the venom of the noxious scorpion *Tityus serrulatus* (TsNTxP)	peptide scans of TsNTxP and alanine substitution scans of selected peptides; peptide scans of active toxins from the venom of *Tityus serrulatus*	SPOT	[78]
anti-tobacco mosaic virus protein (TMVP) Fab 57P	partial peptide scan of tobacco mosaic virus protein (TMVP); substitutional analysis, truncation analysis and alanine walk of a selected peptide	SPOT	[86]
anti-vascular endothelial cadherin (VE-cadherin) mabs BV6 and Cad 5	peptide scans of the extracellular domain of vascular endothelial cadherin (VE-cadherin)	SPOT	[93]
p60^{c-src} protein tyrosine kinase and γ^{32}P-ATP	Ttds-EEIYGEFF	DIPP	[143]
strepavidin-Cy3 conjugate; avidin-Cy5 conjugate	biotin, HPYPP and WSHPQFEK		
anti-human insulin mab HB125 and anti-mouse IgG-Cy5 conjugate	biotin, wGeyidvk, pqrGstG, WSHPQFEK and YGGFL		
WEHI-231 cells and negative control cells	wGeyidvk		
anti-genome-derived Ag 33 (GNA33) mab 25; anti-loop 4 of porin A mab P1.2	peptide scans of genome-derived Ag 33 (GNA33) of *Neisseria meningitidis* and loop 4 of porin A	SPOT	[182]
anti-transforming growth factor α (TGFα) mab Tab2	5 substitutional analyses of linear and cyclic (disulfide cyclization and amide cyclization from the N-α to the C-terminal glutamic acid side chain) analogs of the epitope	SPOT	[191]
scFv-phage library; anti-C3a receptor 2e loop scFvs E10 and E6	peptide scans of the 2e loop of C3a receptor	SPOT	[199]

(cont.)

TABLE 7.4. *Continued*

Screening Molecule	Library	Technology	Reference
sera from children with cow's milk allergy	peptide scans of α-lactalbumin and β-lactalbumin	SPOT	[235]
anti-Ha-*ras* antibody mAb-10	peptide scan of the Ha-ras protein	SPOT	[294]
sera from patients with systemic sclerosis, rheumatoid arthritis and undefined connective tissue disease; control sera	complete substitutional analyses of the centromere-associated protein A (CENP-A) epitope GPRRR; combinatorial library of the type GPB_1RB_2; libraries of mimotopes and homologous epitopes from potential autoantigens	SPOT	[341]
mab R5 recognizing prolamins from wheat (glidins), barley (hordeins) and rye (secalins)	peptide scans from α- and γ-type gliadin	SPOT	[391]
sera from *Trypanosoma cruzi* infected patients (Chagas' disease) and sera from experimentally infected mice and rabbits	peptide scan of *trans*-sialidase from *Trypanosoma cruzi*	SPOT	[424]
anti-transforming growth factor α (TGFα) mab Tab2	peptoidic building block substitutional analyses of the mab Tab2 peptide epitope VVSHFND and of peptomers and peptoids derived thereof	SPOT	[456]
5 anti-CEA (carcinoembryonic antigen) mabs, rabbit and goat anti-CEA sera	peptide scan of CEA (carcinoembryonic antigen)	SPOT	[520]
monoclonal antibodies A, 32F81, B, A6.6A	randomly generated libraries of 3640 or 4550 defined peptides	MPC	[561]
10 sera from children who achieved clinical tolerance against cow's milk allergens; 10 sera from patients with persistent cow's milk allergy (CMA)	6 selected peptides from αs1-casein and β-casein	SPOT	[566]
anti-β-Gal 14-3-3 fusion protein polyclonal antibody	peptide scan of *Chlamydomonas reinhardtii* 14-3-3- protein	SPOT	[570]
anti-phosphotyrosine mabs 4G10, PY20, pTyr100; anti-acetyl-lysine polyclonal antibody	peptides containing phosphor-Tyr or acetyl-Lys	DIPP	[571]
2 rabbit sera against TsNTxP and TsIV that are two components of the venom of scorpion *Tityus serrulatus*)	peptide scans of TsNTxP and TsVI from the venom of scorpion *Tityus serrulatus*	SPOT	[11]
several sera as well as pooled sera from borna disease virus (BDV) infected mice, rats, horses, and humans; 1 anti-nucleoprotein (N) of BDV mab; 1 anti-phosphoprotein (P) of BDV mab	peptide scans of BDV-N and BDV-P proteins; length analysis of a selected epitope	SPOT	[43]
Practical protocols describing the prediction of structural features of peptides using amino acid substitution scans, complete substitutional analyses and double mutation ("dipeptide libraries") exemplified by the PM/Scl-100 epitope		SPOT	[49]

TABLE 7.4. *Continued*

Screening Molecule	Library	Technology	Reference
Practical protocols to optimize the mapping of antibody epitopes and protein-protein interactions with respect to protein concentrations, blocking conditions, detection systems, and the efficiency of the regeneration procedures		SPOT	[59]
sera from 13 cow's milk allergic children	peptide scan of α_{s2}-casein	SPOT	[69]
anti-E6 oncoprotein (papillomvirus type 16) mabs 1F1, 6F4, and scFv 1F4	partial E6 peptide scan; truncation analyses, alanine substitution scans, and complete L- amino acid substitutional analses of selected peptides	SPOT	[87]
mixture of anti-phosphotyrosine antibodies, POD-conjugated anti-mouse antibody	phosphotyrosine-containing peptides derived from human insulin receptor; degenerate SPOT library of general structure AABX$_1$ZX$_2$BAA (Z = phosphotyrosine, B = mixture of 20 natural amino acids, X defined amino acids) and peptides derived from human STAT 5A containing both phosphotyrosine and phosphothreonine residues	SPOT	[140]
sera from older patients with persistent cow's milk allergy; sera from younger children with decreasing milk-specific IgE antibodies	25 peptides representing epitopes on cow's milk proteins: 5 epitopes on α_{s1}-casein, 5 on α_{s2}-casein, 9 on κ-casein, 1 on α-lactalbumin, and 5 on β-lactoglobulin)	SPOT	[236]
Practical protocols for mapping monoclonal and polyclonal antibody epitopes using peptide scans, length analyses, amino acid substitution scans and complete substitutional analyses		SPOT	[261]
anti-p53 mAb DO-1 fluorescence shift following Pb^{2+} or As^{3+} binding	peptides SDLHKL, DSLGKL, and SGLHKL dns (dansyl)-ECEE, dns-CCCC, dns-GGGG, EEEE	LDPS	[265]
5 murine mabs against glycoprotein G-2 (gG-2) of herpes simplex virus type 2 (HSV-2)	peptide scans of gG-2 of HSV-2	SPOT	[322]
Practical protocols for (a) affinity purification of subfractions from polyclonal sera that are specific for a defined linear epitope and (b) competition assays on peptide arrays		SPOT	[342]
12 sera from a patient with systemic sclerosis	peptide scans of centromeric protein A (CENP-A) as well as several mimotopes from potential autoantigens	SPOT	[343]
sera from HCV and EBV patients and negative control sera	peptides from HCV core and NS4 and EBV capsid, negative control peptide	DIPP	[361]
anti-SA-Le (tetrasaccharide) mab NS19-9	L-substitutional analysis of the NS19-9 epitope DLWDWVVGKPAG	SPOT	[386]
anti-p53 antibody (PAb240)	substitutional analysis of the PAb240 peptide epitope including proteinogenic amino acids and unnatural building blocks; length analysis of the epitope	LDPS	[406]

(cont.)

TABLE 7.4. *Continued*

Screening Molecule	Library	Technology	Reference
anti-IL-10 mab CB/RS/13; anti-transforming growth factor α (TGFα) mab Tab2; anti-p24 (HIV-1) mab CB4-1	5520-membered library of randomly generated 15-mer L-peptides; 8 substitutional analyses of selected antibody-binding peptides	SPOT	[458]
different pooled sera from autoimmune patients; 2 sera from single systemic lupus erythematodes (SLE) patients; negative control sera;	immunodominant peptides from snRNP proteins, Sm proteins, myelin basic protein (MBP), poly A ribose polymerase (PARP), and histones H1, H2A, H2B, H3, and H4	DIPP	[472]
sera from mice immunized with myelin oligodendrocyte glycoprotein (MOG) p35–55, proteolipid protein (PLP) p139–151, and negative control serum	selected peptides derived from myelin basis protein (MBP), proteolipid protein (PLP), myelin oligodendrocyte glycoprotein (MOG), oligodendrocyte-specific protein (OSP), αb crystalline and other unspecified myelin proteins	DIPP	[475]
incubation with blood sera followed by incubation with biotinylated anti-human IgE Ab and avidin-Cy3 conjugate	peptide scan through major peanut allergens Ara h1, Ara h2 and Ara h3 (210 peptides)	DIPP	[511]
anti-FVIII mab ESH8	ESH8 binding peptides identified by phage display; peptide scan of the C2 domain of blood coagulation factor VIII	SPOT	[567]
anti-FLAG mab M2	FLAG epitope; 3 single site alanine substitutions of the FLAG epitope; HA epitope	DIPP	[581]
pooled sera from *Parietaria judaica* allergic patients	peptide scans of the Par j 1 and 2 allergens from *Parietaria judaica* pollen	SPOT	[17]
sera from peanut allergy patients and nonatopic control sera (detection of IgE antibodies)	8 peptides representing the immunodominant epitopes of the peanut allergens Ara h 1, Ara h 2, and Ara h 3	SPOT	[39]
individual and pooled sea from cow's milk allergy (CMA) patients	truncation libraries and alanine substitution scans of peptides from the α_{s1}-casein allergen	SPOT	[89]
anti-CD30 mabs Ber-H2, Ki-2, Ki-4, R4-4	peptide scans of the extracellular domain of CD30 and substitutional analyses of selected antibody binding peptides	SPOT	[114]
FITC-labeled anti-RHSVV pAb240	multiple substitutional analysis of RHSVV epitop using 28 different amino acids (2304 peptides)	LDPS	[172]
individual and pooled sera form patients with peach allergy	peptide scan of the peach allergen Pru p 3 and alanine substituted analogs of selected peptides	SPOT	[173]
anti-transforming growth factor α (TGFα) mab Tab2	randomly generated library of 8000 hexapeptoids and -peptomers	SPOT	[202]

TABLE 7.4. *Continued*

Screening Molecule	Library	Technology	Reference
anti-MIP mAb, POD-labeled anti-mouse IgG antibody	peptide scan through MIP, substitutional analysis of antigenic MIP derived peptide AYGPRSVGGPIGPNE	SPOT	[203]
5 anti-Hantaan-G2 protein antibodies from phage display panning against native Hantaan virus	peptide scan of the Hantaan-G2 protein; alanine and glycine substituted analogs of selected peptides	SPOT	[262]
human anti-hantaan virus (HNTV) G1 protein mab AH100IgG	peptide scan of G1 protein from hantaan virus (HNTV, strain 76–118); glycine substitution scan of a selected peptide	SPOT	[318]
13 anti-ribosomal P protein human sera	peptide scans of the human ribosomal proteins P0, P1, and P2; truncation analyses, alanine substitution scans and complete L-amino acid substitutional analyses of selected peptides	SPOT	[344]
sera from macaques vaccinated with 3 different DNA and/or recombinant modified vaccinia virus Ankara (rMVA) vaccines encoding Gag-Pol or Gag-Pol-Env	430 proteins and overlapping peptides spanning the simian-human immunodeficiency virus (SHIV) proteome	DIPP	[380]
activating anti-p53 mabs HR231 and Pab421; inhibition of antibody binding by a p53 C-terminal peptide	peptide scans of p53	SPOT	[427]
Detailed protocols for the synthesis of peptide arrays, library design, and incubation procedures for antibody epitope mapping; experimental example for the epitope mapping of anti-IL-10 mab CB/RS/3 using a peptide scan and a substitutional analyis		SPOT	[459]
4 human patient sera with neurological diseases (2), Raynaud's disease (1), undifferentiated connective tissue disease (1); anti-EEA1 murine mab	peptide scans of the early endosome antigen 1 (EEA1)	SPOT	[506]
anti-FVIII mab Bo2C11	mab Bo2C11 binding peptides identified by phage display	SPOT	[568]
7 anti-ribosomal P protein human sera, 2 anti-ribosomal P protein scFv antibody fragments	complete L-amino acid substitutional analyses of the C-10 peptide from the C-terminal P protein sequence	SPOT	[611]
anti-CD30 mab Ki-4	peptide scan of the extracellular domains of CD30	SPOT	[196]
tyrosine kinase Abl and CKII together with FITC-labeled anti-phospho-tyrosine antibody and γ^{32}P-ATP, respectively	collection of more than 13 000 peptides (13meric) derived from human proteins	DIPP	[395]
FITC-labeled anti-phospho-tyrosine antibody	2923 phospho-peptides (13-mers) derived from human proteins		
phospho-specific antibodies MPM2, 3F3/2, PY20	complete L-amino acid substitutional analysis of AXXXX[pS/pT]XXXXA and AXXXX[pYT]XXXXA	SPOT	[477]

(*cont.*)

TABLE 7.4. Continued

Screening Molecule	Library	Technology	Reference
GST fusions of GRB-SH2, SHP2-CSH2, GRB7-SH2, GRB 10-SH2	complete L-amino acid substitutional analysis of AXXXXX[pYT]XXXXA		
protein kinase A (PKA), Cdc 15	complete L-amino acid substitutional analysis of $AX_1X_1X_1X_1[S/T]X_1X_1X_1X_1AX$ = L-amino acids exept for Cys, X1 = L-amino Acids exept for Cys, Ser, Thr		

Antibody Epitope Mapping: Discontinuous Epitopes

Screening Molecule	Library	Technology	Reference
anti-β-factor XIIa mab 201/9	combinatorial starting library XOXOXXXX and positional scanning deconvolution libraries	SPOT	[169]
3 anti-β-factor XIIa mabs	peptide scan of β-factor XIIa, alanine walks and truncation analyses of selected peptides	SPOT	[170]
anti-IL-10 mab; anti-TNFα mabs; TNFα	peptide scans of IL-10; TNFα; 55-kDa TNFα receptor	SPOT	[444]
anti-prion protein (PrP) mabs 15B3 and 6H4	peptide scans of bovine PrP	SPOT	[267]
anti-complement receptor 2 (CR2) mab FE8	peptide scan of the short consensus repeats 1 and 2 (SCR-1, SCR-2) of the complement receptor 2 (CR2)	SPOT	[429]
anti-interleukin-10 (IL-10) mab CB/RS/5	peptide scan and hybritope scan of IL-10	SPOT	[445]
interleukin 10 (IL-10); interleukin 10 receptor a (IL-10Rα); 3 anti-IL-10 mabs; 1 anti-IL-10Rα mab	peptide scans of IL-10 and IL-10Rα	SPOT	[446]
anti-actin mab 2G2	partial peptide scan of actin	SPOT	[180]
anti-interleukin 10 (IL-10) mab CB/RS/1	peptide scan of IL-10; substitutional analyses; cyclization scan	SPOT	[448]
2 anti-interleukin 10 (IL-10) mabs	peptide scans of IL-10 and substitutional analyses of selected peptides	SPOT	[449]
2 anti-huIgG hinge region scFv	substitutional analyses of dimeric peptides derived from the hinge regions of IgG1, IgG2, and IgG4	SPOT	[588]
3 anti-thyroid peroxidase scFv	peptide scans of thyroid peroxidase (no binding observed indicating a discontinuous epitope)	SPOT	[76]
anti-interleukin 10 (IL-10) mab CB/RS/1	peptide scan of IL-10; substitutional analyses; cyclization scan	SPOT	[451]
anti-hen-eg white lysozyme (HEL) mab D1.3	duotope scan of hen-eg white lysozyme (HEL)		
anti-interleukin 10 (IL-10) mab CB/RS/1	peptide scan of IL-10; substitutional analyses; cyclization scan	SPOT	[452]
4 mabs against *Grapevine virus* A	peptide scans of the capsid protein of *Grapevine virus* A	SPOT	[109]
5 murine mabs against glycoprotein G-2 (gG-2) of herpes simplex virus type 2 (HSV-2)	peptide scans of gG-2 of HSV-2	SPOT	[322]

PEPTIDE ARRAYS IN PROTEOMICS AND DRUG DISCOVERY

TABLE 7.4. *Continued*

Screening Molecule	Library	Technology	Reference
anti-FVIII mab ESH8	ESH8 binding peptides identified by phage display; peptide scan of the C2 domain of blood coagulation factor VIII	SPOT	[567]
anti-CD30 mabs Ber-H2, Ki-2, Ki-4, R4-4	peptide scans of the extracellular domain of CD30 and substitutional analyses of selected antibody binding peptides	SPOT	[114]
Antibody Paratope Mapping			
thyroglobulin; lysozyme; angiotensin II	peptide scans of the variable regions of the anti-thyroglobulin mab Tg10, anti-lysozyme antibody HyHEL-5, and anti-angiotensin II mab 4D8	SPOT	[290]
hen egg-white lysozyme; anti-idiotypic mab AI-10 (anti-thyroglobulin mab Tg10)	peptide scan of the variable regions of anti-lysozyme antibody HyHEL-5 and anti-thyroglobulin mab Tg10	SPOT	[291]
CD4	peptide scan of the variable regions of anti-CD4 mab ST40	SPOT	[374]
anti-idiotypic mab AI-10 (anti-thyroglobulin mab Tg10)	peptide scan of the variable regions of anti-thyroglobulin mab Tg10	SPOT	[292]
CD4	alanine scans of 3 peptides derived from the CDRs of anti-CD4 mab ST40	SPOT	[432]
angiotensin II	peptide scan of the variable domain of anti-angiotensin II mab 4D8	SPOT	[90]
Review article describing the application of peptide arrays to identify critical paratope residues for antigen binding and the identification of antigen binding peptides thereof		SPOT	[293]
CD4	alanine substitution scan of peptides from the anti-CD4 Fab 13B8.2 paratope	SPOT	[38]
Protein-Protein Interactions			
Protein-Protein Interactions in General			
procine transferrin (3 different detection methods: (i) immunologically, (ii) ^{59}Fe-loaded transferring, (iii) biotinylated transferrin and streptavidin-alkaline phosphatase)	peptide scan of TfbA of *Actinobacillus pleuropneumoniae*	SPOT	[531]
^{35}S-Met-labeled protein extract from Jurkat cells containing the novel WD-repeat protein FAN	peptide scans of the TNFR-R55 cytoplasmic domain	SPOT	[2]
calmodulin	peptide scans of the cytoplasmic domains of C-CAM1 and C-CAM2 (rat) and BGP (mouse) and Bgp (man)	SPOT	[125]
anti-IL-10 mab; anti-TNFα mabs; TNFα	peptide scans of IL-10, TNFα, 55-kDa TNFα receptor	SPOT	[444]
S100C (EF-hand-type Ca^{2+} binding protein)	peptide scan of the N-terminal region of annexin I	SPOT	[505]

(*cont.*)

TABLE 7.4. *Continued*

Screening Molecule	Library	Technology	Reference
soluble interleukin 6 receptor (sIL-6R); interleukin 6 (IL-6)	peptide scans of IL-6 and IL-6R; substitutional analyses of IL-6 derived peptides	SPOT	[582]
human transferrin (huTf)	peptide scan of the transferrin-binding protein B (TbpB) from *Neisseria meningitidis* (strain M982)	SPOT	[461]
^{125}I-calmodulin	phosphorylated and unphosphorylated peptides derived from cell-cell adhesion molecule (C-CAM) isoforms S and L and single site substitution analogs	SPOT	[126]
porcine protein kinase C β and γ ^{32}P-ATP	partial peptide scans of the C-CAM isoforms S and L		
^{32}P-labeled CaBP1 and CaBP2; peptides LEKDEL and CYDDQKAVKDEL as competitors	peptide scan of the *erd2* receptor	SPOT	[234]
dimeric hepatitis B virus (HBV) core protein variants c1-149 and c1-124	partial peptide scan of the HBV core protein 1–161	SPOT	[266]
plasma-derived factor VIII (FVIII) and recombinant FVIII	partial peptide scan of von Willebrand factor (vWF)	SPOT	[379]
interleukin 10 (IL-10); interleukin 10 receptor a (IL-10Rα); 3 anti-IL-10 mabs; 1 anti-IL-10Rα mab	peptide scans of IL-10 and IL-10Rα	SPOT	[446]
cytosolic domains of the translocase of the outer mitochondrial membrane (Tom) 20, 22, and 70 *(S. cerevisiae)*	peptide scans of the cytochrome c oxidase subunit IV (CoxIV) and phosphate carrier (P_iC) preproteins *(S. cerevisiae)*	SPOT	[62]
α−actinin	peptide scan of zyxin	SPOT	[120]
p24 (HIV-1)	peptide scan of p24 (HIV-1)	SPOT	[205]
tubulin; γ-tubulin	peptide scans of α-, β-, and γ-tubulin	SPOT	[331]
^{125}I vascular endothelial growth factor (VEGF)	peptide scans of VEGF receptor II; substitutional analysis of a selected peptide	SPOT	[417]
α−actinin	peptide scan of zyxin	SPOT	[460]
N-lobe, C-lobe, and complete transferrin binding protein B (TbpB) from *Neisseria meningitidis*; TbpB from *Moraxella catarrhalis*	peptide scans of human transferrin (hTf)	SPOT	[463]
CD4; anti-gp17 polyclonal antibody	peptide scan of gp17; alanine walks of selected peptides	SPOT	[28]
cytoplasmic and core domain of mitochondrial import receptor Tom70 from Saccaromyces cerevisiae	peptide scan of the phosphate carrier protein	SPOT	[63]
cardiac troponin C	scans of cardiac troponin I	SPOT	[146]
α_2−macroglobulin	peptide scan of β_2-microglobulin	SPOT	[181]

TABLE 7.4. Continued

Screening Molecule	Library	Technology	Reference
secretory immunogobulin A (SIgA) and scretory component (SC)	peptide scans of the SIgA and SC binding domain of *Streptococcus pneumoniae* surface protein SpsA; truncation analyses of selected peptides	SPOT	[194]
ActA (*L. monocytogenes*) and p21-Arc	peptide scans of p21-Arc and ActA	SPOT	[423]
^{125}I-PHF43 fragment of τ protein	partial peptide scan of τ protein	SPOT	[574]
calmodulin	partial peptide scan of human Ca^{2+}-calmodulin-dependent protein kinase II β chain (CaM kinase II); peptide scan of the rat N-STOP protein	SPOT	[55]
anti-α-bungarotoxin scFv C12; nicotinic acetylcholine receptor nAchR	peptide scan of α-bungarotoxin	SPOT	[58]
cpSRP, cpSRP54 and cpSRP43 from *Arabidopsis thaliana* (cpSRP = chloroplast signal recognition particle)	peptide scans of cpSRP54 and preLhcb1 from pea	SPOT	[185]
8 kDa dynein light chain (LC8)	(partial) peptide scans of cellular (nNOS, Kid-1 renal transcription factor, Bim, DNA cytosine methyl transferase, microtubule-associated protein, dynein heavy chain, Drosophila swallow, GKAP) and viral target proteins from Mokola virus, rabies virus and ASF virus; N- and C-terminal extensions of the consensus binding motif KSTQT	SPOT	[478]
Lyn	partial peptide scan of the erythropoietin receptor (EPOR)	SPOT	[543]
Practical protocols for the mapping of protein-protein interactions using peptide arrays prepared by SPOT synthesis		SPOT	[184]
Puumala virus nucleocapsid protein (PUUV-N)	partial peptide scan of the C-terminal 243 amino acids of the human apoptosis enhancer Daxx	SPOT	[315]
human islet amyloid polypeptide (hIAPP) fused with maltose binding protein (MBP)	peptide scan of hIAPP	SPOT	[358]
human fibrinogen	alanine substitution scan and complete L-amino acid substitutional analysis of a selected peptide from FbsA (fibrinogen receptor from group B *Streptococcus*); control peptides	SPOT	[499]

(cont.)

TABLE 7.4. *Continued*

Screening Molecule	Library	Technology	Reference
regulatory subunit RII from protein kinase A	partial peptide scans of 10 different A kinase anchoring proteins (AKAPs); complete L-amino acid substitutional analysis of an AKAP-specific "position-dependent scoring matrix" (PDSM) consensus sequence derived from the AKAP peptide scans	SPOT	[10]
human plasminogen	peptide scan of pneumococcal α-enolase (Eno) and length analysis of the human plasminogen biding sequence	SPOT	[35]
regulatory subunits RIα and RIIα of protein kinase A	L-substitutional analysis of a 27-mer peptide derived from the dual specific A kinase anchoring protein 2 (D-AKAP2)	SPOT	[68]
E. coli YoeB-GST fusion protein	peptide scans of E. coli YefM and single site substitution analogs of a selected peptide	SPOT	[81]
PSK (a protein-bound polysaccharide from *Basidiomycetes*)	peptide scan of chicken gizzard regulatory light chain (RLC) of smooth muscle myosin	SPOT	[165]
^{125}I-human soluble urokinase-type plasminogen activator receptor suPAR	peptide scan of uPAR	SPOT	[317]
Cdc42-maltose binding protein (MBP) fusion protein in the presence of a non-hydrlyzable GTP analog (GMPPMP)	peptide scan of the C-terminal half of IQGAP1	SPOT	[353]
^{125}I vascular endothelial growth factor (VEGF)	complete D-amino acid substitutional analyses of VEGF receptor II-derived peptides	SPOT	[418]
N-terminal clathrin domain (clathrin-TD); α-ear domain of AP-2	partial peptide scans of auxilin	SPOT	[491]
soluble cytosolic domains of Tom20, Tom22, Tom70	peptide scan of the inner membrane mitochondrial protein BCS1	SPOT	[524]
α_MI-domain of integrin $\alpha_M \beta_2$	partial peptide scans of the P2 peptide of the γC domain of fibrinogen	SPOT	[556]
Tim9/Tim10 comlex	peptide scans of Tim17, Tim22, Tim23, Sc AAC2, FLX1, Hs UCP1, and 11 control peptides	SPOT	[564]
Protein Domain—Ligand Interactions			
γ^{32}P-labeled GST-WW-domain of human YAP	substitutional analysis (including phospho amino acids) of WBP-1 derived peptide GTPPPPYTVG	SPOT	[79]
WW domain of human Yes-Associated Protein (hYAP)	substitutional analysis of a peptide derived from the WW binding protein WBP	SPOT	[323]

TABLE 7.4. Continued

Screening Molecule	Library	Technology	Reference
vasodilator-stimulated phosphoprotein (VASP) and Mena (mammalian homolog of *Drosophila* Ena) both containing EVH1 domains (Ena-VASP homology domain 1)	peptide scans of ActA (*Listeria monocytogenes*) and truncation and substitutional analyses of selected peptides; a library of proline-rich sequences from cytoskeletal proteins; a library of SH3 and WW/WWP ligands	SPOT	[382]
syntrophin PDZ domain	combinatorial library of the type XXB_1B_2SXV; substitutional analysis of a peptide ligand for the syntrophin PDZ domain	SPOT	[501]
SH3 domains of amphiphysin 1 and endophilin 2	peptide scan of synaptojanin and substitutional analyses of selected peptides; peptides derived from dynamins from *Drosophila*, rat, humans	SPOT	[73]
α—actinin	peptide scan of zyxin	SPOT	[120]
^{32}P-labeled GST-hYAP WW1 and FE65 WW domain; several mutated WW domains to switch the specificity between hYAPWW1 and FE65	PPPP**X**PAAAA and 19 L-amino acid substitutions at position **X**	SPOT	[138]
syntrophin PDZ domain	library of C-terminal peptides from all known human proteins (3514 spots); substitutional analysis of a selected peptide	SPOT	[211]
phospho-tyrosine –binding domain (PTB) of Dab1	amyloid precursor protein (APP) derived peptides and known ligands of PTB domains; substitutional analyses of selected peptides	SPOT	[220]
tumor necrois factor receptor associated factors (TRAFs) 1, 2, 3, and 6	L- and D-amino acid substitutional analyses of peptides derived from the cytoplasmatic portion of CD40	SPOT	[434]
vasodilator-stimulated phosphoprotein (VASP) EVH1 domain	substitutional analysis of an ActA (*Listeria monocytogenes*) derived peptide	SPOT	[19]
STAT5a and a mutant STAT5a (STAT: signal transducer and activator of transcription)	peptide scan of the intracellular domain of erythropoietin receptor (EPOR)	SPOT	[44]
^{32}P-labeled GST, GST- RSP5-WW1, GST- RSP5-WW2, GST-RSP5-WW3; (RSP = reversion pf Spt phenotype, *Saccharomyces cerevisiae*)	truncation analyses, alanine substitution scans, and phospho amino acid analogs of carboxy-terminal domain (CTD) derived peptides of the RNA polymerase II large subunit	SPOT	[75]
Mena (mammalian homolog of *Drosophila* Enabled (Ena) protein)	peptide scan of zyxin and Phe-Ala substitution analogs of selected peptides	SPOT	[121]
vasodilator-stimulated phosphoprotein (VASP)	peptide scans of Fyb/SLAP1 and 2 (Fyn- and SLP-76-associated protein	SPOT	[277]

(cont.)

TABLE 7.4. *Continued*

Screening Molecule	Library	Technology	Reference
murine EVH1 domain of Vesl 2	partial peptide scan of mGluR5 (metabotropic glutamate receptor 5); length analysis and X-scan of a selected peptide (X = degenerate position)	SPOT	[27]
C-terminal part of p53-binding protein-2 GST fusion protein (including SH3 domain) labeled with ^{32}P	peptide scan of hYAP (Yes-associated protein); alanine scan and progressive alanine substitutions of a selected peptide	SPOT	[139]
several mutated GST-Cterm. Variants	peptide derived from the hYAP binding site and 2 substitution analogs		
^{32}P-labeled GST-hYAP WW1 domain	partial peptide scan of p53-BP-2; alanine scan, progressive alanine substituions, and phospho amino acid and Glu and Asn substitution analogs of a selected peptide		
cell lysates from erythropoietin stimulated and unstimulated cells; JAK2, SHP1, PLCγ	peptide scan of the intracellular domain of the EPO receptor	SPOT	[278]
WW domain of human Yes-Associated Protein (hYAP)	substitutional analyis and truncation analysis of the peptide GTPPPPYTVG	SPOT	[419]
peptide ligand of the human Yes-Associated Protein (hYAP) WW domain	substitutional analysis human Yes-Associated Protein (hYAP) WW domain (44-mers)	SPOT	[546, 547]
Gads C-SH3-GST fusion protein; Grb2 C-SH3-GST fusion protein	partial peptide scan of SLP-76; alanine substitution scan and complete L-amino acid substitutional analysis of a selected SLP-76-derived peptide; peptides derived from the Gads binding partners UBPY, AMSH, BLNK;	SPOT	[37]
human Yes-associated protein (hYAP) WW domain, rFE65 WW domain, and mMena EVH1 domain displayed on phage	peptide ligands for hYAP WW (GTPPPPYTVG), rFE65 WW (PPPPPPPLPAPPPQP), and mMena EVH1 (SFEFPPPPTDEELRL) domains; negative control peptide	SPOT	[42]
2 POD-labeled peptide ligands of WW domains	arrays with 42 WW domains	DIPP	[393]
6 POD labeled WW domains	6 substitutional analyses of peptide ligands	SPOT	
5 POD labeled WW domains	phospho-peptide analogs of the WW ligand GPPPP-*pAA*-P	SPOT	
SH3 domain of ScPex13p GST fusion protein (peroxisomal protein import machinery)	peptide scans of ScPex5p and ScPex14p; L-amino acid substitutional analysis and truncation analysis of a selected ScPex5P peptide	SPOT	[420]

TABLE 7.4. *Continued*

Screening Molecule	Library	Technology	Reference
22 different dye-labeled peptide ligands (*e.g.*, dye-GTPPPPYTVG)	11859 variants with combinatorially designed substitutions by proteinogenic and unnatural amino acids at three positions of the hYAP WW domain	SPOT Ch. Lig.	[548]
VASP-EVH1 domain GST fusion protein	peptoid building block substitutional analysis of a VASP-EVH1 peptide ligand; combinatorial peptoid building block library of two positions within a VASP-EVH1 peptide ligand	SPOT	[617]
phospho-specific antibodies MPM2, 3F3/2 and PY20	complete L-amino acid substitutional analysis of AXXXX[pS/pT]XXXXA and AXXXX[pYT]XXXXA	SPOT	[477]
GST fusions of GRB-SH2, SHP2-CSH2, GRB7-SH2 and GRB 10-SH2	complete L-amino acid substitutional analysis of AXXXX[pYT]XXXXA		
protein kinase A (PKA), Cdc 15	complete L-amino acid substitutional analysis of $AX_1X_1X_1X_1[S/T]X_1X_1X_1X_1A$ X = L-amino acids exept for Cys, X1 = L-amino acids exept for Cys, Ser, Thr		
Protein Domain Arrays			
peptide ligand of the human Yes-Associated Protein (hYAP) WW domain	substitutional analysis human Yes-Associated Protein (hYAP) WW domain (44-mers)	SPOT	[546, 547]
2 POD-labeled peptide ligands of WW domains	arrays with 42 WW domains	DIPP	[393]
6 POD labeled WW domains	6 substitutional analyses of peptide ligands	SPOT	
5 POD labeled WW domains	phospho-peptide analogs of the WW ligand GPPPP-*pAA*-P	SPOT	
22 different dye-labeled peptide ligands (*e.g.*, dye-GTPPPPYTVG)	11859 variants with combinatorially designed substitutions by proteinogenic and unnatural amino acids at three positions of the hYAP WW domain	SPOT Ch. Lig.	[548]
Chaperones			
DnaK *(Escherichia coli)*	peptide scan of the heat shock transcription factor σ^{32} *(E. coli)*	SPOT	[359]
DnaK *(Escherichia coli)*	peptide scans of 37 substrate proteins; combinatorial peptide libraries with two defined positions (XXXXXBXBXXXXX, XXXXXBXXBXXXX)	SPOT	[481]
GroEL *(Escherichia coli)*	peptide scan of Raf-1 protein kinase	SPOT	[209]
SecB *(Escherichia coli)*	peptide scans of 23 substrate proteine	SPOT	[260]

(*cont.*)

TABLE 7.4. Continued

Screening Molecule	Library	Technology	Reference
DnaK (Escherichia coli), 4 mutant analogs and 1 truncation analog	peptide scans of the λ CI protein	SPOT	[482]
Triggerfactor (TF) (Escherichia coli) and TF-fragments	peptide scans of E. coli proteins: EF-Tu (elongation factor Tu), MetE (methionine biosynthesis enzyme), ICDH (isocitrate dehydrogenase) GlnRS (glutamine-tRNA-synthetase), alkaline phosphatase, β-galactosidase, FtsZ (involved in cell division), GBP (galactose binding protein), L2 (a ribosomal protein), lambda cI, pro OmpA (pro-outer membrane protein A), sigma 32, SecA (involved in secretion and murine DHFR (dehydrofolate reductase); yeast cytochrome B2, F1 β- and Su9-ATPase subunits; *Photinus pyralis* luciferase, RNaseT1 from *Aspergillus oryzae*; human PrP (prion protein)	SPOT	[399]
DnaJ and DnaK (Escherichia coli)	peptide scans of 14 substrate proteins	SPOT	[483]
Tim9p-Tim10p complex of the mitochondrial intermembrane space (Tim = translocase of inner membrane)	peptide scan of the ADP/ATP carrier (AAC) and Tim 23p	SPOT	[96]
Hsc66 (*E. coli*); DnaK (*E. coli*); Hsc20 (*E. coli*) as competitor in the presence of ADP	peptide scans of the Fe/S protein *E. coli* IscU	SPOT	[210]
Dnak (*E. coli*)	peptide scans of human and mouse interferon-γ	SPOT	[563]
Trigger Factor and DnaK from *E. coli*	peptides derived from Trigger Factor and DnaK substrate proteins	SPOT	[112]
Enzyme-Substrate and Enzyme-Inhibitor Interactions			
Kinases and Phosphatases			
protein kinase A (PKA); cytoplasmic (kinase) domains of transforming growth factor β (TGFβ) type I and II receptors; γ^{32}P-ATP	several combinatorial libraries with fixed amino acids and randomized positions (17 amino acids excludinf Cys, Ser, Thr) at different positions	SODA	[337]
protein tyrosine kinase Lyn; anti-phoshpo-tyrosine antibody and POD-labeled second antibody for detection	panel of 23 PKCδ-derived 15-mer peptides	SPOT	[534]
catalytic subunit of protein kinase A or cGMP activated protein kinase G and γ^{32}P-ATP	combinatorial library of the type $XXXO_1O_2XXX$ and iterative deconvolution libraries	SPOT	[539]
anti-porcine tubulin-tyrosine ligase (TTL) mab 1D3	substitutional and truncation analysis of a TTL-derived peptide; combinatorial library (XXB_1B_2XX) and deconvolution libraries	SPOT	[159]

TABLE 7.4. Continued

Screening Molecule	Library	Technology	Reference
streptavidin-alkaline phosphatase	combinatorial library (XXB$_1$B$_2$XX) and deconvolution libraries		
cAMP-dependent protein kinase	combinatorial library (XXB$_1$B$_2$XX) and deconvolution libraries		
casein kinase I (CKI) or II (CKII) protein kinase C (PKC) catalytic subunit of protein kinase A (PKA) and γ^{32}P-ATP	peptides derived from the phosphorylation sites RRASVA, QKRPSQRAKYL, DDDDEESITRR, DDDSDDDAAAA	SPOT	[549]
potato acid phosphatase	RRASS*VA, QKRPS*QRAKYL, DDDDEES*ITRR, DDDS*DDDAAAA (S* = phospho-serine)		
protein kinase A (PKA), protein kinase C (PKC), casein kinase I (CKI), casein kinase II (CKII) and γ^{32}P-ATP	panel of human phosphorylation sites	SPOT	[550]
^{125}I-calmodulin	phosphorylated and unphosphorylated peptides derived from cell-cell adhesion molecule (C-CAM) isoforms S and L and single site substitution analogs	SPOT	[126]
porcine protein kinase C β and γ^{32}P-ATP	partial peptide scans of the C-CAM isoforms S and L		
androgen receptor protein kinase and γ^{32}P-ATP	panel of kinase substrates known from literature	SPOT	[370]
enzyme I of bacterial sugar phosphotransferase system and ^{32}P-phospho*enol*pyruvate	combinatorial library of the type XXXXXO$_1$HO$_2$XXXXX and iterative deconvolution libraries	SPOT	[376]
catalytic subunit of protein kinase A or cGMP activated protein kinase G and γ^{32}P-ATP	combinatorial libraries of the type XXXO$_1$O$_2$XXX and XXXRRO$_1$O$_2$X	SPOT	[540]
catalytic subunit of protein kinase A or cGMP activated protein kinase G and γ^{32}P-ATP	combinatorial libraries of the type O$_1$KARKKSNO$_2$, O$_1$O$_2$TQAKRKKSLA, O$_1$O$_2$KATQAKRKKSLA, TQAKRKKSLAO$_1$O$_2$, and TQAKRKKSLAMAO$_1$O$_2$	SPOT	[115]
catalytic subunit of protein kinase A or cGMP activated protein kinase G and γ^{32}P-ATP	combinatorial library of the type XXXO$_1$O$_2$XXX and deconvolution libraries: XXXRKO$_1$O$_2$X, XRKKKO$_1$O$_2$X, O$_1$RKKKKKO$_2$, LRKKKKKHO$_1$O$_2$, and O$_1$O$_2$LRKKKKKH	SPOT	[116]
GST-Dyrk1A-Δ(500–763) and γ^{32}P-ATP	partial substitutional analysis of RRRFRPASPLRGPPK	SPOT	[207]
maize Ca^{2+}-dependent protein kinase (CDPK-1) and γ^{32}P-ATP	partial substitutional analysis of LARLHSVRER	SPOT	[334]
protein kinase A (PKA), casein kinase II (CKII), p42-MAP kinase (Erk2) and γ^{32}P-ATP	kinase substrates Kemptide and Elk1; protein kinase inhibitor 2	CLPP	[338]

(cont.)

TABLE 7.4. Continued

Screening Molecule	Library	Technology	Reference
p60^{c-src} protein tyrosine kinase and γ^{32}P-ATP	Ttds-EEIYGEFF	DIPP	[143]
strepavidin-Cy3 conjugate; avidin-Cy5 conjugate	biotin, HPYPP and WSHPQFEK		
anti-human insulin mab HB125 and anti-mouse IgG-Cy5 conjugate	biotin, wGeyidvk, pqrGstG, WSHPQFEK and YGGFL		
WEHI-231 cells and negative control cells	wGeyidvk		
Practical protocols to measure phosphorylation of peptides on peptide arrays prepared by SPOT synthesis exemplified by the kinase CKII		SPOT	[50]
cyclic GMP-dependent protein kinase (phosphorylation γ^{32}P-ATP)	XXXRKB$_1$B$_2$X (deconvolution libraries described but data not shown)	SPOT	[117]
^{32}P-labeled cyclic GMP-dependent protein kinase (binding)	B$_1$RKKKKKB$_2$ (preceding libraries described but data not shown)		
GST-protein tyrosine phosphatase 1B; mixture of anti-phospho-tyrosine antibodies, POD-conjugated anti-mouse antibody	15-mer phospho-tyrosine-containing peptides derived from human insulin receptor; degenerate SPOT library of general structure AABX$_1$ZX$_2$BAA (Z = phospho-tyrosine, B = mixture of 20 natural amino acids, X defined amino acids) and peptides derived from human STAT 5A containing both phospho-tyrosine and phospho-threonine residues	SPOT	[140]
GST-protein tyrosine phosphatase ß; mixture of anti-phospho-tyrosine antibodies, POD-conjugated anti-mouse antibody	15-mer phospho-tyrosine-containing peptides derived from human Tie2 receptor containing additional phospho-serine or phospho-threonine residues		
GST-fusion protein of substrate trapping mutant of protein tyrosine phosphatase 1B (D181A), radioactively labeled by incubation with protein kinase A in the presence of γ^{32}P-ATP	15-mer phospho-tyrosine-containing peptides derived from human insulin receptor; degenerate SPOT library of general structure AABX$_1$ZX$_2$BAAA (Z = phospho-tyrosine, B = mixture of 20 natural amino acids, X defined amino acids), progressive alanine substitution of MTRDIYETDYZRKGG (Z = phospho-tyrosine), alanine scan of TRDIYETDYZRKGGKGL, substitutional analysis for X in MTRDIYETDXZRKGG (Z = phospho-tyrosine)		
c-Src and γ^{32}P-ATP or anti-phospho-tyrosine antibody	panel of known kinase substrates	DIPP	[218]
c-Src and γ^{32}P-ATP	c-Src substrate	DIPP	[219]

PEPTIDE ARRAYS IN PROTEOMICS AND DRUG DISCOVERY 255

TABLE 7.4. *Continued*

Screening Molecule	Library	Technology	Reference
p60 and FITC-labeled anti-phospho-tyrosine antibody	YIYGSFK, ALRRASLG, KGTGYIKTG and monophosphorylated derivatives	DIPP	[310]
PKA or p60 and FITC-labeled anti-phospho-tyrosine or anti-phospho-serine antibody	YIYGSFK, ALRRASLG, YIYGSFK and monophosphorylated derivatives	DIPP	[311]
NEK6 kinase and γ^{32}P-ATP	human annotated phosphorylation sites and mutational analysis of GLAKSFGSPNRAY	DIPP	[330]
Protein tyrosine kinase ABL and γ^{32}P-ATP or FITC-labeled anti-phospho-tyrosine antibody	collection of 720 peptides (13-mers) derived from human annotated phosphorylation sites (databases SwissProt and Phosphobase)	DIPP	[443]
GST-ERK-2 or GST-MEK-EE, anti-phospho-Elk-1 antibody, POD labeled anti-rabbit antibody	double peptide synthesis of ELK1-substrate peptide FWSTLSPIAPR, D-loop peptide KGRKPRDLELP and control peptide MNGGAANGRIL	SPOT	[141]
GST-fusion protein-protein tyrosine phosphatase 1B; anti-phospho-tyrosine antibody, POD-conjugated anti-mouse antibody	double peptide synthesized SPOTs containing the PTP-1B substrate IYETDYZRKGG (Z = phospho-tyrosine) and on the second site a scan of 11-mers overlapping peptides derived from the cytoplasmic domain of insulin receptor		
GST-fusion protein of substrate trapping mutant of protein tyrosine phosphatase 1B (D181A), radioactively labeled by incubation with protein kinase A in the presence of γ^{32}P-ATP	double peptide synthesis of insulin receptor derived substrate peptide IYETDYZRKGG (Z = phospho-tyrosine) and binding motif for YAP WW1 domain and p53 binding protein-2, respectively (YPPYPPPPYPS)		
PKA and Abl; Pro-Q Diamond phospho-specific stain or FITC-labeled anti-phospho-tyrosine antibody for detection	Kemptide, p60 c-src (521-533), delta sleep inducing peptide (DSIP), phosphoDSIP, CamKII peptide (GS1-10), different proteins	DIPP	[350]
GST-proteins of substrate trapping mutants of tyrosine phosphatases (PTP-H1, SAP-1, TC-PTP, PTP-1B) radioactively labeled by incubation with protein kinase A in the presence of γ^{32}P-ATP	7 peptides (14-mers) derived from human GHR together with the appropriate phospho-tyrosine-containing derivatives	SPOT	[397]
tyrosine kinase p60 c-scr and FITC-labeled anti-phospho-tyrosine antibody	deletion, alanine scanning, positional scanning- and full combinatorial mixture libraries of CGG-YIYGSFK (p60 c-src substrate)	DIPP	[558]

(*cont.*)

TABLE 7.4. *Continued*

Screening Molecule	Library	Technology	Reference
phospho-specific antibodies MPM2, 3F3/2, PY20	complete L-amino acid substitutional analysis of AXXXX[pS/pT]XXXXA and AXXXX[pYT]XXXXA	SPOT synthesis	[477]
GST fusions of GRB-SH2, SHP2-CSH2, GRB7-SH2, GRB 10-SH2 protein kinase A (PKA), Cdc 15	complete L-amino acid substitutional analysis of AXXXX[pYT]XXXXA complete L-amino acid substitutional analysis of $AX_1X_1X_1X_1[S/T]X_1X_1X_1X_1A$ $X =$ L-amino acids exept for Cys, $X1 =$ L-amino Acids exept for Cys, Ser, Thr		
tyrosine kinase Abl and CKII together with FITC-labeled anti-phospho-tyrosine antibody and γ^{32}P-ATP, respectively	collection of more than 13 000 peptides (13-mers) derived from human proteins	DIPP	[395]
FITC-labeled anti-phospho-tyrosine antibody	2923 phosphopeptides (13-mers) derived from human proteins		
tyrosine kinase Abl and γ^{32}P-ATP	1433 randomly generated 15-mers peptides	DIPP	[484]
kinases PDK1, Tie2, CKII, PKA GSK3 and γ^{32}P-ATP	peptide scans trough MBP and Tie2, 720 human annotated phosphorylation sites (13-mers peptides, 2923 phosphopeptides (13-mers) derived from human proteins, 1394 peptides derived from activation loops of human kinases	DIPP	[503]
Proteases			
chymotrysin, papain	combinatorial dipeptide libraries (400 spots) for substrate identification	SPOT	[122]
trypsin	substitution libraries GGRBG and GGKBG (B = 20 amino acids)	SPOT	[275]
trypsin	internally quenched libraries of the type XXBXX and XXB_1B_2XX for substrate identification	SPOT	[450]
porcine pancreatic elastase	subsitutional analyses and truncation analyses of a peptide derived from the 3rd domain of turkey ovomucoid inhibitor (inhibitor identification)	SPOT	[206]
trypsin; *Escherichia coli* outer membrane protease T (OmpT)	internally quenched substitution analog libraries of known substrates (substrate specificity mapping)	SPOT	[102]
porcine pancreatic elastase	combinatorial library of all 400 dipeptide combinations (N-terminally labeled with FTC) from the genetically encoded amino acids	SPOT	[295]
caspase-3	complete L-amino acid substitutional analysis of the caspase-3 substrate Abz-VDQMDGW (Abz = amino benzoic acid	SPOT	[455]

TABLE 7.4. *Continued*

Screening Molecule	Library	Technology	Reference
caspase-3	substitutional analysis of a known substrate; protein sequence derived, combinatorial, positional scanning, and randomly generated peptide libraries (data not shown) for substrate identification	SPOT	[457]
cathepsin C and cathepsin L	peptidic cathepsin inhibitors tethered to fluorescence-labeled peptide nucleic acids	DIPP	[596]
elastase, cathepsin G	partial peptide scan and substitution analogs of the transferrin receptor stalk	SPOT	[254]
trypsin, granzyme B and thrombin	Ac-LGPL-U, Ac-Nle-TPK-U, Ac-IEPD-U and 361 variants of Ac-AOOK-U (Cys omitted); U = alkoxyamino-substituted coumarin derivative	DIPP	[486]
caspase 3, granzyme B, Jurkat cell lysates and granzyme B activated apoptotic Jurkat cell lysates	peptide derivatives tethered to fluoresceine-labeled peptide nucleic acids	DIPP	[597]
lysyl-endopeptidase (LEP), chymotrypsin, V8 protease; BSA and ConA as negative controls; TLCK and Boc-glu as inhibitors	3 substrate peptides with LEP, chymotrypsin or V8 protease specificity		[258]
Isomerases			
human peptidyl-prolyl *cis/trans* isomerase Pin1, sequential electo-blotting to PVDF-membranes, incubation of PVDV-membranes with anti-Pin1 antibody followed by secondary antibody	peptide scan of human Cdc25C including phosphorylated S/T-P moieties	SPOT	[336]
Triggerfactor (TF) (*Escherichia coli*) and TF-fragments	peptide scans of *E. coli* proteins: EF-Tu (elongation factor Tu), MetE (methionine biosynthesis enzyme), ICDH (isocitrate dehydrogenase), GlnRS (glutamine-tRNA-synthetase), alkaline phosphatase, β-galactosidase, FtsZ (involved in cell division), GBP (galactose binding protein), L2 (a ribosomal protein), lambda cI, pro OmpA (pro-outer membrane protein A), sigma 32, SecA (involved in secretion) and murine DHFR (dehydrofolate reductase), yeast cytochrome B2, F1β- and Su9-ATPase subunits, *Photinus pyralis* luciferase, RNaseT1 from *Aspergillus oryzae*, and human PrP (prion protein)	SPOT	[399]

(*cont.*)

TABLE 7.4. Continued

Screening Molecule	Library	Technology	Reference
Trigger Factor and DnaK from *E. coli*	peptides derived from Trigger Factor and DnaK substrate proteins	SPOT	[112]
Diverse Enzymes			
pertussis toxin (*Bordetella pertussis*) and ATP + ^{32}P-NAD: ADP ribosylation	5 to 25-mer C-terminal peptides of the G-Protein α-subunits of G_{i3}, G_i, G_s, G_{o1}, G_{o2}, G_{oX1}, T_{rod}, G_z, $G_{q/11}$ and G_h, partial substitutional analysis of the 16-mer C-terminal peptide of the G_{i3}α-subunit	SPOT	[576]
peptide detection by MALDI-TOF MS concanavalin A (lectin binding) β-1,4-galactosyltransferase galactosidase	Ac-IYAAPKKKC-NH2 α-mannose *N*-acetylglucosamine *N*-acetyllactosamine	DIPP	[532]
Rhodococcus rhodochrous epoxide hydrolase, *Electrophorus electricus* acetylcholine esterase, bovine pancreas trypsin, bovine intestinal mucosa alkaline phosphatase	5 substrates for the 4 hydrolases among them being 2 peptidic substrates for trypsin and caspases	DIPP	[616]

Applications of Peptide Arrays: Miscellaneous
DNA Binding Peptides

oligonucleotide with an endonuclease *Eco*RII specific recognition site	peptide scan of *Eco*RII and substitutional analyses of selected peptides	SPOT	[465]
Practical protocols for the analysis of DNA-protein and DNA-peptide interactions		SPOT	[466]

Metal Ion Binding Peptides

anti-transforming growth factor (TGF) mab Tab2; silver(I); dsDNA	combinatorial libraries of the type XXB_1B_2XX and XB_1XB_2XX	SPOT	[269]
technetium–99m; nickel(II); anti-transforming growth factor α (TGFα) mab Tab2	combinatorial with 2 defined positions (linear, cyclic, L- and D- amino acids) and deconvolution libraries	SPOT	[270]
technetium–99m	combinatorial hexapeptide library $B_1XB_2XB_3X$ (8000 peptide mixtures) and deconvolution libraries	SPOT	[346]
anti-transforming growth factor α (TGFα) mab Tab2; anti-p24 (HIV-1) mab CB4-1	substitutional analyses; combinatorial library of the type $XXXXB_1B_2B_3XXXX$; positional scanning combinatorial library	SPOT	[494]
lead(II); calcium(II); iron(III); silver(I); nickel(II); technetium-99m	combinatorial libraries of the type XB_1XB_2XX; deconvolution libraries; positional scanning library		
anti-p53 mAb DO-1	peptides SDLHKL, DSLGKL, and SGLHKL	LDPS	[265]
fluorescence shift following Pb^{2+} or As^{3+} binding	dns(dansyl)-ECEE, dns-CCCC, dns-GGGG, dns-EEEE		
fluorescence shift following Pb^{2+} binding	dns(dansyl)-ECEE, dns-CCCC, dns-GGGG, dns-EEEE	LDPS	[172]

PEPTIDE ARRAYS IN PROTEOMICS AND DRUG DISCOVERY

TABLE 7.4. Continued

Screening Molecule	Library	Technology	Reference
Arrays of 4 α-Helix Bundle Mini-Proteins			
Fe^{III}-protoporphyrin IX	libraries of antiparallel 4-α-helix bundle hemoproteins	SPOT	[441]
copper(II), cobalt(II)	libraries of antiparallel 4-α-helix bundle metalloproteins	SPOT	[495]
Identification, Characterization, and Optimization of Peptidic Ligands			
anti-porcine tubulin-tyrosine ligase (TTL) mab 1D3	substitutional and truncation analysis of a TTL-derived peptide; combinatorial library (XXB_1B_2XX) and deconvolution libraries	SPOT	[159]
streptavidin-alkaline phosphatase	combinatorial library (XXB_1B_2XX) and deconvolution libraries		
cAMP-dependent protein kinase	combinatorial library (XXB_1B_2XX) and deconvolution libraries		
streptavidin	substitution analogs of Strep-tag II	SPOT	[493]
amyloid β-peptide (Aβ)	peptide scan of amyloid β-peptide (Aβ); truncation analysis and alanine scan of a selected peptide	SPOT	[544]
amyloid β-peptide (Aβ); Aβ-derived peptide LBMP1620	peptide scans of amyloid β-peptide (Aβ) and truncation analyses of selected peptides	SPOT	[545]
anti-cholera toxin mab TE33; anti-p24 (HIV-1) mab CB4-1	D-amino acid substitutional analyses for the transformation of linear wt-epitopes into all-D peptide analogs	SPOT	[274]
strepavidin-FITC conjugate	2888 O_1O_2-P/L-Q/F-F/L derived peptides bound to functionalized teflon membranes	SODA	[298]
plasma-derived factor VIII (pdFVIII), recombinant factor VIII (rFVIII)	dual position scanning library with the sublibraries O_1O_2XXXXXX, XXO_1O_2XXXX, XXXXO_1O_2XX, XXXXXXO_1O_2; combination library of hits from the dual position sublibraries	SPOT	[12]
anti-p24 (HIV-1) mab CB4-1	libraries with all possible single step transition pathways between 3 sequentially and structurally unrelated mab CB4-1 binding peptides; substitutional analyses of peptides from selected transition pathways	SPOT	[212]
^{35}S-labeled calmodulin	mastpoparan (Ac-INLKALAALAKKIL) and truncation and deletion analogs	SPOT	[223]
pIII phage display library of calmodulin variants	mastpoparan (Ac-INLKALAALAKKIL) and truncation and deletion analogs		

(cont.)

TABLE 7.4. *Continued*

Screening Molecule	Library	Technology	Reference
pIII phage display scFv library presenting the antibody repertoire of a mouse immunized with the recombinant extracellular domain of the C3a receptor	partial peptide scan of the extracellular domain of the C3a receptor		
Fe^{III}-*meso*-tetrakis(4-carboxyphenyl)porphyrin (Fe^{III}-TCPP)	alanine walk, partial substitutional anlaysis, and truncation analysis of an anti-porphyrin mab CDR-derived peptide	SPOT	[535]
biotinylated α-bungarotoxin (snake neurotoxin)	combinatorial 14-mer peptide library with 5 invariant, 4 partially variant and 5 totally variant positions	SPOT	[57]
plasma-derived factor VIII (pdFVIII), recombinant factor VIII (rFVIII)	complete substitutional analyses of the factor VIII binding peptide EYKSWEYC		[409]
^{125}I vascular endothelial growth factor (VEGF)	complete D-amino acid substitutional analyses of VEGF receptor II-derived peptides	SPOT	[418]
α-thrombin	peptide scans of clones T10-11 and T10-39 identified by mRNA display with α-thrombin as target; L-substitutional analyses of T10-11 and T10-39	SPOT	[439]
Chemical Transformations			
Dpab-Ala-O-FM = Methyl Red (2{[4-(dimethylamino)phanyl]azo} benzoic acid)-labeled alanine α-formylmethyl (FM) ester aldehydes for the evaluation of amide chemical ligation strategies	combinatorial libraries of all possible dipeptide combinations	SPOT	[538]
combinatorial dipeptide libraries were used to elucidate the reactivity of the amino termini and amino acid side chains to be non-enzymatically modified by reducing sugars leading to advanced glycation end products (AGEs) or crosslinked by BSA-AGE		SPOT	[377]
Cell Binding to Peptide Arrays			
p60$^{c\text{-}src}$ protein tyrosine kinase and γ^{32}P-ATP	Ttds-EEIYGEFF	DIPP	[143]
strepavidin-Cy3 conjugate; avidin-Cy5 conjugate	biotin, HPYPP, and WSHPQFEK		
anti-human insulin mab HB125 and anti-mouse IgG-Cy5 conjugate	biotin, wGeyidvk, pqrGstG, WSHPQFEK, and YGGFL		
WEHI-231 cells and negative control cells	wGeyidvk		
3T3 fibroblasts	Gly-Arg-Gly-Asp-Ser	DIPP	[609]
Review for therapeutic cancer targeting peptides including a short peptide microarray section and experimental data of an array with 42 different peptides that was probed with Jurkat T-lymphoma cells		DIPP	[8]

PEPTIDE ARRAYS IN PROTEOMICS AND DRUG DISCOVERY 261

TABLE 7.4. *Continued*

Screening Molecule	Library	Technology	Reference
T-cell Epitope Mapping			
HLA-DR1 and HLA-DR4 molecules, HLA-DR4 binding self peptide as competitor, malaria-infected human serum	partial peptide scans of the merozoite surface glycoprotein 1 (MSP1) of *plasmodium falciparum*	SPOT	[163]
9C5.D8-H T-cell hybridoma	peptide scan of rabies virus nucleoprotein	SPOT	[394]
Identification of protein-protein interaction partners			
[35]S-Met-labeled protein extract from Jukat cells containing the novel WD-repeat protein FAN	peptide scans of the TNFR-R55 cytoplasmic domain	SPOT	[2]
syntrophin PDZ domain	library of C-terminal peptides from all known human proteins (3514 spots); substitutional analysis of a selected peptide	SPOT	[211]
cell lysates from erythropoietin stimulated and unstimulated cells; JAK2, SHP1, PLCγ	peptide scan of the intracellular domain of the EPO receptor	SPOT	[278]
human Yes-associated protein (hYAP) WW domain, rFE65 WW domain, and mMena EVH1 domain displayed on phage	peptide ligands for hYAP WW (GTPPPPYTVG), rFE65 WW (PPPPPPPLPAPPPQP), and mMena EVH1 (SFEFPPPPTDEELRL) domains; negative control peptide	SPOT	[42]
Peptidomimetics			
synthesis of peptide nucleic acid (PNA) oligomer arrays on polymer membranes and their hybridisation with DNA probes		SPOT	[583]
simutaneous synthesis of peptides, peptomers and peptoids on continuous surfaces; amino funtionalization of cellulose membranes with epibromohydrin and 4,7,10-trioxa-1,13-tridecanediamine		SPOT	[16]
fluorescent-labeled model DNA oligonucletide	peptide nucleic acid (PNA) oligomer arrays on cellulose, polypropylene and PTFE membranes with different spacer molecules	SPOT	[355]
anti-transforming growth factor α (TGFα) mab Tab2	amino and amino-oxy-substituted 1,3,5-triazine arrays (detailed synthesis protocols	SPOT	[488]
Synthesis of 1,3,5-triazines on continuous surfaces		SPOT	[489]
synthesis of 1,3,5-trisubstituted hydantoin arrays on cellulose membranes		SPOT	[201]
anti-transforming growth factor α (TGFα) mab Tab2	peptoidic building block substitutional analyses of the mab Tab2 peptide epitope VVSHFND and of peptomers and peptoids derived thereof	SPOT	[456]
Synthesis of macrocyclic peptidomimetics which incorporate heteroaromatic building blocks such as 2,4,6-trichloro-[1,3,5]triazine, 2,4,6-trichloropyrimidine, 4,6-dichloro-5-nitropyrimidine, 2,6,8-trichloro-7-methylpurine		SPOT	[490]
anti-transforming growth factor α (TGFα) mab Tab2	randomly generated library of 8000 hexapeptoids and -peptomers	SPOT	[202]

(*cont.*)

TABLE 7.4. *Continued*

Screening Molecule	Library	Technology	Reference
VASP-EVH1 domain GST fusion protein	peptoid building block substitutional analysis of a VASP-EVH1 peptide ligand; combinatorial peptoid building block library of two positions within a VASP-EVH1 peptide ligand	SPOT	[617]

Review Articles

Description of the SPOT method in the context of other combinatorial library techniques: practical approaches, library types, binding assays with anti-porcine tubulin-tyrosine ligase (TTL) mab 1D3		SPOT	[157]
Different types of combinatorial peptide libraries are described: XB_1XB_2XX (400 spots), $XXXXB_1B_2B_3XXXX$ (8.000 spots) and a positional scanning combinatorial library XXXX(3B3X)XXXX (68.000 spots) as well as deconvolution libraries and substitutional analyses. These libraries were applied for the identification of linear antibody epitopes (anti-TGFα mab Tab2 and anti-p24 (HIV-1) mab CB4-1) and metal ion binding peptides.		SPOT	[494]
Review describing the technology and applications of peptides on pins, macroscopic DNA arrays, light-directed synthesis (DNA and peptides), peptides on paper (SPOTmethod), and inorganic combinatorial libraries		Diff.	[422]
Solid-Phase synthesis on planar supports including synthesis on glass, cellulose, cotton, and teflon membrane supports)		Diff.	[298]
Personal comments on "classical" papers in combinatorial chemistry (e.g., R. Frank on SPOT synthesis)		Diff.	[299]
Describes the mapping of linear and discontinuous interactions of antibody/antigen and protein/protein complexes using sequence-derived peptide scans, substitutional analyses and combinatorial libraries prepared by SPOT synthesis		SPOT	[447]
Mapping of linear epitopes of 2 anti-interleukin-10 (IL-10) mabs with peptide scans and substitutional analyses, mapping of the IL-10/IL-10R interaction with peptide scans, identification of the anti-TGFα mab Tab2 by combinatorial hexapeptide libraries, identification of the anti-p24 (HIV-1) mab CB4-1 with a positional scanning combinatorial library			
Summary of the SPOT synthesis technique particularly addressing small molecule and peptidomimetic arrays		SPOT	[53]
Review article describing peptide and protein arrays		Diff.	[133]
Synthesis and application of peptide and peptidomimetic/small molecule arrays prepared by SPOT synthesis; brief description of peptide library types		SPOT	[179]
Extensive review describing polymeric solid supports for SPOT synthesis (cellulose and polypropylene) amino functionalization of membrane suports, linker systems		SPOT	[592]
peptide synthesis on solid supports, examples for product quality, compatibility of polymer membranes for peptide-antibody binding assays (p24 (HIV-1) mab CB4-1 and anti-interleukin-10 mab CB/RS/3),			
biological screening (types of peptide libraries, mapping of antibody epitopes and protein-protein interactions, enzymatic modification of cellulose-bound peptides, DNA and metal binding peptides, T-cell epitope mapping, and cellular assays)			
Detailed experimental assay protocols for the mapping of linear and discontinuous antibody epitopes and protein-protein interactions		SPOT	[453]

TABLE 7.4. *Continued*

Screening Molecule	Library	Technology	Reference
Review article describing assay principles and the main applications of peptide arrays prepared by SPOT synthesis		SPOT	[454]
Review for therapeutic cancer targeting peptides including a short peptide microarray section and experimental data of an array with 42 different peptides that was probed with Jurkat T-lymphoma cells		DIPP	[8]
Review of the history of peptide arrays prepared by SPOT synthesis (chemistry, automation, analytical procedures) and comprehensive overview of library types and applications		SPOT	[160]
Short introduction into peptide arrays on membrane supports and a comprehensive directory of peptide array applications		SPOT	[161]
Extensive review of synthetic peptide arrays and microarrays as tools for functional genomics and proteomics		Diff.	[162]
Autoantibody profiling for the study and treatment of autoimmune disease (B-cell responses). The main topic is protein arrays. Short comments on peptide arrays are given		Diff.	[221]
Review describing different types of microarrays (peptides, small molecules, proteins, nucleic acids)		Diff.	[287]
Update of the review article (Reineke, 2001b) describing applications of the SPOT technology and novel developments in the field of microarrays with a special focus and experimental data for kinase substrate micro arrays		SPOT DIPP, CLPP	[443]
Protein and peptide array analysis of autoimmune diseases (peptides derived from putative rheumatoid arthritis autoantigens and myelin basic protein)		DIPP	[473]
Proteomics technologies for the study of autoimmune disease including protein arrays and few examples for peptide arrays		Diff.	[474]
Application of "one bead one compound" combinatorial libraries and chemical microarrays (small molecules and peptides)		Diff.	[288]
Review article describing combinatorial peptide library methods (array and non-array approaches) for immunological research		Diff.	[327]
Protein arrays for autoantibody profiling and fine-specificity mapping including few peptide array examples		Diff.	[476]
Peptide library review including a short peptide arrays section		Diff.	[555]
Protein array review including a short peptide arrays section		Diff.	[516]
Synthesis			
First full paper describing the SPOT synthesis (principle, supports, synthesis, assays, library types)		SPOT	[155]
Brief desciption of SPOT synthesis		SPOT	[156]
Detailed description of synthetic procedures to generate peptide mixtures with randmized positions by SPOT synthesis. The srategy is exemplified by synthesis of combinatorial peptide libraries of the type XXB_1B_2XX and XB_1XB_2XX and their screening with anti-transforming growth factor α (TGFα) mab Tab2, silver(I), and dsDNA.		SPOT	[269]
Detailed description of synthetic procedures to generate peptide mixtures with randmized positions by SPOT synthesis. The srategy is exemplified by a substitutional analysis of the linear epitope for anti-p24 (HIV-1) mab CB4-1.		SPOT	[572]
Detailed laboratory manual for synthesis of SPOT arrays and binding studies with antibodies (library types described: peptide scan, truncation analysis, substitutional analysis, combinatorial library (XXB_1B_2XX) and deconvolution libraries)		SPOT	[158]
Development of a sequence design software, optimization of coupling methods for SPOT synthesis, preparation of peptides up to 29 amino acids in length, incubation of model peptides with their cognate mabs		SPOT	[371]

(*cont.*)

TABLE 7.4. *Continued*

Screening Molecule	Library	Technology	Reference
Attachment of the p-hydroxymethyl-benzoic acid (HMB) linker to continuous cellulose membranes		SPOT	[573]
Detailed laboratory manual for synthesis of SPOT arrays and binding studies with antibodies (library types described: peptide scan, truncation analysis, substitutional analysis, combinatorial library (XXB_1B_2XX))		SPOT	[273]
Description of the BioDisk-Synthesizer for synthesis on non-porous planar surface (Compact Disk format) in a circular r/φ-array format		SPOT	[4]
Optimization of side chain deprotection procedures, optimization of the peptide density for screening purposes, correlation between signal intensity and binding affinity for binding assays (3 different detection methods)		SPOT	[276]
International patent application describing a method for manufacturing a carrier for chemical or biochemical assays. The carrier of e.g., 2.5 × 7.5 cm is structured as an array of e.g., 3, 1, or 0.1 µl wells that can be used to synthesize or immobilize e.g., peptides		MPC	[633]
Spatially addressed SPOT synthesis on polymeric membranes (synthesis and application for antibody epitope mapping)		SPOT	[590]
Detailed practical SPOT synthesis protocols (membrane functionalization; peptides, peptidomimetics, and PNA synthesis)		SPOT	[591]
SPOT synthesis protocols: synthesis on derivatized polypropylene membranes in a 96-well plate format synthesis block to create distinct "microreactors" for each spot; application of peptide arrays for epitope mapping of an anti-Ha-*ras* mab using a Ha-*ras*-derived peptide scan		SPOT	[294]
A fully automated SPOT synthesizer for oligmer synthesis of peptide nucleic acids (PNAs) is described. The key feature is a synthesis block generating reaction chambers on the synthesis membrane. The protocols are also applicable for peptide synthesis.		SPOT	[356]
Synthesis of macrocyclic peptidomimetics which incorporate heteroaromatic building blocks such as 2,4,6-trichloro-[1,3,5]triazine; 2,4,6-trichloropyrimidine; 4,6-dichloro-5-nitropyrimidine; 2,6,8-tricloro-7-methylpurine		SPOT	[490]
Automated synthesis of peptide arrays by SPOT synthesis		SPOT	[175]
Practical protocol for manual SPOT synthesis of peptide arrays		SPOT	[408]
Practical protocols for Fmoc peptide synthesis on membrane supports		SPOT	[612]
Combination of SPOT synthesis and native chemical ligation, application for the synthesis of 11859 variants of the hYAP WW domain		SPOT Ch. Lig.	[548]
Synthesis and screening of a 8000-membered hexapeptoid and –peptomer library		SPOT	[202]
Detailed protocols for the synthesis of peptide arrays, library design, and incubation procedures for antibody epitope mapping		SPOT	[459]

SPOT = spot synthesis of peptides
LDPS = light directed peptide synthesis
DIPP = directed immobilization of pre-synthesized peptides
CLPP = cross linking of pre-synthesized peptides
CPPP = co-polymerization of pre-synthesized peptides
SODA = synthesis on defined areas
MPC = mini-Pepscan cards
Ch. Lig. = chemical ligation
Diff. = different technologies

technology employed. Therefore, the authors appreciate any additional information about relevant publications in order to complete this bibliography and make it available to the public domain, either in further publications or via the World Wide Web.

Overall, the entries in Table 7.4 are arranged according to Section 7.5 "Application of peptide arrays". The Table has some additional headings that do not reflect a particular subsection in Section 7.5 because only a few examples are published. These are protein domain arrays, isomerases, diverse enzymes, and T-cell epitope mapping. Two additional Sections of Table 7.4 summarize review articles and publications describing synthesis procedures for stepwise *in situ* peptide array preparation technologies. Within any one part of Table 7.4 the entries are arranged according to the publication year and then alphabetically by author surnames. The Table lists screening molecules, library types, array preparation technologies, and each publication citation. Publications relating to more than one category are re-listed in each relevant part of the Table.

REFERENCES

[1] M. Adamczyk, J.C. Gebler, R.E. Reddy, and Z. Yu. *Bioconjugate Chem.*, 12:139, 2001.
[2] S. Adam-Klages, D. Adam, K. Wiegmann, S. Struve, W. Kolanus, J. Schneider-Mergener, and M. Krönke. *Cell*, 86:937, 1996.
[3] R.L. Adams and J. Roy. *J. Appl. Mech.*, 53:193, 1986.
[4] F. Adler, G. Türk, R. Frank, N. Zander, W. Wu, R. Volkmer-Engert, J. Schneider-Mergener, and H. Gausepohl, in Proceedings of the International Symposium on Innovation and Perspectives in Solid Phase Synthesis, edited by R. Epton (Mayflower Worldwide Ltd., Kingswinford, 1999), p. 221.
[5] V. Afanassiev, V. Hanemann, and S. Wölfl. *Nucleic Acid Res.*, 28:e66, 2000.
[6] G. Agarwal, R.R. Naik, and M.O. Stone. *J. Am. Chem. Soc.*, 125:7408, 2003a.
[7] G. Agarwal, L.A. Sowards, R.R. Naik, and M.O. Stone. *J. Am. Chem. Soc.*, 125, 580, 2003b.
[8] O.H. Aina, T.C. Sroka, M.-L. Chen, and K.S. Lam. *Biopolymers*, (Peptide Science) 66:184, 2002.
[9] T. Albrecht, A. Koch, A. Lode, B. Greve, J. Schneider-Mergener, and M. Steup. *Planta*, 213:602, 2001.
[10] N.M. Alto, S.H. Soderling, N. Hoshi, L.K. Langeberg, R. Fayos, P.A. Jennings, and J.D. Scott. *Proc. Natl. Acad. Sci. U.S.A.*, 100:4445, 2003.
[11] L.M. Alvarenga, C.R. Diniz, C. Granier, and C. Chávez-Olórtegui. *Toxicon*, 40:89, 2002.
[12] K. Amatschek, R. Necina, R. Hahn, E. Schallaun, H. Schwinn, D. Josić, and A. Jungbauer. *J. High Resol. Chromatogr.*, 23:47, 2000.
[13] C.H. Ang, J.L. Garnett, R. Levot, M.A. Long, and N.T. Yen. *J. Polym. Sci. Polym. Lett. Ed.*, 18:471, 1980.
[14] A. Asai. *Japanese J. Appl. Physics*, 28:909, 1989.
[15] A. Asai. *J. Fluids Eng.*, 114:638, 1992.
[16] T. Ast, N. Heine, L. Germeroth, J. Schneider-Mergener, and H. Wenschuh. *Tetrahedron Lett.*, 40:4317, 1999.
[17] J.A. Asturias, N. Gómez-Bayón, J.L. Eseverri, and A. Martínez. *Clin. Exp. Allergy*, 33:518, 2003.
[18] M. Baensch, R. Frank, and J. Köhl. *Microbiology*, 144:2241, 1998.
[19] L.J. Ball, R. Kühne, B. Hoffmann, A. Häfner, P. Schmieder, R. Volkmer-Engert, M. Hof, M. Wahl, J. Schneider-Mergener, U. Walter, H. Oschkinat, and T. Jarchau. *EMBO J.*, 19:4903, 2000.
[20] B. Banerjee, X. Wang, K.J. Kelly, J.N. Fink, G.L. Sussman, and V.P. Kurup. *J. Immunol.*, 159:5724, 1997.
[21] B. Banerjee, P.A. Greenberger, J.N. Fink, and V.P. Kurup. *Infect. Immun.*, 67:2284, 1999.
[22] B. Banerjee, K. Kanitpong, J.N. Fink, M. Zussman, G.L. Sussman, K.J. Kelly, and V.P. Kurup. *Mol. Immunol.*, 37:789, 2000.
[23] L. Bao, C.E. Varden, W.E. Zimmer, and R. Balczon. *Mol. Biol. Rep.*, 25:111, 1998.
[24] D.J. Barlow, M.S. Edwards, and J.M. Thornton. *Nature*, 322:747, 1986.
[25] M.A. Bartlett and M. Yan. *Polym. Mater. Sci. Eng.*, 83:451, 2000.
[26] M.A. Bartlett and M. Yan. *Adv. Mater.*, 13:1449, 2001.
[27] M. Barzik, U.D. Carl, W.-D. Schubert, R. Frank, J. Wehland, and D.W. Heinz. *J. Mol. Biol.*, 309:155, 2001.

[28] S. Basmaciogullari, M. Auterio, R. Culerrier, J.-C. Mani, M. Gaubin, Z. Mishal, J. Guardiola, C. Granier, and D. Piatier-Tonneau. *Biochemistry*, 39:5332, 2000.
[29] D.H. Beezhold, D.A. Kostyal, and G.L. Sussmann. *Clin. Exp. Immunol.*, 108:114, 1997.
[30] D.H. Beezhold, V.L. Hickey, J.E. Slater, and G.L. Sussmann. *J. Allergy Clin. Immunol.*, 103:1166, 1999.
[31] M. Beier and J.D. Hoheisel. *Nucleic Acid Res.*, 27:1970, 1999.
[32] L. Belov, O. de la Vega, C.G. dos Remedios, S.P. Mulligan, and R.I. Christopherson. *Cancer Res.*, 61:4483, 2001.
[33] R. Benters, C.M. Niemeyer, and D. Wöhrle. *Chembiochem.*, 2:686, 2001.
[34] R.H. Berg, K. Almdal, W. Peterson, A. Holm, J.P. Tam, and R.B. Merrifield, *J. Am. Chem. Soc.*, 111:8024, 1989.
[35] S. Bergmann, D. Wild, O. Diekmann, R. Frank, D. Bracht, G.S. Chhatwal, and S. Hammerschmidt. *Mol. Microbiol.*, 49:411, 2003.
[36] A. Bernhard, E. Delamarche, H. Schmid, B. Michel, H.R. Bosshard, and H. Biebuyck. *Langmuir*, 14:2225, 1998.
[37] D.M. Berry, P. Nash, S.K.-W. Liu, T. Pawson, and C.J. McGlade. *Curr. Biol.*, 12:1336, 2002.
[38] C. Bès, L. Briant-Longuet, M. Cerutti, F. Heitz, S. Troadec, M. Pugnière, F. Roquet, F. Molina, F. Casset, D. Bresson, S. Péraldi-Roux, G. Devauchelle, C. Devaux, C. Granier, and T. Chardès. *J. Biol. Chem.*, 278:14265, 2003.
[39] K. Beyer, L. Ellman-Grunther, K.M. Järvinen, R.A. Wood, J. Hourihane, and H.A. Sampson. *J. Allergy Clin. Immunol.*, 112:202, 2003.
[40] T.N. Bhat, G.A. Bentley, T.O. Fischmann, G. Boulot, and R.J. Poljak. *Nature*, 347:483, 1990.
[41] T.N. Bhat, G.A. Bentley, G. Boulot, M.I. Greene, D. Tello, W. Dell'Acqua, H. Souchon, F.P. Schwarz, R.A. Mariuzza, and R.J. Poljak. *Proc. Natl. Acad. Sci. U.S.A.*, 91:1089, 1994.
[42] K. Bialek, A. Swistowski, and R. Frank. *Anal. Bioanal. Chem.*, 376:1006, 2003.
[43] C. Billich, C. Sauder, R. Frank, S. Herzog, K. Bechter, K. Takahashi, H. Peters, P. Staehli, and M. Schwemmle. *Biol. Psychiatry*, 51:979, 2002.
[44] T. Bittorf, T. Sasse, M. Wright, R. Jaster, L. Otte, J. Schneider-Mergener, and J. Brock. *Cell. Signal.*, 12:721, 2000.
[45] J. Blake and L. Litzi-Davis. *Bioconjugate Chem.*, 3:510, 1992.
[46] A.P. Blanchard, R.J. Kaiser, and L.E. Hood. *Biosens. Bioelectronics*, 11:687, 1996.
[47] M. Blüthner, C. Schäfer, C. Schneider, and F.A. Bautz. *Autoimmunity*, 29:33, 1999.
[48] M. Blüthner, M. Mahler, D.B. Müller, H. Dünzl, and F.A. Bautz. *J. Mol. Med.*, 78:47, 2000.
[49] M. Blüthner, J. Koch, and M. Mahler. In J. Koch and M. Mahler (eds.), *Peptide Arrays on Membrane Supports—Synthesis and Applications*, Springer Verlag, Berlin, Heidelberg, p. 123, 2002.
[50] J. Bodem, and M. Blüthner. In J. Koch and M. Mahler (eds.), *Peptide Arrays on Membrane Supports—Synthesis and Applications*, Springer Verlag, Berlin, Heidelberg, p. 141, 2002.
[51] T. Boland, E.E. Johnston, A. Huber, and B.D. Ratner. In Scanning Probe Microscopy of Polymers. B.D. Ratner and V.V. Tsukruk (eds.), American Chemical Society, Washington, p. 342, 1998.
[52] T. Böldicke, F. Struck, F. Schaper, W. Tegge, H. Sobek, B. Villbrandt, P. Lankenau, and M. Böcher. *J. Immunol. Methods*, 240:165, 2000.
[53] S. Borman. *Chemical & Engineering News*, 78:25, 2000.
[54] C.A. Borrebaeck, S. Ekstrom, A.C. Hager, J. Nilsson, T. Laurell, and G. Marko-Vaga. *Biotechniques*, 30:1126, 2001.
[55] C. Bosc, R. Frank, E. Denarier, M. Ronjat, A. Schweitzer, J. Wehland, and D. Job. *J. Biol. Chem.*, 276:30904, 2001.
[56] V. Böttger. In Antibody Engineering (Springer Lab Manual). R. Kontermann and S. Dübel (eds.), Springer-Verlag, Berlin, Heidelberg, p. 460, 2001.
[57] L. Bracci, L. Lozzi, B. Lelli, A. Pini, and P. Neri. *Biochemistry*, 40:6611, 2001a.
[58] L. Bracci, A. Pini, L. Lozzi, B. Lelli, P. Battestin, A. Spreafico, A. Bernini, N. Niccolai, and P. Neri. *J. Neurochem.*, 78:24, 2001b.
[59] R. Bräuning, M. Mahler, B. Hülge-Dörr, M. Blüthner, J. Koch, and G. Petersen. In J. Koch and M. Mahler (eds.), *Peptide Arrays on Membrane Supports—Synthesis and Applications*, Springer Verlag, Berlin, Heidelberg, p. 153, 2002.
[60] M. Brehmer, L. Conrad, and L. Funk. *J. Dispension Sci. Tech.*, 24:291, 2003.
[61] T.M. Brennan. patent, US 005474796A, 1995.

[62] J. Brix, S. Rüdiger, B. Bukau, J. Schneider-Mergener, and N. Pfanner. *J. Biol. Chem.*, 274:16522, 1999.
[63] J. Brix, G.A. Ziegler, K. Dietmeier, J. Schneider-Mergener, G.E. Schulz, and N. Pfanner. *J. Mol. Biol.*, 303:479, 2000.
[64] M.E. Browning-Kelley, K. Wadu-Mesthrige, V. Hari, and G.Y. Liu. *Langmuir*, 13:343, 1997.
[65] A. Bubert, P. Schubert, S. Köhler, R. Frank, and W. Goebel. *Appl. Environ. Microb.*, 60:3120, 1994.
[66] N. Bugdayci, D.B. Bogy, and F.E. Talke. *IBM J. Res. Dev.*, 27:171, 1983.
[67] M. Burgener, M. Sänger, and U. Candrian. *Bioconjugate Chem.*, 11:749, 2000.
[68] L.L. Burns-Hamuro, Y. Ma, S. Kammerer, U. Reineke, C. Self, C. Cook, G.L. Olson, C. R. Cantor, A. Braun, and S.S. Taylor. *Proc. Natl. Acad. Sci. U.S.A.*, 100:4072, 2003.
[69] P.J. Busse, K.-M. Järvinen, L. Vila, K. Beyer, and H.A. Sampson. *Int. Arch. Allergy Immunol.*, 129:93, 2002.
[70] J.H. Butler, M. Cronin, K.M. Anderson, G.M. Biddison, F. Chatelain, M. Cummer, D.J. Davi, L. Fisher, A.W. Frauendorf, F.W. Frueh, C. Gjerstad, T.F. Harper, S.D. Kernahan, D.Q. Long, M. Pho, J.A. Walker, and T.M. Brennan. *J. Am. Chem. Soc.*, 123:8887, 2001.
[71] I. Caelen, H. Gao, and H. Sigrist. *Langmuir*, 18:2463, 2002.
[72] F. Carbonnet, D. Blanchard, C. Hattab, S. Cochet, Y. Petit-Leroux, M.-J. Loirat, J.-P. Cartron, and O. Bertrand. *Transfusion Med.*, 10:145, 2000.
[73] G. Cestra, L. Castagnoli, L. Dente, O. Minenkova, A. Petrelli, N. Migone, U. Hoffmüller, J. Schneider-Mergener, and G. Cesareni. *J. Biol. Chem.*, 274:32001, 1999.
[74] S.-F. Chang, B. Reich, J.D. Brunzell, and H. Will. *J. Lipid Res.*, 39:2350, 1998.
[75] A. Chang, S. Cheang, X. Espanel, and M. Sudol. *J. Biol. Chem.*, 275:20562, 2000.
[76] N. Chapal, S. Peraldi-Roux, D. Bresson, M. Pugniere, J.-C. Mani, C. Granier, L. Baldet, B. Guerrier, B. Pau, and M. Bouanani. *J. Immunol.*, 114:4162, 2000.
[77] P. Chatchatee, K.M. Järvinen, L. Bardina, K. Beyer, and H.A. Sampson. *J. Allergy Clin. Immunol.*, 107:379, 2001.
[78] C. Chavez-Olortegui, F. Molina, and C. Granier. *Mol. Immunol.*, 38:867, 2001.
[79] H.I. Chen, A. Einbond, S.-J. Kwak, H. Linn, E. Koepf, S. Peterson, J.W. Kelly, and M. Sudol. *J. Biol. Chem.*, 272:17070, 1997.
[80] G.Y.J. Chen, M. Uttamchandani, Q. Zhu, G. Wang, and S.Q. Yao. *Chem. Bio. Chem.*, 4:336, 2003.
[81] I. Cherny and E. Gazit. *J. Biol. Chem.*, 279:8252, 2004.
[82] Y. Chevolot, J. Martins, N. Milosevic, D. Leonard, S. Zeng, M. Malissard, E.G. Berger, P. Maier, H.J. Marthieu, D.H.G. Crout, and H. Sigrist. *Bioorg. Med. Chem.*, 9:2943, 2001.
[83] D.T. Chiu, N.L. Jeon, S. Huang, R.S. Kane, C.J. Wargo, I.S. Choi, D.E. Ingber, and G.M. Whitesides. *Proc. Natl. Acad. Sci. U.S.A.*, 97:2408, 2000.
[84] C.Y. Cho, E.J. Moran, S.R. Cherry, J.C. Stephans, S.P. Fodor, C.L. Adams, A. Sundaram, J.W. Jacobs, and P.G. Schultz. *Science*, 261:1303, 1993.
[85] S.J. Cho, H.J. Kim, and J.W. Park. *Mol. Cryst. & Liq. Cryst.*, 371:71, 2001.
[86] L. Choulier, D. Laune, G. Orfanoudakis, H. Wlad, J.-C. Janson, C. Granier, and D. Altschuh. *J. Immunol. Methods*, 249:253, 2001.
[87] L. Choulier, G. Orfanoudakis, P. Robinson, D. Laune, M.B. Khalifa, C. Granier, E. Weiss, and D. Altschuh. *J. Immunol. Methods*, 259:77, 2002.
[88] R.A. Clark, P.B. Hietpas, and A.G. Ewing. *Anal. Chem.*, 69:259, 1997.
[89] R.R. Cocco, K.-M. Järvinen, H.A. Sampson, and K. Beyer. *J. Allergy Clin. Immunol.*, 112:433, 2003.
[90] P. Cohen, D. Laune, I. Teulon, T. Combes, M. Pugnière, G. Badouaille, C. Granier, J.-C. Mani, and D. Simon. *J. Immunol. Methods*, 254:147, 2001.
[91] A. Collioud, J.-F. Clemence, M. Sänger, and H. Sigrist. *Bioconjugate Chem.*, 4:528, 1993.
[92] U. Commandeur, R. Koenig, R. Manteuffel, L. Torrance, P. Lüddecke, and R. Frank. *Virology*, 198:282, 1994.
[93] M. Corada, F. Liao, M. Lindgren, M.G. Lampugnani, F. Breviario, R. Frank, W.A. Muller, D.J. Hicklin, P. Bohlen, and E. Dejana. *Blood*, 97:1679, 2001.
[94] J. Craft and S. Fatenejad. *Arthritis Rheum.*, 40:1374, 1997.
[95] B.C. Cunningham and J.A. Wells. *Science*, 244:1081, 1989.
[96] S.P. Curran, D. Leuenberger, W. Oppliger, and C.M. Koehler. *EMBO J.*, 21:942, 2002.
[97] S.B. Daniels, M.S. Bernatowicz, J.M. Coull, and H. Köster. *Tetrahedron Lett.*, 30:4345, 1989.
[98] A. Darji, K. Niebuhr, M. Hense, J. Wehland, T. Chakraborty, and S. Weiss. *Infect. Immun.*, 64:2356, 1996.

[99] P.E. Dawson, T.W. Muir, I. Clark-Lewis, and S.B.H. Kent. *Science*, 266:776, 1994.
[100] E. DeGiglio, L. Sabbatini, S. Colucci, and G. Zambonin. *J. Biomater. Sci. Polymer Edition*, 11:1073, 2000.
[101] B. de Heij, C. Steinert, H. Sandmaier, and R. Zengerle. *Proceedings of MEMS 2002: The Fifteenth IEEE International Conference (Las Vegas) on Micro Electro Mechanical Systems (Institue of Electrical and Electronics Engineeres*, New York, N.Y., p. 706, 2002.
[102] N. Dekker, R.C. Cox, R.A. Kramer, and M.R. Egmond. *Biochemistry*, 40:1694, 2001.
[103] E. Delamarche, A. Bernhard, H. Schmid, B. Michael, and H. Biebuyck. *Science*, 276:779, 1997.
[104] E. Delamarche, A. Bernhard, H. Schmid, A. Bietsch, B. Michael, and H. Biebuyck. *J. Am. Chem. Soc.*, 120:500, 1998a.
[105] E. Delamarche, H. Schmid, A. Bietsch, N.B. Larsen, H. Rothuizen, B. Michel, and H. Biebuyck. *J. Phys. Chem. B*, 102:3324, 1998b.
[106] E. Delamarche, M. Geissler, H. Wolf, and B. Michel. *J. Am. Chem. Soc.*, 124:3834, 2002.
[107] E. Delamarche, C. Donzel, F.S. Kamounah, H. Wolf, M. Geissler, R. Stutz, P. Schmidt-Winkler, B. Michel, H.J. Mathieu, and K. Schaumburg. *Langmuir*, 19:8749, 2003.
[108] W.L. DeLano. *Curr. Opin. Struct. Biol.*, 12:14, 2002.
[109] M. Dell'Orco, P. Saldarelli, A. Minafra, D. Boscia, and D. Gallitelli. *Arch. Virol.*, 147:627, 2002.
[110] L.M. Demers, D.S. Ginger, S.-J. Park, Z. Li, S.-W. Chung, and C.A. Mirkin. *Science*, 296:1836, 2002.
[111] B. Déprez, X. Willard, L. Bourel, H. Coste, F. Hyafil, and A. Tartar. *J. Am. Chem. Soc.*, 117:5405, 1995.
[112] E. Deuerling, H. Patzelt, S. Vorderwulbecke, T. Rauch, G. Kramer, E. Schaffitzel, A. Mogk, A. Schulze-Specking, H. Langen, and B. Bukau. *Mol. Microbiol.*, 47:1317, 2003.
[113] L. Dong, J. Schneider-Mergener, and A. Kramer. In Peptides 1998: Proceedings of the Twenty-Fifth European Peptide Symposium, S. Bajusz and F. Hudecz (eds.), Akadémiai Kiadó, Budapest, p. 530, 1999.
[114] L. Dong, M. Hülsmeyer, H. Dürkop, H.P. Hansen, J. Schneider-Mergener, A. Ziegler, and B. Uchanska-Ziegler. *J. Mol. Recognit.*, 16:28, 2003.
[115] W.R.G. Dostmann, C. Nickl, S. Thiel, I. Tsigelny, R. Frank, and W.J. Tegge. *Pharmacol. Ther.*, 82:373, 1999.
[116] W.R.G. Dostmann, M.S. Taylor, C.K. Nickl, J.E. Brayden, R. Frank, and W.J. Tegge. *Proc. Natl. Acad. Sci. U.S.A.*, 97:14772, 2000.
[117] W.R.G. Dostmann, W. Tegge, R. Frank, C.K. Nickl, M.S. Taylor, and J.E. Brayden. *Pharmacol. Ther.*, 93:203, 2002.
[118] D. Dottavio. In G.E. Morris (ed.), Methods in Molecular Biology, Humana Press Inc., Totowa, NJ, p. 181, 1996.
[119] J.-C. Doury, J.-L. Goasdoue, H. Tolou, M. Martelloni, S. Bonnefoy, and O. Mercereau-Puijalon. *Mol. Biochem. Parasitol.*, 85:1997.
[120] B.E. Drees, K.M. Andrews, and M.C. Beckerle. *J. Cell. Biol.*, 147:1999.
[121] B. Drees, E. Friederich, J. Fradelizi, D. Louvard, M.C. Beckerle, and R.M. Golsteyn. *J. Biol. Chem.*, 275:22503, 2000.
[122] Y. Duan and R.A. Laursen. *Anal. Biochem.*, 216:431, 1994.
[123] A. Ducki, O. Grundmann, L. Konermann, F. Mayer, and M. Hoppert. *J. Gen. Appl. Microbiol.*, 44:327, 1998.
[124] J. Ducree, H. Gruhler, N. Hey, M. Müller, S. Bekesi, M. Freygang, H. Sandmaier, and R. Zengerle. *Proceedings of the MEMS 2000: The Thirteenth IEEE International Micro Electro Mechanical Systems Conference*, 2000.
[125] M. Edlund, I. Blikstad, and B. Öbrink. *J. Biol. Chem.*, 271:1393, 1996.
[126] M. Edlund, K. Wikström, R. Toomik, P. Ek, and B. Öbrink. *FEBS Lett.*, 425:166, 1998.
[127] J. Eichler, M. Beyermann, and M. Bienert. *Collect. Czech. Chem. Commun.*, 54:1746, 1989.
[128] J. Eichler, M. Bienert, A. Stierandowa, and M. Lebl. *Peptide Res.*, 4:296, 1991.
[129] J.K. Elder, M. Johnson, N. Milner, K.U. Mir, M. Sohail, and E.M. Southern. In DNA Microarrays, A Practical Approach, M. Schena (ed.), Oxford University Press, New York, p. 77, 1999.
[130] M.S.M. Eldin, M. Portaccio, N. Diano, S. Rossi, U. Bencivenga, A. D'Uva, P. Canciglia, F.S. Gaeta, and D.G. Mita. *J. Mol. Cat. B: Enzymatic*, 7:251, 1999.
[131] R. Elies, L.X.M. Fu, P. Eftekhari, G. Wallukat, W. Schulze, C. Granier, Å. Hjalmarson, and J. Hoebeke. *Eur. J. Biochem.*, 251:659, 1998.

[132] S.A. Elrod, B. Hadimioglu, B.T. Khuri-Yakub, E.G. Rawson, E. Richley, C.F. Quate, N.N. Mansour, and T.S. Lundgren. *J. Appl. Phys.*, 65:3441, 1989.
[133] A.Q. Emili and G. Cagney. *Nat. Biotechnol.*, 18:393, 2000.
[134] D.R. Englebretsen and D.R.K. Harding. *Peptide Res.*, 7:322, 1994.
[135] C. Erck, R. Frank, and J. Wehland. *Neurochem. Res.*, 25:5, 2000.
[136] E. Erdmanntraut, K. Wohlfart, S. Wölfl, T. Schulz, and M. Köhler. In W. Ehrfeld (ed.), *Microreaction Technology-Proceedings of the First International Conference on Microreaction Technology*. Springer, Heidelberg, p. 332, 1997.
[137] E. Erdmanntraut, T. Schulz, J. Tuchscherer, S. Wölfel, H.P. Saluz, E. Thallner, and M.J. Köhler. In Proceedings of μTAS'98, Kluwer Scientific Publishing, Utrecht, p. 217, 1998.
[138] X. Espanel and M. Sudol. *J. Biol. Chem.*, 274:17284, 1999.
[139] X. Espanel and M. Sudol. *J. Biol. Chem.*, 276:14514, 2001.
[140] X. Espanel, M. Huguenin-Reggiani, and R.H. van Huijsduijnen. *Protein Sci.*, 11:2326, 2002.
[141] X. Espanel, S. Wälchli, T. Rückle, A. Harrenga, M. Huguenin-Reggiani, and R.H. van Huijsduijnen. *J. Biol. Chem.*, 278:15162, 2003.
[142] M. Esposito, V. Venkatesh, L. Otvos, Z. Weng, S. Vajda, K. Banki, and A. Perl. *J. Immunol.*, 163:1999.
[143] J.R. Falsey, M. Renil, S. Park, S. Li, and K.S. Lam. *Bioconjugate Chem.*, 12:346, 2001.
[144] F. Fazio, M.C. Bryan, O. Blixt, J.C. Paulson, and C.-H. Wong. *J. Am. Chem. Soc.*, 124:14397, 2002.
[145] G. Ferrières, C. Calzolari, J.-C. Mani, D. Laune, S. Trinquier, M. Laprade, C. Larue, B. Pau, and C. Granier. *Clin. Chem.*, 44:487, 1998.
[146] G. Ferrières, M. Pugnière, J.-C. Mani, S. Villard, M. Laprade, P. Doutre, B. Pau, and C. Granier. *FEBS Lett.*, 479:99, 2000a.
[147] G. Ferrières, S. Villard, M. Pugnière, J.-C. Mani, I. Navarro-Teulon, F. Rharbaoui, D. Laune, E. Loret, B. Pau, and C. Granier. *Eur. J. Biochem.*, 267:1819, 2000b.
[148] H. Fiehn, S. Howitz, and T. Wegener. *Pharmazeutische Industrie*, 59:814, 1997.
[149] G.B. Fields, and R.L. Noble. *Int. J. Peptide Prot. Res.*, 35:161, 1990.
[150] V.L. Filatov, A.G. Katrukha, A.V. Bereznikova, T.V. Esakova, T.V. Bulargina, O.V. Kolosova, E.S. Severin, and N.B. Gusev. *Biochem. Mol. Biol. Int.*, 45:1179, 1998.
[151] S.P.A. Fodor, J.L. Read, M.C. Pirrung, L. Stryer, A.T. Lu, and D. Solas. *Science*, 251:767, 1991.
[152] D. Foedinger, A. Elbe-Bürger, B. Steniczky, M. Lackner, R. Horvat, K. Wolff, and K. Rappersberger. *J. Invest. Dermatol.*, 111:503, 1998.
[153] R. Frank and R. Döring. *Tetrahedron*, 44:6031, 1988.
[154] R. Frank, S. Güler, S. Krause, and W. Lindenmaier. In E. Giralt and D. Andreu (eds.), Peptides 1990 Proceedings of the 21st European Peptide Symposium. ESCOM, Leiden, p. 151, 1991.
[155] R. Frank. *Tetrahedron*, 48:9217, 1992.
[156] R. Frank. *Bioorg. Med. Chem. Lett.*, 3:425, 1993.
[157] R. Frank. *J. Biotechnol.*, 41:259, 1995.
[158] R. Frank, and H. Overwin. In G.E. Morris (ed.), Methods in Molecular Biology. Humana Press, Inc., Totowa, NJ, p. 149, 1996a.
[159] R. Frank, S. Hoffmann, M. Kieß, H. Lahmann, W. Tegge, C. Behn, and H. Gausepohl. In G. Jung (ed.), Combinatorial Peptide and Nonpeptide Libraries, VCH Verlagsgesellschaft, Weinheim, p. 363, 1996b.
[160] R. Frank. *J. Immunol. Methods*, 267:13, 2002a.
[161] R. Frank and J. Schneider-Mergener. In J. Koch and M. Mahler (eds.), Peptide Arrays on Membrane Supports—Synthesis and Applications. Springer Verlag, Berlin, Heidelberg, p. 1, 2002b.
[162] R. Frank. *Comb. Chem. High Throughput Screen.*, 5:429, 2002c.
[163] J. Fu, M. Hato, K. Igarashi, T. Suzuki, H. Matsuoka, A. Ishii, J.L. Leafasia, Y. Chinzei, and N. Ohta. *Microbiol. Immunol.*, 44:249, 2000a.
[164] J. Fu, M. Hato, H. Ohmae, H. Matsuoka, M. Kawabata, K. Tanabe, Y. Miyamoto, J.L. Leafasia, Y. Chinzei, and N. Ohta. *Parasitol. Res.*, 86:345, 2000b.
[165] T. Fujii and M. Kunimatsu. *Biol. Pharm. Bull.*, 26:771, 2003.
[166] Á. Furka. *Drug Discov. Today*, 7:1, 2002.
[167] H.F. Gaertner, K. Rose, R. Cotton, D. Timms, R. Camble, and R.E. Offord. *Bioconjugate Chem.*, 3:262, 1992.
[168] M.A. Gallop, R.W. Barrett, W.J. Dower, S.P.A. Fodor, and E.M. Gordon. *J. Med. Chem.*, 37:1233, 1994.
[169] B. Gao and M.P. Esnouf. *J. Biol. Chem.*, 271:24634, 1996a.

[170] B. Gao and M.P. Esnouf. *J. Immunol.*, 157:183, 1996b.
[171] X. Gao, P. Yu, E. LeProust, L. Sonigo, J.P. Pellois, and H. Zhang. *J. Am. Chem. Soc.*, 120:12698, 1998.
[172] X. Gao, X. Zhou, and E. Gulari. *Proteomics*, 3:2135, 2003.
[173] G. García-Casado, L.F. Pacios, A. Díaz-Perales, R. Sánchez-Monge, M. Lombardero, F.J. García-Selles, F. Polo, D. Barber, and G. Salcedo. *J. Allergy Clin. Immunol.*, 112:599, 2003.
[174] H.R. Garner. *Proceedings Lab Chips and MicroArrays Biotech. Appl.*, Zürich, 1999.
[175] H. Gausepohl and C. Behn. In J. Koch and M. Mahler (eds.), Peptide Arrays on Membrane Supports— Synthesis and Applications. Springer Verlag, Berlin, Heidelberg, p. 55, 2002.
[176] H.M. Geysen, R.H. Meloen, and S.J. Barteling. *Proc. Natl. Acad. Sci. U.S.A.*, 81:3998, 1984.
[177] H.M. Geysen, S.J. Rodda, and T.J. Mason. *Mol. Immunol.*, 23:709, 1986.
[178] H.M. Geysen, S.J. Rodda, T.J. Mason, G. Tribbick, and P.G. Schofs. *J. Immunol. Methods*, 102:259, 1987.
[179] L. Germeroth, U. Reineke, K. Dietmeier, C. Piossek, N. Heine, D. Scharn, T. Ast, M. Schulz, H. Matuschewski, A. Kramer, J. Scheider-Mergener, and H. Wenschuh. In W. Ehrfeld (ed.), Microreaction Technology: Industrial Prospects. Springer Verlag, Berlin, Heidelberg, New York, p. 124, 2000.
[180] S.M. Gonsior, S. Platz, S. Buchmeier, U. Scheer, B.M. Jockusch, and H. Hinssen. *J. Cell Science*, 112:797, 1999.
[181] A. Gouin-Charnet, D. Laune, C. Granier, J.-C. Mani, B. Pau, G. Mourad, and A. Argilés. *Clin. Sci.*, 98:427, 2000.
[182] D.M. Granoff, G.R. Moe, M.M. Giuliani, J. Adu-Bobie, L. Santini, B. Brunelli, F. Piccinetti, P. Zuno-Mitchell, S.S. Lee, P. Neri, L. Bracci, L. Lozzi, and R. Rappuoli. *J. Immunol.*, 167:6487, 2001.
[183] D.E. Gray, S.C. Case-Green, T.S. Fell, P.J. Dobson, and E.M. Southern. *Langmuir*, 13:2833, 1997.
[184] M.R. Grooves and I. Sinning. In J. Koch and M. Mahler (eds.), Peptide Arrays on Membrane Supports— Synthesis and Applications. Springer Verlag, Berlin, Heidelberg, p. 83, 2002.
[185] M.R. Groves, A. Mant, A. Kuhn, J. Koch, S. Dübel, C. Robinson, and I. Sinning. *J. Biol. Chem.*, 276:27778, 2001.
[186] J.T. Groves and S.G. Boxer. *Accounts Chem. Res.*, 35:149, 2002.
[187] S.M. Gruber, P. Yu-Yang, and S.P.A. Fodor. In J.A. Smith and J.E. Rivier (eds.), Peptides—Chemistry and Biology: Proceedings of the 12th American Peptide Symposium 1991. ESCOM, Leiden, p. 489, 1992.
[188] Z.-Z. Gu, A. Fujishima, and O. Sato. *Angew. Chem.*, 144:2172, 2002.
[189] F. Guillaumie, O.R.T. Thomas, and K.J. Jensen. *Bioconjugate Chem.*, 13:285, 2002.
[190] D. Guschin, G. Yershov, A. Zaslavsky, A. Gemmell, V.V. Shick, D. Proudnikov, P. Arenkov, and A.D. Mirzabekov. *Anal. Biochem.*, 250:203, 1997.
[191] M. Hahn, D. Winkler, K. Welfle, R. Misselwitz, H. Welfle, H. Wessner, G. Zahn, C. Scholz, M. Seifert, R. Harkins, J. Schneider-Mergener, and W. Höhne. *J. Mol. Biol.*, 314:293, 2001.
[192] H. Halimi, M. Eisenstein, J.-W. Oh, M. Revel, and J. Chebath. *Eur. Cytokine Netw.*, 6:135, 1995.
[193] H. Halimi, H. Dumortier, J.-P. Briand, and S. Muller. *J. Immunol. Methods*, 199:77, 1996.
[194] S. Hammerschmidt, M.P. Tillig, S. Wolff, J.-P. Vaerman, and G.S. Chhatwal. *Mol. Microbiol.*, 36:726, 2000.
[195] S.H. Hang and C.A. Mirkin. *Science*, 288:1808, 2000.
[196] H.P. Hansen, A. Recke, U. Reineke, B. von Tresckow, P. Borchmann, E.P. von Strandmann, H. Lange, H. Lemke, and A. Engert. *FASEB J.* In press, 2004.
[197] N.C. Hauser, M. Vingron, M. Schneideler, B. Krems, K. Hellmuth, K.-D. Entian, and D.J. Hoheisel. *Yeast*, 14:1209, 1998.
[198] H. Hawlisch, R. Frank, M. Hennecke, M. Baensch, B. Sohns, L. Arseniev, W. Bautsch, A. Kola, A. Klos, and J. Köhl. *J. Immunol.*, 160:2947, 1998.
[199] H. Hawlisch, M. Müller, R. Frank, W. Bautsch, A. Klos, and J. Köhl. *Anal. Biochem.*, 293:142, 2001.
[200] P. Heiduschka, W. Göpel, W. Beck, W. Kraas, S. Kienle, and G. Jung. *Chem. Eur. J.*, 2:667, 1996.
[201] N. Heine, L. Germeroth, J. Schneider-Mergener, and H. Wenschuh. *Tetrahedron Lett.*, 42:227, 2001.
[202] N. Heine, T. Ast, J. Schneider-Mergener, U. Reineke, L. Germeroth, and H. Wenschuh. *Tetrahedron*, 59:9919, 2003.
[203] J.H. Helbig, B. König, H. Knospe, B. Bubert, C. Yu, C.P. Lück, A. Riboldi-Tunnicliffe, R. Hilgenfeld, E. Jacobs, J. Hacker, and G. Fischer. *Biol. Chem.*, 384:125, 2003.
[204] P.J. Hergenrother, K.M. Depew, and S.L. Schreiber. *J. Am. Chem. Soc.*, 122:7849, 2000.
[205] K. Hilpert, J. Behlke, C. Scholz, R. Misselwitz, J. Schneider-Mergener, and W. Höhne. *Virology*, 254:6, 1999.
[206] K. Hilpert, G. Hansen, H. Wessner, J. Schneider-Mergener, and W. Höhne. *J. Biochem.*, 128:1051, 2000.

[207] S. Himpel, W. Tegge, R. Frank, S. Leder, H.-G. Joost, and W. Becker. *J. Biol. Chem.*, 274:2431, 2000.
[208] N. Hird, I. Hughes, D. Hunter, M.G.J.T. Morrison, D.C. Sherrington, and L. Stevenson. *Tetrahedron*, 55:9575, 1999.
[209] M.F. Ho, B.A. Wilson, and J.W. Peterson. *J. Chin. Chem. Soc.*, 46:735, 1999.
[210] K.G. Hoff, D.T. Ta, T.L. Tapley, J.J. Silberg, and L.E. Vickery. *J. Biol. Chem.*, 277:27353, 2002.
[211] U. Hoffmüller, M. Russwurm, F. Kleinjung, J. Ashurst, H. Oschkinat, R. Volkmer-Engert, D. Koesling, and J. Schneider-Mergener. *Angew. Chem. Int. Ed.*, 38:2000, 1999.
[212] U. Hoffmüller, T. Knaute, M. Hahn, W. Höhne, J. Schneider-Mergener, and A. Kramer. *EMBO J.*, 19:4866, 2000.
[213] C.P. Holmes, C.L. Adams, L.M. Kochersperger, R. B. Mortensen, and L.A. Aldwin. *Biopolymers* (Peptide Science), 37:199, 1995.
[214] S.H. Hong, J. Zhu, and C.A. Mirkin. *Science*, 286:523, 1999.
[215] R.A. Houghten, C. Pinilla, S.E. Blondelle, J.R. Appel, C.T. Dooley, and J.H. Cuervo. *Nature*, 354:84, 1991.
[216] R.A. Houghten, J.R. Appel, S.E. Blondelle, J.H. Cuervo, C.T. Dooley, and C. Pinilla. *BioTechniques*, 13:412, 1992.
[217] B.T. Houseman and M. Mrksich. *Angew. Chem.*, 111:876, 1999.
[218] B.T. Houseman, J.H. Huh, S.J. Kron, and M. Mrksich. *Nat. Biotechnol.*, 20:270, 2002a.
[219] B.T. Houseman and M. Mrksich. *Trends Biotechnol.*, 20:279, 2002b.
[220] B.W. Howell, L.M. Lanier, R. Frank, F.B. Gertler, and J.A. Cooper. *Mol. Cell. Biol.*, 19:5179, 1999.
[221] W. Hueber, P.J. Utz, L. Steinman, and W.H. Robinson. *Arthritis Res.*, 4:290, 2002.
[222] T.R. Hughes, M. Mao, A.R. Jones, J. Burchard, M.J. Marton, K.W. Shannon, S.M. Lefkowitz, M. Ziman, J.M. Schelter, M.R. Meyer, S. Kobayashi, C. Davis, H. Dai, Y. D. He, S.B. Stephaniants, G. Cavet, W.L. Walker, A. West, E. Coffey, D.D. Shoemaker, R. Stoughton, A.P. Blanchard, S.H. Friend, and P.S. Linsley. *Nat. Biotechnol.*, 19:342, 2001.
[223] C. Hultschig, M. Baensch, J. Köhl, and R. Frank. In R. Epton (ed.), Solid Phase Synthesis & Combinatorial Libraries. Mayflower Worldwide, Birmingham, p. 7, 2000.
[224] B. Hutter and M. Singh. *Gene*, 212:21, 1998.
[225] J. Hyun and A. Chilkoti. *J. Am. Chem. Soc.*, 123:6943, 2001.
[226] J. Hyun, Y. Zhu, A. Liebmann-Vinson, T.P. Beebe Jr., and A. Chilkoti. *Langmuir*, 17:6358, 2001.
[227] S.K. Ibrahim, C.J. Pickett, and C. Sudbrake. *J. Electroanal. Chem.*, 387:139, 1995.
[228] Y. Ikada. *Biomaterials*, 15:725, 1994.
[229] H.D. Inerowicz, S. Howell, F.E. Regnier, and R. Reifenberger. *Langmuir*, 18:5263, 2002.
[230] R.J. Jackman, D.C. Duffy, E. Ostuni, N.D. Willmore, and G.M. Whitesides. *Anal. Chem.*, 70:2280, 1998.
[231] T. Jacobs, M.D. Cima-Cabal, A. Darji, F.J. Méndez, F. Vázquez, A.A.C. Jacobs, Y. Shimada, Y. Ohno-Iwashita, S. Weiss, and J.R. de los Toyos. *FEBS Lett.*, 459:463, 1999.
[232] J. James and J. Harley. *Immunol. Rev.*, 164:789, 1998.
[233] C.D. James, R. Davis, R. Meyer, A. Turner, S. Turner, G. Withers, L. Kam, G. Banker, H. Craighead, M. Isaacson, J. Turner, and W. Shain. *IEEE Transactions Biomed. Eng.*, 47:17, 2000.
[234] I.M. Janson, R. Toomik, F. O'Farrell, and P. Ek. *Biochem. Biophys. Res. Commun.*, 247:447, 1998.
[235] K.-M. Järvinen, P. Chatchatee, L. Bardina, K. Beyer, and H.A. Sampson. *Int. Arch. Allergy Immunol.*, 126:111, 2001.
[236] K.-M. Järvinen, K. Beyer, L. Vila, P. Chatchatee, P.J. Busse, and H.A. Sampson. *J. Allergy Clin. Immunol.*, 110:293, 2002.
[237] C.L. Jellis, T.J. Cradick, P. Rennert, P. Salinas, J. Boyd, T. Amirault, and G.S. Gray. *Gene*, 137:63, 1993.
[238] M. Jensen, T. Hartmann, B. Engvall, R. Wang, S.N. Uljon, K. Sennvik, J. Näslund, F. Muehlhauser, C. Nordstedt, K. Beyreuther, and L. Lannfelt. *Mol. Med.*, 6:291, 2000.
[239] P.G. Jerlström, S.R. Talay, P. Valentin-Weigand, K.N. Timmis, and G.S. Chhatwal. *Infect. Immun.*, 64:2787, 1996.
[240] P.M. John and H.G. Craighead. *Appl. Phys. Lett.*, 68:1022, 1996.
[241] J.V.St. John and P. Wisian-Neilson. *Mat. Res. Soc. Symp.*, 629:FF9.5.1, 2000.
[242] B. Johnsson, S. Löfas, and G. Lindquist. *Anal. Biochem.*, 198:268, 1991.
[243] P. Joki-Korpela, M. Roivainen, H. Lankinen, T. Pöyry, and T. Hyypiä. *J. Gen. Virol.*, 81:1709, 2000.
[244] S. Jones and J.M. Thornton. *Proc. Natl. Acad. Sci. U.S.A.*, 93:13, 1996.
[245] V.W. Jones, J.R. Kenseth, M.D. Porter, C.L. Mosher, and E. Henderson. *Anal. Chem.*, 70:1233, 1998.
[246] G. Jung and A.G. Beck-Sickinger. *Angew. Chem.*, 4:375, 1992.

[247] G. Jung. Combinatorial Peptide and Nonpeptide Libraries (VCH Verlagsgesellschaft mbH, Weinheim, Germany, 1996).
[248] F. Jung, G. Neuer, and F.A. Bautz. *Arthritis Rheum.*, 40:1803, 1997.
[249] H. Jung, R. Kulkarni, and C.P. Collier. *J. Am. Chem. Soc.*, 125:12096, 2003.
[250] L. Kaikkonen, H. Lankinen, I. Harjunpää, K. Hokynar, M. Söderlund-Venermo, C. Oker-Blom, L. Hedmann, and K. Hedmann. *J. Clin. Microbiol.*, 37:3952, 1999.
[251] T.G. Kang and Y.-H. Cho. *Proceedings of MEMS 2003. The Sixteenth Annual International Conference on Micro Electro Mechanical Systems. Institute of Electrical and Electronics Engineers*, New York, N.Y., p. 690, 2003.
[252] N. Kanoh, S. Kumashiro, S. Simizu, Y. Kondoh, S. Hatakeyama, H. Tashiro, and H. Osada. *Angew. Chem.*, 115:5742, 2003.
[253] H. Kaufmann, J. Bailey, and M. Fussenegger. *Proteomics*, 1:194, 2001.
[254] M. Kaup, K. Dassler, U. Reineke, C. Weise, R. Tauber, and H. Fuchs. *Biol. Chem.*, 383:1011, 2002.
[255] T. Keitel, A. Kramer, H. Wessner, C. Scholz, J. Schneider-Mergener, and W. Höhne. *Cell*, 91:811, 1997.
[256] J.R. Kenseth, J.A. Harnisch, V.W. Jones, and M.D. Porter. *Langmuir*, 17:4105, 2001.
[257] H.J. Kim, J.H. Moon, and J.W. Park. *J. Colloid Interface Sci.*, 227:247, 2000.
[258] S. Kiyonaka, K. Sada, I. Yoshimura, S. Shinkai, N. Kato, and I. Hamachi. *Nat. Mater.*, 3:58, 2003.
[259] S. Kneissel, I. Queitsch, G. Petersen, O. Behrsing, B. Micheel, and S. Dübel. *J. Mol. Biol.*, 288:21, 1999.
[260] N.T.M. Knoblauch, S. Rüdiger, H.-J. Schönfeld, A.J.M. Driessen, J. Schneider-Mergener, and B. Bukau. *J. Biol. Chem.*, 274:34219, 1999.
[261] J. Koch, M. Mahler, and M. Blüthner. In J. Koch and M. Mahler (eds.), Peptide Arrays on Membrane Supports—Synthesis and Applications. Springer Verlag, Berlin, Heidelberg, p. 69, 2002.
[262] J. Koch, M. Liang, I. Queitsch, A.A. Kraus, and E.K. Bautz. *Virology*, 308:64, 2003.
[263] T. Kodadek. *Chem. Biol.*, 8:105, 2001.
[264] N. Kolbus, W. Beuche, K. Felgenhauer, and M. Mäder. *Clin. Immunol. Immunopathol.*, 80:298, 1996.
[265] K. Komolpis, O. Srivannavit, and E. Gulari. *Biotechnol. Prog.*, 18:641, 2002.
[266] S. König, G. Beterams, and M. Nassal. *J. Virol.*, 72:4997, 1998.
[267] C. Korth, C. Stierli, P. Streit, M. Moser, O. Schaller, R. Fischer, W. Schulz-Schaeffer, H. Kretzschmar, A. Raeber, U. Braun, F. Ehrensperger, S. Hornemann, R. Glockshuber, R. Riek, M. Billeter, K. Wüthrich, and B. Oesch. *Nature*, 390:74, 1997.
[268] T. Koyano, M. Saito, Y. Miyamoto, K. Kaifu, and M. Kato. *Biotechnol. Prog.*, 12:141, 1996.
[269] A. Kramer, R. Volkmer-Engert, R. Malin, U. Reineke, and J. Schneider-Mergener. *Peptide Res.*, 6:314, 1993.
[270] A. Kramer, A. Schuster, U. Reineke, R. Malin, R. Volkmer-Engert, C. Landgraf, and J. Schneider-Mergener. *METHODS: A Companion to Methods in Enzymology*, 6:388, 1994.
[271] A. Kramer, E. Vakalopoulou, W.-D. Schleuning, and J. Schneider-Mergener. *Mol. Immunol.*, 32:459, 1995.
[272] A. Kramer, T. Keitel, K. Winkler, W. Stöcklein, W. Höhne, and J. Schneider-Mergener. *Cell*, 91:799, 1997.
[273] A. Kramer, and J. Schneider-Mergener. In S. Cabilly (ed.), Methods in Molecular Biology, Humana Press Inc., Totowa, NJ, p. 25, 1998a.
[274] A. Kramer, R.-D. Stigler, T. Knaute, B. Hoffmann, and J. Schneider-Mergener. *Protein Eng.*, 11:941, 1998b.
[275] A. Kramer, M. Affelt, R. Volkmer-Engert, and J. Schneider-Mergener. Peptides 1998: *Proceedings of the Twenty-Fifth European Peptide Symposium*, S. Bajusz and F. Hudecz (eds.), Akadémiai Kiadó, Budapest, p. 546, 1999a.
[276] A. Kramer, U. Reineke, L. Dong, B. Hoffmann, U. Hoffmüller, D. Winkler, R. Volkmer-Engert, and J. Schneider-Mergener. *J. Pept. Res.*, 54:319, 1999b.
[277] M. Krause, A.S. Sechi, M. Konradt, D. Monner, F.B. Gertler, and J. Wehland. *J. Cell. Biol.*, 149:181, 2000.
[278] E. Krause, T. Bittdorf, and J. Schneider-Mergener. patent, WO 01/18545 A2, 2001.
[279] T. Kubota, T. Watanabe, N. Yokosawa, K. Tsuzuki, T. Indoh, K. Moriishi, K. Sanda, Y. Maki, K. Inoue, and N. Fujii. *Appl. Environ. Microb.*, 63:1214, 1997.
[280] A. Kumar and G.M. Whitesides. *Appl. Phys. Lett.*, 63:2002, 1993.
[281] G. Küttner, A. Kramer, G. Schmidtke, E. Giessmann, L. Dong, D. Roggenbuck, C. Scholz, M. Seifert, R.-D. Stigler, J. Schneider-Mergener, T. Porstmann, and W. Höhne. *Mol. Immunol.*, 36:669, 1999.
[282] J.J. La Clair and M.D. Burkart. *Org. Biomol. Chem.*, 1:3244, 2003.
[283] J. Lahann, H. Höcker, and R. Langer. *Angew. Chem. Int. Ed. Engl.*, 40:726, 2001a.
[284] J. Lahann, D. Klee, W. Pluester, and H. Höcker. *Biomaterials*, 22:817, 2001b.

[285] J. Lahiri, E. Ostuni, and G.M. Whitesides. *J. Am. Chem. Soc.*, 122:6303, 2000.
[286] K.S. Lam, S.E. Salmon, E.M. Hersh, V.J. Hruby, W.M. Kazmierski, and R.J. Knapp. *Nature*, 354:82, 1991.
[287] K.S. Lam and M. Renil. *Curr. Opin. Chem. Biol.*, 6:353, 2002.
[288] K.S. Lam, R. Liu, S. Miyamoto, A.L. Lehman, and J.M. Tuscano. *Accounts Chem. Res.*, 36:370, 2003.
[289] C. Larue, G. Ferrieres, M. Laprade, C. Calzolari, and C. Granier. *Clin. Chem. Lab. Med.*, 36:361, 1998.
[290] D. Laune, F. Molina, G. Ferrieres, J.-C. Mani, P. Cohen, D. Simon, T. Bernardi, M. Piechaczyk, B. Pau, and C. Granier. *J. Biol. Chem.*, 272:30937, 1997.
[291] D. Laune, B. Pau, and C. Granier. *Clin. Chem. Lab. Med.*, 36:367, 1998.
[292] D. Laune, F. Molina, J.-C. Mani, M. Del Rio, M. Bouanani, B. Pau, and C. Granier. *J. Immunol. Methods*, 239:63, 2000.
[293] D. Laune, F. Molina, G. Ferrières, S. Villard, C. Bès, F. Rieunier, T. Chardès, and C. Granier. *J. Immunol. Methods*, 267:53, 2002.
[294] R.A. Laursen, and Z. Wang. In I. Sucholeiki (ed.), High-Throughput Organic Synthesis. Marcel Dekker, New York, p. 117, 2001a.
[295] R.A. Laursen, C. Zhu, and Y. Duan. In I. Sucholeiki (ed.) High-Throughput Organic Synthesis. Marcel Dekker, New York, p. 109, 2001b.
[296] H.P. Le. *J. Imaging Sci. Technol.*, 42:49, 1998.
[297] D. Lebesgue, G. Wallukat, A. Mijares, C. Granier, J. Argibay, and J. Hoebeke. *Eur. J. Pharmacol.*, 348:123, 1998.
[298] M. Lebl. *Biopolymers (Peptide Science)*, 47:397, 1998.
[299] M. Lebl. *J. Comb. Chem.*, 1:3, 1999.
[300] K.B. Lee, S.-J. Park, C.A. Mirkin, J.C. Smith, and M. Mrksich. *Science*, 295:1702, 2002a.
[301] P.H. Lee, S.P. Sawan, Z. Modrusan, L.J. Arnold, and M.A. Reynolds. *Bioconjugate Chem.*, 13:97, 2002b.
[302] S.-W. Lee, H.-C. Kim, K. Kuk, and Y.-S. Oh. *Sens. Actuators*, A95:114, 2002c.
[303] K.-N. Lee, D.-S. Shin, Y.-S. Lee, and Y.K. Kim. *J. Micromech. Microeng.*, 13:18, 2003a.
[304] K.-N. Lee, D.-S. Shin, Y.-S. Lee, and Y.K. Kim. *J. Micromech. Microeng.*, 13:474, 2003b.
[305] J.P. Lee and M.M. Sung. *J. Am. Chem. Soc.*, 126:28, 2003.
[306] V. Lehmann and U. Gruning. *Thin Solid Films*, 297:13, 1997.
[307] G.A. Lemieux and C.R. Bertozzi. *TIBTECH*, 16:506, 1998.
[308] R. Lenigk, M. Carles, N.Y. Ip, and N.J. Sucher. *Langmuir*, 17:2497, 2001.
[309] E. LeProust, J.P. Pellois, P. Yu, H. Zhang, X. Gao, O. Srivannavit, E. Gulari, and X. Zhou. *J. Comb. Chem.*, 2:349, 2000.
[310] M.-L. Lesaicherre, M. Uttamchandani, G.Y.J. Chen, and S.Q. Yao. *Bioorg. Med. Chem. Lett.*, 12:2079, 2002a.
[311] M.-L. Lesaicherre, M. Uttamchandani, G.Y.J. Chen, and S.Q. Yao. *Bioorg. Med. Chem. Lett.*, 12:2085, 2002b.
[312] J.B. Levy, A. Coulthart, and C.D. Pusey. *J. Am. Soc. Nephrol.*, 8:1698, 1997.
[313] Y. Li, J.M. Desimone, C.-D. Poon, and E.T. Samulski. *J. Appl. Polym. Sci.*, 64:883, 1997.
[314] H. Li, D.-J. Kang, M.G. Blamire, and W.T.S. Huck. *Nano Lett.*, 2:347, 2002a.
[315] X.-D. Li, T.P. Mäkelä, D. Guo, R. Soliymani, V. Koistinen, O. Vapalathi, A. Vaheri, and H. Lankinen. *J. Gen. Virol.*, 83:759, 2002b.
[316] O.D. Liang, T. Chavakis, S.M. Kanse, and K.T. Preissner. *J. Biol. Chem.*, 276:28946, 2001.
[317] O.D. Liang, K. Bdeir, R.L. Matz, T. Chavakis, and K.T. Preissner. *J. Biochem.*, 134:661, 2003a.
[318] M. Liang, M. Mahler, J. Koch, Y. Ji, D. Li, C. Schmaljohn, and E.K.F. Bautz. *J. Med. Virol.*, 69:99, 2003b.
[319] L. Libioulle, A. Bietsch, H. Schmid, B. Michel, and E. Delamarche. *Langmuir*, 15:300, 1999.
[320] K. Licha, S. Bhargava, C. Rheinländer, A. Becker, J. Schneider-Mergener, and R. Volkmer-Engert. *Tetrahedron Lett.*, 41:1711, 2000.
[321] J.-Å. Liljeqvist, E. Trybala, B. Svennerholm, S. Jeansson, E. Sjögren-Jansson, and T. Bergström. *J. Gen. Virol.*, 79:1215, 1998.
[322] J.-Å. Liljeqvist, E. Trybala, J. Hoebeke, B. Svennerholm, and T. Bergström. *J. Gen. Virol.*, 83:157, 2002.
[323] H. Linn, K.S. Ermekova, S. Rentschler, A.B. Sparks, B.K. Kay, and M. Sudol. *Biol. Chem.*, 378:531, 1997.
[324] R.J. Lipshutz, S.P.A. Fodor, T.R. Gingeras, and D.J. Lockart. *Nature Genet.*, 21:20, 1999.
[325] C.-F. Liu, C. Rao, and J.P. Tam. *J. Am. Chem. Soc.*, 118:307, 1996.
[326] Z. Liu, D. Song, A. Kramer, A.C.R. Martin, T. Dandekar, J. Schneider-Mergener, E.K.F. Bautz, and S. Dübel. *J. Mol. Recognit.*, 12:103, 1999.

[327] R. Liu, A.M. Enstrom, and K.S. Lam. *Exp. Hematol.*, 31:11, 2003.
[328] T. Livache, H. Bazin, P. Caillat, and A. Roget. *Biosens. Bioelectron.*, 13:629, 1998.
[329] M.A. Livshits and A.D. Mirzabekov. *Biophys. J.*, 71:2795, 1996.
[330] J.M. Lizcano, M. Deak, N. Morrice, A. Kieloch, C.J. Hastie, L. Dong, M. Schutkowski, U. Reimer, and D.R. Alessi. *J. Biol. Chem.*, 277:27839, 2002.
[331] R. Llanos, V. Chevrier, M. Ronjat, P. Meurer-Grob, P. Martinez, R. Frank, M. Bornens, R.H. Wade, J. Wehland, and D. Job. *Biochemistry*, 38:15712, 1999.
[332] J.R. Long, N. Oyler, G.P. Drobny, and P.S. Stayton. *J. Am. Chem. Soc.*, 124:6297, 2002.
[333] Y.-L. Loo, R.L. Willett, K.W. Baldwin, and J.A. Rogers. *J. Am. Chem. Soc.*, 124:7654, 2002.
[334] M. Loog, R. Toomik, K. Sak, G. Muszynska, J. Järv, and P. Ek. *Eur. J. Biochem.*, 267:337, 2000.
[335] W. Lu, M.A. Qasim, and S.B.H. Kent. *J. Am. Chem. Soc.*, 118:8518, 1996.
[336] P.J. Lu, X.Z. Zhou, M. Shen, and K.P. Lu. *Science*, 283:1325, 1999.
[337] K. Luo, P. Zhou, and H.F. Lodish. *Proc. Natl. Acad. Sci. U.S.A.*, 92:11761, 1995.
[338] G. MacBeath and S.L. Schreiber. *Science*, 289:1760, 2000.
[339] D. Mahé, P. Blanchard, C. Truong, C. Arnauld, P. Le Cann, R. Cariolet, F. Madec, E. Albina, and A. Jestin. *J. Gen. Virol.*, 81:1815, 2000.
[340] M. Mahler, R. Mierau, and M. Blüthner. *J. Mol. Med.*, 78:500, 2000.
[341] M. Mahler, R. Mierau, W. Schlumberger, and M. Blüthner. *J. Mol. Med.*, 79:722, 2001.
[342] M. Mahler, M. Blüthner, and J. Koch. In J. Koch and M. Mahler (eds.), Peptide Arrays on Membrane Supports—Synthesis and Applications. Springer Verlag, Berlin, Heidelberg, p. 107, 2002a.
[343] M. Mahler, R. Mierau, E. Genth, and M. Blüthner. *Arthritis Rheum.*, 46:1866, 2002b.
[344] M. Mahler, K. Kessenbrock, J. Raats, R. Williams, M.J. Fritzler, and M. Blüthner. *J. Mol. Med.*, 81:194, 2003.
[345] S.A. Makohliso, D. Leonard, L. Giovangrandi, H.J. Mathieu, M. Ilegems, and P. Aebischer. *Langmuir*, 15:2940, 1999.
[346] R. Malin, R. Steinbrecher, J. Jannsen, W. Semmler, B. Noll, B. Johannsen, C. Frömmel, W. Höhne, and J. Schneider-Mergener. *J. Am. Chem. Soc.*, 117:11821, 1995.
[347] G. Marko-Varga, J. Nilsson, and T. Laurell. *Electrophoresis*, 24:3521, 2003.
[348] W. Martens, I. Greiser-Wilke, T.C. Harder, K. Dittmar, R. Frank, C. Örvell, V. Moennig, and B. Liess. *Vet. Microbiol.*, 44:289, 1995.
[349] B.D. Martin, B.P. Gaber, C.H. Patterson, and D.C. Turner. *Langmuir*, 14:3971, 1998.
[350] K. Martin, T.H. Steinberg, L.A. Cooley, K.R. Gee, J.M. Beechem, and W.F. Patton. *Proteomics*, 3:1244, 2003.
[351] U. Maskos and E.M. Southern. *Nucl. Acid Res.*, 21:2267, 1993a.
[352] U. Maskos and E.M. Southern. *Nucl. Acid Res.*, 21:4663, 1993b.
[353] J.M. Mataraza, M.W. Briggs, Z. Li, R. Frank, and D.B. Sacks. *Biochem. Bioph. Res. Com.*, 305:315, 2003.
[354] R.S. Matson, J.B. Rampal, and P.J. Coassin. *Anal. Biochem.*, 217:306, 1994.
[355] S. Matysiak, N.C. Hauser, S. Würtz, and J.D. Hoheisel. *Nucleos. Nucleot.*, 18:1289, 1999.
[356] S. Matysiak, F. Reuthner, and J.D. Hoheisel. *BioTechniques*, 31:896, 2001.
[357] O. Mayboroda, K. Schlüter, and B.M. Jockusch. *Cell Motil. Cytoskel.*, 37:166, 1997.
[358] Y. Mazor, S. Gilead, I. Benhar, and E. Gazit. *J. Mol. Biol.*, 322:1013, 2002.
[359] J.S. McCarthy, S. Rüdiger, H.-J. Schönfeld, J. Schneider-Mergener, K. Nakahigashi, T. Yura, and B. Bukau. *J. Mol. Biol.*, 256:829, 1996.
[360] M.E. McGovern and M. Thompson. *Can. J. Chem.*, 77:1678, 1999.
[361] O. Melnyk, X. Duburcq, C. Olivier, F. Urbès, C. Auriault, and H. Gras-Masse. *Bioconjugate Chem.*, 13:713, 2002.
[362] R.H. Meloen, W.C. Puijk, J.P.M. Langefeld, J.P.M. Langedijk, A. van Amerongen, and W.M.M. Schaaper. In N.D. Zegers, W.J.A. Boersma, and E. Claassen (eds.), Immunological Recognition of Peptides in Medicine and Biology. CRC Press, Boca Raton, FL, p. 15, 1995.
[363] R.H. Meloen, W.C. Puijk, and J.W. Slootstra. *J. Mol. Recognit.*, 13:352, 2000.
[364] R.H. Meloen, J.P.M. Langeveld, W.M.M. Schaaper, and J.W. Slootstra. *Biologicals*, 29:233, 2001.
[365] B. Michel, A. Bernard, A. Bietsch, E. Delamarche, M. Geissler, D. Junker, H. Kind, J.P. Renault, H. Rothuizen, H. Schmid, P. Schmidt-Winkler, R. Stutz, and H. Wolf. *IBM J. Res. Dev.*, 45:697, 2001.
[366] B. Michel, A. Bernard, A. Bietsch, E. Delamarche, M. Geissler, D. Junker, H. Kind, J.P. Renault, H. Rothuizen, H. Schmid, P. Schmidt-Winkler, R. Stutz, and H. Wolf. *Chimia.*, 56:527, 2002.

[367] A.D. Mirzabekov. *Trends Biotechnol.*, 12:27, 1994.
[368] M.M. Mohebi and J.R.G. Evans. *J. Comb. Chem.*, 4:267, 2002.
[369] M.M. Mohebi and J.R.G. Evans. *J. Am. Ceramic Soc.*, 86:1654, 2003.
[370] A.-M. Moilanen, U. Karvonen, H. Poukka, O.A. Jänne, and J.J. Palvimo. *Mol. Biol. Cell*, 9:2527, 1998.
[371] F. Molina, D. Laune, C. Gougat, B. Pau, and C. Granier. *Peptide Res.*, 9:151, 1996.
[372] F. Molina, B. Pau, and C. Granier. *Lett. Pept. Sci.*, 4:201, 1997.
[373] D. Moll, C. Huber, B. Schlegel, D. Pum, U.B. Sleytr, and M. Sára. *Proc. Natl. Acad. Sci. U.S.A.*, 99:14646, 2002.
[374] C. Monnet, D. Laune, J. Laroche-Traineau, M. Biard-Piechaczyk, L. Briant, C. Bès, M. Pugnière, J.-C. Mani, B. Pau, M. Cerutti, G. Devauchelle, C. Devaux, C. Granier, and T. Chardès. *J. Biol. Chem.*, 274:3789, 1999.
[375] M. Mrksich. *Chem. Soc. Rev.*, 29:267, 2000.
[376] S. Mukhija, L. Germeroth, J. Schneider-Mergener, and B. Erni. *Eur. J. Biochem.*, 254:433, 1998.
[377] G. Münch, D. Schicktanz, A. Behme, M. Gerlach, P. Riederer, D. Palm, and R. Schinzel. *Nat. Biotechnol.*, 17:1006, 1999.
[378] Y. Muro, N. Azuma, H. Onouchi, M. Kunimatsu, Y. Tomita, M. Sasaki, and K. Sugimoto. *Clin. Exp. Immunol.*, 120:218, 2000.
[379] R. Necina, K. Amatschek, E. Schallaun, H. Schwinn, D. Josic, and A. Jungbauer. *J. Chromatogr. B*, 715:191, 1998.
[380] H.E. Neuman de Vegvar, R.R. Amara, L. Steinman, P.J. Utz, H.L. Robinson, and W.H. Robinson. *J. Virol.*, 77:11125, 2003.
[381] D.V. Nicolau, T. Taguchi, H. Taniguchi, and S. Yoshikawa. *Langmuir*, 14:1927, 1998.
[382] K. Niebuhr, F. Ebel, R. Frank, M. Reinhard, E. Domann, U.D. Carl, U. Walter, F.B. Gertler, J. Wehland, and T. Chakraborty. *EMBO J.*, 16:5433, 1997.
[383] K. Nito, S. Suzuki, K. Miyasaki, and K. Ishikawa. *Polym. Prep. Jpn.*, 29:171, 1980.
[384] D.A. Nivens and D.W. Conrad. *Langmuir*, 18:499, 2002.
[385] W.B. Nowall, D.O. Wipf, and W.G. Kuhr. *Anal. Chem.*, 70:2601, 1998.
[386] I.O.L. Otvos, T. Kieber-Emmons, and M. Blaszczyk-Thurin. *Peptides*, 23:999, 2002.
[387] Y. Ogiwana and M. Kanda. *J. Polym. Sci. Polym. Lett. Ed.*, 19:457, 1981.
[388] S.J. Oh, S.J. Cho, C.O. Kim, and J.W. Park. *Langmuir*, 18:1764, 2002.
[389] A.A. Osman, H. Uhlig, B. Thamm, J. Schneider-Mergener, and T. Mothes. *FEBS Lett.*, 433:103, 1998.
[390] A.A. Osman, T. Günnel, A. Dietl, H.H. Uhlig, M. Amin, B. Fleckenstein, T. Richter, and T. Mothes. *Clin. Exp. Immunol.*, 121:248, 2000.
[391] A.A. Osman, H.H. Uhlig, I. Valdes, M. Amin, E. Méndez, and T. Mothes. *Eur. J. Gastroenterol. Hepat.*, 13:1189, 2001.
[392] E. Ostuni, C.S. Chen, D.E. Ingber, and G.M. Whitesides. *Langmuir*, 17:2828, 2001.
[393] L. Otte, U. Wiedemann, B. Schlegel, J.R. Pires, M. Beyermann, P. Schmieder, G. Krause, R. Volkmer-Engert, J. Schneider-Mergener, and H. Oschkinat. *Protein Sci.*, 12:491, 2003.
[394] L. Otvos Jr., A.M. Pease, K. Bokonyi, W. Giles-Davis, M.E. Rogers, P.A. Hintz, R. Hoffmann, and H.C.J. Ertl. *J. Immunol. Methods*, 233:95, 2000.
[395] S. Panse, L. Dong, A. Burian, R. Carus, M. Schutkowski, U. Reimer, and J. Schneider-Mergener. *Mol. Divers.*, 8:291, 2004.
[396] C.D. Partidos, F.B. Salani, J. Ripley, and M.W. Steward. *Vaccine*, 18:321, 2000.
[397] C. Pasquali, M.-L. Curchod, S. Wälchli, X. Espanel, M. Guerrier, F. Arigoni, G. Strous, and R.H. van Huijsduijnen. *Molec. Endocrinol.*, 17:2228, 2003.
[398] N. Patel, G.H.W. Sanders, K.M. Shakesheff, S.M. Cannizzaro, M.C. Davies, R. Langer, C.J. Roberts, S.J.B. Tendlers, and P.M. Williams. *Langmuir*, 15:7252, 1999.
[399] H. Patzelt, S. Rüdiger, D. Brehmer, G. Kramer, S. Vorderwülbecke, E. Schaffitzel, A. Waitz, T. Hesterkamp, L. Dong, J. Schneider-Mergener, B. Bukau, and E. Deuerling. *Proc. Natl. Acad. Sci. U.S.A.*, 98:14244, 2001.
[400] M. Pawlak, E. Grell, E. Schick, D. Anselmetti, and M. Ehrat. *Faraday Discuss.*, 111:273, 1998.
[401] M. Pawlak, E. Schick, M.A. Bopp, M.J. Schneider, P. Oroszlan, and M. Ehrat. *Proteomics*, 2:383, 2002.
[402] T. Pawson and P. Nash. *Gene. Dev.*, 14:1027, 2000.
[403] T. Pawson and P. Nash. *Science*, 300:455, 2003.
[404] E. Peeters and S. Verdonckt-Vandebrock. *IEEE Circuits and Devices*, 13(4):19, 1997.
[405] J.P. Pellois, W. Wang, and X. Gao. *J. Comb. Chem.*, 2:355, 2000.

[406] J.P. Pellois, X. Zhou, O. Srivannavit, T. Zhou, E. Gulari, and X. Gao. *Nat. Biotechnol.*, 20:922, 2002.
[407] L.G. Pereira, L. Torrance, I.M. Roberts, and B.D. Harrison. *Virology*, 203:277, 1994.
[408] G. Petersen. In J. Koch and M. Mahler (eds.), Peptide Arrays on Membrane Supports—Synthesis and Applications. Springer Verlag, Berlin, Heidelberg, p. 41, 2002.
[409] K. Pflegerl, R. Hahn, E. Berger, and A. Jungbauer. *J. Peptide Res.*, 59:174, 2002.
[410] J. Piehler, A. Brecht, K.E. Geckler, and G. Gauglitz. *Biosens. Bioelectron.*, 11:579, 1996.
[411] J. Piehler, A. Brecht, K. Hehl, and G. Gauglitz. *Colloids Surfaces B: Biointerfaces*, 13:325, 1999.
[412] J. Piehler, A. Brecht, R. Valiokas, B. Liedberg, and G. Gauglitz. *Biosens. Bioelectron.*, 15:473, 2000.
[413] R.D. Piner, J. Zhu, F. Xu, S. Hong, and C.A. Mirkin. *Science*, 283:661, 1999.
[414] C. Pinilla, J.R. Appel, P. Blanc, and R.A. Houghten. *BioTechniques*, 13:901, 1992.
[415] C. Pinilla, J.R. Appel, and R.A. Houghten. *Mol. Immunol.*, 30:577, 1993.
[416] C. Pinilla, J.R. Appel, C. Dooley, S. Blondelle, J. Eichler, B. Dörner, J. Ostresh, and R.A. Houghten. In G. Jung (ed.) Combinatorial Peptide and Nonpeptide Libraries. VCH Verlagsgesellschaft, Weinheim, p. 139, 1996.
[417] C. Piossek, J. Schneider-Mergener, M. Schirner, E. Vakalopoulou, L. Germeroth, and K.-H. Thierauch. *J. Biol. Chem.*, 274:5612, 1999.
[418] C. Piossek, K.-H. Thierauch, J. Schneider-Mergener, R. Volkmer-Engert, M.F. Bachmann, T. Korff, H.G. Augustin, and L. Germeroth. *Thromb. Haemost.*, 90:501, 2003.
[419] J.R. Pires, F. Taha-Nejad, F. Toepert, T. Ast, U. Hoffmüller, J. Schneider-Mergener, R. Kühne, M.J. Macias, and H. Oschkinat. *J. Mol. Biol.*, 314:1147, 2001.
[420] J.R. Pires, X. Hong, C. Brockmann, R. Volkmer-Engert, J. Schneider-Mergener, H. Oschkinat, and R. Erdmann. *J. Mol. Biol.*, 326:1427, 2003.
[421] M.C. Pirrung and J. Chen. *J. Am. Chem. Soc.*, 117:1240, 1995.
[422] M.C. Pirrung. *Chem. Rev.*, 97:473, 1997.
[423] S. Pistor, L. Gröbe, A.S. Sechi, E. Domann, B. Gerstel, L.M. Machesky, T. Chakraborty, and J. Wehland. *J. Cell Sci.*, 113:3277, 2000.
[424] T.A. Pitcovsky, J. Mucci, P. Alvarez, M.S. Leguizamón, O. Burrone, P.M. Alzari, and O. Campetella. *Infect. Immun.*, 69:1869, 2001.
[425] M.A. Podyminogin, E.A. Lukhtanov, and M.W. Reed. *Nucleic Acid Res.*, 29:5090, 2001.
[426] J.-M. Portefaix, S. Thebault, F. Bourgain-Guglielmetti, M. Del Rio, C. Granier, J.-C. Mani, I. Navarro-Teulon, M. Nicolas, T. Soussi, and B. Pau. *J. Immunol. Methods*, 244:17, 2000.
[427] J.-M. Portefaix, M. Del Rio, C. Granier, F. Roquet, B. Pau, and I. Navarro-Teulon. *Peptides*, 24:339, 2003.
[428] A. Poustka, F. Breitling, K.-H. Gross, S. Duebel, and R. Saffrich. patent, EP 1140977 A, 2001.
[429] W.M. Prodinger, M.G. Schwendinger, J. Schoch, M. Köchle, C. Larcher, and M.P. Dierich. *J. Immunol.*, 161:4604, 1998.
[430] D. Proudnikov, E. Timofeev, and A.D. Mirzabekov. *Anal. Biochem.*, 259:34, 1998.
[431] M. Ptushkina, T. von der Haar, S. Vasilescu, R. Frank, R. Birkenhäger, and J.E.G. McCarthy. *EMBO J.*, 17:4798, 1998.
[432] M. Pugnière, T. Chardès, C. Bès, C. Monnet, D. Laune, C. Granier, and J.-C. Mani. *Int. J. Bio-Chromatogr.*, 5:187, 2000.
[433] W.C. Puijk, patent, WO 00/05584, 2000.
[434] S.S. Pullen, T.T.A. Dang, J.J. Crute, and M.R. Kehry. *J. Biol. Chem.*, 274:14246, 1999.
[435] T. Pulli, H. Lankinen, M. Roivainen, and T. Hyypiä. *Virology*, 240:202, 1998.
[436] M. Purtscher, A. Trkola, A. Grassauer, P.M. Schulz, A. Klima, S. Döpper, G. Gruber, A. Buchacher, T. Muster, and H. Katinger. *Aids*, 10:587, 1996.
[437] D. Qin, Y.N. Xia, J.A. Rogers, R.J. Jackman, X.-M. Zhao, and G.M. Whitesides. *Top. Curr. Chem.*, 194:1, 1998.
[438] A.P. Quist, A.A. Bergman, C.T. Reimann, S.O. Oscarsson, and B.U.R. Sundquist. *Scanning Microsc.*, 9:395, 1995.
[439] N. Raffler, J. Schneider-Mergener, and M. Famulok. *Chem. Biol.*, 10:69, 2003.
[440] D. Rama, C. Calzolari, C. Granier, and B. Pau. *Hybridoma*, 16:153, 1997.
[441] K.H. Rau, N. DeJonge, and W. Haehnel. *Angew. Chem. Int. Ed.*, 112:256, 2000.
[442] P. Reichmuth, H. Sigrist, M. Badertscher, W.E. Morf, N.F. de Rooij, and E. Pretsch. *Bioconjugate Chem.*, 13:90, 2002.
[443] U. Reimer, U. Reineke, and J. Schneider-Mergener. *Curr. Opin. Biotech.*, 13:315, 2002.

[444] U. Reineke, R. Sabat, A. Kramer, R.-D. Stigler, M. Seifert, T. Michel, H.-D. Volk, and J. Schneider-Mergener. *Mol. Divers.*, 1:141, 1996.
[445] U. Reineke, B. Ehrhard, R. Sabat, H.-D. Volk, and J. Schneider-Mergener. In R. Ramage and R. Epton (eds.), Peptides 1996: Proceedings of the Twenty-Fourth European Peptide Symposium. Mayflower Scientific Ltd., Kingswinford, p. 751, 1998a.
[446] U. Reineke, R. Sabat, H.-D. Volk, and J. Schneider-Mergener. *Protein Sci.*, 7:951, 1998b.
[447] U. Reineke, A. Kramer, and J. Schneider-Mergener. *Curr. Top. Microbiol. Immunol.*, 243:23, 1999a.
[448] U. Reineke, R. Sabat, R. Misselwitz, H. Welfle, H.-D. Volk, and J. Schneider-Mergener. *Nat. Biotechnol.*, 17:271, 1999b.
[449] U. Reineke, J. Schneider-Mergener, R. W. Glaser, R.-D. Stigler, M. Seifert, H.-D. Volk, and R. Sabat. *J. Mol. Recognit.*, 12:242, 1999c.
[450] U. Reineke, S. Bhargava, M. Schutkowski, C. Landgraf, L. Germeroth, G. Fischer, and J. Schneider-Mergener. In S. Bajusz and F. Hudecz (eds.), Peptides 1998: Proceedings of the Twenty-Fifth European Peptide Symposium. Akadémiai Kiadó, Budapest, p. 562, 1999d.
[451] U. Reineke, U. Hoffmüller, and J. Schneider-Mergener. patent, DE 198 31 429 A1, 2000a.
[452] U. Reineke, R. Sabat, U. Hoffmüller, M. Schmidt, D. Kurzhals, H. Wenschuh, H.D. Volk, L. Germeroth, and J. Schneider-Mergener. In G.B. Fields, J.P. Tam, and G. Barany (eds.), Peptides for the New Millennium: Proceeding of the Sixteenth American Peptide Symposium 1999. Kluwer Academic Publishers, Dordrecht, p. 167, 2000b.
[453] U. Reineke, A. Kramer, and J. Schneider-Mergener. In R. Kontermann and S. Dübel (eds.), Antibody Engineering (Springer Lab Manual). Springer-Verlag, Berlin, Heidelberg, p. 443, 2001a.
[454] U. Reineke, R. Volkmer-Engert, and J. Schneider-Mergener. *Curr. Opin. Biotech.*, 12:59, 2001b.
[455] U. Reineke, D. Kurzhals, A. Köhler, C. Blex, J.E.G. McCarthy, P. Li, L. Germeroth, and J. Schneider-Mergener. In J. Martinez and J. A. Fehrentz (eds.), Peptides 2000: Proceedings of the Twenty-Sixth European Peptide Symposium. Éditions EDK, Paris, p. 721, 2001c.
[456] U. Reineke, B. Hoffmann, T. Ast, T. Polakowski, J. Schneider-Mergener, and R. Volkmer-Engert. In M. Lebl and R.A. Houghten (eds.), Peptides: The Wave of the Future; Proceedings of the Second International and the Seventeenth American Peptide Symposium. American Peptide Society, San Diego, CA, p. 577, 2001d.
[457] U. Reineke and U. Hoffmüller. American Biotechnology Laboratory, p. 50, 2001e.
[458] U. Reineke, C. Ivascu, M. Schlief, C. Landgraf, S. Gericke, G. Zahn, H. Herzel, R. Volkmer-Engert, and J. Schneider-Mergener. *J. Immunol. Methods*, 267:37, 2002.
[459] U. Reineke. In B.K.C. Lo (ed.), Antibody Engineering: Methods and Protocols. Humana Press Inc., Totowa, NJ, p. 443, 2003.
[460] M. Reinhard, J. Zumbrunn, D. Jaquemar, M. Kuhn, U. Walter, and B. Trueb. *J. Biol. Chem.*, 274:13410, 1999.
[461] G. Renauld-Mongénie, D. Poncet, L. von Olleschik-Elbheim, T. Cournez, M. Mignon, M. A. Schmidt, and M.-J. Quentin-Millet. *J. Bacteriol.*, 179:6400, 1997.
[462] A. Ressine, S. Ekstrom, G. Marko-Varga, and T. Laurell. *Anal. Chem.*, 75:6968, 2003.
[463] M.D. Retzer, R.H. Yu, and A.B. Schryvers. *Mol. Microbiol.*, 32:111, 1999.
[464] P. Reusch, S. Arnold, C. Heusser, K. Wagner, B. Weston, and W. Sebald. *Eur. J. Biochem.*, 222:491, 1994.
[465] M. Reuter, J. Schneider-Mergener, D. Kupper, A. Meisel, P. Mackeldanz, D.H. Krüger, and C. Schroeder. *J. Biol. Chem.*, 274:5213, 1999.
[466] M. Reuter and E. Möncke-Buchner. In J. Koch and M. Mahler (eds.), Peptide Arrays on Membrane Supports—Synthesis and Applications. Springer Verlag, Berlin, Heidelberg, p. 97, 2002.
[467] A. Rezania and K.E. Healy. *Biotechnol. Prog.*, 15:19, 1999.
[468] F. Rharbaoui, C. Granier, M. Kellou, J.-C. Mani, P. van Endert, A.-M. Madec, C. Boitard, B. Pau, and M. Bouanani. *Immunol. Lett.*, 62:123, 1998.
[469] G. Riemekasten, J. Marell, G. Trebeljahr, R. Klein, G. Hausdorf, T. Häupl, J. Schneider-Mergener, G.R. Burmester, and F. Hiepe. *J. Clin. Invest.*, 102:754, 1998.
[470] J.M. Rini, R.L. Stanfield, E.A. Stura, P.A. Salinas, A.T. Profy, and I.A. Wilson. *Proc. Natl. Acad. Sci. U.S.A.*, 90:6325, 1993.
[471] H.A. Rinia, R.A. Kik, R.A. Demel, M.M.E. Snel, J.A. Killian, J.P.J.M. van der Eerden, and B. de Kruijff. *Biochemistry*, 39:5852, 2000.

[472] W.H. Robinson, C. DiGennaro, W. Hueber, B.B. Haab, M. Kamachi, E.J. Dean, S. Fournel, D. Fong, M.C. Genovese, H.E. Neuman de Vegvar, K. Skriner, D.L. Hirschberg, R.I. Morris, S. Muller, G.J. Pruijn, W.J. van Venrooij, J.S. Smolen, P.O. Brown, L. Steinman, and P.J. Utz. *Nat. Med.*, 8:295, 2002a.
[473] W.H. Robinson, L. Steinman, and P.J. Utz. *BioTechniques*, 33:S66, 2002b.
[474] W.H. Robinson, L. Steinman, and P.J. Utz. *Arthritis Rheum.*, 46:885, 2002c.
[475] W.H. Robinson, H. Garren, P.J. Utz, and L. Steinman. *Clin. Immunol.*, 103:7, 2002d.
[476] W.H. Robinson, L. Steinman, and P.J. Utz. *Proteomics*, 3:2077, 2003.
[477] M. Rodriguez, S.S.-C. Li, J. W. Harper, and Z. Songyang. *J. Biol. Chem.*, 279:8802, 2004.
[478] I. Rodríguez-Crespo, B. Yélamos, F. Roncal, J.P. Albar, P.R.O. de Montellano, and F. Gavilanes. *FEBS Lett.*, 503:135, 2001.
[479] S.D. Rose. *JALA*, 3:53, 1998.
[480] L.F. Rozsnyai, D.R. Benson, S.P.A. Fodor, and P. Schultz. *Angew. Chem.*, 104:801, 1992.
[481] S. Rüdiger, L. Germeroth, J. Schneider-Mergener, and B. Bukau. *EMBO J.*, 16:1501, 1997.
[482] S. Rüdiger, M.P. Mayer, J. Schneider-Mergener, and B. Bukau. *J. Mol. Biol.*, 304:245, 2000.
[483] S. Rüdiger, J. Schneider-Mergener, and B. Bukau. *EMBO J.*, 20:1042, 2001.
[484] L. Rychlewski, M. Kschischo, L. Dong, M. Schutkowski, and U. Reimer. *J. Mol. Biol.*, 336:307, 2004.
[485] K. Sachse, J.H. Helbig, I. Lysnyansky, C. Grajetzki, W. Müller, E. Jacobs, and D. Yogev. *Infect. Immun.*, 68:680, 2000.
[486] C.M. Salisbury, D.J. Maly, and J.A. Ellman. *J. Am. Chem. Soc.*, 124:14868, 2002.
[487] F. Schaper, S. Kirchhoff, G. Posern, M. Köster, A. Oumard, R. Sharf, B.-Z. Levi, and H. Hauser. *Biochem. J.*, 335:147, 1998.
[488] D. Scharn, H. Wenschuh, U. Reineke, J. Schneider-Mergener, and L. Germeroth. *J. Comb. Chem.*, 2:361, 2000a.
[489] D. Scharn, H. Wenschuh, J. Schneider-Mergener, and L. Germeroth. In R. Epton (ed.), Innovation and Perspectives in Solid Phase Synthesis and Combinatorial Libraries. Mayflower Worldwide Ltd., Birmingham, p. 157, 2000b.
[490] D. Scharn, L. Germeroth, J. Schneider-Mergener, and H. Wenschuh. *J. Org. Chem.*, 66:507, 2001.
[491] U. Scheele, J. Alves, R. Frank, M. Düwel, C. Kalthoff, and E. Ungewickell. *J. Biol. Chem.*, 278:25357, 2003.
[492] M. Schmidt, J. Eichler, J. Odarjuk, E. Krause, M. Beyermann, and M. Bienert. *Bioorg. Med. Chem. Lett.*, 3:441, 1993.
[493] T.G.M. Schmidt, J. Koepke, R. Frank, and A. Skerra. *J. Mol. Biol.*, 255:1996.
[494] J. Schneider-Mergener, A. Kramer, and U. Reineke. In R. Cortese (ed.), Combinatorial Libraries—Synthesis, Screening and Application Potential, Walter de Gruyter, Berlin, New York, p. 53, 1996.
[495] R. Schnepf, P. Hörth, E. Bill, K. Wieghardt, P. Hildebrandt, and W. Haehnel, *J. Am. Chem. Soc.* 123:2186, 2001.
[496] M. Schnölzer and S.B.H. Kent. *Science*, 256:221, 1992.
[497] M. Schreiber, C. Wachsmuth, H. Müller, S. Odemuyiwa, H. Schmitz, S. Meyer, B. Meyer, and J. Schneider-Mergener. *J. Virol.*, 71:9198, 1997.
[498] A.B. Schroeijers, A.C. Sira, G.L. Scheffer, M.C. de Jong, S.C.E. Bolick, D.F. Dukers, J.W. Slootstra, R.H. Mcloen, E. Wiemer, V.A. Kickhoefer, L.H. Rome, and R.J. Scheper. *Cancer Res.*, 60:1104, 2000.
[499] A. Schubert, K. Zakikhany, M. Schreiner, R. Frank, B. Spellerberg, B.J. Eikmanns, and D.J. Reinscheid. *Mol. Microbiol.*, 2:557, 2002.
[500] A. Schultz, V. Hoffacker, A. Wilisch, W. Nix, R. Gold, B. Schalke, S. Tzartos, H.K. Müller-Hermelink, and A. Marx. *Ann. Neurol.*, 46:167, 1999.
[501] J. Schultz, U. Hoffmüller, G. Krause, J. Ashurst, M.J. Macias, P. Schmieder, J. Schneider-Mergener, and H. Oschkinat. *Nat. Struct. Biol.*, 5:19, 1998.
[502] M. Schuerenberg, C. Luebbert, H. Eickhoff, M. Kalkum, H. Lehrach, and E. Nordhoff. *Anal. Chem.*, 72:3436, 2000.
[503] M. Schutkowski, U. Reimer, S. Panse, L. Dong, J.M. Lizcano, D.R. Alessi, and J. Schneider-Mergener. *Angew. Chem. Int. Ed. Engl.*, 43:2671, 2004.
[504] J.K. Scott and G.P. Smith. *Science*, 249:386, 1990.
[505] J. Seemann, K. Weber, and V. Gerke. *Biochem. J.*, 319:123, 1996.
[506] S. Selak, M. Mahler, K. Miyachi, M.L. Fritzler, and M.J. Fritzler. *Clin. Immunol.*, 109:154, 2003.

[507] R.R. Shah, D. Merreceyes, M. Husemann, I. Rees, N.L. Abbott, C.J. Hawker, and J.L. Hedrick. *Macromolecules*, 33:597, 2000.
[508] J. Shieh, C. To, J. Carramao, N. Nishimura, Y. Maruta, Y. Hashimoto, D. Wright, H.C. Wu, and A. Azarani. *BioTechniques*, 32:1360, 2002.
[509] T.W. Shield, D.B. Bogy, and F.E. Talke. *IBM J. Res. Dev.*, 31:96, 1987.
[510] D.S. Shin, K.N. Lee, K.H. Jang, J.K. Kim, W.J. Chung, Y.K. Kim, and Y.S. Lee. *Biosens. Bioelectron.*, 19:485, 2003.
[511] W.G. Shreffler, K. Beyer, T. Chu, L. Ellman, S. Stanley, G.A. Bannon, W. Burks, and H.A. Sampson. *J. Allergy Clin. Immunol.*, 109:876, 2002.
[512] S.K. Sia and G.M. Whitesides. *Electrophoresis*, 24:3563, 2003.
[513] F.-Z. Sidouni, N. Nurdin, P. Chabrecek, D. Lohmann, J. Vogt, N. Xanthopoulos, H.J. Mathieu, P. Francios, P. Vaudaux, and P. Descouts. *Surface Science*, 481:355, 2001.
[514] S. Singh-Gasson, R.D. Green, Y. Yue, C. Nelson, F. Blattner, M.R. Sussman, and F. Cerrina. *Nat. Biotechnol.*, 17:974, 1999.
[515] H. Siray, C. Frömmel, T. Voronkova, S. Hahn, W. Arnold, J. Schneider-Mergener, S. Scherneck, and R. Ulrich. *Viral Immunol.*, 13:533, 2000.
[516] J.W. Slootstra, W.C. Puijk, G.J. Ligtvoet, J.P.M. Langeveld, and R.H. Meloen. *Mol. Divers.*, 1:87, 1995.
[517] J.W. Slootstra, W.C. Puijk, G.J. Ligtvoet, D. Kuperus, W.M.M. Schaaper, and R.H. Meloen. *J. Mol. Recognit.*, 10:217, 1997.
[518] J.W. Slootstra and W.C. Puijk, patent, WO 0029851, 2000.
[519] J.C. Smith, K.B. Lee, Q. Wang, M.G. Finn, J.E. Johnson, M. Mrksich, and C.A. Mirkin. *Nano Lett.*, 3:883, 2003.
[520] I. Solassol, C. Granier, and A. Pèlegrin. *Tumor Biol.*, 22:184, 2001.
[521] E.M. Southern, U. Maskos, and J.K. Elder. *Genomics*, 13:1008, 1992.
[522] E.M. Southern, S.C. Case-Green, J.K. Elder, M., Johnson, K.U. Mir, L. Wang, and J.C. Williams. *Nucl. Acid Res.*, 22:1368, 1994.
[523] E.M. Southern, K. Mir, and M. Shchepinov. *Nature Genet. Suppl.*, 21:5, 1999.
[524] T. Stan, J. Brix, J. Schneider-Mergener, N. Pfanner, W. Neupert, and D. Rapaport. *Mol. Cell. Biol.*, 23:2239, 2003.
[525] G.C.M. Steffens, L. Nothdurft, G. Buse, H. Thissen, H. Hocker, and D. Klee. *Biomaterials*, 23:3523, 2002.
[526] T.H. Steinberg, B.J. Agnew, K.R. Gee, W.Y. Leong, T. Goodman, B. Schulenberg, J. Hendrickson, J.M. Beechem, R.P. Hougland, and W.F. Patton. *Proteomics*, 3:270, 2003.
[527] M.J. Stevens. *Langmuir*, 15:2773, 1999.
[528] R.-D. Stigler, F. Rüker, D. Katinger, G. Elliott, W. Höhne, P. Henklein, J.X. Ho, K. Keeling, D.C. Carter, E. Nugel, A. Kramer, T. Porstmann, and J. Schneider-Mergener. *Protein Eng.*, 8:471, 1995.
[529] R.-D. Stigler, B. Hoffmann, R. Abagyan, and J. Schneider-Mergener. *Structure*, 7:663, 1999.
[530] M.L. Stolowitz, C. Ahlem, K.A. Hughes, R.J. Kaiser, E.A. Kesicki, G. Li, K.P. Lund, S.M. Torkelson, and J.P. Wiley. *Bioconjugate Chem.*, 12:229, 2001.
[531] K. Strutzberg, L. von Olleschik, B. Franz, C. Pyne, M. A. Schmidt, and G.-F. Gerlach. *Infect. Immun.*, 63:3846, 1995.
[532] J. Su and M. Mrksich. *Angew. Chem.*, 114:4909, 2002.
[533] J. Su and M. Mrksich. *Langmuir*, 19:4867, 2003.
[534] Z. Szallasi, M.F. Denning, E.-Y. Chang, J. Rivera, S.H. Yuspa, C. Lehel, Z. Ohla, W.B. Anderson, and P.M. Blumberg. *Biochem. Biophys. Res. Commun.*, 214:888, 1995.
[535] M. Takahashi, A. Ueno, and H. Mihara. *Chem. Eur. J.*, 6:3196, 2000.
[536] A.C. Tam and W.D. Gill. *Appl. Opt.*, 21:1891, 1982.
[537] J.P. Tam, Y.-A. Lu, C.-F. Liu, and J. Shao. *Proc. Natl. Acad. Sci. U.S.A.*, 92:12485, 1995a.
[538] J.P. Tam, C. Rao, C.-F. Liu, and J. Shao. *Int. J. Pept. Protein Res.*, 45:209, 1995b.
[539] W. Tegge, R. Frank, F. Hofmann, and W.R.G. Dostmann. *Biochemistry*, 34:10569, 1995.
[540] W. Tegge and R. Frank. In S. Cabilly (ed.), Methods in Molecular Biology. Humana Press Inc., Totowa, NJ, p. 99, 1998.
[541] T.P. Theriault, S.C. Winder, and R.C. Gamble. In M. Schena (eds.), DNA Microarrays: A Practical Approach. Oxford University Press, New York, p. 101, 1999.

[542] G. Thinakaran, D.R. Borchelt, M.K. Lee, H.H. Slunt, L. Spitzer, G. Kim, T. Ratovitsky, F. Davenport, C. Nordstedt, M. Seeger, J. Hardy, A.I. Levey, S.E. Gandy, N.A. Jenkins, N.G. Copeland, D.L. Price, and S.S. Sisodia. *Neuron*, 17:181, 1996.
[543] P.A. Tilbrook, G.A. Palmer, T. Bittorf, D.J. McCarthy, M.J. Wright, M.K. Sarna, D. Linnekin, V.S. Cull, J.H. Williams, E. Ingley, J. Schneider-Mergener, G. Krystal, and S. P. Klinken. *Cancer Res.*, 61:2453, 2001.
[544] L.O. Tjernberg, J. Näslund, F. Lindqvist, J. Johansson, A.R. Karlström, J. Thyberg, L. Terenius, and C. Nordstedt. *J. Biol. Chem.*, 271:8545, 1996.
[545] L.O. Tjernberg, C. Lilliehöök, D.J.E. Callaway, J. Näslund, S. Hahne, J. Thyberg, L. Terenius, and C. Nordstedt. *J. Biol. Chem.*, 272: 1997.
[546] F. Toepert, T. Kaute, S. Guffler, and J. Schneider-Mergener. In M. Lebl and R.A. Houghten (eds.), Peptides: Proceedings of the Second International and the Seventeenth American Peptide Symposium. Kluwer Academic Publishers, Dordrecht, p. 212, 2001a.
[547] F. Toepert, J. R. Pires, C. Landgraf, H. Oschkinat, and J. Schneider-Mergener. *Angew. Chem. Int. Ed.*, 40:897, 2001b.
[548] F. Toepert, T. Knaute, S. Guffler, J.R. Pirés, T. Matzdorf, H. Oschkinat, and J. Schneider-Mergener. *Angew. Chem. Int. Ed.*, 42:1136, 2003.
[549] R. Toomik, M. Edlund, P. Ek, B. Öbrink, and L. Engström. *Peptide Research*, 9:6, 1996.
[550] R. Toomik and P. Ek. *Biochem. J.*, 322:455, 1997.
[551] F. Torreilles, F. Roquet, C. Granier, B. Pau, and C. Mourton-Gilles. *Mol. Brain Res.*, 78:181, 2000.
[552] I. Torréns, O. Reyes, A.G. Ojalvo, A. Seralena, G. Chinea, L.J. Cruz, and J. de la Fuente. *Biochem. Bioph. Res. Com.*, 259:162, 1999.
[553] V.V. Tsukruk, I. Luzinov, and D. Julthongpiput. *Langmuir*, 15:3029, 1999.
[554] P. Tunbäck, J.-Å. Liljeqvist, G.-B. Löwhagen, and T. Bergström. *J. Gen. Virol.*, 81:1033, 2000.
[555] B.E. Turk and L.C. Cantley. *Curr. Opin. Chem. Biol.*, 7:84, 2003.
[556] T.P. Ugarova, V.K. Lishko, N.P. Podolnikova, N. Okumura, S.M. Merkulov, V.P. Yakubenko, V.C. Yee, S.T. Lord, and T.A. Haas. *Biochemistry*, 42:9365, 2003.
[557] M. Ulbricht, K. Richau, and H. Kamusewitz. *Colloids & Surfaces A—Physicochem. Eng. Aspects*, 138:353, 1998.
[558] M. Uttamchandani, E.W.S. Chan, G.Y.J. Chen, and S.Q. Yao. *Bioorg. Med. Chem. Lett.*, 13:2997, 2003.
[559] M. Valle, L. Kremer, C. Martínez-A., F. Roncal, J.M. Valpuesta, J.P. Albar, and J.L. Carrascosa. *J. Mol. Biol.*, 288:899, 1999a.
[560] M. Valle, N. Muñoz, L. Kremer, J.M. Valpuesta, C. Martínez-A., J.L. Carrascosa, and J.P. Albar. *Protein Sci.*, 8:883, 1999b.
[561] P.J. van der Veen, L.F.A. Wessels, J.W. Slootstra, R.H. Meloen, M.J.T. Reinders, and J. Hellendoorn. In O. Gascuel and B.M.E. Moret (eds.), Lecture Notes in Computer Science, Workshop on Algorithms in Bioinformatics (WABI 2001). Springer-Verlag, Berlin, Heidelberg, p. 264, 2001.
[562] M.H.V. van Regenmortel. In C.J. Van Oss and M.H.V. Regenmortel (eds.), Immunochemistry. Marcel Dekker, Inc., New York, p. 277, 1994.
[563] K. Vandenbroeck, I. Alloza, D. Brehmer, A. Billiau, P. Proost, N. MsFerran, S. Rüdiger, and B. Walker. *J. Biol. Chem.*, 277:25668, 2002.
[564] A. Vasilijev, U. Ahting, F.E. Nargang, N.E. Go, S.J. Habib, C. Kozany, V. Panneels, I. Sinning, H. Prokisch, W. Neupert, S. Nussberger, and D. Rapaport. *Mol. Cell. Biol.* In press Epub ahead of print Dec 10 (2003).
[565] A.V. Vasilikov, E.N. Timofeev, S.A. Surzhikov, A.L. Drobyshev, V.V. Shick, and A.D. Mirzabekov. *Biotechniques*, 27:592, 1999.
[566] L. Vila, K. Beyer, K.-M. Jaärvinen, P. Chatchatee, L. Bardina, and H.A. Sampson. *Clin. Exp. Allergy*, 31:1599, 2001.
[567] S. Villard, D. Piquer, S. Raut, J.-P. Léonetti, J.-M. Saint-Remy, and C. Granier. *J. Biol. Chem.*, 277:27232, 2002.
[568] S. Villard, S. Lacroix-Desmazes, T. Kieber-Emmons, D. Piquer, S. Grailly, A. Benhida, S.V. Kaveri, J.M. Saint-Remy, and C. Granier. *Blood*, 102:949, 2003.
[569] U. Vogel, A. Weinberger, R. Frank, A. Müller, J. Köhl, J.P. Atkinson, and M. Frosch. *Infect. Immun.*, 65:4022, 1997.
[570] J. Voigt, I. Liebich, M. Kieß, and R. Frank. *Eur. J. Biochem.*, 268:6449, 2001.
[571] S. Volinia, F. Francioso, L. Venturoli, L. Tosi, M. Marastoni, P. Carinci, M. Carella, P. Stanziale, and R. Evangelisti. *Minerva Biotec.*, 13:281, 2001.

[572] R. Volkmer-Engert, B. Ehrhard, J. Hellwig, A. Kramer, W. Höhne, and J. Schneider-Mergener. *Lett. Pept. Sci.*, 1:243, 1994.
[573] R. Volkmer-Engert, B. Hoffmann, and J. Schneider-Mergener. *Tetrahedron Lett.*, 38:1029, 1997.
[574] M. von Bergen, P. Friedhoff, J. Biernat, J. Heberle, E.-M. Mandelkow, and E. Mandelkow. *Proc. Natl. Acad. Sci. U.S.A.*, 97:5129, 2000.
[575] L. von Olleschik-Elbheim, A. el Bayâ, and M.A. Schmidt. *J. Immunol. Methods*, 197:181, 1996.
[576] L. von Olleschik-Elbheim, A. el Bayâ, and M.A. Schmidt. In Haag and Koch-Nolte (ed.), ADP-Ribosylation in Animal Tissue. Plenum Press, New York, p. 87, 1997.
[577] K. Wadu-Mesthrige, S. Xu, N.A. Amro, and G.-Y. Liu. *Langmuir*, 15:8580, 1999.
[578] Z. Wang and R.A. Laursen. *Peptide Res.*, 5:275, 1992.
[579] Z. Wang, W.P. Carney, and R.A. Laursen. *J. Peptide Res.*, 50:483, 1997.
[580] X.F. Wang, K.S. Ryu, D.A. Bullen, J. Zou, H. Zhang, C.A. Mirkin, and C. Liu. *Langmuir*, 19:8951, 2003.
[581] G.J. Wegner, H.J. Lee, and R.M. Corn. *Anal. Chem.*, 74:5161, 2002.
[582] O. Weiergräber, J. Schneider-Mergener, J. Grötzinger, A. Wollmer, A. Küster, M. Exner, and P.C. Heinrich. *FEBS Lett.*, 379:122, 1996.
[583] J. Weiler, H. Gausepohl, N. Hauser, O.N. Jensen, and J.D. Hoheisel. *Nucleic Acids Res.*, 25:2792, 1997.
[584] S.R. Weinberger, T.S. Morris, and M. Pawlak. *Pharmacogenomics*, 1:395, 2000.
[585] S.R. Weinberger, E.A. Dalmasso, and E.T. Fung. *Curr. Opin. Chem. Biol.*, 6:86, 2001.
[586] V. Weinrich, P. Sondermann, N. Bewarder, K. Wissel, and J. Frey. *Hybridoma*, 15:109, 1996.
[587] K. Welfle, R. Misselwitz, R. Sabat, H.-D. Volk, J. Schneider-Mergener, U. Reineke, and H. Welfle. *J. Mol. Recognit.*, 14:89, 2001.
[588] M. Welschof, U. Reineke, C. Kleist, S. Kipriyanov, M. Little, R. Volkmer-Engert, J. Schneider-Mergener, G. Opelz, and P. Terness. *Hum. Immunol.*, 60:282, 1999.
[589] H. Wenschuh, B. Hoffmann, S. Schaller, L. Germeroth, J. Schneider-Mergener, and R. Volkmer-Engert. In S. Bajusz and F. Hudecz (eds.), Peptides 1998: Proceedings of the Twenty-Fifth European Peptide Symposium. Akademiai Kiado, Budapest, p. 772, 1999.
[590] H. Wenschuh, M. Schmidt, L. Germeroth, U. Reineke, D. Scharn, N. Heine, G. Hummel, L. Jobron, J. Schneider-Mergener, H. Matuschewski, M. Ulbricht, U. Schedler, and M. Schulz. In R. Epton (ed.), Innovation and Perspectives in Solid Phase Synthesis and Combinatorial Libraries. Mayflower Worldwide Ltd., Birmingham, p. 383, 2000a.
[591] H. Wenschuh, H. Gausepohl, L. Germeroth, M. Ulbricht, H. Matuschewski, A. Kramer, R. Volkmer-Engert, N. Heine, T. Ast, D. Scharn, and J. Schneider-Mergener. In H. Fenniri (ed.), Combinatorial Chemistry: A Practical Approach. Oxford University Press, Oxford, p. 95, 2000b.
[592] H. Wenschuh, R. Volkmer-Engert, M. Schmidt, M. Schulz, J. Schneider-Mergener, and U. Reineke. *Biopolymers (Peptide Science)* 55:188, 2000c.
[593] M.B. Wilson, and P.K. Nakane. In W. Knapp, K. Holubar, and G. Wick (eds.), Immunofluorescence Related Staining Techniques. Elsevier, Amsterdam, p. 215, 1978.
[594] K. Winkler, H. Scharnagl, U. Tisljar, H. Hoschützky, I. Friedrich, M.M. Hoffmann, M. Hüttinger, H. Wieland, and W. März. *J. Lipid Res.*, 40:447, 1999.
[595] K. Winkler, A. Kramer, G. Küttner, M. Seifert, C. Scholz, H. Wessner, J. Schneider-Mergener, and W. Höhne. *J. Immunol.*, 165:4505, 2000.
[596] N. Winssinger, J.L. Harris, B.J. Backes, and P.G. Schultz. *Angew. Chem.*, 113:3254, 2001.
[597] N. Winssinger, S. Ficarro, P.G. Schultz, and J.L. Harris. *Proc. Natl. Acad. Sci. U.S.A.*, 99:11139, 2002.
[598] G. Wittstock, R. Hesse, and W. Schumann. *Electroanalysis*, 9:746, 1997.
[599] B. Wu, G. Toussaint, L. Vander Elst, C. Granier, M.G. Jacquemin, and J.-M.R. Saint-Remy. *Eur. J. Immunol.*, 30:291, 2000.
[600] Y.N. Xia and G.M. Whitesides. *Langmuir*, 13:2059, 1997.
[601] Y.N. Xia and G.M. Whitesides. *Angew. Chem. Int. Ed. Engl.*, 37:550, 1998.
[602] Y.N. Xia, J.A. Rogers, K.E. Paul, and G.M. Whitesides. *Chem. Rev.*, 99:1823, 1999.
[603] S. Xiao, M. Textor, and N.D. Spencer. *Langmuir*, 14:5507, 1998.
[604] M. Yan and M.A. Bartlett. *Nano Lett.*, 2:275, 2002.
[605] Z. Yang, W. Frey, T. Oliver, and A. Chilkoti. *Langmuir*, 16:1751, 2000.
[606] S.Y. Yang and M.F. Rubner. *J. Am. Chem. Soc.*, 124: 2100, 2002.
[607] M.N. Yousaf, E.W.L. Chan, and M. Mrksich. *Angew. Chem.*, 112:2019, 2000.
[608] M.N. Yousaf, B.T. Houseman, and M. Mrksich. *Angew. Chem. Int. Ed. Engl.*, 40:1093, 2001a.

[609] M.N. Yousaf, B.T. Housman, and M. Mrksich. *Proc. Natl. Acad. Sci. U.S.A.*, 98:5992, 2001b.
[610] A. Yu, E.I. Dementieva, A.A. Stomakhin, E.L. Dari, S.V. Pankov, V.E. Barsky, S.M. Ivanov, E.V. Konovalova, and A.D. Mirzabekov. *BioTechniques*, 34:1008, 2003.
[611] S. Zampieri, M. Mahler, M. Blüthner, Z. Qiu, K. Malmegrim, A. Ghirardello, A. Doria, W.J. van Venrooij, and J.M.H. Raats. *Cell. Mol. Life Sci.*, 60:588, 2003.
[612] N. Zander and H. Gausepohl. In J. Koch and M. Mahler (eds.), Peptide Arrays on Membrane Supports—Synthesis and Applications. Springer Verlag, Berlin, Heidelberg, p. 23, 2002.
[613] X.-M. Zhao, Y.N. Xia, and G.M. Whitesides. *J. Mat. Sci.*, 7:1069, 1997.
[614] H. Zhu and M. Snyder. *Curr. Opin. Chem. Biol.*, 5:40, 2001.
[615] H. Zhu and M. Snyder. *Curr. Opin. Chem. Biol.*, 7:55, 2003.
[616] Q. Zhu, M. Uttamchandani, D. Li, M.L. Lesaicherre, and S.Q. Yao. *Org. Lett.*, 5:1257, 2003.
[617] J. Zimmermann, R. Kühne, R. Volkmer-Engert, T. Jarchau, U. Walter, H. Oschkinat, and L.J. Ball. *J. Biol. Chem.*, 278:36810, 2003.

8

From One-Bead One-Compound Combinatorial Libraries to Chemical Microarrays

Kit S. Lam, Ruiwu Liu, Jan Marik, and Pappanaicken R. Kumaresan
University of California Davis Cancer Center, Division of Hematology/Oncology and Department of Internal Medicine, University of California Davis, 4501 X Street, Sacramento, CA 95817

8.1. INTRODUCTION

In 1984, Geysen *et al.* reported the synthesis of peptides on polyethylene pins in a 96-well foot-print (the multi-pin system) [45]. In 1992, Frank *et al.* described the synthesis of multiple different peptides on cellulose paper [40]. The number of compounds prepared by these parallel synthesis techniques is limited, but they represent the early development of synthetic combinatorial chemistry and chemical arrays. Geysen *et al.* applied the multi-pin system to synthesize peptide mixtures on individual pins, and applied iterative screening (enzyme-linked immunoabsorbant assay) and synthesis approaches to elucidate the chemical structure of the biologically active peptides [46]. Fodor *et al.* first reported the minaturization of the chemical arrays by using light-directed photolithographic chemical synthesis techniques to construct 1042 peptides on a glass chip [38]. Such spatially-addressable peptide microarrays were probed with fluorescent-labeled antibodies, and quantitated with a fluorescent scanner. About the same time, we described the use of split-mix synthesis method to generate millions of random peptide-beads such that each bead displayed only one peptide entity [84]. These "one-bead one-compound" (OBOC) peptide libraries were then screened with an enzyme-linked colorimetric assay, and positive peptide-beads were physically isolated for structural analysis. Such bead libraries can be considered as chemical microarrays that are not addressable but spatially separable. In this mini-review, the development and applications of the OBOC combinatorial library methods and chemical microarray techniques will be reviewed. For chemical microarrays, we shall focus our attention on the various peptide and small molecule microarray techniques. Protein microarrays will only be briefly addressed. Other combinatorial library methods, such as the phage-displayed peptide libraries [24, 28, 148], positional scanning library methods [132], and affinity column library methods [186] will not be discussed here.

8.2. OBOC PEPTIDE LIBRARIES

In 1991, we recognized that by using a "split-mix" synthesis method (Figure 8.1A), we could generate huge peptide libraries on beads [84]. Since each bead is exposed to only one building block at each coupling cycle and the reaction is driven to completion, each bead displays only one peptide-entity. Each 80–100 μm bead contains approximately 100 pmole or 10^{13} copies of the same peptide. The synthesis of such OBOC combinatorial libraries is highly efficient. For example, a heptapeptide bead library consisting of 20 amino acids can be synthesized within 2 to 4 days, which has 1.28 billion permutations (Figure 8.1B). TentaGel resin (Rapp Polymere, Tubingen, Germany) is a good choice for OBOC library. This resin consists of a cross-linked polystyrene core grafted with 3000–4000 Dalton amino-polyethyleneglycol (PEG) chains. These beads can be swollen in water as well as a wide range of organic solvents such as dimethyl formamide, dichloromethane, and toluene. Therefore, TentaGel resin is compatible with both organic synthesis and biological screening. For peptide library construction, Fmoc/*t*-But (9-fluorenylmethoxycarbonyl/*tert*-butyl) chemistry is preferred because unlike Boc/Bn (*tert*-butoxycarbonyl/benzyl) chemistry, hydrofluoric acid (HF) is not needed. HF is highly toxic, requires a special apparatus, and partially degrades the PEG chain on the resin.

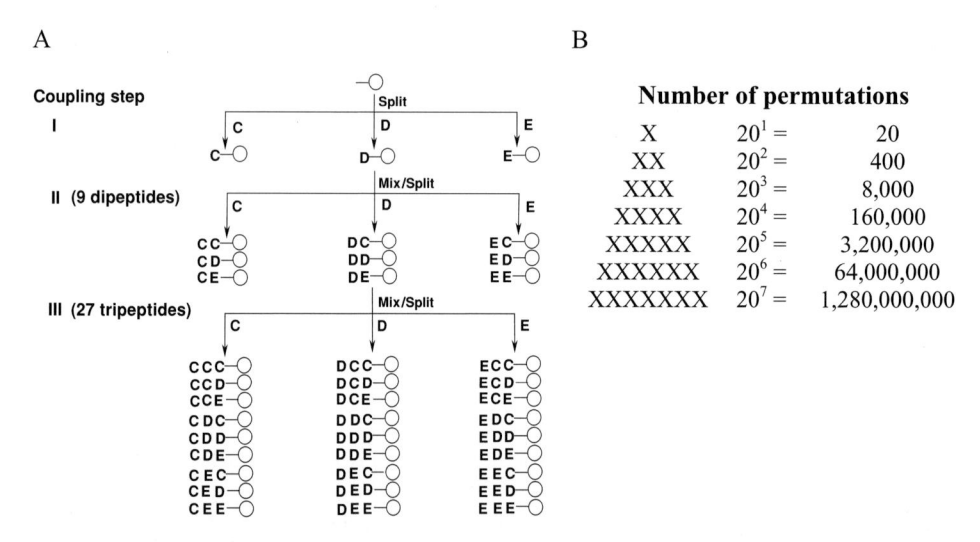

Number of permutations		
X	$20^1 =$	20
XX	$20^2 =$	400
XXX	$20^3 =$	8,000
XXXX	$20^4 =$	160,000
XXXXX	$20^5 =$	3,200,000
XXXXXX	$20^6 =$	64,000,000
XXXXXXX	$20^7 =$	1,280,000,000

FIGURE 8.1. A: Synthetic scheme of the "split synthesis" method to generate a one-bead one-compound combinatorial library; B: Number of permutations for random peptide libraries constructed with 20 amino acids per coupling cycle.

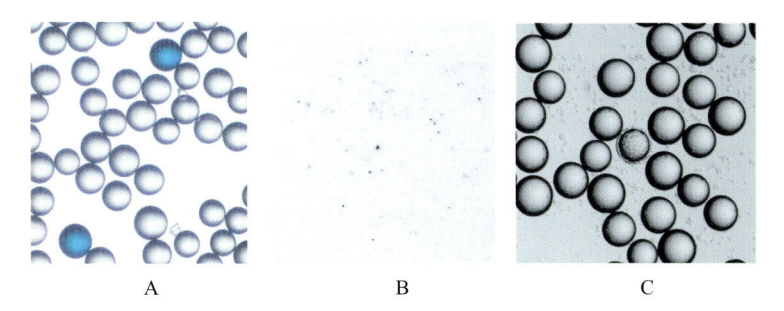

FIGURE 8.2. Various approaches to screen OBOC combinatorial libraries: (A) enzyme-linked colorimetric assay for target protein binding, (B) ^{33}P phosphorylation functional assay for protein kinase substrates (autoradiogram, low power), and (C) whole cell binding assay for cell surface binding ligands.

The well-established on-bead screening assays used in OBOC libraries are highly efficient. Literally millions of compounds can be screened in parallel for a specific acceptor molecule (receptor, antibody, enzyme, virus, whole cell, etc) within a day or two. With various screening techniques, compound-beads with a specific biological, chemical, or physical property can be rapidly identified. We have employed an enzyme-linked colorimetric assay, similar to Western blot, to identify ligands for an anti-β-endorphin antibody [77], streptavidin [81] (Figure 8.2A), avidin [79], an anti-insulin monoclonal antibody that recognizes a discontinuous epitope [78], surface idiotype of B-cell lymphoma cell lines [83], and MHC-Class I molecules [153]. We reported the use of a ^{32}P or ^{33}P phosphorylation assay and an autoradiographic method to identify specific and efficient peptide substrates for protein kinases (Figure 8.2B) [85, 172, 173]. We have also described the use of a whole cell binding assays in which bead-libraries are mixed with live cells (Figure 8.2C) to identify cell surface binding peptide ligands specific for prostate cancer, non-small cell lung cancer, and lymphoma cells [27, 86, 124, 129]. Meldal et al. reported using a fluorogenic quench screening method to identify protease substrates from an OBOC library [106]. Highly porous PEG-based resin (PEGA beads) was used for the peptide library construction because it allowed the enzyme to gain access to the bead interior. Those beads with the active substrates were cleaved by specific protease, resulting in the removal of the quencher and therefore appearance of fluorescent signals. These fluorescent-labeled beads were detected and isolated under a fluorescent microscope, or isolated by a fluorescent activated bead sorter (COPAS™ BIOBEAD, Union Biometrica, Inc, Somerville, MA). Edman sequence analysis of positive beads reveals the substrate sequence, the cleavage point, and the degree of cleavage. This method is a powerful tool to investigate the protease activity and specificity since it gives a complete map of the substrate specificity and affinity for the cleavage site. Combining the OBOC concept and fluorogenic quench screening strategy, Meldal et al. later reported a so-called one-bead two-compounds approach to directly identify protease inhibitors. In this method, each bead contains a putative protease inhibitor along with a fluorescence-quenched substrate for the protease [50, 105]. Very recently, Juskowiak et al. employed a novel on-bead screening method to identify short peptides that were capable of converting to fluorgenic compounds under an ambient photooxidative condition [63]. In this method, a random peptide bead library was illuminated with a tungsten-halogen lamp while the bead mixture was agitated with a stream of air bubbles. After illumination at 40–50 °C for 2 days, the resulting fluorescent beads were detected under a fluorescent microscope. Kodadek and co-workers recently reported the use of fluorescent-labeled proteins as probes to screen OBOC diverse peptoid (N-substituted oligoglycine) libraries [4]. They also reported the use of quantum dots, rather than organic fluorescent dye, as a fluorophor to label the target protein [121]. The main advantage of this latter library screening approach is that the signal to noise ratio is high.

Thus far, only a few groups have reported on the release of compounds from OBOC libraries for solution phase assays. We have described the use of the OBOC combinatorial libraries in a solution phase screening assay in which the compound-beads were immobilized in a thin layer of soft agar together with cancer cells [145]. After cleavage from the linker, compounds from each bead were released and diffused outward. A clear zone of growth inhibition around the positive bead was detected by adding 3-(4,5-dimethylthazol-2-yl)-2,5-diphenyltetrazolium bromide (MTT) to the culture plate, in which only the live cells were stained purple. Jayawickreme *et al.* reported a similar ultra-high throughput assay approach termed "cell-based lawn format" utilizing an *in situ* photocleavage method to release the compound [59, 60]. They first grew a confluent monolayer of indicator cells (frog melanophore cells that had been transfected with a specific receptor) in a petri dish. The cells were then covered with agarose that contained OBOC combinatorial peptide-beads. In response to an agonist released from a specific bead, the frog cells surrounding that bead would turn dark. Silen *et al.* has also reported the use of similar lawn-based *in situ* photocleavage assay to screen an OBOC small molecule library (triazines) to identify novel antimicrobial agents [151]. The sensitivity of this lawn-based screen method has been modified to detect antibacterial compounds with modest potency. These *in situ* releasable solution phase assays have great potential but will require further development before they can be reliably used for drug screening. For instance, special solid supports need to be developed such that all compounds will diffuse freely out of all beads into the surrounding media. An alternative approach to using solution phase assays to screen OBOC library is to release compounds from an individual or small collection of compound beads in a microtiter plate. The released compounds were then subjected to standard solution phase assays [144]. However, to retrieve minute amounts of soluble compounds (100 pmol) from one single 80–100 µm bead is not easy, and often there is not enough material from one single bead for biological assays. Schreiber and his co-workers expanded on the OBOC library methods by developing releasable OBOC libraries in microtiter plates using bigger beads (500 µm macrobeads) [14, 15, 22]. However, the capacity of these macrobeads is still rather limited (<0.1 µmol/bead). Further increases in bead size leads to incomplete coupling because reagent diffusion will become a limiting factor. We have addressed this problem by developing a "one-aggregate one-compound" method [102]. The loading capacity of each bead aggregate ranges between 1–10 µmol, i.e., 10–100 fold more than that of the macrobeads used by Schreiber's group. Bead aggregates were prepared by cross-linking the TentaGel resin beads with glutaraldehyde. Each bead aggregate contains some colored beads with an orthogonal protecting group that can be chemically encoded during the library synthesis and retrieved for decoding after biological screening. Taking full advantage of the OBOC concept, this "one-aggregate one-compound" method is highly efficient and much more economical than the IRORI Nanokan® system (Discovery Partners International Inc) [116]. Such diverse solution phase libraries can be screened with the standard solution phase assay or used to print multiple replicates of chemical microarrays (see below).

As mentioned above, positive compound-beads from screening are physically isolated under a microscope. For peptide libraries that consist of natural amino acids, the straightforward approach to determine the amino acid sequence of the positive bead is to use Edman Chemistry with an automatic protein sequencer (e.g., Procise 494, Perkin-Elmer/Applied Biosystems). The inclusion of unnatural amino acids in the construction of peptide libraries not only greatly increase the diversity of the library but also renders some of the peptide library members more resistant to proteolysis. We have developed methods to sequence peptides containing many unnatural α-amino acids [93]. However, Edman sequencing is time-consuming and expensive compared with mass spectrometry analysis. Youngquist *et al.* introduced 'ladder sequencing' to determine the full peptide sequence of single beads with matrix-assisted laser desorption/ionization time-of-flight mass spectrometry (MALDI-TOF MS) [180]. Later, several groups have modified this method with either replacement of the capping step by the partial incorporation of a methionine residue at each coupling stage [26] or use of the dual capping groups and analysis of molecular ion redundancy to directly elucidate the structure by mass spectrometry analysis [52]. Very recently, this method has been further improved by using an isotope-labeling strategy to terminate the N terminus, and analyzing the generated ladders with ion-extraction mass spectrometry [125].

8.3. ENCODED OBOC SMALL MOLECULE COMBINATORIAL LIBRARIES

Although peptides are useful for targeting extracellular macromolecules or cell surface receptors, they often cannot penetrate cell membranes and therefore are not good drug candidates for intracellular targets. Moreover, the diversity of peptides is limited by the peptide backbone. Therefore, the major effort in combinatorial chemistry has been on small molecule libraries. In OBOC small molecule combinatorial libraries, the chemical structure of the compound on a positive bead has to be determined either directly or via an encoding strategy. Currently, there is no reliable method to directly determine the chemical structure of small molecule on one single bead (100 pmol) isolated from a huge diversity library with, e.g., 150,000 members. Several research groups have developed various physical and chemical encoding methods to encode OBOC small molecule libraries. This subject has been reviewed [2, 8, 26, 80, 149, 175].

Chemical encoding is ideal for OBOC small molecule libraries. Chemical tags are added to the bead during the synthetic steps so that the synthetic history of each compound bead in the chemical library can be recorded. These chemical codes can then be decoded by spectroscopic or chromatographic methods such as HPLC (high performance liquid chromatography), GC (gas chromatography), MS, fluorescence, IR (infrared), NMR (nuclear magnetic resonance) spectroscopy, and electron capture. Several chemical encoding methods have been reported, such as using oligonucleotide tags [19], fluorophenyl ether tags [114, 119], secondary amine tags [37, 115], peptide tags [66, 118], and trityl-based mass tags [150]. However, these methods suffer two disadvantages. First, the coding structure may interfere with the screening assay involving the testing compound. Second, these chemical encoding methods require that the chemistry of adding the tag and synthesizing the library be orthogonal, resulting in nearly doubling the number of synthetic steps. To solve these two problems, we have recently developed a novel and robust peptide-based encoding strategy for OBOC small molecule combinatorial libraries [94]. In this method,

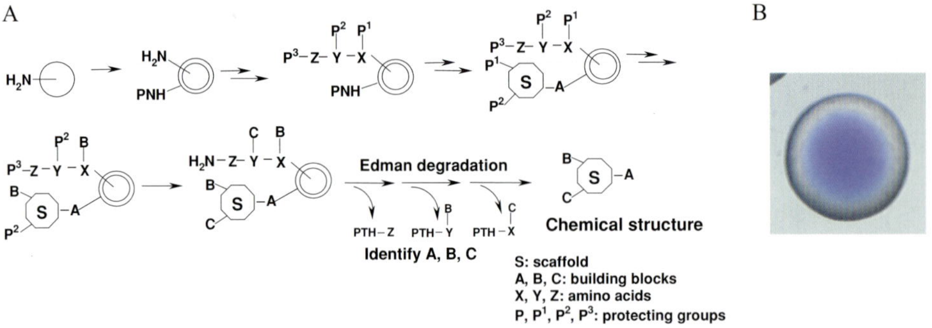

FIGURE 8.3. (A) General synthetic and decoding scheme of peptide-encoded library, (B) Photomicrograph of the topographically segregated bifunctional bead. Free amine at the core reacted with bromophenol blue.

the testing molecule is on the bead surface, and the coding tag is in the interior of the beads. Therefore, the coding tags will not interfere with screening (Figure 8.3). This encoding method is highly efficient as each of the building blocks is incorporated into the testing arm (bead surface) and coding peptide backbone (bead interior) simultaneously. Consequently, no additional synthetic steps are needed. After screening, the positive beads can be isolated, and the peptide coding tags, which consist of α-amino acids with side chains derivatized by the building blocks, can be readily decoded by Edman microsequencing.

Recently, we have further improved our encoding method by incorporating triple or quadruple cleavable coding arms in the bead interior so that mass spectrometry can be used for the decoding process (Figure 8.4) [155]. Prior to library synthesis, the inner core of each bead is derivatized with three or four different coding arms on a cleavable linker. Each of these coding arms contains a functional group that is identical or related to the functional groups on the scaffold of the testing compound to be synthesized. Similar to the above-mentioned peptide-based encoding method, each building block will react with the testing and encoding arms simultaneously, thus eliminating many synthetic steps. After screening, the coding tags in the positive beads are released, followed by molecular mass determination using matrix-assisted laser desorption/ionization-Fourier transform mass spectrometry (MALDI-FTMS). This MS-decoding method may have broader applications than the

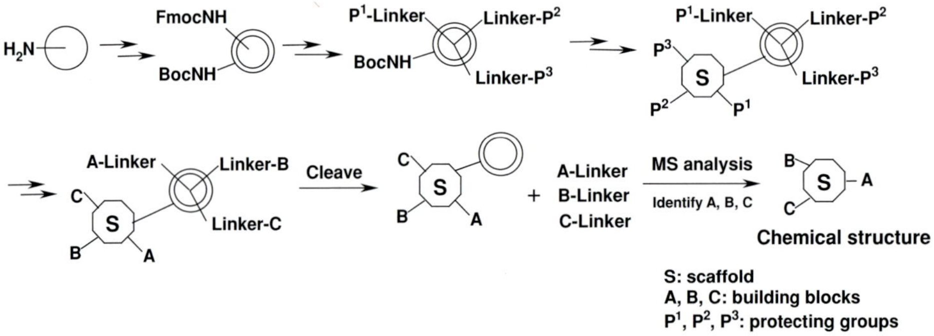

FIGURE 8.4. General synthetic and encoding strategy of the MS-encoded library.

peptide-based encoding method due to the following reasons: (i) MALDI mass spectrometer is readily available in many chemical laboratories, (ii) MS offers rapid analysis (e.g., over 100 compounds per day), and (iii) MS requires minute quantity of material for analysis.

8.4. PEPTIDE AND CHEMICAL MICROARRAYS

As mentioned in the introduction, peptide microarrays [38, 40, 45] preceded DNA microarrays [134]. With the tremendous success of DNA microarrays in the field of genomics in the last decade, the field of chemical microarrays has re-emerged and many researchers have developed methods to prepare microarrays composed of proteins [69], peptides [31], carbohydrates [68, 97, 107], small molecules and other biological molecules. These new techniques enable investigators to rapidly analyze, in parallel, molecular interactions between immobilized molecules and complex biological mixtures. They have been widely applied in the field of diagnostics and proteomics. The immobilized molecules on chemical microarrays are generally addressable, which means that the chemical identity of each immobilized molecule is known. The molecules can be immobilized by *in situ* synthesis [38, 40], chemical ligation, or non-specific adsorption. Proteins can be readily immobilized on polystyrene surfaces, and nitrocellulose or PVDF membranes through non-specific adsorption. However, immobilization of small molecules, carbohydrates, or short peptides often requires covalent ligation to the solid surface, unless these molecules are first ligated to a macromolecule prior to adsorption onto the solid support [102, 178].

An OBOC combinatorial bead library, with one compound entity expressed on each individual bead, can also be considered as a spatially separable but non-addressable chemical microarray [84]. These compounds are synthesized *in situ* using a "split-mix synthesis" approach. Various applications of such libraries have already been addressed in the earlier section of this review.

For most of the peptide and small molecule chemical microarrays, the compounds are synthesized as stock solutions prior to immobilization. In the past decade, several combinatorial chemistry techniques have been developed for parallel synthesis of peptides or small molecules. These include the Multipin system [45], the SynPhase lantern system [72, 126, 174], multi-syringe system [73] and the 96 deep-well plate system [147]. These methods are labor-intensive but can be facilitated by robotics. An alternative but more efficient approach to synthesizing large number of compounds is to exploit the "split-mix" synthesis method as described in the OBOC combinatorial library approach. In this case, one may use the following solid supports to generate large amount of compounds: macrobeads [15, 22] that contain up to 0.1 µmol per bead, bead aggregates with a loading capacity of 1–10 µmol per aggregate [102], or the commercially available IRORI Nanokans [116] ranging from 1–10 µmol. However, these "split-mix" synthesis approaches require a built-in encoding system, so that the chemical structure of the compounds can be elucidated by decoding.

8.4.1. Immobilization Methods for Pre-Synthesized Libraries

The most commonly used solid support for microarray printing is a standard microscope glass slide, but other supports have also been used such as polystyrene [3], nitrocellulose membranes [44], PVDF membrane [102], Hybond ECL membranes [57, 58], gold surfaces [54], and chemical vapor deposited diamond films [168]. For silicon oxide surfaces such as glass slides, the surface is first cleaned with strong oxidizers (H_2SO_4/H_2O_2, oxygen plasma or NH_3/H_2O_2), followed by the coating of organosilane bearing the functionality that will be used for the attachment of small molecules. It is also advantageous to include a hydrophilic linker between the solid surface and the small molecule to minimize steric hindrance caused by the solid support. In some applications, particularly in the drug discovery field where many compounds will be evaluated against limited sets of immobilized molecules, replicates of microarrays can be printed in the bottom of each well of the 96-well plate.

Many automatic arrayers have been developed to print DNA microarrays. Most of these commercially available arrayers can also be used to print peptide and chemical microarrays. The most common mechanical micro-spotting method involves surface contact between the solid support and the tip of the needle or pin. The amount of liquid delivered of by this method ranges between 50 pL–100 nL, resulting in spot size ranges between 75–500 μm. The non-contact electrospraying technique in cone-jet mode can deposit minute but reproducible amounts of enzyme solutions on solid surface, resulting in spots with 130–350 μm in diameter [111]. Avseenko et al. described the further development of the electrospray deposition method for fabricating protein microarrays [7]. They use an electrospray technique to deposit protein solution on an oxidized dextran-grafted surface to form a Shiff base which was subsequently reduced by sodium cyanoborohydride solution. The deposited spots were 30–40 μm in diameter. The laser-based printing method, or so called MAPLE DW (matrix assisted pulsed laser evaporation direct writing) method, has been described for efficient dispension of picoliter volumes of protein solutions and antibodies onto standard solid phase supports [138] to generate spots of 50 μm in diameter. Recently, Ouyang et al. reported a protein microarray preparation method called "soft-landing of mass-selected ions". This method employs electrospray ionization (ESI) followed by individual selection in modified quadruple mass spectrometer (according to mass charge ratio m/z) and subsequent deposition onto solid support [122].

Large macromolecules such as proteins can be easily immobilized via non-specific adsorption. This approach has been used in standard ELISA, dot blot, and Western blot for many years [170]. Specific interactions between chemical tags and macromolecules can also be used for immobilization of tagged proteins. For example, the biotin-streptavidin system was used for immobilization of biotinylated proteins onto streptavidin coated surfaces [142]. The poly-His-Ni^{2+} specific interaction was used for immobilization of poly His tagged proteins onto a Ni^{2+} chelating surface [181]. Anti GST antibody or glutathione coated slides have been reported for GST-fusion proteins immobilization. Unlike plastic and PVDF or nitrocellulose membranes, glass surfaces have low capacity for protein adsorption. To immobilize proteins on glass slides, the glass surface needs to be activated. For example, aldehyde derivatized glass slides have been used for covalent protein immobilization via Shiff base formation of ε-lysine amine groups [100]. Glass surfaces coated with succinimidyl ester or isocyanate functionalized dendrimer have been used for immobilization of proteins and nucleic acids [11], and bisulfosuccinimidyl suberate derivatized glass slides have been used to immobilize proteins [91]. Recently, the use of photoinduced methods for immobilization of proteins using an aryl nitrene [110] or aryl carbene [21] have also been reported. Fang et al. described the fabrication of a novel membrane protein microarray [32, 33]. They first derivatized the glass or gold-coated glass with γ-aminopropylsilane.

Cell membrane preparation containing G-protein-coupled receptors was then spotted with a quill-pin printer. Such microarrays can be use to analyze ligand binding to membrane receptors.

Like proteins and large carbohydrates, native DNA macromolecules can adsorb onto solid support [134]. However, shorter oligonucleotides, small molecules and short peptides have to be immobilized by covalent attachment. We have recently reviewed the methodologies for immobilization of peptides and small molecules on solid support [177]. Rogers *et al.* reported the use of disulfide bond formation to immobilize oligonucleotides on thiol derivatized glass surface [140]. Using Cu-catalyzed [3 + 2] cycloaddition reaction, azide-derivatized oligosaccharides were conjugated to alkynes that have been immobilized to the solid support via the C_{14} hydrocarbon tail [34]. Houseman *et al.* described the covalent attachment of carbohydrates to gold surface [53]. In this method, the gold surface was first derivatized with thiol-polyoxyethylene-benzoquinone conjugate. The cyclopentadiene derivatized carbohydrate was then attached to the benzoquinone group by Diels-Alder reaction. Park *et al.* reported the ligation of maleimide-modified mono- and di-saccharides to glass surface that has been derivatized with thiol group [123]. Zhu *et al.* used TBTU/HOBt/DIEA [2-(1H-benzotriazole-1-yl)-1,1,3,3-tetramethyluronium tetrafluoroborate/*N*-hydroxybenzotriazole/ diisopropylethylamine] as the coupling reagent to immobilize coumarin-4-carboxylic acid derivatives to the amine modified glass surface and employed such arrays to profile enzyme activities [185]. The mild reaction of an aminooxy group with ketone and aldehyde has been widely used for chemselective ligation of various molecules or unprotected peptide fragments [101]. This chemistry has also been applied to the preparation of chemical microarrays. We modified the glass surface with a glyoxylyl group which could then react with peptides or small molecules bearing an aminooxy moiety to form an oxime bond [31]. In principle, proteins with a N-terminal cysteine can also be ligated, chemoselectively, to such surface to form a stable thiazolidine ring [133]. We subsequently decribed a more efficient approach to prepare glyoxylyl coated glass surfaces by first coupling acrylic acid to the amino glass surface, followed by oxidation with $NaIO_4/OsO_4$ to form a glyoxylyl group [176]. Salisbury *et al.* used a similar approach to immobilize aminooxy derivatized peptides to aldehyde modified glass surface [143]. Similar method of immobilization of peptides or oligonucleotides involves the formation of alpha-oxo semicarbazone has also been reported by Olivier *et al.* [121]. They first functionalized the glass surface with semicarbazide and then ligated the glyoxylyl derivatized oligonucleotides or peptides to the semicarbazide group. Lesaicherre *et al.* employed a different chemoselective ligation approach to immobilize C-terminal thioester modified peptides onto an amino modified glass [89]. This microarray was subsequently used for antibody based fluorescence detection of kinase activity [88]. The Staudinger reaction has been recently used for immobilization of proteins and peptides. In this method, Soellner *et al.* ligated the azide modified peptide to phospinothioester-functionalized glass surface to form an amide bond [154]. Kohn *et al.* used Staudinger reaction to immobilize small molecules onto a phosphane-decorated glass slide [71]. The formation of an ether bond between chlorinated glass and a hydroxy group of small molecules has been used for preparation of microarrays by Schreiber *et al.* [49, 73, 75]. The same group has also reported the immobilization of thiol group containing small molecules by Michael addition to maleimide-functionalized glass surface [99]. Recently, they reported the use of diazobenzilidene-functionalized glass slides for immobilization of compounds containing acidic protons, such as phenols, carboxylic acids, and sulfonamides [9]. Kanoh *et al.* recently reported a "non selective" immobilization method using the chemistry developed for photoaffinity labeling [64]. In this method, they first functionalized the glass surface with a photoreactive group, such as diazirine. A small molecule was then spotted and the coupling reaction was initiated by UV irradiation. The chemistries of many of the above mentioned immobilization methods are summarized in Table 8.1.

Macromolecules like proteins, native DNAs or high molecular weight carbohydrates [68, 165] can be immobilized onto a surface, such as plastic and PVDF or nitrocellulose membranes by simple adsorption. Hence, small molecules or peptides could first be conjugated to a macromolecule prior to spotting. Adams *et al.* reported the ligation of low molecular weight carbohydrates to proteins followed by non-covalent immobilization of the carbohydrate-protein conjugates onto solid support [1]. Fukui *et al.* conjugated carbohydrates to lipids, which were subsequently immobilized non-covalently to PVDF or nitrocellulose membranes [41]. Such arrays have been used for studying protein carbohydrate interactions. We spotted peptides conjugated to human serum albumin on polystyrene slides or PVDF membranes as a way to display these peptides on chemical microarrays [178]. Another approach that has been widely used in our laboratory is to chemoselectively conjugate small molecules or peptides bearing aminooxy groups to ketone-modified agarose [102]. Agarose not only serves as a carrier for the small molecule or peptide ligand, but it also provides a highly hydrophilic environment for the analyte to interact with the immobilized compounds. This two-step approach has several advantages. First, the chemical ligation reaction between ligand and polymer scaffold is site-specific and occurs in solution, therefore the yield is high. Second, the concentration of ligand bound to the scaffold is identical among different samples, if excess ligands and the same batch of scaffold are used in the coupling reaction. Third, once the ligand-scaffold conjugate is made, it can be stored indefinitely and used for subsequent spotting. These unique features enable one to generate highly reproducible chemical microarrays with uniform ligand concentration among spots and between slides, which is often difficult with microarrays that are prepared by direct chemical ligation of ligands to the functionalized surface of the solid support.

8.4.2. In Situ Synthesis of Microarrays

As indicated earlier, the synthesis of peptide arrays on paper or cellulose membrane was first described by Frank *et al.* and this method was referred as SPOT synthesis [40]. However, the array generated by this method is generally low density, even with the commercially available automatic SPOT synthesizer [136]. Foder *et al.* [38] first described the high density peptide microarray using the photolithographic light directed parallel synthesis method. This method represents the basic technique that has been used for the generation of the commercially available Affymetrix DNA chips. This method, however, requires building blocks with photolabile protecting groups (e.g., NVoc, MeNPoc, NNeoc, DMBoc, NPPoc, PYMoc, Figure 8.5) which need to be synthesized by the investigator because they are not yet commercially available. To address this problem, Gao *et al.* [42, 43, 87] combined light directed synthesis with microfluidics so that acid can be generated *in situ* (with light) to remove the protecting groups of standard commercially available building blocks such as Boc-protected amino acids [127, 128]. In addition, Gao and others described the use of digital micromirror devices, consisting of 600×800 array of micromirrors to form a virtual

TABLE 8.1. Immobilization Methods

Solid Support	Functionalized Linker	Small molecule or Peptide (R)	Product	Ref.
glass		R-OH		[49, 70, 75]
glass		R-SH		[99]
glass		ArOH, RSO$_2$NH, RCOOH		[9]
glass		R-N$_3$		[71]
glass		R-OSu		[11]
gold		R-Cp		[53, 55]
glass		Cys-R		[89]
glass		R-NH$_2$		[183]

(Cont.)

TABLE 8.2. Continued

Solid Support	Functionalized Linker	Small molecule or Peptide (R)	Product	Ref.
glass		R-O-NH$_2$		[143]
glass, plastic, PVDF		R-O-NH$_2$		[102, 178]
glass		Cys-R		[88]
glass		RCOCHO		[120]
glass		UV		[71]
glass		R-NH$_2$		[11, 91]
glass		R-NH$_2$		[100]
glass		R-NH$_2$		[11]
glass		R-Aoa		[31, 176]

Su = succinimidyl; Cp = cyclopentadienyl; Aoa = aminooxyacetyl, PAMAM = polyamidoamine dendrimer.

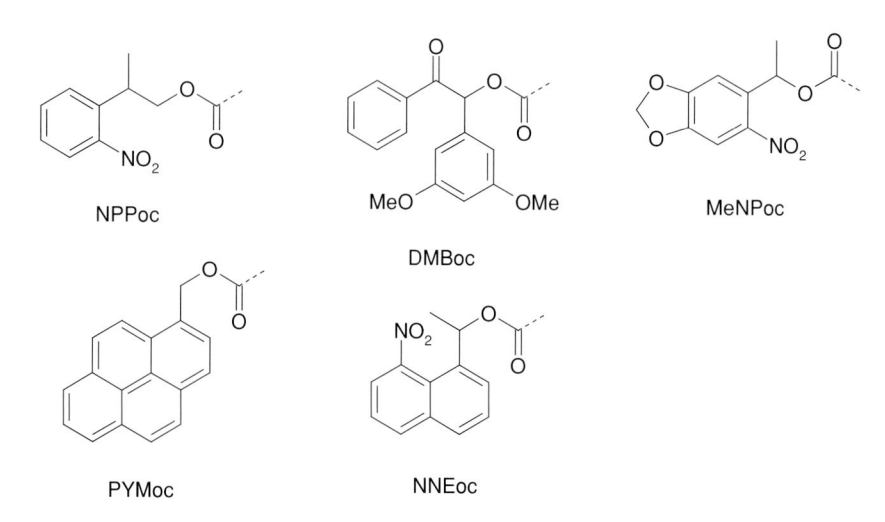

FIGURE 8.5. Photolabile protecting groups used in light-directed parallel synthesis of microarrays Nvoc: 6-nitroveratryloxycarbonyl, MeNPoc: 5'-(α-methyl-2-nitropiperonyl)oxycarbonyl, Nneoc: 8-nitronaphtylethyloxycarbonyl, DMBoc: 5'-O-dimethoxybenzoincarbonyl, NPPoc: 2-(2-nitrophenyl)-propylxycarbonyl, PYMoc: pyrenylmethyloxycarbonyl.

mask [152], to facilitate the *in situ* synthesis of peptide [128] or oligonucleotide [87, 42] microarrays on glass surfaces.

8.4.3. CD, Microfluidics, Fiber Optic Microarray, Multiplex Beads

The compact disc-based microarray system was developed by Kido *et al.* and used for immunoassays [67]. They used the piezoelectric inkjet applicator to deposit the proteins onto a polycarbonate disc. Recently, Clair *et al.* reported the attachment of small molecules to a polycarbonate compact disc (CD) surface via phosphodiester bond. Molecular interactions between analyte and some of these molecules could be detected with a conventional CD player [76]. Walt *et al.* developed the fiber-optic microarray biosensor technology that has been commercialized, and is now referred as the Illumina BeadArray. This microarray is composed of bundles of selectively etched glass fibers which are dipped into the OBOC oligonucleotide library of microspheres (3 or 5 μm). For some reason, these microspheres are captured randomly and spontaneously at the end of each optical fiber [29, 35, 108, 163]. The array is reusable, allows a rapid response, and has an extremely low detection limits. However, because the microarray is spatially non-addressable, an encoding method is needed. The earlier encoding method uses unique ratio of orthogonal fluorescent dyes to encode the oligonucleotides on each bead [35, 108, 109]. An alternative decoding method uses a unique sequence on each bead of the randomly formed microsphere array as an address. To determine the structure, the microsphere bound oligonucleotide is hybridized to a series of biotinylated complementary oligonucleotides and subsequently visualized by interactions with labeled streptavidin [179]. Epstein *et al.* described a similar encoding methodology using the displayed oligonucleotide itself to identify its sequence. But, to determine the sequence, the arrays are hybridized to a series of combinatorial decoding libraries. Four orthogonally labeled decoding libraries are used for each position whereas the screened position in the library is not randomized therefore the affinity of one library to particular bead is higher hence the identity can be determined from level of fluorescence [29]. Recently, encoded fiber-optic microsphere arrays of carbohydrates have been reported [1]. The saccharides were prepared with thiol-terminated ethylenedioxy linker attached to the anomeric center and chemoselectively ligated to commercially available maleimide-activated bovine serum albumin (BSA). The saccharide-BSA conjugate was then coupled to solid support by using water soluble carbodiimide. The BSA molecule serves here as a spacer between the solid support and carbohydrates. The above-mentioned fluorescent dye spectral signature method was used to encode such bead libraries.

8.5. DETECTION METHODS IN CHEMICAL MICROARRAYS

Several methods have been developed for the detection of interactions between the immobilized molecules on the microarrays and the added complex biological mixtures, or analytes. These methods can be broadly classified into two categories: (i) those that involve the detection, identification and characterization of biomolecules (e.g., proteins) present in the analytes that bind to or covalently link to individual chemical spots on the microarrays, and (ii) those that involve the detection and identification of the individual chemical spot that have been enzymatically modified by the enzyme(s), or chemically modified by component(s) present in the analytes.

8.5.1. Identification and Characterization of Bound Proteins

Proteins can bind to peptides or small molecules either through covalent, or non-covalent interactions such as hydrophobic interaction, Van der Waals forces, salt bridges, and hydrogen bonds etc. Common detection methods for protein binding include enzyme-linked colorimetric, fluorescence, luminescence, and radioisotope methods. The fluorescence method is often preferred because it is simple, safe, extremely sensitive, and compatible with the commercially available microarray scanners. The most widely used fluorescent techniques include laser-induced fluorescence, fluorescence resonance energy transfer (FRET), fluorescence polarization, homogenous time resolved fluorescence, and fluorescence correlation spectroscopy [65]. In the standard fluorescence methods, analytes can be directly labeled with fluorescent probe or indirectly with biotin, followed by fluorescent-labeled streptavidin. However, most of the chemical derivatization methods are not site-specific and there is always a concern that chemical labeling of a protein may negatively affect its binding activity to the ligand. An alternative method to fluorescently label a protein is to construct and express the protein as either a fluorophore-puromycin conjugate or red/green fluorescent fusion protein, which can be detected by a conventional fluorescence slide scanner [65, 74]. These methods, while useful, require additional steps, work in some assay systems, but do not work in all proteins. Therefore, alternative methods to detect the bound proteins in their native form are sometimes preferred. One approach is to use fluorescent labeled-antibodies to detect specific proteins bound to the microarrays.

Label-free optical techniques for detecting bound proteins on microarrays have been recently reviewed [23, 90, 162]. These methods include surface plasmon resonance (SPR) [62], grating couplers [13, 112, 169] and reflectometry [18]. SPR has now matured as a versatile detection tool to analyze the kinetics of protein-ligand interactions over a wide range of molecular weights, affinities, and binding rates [104, 113]. Houseman and Mrksich reported the use of SPR as a detection method to profile carbohydrate-lectin interactions on a carbohydrate arrays [54]. As an alternative detection method, Sapsford *et al.* used a planar waveguide to develop an antibody array biosensor and studied the kinetics of antigen-antibody interactions in parallel [146]. The use of atomic force microscopy (AFM) method to detect the surface topological changes of the microarrays due to bound proteins has been reported [61]. However, AFM is very slow, and the method is impractical unless the chip can be scanned with a large array of AFM probes in parallel.

In recent years, various mass spectrometry technologies have evolved as the dominant tools for identification and characterization of bound proteins [36]. Surface-enhanced laser desorption/ionization (SELDI) mass spectrometry has been developed to detect proteins captured by the very low-density affinity arrays [166]. In this method, captured proteins bound to the metal surface (SELDI protein array) are vaporized using a laser beam, followed by identification of these proteins by mass spectrometry. The detection of a minute amount of bound proteins on high density microarrays by this approach, however, remain to be developed.

8.5.2. Detection Methods to Identify Post-Translational Modification of Proteins or to Quantitate Enzyme Activity in Analytes

Peptide microarrays have been used successfully to probe the activities of specific post-translational modification of enzymes (e.g., proteases, protein kinases, esterases, glycosyl transferases, and acetylase) present in an analyte. Modification of protein, peptide or small molecule spots by these enzymes can be detected by lectins, antibodies, fluorescent or radioactive probes [31, 141, 182]. Zhu *et al.* analyzed the kinase-substrate specificity of almost all (119 of 122) yeast kinases using 17 different protein substrates [183]. The substrates were first covalently immobilized on the surface of individual nanowells, and individual protein kinases in kinase buffer with [γ^{33}P] ATP were incubated with the substrates. After washing, the nanowell chips were analyzed for ^{33}P-labeled substrates using a phosphoimager (Molecular Dynamics, Inc.) [183]. Recently, a small molecule fluorophore phosphosensor technology referred as Pro-Q Diamond dye has been developed to detect and quantitate phosphorylated amino acids of peptides and proteins in microarrays [103]. To determine the protease substrate specificity, several groups have developed fluorescent-quenching methods [20, 143, 185]. These methods are very similar to the on-bead assay for OBOC libraries developed by Meldal's group described earlier [106]. In this method, quenched fluorescent substrates were prepared by coupling the peptide substrate to coumarin or 2-aminobenzoic acid. These peptide substrates were then spotted onto the solid support and incubated with proteases such as caspase, MMP-2, and trypsin. Peptides that were susceptible to proteolysis fluoresced while others did not.

8.6. APPLICATION OF CHEMICAL MICROARRAY

Chemical genomics is a highly interdisciplinary approach that integrates chemistry and cell biology [157, 158]. Chemical microarray represents an important tool to exploit this emerging field. Chemical microarrays allow investigators to perform many different assays in parallel using minimal amounts of analytes. Large numbers of biomolecular interactions like protein-protein, protein-ligand, protein-lipid, protein-carbohydrate, and peptide/small molecule-DNA interactions and post-translational modifications can be studied simultaneously by this approach. Information obtained from such studies will facilitate our understanding of cell signaling and function. Like DNA microarrays, the overall pattern of interactions between the analyte and a large number of chemical spots is very informative. Such interaction profiles will allow the investigator to generate conclusions that otherwise would not be possible by only examining a limited number of molecular interactions. Similar to DNA microarrays, bioinformatics and related analytical tools are needed to successfully analyze the data obtained from chemical microarrays. Various aspects of protein microarray technology and antibody microarrays have been recently reviewed, and will only be briefly discussed here [47, 98, 160, 182, 184]. Below is a description on the various biological applications of chemical microarrays, with focus on peptide and small molecule microarrays.

8.6.1. Protein Binding Studies

Protein chips can be used to study protein-protein, protein-nucleic acid, protein-small molecule, and protein-drug interactions [53]. However protein production, purification, and stability are some of the major limiting factors in functional protein microarrays. Protein-protein interactions often occur between specific protein domains that involve short peptides. For example the SH_2 domain binds to a phosphotyrosyl peptide, the SH_3 domain binds to a polyproline helix, and integrins bind short peptides such as Arg-Gly-Asp. Therefore, peptide microarrays are useful tools to study protein-protein interactions. Peptides, unlike proteins, are relatively stable and can be readily synthesized by standard solid phase peptide synthesis methods [131]. Espejo *et al.* reported the development of peptide microarrays with peptide ligands to various known regulatory domain of proteins, such as SH_2, SH_3, PH, EVHI, PZ and WW. By incubating the peptide microarrays with whole cell lysates, they were able to identify new signaling and associated proteins [6, 30].

Peptide arrays have been used successfully by many investigators for B-cell epitope mapping. As early as 1984, Geysen used multipin technology (peptides immobilized on pins in a 96-well footprint) to map epitopes for monoclonal antibodies [45]. Frank *et al.* used SPOT synthesis method to prepare a peptide array on a cellulose membrane for epitope mapping [40]. Similarly, the initial application of the first light-directed synthesis of a high-density peptide microarray was to map the B-cell epitope of anti-β endorphin monoclonal antibody [38].

Recently, Frank and his coworkers have proposed the use of low-density peptide microarrays, which are *in situ* synthesized on membrane by SPOT synthesis, to pan a phage-display protein library that is derived from a randomly fragmented and cloned cDNA library [12]. After affinity enrichment, peptide specific phage populations will be eluted, propagated, labeled, and the identity of the displayed protein determined. This approach is similar to the approach that we have used for studying interactions between OBOC combinatorial small molecule libraries and whole cell extracts, in which a library of the immobilized compounds were screened against a library of target proteins [82]. Reuter *et al.* reported the use of a peptide array to identify the binding sites of DNA to endonuclease *Eco*RII. In this study, [^{32}P]-labeled DNA was used to probe a peptide array derived from endonuclease *Eco*RII sequence [137].

Although there are many reports on using peptide microarrays to probe interactions between peptides and cellular proteins, there have been a limited number of reports on the use of small molecule microarrays to probe cellular proteins. Barnes-Seeman et al. described the identification of new calmodulin-binding small molecules by screening a 6336-member phenol containing fused heterocyclic molecule microarrays [9]. Winssinger et al. used spatially addressable small molecule arrays to study the activity-based profile of proteases in crude cell extracts. In this method a small molecule was covalently tethered to a peptide nucleic acid (PNA) tag, whose sequence, when hybridized to an oligonucleotide microarray, could be used to decode the chemical identity of the small molecule. Using this method, they were able to isolate a small molecule that bound to caspase-3 [171].

8.6.2. Post-Translational Modification, Enzyme-Substrate and Inhibitor Studies

Some of the most important post-translational modifications in cell regulations include phosphorylation, glycosylation, acetylation, and proteolysis. Protein microarrays can be used to identify native substrates for such post-translational modifications. As indicated earlier, protein spots modified by the enzymes can be detected by a radiolabeled substrate (e.g. [γ^{32}P] ATP for protein kinases) and by antibodies or lectins against specific post-translational sites. Zhu et al. analyzed the ability of 119 different yeast kinases to phosphorylate 17 different proteins. They found that members of the yeast Ser-Thr family protein kinases were capable of phosphorylating tyrosine residues of some their substrates [183]. Peptide microarrays have also been a useful tool to study protein phosphorylation. For example, about 10 years ago, we reported on the use of a random OBOC combinatorial library to identify substrate motifs for protein kinase A and c-src protein tyrosine kinase [85, 96, 172]. We have also developed peptide microarray methods to profile protein kinase activities [31]. Peptide substrates were first immobilized on a glass slide through a long hydrophilic linker. After incubation with a protein kinase and [γ^{33}P]-ATP, the phosphorylated peptides were detected by autoradiography or phosphorimager. This functional approach can be used to determine the substrate-specificity of a specific protein kinase, or to profile protein kinase activities in a complex biological sample such as cell extract or serum. Lizcano et al. studied the molecular basis for the substrate specificity of a human protein kinase, Nek6, using peptide microarrays containing more than 1000 different peptides [95]. They observed that protein kinase Nek6 required presence of leucine at the third position on the N-terminal side of the phosphorylation site. Recently, Houseman et al. and others reported the development of a peptide chip that could be phosphorylated by c-src protein kinase, and the level of phosphorylation could be determined by surface plasmon resonance, fluorescence and phosphorimaging [53, 161].

Like phosphorylation and dephosphorylation processes for signaling cascades, glycosylation of extracellular proteins and lipids are critical for the recognition of ligands and cell-cell interactions [48, 92, 167]. Houseman developed monosaccharide arrays and demonstrated that N-acetylglucosamine could be glycosylated by β-1,4-galactosyltransferase in the presence of the donor substrate UDPgalactose [34, 55].

Peptide substrates can also be used to profile protease activities. Salisbury et al. described a protease substrate microarray in which the carboxyl end of the peptide substrates was conjugated to 7-amino-4-carbamoylmethyl coumarin, a fluorogenic compound [143]. The conjugate was non-fluorescent when the electron-donating group on the coumarin was attached to the peptide. Upon proteolysis, the peptide was released and the microarray spot fluoresced. In principle, substrate microarrays consisting of peptides or small molecules can be used as a valuable tool to profile many other enzymes. For example, Zhu et al. demonstrated that the small molecule microarrays could be used to detect enzyme activities of epoxide hydrolases and phosphatases [185].

8.6.3. Cell-Binding Studies

Protein or peptide microarrays can be used to profile the surface receptors or to study the biological function of a live cell. Belov et al. immobilized a series of cell surface marker specific antibodies to form an antibody microarrays, which was then used to profile cells present in the peripheral blood [10]. We have used OBOC combinatorial library methods to identify peptide ligands that bind the surface of intact cells [129]. We plan to develop these cell surface binding peptides into targeting agents for cancer [3]. Peptide microarrays can potentially be used as a diagnostic tool to profile patient cancer cells, allowing the physician can tailor an appropriate peptide cocktail for targeted therapy [3, 31, 83]. To test this concept, we immobilized 44 different cell-binding peptides onto polystyrene slides, and used intact Jurkat human T-lymphoma cells to probe the peptide microarray. The bound cells were stained with crystal violet. This micro cell adhesion assay enables us to identify those ligands that bind to live cancer cells [3]. Furthermore, the cell-binding assay, when used in conjunction with appropriate fluorescence labeled antibodies and confocal microscopy, will enable us to detect cell signaling or morphological changes of cells at spots where the cell attachment occurs [31].

8.6.4. Drug Discovery and Cell Signaling

The various approaches to prepare chemical libraries have been discussed in detail earlier in this review. Often, a combination of combinatorial techniques, in conjunction with standard medicinal chemistry, biophysical methods and molecular modeling methods are needed to develop a drug. Chemical libraries can also be prepared in a microarray format. Such libraries can be used for target validation and drug screening [16, 56, 159]. Although the number of chemical compounds one can generate in a chemical microarray format is limited (e.g., 5–10,000 compounds per slide as compared to 100,000–1,000,000 compounds per OBOC chemical library), replicates of chemical microarrays can be prepared and probed with a number of different target proteins. Schreiber et al. prepared OBOC macrobead chemical libraries, released the compound from each bead into micro wells and then printed chemical microarrays [15, 22]. In order to prepare enough compounds for a large number of microarray replicates, we used encoded bead-aggregates to prepare a "one-aggregate one-compound library" and released the compound from each aggregate into a 96-well plate, ligated the compound to agarose, and then printed the chemical microarray [102]. Such microarrays can then be screened with any of the detection methods discussed above.

Signal transduction in mammalian cells is mediated by complex networks of interacting proteins. Elucidating these pathways requires methods to quantitate the activities of multiple proteins in a rapid and accurate manner. Multiplex antibody microarrays have been used to study the receptor tyrosine kinase signaling cascade in crude cell lysates [117]. Similarly, chemical microarrays have been used to identify the small molecule ligands that can affect specific signaling proteins or pathways. For example, Schreiber and his co-workers reported the use of high-density small molecule arrays of 1, 3 dioxane small molecule library to identify the chemical ligands for the Ure2p transcriptional repressor in yeast [75]. The identified ligand, Uretupamine, was capable of modulating Ure2p signaling function inside the yeast cells. They also identified the small molecules that can interact directly to a signaling protein calmodulin and a yeast transcriptional factor Hap3p and demonstrated that the haptamide A inhibited the Hap3p functions in a dose dependent manner [70]. Housman *et al.* reported the use of peptide microarrays as a screening tool to simultaneously evaluate different substrates and inhibitors for c-src protein tyrosine kinase [54].

8.6.5. Diagnostic Studies

Diagnostic tests for determining serum antibody titers to a number of autoantigens, infectious agents, or other exogenous molecules have been used in clinical medicine for many years. Often these tests are performed in the clinical laboratory, one at a time, and require a large quantity of serum and reagents. In principle, all these tests could be miniaturized by immobilizing the antigens (proteins or peptides) in a microarray format. Similarly, antibodies specific to cytokines or other biological molecules can be immobilized on chips, which can be used either as diagnostic tools to evaluate serum levels of these biological molecules in patients or as research tools in proteomics [17, 156]. It is conceivable that within a decade, biochips will be available for clinical diagnosis, where hundreds to thousands of blood tests can be performed simultaneously and economically on each patient using only a minute amount of blood and analytes. Recent studies also suggest that peptide or small molecule microarrays are useful in the discovery of biomarkers for various diseases, such as autoimmune diseases and cancers [39, 130, 157]. Autoantibody profiles and IgE reactivity profiles have been created by arraying hundreds of autoantigens, including peptides, proteins, and other biomolecules and probing with normal and patients blood samples [51, 139]. Wang and co-workers used carbohydrate-based microarrays to analyze the different types of anticarbohydrate antibodies in human and mammalian sera [164, 165]. Interestingly, many of the carbohydrates that react with the sera are normally present in pathogenic microbes, suggesting that the individuals may have acquired these antibodies during a microbial infection. Very recently Amano *et al.* prepared peptide antigens decorated with various xenobiotics to evaluate their reactivities to sera derived from patients with primary biliary cirrhosis [5].

8.6.6. Non-Biological Applications

The non-biological applications of chemical microarrays, similar to that of combinatorial chemistry, have been lacking behind biological applications. There have been only a few reports on applying chemical microarrays to non-biological systems. Rakow *et al.* reported the development of a low-density chemical microarray of a limited number of compounds that can detect selected organic molecules [135]. In this method, a library of vapor-sensing metalloporphyrins dyes were immobilized on solid support. Visual identification of color change was easily achieved while a ligand was bound to the metalloporphyrins dyes. Using this method, a wide range of ligating vapor can be detected even weakly ligating vapor such as halocarbons and ketones. This type of sensing array is of practical importance for general-purpose vapor dosimeters and analyte-specific detectors.

8.7. FUTURE DIRECTIONS

Since the early 1990s, the field of combinatorial chemistry has progressed rapidly and it has now become an indispensable tool for basic research and drug discovery. Microarrays, initially started as a form of a combinatorial peptide library, has evolved to DNA microarrays, protein microarrays, small molecule microarrays, and microarrays of many other biomolecules. These methods enable one to examine thousands of molecular interactions simultaneously. As a result, biological systems can be studied globally and efficiently. However, to fully exploit the potential of microarrays, more efficient and reproducible methods for immobilizing a uniform amount of chemical compounds or biomolecules onto solid surfaces need to be developed. Many effective detection systems for microarrays have already been described. The next challenge will be to develop mass spectroscopic methods, so that bound biomolecules to each microarray spot can be efficiently identified and quantified. DNA microarrays have already made a great impact in the field of genomics, and have begun to provide prognostic information for cancer patients. Protein, peptide, carbohydrate, and small molecule microarrays will continue to play an increasingly important role in the fields of proteomics, diagnostics, and drug development. We anticipate that in a decade, blood tests will no longer be performed one at a time. Instead, biochips that can perform hundreds to thousands of blood tests for each patient will become commonplace in modern medicine. As more proteins are cloned and expressed, proteome chips containing thousands of proteins will become available for probing protein-protein interactions, signaling pathways, and drug target identification. Human sub-proteome arrays, such as tissue specific and disease-specific protein collections, will enable researchers to rapidly characterize disease pathways for identification and validation of drug targets and biomarkers, as well as drug lead identification and global profiling of drug-protein interactions. Proteome microarrays derived from pathogens is expected to contribute greatly to the development of anti-infectious agents. Even though only a limited number of laboratories are currently working on small molecule microarrays, we expect research activities on this area will grow rapidly in the next few years. When probed with cell extracts obtained from normal and diseased tissues, these small molecule microarrays have the potential of generating drug leads, imaging agents, and drug targets at the same time. Furthermore, such microarrays may enable the researcher to isolate and identify protein complexes that are very important in cell signaling.

Combinatorial chemistry and chemical microarray techniques have already proven to be invaluable tools for biomedical research. Other areas that have and will continue to benefit from these new techniques are material science and sensor development. We anticipate that sensor chips based on chemical microarrays that can detect large numbers of environmental chemicals or biologicals will be developed. Material scientists will be developing methods to combinatorially generate large arrays of new material for rapid analysis. It is expected new materials will be discovered through this high-throughput approach.

ACKNOWLEDGEMENTS

The authors would like to thank Amanda Enstrom for the assistance with the manuscript. This work was supported by NSF Grant MCB9728399, NIH Grants R33CA-86364, R33CA-89706, and R01CA-098116. Ruiwu Liu is supported in part by the University of California System wide Biotechnology Research Program, grant number: 2001–07.

ABBREVIATIONS

AFM	atomic force microscopy
ATP	adenosine-5'- tri phosphate
Bn	benzyl
Boc	*tert*-butoxycarbonyl
BSA	bovine serum albumin
t-But	*tert*-butyl
DIEA	N,N-diisopropylethylamine
DNA	deoxyribonucleic acid
ESI	electrospray ionization
EVHI	enabled/vasodilator-stimulated phosphoprotein homology I
Fmoc	9-fluorenylmethoxycarbonyl
FRET	fluorescence resonance energy transfer
GC	gas chromatography
HF	hydrofluoric acid
HPLC	high performance liquid chromatography
HTS	high throughput screening
IR	infrared
MALDI-FTMS	matrix-assisted laser desorption/ionization-Fourier transform mass spectrometry
MALDI-TOF MS	matrix-assisted laser desorption/ionization time-of-flight mass spectrometry
MMP-2	matrix Metalloproteinasese-2
MTT	3-(4,5-dimethylthazol-2-yl)-2,5-diphenyltetrazolium bromide
NMR	nuclear magnetic resonance
OBOC	one-bead one-compound
PEG	polyethyleneglycol
PNA	peptide nucleic acid
PVDF	polyvinylidene fluoride
SELDI	surface-enhanced laser desorption/ionization
SH_2	the Src homology 2 domain
SH_3	the Src homology 3 domain
SPR	surface plasmon resonance
TBTU	O-(benzotriazol-1-yl)-N,N,N',N'-tetramethyluronium tetrafluoroborate tetrafluoroborate
UDP-galactose	uridine-5'-diphospho galactose

REFERENCES

[1] E.W. Adams, J. Ueberfeld, D.M. Ratner, B.R. O'Keefe, D.R. Walt, and P.H. Seeberger. *Angew. Chem. Int. Ed. Engl.*, 42:5317–5320, 2003.
[2] R.L. Affleck. *Curr. Opi. Chem. Biol.*, 5:257–263, 2001.
[3] O.H. Aina, T.C. Sroka, C. Man-Ling, and K.S. Lam. *Biopolymers (Peptide science)*, 66:184–199, 2002.
[4] P.G. Alluri, M.M. Reddy, K. Bachhawat-Sikder, H.J. Olivos, and T. Kodadek. *J. Am. Chem. Soc.*, 125:13995–14004, 2003.
[5] K. Amano, P.S.C. Leung, Q. Xu, J. Marik, C. Quan, M.J. Kurth, M.H. Nantz, A.A. Ansari, K.S. Lam, M. Zeniya, R.L. Coppel, and E. Gershwin. *J. Immunol.*, (In press) 2004.
[6] R. Apweiler, T.K. Attwood, A. Bairoch, A. Bateman, E. Birney, M. Biswas, P. Bucher, L. Cerutti, F. Corpet, M.D. Croning, R. Durbin, L. Falquet, W. Fleischmann, J. Gouzy, H. Hermjakob, N. Hulo, I. Jonassen, D. Kahn, A. Kanapin, Y. Karavidopoulou, R. Lopez, B. Marx, N.J. Mulder, T.M. Oinn, M. Pagni, F. Servant, C.J. Sigrist, and E.M. Zdobnov. *Nucleic Acids Res.*, 29:37–40, 2001.
[7] N.T. Avseenko, T.Y. Morozova, F.I. Ataullakhanov, and V.N. Morozov. *Anal. Chem.*, 74:927–933, 2002.
[8] C. Barnes and S. Balasubramanian. *Curr. Opin. Chem. Biol.*, 4:346–350, 2000.
[9] D. Barnes-Seeman, S.B. Park, A.N. Koehler, and S.L. Schreiber. *Angew. Chem. Int. Ed. Engl.*, 42:2376–2379, 2003.
[10] L. Belov, O. de la Vega, C.G. dos Remedios, S.P. Mulligan, and R.I. Christopherson. *Cancer Res.*, 61:4483–4489, 2001.
[11] R. Benters, C.M. Niemeyer, and D. Wohrle. *Chembiochem*, 2:686–694, 2001.
[12] K. Bialek, A. Swistowski, and R. Frank. *Anal. Bioanal. Chem.*, 376:1006–1013, 2003.
[13] F.F. Bier and F.W. Scheller. *Biosens. Bioelectron.*, 11:669–674, 1996.
[14] H.E. Blackwell, L. Perez, and S.L. Schreiber. *Angew. Chem. Int. Ed. Engl.*, 40:3421–3425, 2001a.
[15] H.E. Blackwell, L. Perez, R.A. Stavenger, J.A. Tallarico, E.C. Eatough, M.A. Foley, and S.L. Schreiber. *Chem. Biol.*, 8:1167–1182, 2001b.
[16] J. Boguslavsky. *Genom. Proteom.*, 1:44–46, 2001.
[17] C.A. Borrebaeck. *Immunol. Today*, 21:379–382, 2000.
[18] A. Brecht and G. Gauglitz. *Biosens. Bioelectron.*, 10:923–936, 1995.
[19] S. Brenner and R.A. Lerner. *Proc. Natl. Acad. Sci. U.S.A.*, 89:5381–5383, 1992.
[20] J. Buchardt, C.B. Schiodt, C. Krog-Jensen, J.M. Delaisse, N.T. Foged, and M. Meldal. *J. Comb. Chem.*, 2:624–638, 2000.
[21] I. Caelen, H. Gao, and H. Sigrist. *Langmuir*, 18:2463–2467, 2002.
[22] P.A. Clemons, A.N. Koehler, B.K. Wagner, T.G. Sprigings, D.R. Spring, R.W. King, S.L. Schreiber, and M.A. Foley. *Chem. Biol.*, 8:1183–1195, 2001.
[23] M.A. Cooper. *Nat. Rev. Drug. Discov.*, 1:515–528, 2002.
[24] S.E. Cwirla, E.A. Peters, R.W. Barrett, and W.J. Dower. *Proc. Natl. Acad. Sci. U.S.A.*, 87:6378–6382, 1990.
[25] A.W. Czarnik. *Curr. Opin. Chem. Biol.*, 1:60–66, 1997.
[26] M. Davies and M. Bradley. *Tetrahedron Lett.*, 38:8565–8568, 1997.
[27] I.B. DeRoock, M.E. Pennington, T.C. Sroka, R.S. Lam, G.T. Bowden, E.L. Bair, and A.E. Cress. *Can. Res.*, 61:3308–3313, 2001.
[28] J.J. Devlin, L.C. Panganiban, and P.E. Devlin. *Science*, 249:404–406, 1990.
[29] J.R. Epstein, J.A. Ferguson, K.H. Lee, and D.R. Walt. *J. Am. Chem. Soc.*, 125:13753–13759, 2003.
[30] A. Espejo, J. Cote, A. Bednarek, S. Richard, and M.T. Bedford. *Biochem. J.*, 367:697–702, 2002.
[31] J.R. Falsey, M. Renil, S. Park, S. Li, and K.S. Lam. *Bioconjug. Chem.*, 12:346–353, 2001.
[32] Y. Fang, A.G. Frutos, and J. Lahiri. *Chembiochem.*, 3:987–991, 2002a.
[33] Y. Fang, A.G. Frutos, and J. Lahiri. *J. Am. Chem. Soc.*, 124:2394–2395, 2002b.
[34] F. Fazio, M.C. Bryan, O. Blixt, J.C. Paulson, and C.H. Wong. *J. Am. Chem. Soc.*, 124:14397–14402, 2002.
[35] J.A. Ferguson, F.J. Steemers, and D.R. Walt. *Anal. Chem.*, 72:5618–5624, 2000.
[36] D. Figeys, L.D. McBroom, and M.F. Moran. *Methods*, 24:230–239, 2001.
[37] W.L. Fitch, T.A. Baer, W. Chen, F. Holden, C. Holmes, D. Maclean, N. Shah, E. Sullivan, M. Tang, and P. Waybourn. *J. Comb. Chem.*, 1:188–194, 1999.
[38] S.P. Fodor, J.L. Read, M.C. Pirrung, L. Stryer, A.T. Lu, and D. Solas. *Science*, 251:767–773, 1991.
[39] S. Fournel and S. Muller. *Curr. Protein Pept. Sci.*, 4:261–274, 2003.

[40] R. Frank. *Tetrahedron*, 48:9217–9232, 1992.
[41] S. Fukui, T. Feizi, C. Galustian, A.M. Lawson, and W. Chai. *Nat. Biotechnol.*, 20:1011–1017, 2002.
[42] X. Gao, E. LeProust, H. Zhang, O. Srivannavit, E. Gulari, P. Yu, C. Nishiguchi, Q. Xiang, and X. Zhou. *Nucleic Acids Res.*, 29:4744–4750, 2001.
[43] X. Gao, P. Yu, E. LeProust, L. Sonigo, P. Pellois, and H. Zhang. *J. Am. Chem. Soc.*, 120:12698–12699, 1998.
[44] H. Ge. *Nucleic Acids Res.*, 28:e3, 2000.
[45] H.M. Geysen, R.H. Meloen, and S.J. Barteling. *Proceedings of the National Academy of Sciences of the United States of America*. Vol. 81, pp. 3998–4002, 1984.
[46] H.M. Geysen, S.J. Rodda, and T.J. Mason. *Mol. Immunol.*, 23:709–715, 1986.
[47] B.B. Haab. *Curr. Opin. Drug Discov. Devel.*, 4:116–123, 2001.
[48] A. Helenius and M. Aebi. *Science*, 291:2364–2369, 2001.
[49] P.J. Hergenrother, K.M. Depew, and S.L. Schreiber. *J. Am. Chem. Soc.*, 122:7849–7850, 2000.
[50] P.M. Hilaire, S. Sanderson, M.A. Juliano, M. Willert, J. Mottram, G. Coombs, L. Juliano, and M. Meldal. Peptides for the New Millennium, 16th Proc. Am. Pept. Symp., Kluwer Academic Publishers (eds.), Minneapolis, pp. 456–458, 1999.
[51] R. Hiller, S. Laffer, C. Harwanegg, M. Huber, W.M. Schmidt, A. Twardosz, B. Barletta, W.M. Becker, K. Blaser, H. Breiteneder, M. Chapman, R. Crameri, M. Duchene, F. Ferreira, H. Fiebig, K. Hoffmann-Sommergruber, T.P. King, T. Kleber-Janke, V.P. Kurup, S.B. Lehrer, J. Lidholm, U. Muller, C. Pini, G. Reese, O. Scheiner, A. Scheynius, H.D. Shen, S. Spitzauer, R. Suck, I. Swoboda, W. Thomas, R. Tinghino, M. Van Hage-Hamsten, T. Virtanen, D. Kraft, M.W. Muller, and R. Valenta. *Faseb J.*, 16:414–416, 2002.
[52] C. Hoffmann, D. Blechschmidt, R. Kruger, M. Karas, and C. Griesinger. *J. Comb. Chem.*, 4:79–86, 2002.
[53] B.T. Houseman, J.H. Huh, S.J. Kron, and M. Mrksich. *Nat. Biotechnol.*, 20:270–274, 2002a.
[54] B.T. Houseman and M. Mrksich. *Chem. Biol.*, 9:443–454, 2002b.
[55] B.T. Houseman and M. Mrksich. *Trends Biotechnol.*, 20:279–281, 2002c.
[56] D.N. Howbrook, A.M. van der Valk, M.C. O'Shaughnessy, D.K. Sarker, S.C. Baker, and A.W. Lloyd. *Drug Discov. Today*, 8:642–651, 2003.
[57] R.P. Huang. *J. Immunol. Methods*, 255:1–13, 2001.
[58] R.P. Huang, R. Huang, Y. Fan, and Y. Lin. *Anal. Biochem.*, 294:55–62, 2001.
[59] C.K. Jayawickreme, G.F. Graminski, J.M. Quillan, and M.R. Lerner. *Proceedings of the National Academy of Sciences of the United States of America*. Vol. 91, pp. 1614–1618, 1994.
[60] C.K. Jayawickreme, H. Sauls, N. Bolio, J. Ruan, M. Moyer, W. Burkhart, B. Marron, T. Rimele, and J. Shaffer. *J. Pharmacol. Toxicol. Methods*, 42:189–197, 1999.
[61] V.W. Jones, J.R. Kenseth, M.D. Porter, C.L. Mosher, and E. Henderson. *Anal. Chem.*, 70:1233–1241, 1998.
[62] C.E. Jordan and R.M. Corn. *Anal. Chem.*, 69:1449–1456, 1997.
[63] G.L. Juskowiak, S.J. Stachel, P. Tivitmahaisoon, and D.L. Van Vranken. *J. Am. Chem. Soc.*, 126:550–556, 2004.
[64] N. Kanoh, S. Kumashiro, S. Simizu, Y. Kondoh, S. Hatakeyama, H. Tashiro, and H. Osada. *Angew. Chem. Int. Ed. Engl.*, 42:5584–5587, 2003.
[65] Y. Kawahashi, N. Doi, H. Takashima, C. Tsuda, Y. Oishi, R. Oyama, M. Yonezawa, E. Miyamoto-Sato, and H. Yanagawa. *Proteomics*, 3:1236–1243, 2003.
[66] J.M. Kerr, S.C. Banville, and R.N. Zuckerman. *J. Am. Chem. Soc.*, 115:2529–2531, 1993.
[67] H. Kido, A. Maquieira, and B.D. Hammock. *Anal. Chim. Acta.*, 411:1–11, 2000.
[68] L.L. Kiessling and C.W. Cairo. *Nat. Biotechnol.*, 20:234–235, 2002.
[69] T. Kodadek. *Chem. Biol.*, 8:105–115, 2001.
[70] A.N. Koehler, A.F. Shamji, and S.L. Schreiber. *J. Am. Chem. Soc.*, 125:8420–8421, 2003.
[71] M. Kohn, R. Wacker, C. Peters, H. Schroder, L. Soulere, R. Breinbauer, C. M. Niemeyer, and H. Waldman. *Angew. Chem. Int. Ed. Engl.*, 42:5830–5834, 2003.
[72] V. Krchnak and A. Burritt. *Methods Mol. Biol.*, 201:41–59, 2002.
[73] V. Krchnak and J. Vagner. *Pept. Res.*, 3:182–193, 1990.
[74] T. Kukar, S. Eckenrode, Y. Gu, W. Lian, M. Megginson, J.X. She, and D. Wu. *Anal. Biochem.*, 306:50–54, 2002.
[75] F.G. Kuruvilla, A.F. Shamji, S.M. Sternson, P.J. Hergenrother, and S.L. Schreiber. *Nature*, 416:653–657, 2002.
[76] J.J. La Clair and M.D. Burkart. *Org. Biomol. Chem.*, 1:3244–3249, 2003.

[77] K.S. Lam, V.J. Hruby, M. Lebl, R.J. Knapp, W.M. Kazmierski, E.M. Hersh, and S.E. Salmon. *Bioorg. Medicinal Chem. Lett.*, 3:419–424, 1993.
[78] K.S. Lam, D. Lake, S.E. Salmon, J. Smith, M.L. Chen, S. Wade, F. Abdul-Latif, R.J. Knapp, Z. Leblova, R.D. Ferguson, V.V. Krchnak, N.F. Sepetov, and M. Lebl. *Methods*, 9:482–493, 1996.
[79] K.S. Lam and M. Lebl. *Immunological Methods*, 1:11–15, 1992.
[80] K.S. Lam, M. Lebl, and V. Krchnak. *Chem. Rev.*, 97:411–448, 1997.
[81] K.S. Lam, M. Lebl, S. Wade, A. Stierandova, P.S. Khattri, N. Collin, and V.J. Hruby.. Streptavidin-peptide interaction as a model system for molecular recognition. R. S. Hodges. *Peptides: Chemistry, Structure and Biology*. Leiden, Escom, 1005–1006, 1994.
[82] K.S. Lam, R. Liu, S. Miyamoto, A.L. Lehman, and J.M. Tuscano. *Acc. Chem. Res.*, 36:370–377, 2003.
[83] K.S. Lam, Q. Lou, Z.G. Zhao, J. Smith, M.L. Chen, E. Pleshko, and S.E. Salmon. *Biomed. Pept. Proteins Nucleic Acids*, 1:205–210, 1995a.
[84] K.S. Lam, S.E. Salmon, E.M. Hersh, V.J. Hruby, W.M. Kazmierski, and R.J. Knapp. *Nature*, 354:82–84, 1991.
[85] K.S. Lam, J. Wu, and Q. Lou. *Int. J. Pept. Protein Res.*, 45:587–592, 1995b.
[86] D.H. Lau, L.L. Guo, R.W. Liu, A.M. Song, C.K. Shao, and K.S. Lam. *Biotechnol. Lett.*, 24:497–500, 2002.
[87] E. LeProust, J.P. Pellois, P. Yu, H. Zhang, X. Gao, O. Srivannavit, E. Gulari, and X. Zhou. *J. Comb. Chem.*, 2:349–354, 2000.
[88] M.L. Lesaicherre, M. Uttamchandani, G.Y. Chen, and S.Q. Yao. *Bioorg. Med. Chem. Lett.*, 12:2085–2088, 2002a.
[89] M.L. Lesaicherre, M. Uttamchandani, G.Y. Chen, and S.Q. Yao. *Bioorg. Med. Chem. Lett.*, 12:2079–2083, 2002b.
[90] B. Lin, J. Qiu, J. Gerstenmeier, P. Li, H. Pien, J. Pepper, and B. Cunningham. *Biosens. Bioelectron.*, 17:827–834, 2002.
[91] S.C. Lin, F.G. Tseng, H.M. Huang, C.Y. Huang, and C.C. Chieng. *Fresenius J. Anal. Chem.*, 371:202–208, 2001.
[92] H. Lis and N. Sharon. *Eur. J. Biochem.*, 218:1–27, 1993.
[93] R. Liu and K.S. Lam. *Anal. Biochem.*, 295:9–16, 2001.
[94] R. Liu, J. Marik, and K.S. Lam, *J. Am. Chem. Soc.* (In press), 2002.
[95] J.M. Lizcano, M. Deak, N. Morrice, A. Kieloch, C.J. Hastie, L. Dong, M. Schutkowski, U. Reimer, and D.R. Alessi. *J. Biol. Chem.*, 277:27839–27849, 2002.
[96] Q. Lou, M.E. Leftwich, and K.S. Lam. *Bioorg. Med. Chem.*, 4:677–682, 1996.
[97] K.R. Love and P.H. Seeberger. *Angew. Chem. Int. Ed. Engl.*, 41:3583–3586, 3513, 2002.
[98] G. MacBeath. *Nat. Genet.*, 32 Suppl:526–532, 2002.
[99] G. MacBeath, A.N. Koehler, and S.L. Schreiber. *J. Am. Chem. Soc.*, 121:7967–7968, 1999.
[100] G. MacBeath and S.L. Schreiber. *Science*, 289:1760–1763, 2000.
[101] L.A. Marcaurelle, Y. Shin, S. Goon, and C.R. Bertozzi. *Org. Lett.*, 3:3691–3694, 2001.
[102] J. Marik, Q. Xu, X. Wang, L. Peng, and K.S. Lam. T. K. Sawyer and M. Chorev (eds.), *Peptides; Peptide Revolution: Genomics, Proteomics & Therapeutics*, American Peptide Society, Boston, Eighteenth American Peptide Symposium, in press, 2004.
[103] K. Martin, T.H. Steinberg, L.A. Cooley, K.R. Gee, J.M. Beechem, and W.F. Patton. *Proteomics*, 3:1244–1255, 2003.
[104] J.M. McDonnell. *Curr. Opin. Chem. Biol.*, 5:572–577, 2001.
[105] M. Meldal. *Biopolymers*, 66:93–100, 2002.
[106] M. Meldal, I. Svendsen, K. Breddam, and F.I. Auzanneau. *Proceedings of the National Academy of Sciences of the United States of America*. Vol. 91, pp. 3314–3318, 1994.
[107] C.O. Mellet and J.M.G. Fernandez. *ChemBioChem*, 3:819–822, 2002.
[108] K.L. Michael, L.C. Taylor, S.L. Schultz, and D.R. Walt. *Anal. Chem.*, 70:1242–1248, 1998.
[109] K.L. Michael, L.C. Taylor, and D.R. Walt. *Anal. Chem.*, 71:2766–2773, 1999.
[110] J.C. Miller, H. Zhou, J. Kwekel, R. Cavallo, J. Burke, E.B. Butler, B.S. Teh, and B.B. Haab. *Proteomics*, 3:56–63, 2003.
[111] R. Moerman, J. Frank, J.C.M. Marjnissen, T.G.M. Schalkhammer, and G.W.K. van Dedem. *Anal. Chem.*, 73:2183–2189, 2001.
[112] F. Morhard, J. Pipper, R. Dahint, and M. Grunze. *Sens. Actua.*, B70:232–242, 2000.
[113] D.G. Myszka and R.L. Rich. 3:310–317, 2000.

[114] H.P. Nestler, P.A. Bartlett, and W.C. Still. *J. Org. Chem.*, 58:4723–4724, 1994.
[115] Z.J. Ni, D. Maclean, C.P. Holmes, M.M. Murphy, B. Ruhland, J.W. Jacobs, E.M. Gordon, and M.A. Gallop. *J. Med. Chem.*, 39:1601–1608, 1996.
[116] K.C. Nicolaou, K.C. Xiao, Z. Parandoosh, Z. Seneyi, and M. Nova. *Angew. Chem. Int. Ed. Engl.*, 34:1995.
[117] U.B. Nielsen, M.H. Cardone, A.J. Sinskey, G. MacBeath, and P.K. Sorger. *Proc. Natl. Acad. Sci. U.S.A.*, 100:9330–9335, 2003.
[118] V. Nikolaiev, A. Stierandova, V. Krchnak, B. Seligmann, K.S. Lam, S.E. Salmon, and M. Lebl. *Peptide Res.*, 6:161–170, 1993.
[119] M.H. Ohlmeyer, R.N. Swanson, L.W. Dillard, J.C. Reader, G. Asouline, R. Kobayashi, M. Wigler, and W.C. Still. *Proc. Natl. Acad. Sci. U.S.A.*, 90:10922–10926, 1993.
[120] C. Olivier, D. Hot, L. Huot, N. Ollivier, O. El-Mahdi, C. Gouyette, T. Huynh-Dinh, H. Gras-Masse, Y. Lemoine, and O. Melnyk. *Bioconjug. Chem.*, 14:430–439, 2003.
[121] H.J. Olivos, K. Bachhawat-Sikder, and T. Kodadek. *Chembiochem*, 4:1242–1245, 2003.
[122] Z. Ouyang, Z. Takats, T.A. Blake, B. Gologan, A.J. Guymon, J.M. Wiseman, J.C. Oliver, V.J. Davisson, and R.G. Cooks. *Science*, 301:1351–1354, 2003.
[123] S. Park and I. Shin. *Angew. Chem. Int. Ed. Engl.*, 41:3180–3182, 2002a.
[124] S.I. Park, R. Manat, B. Vikstrom, N. Amro, L.W. Song, and K.S. Lam. *Lett. Pept. Sci.*, 8:171–178, 2002b.
[125] J.J. Pastor, I. Lingard, G. Bhalay, and M. Bradley. *J. Comb. Chem.*, 5:85–90, 2003.
[126] M. Patek, P. Safar, M. Smrcina, E. Wegrzyniak, K. Bjegrade, A. Weichsel, and P. Strop. *J. Comb. Chem.*, 6:43–49, 2003.
[127] J.P. Pellois, W. Wang, and X. Gao. *J. Comb. Chem.*, 2:355–360, 2000.
[128] J.P. Pellois, X. Zhou, O. Srivannavit, T. Zhou, E. Gulari, and X. Gao. *Nat. Biotechnol.*, 20:922–926, 2002.
[129] M.E. Pennington, K.S. Lam, and A.E. Cress. *Molecul. Diversity*, 2:19–28, 1996.
[130] E.F. Petricoin, A.M. Ardekani, B.A. Hitt, P.J. Levine, V.A. Fusaro, S.M. Steinberg, G.B. Mills, C. Simone, D. A. Fishman, E.C. Kohn, and L.A. Liotta. *Lancet*, 359:572–577, 2002.
[131] E. Phizicky, P.I. Bastiaens, H. Zhu, M. Snyder, and S. Fields. *Nature*, 422:208–215, 2003.
[132] C. Pinilla, J.R. Appel, P. Blanc, and R.A. Houghten. *Biotechniques*, 13:901–905, 1992.
[133] C. Pinilla, V. Rubio-Godoy, V. Dutoit, P. Guillaume, R. Simon, Y.D. Zhao, R.A. Houghten, J.C. Cerottini, P. Romero, and D. Valmori. *Can. Res.*, 61:5153–5160, 2001.
[134] M. Pirrung. *Angew. Che. Int. Ed.*, 41:1276–1289, 2002.
[135] N.A. Rakow and S.K.S. *Nature*, 406:710–713, 2000.
[136] U. Reineke, R. Volkmer-Engert, and J. Schneider-Mergener. *Curr. Opin. Biotechnol.*, 12:59–64, 2001.
[137] M. Reuter, J. Schneider-Mergener, D. Kupper, A. Meisel, P. Mackeldanz, D. H. Kruger, and C. Schroeder. *J. Biol. Chem.*, 274:5213–5221, 1999.
[138] B.R. Ringeisen, P.K. Wu, H. Kim, A. Pique, R.Y. Auyeung, H.D. Young, D.B. Chrisey, and D.B. Krizman. *Biotechnol. Prog.*, 18:1126–1129, 2002.
[139] W.H. Robinson, C. DiGennaro, W. Hueber, B.B. Haab, M. Kamachi, E.J. Dean, S. Fournel, D. Fong, M.C. Genovese, H.E. de Vegvar, K. Skriner, D.L. Hirschberg, R.I. Morris, S. Muller, G.J. Pruijn, W.J. van Venrooij, J.S. Smolen, P.O. Brown, L. Steinman, and P.J. Utz. *Nat. Med.*, 8:295–301, 2002.
[140] Y.-H. Rogers, P. Jiang-Baucom, Z.-J. Huang, V. Bogdanov, S. Anderson, and M. T. Boyce-jacino. *Anal. Biochem.*, 266:23–30, 1999.
[141] C.A. Rowe, S.B. Scruggs, M.J. Feldstein, J.P. Golden, and F.S. Ligler. *Anal. Chem.*, 71:433–439, 1999.
[142] L.A. Ruiz-Taylor, T.L. Martin, F.G. Zaugg, K. Witte, P. Indermuhle, S. Nock, and P. Wagner. *Proc. Natl. Acad. Sci. U.S.A.*, 98:852–857, 2001.
[143] C.M. Salisbury, D.J. Maly, and J. Ellman. *J. Am. Chem. Soc.*, 124:14868–14870, 2002.
[144] S.E. Salmon, K.S. Lam, M. Lebl, A. Kandola, P.S. Khattri, S. Wade, M. Patek, P. Kocis, V. Krchnak, D. Thorpe, and S. Felder. *Proceedings of the National Academy of Sciences of the United States of America*. Vol. 90, pp. 11708–11712, 1993.
[145] S.E. Salmon, R.H. Liu-Stevens, Y. Zhao, M. Lebl, V. Krchnak, K. Wertman, N. Sepetov, and K.S. Lam. *Mol. Divers.*, 2:57–63, 1996.
[146] K.E. Sapsford, Z. Liron, Y.S. Shubin, and F.S. Ligler. *Anal. Chem.*, 73:5518–5524, 2001.
[147] G. Scnorrenberg and H. Gerhardt. *Tetrahedron*, 45:7759–7764, 1989.
[148] J.K. Scott and G.P. Smith. *Science*, 249:386–390, 1990.
[149] P. Seneci. *J. Recept. Signal. Transduct. Res.*, 21:409–445, 2001.
[150] M.S. Shchepinov, R. Chalk, and E.M. Southern. *Tetrahedron*, 56:2712–2724, 2000.

[151] J.L. Silen, A.T. Lu, D.W. Solas, M.A. Gore, D. MacLean, N.H. Shah, J.M. Coffin, N.S. Bhinderwala, Y. Wang, K.T. Tsutsui, G.C. Look, D.A. Campbell, R.L. Hale, M. Navre, and C.R. DeLuca-Flaherty. *Antimicrob. Agents Chemother.*, 42:1447–1453, 1998.
[152] S. Singh-Gasson, R.D. Green, Y. Yue, C. Nelson, F. Blattner, M. R. Sussman, and F. Cerrina. *Nat. Biotechnol.*, 17:974–978, 1999.
[153] M.H. Smith, K.S. Lam, E.M. Hersh, M. Lebl, and W.J. Grimes. *Mol. Immunol.*, 31:1431–1437, 1994.
[154] M.B. Soellner, K.A. Dickson, B.L. Nilsson, and R.T. Raines. *J. Am. Chem. Soc.*, 125:11790–11791, 2003.
[155] A. Song, J. Zhang, C.B. Lebrilla, and K.S. Lam. *J. Am. Chem. Soc.*, 125:6180–6188, 2003.
[156] A. Sreekumar and A.M. Chinnaiyan. *Curr. Opin. Mol. Ther.*, 4:587–593, 2002.
[157] R.L. Strausberg and S.L. Schreiber. *Science*, 300:294–295, 2003a.
[158] R.L. Strausberg, A.J. Simpson, and R. Wooster. *Nat. Rev. Genet.*, 4:409–418, 2003b.
[159] S.A. Sundberg, A. Chow, T. Nikiforov, and H.G. Wada. *Drug Discov. Today*, 5:92–103, 2000.
[160] M.F. Templin, D. Stoll, M. Schrenk, P.C. Traub, C.F. Vohringer, and T.O. Joos. *Drug Discov. Today*, 7:815–822, 2002.
[161] M. Uttamchandani, E.W. Chan, G.Y. Chen, and S.Q. Yao. *Bioorg. Med. Chem. Lett.*, 13:2997–3000, 2003.
[162] D. Vetter. *J. Cell. Biochem. Suppl.*, 39:79–84, 2002.
[163] D.R. Walt. *Science*, 287:451–452, 2000.
[164] D. Wang. *Proteomics*, 3:2167–2175, 2003.
[165] D. Wang, S. Liu, B.J. Trummer, C. Deng, and A. Wang. *Nat. Biotechnol.*, 20:275–281, 2002.
[166] S.R. Weinberger, T.S. Morris, and M. Pawlak. *Pharmacogenomics*, 1:395–416, 2000.
[167] W.I. Weis, M.E. Taylor, and K. Drickamer. *Immunol. Rev.*, 163:19–34, 1998.
[168] S. Wenmackers, K. Haenen, M. Nesladek, P. Wagner, L. Michiels, M. VanDeVen, and M. Ameloot. *Phys. Stat. Sol.*, 199:44–48, 2003.
[169] M. Wiki, R.E. Kunz, G. Voirin, K. Tiefenthaler, and A. Bernard. *Biosens. Bioelectron.*, 13:1181–1185, 1998.
[170] D.S. Wilson and S. Nock. *Angew. Chem. Int. Ed. Engl.*, 42:494–500, 2003.
[171] N. Winssinger, S. Ficarro, P.G. Schultz, and J.L. Harris. *Proc. Natl. Acad. Sci. U.S.A.*, 99:11139–11144, 2002.
[172] J. Wu, Q.N. Ma, and K.S. Lam. *Biochemistry*, 33:14825–14833, 1994.
[173] J.J. Wu, D.E.H. Afar, H. Phan, O.N. Witte, and K.S. Lam. *Combinat. Chem. High Throughput Screen.*, 5:83–91, 2002.
[174] Z. Wu and N.J. Ede. *Org. Lett.*, 5:2935–2938, 2003.
[175] X.-Y. Xiao. *Front. Biotechnol. Pharm.*, 1:114–149, 2000.
[176] Q. Xu and K.S. Lam. *Tetrahed. Lett.*, 43:4435–4437, 2002.
[177] Q. Xu and K.S. Lam. *J. Biomed. Biotechnol.*, 2003:257–266, 2003.
[178] Q. Xu, S. Miyamoto, and K.S. Lam. *Molecul. Diver.* (In press), 2004.
[179] J.M. Yeakley, J.B. Fan, D. Doucet, L. Luo, E. Wickham, Z. Ye, M.S. Chee, and X.D. Fu. *Nat. Biotechnol.*, 20:353–358, 2002.
[180] R.S. Youngquist, G.R. Fuentes, M.P. Lacey, and T. Keough. *J. Am. Chem. Soc.*, 117:3900–3906, 1995.
[181] H. Zhu, M. Bilgin, R. Bangham, D. Hall, A. Casamayor, P. Bertone, N. Lan, R. Jansen, S. Bidlingmaier, T. Houfek, T. Mitchell, P. Miller, R.A. Dean, M. Gerstein, and M. Snyder. *Science*, 293:2101–2105, 2001a.
[182] H. Zhu, M. Bilgin, and M. Snyder. *Annu. Rev. Biochem.*, 72:783–812, 2003a.
[183] H. Zhu, J.F. Klemic, S. Chang, P. Bertone, A. Casamayor, K.G. Klemic, D. Smith, M. Gerstein, M.A. Reed, and M. Snyder. *Nat. Genet.*, 26:283–289, 2000.
[184] H. Zhu and M. Snyder. *Curr. Opin. Chem. Biol.*, 5:40–45, 2001b.
[185] Q. Zhu, M. Uttamchandani, D. Li, M.L. Lesaicherre, and S.Q. Yao. *Org. Lett.*, 5:1257–1260, 2003b.
[186] R.N. Zuckermann, J.M. Kerr, M.A. Siani, S.C. Banville, and D.V. Santi. *Proceedings of the National Academy of Sciences of the United States of America.* Vol. 89, pp. 4505–4509, 1992.

II

Advanced Microfluidic Devices
and Human Genome Project

9

Plastic Microfluidic Devices for DNA and Protein Analyses

Z. Hugh Fan[1] and Antonio J. Ricco[2]
[1]Department of Mechanical and Aerospace Engineering, Department of Biomedical Engineering, and McKnight Brain Institute, University of Florida, PO Box 116250, Gainesville, FL 32611-6250
[2]NASA Ames Research Center, Mountain View, CA 94035 and Dept. of Electrical Engineering, Stanford University, Stanford, CA 94305

9.1. INTRODUCTION

Microfluidics and integrated microsystems are the current focus of unusually intense interest and activity on the part of academia, industry, and governmental agencies, an assertion substantiated by the solid attendance at the 2003 International Conference on Micro Total Analysis Systems (μTAS) [45] and by the recent publication of various books on this topic [46], both at a time when the economy in general, and the technology sector in particular, are at relative low points. Significant advances have been realized since the concept of μTAS was developed more than a decade ago as a means to enhance versatility and functionality relative to discrete chemical sensor devices [41]. Functional examples of the μTAS concept include on-chip PCR (polymerase chain reaction) [28, 44, 72], DNA analysis and sequencing [14, 58, 76], immunoassays [10, 19, 57], protein separations [23, 48, 64], and intra- and inter-cellular analysis [29, 53, 63]. Advantages of these μ-TAS over bench-top instruments include low reagent consumption, small sample volumes, high separation efficiencies, short reaction times, ease of automation, and potential for mass-production with low cost [41].

9.1.1. Detection

In general, miniaturization of analytical instrumentation requires that detection systems detect ever-decreasing numbers of molecules. A variety of detection methods have been incorporated or appended ("hyphenated") to μTAS devices, including laser-induced fluorescence [18], ultraviolet (UV) absorbance [30], electrochemistry [68, 54], and mass spectrometry [9, 26]. Due to its sensitivity, laser-induced fluorescence (LIF) is most often employed in applications with challenging limits of detection. Using confocal detection, LIF now routinely enables detection of fluorescently labeled molecules at concentrations around 10 pM; with sub-nanoliter detection volumes, the number of molecules detected is in the hundreds to thousands (, ~1 zeptomol). This detection limit is acceptable for most applications that require high sensitivity, such as DNA sequencing and many immunoassays. Though less sensitive in absolute terms, electrochemical detection is low in cost and readily integrated onto fluidic chips: Girault and colleagues integrated carbon microband electrodes in the bottoms or sidewalls of rectangular microchannels in plastics, demonstrating a detection limit of ~1 fmol of ferrocenecarboxylic acid [54].

9.1.2. Materials

Several types of materials have been used for making μTAS devices, including silicon [21, 72], glass [13, 25], and plastics [3, 51, 61]. The primary motivation for the construction of μTAS from plastics is cost. While microfabrication of silicon, glass, and similar materials can be accomplished cheaply when chip sizes are small (a few mm^2 in area), fluidic devices occupying many cm^2 are often dictated by applications (*vide infra*). For diagnostic devices and other contamination-sensitive applications, costs must often be low enough for single-use disposability. In addition, the biocompatibility of various plastics is well documented, allowing plastic devices to be adapted to a wide range of applications, including DNA and protein analyses. For these reasons, this chapter focuses on plastic microfluidic systems.

9.2. ELECTROKINETIC PUMPING

Manipulation of fluids in a microfabricated device requires pumps that are compatible with the typical flow rates used in the microfluidic system. Pumps using electronic and mechanical means have been most often employed. Electronic pumps include electrokinetic (*vide infra*), electrohydrodynamic [37], and dielectrophoretic [17], while mechanical pumps include those using pneumatic pressure, syringe drive, bubble generation, thermal expansion, osmotic pressure, and other transducer-induced motions/forces. Other pumping mechanisms involve thermal gradients, magnetic force, and magnetohydrodynamic flow. In this chapter, we will center on electrokinetic means to move and separate analytes.

Using electrokinetic phenomena as a means to pump and separate analytes was a major catalyst for the rapid growth of μ-TAS during the 1990's. Its utility was demonstrated by Manz and Harrison early in the 1990's [20, 40]. However, this seminal achievement came more than a decade after the first realization of integrated components for miniaturized analytical instrumentation, namely a microfabricated gas chromatography system on a single silicon wafer [66]. The long time delay was due primarily to the lack of an appropriate pumping mechanism suited to micro-scale devices.

The use of electrokinetic pumping was accompanied by widespread acceptance of capillary electrophoresis (CE) as a chip-appropriate separation method. CE was largely developed using fused silica capillaries, their diameter being in the same range as the

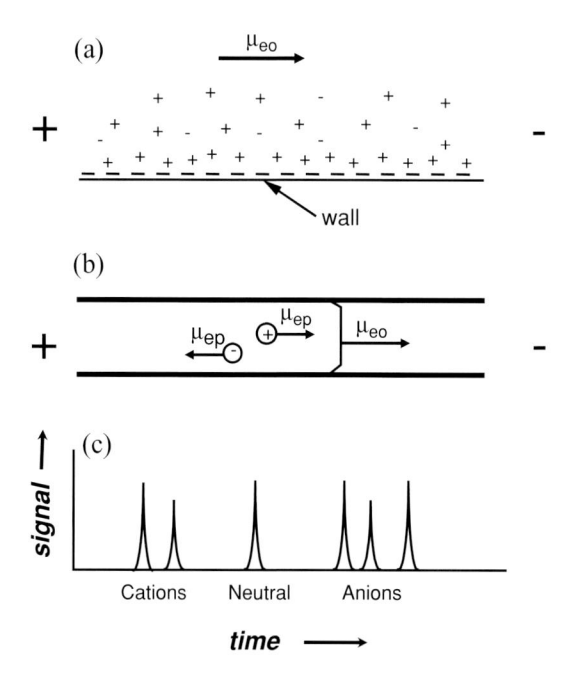

FIGURE 9.1. (a) Formation of the double-layer at a channel wall/solution interface and the generation of electroosmosis under an applied electric field; μ_{eo} is EO mobility. (b) The nearly flat electroosmotic flow profile. In addition to EO, both cations and anions experience electrophoretic flow; μ_{ep} is electrophoretic mobility. (c) A generic electropherogram showing cations, neutral molecules, and anions.

microfabricated channels that are a principal feature of fluidic devices, making its adaptation to chip-based separations straightforward.

The fundamentals of CE, found in textbooks [3] and reviews [12], are discussed briefly here. When a glass channel or a fused silica capillary is filled with an appropriate solution, a phenomenon called electroosmosis (EO) occurs when an electric field is applied to the ends. The generation of EO is schematically illustrated in Figure 9.1a. The walls of the channel or capillary are negatively charged in an aqueous solution at pH > 3 due to the ionization of surface silanol groups [35]. The negative charges on the wall surface attract positive ions from the buffer solution, thus giving a typical ionic double-layer structure. The mobile positive ions in the diffuse layer carry several solvent molecules each and these solvated ions are attracted to the cathode (negative electrode) by the electric field, thus moving parallel to the channel or capillary walls. The movement of this sheath of solvated ions drags with it the solution in the rest of the channel, and the resulting electroosmosis has a unique feature: a nearly flat flow profile across the entire diameter of the capillary, as shown in Figure 9.1b.

While electroosmosis moves both solutes and solvents in the same direction and at the same speed, the separation of solutes results from electrophoresis, which occurs at the same time as EO and is caused by the difference in ionic mobilities in the applied electric field. Although electrophoresis simultaneously drives cations (positive ions) to the cathode and anions (negative ions) to the anode, for glass channels it is often the case that electroosmotic mobility is larger than electrophoretic mobility, hence both cations and anions exhibit net migration toward the cathode and are detected at one end of the channel or capillary. A generic electropherogram is shown in Figure 9.1c, in which cations migrate first to the detector, and then neutral molecules, with anions last. This pattern results from the fact that cations experience both electrophoretic and electroosmotic movement in the same direction, while anions experience them in opposite directions, and neutral molecules experience only electroosmosis.

9.3. PLASTIC DEVICES

Although much of the work on microfluidic devices in the 1990's utilized glass or silicon substrates, plastics have been increasingly studied in the past few years. The plastic materials investigated for microfabrication and microfluidics include polystyrene [51], polymethylmethacrylate (PMMA) [42, 49, 69], polycarbonate [32, 77], polydimethylsiloxane (PDMS) [11, 43], polyolefins [7, 26], polyester [16], fluorinated ethylene propylene [55], and poly(ethylene terephthalate) [51]. The scientists of ACLARA BioSciences, Inc. used several of these plastics for a wide range of applications including DNA sequencing, pharmaceutical candidate screening, miniaturized PCR, and protein separations [27, 42, 56, 64, 78]. Part of their work was recently reviewed [5, 7]. PDMS has been used by several research groups to make microfluidic devices; some of this work was reviewed recently by Whitesides and colleagues [43]. Very complicated devices with large-scale integration have also been fabricated using PDMS for applications such as cell sorting and combinatorial synthesis [67]. Fabrication methods for making plastic microfluidic devices have been reviewed by Soper et al. [61].

Plastic microfluidic devices range in size from that of a thumbnail to that of a microtiter plate (in excess of 100 cm^2). The microchannels are typically tens to hundreds of micrometers in depth and width while circuit-like channels can be centimeters in length. Reagents are fed to these channels via reservoirs, which are typically 0.5–3 mm in diameter and penetrate the depth of the device ($\sim$ 0.5–2 mm), resulting in total reagent volumes of 0.1–15 µL per reservoir. Hundreds of reservoirs and meters of channels can be densely packed into multiplexed designs that form an array of patterns on a single disposable device.

Key differences between silicon, glass, and plastics are compiled in Table 9.1. Advantages of plastics include vast experience in manufacturing low-cost, high-volume labware items such as microcentrifuge tubes and microtiter plates, as well as many plastics' compatibility with biological and chemical reagents. While microfabrication of silicon or glass can be accomplished quite cheaply when chip sizes are small ($\sim$ 1 cm^2), larger fluidic devices ($\sim$100 cm^2) are often required for applications such as DNA sequencing [2] or the parallel screening of large numbers of pharmaceutical candidates [7]. In addition, the drilling of reservoirs and sealing of a large glass cover layer using high-temperature fusion bonding are relatively cumbersome processes [2]. In contrast, plastic parts made by techniques such as injection molding, casting, or embossing, followed by sealing using thermal or adhesive bonding, can be quite inexpensive: the manufacturing cost of an injection-molded compact disc, a two-layer structure made from acrylic or polycarbonate and containing micron-scale features, is presently less than 20 ¢.

TABLE 9.1. Materials of construction for electrokinetic microfluidic devices[a]

	Silicon	Glass/fused silica	Plastics[b]
thermal conductivity (cal/cm-s-°C)	0.35	$\sim 2 \times 10^{-3}$	4.5×10^{-4} [c]
bioassay compatibility	fair (oxide/nitride surface layer)	fair	very good
optical detection	visible/UV: strong absorbance IR: transparent	glass: very good fused silica: excellent	poor–v. good (varies according to polymer choice and wavelength)
microfabrication	many well-developed approaches	isotropic wet etching only	Si or glass mastering plus replication techniques; direct methods (ablation, dry etching)
feature aspect ratio (depth:width of microchannels)	<0.1–40	<0.5	dependent on master for replication methods
manufacturing methods	well developed	need development	well developed
cost	inexpensive (small single devices) to expensive (large-area device arrays)	moderately expensive	inexpensive

[a] Harrison, D. J.; Glavina, P. G.; Manz, A. *Sensors and Actuators*, **1993**, *B10*, 107–116; Weast, R. C. *CRC Handbook of Chemistry and Physics*, 59th edition, CRC Press, Florida, **1979**.
[b] acrylics, polycarbonates, polyolefins, and polydimethylsiloxane have been most utilized and studied.
[c] value for polymethylmethacrylate.

Thus, plastic microfluidic devices might be made so cheaply as to be disposable after a single use. This could have tremendous impact in applications where cross-contamination of sequential samples is of concern. Point-of-care clinical diagnostics, high-throughput screening of candidate pharmaceutical libraries, and genotyping for forensic identification are three instances where contamination of one sample by another is most undesirable.

9.3.1. Pumping and Detection

With regard to electrokinetic pumping, most plastics, including PMMA, polycarbonate, PDMS, and even Teflon® exhibit a degree of EOF—typically a factor of five or so smaller than silica/glass—which results from either a low level of fixed, negatively charged surface functional groups, or ions adsorbed from the aqueous buffer in the channels [6, 35]. The surface properties of plastics usually have no direct effect on electrophoretic separations (so long as there is no unwanted nonspecific adsorption), which depend on the properties of analytes and separation medium.

From a detection standpoint, the most obvious concern in working with plastic microfluidic devices is background fluorescence, especially using short-wavelength excitation. The fluorescence of a plastic device depends not only on the bulk properties of the plastic but can be influenced by the chip manufacturing process and/or adhesives used for bonding a cover onto the device. An approach to increasing signal-to-noise ratio by rejecting background is confocal epifluorescence detection. The use of high numerical aperture objectives allows efficient collection of light while a pinhole positioned in the image plane successfully minimizes background fluorescence from the chip. An alternative strategy is to use red- or near-infrared-absorbing fluorophores, since background fluorescence is typically much reduced in this wavelength region. Experimental results indicate that, under conditions of photobleaching, the background fluorescence from several plastic (e.g., PMMA) chips, when excited by 488 nm laser light, is just a few times that of a glass microfluidic device of similar thickness [7].

9.3.2. Device Fabrication

The numerous methods for manufacturing plastic devices can be divided into two classes, direct fabrication and replication. Direct techniques such as mechanical machining or laser ablation are usually not the most economical, since the fabrication operation must be conducted on each and every part; in some cases, these methods produce surfaces too rough for high-resolution electrophoresis [51]. Replication methods shift the burden of creating a high-quality surface to the creation of the template, master, or mold tool from which the polymer part is to be formed. Replication methods include injection molding, compression molding, embossing, and casting; their definitions are given in Table 9.2. These techniques are demonstrably economical when large volumes of identical plastic parts or devices are formed using existing manufacturing methods and machinery.

One approach to mass production of plastic microfluidic devices, similar to that used in the commercial manufacture of compact discs, involves two primary steps: (1) formation of "open" microstructures such as channels and reservoirs on one surface of a base layer of polymer, and (2) sealing of the features in the base layer by a plastic cover layer [7, 42]. The overall process is diagrammed in Figure 9.2. To form the base layer, fluidic features are first microfabricated on a glass plate, fused silica plate, or silicon wafer using standard photolithographic patterning and etching techniques; requirements for surface smoothness (sub-micron) as well as lateral structure resolution (tens of microns) are readily attained with such methods. Next, a metal mold or "master" is created by electroplating from aqueous solution to deposit anywhere from a few hundred microns to a few millimeters of metal onto the surface of the master, creating a precise replicate "electroform" with inverse topology:

TABLE 9.2. Glossary of plastics processing[a]

Injection Molding	A molding procedure whereby a heat-softened plastic material is forced from a cylinder into a cavity that gives the article the desired shape
Compression Molding	A molding technique whereby molding compound is introduced into an open mold and formed under heat and pressure
Extrusion	Process of compacting and melting a plastic material and forcing it through an orifice in a continuous fashion
Embossing	Techniques used to create depressions of a specific pattern in plastic films or sheets
Electroforming	Moldmaking method whereby a thin layer of metal is deposited onto a patterned substrate, then removed for use as a tool

[a] Berins, M.L. *Plastics Engineering Handbook*, 5th edition, Van Nostrand Reinhold, NY, **1991**.

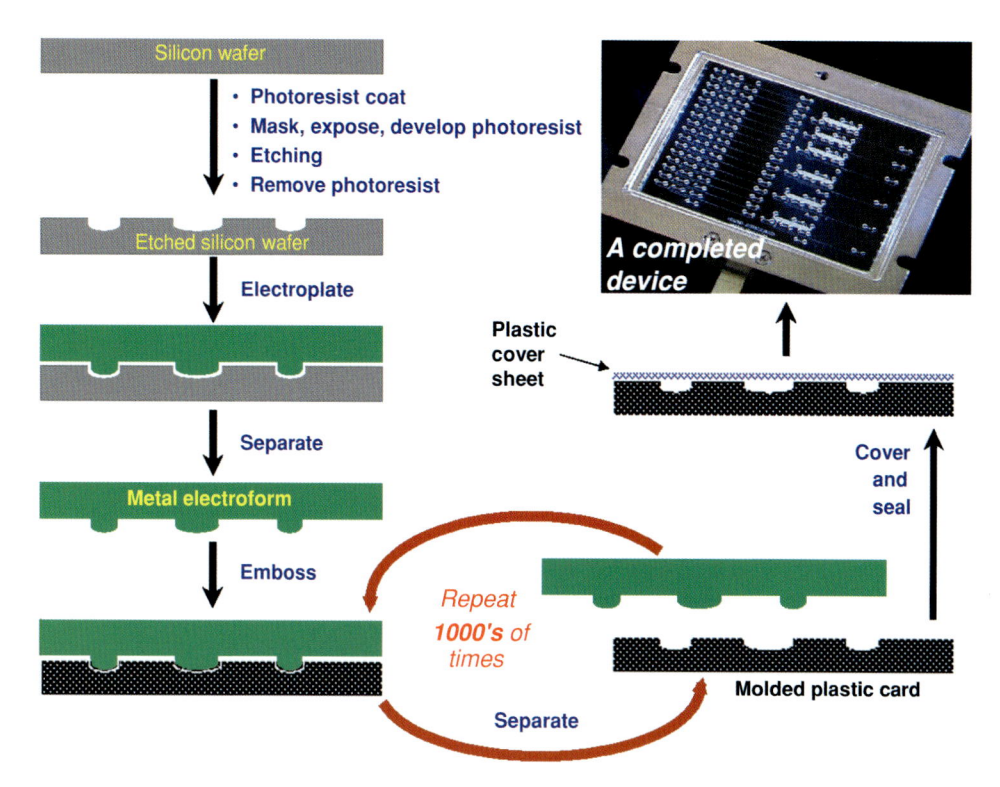

FIGURE 9.2. Process overview for mass manufacturing of plastic microfluidic systems. A microfluidic design is patterned and etched on a glass or silicon substrate using standard micromachining techniques to form a positive "master." A metal mold or "electroform" is formed via electroplating onto the surface of the master, creating a precise negative replica of the pattern. Thousands to millions of plastic parts with channeled structures are thermally formed against the metal mold tool. Each molded device is then sealed with a plastic layer that encloses the microchannels [7].

a channel in the silicon or glass becomes a ridge in the electroform. Such a metal tool can be mounted as the molding tool on an embossing machine or an injection- or compression-molding system. Polymer base layers are then formed in volume from melted or softened polymer resin or sheet to create smooth and precise channels in the finished fluidic device (Figure 9.2). A master or metal tool can also be used as the template for the casting of *in-situ*-polymerized substrates.

The open microchannels and reservoirs of a molded or embossed fluidic base layer are sealed to form closed channels by bonding to a smooth plaque of rigid polymer material or a thin polymer film. In either case, the covering material can be sealed to the fluidic base layer using an adhesive interlayer (e.g., films coated with pressure-sensitive adhesive) or by thermal techniques that fuse the two layers to one another. The chemical, optical, electrical, and mechanical properties of both the base material and the sealing layer must be compatible with the chemical reagents and procedures, the chosen method of motivation of solvent and solute through the microchannels, and the technology selected to detect the analytical targets.

9.4. DNA ANALYSES

The full utility of a microfabricated analysis system can be exploited by integrating sample preparation steps as part of a complete assay. This has the potential to minimize user intervention, automate all steps in a process, and reduce sample-to-answer times. However, integration of operations such as sample extraction and sample amplification with electrophoretic analysis poses significant challenges. For instance, samples need to remain confined within specific zones on an integrated device to allow the necessary reactions and chemistries to occur, but individual zones also need to be fluidically linked to allow transfer of a sample from one processing step to the next. In addition, different sample preparation steps may occur at different temperatures and therefore zones need to be thermally isolated from one another. Elevated temperatures required in applications such as PCR complicate the confinement of fluids to defined zones because of thermal-expansion-driven convection and flow. The accurate metering and positioning of fluids are also critical in an integrated device.

To date, there have been few published examples of the integration of sample processing and analysis on a single device. We choose here a couple of examples that illustrate the integration of microfluidic components and their application in DNA analyses.

9.4.1. Integrating PCR and DNA Fragment Separations

Anderson et al. showed nucleic acid extraction, PCR amplification or reverse transcription, and hybridization analysis integrated onto a single microfluidic device fabricated from polycarbonate or polypropylene [1]. This device had reactor volumes in the 5–20 µL range. Several groups have reported integrating PCR amplification and electrophoresis on silicon [8], glass [70], and silicon/glass hybrid microchips [74]. Soper et al. coupled nanoliter-scale amplification in fused-silica capillaries with electrophoretic DNA separation in PMMA microchips [62]. These examples represent impressive advances in integrating sample preparation with other analyses on a single device.

Koh et al. recently designed and fabricated an integrated plastic microfluidic device for bacterial detection and identification [27]. The device, made from poly (cyclic olefin) with integrated graphite ink electrodes and photopatterned gel domains, accomplished DNA amplification, microfluidic valving, sample injection, on-column labeling, and separation. The layout of the device is shown in Figure 9.3. Wells 5 and 6 are for introducing a sample into the PCR channel reactor between them. Well 7 is for the introduction of a DNA sizing ladder, which serves as a calibration standard to accurately identify the amplicons according to their sizes. Wells 8 and 4 are for loading a sample plug into the separation channel, while Wells 1 and 3 are for running the separation. Well 2 was not used.

PCR was conducted in the channel reactor containing a volume of 84 nL; thermal cycling utilized screen-printed graphite ink resistors. Integrated heaters provided rapid thermal cycling due to localized heating combined with low thermal mass. The ink heater was compared with a commercial thermal cycler, and the results showed that the temperature ramp-up speed using the ink heater was roughly 4 times faster than that of the commercial thermal cycler. The higher speed resulted in rapid analysis as well as fewer false positive results due to less time for mismatching that can occur between primers and targets [73].

PLASTIC MICROFLUIDIC DEVICES FOR DNA AND PROTEIN ANALYSES 319

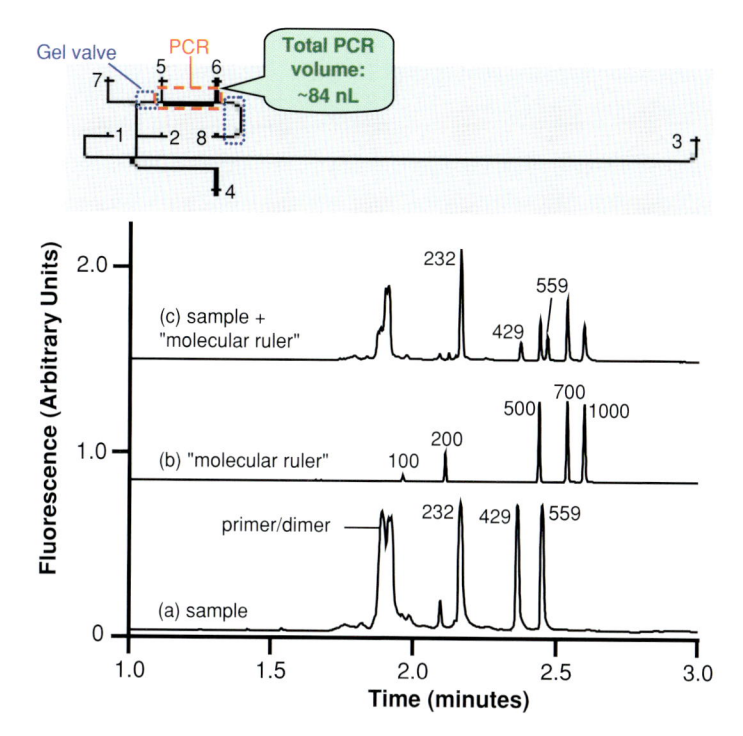

FIGURE 9.3. (Top) Layout of a plastic microfluidic device; the heated region for PCR thermal cycling is indicated by long dotted lines whereas the gel valve areas are designated by short dotted lines. (Bottom) Electropherograms of (a) amplicons, (b) DNA sizing ladder, and (c) the mixture formed by simultaneously injecting amplicons and the ladder. The separation medium consisted of 1.5% hydroxypropylcellulose, 0.4% hydroxyethylcellulose, 4 µM thiazole orange in 1X TBE buffer, pH 8.4. The field strength was 193 V/cm and the effective separation distance, from the injection point to the detection point was 3 cm. (Adapted from [27] published with permission.)

In-situ gel polymerization was employed to form local microfluidic valves that minimize convective flow of the PCR mixture into other regions. To make a gel valve at a precise location, acrylamide monomer solution containing a photosensitive initiator was first filled into all channels and wells. After the device had been covered by a pre-defined photomask, photopolymerization was conducted by exposure to UV irradiation. The solution in the exposed regions polymerized to form plugs of gel. The solution in the regions blocked by the photomask did not polymerize; non-polymerized solution was removed by flushing the channels at the end of the process and the channels were then filled with the appropriate reagents.

After PCR, amplicons (PCR products) were electrokinetically injected *through* the gel valve (the gel readily passes charged ions under electrophoretic drive), followed by on-chip electrophoretic separation. This is an excellent feature of the gel valve, which is able to confine the PCR mixture during thermal cycling while allowing amplicons to be transferred after PCR. To correctly size the amplicons, a DNA sizing ladder was mixed with the sample during injection as in traditional slab-gel measurements. From the migration times of the sizing ladder, the sizes of the amplicons were calculated by interpolation. To facilitate detection, an intercalating dye was admixed in the separation medium to label the amplicons, which were then detected using LIF.

Two model bacteria, *Escherichia coli* O157 and *Salmonella typhimurium*, were chosen to demonstrate bacterial detection and identification using this device. One primer set was designed for *E. coli* according to the literature [50], resulting in one amplicon at 232 bp, whereas two primer sets were designed for *salmonella*, producing two amplicons at 559 bp and 429 bp. Figure 9.3 shows the electropherograms obtained for amplicons from these bacteria; all expected amplicons were produced with a gain ranging from 8.5×10^5 to 8.0×10^6. The limit of detection was measured to be about 6 copies of target DNA.

9.4.2. DNA Sequencing

The sequencing of the human genome was accomplished at a remarkable pace, due in large part to the utilization of 96-capillary-array electrophoresis instruments, which provide increased efficiency and reduced costs compared to slab-gel electrophoresis systems. Capillary bundles filled with high-molecular-weight sieving polymers, such as linear polyacrylamide, allow rapid, high-resolution separations and automated capillary filling. Replacing capillary arrays with an array of microchannels fabricated in planar substrates may, in the future, lead to highly multiplexed microfabricated DNA sequencers [2, 33]. The potential to integrate additional functionalities into such systems—such as on-chip means to purify the sample prior to CE sequencing—offers the hope that many of the manual operations that currently account for over half of the cost of obtaining a DNA sequence can be automated.

Fundamentally, sequencing separations in microchannels are quite similar to those in CE. The speed and resolution of a separation is strongly influenced by the choice and formulation of the separation matrix, separation temperature, electric field strength, and separation length. However, there are also significant differences between the two techniques. First, the method of sample introduction differs: in CE with gel-filled capillaries, samples are electrokinetically injected directly into one end of the separation capillary. The quantity of sample injected depends on the field strength and time of injection, as well as the electrophoretic mobility of the DNA fragments, which varies according to their size, resulting in so-called "injection bias": shorter fragments are injected in higher concentrations than longer ones. By contrast, the microchannel format employs a intersecting-channels injection format that allows a controlled, reproducible injection-plug size that is independent of injection time [13, 25] and minimizes electrophoretic injection bias. Secondly, as the number of capillaries in an array increases, the array becomes increasingly difficult to manufacture and optical alignment for detection is more challenging. In contrast, planar microfluidic channel arrays should be simpler to manufacture in large volumes, and channel-to-channel alignment is fixed, making multiplexed detection strategies more robust.

There are several examples of DNA sequencing in microchannels etched in glass [2, 33, 58, 75], including so-called four-color sequencing, in which all four base types (A, C, G, and T) are analyzed simultaneously using a corresponding set of four spectrally distinguishable dyes to label each DNA fragment according to its terminal base; 500 bases can be sequenced in 20 minutes at 99.4% accuracy [34]. Because glass surfaces are charged at sequencing buffer pH values, they must be derivatized prior to separation to suppress EO, which otherwise would degrade the resolution of the electrophoretic separation process.

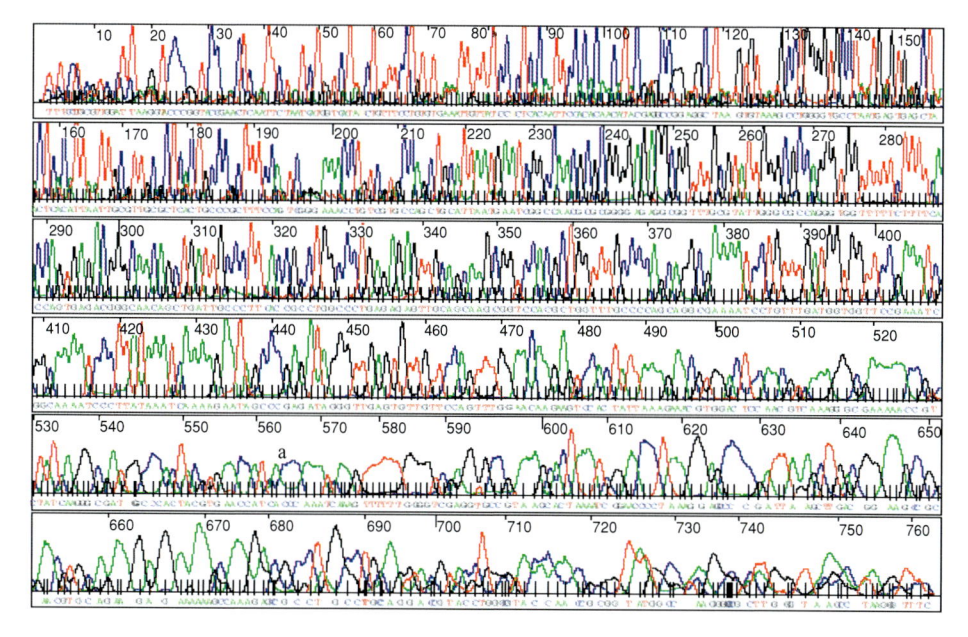

FIGURE 9.4. DNA separation data obtained using plastic microfluidic devices. Four-color electrophoretic separation of single-stranded DNA sequencing fragments generated from a M13mp18 template using BigDye™-labeled primers; the four colors black, red, green, and blue correspond to terminal bases G, T, A, and C, respectively. The separation channel is filled with a linear polyacrylamide matrix to enhance resolution. The total electrophoresis time is approximately 30 minutes for this particular run and base-calling accuracy is 98% up to a read length of 640 bases. (Adapted from [7].)

The stability and lifetime of covalently attached surface coatings has been an issue in CE [39] and is likely to pose similar concerns in microchannel separations.

DNA sequencing separations in plastic (acrylic and polyolefin) microchannels has been accomplished using devices produced via hot embossing from the sort of metal mold tool described above [15]. Altering the substrate material (and therefore the chemical nature of the channel surface) has implications in the choice and performance optimization of the separation matrix, as well as requirements for surface modification. Figure 9.4 shows a typical four-color DNA sequencing electropherogram of a sample run in an 18-cm-long acrylic microchannel. These data correspond to the correct identification of 640 bases with 98% accuracy; such analyses require about 30 min [15].

9.4.3. DNA Sample Purification

As discussed above, integration of sample purification prior to DNA sequencing could generate high-quality data while improving sequencing accuracy. There exists the potential to integrate DNA sequencing with sample purification to take full advantage of the power of microfluidics. Recently, Soper's research group coupled a plastic device for sample purification with a capillary for DNA sequencing [77]. Although the effort failed to implement both procedures in a single planar device, it demonstrated the capability of solid-phase

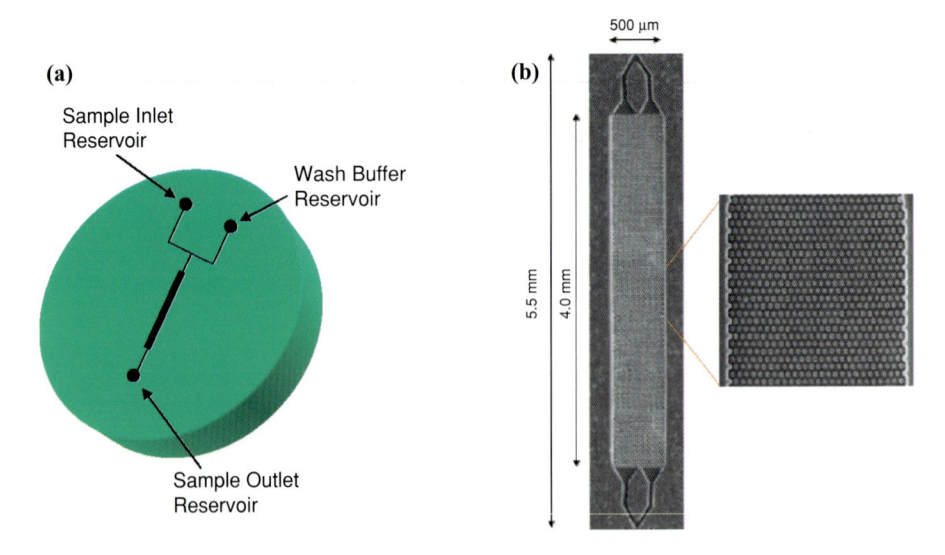

FIGURE 9.5. (a) Schematic diagram of the microfluidic device used for solid-phase reversible immobilization prior to gel electrophoretic sorting of DNA fragments. The channel was 500 μm in width, 50 μm in depth, and 4.0 mm in length. Reservoirs were formed from 500-μm i.d. holes drilled through the chip. (b) Optical micrograph of polycarbonate SPRI capture bed and its dimensions. Shown is the embossed piece in polycarbonate fabricated using a metal master. (Adapted from [77]; published with permission.)

reversible immobilization (SPRI) in a microfabricated plastic device for purification of dye-terminator-labeled DNA fragments prior to DNA sequencing.

In SPRI, DNA purification is achieved by an immobilization bed, which was produced in Soper's work by exposing a "posted" (see below) microchannel to UV irradiation. The radiation induced a surface photooxidation of the polycarbonate (PC), resulting in the production of carboxylate groups. The immobilization bed contained microposts to increase the loading level of surface-captured DNA fragments to improve signal intensity without the need for preconcentration. Figure 9.5 shows a schematic diagram of the microfluidic device and a micrograph of the polycarbonate SPRI capture bed. When the sequencing mixtures were mixed with an immobilization buffer in the bed, the DNA fragments demonstrated a high affinity for the photo-activated carboxylated surface. The loading density of DNA fragments on this activated surface was determined to be 3.9 pmol/cm^2.

To release the captured DNA, deionized water was subsequently introduced to the capture bed. After incubation, the released DNA fragments were pumped into a gel-filled capillary or a slab-gel for DNA sequencing. It was found that SPRI cleanup of dye-terminator DNA fragments using the photoactivated PC chip are comparable in terms of read length to the conventional SPRI format utilizing carboxylated magnetic beads and a magnetic field for capture. The read length for the SPRI chip format was found to be 620 bases with a calling accuracy of 98.9%.

9.5. PROTEIN ANALYSES

Completion of the mapping of the human genome has prompted strong interest in deciphering the human proteome. However, it is a daunting task to analyze the complete complement of proteins present in even a single type of cell or tissue and to study their variation in space and time. Therefore, it is likely that microfluidics can be exploited for niche applications to accelerate proteomic studies due to its unique features and advantages. The examples below illustrate possible connections between microfluidics and protein analyses.

9.5.1. Isoelectric Focusing for Studying Protein Interactions

Among many approaches being developed for protein analysis, two-dimensional gel electrophoresis (2-DE) and mass spectroscopy (MS) are becoming the primary tools for proteomics [31, 79]. Traditional 2-DE consists of isoelectric focusing (IEF) as the first dimension and slab-gel electrophoresis as the second dimension. However, efforts have been made to interface directly capillary IEF (CIEF) with MS, so that the mass information is obtained from MS without the second dimension of 2-D gel electrophoresis. Attempts by a few research groups [60, 65] suggest that this approach is sufficient to generate protein maps from microorganisms. CIEF maintains the high resolving power of traditional IEF, and utilizes CE's advantages such as rapid separation under high electric fields [52, 60, 65].

CIEF has also been explored in microfluidic devices by several research groups. Santiago's group investigated the possibility of having a portable CIEF system [22], while Hofmann et al. evaluated three mobilization methods in conjunction with IEF in a glass device [24]. Smith's group developed a microfabricated IEF device as a direct interface to electrospray MS [71] and Yager et al. electrochemically generated pH gradients for IEF in a microchannel [38].

Tan et al. demonstrated miniaturized capillary isoelectric focusing (CIEF) in a plastic microfluidic device [64]. Conventional CIEF techniques were adapted to the microfluidic devices to separate proteins and to detect protein-protein interactions. Both acidic and basic proteins with isoelectric point (pI) ranging from 5.4 to 11.0 were rapidly focused and detected in a 1.2 cm-long channel with a total analysis time of 150 seconds. The work also experimentally confirmed that IEF resolution is essentially independent of focusing channel length when the applied voltage is kept the same and within a range that does not cause Joule heating.

Understanding protein-protein interactions is one of the major tasks of deciphering the human proteome, since most cellular processes involve the interaction of protein-to-protein, protein-to-small molecule, and/or protein-to-nucleic acid [47]. Protein arrays, yeast-two-hybrid systems, and other methods are poised to play a major role in this aspect of proteomics. IEF makes a unique contribution to this field because of its ability to resolve a wide range of proteins and protein complexes that have small differences in pI [52]. This property becomes very useful when one studies protein-protein interactions and there is a difference in pI between a protein-protein complex and its constituent proteins.

Tan et al. further demonstrated the use of miniaturized CIEF to study interactions between proteins [64]. One pair of proteins studied was immunoglobulin G (IgG) and protein G, which are known to interact with one another. Protein G binds to IgG's Fc region, hence protein G separation columns are often used to purify immunoglobulins. MacBeath and Schreiber used this pair of proteins to demonstrate the concept of protein arrays [36]. The detection of the interaction between these two proteins using miniaturized CIEF was demonstrated, as shown in Figure 9.6. The experiment involved the comparison of CIEF electropherograms of protein G only, and of a mixture of protein G and IgG. Alexa-labeled protein G (APG) fluoresces upon excitation and thus was detected by LIF,

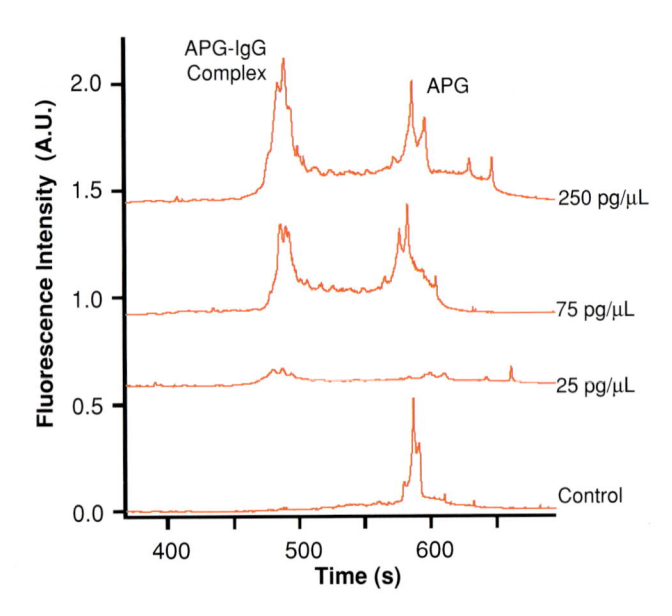

FIGURE 9.6. Detection of protein-protein interactions using CIEF in a microfabricated plastic device. Alexa-labeled protein G (APG) and IgG were incubated at room temperature for 30 minutes. The mixture was diluted in 1:20 with ampholyte buffer before being subjected to miniaturized CIEF. The concentration of IgG was fixed at 250 pg/μL and that of APG is shown next to the electropherograms. The control sample has APG only. (Adapted from [64].)

whereas unlabeled IgG was not detectable. The IgG-APG complex was detectable as well. When the concentration of APG was increased from 25 pg/μL to 250 pg/μL while IgG was fixed at 250 pg/μL, the signal of the formed complex increased accordingly. The multiple peaks were explained by the heterogeneity of the APG (which may result from multiple-dye labeling). The results suggested APG was detectable at 25 pg/μL, corresponding to 50 fmol (5×10^{-14} mol) of protein when a sample of 40 μL was prepared.

9.5.2. Enzymatic Digestion for Protein Mapping

To determine protein identity and post-translational modifications, protein digestion with a proteolytic enzyme (e.g., trypsin) is typically carried out prior to mapping the masses of peptides. This enzymatic digestion preferably takes place on a solid support to eliminate the undesired autodigestion that would occur in solution. In addition, the solid support may function as a pre-concentration step (e.g., solid-phase extraction), which is very useful for those proteins present at very low concentrations.

A microfluidic device demonstrated recently contains both the functions of a solid-phase extractor (SPE) and an enzymatic microreactor [48]. The device was fabricated from a porous poly (butyl methacrylate-co-ethylene dimethacrylate) monolith prepared within a capillary. Photografting with irradiation through a mask was then used to selectively functionalize a portion of the monolith, introducing reactive poly (2-vinyl-4,4-dimethylazlactone) chains to enable the subsequent attachment of trypsin, thereby creating an enzymatic microreactor with high proteolytic activity. The remaining

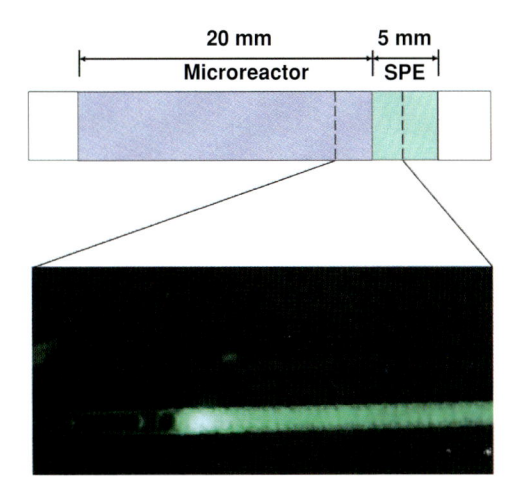

FIGURE 9.7. Scheme and fluorescence microscope image of the monolithic dual-function device used for the digestion of labeled casein and capture of fluorescent peptides. When 5 μL of 10 μg/mL BODIPY-labeled casein solution in 50 mM Tris buffer (pH 8.0) is pumped at a flow rate of 0.5 μL/min through the enzyme reactor on the left, digestion takes place, producing peptides. The resulting peptides are collected in the hydrophobic SPE on the right. After washing the monolith with 10 μL of Tris buffer to remove all unbound species, the enzyme reactor (left) exhibits no fluorescence while the SPE (right) shows strong fluorescence, indicating successful digestion and extraction. (Adapted from [48]; published with permission.)

unmodified hydrophobic monolith served as a micro SPE. Figure 9.7 shows schematically the portion of the device containing a SPE and a microreactor with immobilized enzyme.

The dual function of the device was demonstrated using BODIPY-labeled casein. Casein consists of intra-molecularly quenched fluorescing moieties that are unquenched only by proteolytic digestion. When a casein solution was pumped through the enzyme reactor on the left in Figure 9.7, digestion occurred, producing peptides. The resulting peptides were collected in the hydrophobic SPE on the right. After washing the monolith with a buffer solution to remove all unbound species, the left part of the monolith containing immobilized trypsin exhibited no fluorescence while the right portion containing SPE with extracted, fluorescently labeled peptides showed strong fluorescence. The presence of captured fluorescent species confirmed the digestive function of the device and the high fluorescence of peptides retained in the hydrophobic part of the monolith. The dual function of the device was further demonstrated by concentrating myoglobin, followed by elution and digestion, prior to coupling with MS for peptide mass mapping [48].

In addition, several research groups reported using microfluidic devices to immobilize enzymes for a variety of other applications. For instance, Crooks' research group developed a method for determining enzyme kinetics using a continuous-flow microfluidic system [59]. The method involves immobilizing the enzyme on microbeads, then packing the microbeads into a chip-based microreactor. When the substrate flows over the packed bed, the signal is collected and data are analyzed in a way similar to conventional measurements based on the Michaelis-Menten equation. They studied the horseradish peroxidase-catalyzed reaction between hydrogen peroxide and N-acetyl-3,7-dihydroxyphenoxazine to yield fluorescent resorufin. The experimental results indicated the microfluidics-based method yielded the same result as conventional methods. The advantages of the continuous-flow microfluidic system include rapid determination of enzyme kinetics and less consumption of reagents and enzymes.

CONCLUDING REMARKS

The examples in this chapter exemplify the sort of DNA and protein analyses that can be accomplished in plastic microfluidic devices. Sample preparation, including PCR, purification of DNA fragments, and enzymatic digestion of proteins have been demonstrated in custom-designed devices; some of them have been integrated with separations. In addition, plastic microfluidic devices were illustrated to be capable of performing DNA sequencing based on electrophoretic separation, and isoelectric focusing was shown as a means to study protein-protein interactions. To realize the "lab-on-a-chip" vision articulated by Manz and Harrison in the early 1990's [20, 40], further integration of these devices with various hardware and accessories must be achieved, in addition to addressing the challenges associated with macro-micro interfaces.

ACKNOWLEDGEMENTS

The authors gratefully acknowledge contributions from our former coworkers at ACLARA BioSciences, Inc., particularly our co-authors on the several ACLARA papers cited here. We thank Professors Soper and Frechet for providing Figures 9.5 and 9.7, respectively. ZHF acknowledges the financial support from McKnight Brain Institute, the startup fund from the University of Florida, and the grants from National Aeronautics and Space Administration (NASA) via UF Space Biotechnology and Commercial Applications Program, UCF-UF Space Research Initiative, and Hydrogen Research for Spaceport and Space Based Applications.

REFERENCES

[1] R.C. Anderson, G.J. Bogdan, Z. Barniv, T.D. Dawes, J. Winkler, and K. Roy. *Proc. 1997 International Conf. Solid-State Sensors and Actuators*, p. 477–480, 1997.
[2] C. Backhouse, M. Caamano, F. Oaks, E. Nordman, A. Carrillo, B. Johnson, and S. Bay. *Electrophoresis*, 21:50–156, 2000.
[3] D.R. Baker. *Capillary Electrophoresis*, Chapter 2, John Wiley & Sons Inc., New York, 1995.
[4] S.L. Barker, M.L. Tarlov, H. Canavan, J.J. Hickman, and L.E. Locascio. *Anal. Chem.* 72:4899–4903, 2000.
[5] G. Binyamin, T.D. Boone, H.S. Lackritz, A.J. Ricco, A.P. Sassi, and S.J. Williams. Plastic microfluidic devices: electrokinetic manipulations, life science applications, and production technologies, In R.E. Oosterbroek and A. van den Berg (eds.) *"Lab-on-a-Chip: Miniaturized Systems for (Bio)Chemical Analysis and Synthesis, "'* Elsevier, Amsterdam, p. 83–112, 2003.
[6] T.D. Boone, H.H. Hooper, and D.S. Soane. *Technical Digest of the 1998 Solid-State Sensor and Actuator Workshop*, Transducers Research Foundation, Cleveland, p. 87–92, 1998.
[7] T.D. Boone, Z.H. Fan, H.H. Hooper, A.J. Ricco, H. Tan, and S.J. Williams. *Anal. Chem.* 74:78A–86A, 2002.
[8] M.A. Burns, B.N. Johnson, S.N. Brahmasandra, K. Handique, J.R. Webster, M. Krishnan, T.S. Sammarco, P.M. Man, D. Jones, D. Heldsinger, C.H. Mastrangelo, and D.T. Burke. *Science*, 282:484–487, 1998.

[9] J. Chan, A.T. Timperman, T. Qin, and R. Aebersold. *Anal. Chem.*, 71:4437–4444, 1999.
[10] N. Chiem and D.J. Harrison. *Anal. Chem.*, 69:373–378, 1997.
[11] C.S. Effenhauser, G.J.M. Bruin, A. Paulus, and M. Ehrat. *Anal. Chem.* 69:3451–3457, 1997.
[12] A.G. Ewing, R.A. Wallingford, and T.M. Olefirowicz. *Anal. Chem.*, 61:292A–303A. 1989.
[13] Z.H Fan and D.J. Harrison. *Anal. Chem.*, 66:177–184, 1994
[14] Z.H. Fan, S. Mangru, R. Granzow, Ho, W Heaney, Q. Dong, and R. Kumar, *Anal. Chem.*, 71:4851–4859, 1999.
[15] Z.H. Fan, W. Tan, H. Tan, X.C. Qiu, T.D. Boone, P. Kao, A.J. Ricco, M. Desmond, S. Bay, and K. Hennessy. *Plastic Microfluidic Devices for DNA Sequencing and Protein Separations, Micro Total Analysis Systems 2001*, Kluwer Academic Publishers, Boston, pp. 19–21, 2001.
[16] G.S. Fiorini, G.D.M. Jeffries, D.S.W. Lim, C.L. Kuyper, and D.T. Chiu. *Lab on a chip*, 3:158–163, 2003.
[17] P.R. Gascoyne and J. Vykoukal. *Electrophoresis*, 23:1973–83, 2002.
[18] B.B. Haab and R.A. Mathies. *Anal. Chem.*, 71:5137–5145, 1999.
[19] A.G. Hadd, D.E. Raymond, J.W. Halliwell, S.C. Jacobson, and J.M. Ramsey. *Anal. Chem.*, 69:3407–3412, 1997.
[20] D.J. Harrison, A. Manz, Z. Fan, H. Lüdi, and H.M. Widmer. *Anal. Chem.*, 64:1926–1932, 1992.
[21] D.J. Harrison, P.G. Glavina, and A. Manz. *Sensors and Actuators B*, 10:107–116, 1993.
[22] A.E. Herr, J.I. Molho, J.G. Santiago, T.W. Kenny, D.A. Borkholder, G.J. Kintz, P. Belgrader, and M.A. Northrup. *Electrophoresis*, 22:2291–2295, 2001.
[23] A.E. Herr, J.I. Molho, K.A. Drouvalakis, J.C. Mikkelsen, P.J. Utz, J.G. Santiago, and T.W. Kenny. *Anal. Chem.*, 75:1180–1187, 2003.
[24] O. Hofmann, D. Che, K.A. Cruickshank, and U.R. Muller. *Anal. Chem.*, 71:678–686, 1999.
[25] S.C. Jacobson, R. Hergenroder, L.B. Koutny, R.J. Warmack, and J.M. Ramsey. *Anal. Chem.*, 66:1107–1113, 1994.
[26] J. Kameoka, H.G Craighead, H. Zhang, and J. Henion. *Anal. Chem.*, 73:1935–1941, 2001.
[27] C.G. Koh, W. Tan, M. Zhao, A.J. Ricco, and Z.H. Fan. *Anal. Chem.*, 75:4591–4598, 2003.
[28] M.U. Kopp, A.J. de Mello, and A. Manz. *Science*, 280:1046–1048, 1998.
[29] P.C.H Li and D.J. Harrison. *Anal. Chem.*, 69:1564–1568, 1997.
[30] Z. Liang, N. Chiem, G. Ocvrik, T. Tang, K. Fluri, and D.J. Harrison. *Anal. Chem.*, 68:1040–1046, 1996.
[31] A.J. Link, J. Eng, D.M. Schieltz, E. Carmack, G.J. Mize, D.R. Morris, B.M. Garvik, and J.R. III Yates. *Nat. BioTechnol.*, 17:676–682, 1999.
[32] Y. Liu, C.B. Rauch, R.L. Stevens, R. Lenigk, J. Yang, D.B. Rhine, and P. Grodzinski. *Anal. Chem.*, 74:3063–3070, 2002.
[33] S. Liu, H. Ren, Q. Gao, D.J. Roach, R.T. Loder, T.M. Armstrong, Q. Mao, L. Blaga, D.L. Barker, and S.B. Jovanovich. *Proc. Natl. Acad. Sci. USA*, 97:5369–5374, 2000.
[34] S. Liu, Y. Shi, W.W. Ja, and R.A. Mathies. *Anal. Chem.*, 71:566–573, 1999.
[35] K.D. Lukacs and J.W. Jorgenson. *J. High Res. Chromatogr. & Chromatogr. Commun.*, 8:407–411, 1985.
[36] G. MacBeath and S.L. Schreiber. *Science*, 289:1760–1763, 2000.
[37] S.E. McBride, R.M. Moroney, and W. Chiang. 'Electrohydrodynamic pumps for high-density microfluidic arrays. In D.J. Harrison and A van den Berg (eds.), Micro Total Analysis System', Kluwer Academic Publishers, pp. 45–48, 1998.
[38] K. Macounova, C.R. Cabrera, M.R. Holl, and P. Yager. *Anal. Chem.*, 72:3745–3751, 2000.
[39] R.S. Madabhushi. *Electrophoresis*, 19:224–230, 1998.
[40] A. Manz, J.C. Fettinger, E. Verpoorte, H. Lude, H. M. Widmer, and D.J. Harrison. *Trends in Anal. Chem.*, 10:144–149, 1991.
[41] A. Manz, Y. Miyahara, J. Miura, Y. Watanabe, H. Miyagi, and K. Sato. *Sensors and Actuators*, B1:249–255, 1990.
[42] R.M. McCormick, R.J. Nelson, M.G. Alonso-Amigo, D.J. Benvegnu, and H.H. Hooper. *Anal. Chem.*, 69:2626–2630, 1997.
[43] J.C. McDonald, D.C. Duffy, J.A. Anderson, D.T. Chiu, H. Wu, O.J. Schueller, and G.M. Whitesides. *Electrophoresis*, 21:27–40, 2000.
[44] M.A. Northrup, B. Benett, D. Hadley, P. Landre, S. Lehew, J. Richards, and P. Stratton. *Anal. Chem.*, 70:918–922, 1998.
[45] M.A. Northrup, K.F. Jensen, and D.J. Harrison. *Micro Total Analysis System'2003*, The Transducer Research Foundation, 2003.

[46] R.E. Oosterbroek and A. van den Berg. '*Lab-on-a-Chip: Miniaturized Systems for (Bio)Chemical Analysis and Synthesis*', Elsevier, Amsterdam, 2003.
[47] S. Park and R. Raines. *Nature Biotech.*, 18:847–851, 2000.
[48] D.S. Peterson, T. Rohr, F. Svec, and J.M.J. Frechet. *Anal. Chem.* 75:5328–5335, 2003.
[49] D.L. Pugmire, E.A. Waddell, R. Haasch, M.J. Tarlov, and L.E. Locascio. *Anal. Chem.*, 74:871–878, 2002.
[50] R. Riffon, K. Sayasith, H. Khalil, P. Dubreuil, M. Drolet, and J.J. Lagace. *Clin. Microbiol.*, 39:2584–2589, 2001.
[51] M.A. Roberts, J.S. Rossier, P. Bercier, and H.H. Girault. *Anal. Chem.* 69:2035–2042, 1997.
[52] R. Rodriguez-Diaz, T. Wehr, and M. Zhu. *Electrophoresis*, 18:2134–2144, 1997.
[53] M.G. Roper, J.G. Shackman, G.M. Dahlgren, and R.T. Kennedy. *Anal. Chem.* 75:4711–4717, 2003.
[54] J.S. Rossier, M.A. Roberts, R. Ferrigno, and H.H. Girault. *Anal. Chem.*, 71:4294–4299, 1999.
[55] E. Sahlin, A.T. Beisler, S.J. Woltman, and S.G. Weber. *Anal. Chem.*, 74:4566–4569, 2002.
[56] A.P. Sassi, Q. Xue, and H.H. Hooper. *Amer. Lab.* Oct.:36–41, 2000.
[57] K. Sato, M. Yamanaka, H. Takahashi, M. Tokeshi, H. Kimura, and T. Kitamori. *Electrophoresis*, 23:734–739, 2002.
[58] D. Schmalzing, A. Adourian, L. Koutny, L. Ziaugra, P. Matsudaira, and D. Ehrlich. *Anal. Chem.*, 70:2303–2310, 1998.
[59] G.H. Seong, J. Heo, and R.M. Crooks. *Anal. Chem.*, 75(13):3161–3167, 2003.
[60] Y. Shen, S.J. Berger, and R.D. Smith. *Anal. Chem.*, 72:4603–4607, 2000.
[61] S.A. Soper, S.M. Ford, S. Qi, R.L. McCarley, K. Kelly, and M.C. Murphy. *Anal. Chem.*, 72:643A–651A, 2000.
[62] S.A. Soper, S.M. Ford, Y. Xu, S. Qi, S. McWhorter, S. Lassiter, D. Patterson, and R.C.J. Bruch. *Chromatogr. A.*, 853:107–120, 1999.
[63] E. Tamaki, K. Sato, M. Tokeshi, K. Sato, M. Aihara, and T. Kitamori. *Anal. Chem.*, 74:1560–1564, 2002.
[64] W. Tan, Z.H. Fan, C.X. Qiu, A.J. Ricco, and I. Gibbons. *Electrophoresis*, 23:3638–3645, 2002.
[65] Q. Tang, A.K. Harrata, and C.S. Lee. *Anal. Chem.*, 69:3177–3182, 1997
[66] S.C. Terry, J.H. Jerman, and J.B. Angell. *IEEE Trans. Electron. Devices*, ED-26:1880–1886, 1979.
[67] T. Thorsen, S.J. Maerkl, and S.R. Quake. *Science*, 298:580–584, 2002.
[68] J. Wang and M. Pumera. *Anal Chem.*, 74:5919–23, 2002.
[69] Y. Wang, B. Vaidya, H.D. Farquar, W. Stryjewski, R.P. Hammer, R.L. McCarley, S.A. Soper, Y.W. Cheng, and F. Barany. *Anal. Chem.*, 75:1130–40, 2003.
[70] L.C. Waters, S.C. Jacobson, N. Kroutchinina, J. Khandurina, R.S. Foote, and J.M. Ramsey. *Anal. Chem.*, 70:5172–5176, 1998.
[71] J. Wen, Y. Lin, F. Xiang, D.W. Matson, H.R. Udseth, and R.D. Smith, 21:191–197, 2000.
[72] P. Wilding, M.A. Shoffner, and L. Kricka. *J. Clin. Chem.*, 40:1815–1818, 1994.
[73] C.T. Wittwer, D.J. Garling. *Biotechniques*, 10:76–83, 1991.
[74] A.T. Woolley, D. Hadley, P. Landre, A.J. deMello, R.A. Mathies, and M.A. Northrup. *Anal. Chem.*, 23:4081–4086, 1996.
[75] A.T. Woolley and R.A. Mathies. *Anal. Chem.*, 67:3676–3680, 1995.
[76] A.T. Woolley and R.A. Mathies. *Proc. Natl. Acad. Sci. USA*, 91:11348–11352, 1994.
[77] Y. Xu, B. Vaidya, A.B. Patel, S.M. Ford, R.L. McCarley, and S.A. Soper, *Anal. Chem.* 75:2975–2984, 2003.
[78] Q. Xue, A. Wainright, S. Ganakhedkar, and I. Gibbons. *Electrophoresis*, 18:4000–4007, 2001.
[79] X. Zuo and D.W Speicher. *Proteomics*, 2:58–68, 2002.

10

Centrifuge Based Fluidic Platforms

Jim V. Zoval[1] and M.J. Madou[2]
[1] University of California, Irvine Department of Mechanical and Aerospace Engineering
[2] University of California, Irvine Department of Mechanical and Aerospace Engineering and Department of Biomedical Engineering

10.1. INTRODUCTION

Once it became apparent that individual chemical or biological sensors used in complex samples would not attain the hoped for sensitivity or selectivity, wide commercial use became severely hampered and sensor arrays and sensor instrumentation were proposed instead. It was projected that by using orthogonal sensor array elements (e.g., in electronic noses and tongues) selectivity would be improved dramatically [1]. Instrumentation—it was envisioned—would reduce matrix complexities through filtration, separation, and concentration of the target compound, while, at the same time, ameliorating selectivity and sensitivity of the overall system by frequent recalibration and washing of the sensors. Through miniaturization of analytical equipment (using microfluidics), shortcomings associated with large and expensive instrumentation may potentially be overcome: reduction in reagent volumes, favorable scaling properties of several important instrument processes (basic theory of hydrodynamics and diffusion predicts faster heating and cooling and more efficient chromatographic and electrophoretic separations in miniaturized equipment) and batch-fabrication which may enable low cost, disposable instruments to be used once and then thrown away to prevent sample contamination [2]. Micromachining (MEMS) might also allow co-fabrication of many integrated functional instrument blocks. Tasks that are now performed in a series of conventional bench top instruments could then be combined into one unit, reducing labor and minimizing the risk of sample contamination.

Today it appears that sensor array development in electronic noses and tongues has slowed down because of the lack of highly stable chemical and biological sensors: too frequent recalibration of the sensors and relearning of the pattern recognition software is putting a damper on the original enthusiasm for this sensor approach. In the case of miniaturization of instrumentation through the application of microfluidics, progress was made in the development of platforms for high-throughput screening (HTS) as evidenced by new products introduced, by, for example, Caliper and Tecan Boston [3, 4]. In contrast, for sensing and diagnostic applications, not much progress was made using miniaturized analytical equipment. There have been platforms developed for a limited amount of human and veterinary diagnostic test that do not require complex fluidic design, for example, Abaxis [5]. In this review paper we are, in a narrow sense, summarizing the state of the art of compact disc (CD) based microfluidics and in a broader sense we are comparing the technical barriers involved in applying microfluidics to sensing and diagnostics as opposed to applying such techniques to high throughput screening (HTS). It will quickly become apparent that the former poses the more severe technical challenges and as a result the promise of lab-on-a-chip has not been fulfilled yet.

10.2. WHY CENTRIFUGE AS FLUID PROPULSION FORCE?

There are various technologies for moving small quantities of fluids or suspended particles from reservoirs to mixing and reaction sites, to detectors, and eventually to waste or to a next instrument. Methods to accomplish this include syringe and peristaltic pumps, electrochemical bubble generation, acoustics, magnetics, DC and AC electrokinetics, centrifuge, etc. In Table 10.1 we compare four of the more important and promising fluid propulsion means [6]. The pressure that mechanical pumps have to generate to propel fluids through capillaries is higher the narrower the conduit. Pressure and centrifugal force are both volume-dependent forces, which scale as L^3(in this case L is the characteristic length corresponding to the capillary diameter). Piezoelectric, electroosmotic, electrowetting and electrohydrodynamic (EHD) pumping (the latter two are not shown in Table 10.1) all scale as surface forces (L^2), which represent more favorable scaling behavior in the micro-domain (propulsion forces scaling with a lower power of the critical dimension become more attractive in the micro-domain) and lend themselves better to pumping in smaller and longer channels. In principle, this should make pressure- and centrifuge-based systems less favorable but other factors turn out to be more decisive; despite better scaling of the non-mechanical pumping approaches in Table 10.1, almost all biotechnology equipment today remain based on traditional external syringe or peristaltic pumps. The advantages of this approach are that it relies on well-developed, commercially available components and that a very wide range of flow rates is attainable. Although integrated micromachined pumps based on two one-way valves may achieve a precise flow control on the order of 1 μl/min with fast response, high sensitivity, and negligible dead volume, these pumps generate only modest flow rates and low pressures, and consume a large amount of chip area and considerable power.

Acoustic streaming is a constant (DC) fluid motion induced by an oscillating sound field at a solid/fluid boundary. A disposable fluidic manifold with capillary flow channels can simply be laid on top of the acoustic pump network in the reader instrument. The method

TABLE 10.1. Comparison of microfluidics propulsion techniques

Comparison	Fluid Propulsion Mechanism			
	Centrifuge	Pressure	Acoustic	Electrokinetic
Valving solved ?	Yes for liquids no for vapor	Yes for liquids and vapor	No solution shown yet for liquid or vapor	Yes for liquids, no for vapor
Maturity	R&D stage	Products available	Research	Products available
Propulsion force influenced by	Density and viscosity	Generic	Generic	pH, ionic strength
Power source	Rotary motor	Pump, Mechanical roller	5 to 40 V	10 kV
Materials	Plastics	Plastics	Piezoelectrics	Glass, plastics
Scaling	L^3	L^3	L^2	L^2
Flow rate	From less than 1nL/sec to greater than 100 μl/sec	Very wide range (less than nL/sec to L/sec)	20 μl/sec	0.001–1 μl/sec
General remarks	Inexpensive CD drive, mixing is easy, most samples possible (including cells). Better for diagnostics.	Standard technique. Difficult to miniaturize and multiplex.	Least mature of the four techniques. Might be too expensive. Better for smallest samples.	Mixing difficult. High voltage source is dangerous and many parameters influence propulsion, better for smallest samples (HTS)

is considerably more complex to implement than electro-osmosis (described next) but the insensitivity of acoustic streaming to the chemical nature of the fluids inside the fluidic channels and its ability to mix fluids make it a potentially viable approach. A typical flow rate measured for water in a small metal pipe lying on a piezoelectric plate is 0.02 cc/s at 40 V, peak to peak [7]. Today acoustic streaming as a propulsion mechanism remains in the research stage.

Electro-osmotic pumping (DC electrokinetics) in a capillary does not involve any moving parts and is easily implemented. All that is needed is a metal electrode in some type of a reservoir at each end of a small flow channel. Typical electro-osmotic flow velocities are on the order of 1 mm/sec with a 1200 V/cm applied electric field. For example, in free-flow capillary electrophoresis work by Jorgenson, electro-osmotic flow of 1.7 mm/sec was reported [8]. This is fast enough for most analytical purposes. Harrison et al. achieved electroosmotic pumping with flow rates up to 1 cm/sec in 20 μm capillaries that were micromachined in glass [9]. They also demonstrated the injection, mixing and reaction of fluids in a manifold of micromachined flow channels without the use of valves. The key aspect for tight valving of liquids at intersecting capillaries in such a manifold is the suppression of convective and diffusion effects. The authors demonstrated that these effects can be controlled by the appropriate application of voltages to the intersecting channels simultaneously. Some disadvantages of electro-osmosis are the required high voltage (1–30 kV power supply) and direct electrical-to-fluid contact with resulting sensitivity of

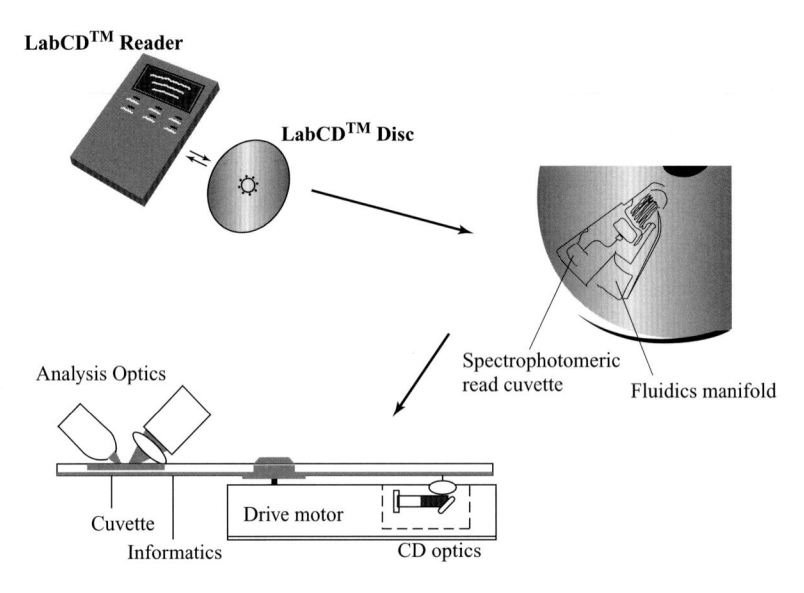

FIGURE 10.1. LabCD™ Instrument and disposable disc. Here, the analytical result is obtained through reflection spectrophotometry.

flow rate to the charge of the capillary wall and to the ionic strength and pH of the solution. It is consequently more difficult to make it into a generic propulsion method. For example, liquids with high ionic strength cause excessive Joule heating; it is therefore difficult or impossible to pump biological fluids such as blood and urine.

Using a rotating disc, centrifugal pumping provides flow rates ranging from less than 10 nL/s to greater than 100 µL/s depending on disc geometry, rotational rate (RPM), and fluid properties (see Figure 10.1)[10]. Pumping is relatively insensitive to physicochemical properties such as pH, ionic strength, or chemical composition (in contrast to AC and DC electrokinetic means of pumping). Aqueous solutions, solvents (e.g., DMSO), surfactants, and biological fluids (blood, milk, and urine) have all been pumped successfully. Fluid gating, as we will describe in more detail further below, is accomplished using "capillary" valves in which capillary forces pin fluids at an enlargement in a channel until rotationally induced pressure is sufficient to overcome the capillary pressure (at the so-called burst frequency) or by hydrophobic methods. Since the types and the amounts of fluids one can pump on a centrifugal platform spans a greater dynamic range than for electrokinetic and acoustic pumps, this approach seems more amenable to sample preparation tasks than electrokinetic and acoustic approaches. Moreover miniaturization and multiplexing are quite easily implemented. A whole range of fluidic functions including valving, decanting, calibration, mixing, metering, sample splitting, and separation can be implemented on this platform and analytical measurements may be electrochemical, fluorescent or absorption based and informatics embedded on the same disc could provide test-specific information.

A most important deciding factor in choosing a fluidic systems is the ease of implementing valves; the method that most elegantly solves the valving issue is already commercially accepted, even if the scaling is not the most favorable namely in the use of traditional pumps. In traditional pumps two one-way valves form a barrier for both liquids and vapors. In the case of the micro-centrifuge, valving is accomplished by varying rotation speed and capillary diameter. Thus, there is no real physical valve required for stopping water flow, but as in the case of acoustic and electrokinetic pumping there is no simple means to stop vapors from spreading over the whole fluidic platform. If liquids need to be stored for a long time on the disposable, as often is the case for use in sensing and diagnostics, valves must be barriers for both liquid and vapor. Some timid attempts at implementing vapor barriers on the CD will be reported in this review.

From the preceding comparison of fluidic propulsion methods for sensing and diagnostic applications, centrifugation in fluidic channels and reservoirs crafted in a CD-like plastic substrate as shown in Figure 10.1 constitutes an attractive fluidic platform.

10.3. COMPACT DISC OR MICRO-CENTRIFUGE FLUIDICS

10.3.1. How it Works

CD fluid propulsion is achieved through centrifugally induced pressure and depends on rotation rate, geometry and location of channels and reservoirs, and fluid properties. Madou et al [11] and Duffy et al [9] characterized the flow rate of aqueous solutions in fluidic CD structures and compared the results to simple centrifuge theory. The average velocity of the liquid (U) from centrifugal theory is given as:

$$U = D_h^2 \rho \omega^2 \bar{r} \Delta r / 32 \mu L, \qquad (10.1)$$

and the volumetric flow rate (Q) as:

$$Q = UA \qquad (10.2)$$

where D_h is the hydraulic diameter of the channel (defined as 4A/P, where A is the cross-sectional area and P is the wetted perimeter of the channel), ρ is the density of the liquid, ω is the angular velocity of the CD, $\bar{r}$ is the average distance of the liquid in the channels to the center of the disc, Δr is the radial extent of the fluid, μ is the viscosity of the solution, and L is the length of the liquid in the capillary channel (see also Figure 10.2A). Flow rates, ranging from 5nL/s to > 0.1mL/s, have been achieved by various combinations of rotational speeds (from 400 to 1600rpm), channel widths (20–500μm), and channel depths (16–340μm). The experimental flow rates were compared to rates predicted by the theoretical model and exhibited an 18.5% coefficient of variation. The authors note that experimental errors in measuring the highest and lowest flow rates made for the largest contribution to this coefficient of variation. The absence of systematic deviation from theory validates the model for describing flow in microfluidic channels under centripetal force. Duffy et al [10] measured flow rates of water, plasma, bovine blood, three concentrations of hematocrit, urine, dimethyl sulfoxide (DMSO), and PCR products and report that centrifugal pumping is relatively insensitive to such physiochemical properties as ionic strength, pH, conductivity,

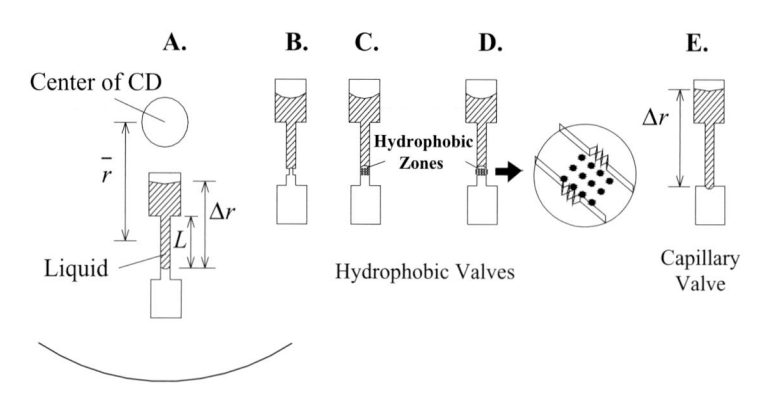

FIGURE 10.2. Schematic illustrations for the description of CD microfluidics. (A.) Two reservoirs connected by a microfluidic chamber. (B.) Hydrophobic valve made by a constriction in a chamber made of hydrophobic material. (C.) Hydrophobic valve made by the application of hydrophobic material to a zone in the channel. (D.) Hydrophobic channel made by the application of hydrophobic material to a zone in a channel made with structured vertical walls (see inset). (E.) Capillary valve made by a sudden expansion in channel diameter such as when a channel meets a reservoir.

and the presence of various analytes, noting good agreement between experiment and theory for all the liquids.

10.4. SOME SIMPLE FLUIDIC FUNCTION DEMONSTRATED ON A CD

10.4.1. Mixing of Fluid

In the work by Madou et al [11] and Duffy et al [10], different means to mix liquids were designed, implemented, and tested. Observations of flow velocities in narrow channels on the CD enabled Reynolds numbers (R_e) calculations establishing that the flow remained laminar in all cases. Even in the largest fluidic channels tested R_e was <100, well below the transition regime from laminar to turbulent flow ($R_e \sim 2300$) [12]. The laminar flow condition necessitates mixing by simple diffusion or by creating special features on the CD that do enable advection or turbulence. In one scenario, fluidic diffusional mixing was implemented by emptying two microfluidic channels together into a single long meandering fluidic channel. Proper design of channel length and reagent reservoirs allowed for stoichiometric mixing in the meandering channel by maintaining equal flow rates of the two streams joining into the mixing channel. Concentration profiles may be calculated from the diffusion rates of the reagents and the time required for the liquids to flow through the tortuous path. Mixing can also be achieved by chaotic advection [6]. Chaotic advection is a result of the rapid distortion and elongation of the fluid/fluid interface, increasing the interfacial area where diffusion occurs, which increases the mean values of the diffusion gradients that drive the diffusion process, one may call this process an enhanced diffusional process. In addition to simple and enhanced diffusional processes, one can create turbulence on the CD by emptying two narrow streams to be mixed into a common chamber. The streams violently splash against a common chamber wall causing their effective mixing (no continuity of the liquid columns is required on the CD as opposed to the case of electrokinetics platforms where a broken up liquid column would cause a voltage overload).

10.4.2. Valving

Valving is a most important function in any type of fluidic platform. Both hydrophobic and capillary valves have been integrated into the CD platform [10, 11, 13–23]. Hydrophobic valves feature an abrupt decrease in the hydrophobic channel cross-section, i.e., a hydrophobic surface prevents further fluid flow (Figure 10.2B–D). In contrast, in capillary valves (Figure 10.2E), liquid flow is stopped by a capillary pressure barrier at junctions where the channel diameter suddenly expands.

10.4.2.1. Hydrophobic Valving

The pressure drop in a channel with laminar flow is given by the Hagen-Poiseuille equation [12]

$$\Delta P = \frac{12 L \mu Q}{w h^3} \quad (10.3)$$

where L is the microchannel length, μ is the dynamic viscosity, Q is the flow rate, and w and h are the channel width and height. The required pressure to overcome a sudden narrowing in a rectangular channel is given by [6]:

$$\Delta P = 2\sigma_l \cos(\theta_c) \left[\left(\frac{1}{w_1}\right) + \left(\frac{1}{h_1}\right) \right] - \left[\left(\frac{1}{w_2}\right) + \left(\frac{1}{h_2}\right) \right] \quad (10.4)$$

where σ_l is the liquid's surface tension, θ_c is the contact angle, w_1 h_1 are the width and height of the channel before the restriction, and w_2 h_2 are the width and height after the restriction. In hydrophobic valving, in order for liquid to move beyond these pressure barriers, the CD must be rotated above a critical speed, at which point the centripetal forces exerted on the liquid column overcome the pressure needed to move past the valve.

Ekstrand et al [13] used hydrophobic valving on a CD to control discrete sample volumes in the nanoliter range with centripetal force. Capillary forces draw liquid into the fluidic channel until there is a change in the surface properties at the hydrophobic valve region. The valving was implemented as described schematically in figure 10.2C. Tiensuu et al [14] introduced localized hydrophobic areas in CD microfluidic channels by ink-jet printing of hydrophobic polymers onto hydrophilic channels. In this work, hydrophobic lines were printed onto the bottom wall of channels with both un-structured (Figure 10.2C) and structured (Figure 10.2D) vertical channel walls. Several channel width to depth ratios were investigated. The CDs were made by injection molding of polycarbonate. The CDs were subsequently rendered hydrophilic by oxygen plasma treatment, and then ink-jet printing was used for the introduction of the hydrophobic polymeric material at the valve position. The parts were capped with polydimethylsiloxane (PDMS) to form the fourth wall of the channel. In testing of non-structured channels (without saw tooth pattern) there were no valve failures for 300 and 500 µm wide channels but some failures for the 100 µm channels, however, in structured vertical walls (with saw tooth patterns), there were no valve failures. The authors attribute the better results of the structured vertical walls to both the favorable distribution of hydrophobic polymer within the channel and the sharper sidewall geometry to be wetted (the side walls are hydrophilic since the printed hydrophobic material is only on the bottom of the channel) compared to the non-structured vertical channel walls.

10.4.2.2. Capillary Valving

Capillary valves have been implemented frequently on CD fluidic platforms [10, 11, 15, 16, 17, 18, 21, 22]. The physical principle involved is based on the surface tension, which develops when the cross section of a hydrophilic capillary expands abruptly as illustrated in Figure 10.2E. As shown in this Figure, a capillary channel connects two reservoirs, and the top reservoir (the one closest to the center of the CD) and the connecting capillary is filled with liquid. For capillaries with axisymetric cross-sections, the maximum pressure at the capillary barrier expressed in terms of the interfacial free energy [16] is given by:

$$P_{cb} = 4\gamma_{al}\sin\theta_c/D_h \tag{10.5}$$

where γ_{al} is the surface energy per unit area of the liquid-air interface, θ_c is the equilibrium contact angle, and D_h is the hydraulic diameter. Assuming low liquid velocities, the flow dynamics may be modeled by balancing the centrifugal force and the capillary barrier pressure (see Equation 10.5). The liquid pressure at the meniscus, from the centripetal force acting on the liquid, can be described as follows:

$$P_m = \rho\omega^2\bar{r}\Delta r, \tag{10.6}$$

where, ρ is the density of the liquid, ω is the angular velocity, $\bar{r}$ is the average distance from the liquid element to the center of the CD, and Δr is the radial length of the liquid sample (Figure 10.2A & 10.2E). Liquid will not pass a capillary valve as long as the pressure at the meniscus (P_m) is less than or equal to the capillary barrier pressure (P_{cb}). Kellogg and coworkers [16] named the point at which P_m equals P_{cb}, the *critical burst condition* and the rotational frequency at which it occurs they called the *burst frequency*. Experimental values of critical burst frequencies versus channel geometry, for rectangular cross-sections over a range of channel sizes, show good agreement with simulation over the entire range of diameters studied. Since these simulations did not assume an axisymetric capillary with a circular contact line and a diameter D_h, the meniscus contact line may be a complex shape. Burst frequencies were shown to be cross-section dependent for equal hydraulic diameters. The theoretical burst frequency equation was modified as follows to account for variation of channel cross section:

$$\rho\omega^2\bar{r}\Delta r < 4\gamma_{al}\sin\theta_c/(D_h)^n, \tag{10.7}$$

where n = 1.08 for an equilateral triangular cross section and, n = 1.14 for a rectangular cross-section. For "pipe flow" (circular cross-section) an additional term is used in the burst frequency expression:

$$\rho\omega^2\bar{r}\Delta r < 4\gamma_{al}\sin\theta_c/(D_h) + \gamma_{al}\sin\theta_c(1/D_h - 1/D_0), \tag{10.8}$$

where the empirically determined constant $D_0 = 40\mu m$. The physical reason for the additional "pipe flow" term, used to get a fit to the simulation results, is not well understood at this time.

Duffy et al [10] modeled capillary valving by balancing the pressure induced by the centripetal force ($\rho\omega^2 \bar{r} \Delta r$) at the exit of the capillary with the pressure inside the liquid droplet being formed at the capillary outlet and the pressure required to wet the chamber beyond the valve. The pressure inside a droplet is given by the Young-Laplace equation [24];

$$\Delta P = \gamma(1/R_1 + 1/R_2) \qquad (10.9)$$

where γ is the surface tension of the liquid and R_1 & R_2 are the meniscus radii of curvature in the x and y dimensions of the capillary cross section. In the case of small circular capillary cross sections with spherical droplet shapes, $R_1 = R_2 \cong$ channel cross section radius and equation (9) can be rewritten as:

$$\Delta P = 4\gamma/D_h. \qquad (10.10)$$

On this basis Duffy et al [10] derived a simplified expression for the critical burst frequency (ω_c) as;

$$\rho\omega_c^2 \bar{r} \Delta r = a(4\gamma/D_h) + b \qquad (10.11)$$

with the first term on the right representing the pressure inside the liquid droplet being formed at the capillary outlet scaled by a factor a (for non spherical droplet shapes) and the second term on the right, b, representing the pressure required to wet the chamber beyond the valve. The b term depends on the geometry of the chamber to be filled and the wettability of its walls.

A plot of the centripetal pressure ($\rho\omega_c^2 \bar{r} \Delta r$) at which the burst occurs verses $1/D_h$ was linear, as expected from equation (10.11), with a 4.3% coefficient of variation. The authors note a potential limitation with capillary valves due to the fact that liquids with low surface tension tend to wet the walls of the chamber at the capillary valve opening resulting in the inability to gate the flow. The b term in equation (10.11) is beneficial in gating flow unless the surface walls at the abrupt enlargement of the capillary valve are so hydrophilic that the liquid is drawn past the valve and into the reservoir.

Madou et al [17, 18] have designed a CD to sequentially valve fluids through a monotonic increase of rotational rate with progressively higher "burst" frequencies. The CD, shown in Figure 10.3, was designed to carry out an assay for ions based on an optode-based detection scheme. The CD design employed 5 serial capillary valves opening at different times as actuated by rotational speed. Results show good agreement between the observed and the calculated burst frequencies (see further below).

It is very important to realize that the valves we mentioned thus far constitute liquid barriers and that they are not barriers for vapors. Vapor barriers must be implemented in any fluidic platform where reagents need to be stored for long periods of time. This is especially important for a disposable diagnostic assay platform. A multi-month, perhaps

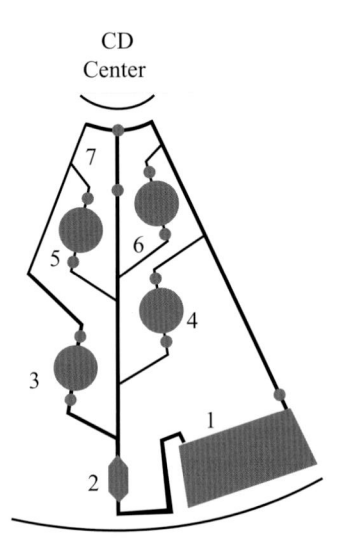

FIGURE 10.3. Schematic illustration of the microfluidic structure employed for the ion selective optode CD platform. The fluidic structure contains five solution reservoirs (numbered 1–5), a detection chamber (6), and a waste reservoir (7). Reservoir (1) and (3) contain the first and second calibrant respectively, reservoirs (2) and (4) contain wash solutions, and reservoir (5) contains the sample. Upon increasing rotation rates, calibrant 1, wash 1, calibrant 2, rinse 2, and then sample were serially gated into the optical detection chamber. Absorption of the calibrants and sample was measured.

multi-year, shelf life would require vapor locks in order to prevent reagent solutions from drying or liquid evaporation and condensation in undesirable areas of the fluidic pathway. Tecan-Boston did investigate vapor resistant valves made of wax that was melted to actuate valve opening [25].

10.4.3. Volume Definition (Metering) and Common Distribution Channels

The CD centrifugal microfluidic platform enables very fine volume control (or metering) of liquids. Precise volume definition is one of the important functions, necessary in many analytical sample processing protocols, which has been added, for example, to the fluidic design in the Gyrolab MALDI SP1 CD [20]. In this CD, developed for matrix assisted laser desorption ionization (MALDI) sample preparation, a common distribution channel feeds several parallel individual sample preparation fluidic structures (Figure 10.4). Reagents are introduced by the capillary force exerted by the hydrophilic surfaces into the common channel and defined volume (200nl) chambers until a hydrophobic valve stops the flow. When all of the defined-volume-chambers are filled, the CD is spun at a velocity large enough to move the excess liquid from the common channel into the waste. Although there is sufficient g-force to empty the common channel, the velocity is not so high as to allow liquid to move past the hydrophobic valve and the well-defined volume chambers remain filled. These precisely defined volumes can be introduced into the subsequent fluidic structures by increasing the CD angular momentum until the centripetal force allows the liquid to move past the hydrophobic barriers.

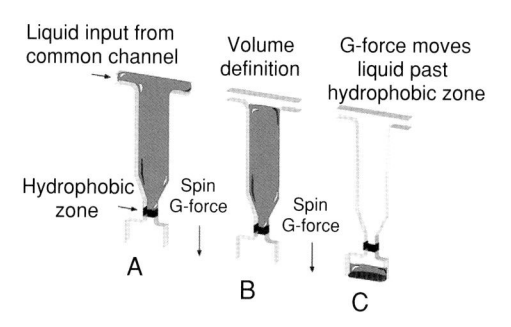

FIGURE 10.4. Schematic illustration of liquid metering. (A) The common distribution channel and liquid metering reservoirs are filled (by capillary forces) with a reagent to be metered. Liquid entering the reservoir does not pass the hydrophobic zone (valve) because of surface tension forces. (B) The CD is rotated at a rate that supplies enough centripetal force to empty the common distribution but not enough to force the liquid through the hydrophobic zone. The volume of the fluid metered is determined by the volume of the reservoir. (C) A further increase in the rotational speed provides enough force to move the well defined volume of solution past the hydrophobic valve. (With kind permission; Application Report 101, Gyrolab MALDA SP1, Gyros AB, Uppsala, Sweden)

10.4.4. Packed Columns

Many commercial products are now available that use conventional centrifuges to move liquid, in a controlled manner, through a chromatographic column. One example is the Quick Spin™ protein desalting column (Roche Diagnostics Corp, Indianapolis, IN), based on the size exclusion principle. There is an obvious fit for this same type of separation experiment to be carried out on a CD fluidic device (we sometimes refer to the CD platform as a smart, miniaturized centrifuge). Affinity chromatography has been implemented in the fluidic design of the Gyrolab MALDI SP1 CD [20]. A reverse phase chromatography column material (SOURCE™ 15 RPC) is packed in a microfluidic channel and protein is adsorbed on the column from an aqueous sample as it passes through the column under centrifugally controlled flow rates. A rinse solution is subsequently passed through the column and finally an elution buffer is flowed through to remove the protein and carry it into the fluidic system for further processing. The complete Gyrolab MALDI SP1 CD is discussed in a later section of this review.

10.5. CD APPLICATIONS

10.5.1. Two-Point Calibration of an Optode-Based Detection System

A CD based system with ion-selective optode detection and a two-point-calibration structure for the accurate detection of a wide variety of ions was developed [15, 17, 18]. The microfluidic architecture, depicted in Figure 10.3, is comprised of channels, five solution reservoirs, a chamber for colorimetric measurement of the optode membrane, and a waste reservoir, all manufactured onto a poly(methyl methacrylate) disc. Ion-selective optode membranes, composed of plasticized poly(vinyl chloride) impregnated with an ionophore, a chromoionophore, and a lipophilic anionic additive, were cast, with a spin-on device, onto a support layer and then immobilized on the disc. With this system, it is possible to deliver calibrant solutions, washing buffers, and "unknown" solutions (e.g., saliva, blood, urine, etc) to the measuring chamber where the optode membrane is located. Absorbance measurements on a potassium optode indicate that optodes immobilized on the platform exhibit the theoretical absorbance response. Samples of unknown concentration can be quantified to within 3% error by fitting the response curve for a given optode membrane using an acid (for measuring the signal for a fully protonated chromoionophore), a base (for fully deprotonated chromoionophore), and two standard solutions. Further, the ability to measure ion concentrations employing one standard solution in conjunction with acid and base, and with two standards alone were studied to delineate whether the current architecture could be simplified. Finally, the efficacy of incorporating washing steps into the calibration protocol was investigated.

This work was further extended to include anion-selective optodes and fluorescence rather than absorbance detection [17]. Furthermore, in addition to employing a standard excitation source where a fiber optic probe is coupled to a lamp, LASER diodes were evaluated as excitation sources to enhance the fluorescence signal.

10.5.2. CD Platform for Enzyme-Linked Immunosorbant Assays (ELISA)

The automation of immunoassays on microfluidic platforms presents multiple challenges because of the high number of fluidic processes and the many different liquid reagents involved. Often there is also the need for highly accurate quantitative results at extremely low concentration and care must be taken to prevent non-specific binding of reporter enzymes and to deliver well defined volumes of reagents consistently. An enzyme-linked immunosorbant assay (i.e., ELISA) is one of the most common immunoassay methods and is often carried out in microtiter plates using labor-intensive manual pippetting techniques. Recently, Lee and et al [21] have implemented an automated enzyme-linked immunosorbant assay on the CD platform. This group used a five-step flow sequence in the same CD design illustrated in Figure 10.3. A capture antibody (anti rat IgG) was applied to the detection reservoir (reservoir 2 in Figure 10.3) by adsorption to the PMMA CD surface, then the surface was blocked to prevent non-specific binding. Antigen/sample (rat IgG), wash solution, 2nd antibody, and substrate solutions were loaded into reservoirs 3–7 (Figure 10.3) respectively. Using capillary valving techniques, the sample and reagents were pumped, one at a time, through the detection chamber. First, the sample was introduced for antibody antigen binding (reservoir 3), then a wash solution (reservoir 4), then an enzyme labeled secondary antibody (reservoir 5), then another wash solution (reservoir 6), and finally the substrate was added (reservoir 7). The U-shaped bend in the fluidic path allows the solutions to incubate in the capture zone/detection chamber until the next solution is released into the chamber. Detection of the fluorescence is performed after the substrate is introduced into the detection reservoir (reservoir 2). Endpoint measurements (completion of enzyme-substrate reaction) were made and compared to conventional microtiter plate methods using similar protocols. The CD ELISA platform was shown to have advantages such as lower reagent consumption, and shorter assay times, explained in terms of larger surface to volume ratio's which favor diffusion limited processes. Since the reagents were all loaded into the CD at the same time, there was no need for manual operator interventions in between fluidic assay steps. The consistent control and repeatability of liquid propulsion removes experimental errors associated with inconsistent manual pippetting methods, for

10.5.3. Multiple Parallel Assays

The ability to obtain simultaneous and identical flow rates, incubation times, mixing dynamics, and detection makes the CD an attractive platform for multiple parallel assays. Kellogg et al [10] have reported on a CD system that performs multiple (48) enzymatic assays simultaneously by combining centrifugal pumping in microfluidic channels with capillary valving and colorimetric detection. The investigation of multiplexed parallel enzyme inhibitor assays are needed for high-throughput screening in diagnostics and in screening of drug libraries. For example, enzymatic dephosphorylation of colorless p-nitrophenol phosphate by alkaline phosphatase results in the formation of the yellow colored p-nitrophenol and inhibition of this reaction may be quantified by light absorption measurement. Theophylline, a known inhibitor of the reaction, was used as the model inhibitory compound in Kellogg et al's [10] feasibility study. A single assay element on the CD contains 3 reservoirs: one for the enzyme, one for the inhibitor, and one for the substrate. Rotation of the CD allows the enzyme and inhibitor to pass capillary valves, mix in a meandering 100-μm-wide channel, and then move to a point where flow is stopped by another capillary valve. A further increase in the rotational speed allows the enzyme/inhibitor mixture and substrate to pass through the next set of capillary valves where they are mixed in a second meandering channel and emptied into an on-disc planar cuvette. The CD is slowed and absorption through each of the 48 parallel assay cuvettes is measured by reflectance, all in a period of 60s, the entire fluidic process including measurement took about 3 minutes. The CDs were fabricated using PDMS replication techniques [26], with the addition of a white pigment to the PDMS polymerization for enhanced reflectivity in the colorimetric measurements. The flow rates and meandering channel widths were selected such that the diffusion rate would allow 90% mixing of the solutions.

The variation in performance between the individual fluidic CD structures was quantified by carrying out the same assay 45 times simultaneously on a CD. The background corrected absorbance was measured and the coefficient of variation in the assay was ~3.2% and when the experiment was repeated on different discs the coefficient of variation ranged from 3–3.5%. Furthermore, variation of absorption across a single cuvette was less than 1%, confirming complete mixing. In experiments to show enzyme inhibition, 45 simultaneous reactions were carried out on the CD using fixed concentrations of enzyme and substrate and 15 concentrations of theophylline in triplicate and a complete isotherm was generated for the inhibition of alkaline phosphatase. The 3 remaining structures were used for calibration with known concentrations of p-nitrophenol. A dose response was seen over 3 logs of theophylline concentration in the range of 0.1mM to 100mM. The authors conclude that a large number of identical assays, with applications in rapid, high-throughput screening, can be carried out on the CD platform simultaneously because of the symmetric force acting on the fluids in high-quality identical microfluidic structures and that detection was simplified by rotating all the reaction mixtures under a fixed detector. In later work [22], the same group has extended the number of assays to 96 per CD and has investigated fluorescent enzymatic assays.

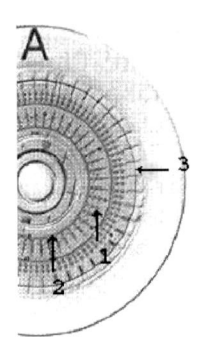

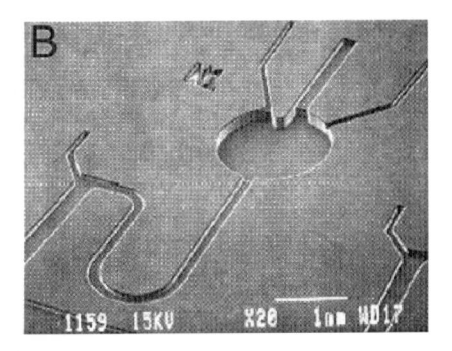

FIGURE 10.5. Microfabricated cell culture CD. (**A**) The CD caries a number of cell growth chambers (1) radially arranged around a common distribution channel (2) and is sealed with a silicone cover (10.3). (**B**) SEM close-up of an individual cell growth chamber and microfluidic connections. (Micro Total Analysis Systems 2000 ©, (2000) pp. 249–252, Nick Thomas, Anette Ocklind, Ingrid Blikstad, Suzanne Griffiths, Michael Kenrick, Helene Derand, Gunnar Ekstrand, Christel Ellström, Anders Larsson and Per Anderson "Integrated Cell Based Assays in Microfabricated Disposable CD Devices," with kind permission of Kluwer Academic Publishers)

10.5.4. Cellular Based Assays on CD Platform

Cell based assays are often used in drug screening [27] and rely on labor intensive microtiter plate technologies. Microtiter plate methods may be difficult to automate without the use of large and expensive liquid handling systems and they present problems with evaporation when scaled down to small volumes. Thomas and et al [23] have reported on a CD platform-based automated adherent cell system. This adherent cell assays involved introducing the compounds to be screened to a cell culture, then determining if the cells were killed (cell viability assay).

Reagents for cell growth, rinsing and viability staining were serially loaded into an annular common distribution chamber and centripetal force was used for reagent loading, exchange, and rinsing of the cell growth chamber (Figure 10.5). Individual inlets are for the addition of compounds to be screened. The plastic channels, Figure 10.5B, were capped with a Polydimethylsiloxane (PDMS) sheet capable of fast gas transport in and out of the culture reservoirs.

HeLA, L929, CHO-M1, and MRC-5 cell lines were cultivated on the CD device. Cell viability assays were performed, on the CD, by removing the growth medium from the cells, washing the cells with PBS, and introducing a solution of the fluorescence assay reagents into the growth chamber. The LIVE/DEAD® Viability Assay (Molecular Probes, Inc., Eugene, OR) uses a mixture of calcein green-fluorescent nucleic acid stain and the red-fluorescent nucleic acid stain, ethidium. The assay performance is based on the differing abilities of the stains to penetrate healthy bacterial cells. The calcein green-fluorescent dye will label all cells, live or dead. The red red-fluorescent ethidium stain will only label cells with damaged membranes. The red stain causes a reduction in the green stain fluorescence when both dyes are present. When the appropriate mixture of green and red stains is used, cells with intact membranes will have a green fluorescence and cells with damaged membranes will have a red fluorescence. The background remains almost completely non-fluorescent (Figure 10.6). All liquid transfers were carried out using centripetal force from CD rotation with angular frequencies between 200 and 600 rpm. Quantitative detection of multiple cell

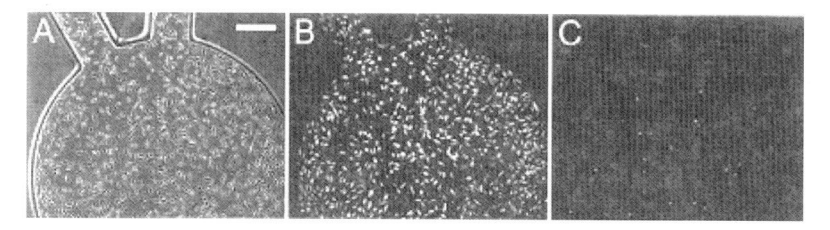

FIGURE 10.6. L929 fibroblasts cultured for 48 hours in CD growth chambers. (**A**) phase contrast (scale bar 100 μm), (**B**) epifluorescence image of calcein stained viable cells, (B) epifluorescence image of ethidium stained non-viable cells. (Micro Total Analysis Systems 2000 ©, (2000) pp. 249–252, Nick Thomas, Anette Ocklind, Ingrid Blikstad, Suzanne Griffiths, Michael Kenrick, Helene Derand, Gunnar Ekstrand, Christel Ellström, Anders Larsson and Per Anderson "Integrated Cell Based Assays in Microfabricated Disposable CD Devices," with kind permission of Kluwer Academic Publishers)

viability assays, within 30 seconds, was carried out by measurement of calcien fluorescence with a CCD based fluorescence imaging system. These experimental results show linear fluorescence intensity across the range of 200–4000 cells and give an indication of the potential of this platform for miniaturized quantitative cell based assays.

In the same work, the authors reported the results of experiments designed to investigate the effect on cells of using centripetal force to move liquids. The cells tested were shown to be compatible with centripetal forges of at least 600 × g, much larger than the 50–100 × g needed for filling and emptying cell chambers. Furthermore, it was reported that cells grown in such devices appear to show the same cell morphology as cells grown under standard conditions.

In separate work done by our group in collaboration with NASA Ames et al [28] the LIVE/DEAD®BacLight™ Bacterial Viability Kit (Molecular Probes, Inc., Eugene, OR) has been integrated to a completely automated process on CD. Disposable and reusable CD structures, hardware, and software were developed for the LIVE/DEAD assay.

The CD design for assay automation must have the following functions or properties: contain separate reservoirs for each dye and the sample, retain those solutions in the reservoir until the disc is rotated at a certain velocity, evenly and completely mix the two dyes, evenly and completely mix the dye-mixture with the sample containing the cells, collect this final mixture in a reservoir with good optical properties. Two methods for quick fabrication of prototype CDs were used. One method used molded PDMS structures. In a second method, a dry film photoresist (DF 8130, Think & Tinker, Palmer Lake, CO) was laminated onto a 1 mm thick polycarbonate disc with pre-drilled holes for sample introduction. The microfluidic pattern was made using a photolithographic pattern on the negative photoresist. The fluidic system was capped with a polycarbonate disc that had been laminated with an optical quality pressure sensitive adhesive (3M 8142, 3M, Minneapolis, MN). Figure 10.7 shows the fluidic pattern for this assay. This pattern is based on the structure developed in a similar approach used to demonstrate multiple enzymatic assays on CD [10].

The dyes and sample were introduced into reservoir chambers using a pipette. The dyes fill the chamber stopping at a capillary valve (valve 1 in figure 10.7). Similarly, the sample containing cells was introduced into the sample reservoir. Upon rotation, the dyes were forced through the capillary valves and were mixed as they flowed through the switchback turns of the microfluidic channels. Simultaneously, the sample passed from its reservoir

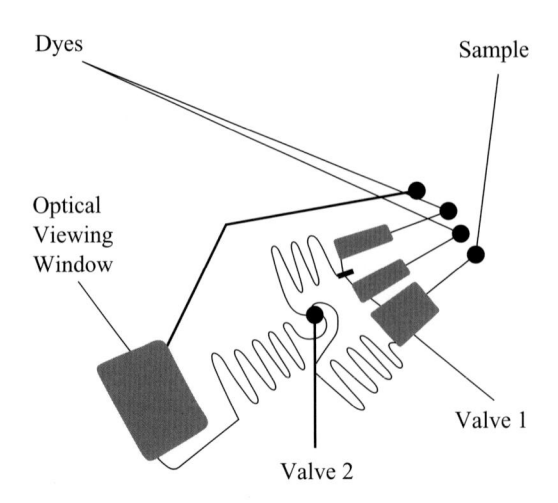

FIGURE 10.7. Microfluidic pattern for LIVE/DEAD®BacLight™ Bacterial Viability Assay. The dyes and sample are introduced into the reservoir chambers using a pipette. The dyes fill the chamber stopping at a capillary valve (valve 1). Similarly, the sample containing cells is introduced into the sample reservoir. The disc is rotated to a velocity of 800 rpm, the dyes are forced through the capillary valves and they are mixed as they flow through the switchback turns of the microfluidic channels. Simultaneously, the sample passes from the reservoir into a fluid channel where it meets the dye mixture at valve 2. The velocity of the disc is increased to 1600 rpm and the dye mixture and sample combine and mix in the switchback microfluidic path leading to the optical viewing window.

into a fluid channel where it met the dye mixture at valve 2 of Figure 10.7. The velocity of the disc was increased and the dye mixture and sample combine and mix in the switchback microfluidic path leading to the optical viewing window. The dye-sample mixture is allowed to incubate in the dark at room temperature for 5 minutes. The optical viewing chamber was imaged twice, once with optics for the green signal and then with optics for the red signal. A typical fluorescence microscopy image of an overlay of the red and green images of stained *E. coli* is shown in Figure 10.8.

The instrument for disc rotation and fluorescence imaging (Figure 10.9) used a programmable rotational motor for various velocities and acceleration/deceleration rates. The use of standard microscope objectives enabled the selection of magnification. An automatic focusing system was used. The light source was a mercury lamp, which used standard low-pass excitation filters for fluorescent excitation. A CCD camera was combined with standard emission filter cubes for imaging.

10.5.5. Automated Cell Lysis on a CD

In work done by our group [29], cell lysis was demonstrated on a microfluidic CD (Compact Disc) platform. In this purely mechanical lysis method, spherical particles (beads) embedded in the CD cause shear-induced disruption of mammalian cells, *E. coli*, and yeast. Interactions between beads and cells were generated in the rimming flow established inside a partially filled annulus chamber in the CD rotating around a horizontal axis of rotation. To maximize bead-cell interactions in the lysis chamber, the CD was spun forward and backwards at high acceleration for up to 5 to 7 minutes. For this novel lysis method we

CENTRIFUGE BASED FLUIDIC PLATFORMS 345

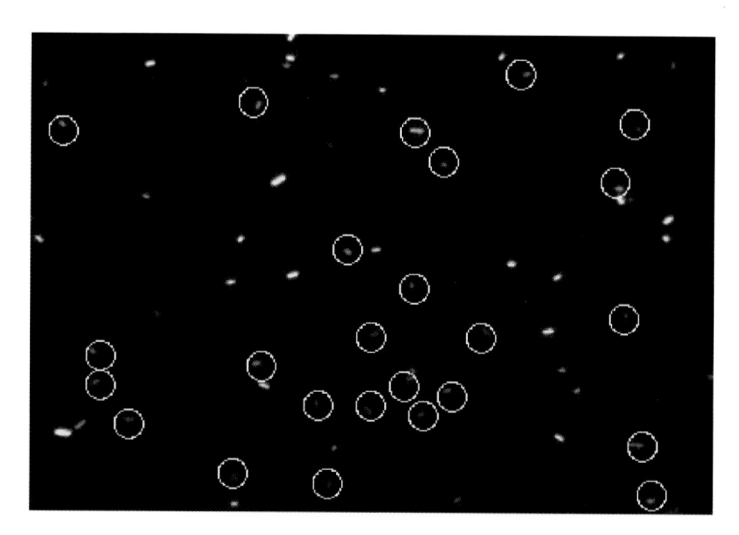

FIGURE 10.8. Fluorescent microscopy over-laid images of red (circled cells) and green (un-circled cells) stained E. coli on CD from LIVE/DEAD®BacLight™ Bacterial Viability Assay.

investigated which of the parameters such as bead density, angular velocity, acceleration rate, and solid volume fraction are the most important for efficient cell lysis. Cell disruption was verified either through direct microscopic viewing or measurement of the DNA concentration extracted upon lysis using the PicoGreen dsDNA Quantitation Kit (Molecular Probes, OR). Lysis efficiency was also compared with a conventional lysis protocol. In the long term, this work is geared towards CD-based sample to answer nucleic acid analysis which will perform cell lysis, DNA purification, DNA amplification, and detection.

Although extensive research on microfluidic devices[6] for biomedical and clinical diagnostic applications has been conducted in the past decade, a majority of the miniaturized devices developed has been limited to single purpose applications due to the complexity and difficulty of integrating various microfluidic components into a single small unit. Compared to other microfluidic technologies [2, 6, 12, 30] for moving small amounts of fluid or

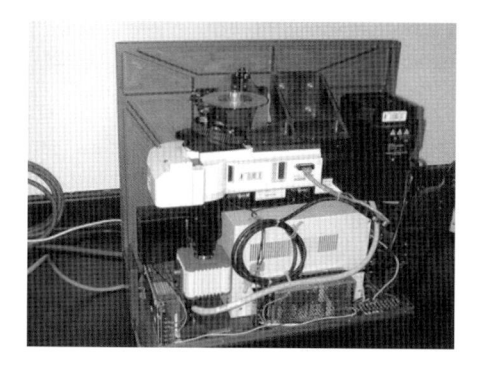

FIGURE 10.9. Left: Optical disc drive/imager with cover removed. Size of unit is made to fit in specific cargo bay of Space-Lab. Right: Zoom of microscope objectives and a disc loaded in the drive.

suspended particles from site to site, a centrifuge-based system is well suited for various crucial microfluidic functions such as flow sequencing, cascade mixing, capillary metering and flow switching.[11, 31, 32] Those functions can be implemented through the exploitation of (1) centrifugal and coriolis forces induced by spinning the CD for sample propulsion, (2) capillary force due to interfacial tension for stopping the flow and (3) microfluidic designs. In this regard, a polymer based microfluidic CD platform is a highly promising approach, not only for the integration of multi-microfluidic functions for diagnostic applications but also as a platform enabling the automated, multiple parallel processing needs in high-throughput screening (HTS). In this paper, we investigate the achievability of mechanical cell lysis on a CD platform and compare its performance with a traditional lysis protocol.

There are many types of cell lysis methods used today that are based on mechanical [33], physico-chemical [34], chemical [35], and enzymatic [35] principles. The most commonly used methods in biology research labs rely on chemical and enzymatic principles. The main drawbacks of those procedures include intensive labor, adulteration of cell lysate, and the need for additional purification steps. In order to minimize the required steps for cell lysis, a rapid and reagentless cell lysis method [33] would be greatly appreciated especially in sample preparation on a microfluidic CD. This would also help avoid the necessity of storing chemicals in chambers and of intensive mixing which is generally considered difficult in the micro domain. In this work, a purely mechanical lysis method was investigated based on rapid granular shear flow [36, 37] in an annulus chamber which spins about a horizontal axis of rotation alternating between the forward and reverse directions. When the CD is at rest, spherical particles, such as glass beads, suspended in an aqueous medium containing the cells, lie in a pool at the bottom of the chamber. However, during spinning all particles are dragged up by the shear of the surrounding medium and uniformly coat the outer wall of the chamber in response to both shear and centrifugal forces as shown in Figure 10.10. This flow is often referred to as "rimming flow" [38, 39]. In order for a cell to be disrupted in a rapid granular flow, the cell should be brought into physical contact with the colliding particles. The bead interactions with the cells consist of two main types: impulsive contact (collision)

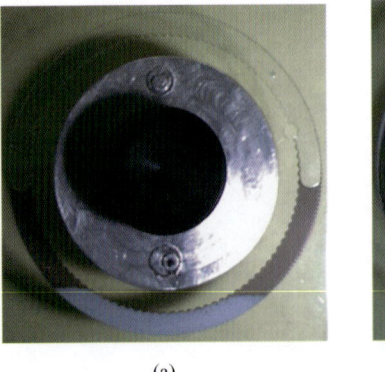

(a)   (b)

FIGURE 10.10. Flow patterns for two rotational states (a) At rest: fluidized beads are sedimented at the bottom of chamber (b) At spinning: two circumferential bands of beads and liquid are observed for a constant velocity.

CENTRIFUGE BASED FLUIDIC PLATFORMS

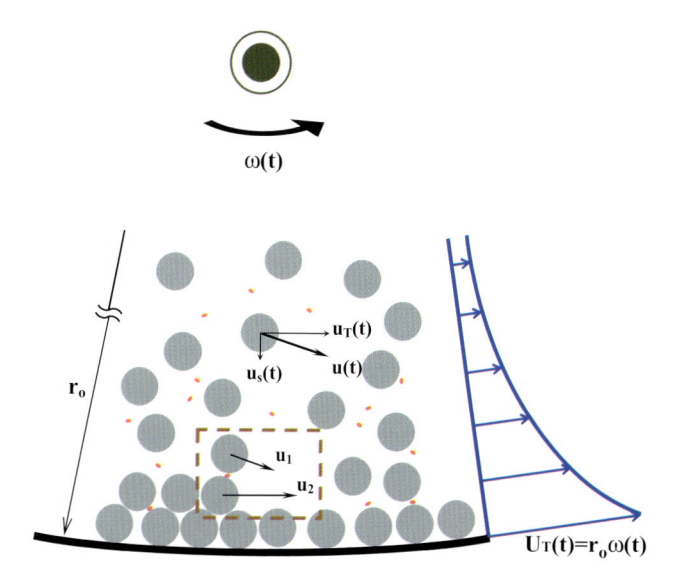

FIGURE 10.11. Illustrative views of particle interactions (collision and friction) that are generated in colliding particles in acceleration of a CD. A tangential velocity profile of fluid in the chamber is depicted on the right and varies with radial locations and time. $u_T(t)$ represents tangential velocity of a particle and follows with the flow passing around it. $u_s(t)$ indicates settling velocity under a centrifugal field.

due to the beads responding to the centrifugal field and sustained contact (friction) due to shearing.

We first discuss inter-particle forces involved in the rimming flow and identify those parameters that may have the largest contribution to the inter-particle forces and the particle interaction frequency. The influence of these parameters on cell lysis is then verified experimentally by microscopy or by measuring the double stranded DNA concentration extracted from the cells upon cell lysis.

Rapid granular rimming flow in a horizontally rotating annulus chamber features two types of particle interactions: collision and friction (grinding). In Fig. 10.11 we schematically depict these particle interactions generated during an acceleration period in a partially-filled lysis chamber. Collisions occur when suspended particles are forced to move outwards radially (vector $u_s(t)$) responding to the centrifugal force and impact upon particles closer to the outer wall. Friction between particles is generated in the process of rearranging the impacting particles into a homogeneous granule band along the outer wall. The impacting particles (vector u_1) travel slower than the particles close to the chamber outer wall (vector u_2) due to differential shear rates imposed on them. Particles accumulated against the chamber outer wall move almost at the same tangential velocity (see $U_T(t)$) as the outer chamber wall as they more or less stick to it. It is actually the velocity differential between colliding particles moving in the same tangential direction that leads to the friction (grinding) between those particles.

The particles make a complete circle on a typical time scale of 0.1 sec at angular velocities ranging from 10 to 20 rev/sec. This velocity (vector $u_T(t)$) is approximately three orders of magnitude larger than the rate of radial motion of particles (vector $u_s(t)$) i.e., particle settling velocity (see also below). On the basis of this consideration, one can expect that the suspended particles, upon spinning the CD, uniformly coat the chamber wall while undergoing both collision and friction.

We discuss the grinding force here as a compressive pressure P_g exerted between the particles in a shear flow. It was empirically found [40] that the compressive pressure is proportional to a shear stress τ which has the following relation

$$\tau \propto \rho_p \cdot D_p^2 \cdot \gamma^2 \cdot f(\nu) \tag{10.12}$$

where ρ_p is particle density, D_p is particle diameter, γ is shear rate, and $f(\nu)$ is a function of the volumetric concentration of particles.

Frictional contact of particles depends on the shear stress τ which is proportional to the density of particles ρ_p and the square of local shear rate γ. This is similar to a plane Couette flow where the shear rate (dU_T/dy) in steady-state is proportional to the angular velocity of the CD. During the steady-state many particles will, however, have already reached the mound of particles at the outer wall of the chamber, and therefore make no relative motions contributing to lysing. Hence, an acceleration period is more contributive to the shear rate than a steady state mode. Accordingly, higher acceleration rate will lead to larger magnitude of frictional contact of particles.

Most of the theories on particle collisions [41, 42] in granular flow are based on models describing the individual collisions of spheres such that the impact properties employed in the theories should adequately describe the average properties of the flow. However, the measurement of impact coefficients to test these theories is extremely challenging in the current application and is beyond the scope of this treatise. Therefore, we have attempted to evaluate the parameters associated with the particle collisions experimentally to predict the magnitude of collisions qualitatively.

In the work by Batchelor [43] and Zenit [44], a collisional particle pressure in a solid-liquid mixture is represented as

$$P_c = \nu \cdot \rho_p \cdot \bar{u}^2 \cdot F(\nu) \tag{10.13}$$

where ν is solid volume fraction (SVF), ρ_p is particle density, and $\bar{u}$ is local mean particle velocity. Batchelor suggests the following relation for a function of solid volume fraction ν

$$F(\nu) \approx \frac{\nu}{\nu_{cp}} \left(1 - \frac{\nu}{\nu_{cp}}\right) \tag{10.14}$$

where ν_{cp} is the closed-packed solid fraction ($\nu_{cp} \approx 0.62$) for a random packed bed of uniform sized particles. Here, considering a solid-liquid suspension in a centrifugal field, it is appropriate that we substitute terminal settling velocity u_t with the local mean particle velocity $\bar{u}$ in Eqn. (10.13). From a balance of the centrifugal, buoyancy and drag force, the motion of a spherical particle is expressed as

$$m\frac{du}{dt} = mr\omega^2 - m\frac{\rho}{\rho_p}r\omega^2 - \frac{C_D}{2}u^2 A_p \rho \tag{10.15}$$

which can be rewritten as

$$\frac{du}{dt} = r\omega^2 \frac{\rho_p - \rho}{\rho_p} - \frac{C_D A_p \rho}{2m} u^2 \qquad (10.16)$$

where ω is the angular frequency, r is distance of particle from the center of the disc, C_D is drag coefficient, and A_p is projected area of particle. In Eqn. (10.16), u is the velocity of the particle relative to the fluid and is directed outwardly along a radius. As a particle travels, the drag quadratically increases with the velocity. Therefore, the particle acceleration quickly decreases and approaches zero. The particle then reaches a maximum constant velocity, terminal settling velocity, which is represented as

$$u_t = \sqrt{\frac{4 \cdot r\omega^2 \cdot D_p \cdot (\rho_p - \rho)}{3 \cdot C_D \cdot \rho}} \qquad (10.17)$$

Since the particle Reynolds number, $\text{Re}_p = (D_p u_t \rho_f)/\eta$, is calculated to be slightly larger than 1 for the current experiments, falling into a transition region very close to Stokes regime, the drag coefficient [45] is found as

$$C_D = \frac{24}{\text{Re}_p} + \frac{3}{\sqrt{\text{Re}_p}} + 0.34 \qquad (10.18)$$

Thus, the collisional pressure in Eqn. (10.13) can now be rewritten as

$$P_c = \left\{\frac{4}{3} \cdot \frac{r D_p}{C_D}\right\} \cdot \left\{\rho_p \left(\frac{\rho_p}{\rho} - 1\right) \cdot \omega^2 \cdot vF(v)\right\} \qquad (10.19)$$

From equations 10.12 and 10.19, the important parameters for lysis appear to be acceleration rate, particle density, angular velocity, and solid volume fraction. The particle size is excluded from the important influencing parameters here because increasing particle size conversely creates larger interparticle voids, which will reduce the number of particle collisions with cells squeezed in-between (i.e., the effective impact frequency). The particle size is expected to be more of an important parameter for the effective impact frequency. Concerning the volume fraction of particles, it plays a binary role in both the magnitude of particle interactions and the effective impact frequency. If only the collisional pressure in Eqn. (10.13) is considered for the inter-particle force, the highest collisional pressure will lie at the SVF of 0.4 as shown in Fig. 10.12.

The CDs were fabricated using the Polydimethylsiloxane (PDMS) replication techniques described earlier [26]. For the purpose of mechanical cell disruption, an ultra-thick SU-8 process was developed to fabricate a mold featuring extra high structures (thickness of a 1mm) so that sufficiently high spaces could be formed in the PDMS for the interacting beads. The rotating CD platform was completed by sandwiching the micromachined PDMS sheet between two polycarbonate discs. Fig 10.13 shows a CD with an annulus chamber designed for cell lysis.

As shown in Fig. 10.14, a custom CD spin-stand is equipped with a motor and an amplifier/controller to allow for various rotational profiles such as RPM-specified rotation

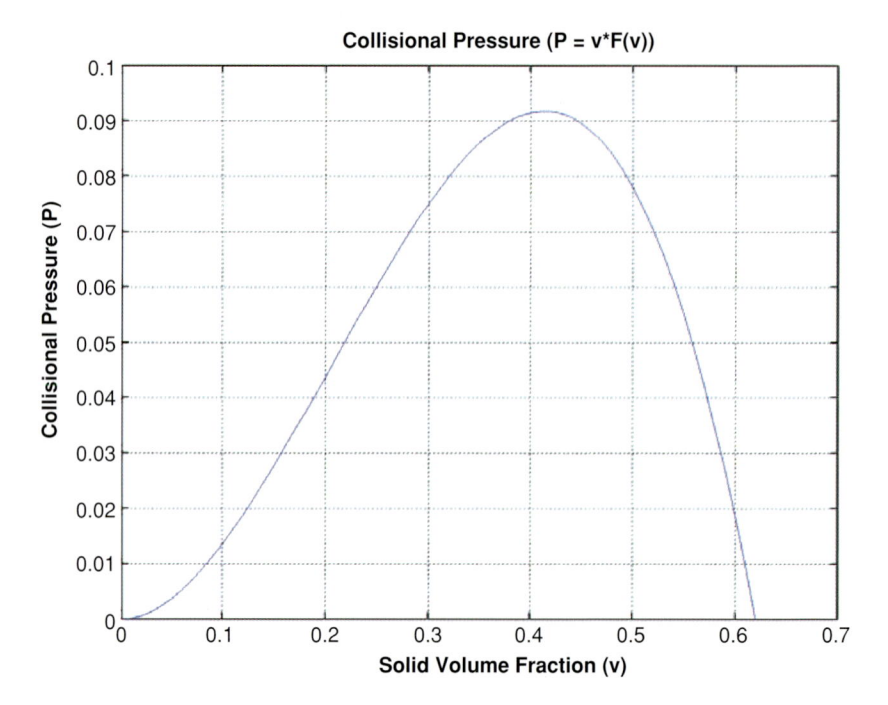

FIGURE 10.12. Plot of collisional pressure with respect to solid volume fraction.

and precise positioning. A centrifuge vision system was used to view a sequence of color images of the area of interest on the CD in real time while it is running and store the captured frames on a computer.

Periodic granular rimming flow is induced in a partially filled circular chamber by spinning a CD around a horizontal axis of symmetry with high acceleration (e.g., 200 rev/sec^2) and changing rotational directions with a tiny rest period in between (e.g., 150 msec) to

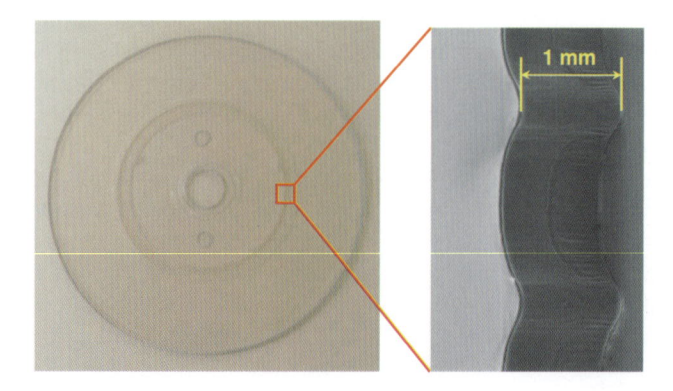

FIGURE 10.13. Microfluidic view of a CD (a) CD with a circular chamber (1mL) (b) SEM photo showing the side view of inner wall of the chamber.

CENTRIFUGE BASED FLUIDIC PLATFORMS

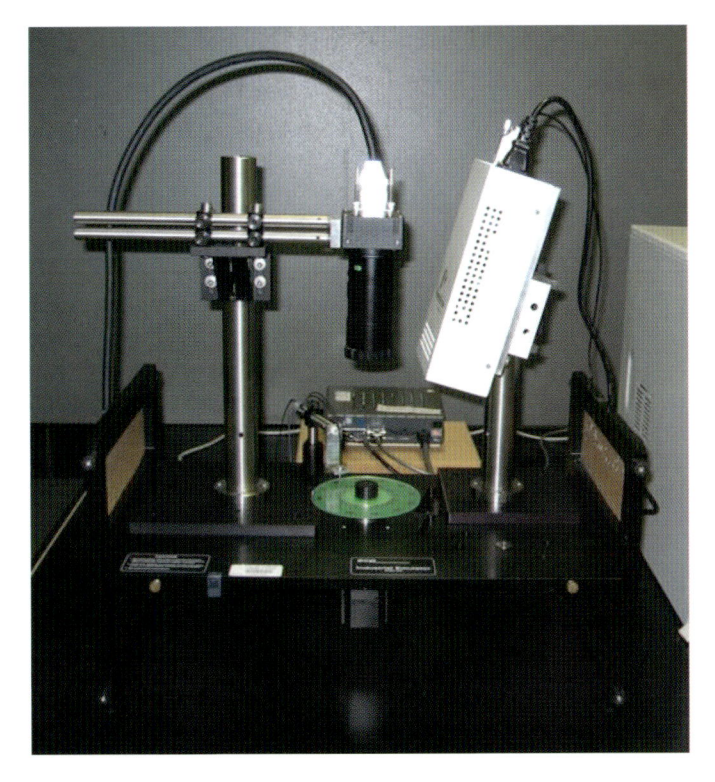

FIGURE 10.14. Spin-stand and centrifuge vision system. the CD spin-stand is equipped with a motor (Pacific Scientific Servo Motor, 9000rpm) and an amplifier/controller (PAC SCI Programmable Servo Drive) to allow for various rotational profiles such as RPM-specified rotation and precise positioning. The servo drive uses a graphical user interface program, ToolPAC, to easily configure and program the motor to specific applications. A microfluidic CD is placed on the aluminum saddle coupled with the motor shaft. The digital video recording system is composed of: a camera (Basler A301bc, 640 × 480 pixels, 80 fps max., 10× zoom lens mounted), a strobe light (PerkinElmer MVS-4200, 6µs duration), and a retro-reflective sensor (Banner D10 Expert Fiber-Optic Sensor). The strobe light with a 100 Hz maximum repetition frequency is employed to reduce blurry images of a fast moving object.

accumulate a solid-liquid mixture at the bottom of the chamber. The CD makes 3 or 4 revolutions depending on the chosen angular velocity (e.g., 10–20 rev/sec) and returns to the original position with the same angular velocity and the same number of revolutions—a typical cycle profile is shown in Fig. 10.15. The CD then runs for hundreds of such cycles to ensure that a large number of effective particle collisions occur. The angular acceleration rate is so high that a circumferential particle band begins to be observed in one revolution corresponding to about 0.1 sec. Fig. 10.16 shows a sequence of flow patterns to illustrate how a solid-liquid mixture forms into rimming flow.

CHO-K1 (Clontech, CA) was used as a model cell line for mammalian cell lysis. The cell density was 100,000 cells/mL. *Saccharomyces cerevisiae* strain TMy16 (*MAT*a *trp1*- Δ901 *ura3*-52 *his3*-Δ201 *ade2*-101 *lys2*-1 *leu1*-12 *can1*-100 *GAL*$^+$) [46] and *Escherichia coli* XL1-Blue (Stratagene, CA) were used for yeast and bacteria lysis. The cell concentration was measured at O.D$_{600}$ with a spectrophotometer.

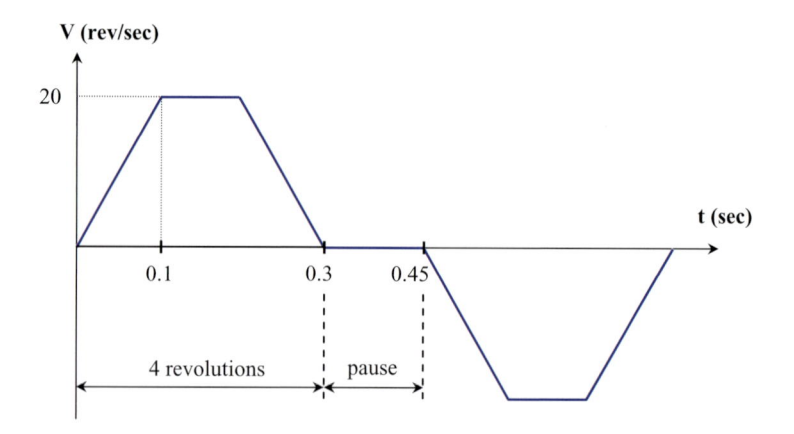

FIGURE 10.15. Schematic rotational profiles for one cycle: A CD ramps up to 20 rev/sec with 200 rev/sec^2 making 4 revolutions and spins back to the original position.

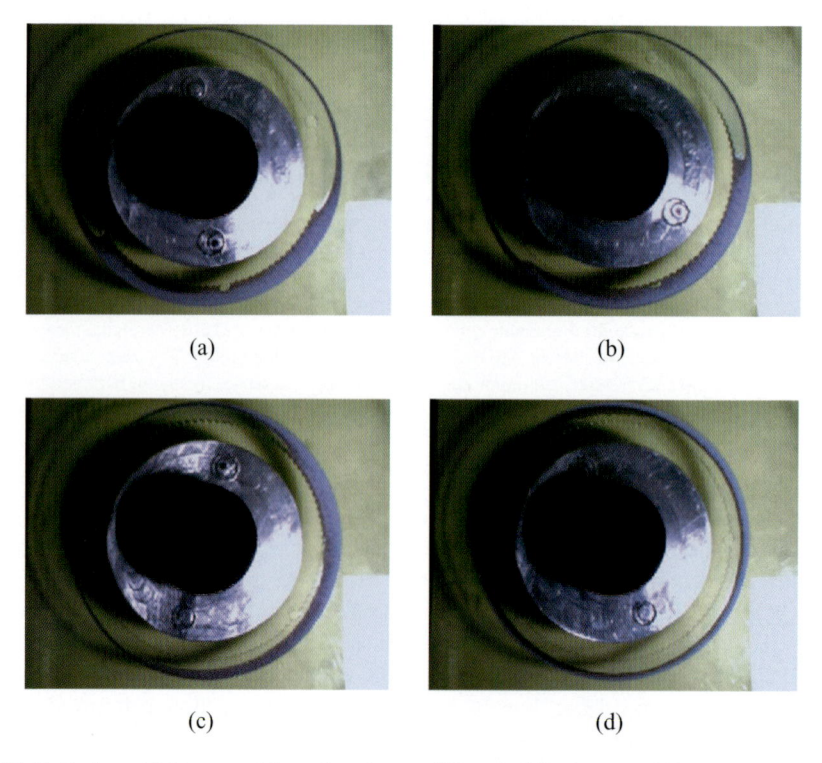

FIGURE 10.16. Sequential images of flow of a mixture of Zirconia-Silica beads and blue-dyed liquid forming into circumferential bands (rimming flow): solid volume fraction of 1/3 and angular velocity of 20 rev/sec. The frames are captured with a continuous pulses generation at a rate of 60 Hz without using the fiber optical sensor.

The protocol we compared our CD based lysis with will now be described. A volume of 0.3 mL of Yeast cells (O.D.$_{600}$ = 2.0) was transferred to a microcentrifuge tube and centrifuged for 1 minute to pellet the cells. The supernatant was discarded and 200μL of lysis buffer (1%SDS, 1mM EDTA, 100mM NaCl, 10mM Tris (pH 8), 2% Triton X-100), phenol-chloroform:isoamyl alcohol and 0.3g of glass beads were added. Using a vortex placed in the refrigerator, we vortexed the microtube for 5 min. After vortexing we added 200μL of water, briefly mixed manually, spun the tube for 10 more minutes and finally transferred the top aqueous layer to a new microcentrifuge tube to be analyzed.

CHO-K1 cells were observed by a microscope and photographed before and after CD spinning to verify the cell disruption. For *E. coli* and yeast, the amount of DNA extracted from lysis was measured by PicoGreen Quantitation Kit (Molecular Probes, Eugene, OR) that exploits an affinity property of PicoGreen reagent to double stranded DNA and its fluorescence intensity dependence on the dsDNA concentration. The samples were excited at 480 nm and the fluorescence emission intensity measured at 520 nm. The DNA concentration of each sample was extrapolated from a standard curve of fluorescence emission intensity versus DNA concentration.

A volume of 0.3 mL Chinese Hamster Ovary (CHO-K1) cells were loaded into a CD lysis chamber along with 0.25 g of acid-washed glass beads (density of 2.5 g/cc). Two sizes of glass beads were used, the first is around 106 μm in diameter and the second ranges from 150 to 212 μm. In the test of the mammalian cells, the two types of beads were mixed in a 7:3 ratio. A mixture of medium and beads approximately filled the lysis chamber halfway with a solid volume fraction of 0.33. The CD was then spun up to 10 rev/sec for a period of 300 cycles. Mammalian cells are, in general, easy to disrupt and thus may not be the optimal choice to test the important parameters for lysis discussed before, so we only checked mammalian cell lysing using simple microscopy. A microscope view before and after spinning the CD in Fig. 10.17 shows that almost all the cells are lysed leaving large amounts of cell debris.

For *E. coli* lysis tests, the same glass beads used in the mammalian test were exploited to study the effect of parameters such as solid volume fractions (SVF) and angular velocity. The two sizes of glass beads were also mixed at the same ratio (i.e., 7:3). The total volume

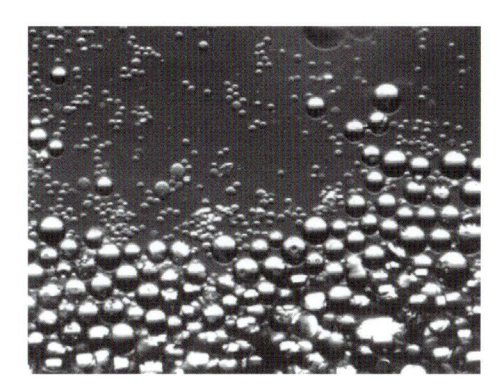

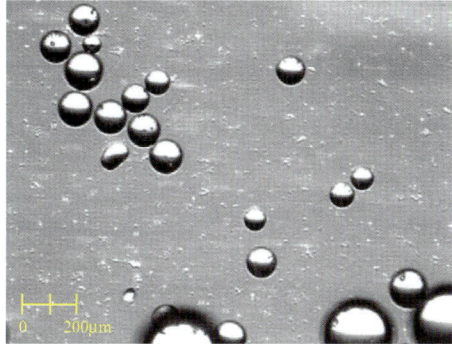

FIGURE 10.17. Comparison of *before (left)* and *after (right)* spinning for mammalian cells. The large spheres are glass beads and the small semicircular spot (~10 μm) on the left are CHO-K1. The image on the right shows cell debris from the lysis of the CHO-K1 cells.

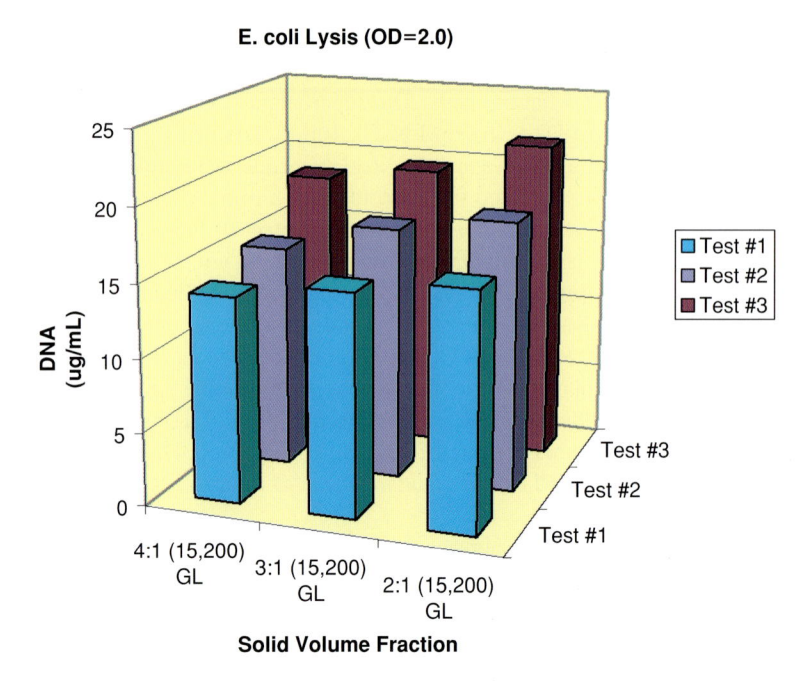

FIGURE 10.18. Plot of dsDNA concentrations of *E. coli* lysis with respect to volumetric ratio of solution to glass beads. The numbers separated by a comma in parenthesis represent angular velocity (rev/sec) and acceleration rate (rev/sec^2). All CDs were spun for a period of 600 cycles. The concentrations for each column were extrapolated from a standard dsDNA calibration curve.

of a mixture was fixed to 0.45 mL for each volumetric ratio of solution to beads of 2:1, 3:1, and 4:1. A CD was ramped up to 15 rev/sec for the intended spin profile and continued for 600 cycles. As shown in Fig. 10.18, a higher solid volume fraction resulted in larger amount of DNA extracted, reaching a maximum at a ratio of 2:1. Above the 2:1 ratio, no more rimming flow was observed, but the bulk of accumulated beads rotated at the same speed of the disc, e.g., a solid body rotation resulted instead. We also tested a control CD containing sample only to determine the contribution of the beads for lysis and verified that very little DNA was extracted from the control CD.

Yeast constitutes the hardest cell to disrupt among our choice of test cells and was tested for all specified control parameters to find the contribution of each parameter. For the solid volume fraction (SVF) test Zirconia-Silica beads (density of 3.7 g/cc) with a particle size ranging from 150 to 250 μm were exploited at an angular velocity of 20 rev/sec for a period of 500 cycles. As shown in Fig 10.19(a), the amount of extracted dsDNA increases with higher solid volume fraction, reaching a peak at 2:1 ratio—the same trend we observed in the *E. coli* tests. Additional yeast tests with the 2:1 ratio were performed while varying other parameters such as acceleration rate, angular velocity, and material density of beads. From these test sets, those control parameters were compared, in terms of DNA concentration extracted from lysis, to find the factors contributing the most to cell lysis efficiency. The use of different beads but of the same size (e.g., glass bead and Zirconia-Silica bead) resulted in the largest difference in DNA concentration as shown in Fig 10.19(b). Therefore the bead

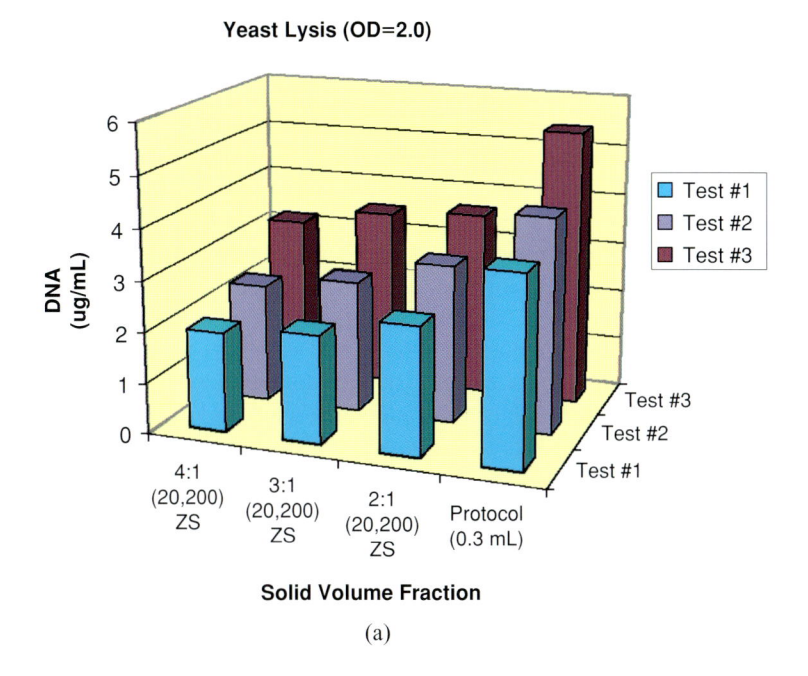

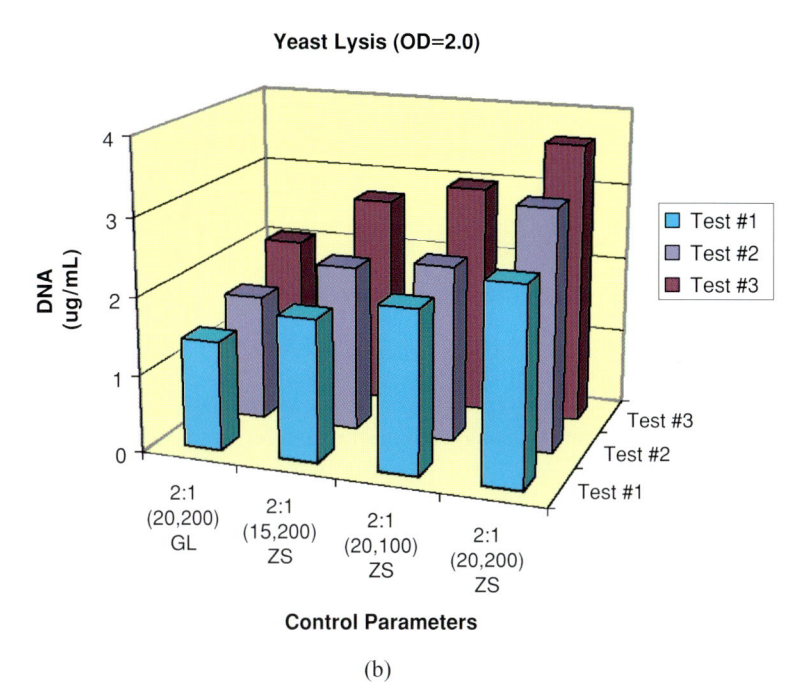

FIGURE 10.19. Plot of dsDNA concentrations of yeast lysis (a) With respect to volumetric ratio of solution to Zirconia-Silica beads and from a common lysis protocol with amount of 0.3 mL (b) For the control parameters (from left): bead material (density), angular velocity, and acceleration rate. GL and ZS stand for glass bead and Zirconia-Silica bead, respectively and are of the same size ranging 150 ∼ 250 μm. The numbers separated by a comma in parenthesis represent angular velocity (rev/sec) and acceleration rate (rev/sec^2). A period of 500 cycles was applied to all test sets for yeast lysis.

density was concluded to contribute the most to the magnitude of interactions in the present model followed by angular velocity as the second most important parameter. In practice, the actual impact velocity is thought to be slower than the terminal settling velocity in Eqn.10.17 due to the fact that a majority of particle impacts may occur within the acceleration period and thus the particles could not fully developed to their terminal settling velocities for a intended angular velocity. A common yeast lysis method (see above) used in biological laboratories was also compared with our model, and its efficiency was found to be a little higher than our optimally conditioned model as shown in Figure 10.19(a).

We have investigated a mechanical method for rapid cell lysis in the rimming flow established inside a partially solid-liquid mixture-filled annulus chamber in a microfluidic CD which rotates around a horizontal axis of rotation. Two types of inter-particle forces are at play i.e., collision and friction (grinding). These forces simultaneously apply to enhance the disruption of biological cells in the rapid granular shear flow.

Our experiments identified the relative importance of control parameters such as bead density, angular frequency, acceleration rate, and solid volume fraction. We concluded that the most important parameter among them was the bead density followed by angular velocity. In addition, the experiments showed the dependence of lysis yield on acceleration rate of a disc, indicating that higher acceleration rate would lead to larger magnitude of frictional contact of particles. For higher particle concentrations, more cells were disrupted due to larger inter-particle force as well as increased number of effective impacts. In practice, a SVF of 0.33 (e.g., 2:1 ratio) was found to be a peak for the maximum lysis efficiency since the rimming flow was hardly observed beyond it. Moreover, compared to a yeast lysis method frequently used in biological laboratory, our novel design demonstrated almost comparable lysis capability without the use of any chemical lysis agent.

We conclude this section with a brief comment that our experiments successfully demonstrated cell lysis on a CD platform through a mechanical method, and this work will be an integral part for our long-term goal—automated CD-based sample preparation platform.

10.5.6. Integrated Nucleic Acid Sample Preparation and PCR Amplification

Nucleic acid analysis is often facilitated by Polymerase Chain Reaction (PCR) and requires substantial sample preparation that, unless automated, is labor extensive. After the initial sample preparation step of cell lyses to release the DNA/RNA, a step must be taken to prevent PCR inhibitors, usually certain proteins such as hemoglobin, from entering into the PCR thermocyle reaction. This can be done by further purification methods such as precipitation and centrifugation, solid phase extraction, or by denaturing the inhibitory proteins. Finally, the sample must be mixed with the PCR reagents followed by thermocyling, a process that presents difficulty in a microfluidic environment because of the relatively high temperatures (up to 95C) required. In a small volume microfluidic reaction chamber, the liquid will be easily evaporated unless care is taken to prevent vapor from escaping.

Kellogg et al [22] combine sample preparation with PCR on the CD. The protocol involves the following steps: (1) mixing raw sample (5 μL of dilute whole bovine blood or *E. coli* suspension) with 5μL of 10mM NaOH; (2) heating to 95C for 1–2 minutes (cell lyses and inhibitory protein denaturization); (3) neutralization of basic lysate by mixing

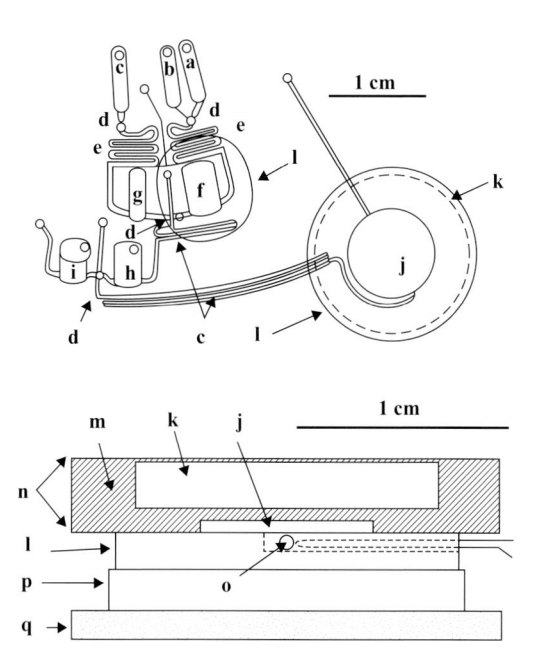

FIGURE 10.20. Schematic illustration of the CD microfluidic PCR structure. The center of the disc is above the figure. The elements are (a) sample, (b) NaOH, (c) tris-HCL, (d) capillary valves, (e) mixing channels, (f) lysis chamber, (g) tris-HCl holding chamber, (h) neutralization lysate holding chamber, (i) PCR reagents, (j) thermal cycling chamber, (k) air gap. Fluids loaded in (a), (b), and (c) are driven at a first RPM into reservoirs (g) and (f), at which time (g) is heated to 95C. The RPM is increased and the fluids are driven into (h). The RPM is increased and fluids in (h) and (i) flow into (j). On the right, the cross-section shows the disc body (m), air gap (k), sealing layers (n), heat sink (l), thermoelectric (p), PC-board (q) and thermistor (o). (Micro Total Analysis Systems 2000 ©, Gregory J. Kellogg, Todd E. Arnold, Bruce L. Carvalho, David C. Duffy, Norman F. Sheppard, "Centrifugal Microfluidics: Applications," (2000) pp. 239-242, with kind permission of Kluwer Academic Publishers)

with 5μL of 16mM tris-HCl (pH = 7.5); (4) neutralized lysate is mixed with 8–10μL of liquid PCR reagents and primers of interest; and (5) thermal cycling. The CD fluidic design is shown schematically in Figure 10.20. Three mixing channels are used in series to mix small volumes. A spinning "platen" allows control of the temperature by positioning thermoelectric devices against the appropriate fluidic chambers. The CD contacts the PC board platen on the spindle of a rotary motor, with the correct angular alignment, which is connected by a slip ring to stationary power supplies and a temperature controller. Thermocouples are used for closed-loop temperature control and air sockets are used as insulators to isolate heating to reservoirs of interest. The thermoelectric at the PCR chamber both heats and cools and since the PCR reaction chamber is thin, 0.5mm, fast thermocyling is achieved. Slew rates of $\pm 2Cs^{-1}$ with fluid volumes of 25μL and thermal gradients across the liquid of 0.5C are reported. It is important to note here that the PCR chambers were not sealed, vapor generated inside the PCR chamber is condensed on the cooler surfaces of the connecting microfluidic chamber and since the CD is rotating, the condensed drops are centrifuged back into the hot PCR chamber. This micro condensation apparatus is unique for the centrifugal CD platform. Details of the experimental parameters used can be found in the original reference [22], but to summarized, sample preparation and PCR amplification for two types of samples, whole blood and *E. coli*, was demonstrated on the CD platform and shown to be comparable to conventional methods.

10.5.7. Sample Preparation for MALDI MS Analysis

MALDI MS peptide mapping is a commonly used method for protein identification. Correct identification and highly sensitive MS analysis require careful sample preparation. Manual sample preparation is quite tedious, time consuming, and can introduce errors common to multi-step pippetting. MALDI MS sample preparation protocols employ a protein digest followed by sample concentration, purification, and recrystallization with minimal loss of protein. Automation of the sample preparation process, without sample loss or contamination, has been enabled on the CD platform by the Gyrolab MALDI SP1 CD and the Gyrolab Workstation (Gyros AB, Sweden) [20].

The Gyrolab MALDI SP1 sample preparation CD will process up to 96 samples simultaneously using separate microfluidic structures. Protein digest from gels or solutions are concentrated, desalted, and eluted with matrix onto a MALDI target area. The CD is then transferred to a MALDI instrument for analysis without the need for further transfer to a separate target plate. The CD fluidic structure contains functions for common reagent distribution, volume definition (metering), valving, reverse phase column (RPC) for concentration and desalting, washing, and target areas for external calibrants. Figure 10.21

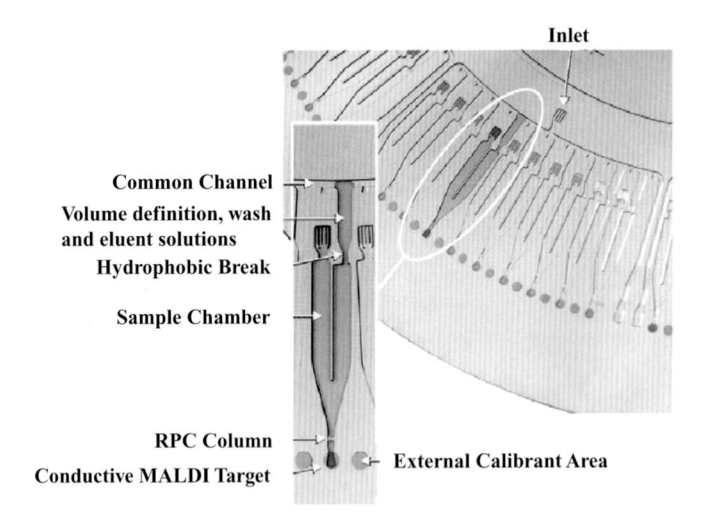

FIGURE 10.21. Image of Gyrolab MALDI SP1 sample preparation CD. The protein digest samples are loaded into the sample reservoir (see inset) by capillary action. Upon rotation, the sample passes through the RPC column. The peptides are bound to the column and the liquid goes out of the system into the waste. A wash buffer is loaded into the common distribution channel and volume definition chamber. The disc is rotated at a PRM that will empty the common distribution channel but not allow the wash solution to pass through the hydrophobic zone. A further increase of the RPM allows the well defined volume of wash solution to pass the hydrophobic break and wash the RPC column then be discarded as waste. Next, a well defined volume of the elution/matrix solution is loaded and passed through the column, taking the peptides to the MALDI target zone. The flow rate is controlled to optimize the evaporation of the solvent crystallization of the protein and matrix at the target zone. (With kind permission; Application Report 101, Gyrolab MALDA SP1, Gyros AB, Uppsala, Sweden)

shows the Gyrolab MALDI SP1 sample preparation CD. The CDs are loaded with reagents and processed in a completely automated, custom workstation capable of holding up to 5 microtiter plates containing samples and reagents and up to 5 CD micro-laboratories. The reagents are taken from the microplates to the CD inlets using a precision robotic arm fitted with multiple needles, the liquid is drawn into specific inlets by capillary forces, and then the needles are cleaned by rinsing at a wash station. Samples are applied in aliquots from 200nl up to 5µl sequentially to each channel where it is contained using hydrophobic surface valves. The CD is then rotated, at an optimized rate, causing the sample to flow through an imbedded reverse phase chromatography column and liquid that passes through the column is collected in a waste container. Controlling the angular velocity dependent liquid flow rate maximizes protein binding to the column. A wash solution is introduced, by capillary action, into common distribution channels connected to groups of microstructures. The wash solution fills a volume definition chamber (200nl) until it reaches a hydrophobic valve and the CD is rotated to clear the excess liquid in the distribution channel. Not until the rotational velocity is further increased is the defined wash volume able to pass through the hydrophobic valve and into the RPC column (SOURCE^{TM} 15 RPC). The peptides are eluted from the column and directly on to the MALDI target area using a solution that contains α-cyano-4-hydroxycinnamic acid and acetonitrile using the same common distribution channel and defined volume as the previous wash step. Optimization of rotational velocity during elution enables maximum recovery and balances the rate of elution with the rate of solvent evaporation from the target surface. Areas in and around the targets are gold plated to prevent charging of the surface that would cause spectral mass shift and ensures uniform field strength. Well-defined matrix/peptide crystals form in the CD MALDI target area. Gyros reports high reproducibility, high sensitivity, and improved performances when compared to conventional pipette tip technologies. Data was shown that includes: comparison of 23 identical samples, processed in parallel on the same CD, from a BSA tryptic digest and analysis of identical samples processed on different CDs, run on different days. Sensitivities were shown in the attomole to femtomole range indicating the ability for identification of low abundance proteins. The report attributed the superior performance of this platform to pretreatment of the CD surface to minimized non-specific adsorption of peptides, reproducible wash volume and flow, and reproducible elution (volume, flow, and evaporation) and crystallization.

10.5.8. Modified Commercial CD/DVD Drives in Analytical Measurements

The commercial CD/DVD drive, commonly used for data storage and retrieval, can be thought of as a laser scanning imager. The CD drive retrieves optically generated electrical signals from the reflection of a highly focused laser light (spot size: full width at half maximum ~1µm), from a 1.2 mm thick polycarbonate disc that contains a spiral optical track feature. The track is fabricated by injection molding and is composed of a series of pits that are 1–4 µm long, 0.15 µm deep, and about 0.5 µm wide. The upper surface of a CD is made reflective by gold or aluminum metallization and protected with a thin plastic coating. Information is generated, as the focused laser follows the spiral track, by converting the reflected light signal into digital information. A flat surface gives a value of zero, an edge of a pit gives a value of one. The data is retrieved at a constant acquisition rate and the serial values (0/1) are converted to data of different kinds for various applications (music, data, etc...). In addition to the code generated by the spacing of the pits, optical signals necessary for focusing, laser tracking of the spiral track, and radial position determination of the read head are monitored and used in feedback loops for proper CD operation. The laser is scanned in a radial direction toward the outer diameter of the disc with an elaborate servo that maintains both lateral tracking and vertical focusing.

Researchers [47, 48] have taken advantage of this low-cost, high-resolution optical platform in analytical DNA array applications. Barathur et al [47] from Burstein Technologies (Irvine, CA), for example, have modified the normal CD drive for use as a sophisticated laser-scanning microscope for analysis of a Bio Compact Disk^{TM} assay, where all analysis is carried out in microfluidic chambers on the CD. The assay is carried out concurrently with the normal optical scanning capabilities of a regular CD drive. The authors report on the application of this device for DNA micro spot-array hybridization assays and comment on its use in other diagnostic and clinical research applications. For the DNA spot-array application, arrays of capture probes for specific DNA sequences are immobilized on the surface of the CD in microfluidic chambers. Sample preparation and multiplexed PCR, using biotinylated primers, are carried out off-disc, then the biotinylated amplicons are introduced into the array chamber and hybridization occurs if amplicons with the correct sequence are present. Hybridization detection is achieved by monitoring the optical signal from the CD photo detector, while the CD is rotating. To generate an optical signal when hybridization has occurred a reporter is used, for the Bio Compact Disk^{TM} assay the reporter is a streptavidin labeled microsphere that will bind only to the array spots which have successfully captured biotinylated amplicons. The unbound microsphere reporters are removed from the array using simple centrifugation and no further rinsing is needed. As the laser is scanned across the CD surface, the microparticle scatters light that would have normally been reflected to the photodetector resulting in less light on the detector (bright field microscopy) and a distinctive electronic signal is generated. The electronic signal intensity data can be stored in memory then de-convoluted into an image. A 1 cm^2 micro-array can be scanned in 20–30 seconds with a data reduction time of 5 minutes and custom algorithms that perform the interpretations in real time. Data was shown for identification of three different species of the *Brucella* coccobacilli on the CD platform. Human infection occurs by transmission from animals by ingestion of infected food products, contact with an infected animal, or inhalation of aerosols. Multiplex PCR-amplified DNA from all three species (common forward primers and specific reverse primers resulting in amplicons of different length for various species, were used for verification of PCR on external gels) were incubated on arrays with species-specific capture probes. Removal of one of the species in the sample resulted in no probes present on that specific array spot verifying the specificity of the assay.

Alexandre et al [48] at Advanced Array Technology (Namur, Belgium), utilize the inner diameter area of a CD and standard servo optics for numerical information and operational control and employ a second scanning laser system to image DNA arrays on transparent surfaces at the outer perimeter of a CD. The second laser system, consisting of a laser diode module that illuminates a 50 micron spot on the CD surface, is scanned radially at a constant linear velocity of 20mm/min while the CD is rotating. Each CD contains 15 arrays arranged in a single ring on the CD perimeter that extends in the radial direction for 15mm. The arrays are rectangular and consist of 4 rows and 11 columns of 300μm spots. The normal CD servo optics are located below the disc and the added imaging optics are above the disc. A photodiode head follows the imaging laser and the refracted light intensity is stored digitally at a high sampling rate. An image of each array on the disc is reconstructed by deconvolution of the light intensity data. The entire CD can be scanned in less than one minute producing a total of 6 MB of information. Sample preparation and PCR amplification was carried out off-disc. Specific DNA capture probes were spotted on the surface of the CD using a custom arrayer that transfers the probes from a multi-well plate on to the surface of up to 12 discs using a robotic arm. Biotinylated amplicons are introduced on to the array chambers (one chamber for each array) and hybridization occurs if amplicons with the correct sequence are present. In order to get an optical signal that can be detected, after a rinse step, a solution of streptavidin labeled colloidal gold particles is applied to the array followed by a Silver Blue solution (AAT, Namur, Belgium). The silver solution causes silver metal to grow on the gold particles thereby making the hybridization-positive micro array spots refractive to the incident laser light. Results were shown for the detection of the 5 most common species of *Staphylococci* and an antibiotic resistant strain. The *fem A* and *mec A* genes of the various species of *Staphylococci* were amplified by primers common to all *Staphylococci* species then hybridized to a micro array containing spots with probes specific for the different *Staphylococci* species. The array also included a capture probe for the genus *Staphylococci* and a probe for the *mec A* gene that is associated with methicillin resistance of the *Staphylococci* species. The results were digitized and quantified with software that is part of the custom Bio-CD™ workstation. Signal to noise ratios were above 50 for all positive signals.

CONCLUSION

In comparing miniaturized centrifugal fluidic platforms to other available micro fluid propulsion methods we have demonstrated how CD based centrifugal methods are advantageous in many analytical situations because of their versatility in handling a wide variety of sample types, ability to gate the flow of liquids (valving), simple rotational motor requirements, ease and economic fabrication methods, and large range of flow rates attainable. Most analytical functions required for a lab-on-a-disc, including; metering, dilution, mixing, calibration, separation, etc. have all been successfully demonstrated in the laboratory. Moreover, the possibility of maintaining simultaneous and identical flow rates, to perform identical volume additions, to establish identical incubation times, mixing dynamics, and detection in a multitude of parallel CD assay elements makes the CD an attractive platform for multiple parallel assays. The platform has been commercialized by Tecan Boston for high throughput screening (HTS) [4], by Gyros AB for sample preparation techniques for MALDI [20] and by Abaxis (in a somewhat larger and less integrated rotor format compared to the CD format) for human and veterinary diagnostic blood analysis [5]. The Abaxis system for human and veterinary medicine uses only dry reagents, but for many diagnostic assays, requiring more fluidic steps, there are severe limitations in progressing toward the "lab-on-a-disc" goal as liquid storage on the disc becomes necessary. In high throughput screening (HTS) situations, the CD platform is being coupled to automated liquid reagent loading systems and no liquids/reagents need to be stored on the disc. The latter made the commercial introduction of the CD platform for HTS somewhat simpler [4, 20, 48]. There is an urgent need though for the development of methods for long term reagent storage that incorporate both liquid and vapor barriers to enable the introduction of lab-on-a-disc platforms for a wide variety of fast diagnostic tests. One possible solution to this problem involves the use of lyophilized reagents with common hydration reservoir feeds, but the issue in this situation becomes the speed of the test as the time required for re-dissolving the lyophilized reagents is often substantial.

The CD platform is easily adapted to optical detection methods because it is manufactured with high optical quality plastics enabling absorption, fluorescence, and microscopy techniques. Additionally, the technology developed by the optical disc industry is being used to image the CD at the micron resolution and move to DVD and HD DVD will allow sub micron resolution. The latter evolution will continue to open up new applications for the CD based fluid platform. Whereas today the CD fluidic platform may be considered a smart micro-centrifuge, we believe that in the future the integration of fluidics and informatics on the DVD and high definition (HD) DVD may lead to a merging of informatics and fluidics on the same disc. One can then envision making very sharp images of the bacteria under test and correlate both test and images with library data on the disc.

ACKNOWLEDGEMENT

The authors thank Sue Cresswell, Gyros AB; Gregory J. Kellogg, Tecan Boston; Horacio Kido and Raj Barathur, Burstein Technologies; Jim Lee, Ohio State University, and Michael Flynn, NASA Ames.

REFERENCES

[1] For commercial electronic noses and tongues see for example ALPHA-MOS: http://www.alpha-mos.com/newframe.htm
[2] E. Verpoorte, Manz, C.S. Effenhauser, N. Burggraf, D.E. Raymond, D.J. Harrison, and H.M. Widmer. Miniaturization of separation techniques using planar chip technology, *HRC High Resol. Chromatogra.*, 16:433–36, 1993.
[3] Caliper Home Page: http://www.calipertech.com/
[4] Tecan-Boston Home Page: Look for LabCD-ADMET system: http://www.tecan-us.com/us-index.htm.
[5] Abaxis Home Page: http://www.abaxis.com/
[6] Marc J. Madou. *Fundamentals of Microfabrication*, 2nd Ed., CRC Press, Boca Raton, London, New York, Washington D.C., 2002.
[7] S. Miyazaki, T. Kawai, and M. Araragi. A Piezo-Electric Pump Driven by a Flexural Progressive Wave. In *Proceedings: IEEE Micro Electro Mechanical Systems (MEMS '91)*, Nara, Japan, pp. 283–88, 1991.
[8] J.W. Jorgenson and E.J. Guthrie. Liquid chromatography in open-tubular columns, *J. Chromatogr.*, 255:335–48, 1983.
[9] D.J. Harrison, Z. Fan, K. Fluri, and K. Seiler. Integrated Electrophoresis Systems for Biochemical Analyses. In *Technical Digest: 1994 Solid State Sensor and Actuator Workshop*, Hilton Head Island, S.C., pp. 21–24, 1994.
[10] D.C. Duffy, H.L. Gills, J. Lin, N.F. Sheppard, and G.J. Kellogg. Microfabicated centrifugal microfluidic systems: characterization and multiple enzymatic assays, *Anal. Chemi.*, 71(20):4669–4678, 1999.
[11] M.J. Madou and G.J. Kellogg. The LabCDTM: a centrifuge-based microfluidic platform for diagnostics. In G.E. Cohn and A. Katzir (eds.), *Systems and Technologies for Clinical Diagnostics and Drug Discovery*, Vol. 3259, San Jose, Calif.: SPIE, pp. 80–93, 1998.
[12] G.T.A. Kovacs. *Micromachined Transducers Sourcebook*, WCB/McGraw-Hill, Boston, chapter 9, pp. 787–793, 1998.

[13] Gunnar Ekstrand, Claes Holmquist, Anna Edman Örlefors, Bo Hellman, Anders Larsson, and Per Anderson, Microfluidics in a rotating CD. In A. van den Berg, W. Olthuis, and P. Bergveld (eds.), *Micro Total Analysis Systems 2000*, Kluwer Academic Publishers, pp. 311–314, 2000.

[14] Anna-Lisa Tiensuu, Ove Öhman, Lars Lundbladh, and Olle Larsson, Hydrophobic valves by ink-jet printing on plastic CDs with integrated microfluidics, In A. van den Berg, W. Olthuis, P. Bergveld (eds.), *Micro Total Analysis Systems 2000*, Kluwer Academic Publishers, pp. 575–578, 2000.

[15] Marc J. Madou, Yumin Lu, Siyi Lai, Jim Lee, and Sylvia Daunert. A centrifugal microfluidic platform—a comparison, In A. van den Berg, W. Olthuis, and P. Bergveld. *Micro Total Analysis Systems 2000*, Kluwer Academic Publishers, pp. 565–570, 2000.

[16] Jun Zeng, Deb Banerjee, Manish Deshpande, John R. Gilbert, David C. Duffy, and Gregory J. Kellogg. Design analysis of capillary burst valves in centrifugal microfluidics. In A. van den Berg, W. Olthuis, P. Bergveld (eds.), *Micro Total Analysis Systems 2000*, Kluwer Academic Publishers, pp. 579–582, 2000.

[17] I.H.A. Badr, R.D. Johnson, M.J. Madou, and L.G. Bachas. Fluorescent ion-selective optode membranes incorporated onto a centrifugal microfluidics platform, *Analytical Chemistry*, 74(21):5569–5575, Nov 2002.

[18] R.D. Johnson, I.H.A. Badr, Gary Barrett, Siyi Lai, Yumin Lu, Marc J. Madou, and Leonidas G. Bachas. Development of a fully integrated analysis system for ions based on ion-selective optodes and centrifugal microfluidics. *Anal. Chem.*, 73(16):3940–3946, Aug 2001.

[19] M. McNeely, M. Spute, N. Tusneem, and A. Oliphant. Hydrophobic microfluidics. *Proceedings Microfluidic Devices and Systems, SPIE*, Vol. 3877, pp. 210–220, 1999.

[20] Application Report 101, Gyrolab MALDI SP1, Gyros AB, Uppsala, Sweden.

[21] S. Lai, S. Wang, J. Luo, J. Lee, S. Yang, and M.J. Madou. *Compact Disc (CD) Platform for Enzyme-Linked Immunosorbant Assays* Manuscript submitted to Analytical Chemistry.

[22] Gregory J. Kellogg, Todd E. Arnold, Bruce L. Carvalho, David C. Duffy, and Norman F. Sheppard, Centrifugal microfluidics: applications. In A. van den Berg, W. Olthuis, and P. Bergveld (eds.), *Micro Total Analysis Systems 2000*, Kluwer Academic Publishers, pp. 239–242, 2000.

[23] Nick Thomas, Anette Ocklind, Ingrid Bliksstad, Suzanne Griffiths, Michael Kenrick, Helene Derand, Gunnar Ekstrand, Christel Ellström, Anders Larsson, and Per Anderson Integrated cell based assays in microfabricated disposable CD devices. In A. van den Berg, W. Olthuis, and P. Bergveld (eds.), *Micro Total Analysis Systems 2000*, Kluwer Academic Publishers, pp. 249–252, 2000.

[24] A.W. Anderson. *Physical Chemistry of Surfaces*, John Wiley & Sons, New York, London, Sidney, Chapter1, pp. 5–6, 1960.

[25] Private communication with Gregory J. Kellogg of Tecan Boston.

[26] D.C. Duffy, J.C. McDonald, O.J.A. Schueller, and G.M. Whitesides. Rapid prototyping of microfluidic systems in Poly(dimethylsiloxane), *Anal. Chem.*, 70:4974–4984, 1998.

[27] J. Burbaum, Minitaturization technologies in HTS: how fast, how small, how soon?, *Drug Discov Today*, 3(7):313–312, 1998.

[28] Jim V. Zoval, Richard Boulanger, Charles Blackwell, Bruce Borchers, Michael Flynn, David Smernoff, Ragnhild Landheim, Rocco Mancinelli Marc J. Madou. *Cell Viability Assay on a Rotating Disc Analytical System*, Manuscript of paper in preparation.

[29] To be submitted for publication.

[30] M.G. Pollack, A.D. Shenderov, and R.B. Fair, Electrowetting-based actuation of droplets for integrated microfluidics, *Lab on a chip*, 2:96–101, 2002.

[31] M.J. Madou, L. James Lee, S. Daunert, S. Lai, and C.-H. Shih. Design and fabrication of CD-like microfluidic platforms for diagnostics: microfluidic functions. *Biomed. Microdevi.*, 3(3):245–254, 2001.

[32] T. Brenner, T. Glatzel, R. Zengerle, and J. Ducree, A Flow Switch Based on Coriolis Force, *7th International Conference on Miniaturized Chemical Biochemical Analysis Systems*.

[33] D.D. Carlo, K-H Jeong, and L. P. Lee, Reagentless mechanical cell lysis by nanoscale barbs in microchannels for sample preparation, *Lab on a Chip*, 3:287–291, 2003.

[34] S.W. Lee, H. Yowanto, and Y.C. Tai. A Micro Cell Lysis Device, *The 11th Annual International Workshop on Micro Electro Mechanical Systems, MEMS '98*.

[35] J. Sambrook and D. W. Russell. *Molecular Cloning*, Cold Spring Harbor Laboratory press, Cold spring harbor, New York, 2001.

[36] Amy Q. Shen, Granular fingering patterns in horizontal rotating cylinders. *Physics of Fluids*, 14(2)462–470, 2002.

[37] L. Bocquet, W. Losert, D, Schalk, T.C. Lubensky, and J.P. Gollub. Granular shear flow dynamics and forces: experiment and continuum theory, *Physical Review E* (65), 011307–1.
[38] K.J. Ruschak and L.E. Scriven, Rimming flow of liquid in a rotating horizontal cylinder, *Fluid Mech.*, 76:113–127, 1976.
[39] S.T. Thoroddsen and L. Mahadevan. Experimental Study of coating flows in a partially-filled horizontally rotating cylinder, *Exper. Fluids*, 23:1–13, 1997.
[40] R.A. Bagnold. Experiments on a gravity-free dispersion of large solid spheres in a Newtonian fluid under shear," *Proceedings of the Royal Society of London. Series A, Mathematical and Physical Sciences*, 225(1160):49–63, 1954.
[41] Cliff K. K. Lun, Granular dynamics of inelastic spheres in Couette flow, *Phys. Fluids*, 8(11), 1996.
[42] S.F. Foerster, M.Y. Louge, H. Chang, and K. Allia. Measurements of the collision properties of small spheres, *Phy. Fluids* 6(3), 1994.
[43] G.K. Batchelor. A new theory of the instability of a uniform fluidized bed, *J. Fluid Mech.*, 193:75–110, 1988.
[44] R. Zenit, M.L. Hunt, and C.E. Brennen. "Collisional particle pressure measurements in solid-liquid flows, *Fluid Mech.* 353:261–283, 1997.
[45] S.A. Morsi and A.J. Alexander. An investigation of particle trajectories in two-phase flow systems, *Fluid Mech.*, 55(2):193–208, 1972.
[46] J. Kirchner J, S.B. Sandmeyer, and D.B. Forrest. Transposition of a Ty3 GAG3-POL3 fusion mutant is limited by availability of capsid protein, *Virol.*, 66(10):6081–92, 1992.
[47] R. Barathur, J. Bookout, S. Sreevatsan, J. Gordon, M. Werner, G. Thor, and M. Worthington. New disc-based technologies for diagnostic and research applications, *Psychiat. Geneti.*, (12)4:193–206, 2002.
[48] I. Alexandre, Y.Houbion, J. Collet, S. Hamels, J. Demarteau, J.-L. Gala, and J. Remacle. Compact disc with both numeric and genomic information as DNA microarray platform, *BioTechniques*, (33)2:435, 2002.

11

Sequencing the Human Genome: A Historical Perspective on Challenges for Systems Integration

Lee Rowen
Multimegabase Sequencing Center, Institute for Systems Biology, 1441 N. 34th Street, Seattle WA 98103

11.1. OVERVIEW

The sequence of the human genome was declared finished on April 14, 2003[1]. Analyses have been published in the journal *Nature* for chromosomes 6, 7, 14, 20, 21, 22 and Y, with the other chromosomes to follow in 2004. Although the Human Genome Project officially began in 1990, most of the publicly accessible sequence data were produced by 20 genome centers in six countries between 1999 and 2002. This group of centers, called the International Human Genome Sequencing Consortium, coordinated their mapping and sequencing efforts and freely shared materials, data and procedures [23]. The International Consortium in turn was supported by a network of funding agency program directors, database managers, resource providers, instrumentation/protocol developers, and conference organizers. In all, several thousand people made sure that the human genome got sequenced, and the world is rightly celebrating their accomplishment.

This being said, it is unlikely that anyone would claim that the effort to obtain a complete sequence of the human genome was efficient. Genomes are being sequenced now at significantly lower cost and higher efficiency thanks to strategies that matured during the course of the human genome project. Efficiency is possible when the requirements for coordination of personnel are low; techniques are robust, automated, and scaleable;

---

[1] See "International consortium completes human genome project" http://www.genome.gov/11006929.

and the integration of cost-effective procedures into a high-throughput and streamlined system has already been achieved. When the project began in 1990, however, the most cost-effective paths to a finished sequence of the human genome were unclear and remained to be determined.

The following historical perspective on how the human genome got sequenced is "internal," meaning that it is written from a participant's point of view. As the coordinator of a mid-sized genome center, the author personally experienced many of the developments that occurred during the course of the human genome project from 1990 to 2003. Like any good story now passing into legend, there are different ways to tell the tale, and there may not be universal agreement on what really happened in terms of facts and rationales for decisions. As for heroes and villains, it is this author's charitable belief that the many characters engaged in the sequencing of the human genome were doing the best they could as they navigated the technological and political twists and turns of the undertaking. The story's ultimate moral is that the genome project embodies a triumph of the human spirit along with a testimony to technological ingenuity and persistence.

In order to provide background for a subsequent discussion of the issues faced by the International Consortium, this review begins with a generic description of the approach used to sequence the human genome. A discussion of specific challenges for systems integration follows, using examples from various phases of the effort. Finally, a brief retrospective consideration of lessons learned that might be applicable to other large-scale technology development endeavors is offered.

11.2. APPROACHES USED TO SEQUENCE THE HUMAN GENOME

11.2.1. Overview

Looking at the big picture, the overarching design for sequencing the human genome entailed dividing individual donors' genomic DNA (genomes) into manageable pieces (cloned genomic inserts; source clones), determining the sequences of the pieces (sequence reads; contigs; source clone sequences), and then reconstructing the sequence of an entire representative human genome from the sequences of overlapping pieces (overlapping source clone sequences), creating at the end one master sequence for each of the 24 chromosomes (Figure 11.1). The International Consortium used a hierarchical sequencing strategy whereby the genome was fragmented into source clones around 150 kilobases (kb) in size which were sequenced using a set of procedures to be described below [23]. The privately funded genome project led by Celera Genomics fragmented the genome into smaller 2 kb, 10 kb, and 50 kb cloned inserts and the overall sequence was assembled from sequence reads derived from the two ends of the cloned inserts using a "whole genome shotgun" approach [48]. For reasons of space, this review primarily covers the strategies used by the publicly funded human genome project.

In the hierarchical approach, the acquisition of sequence data encompassed four major processes:

- mapping–determining the chromosomal location of source clones;
- sequencing–obtaining raw sequence data for source clones;

SEQUENCING THE HUMAN GENOME

A.

B.

genomic DNA

DNA fragmentation and subcloning into suitable source clone vector

source clone DNA

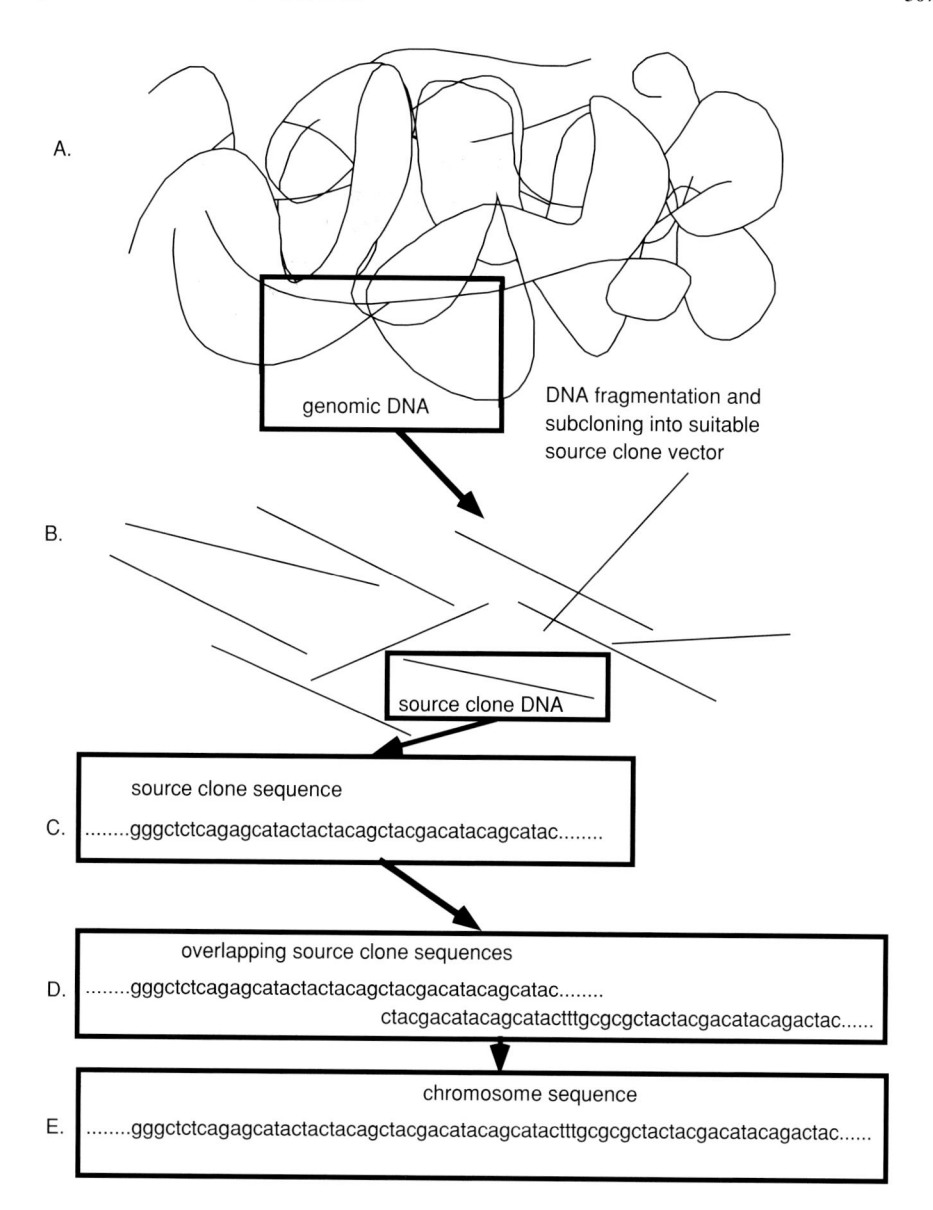

C. source clone sequence
........gggctctcagagcatactactacagctacgacatacagcatac........

D. overlapping source clone sequences
........gggctctcagagcatactactacagctacgacatacagcatac........
          ctacgacatacagcatactttgcgcgctactacgacatacagactac......

E. chromosome sequence
........gggctctcagagcatactactacagctacgacatacagcatactttgcgcgctactacgacatacagactac......

FIGURE 11.1. Strategy for sequencing the human genome. A. Total genomic DNA is obtained from the sperm or cells of an individual, fragmented, subcloned into a suitable cloning vector, and stored as libraries of random source clones. B. A source clone is chosen for sequencing and, C, the sequence of a source clone is obtained. D. Overlapping source clone sequences are identified and, E, merged to create a chromosomal sequence.

- assembly–reconstructing sequences of source clones from raw sequence data and the sequences of chromosomes from source clone sequences;
- finishing–filling gaps, resolving assembly errors, obtaining high quality contiguous sequence over long stretches of genomic terrain.

While the substrate for mapping and sequencing was a source clone (to be described further below), assembly and finishing pertained both to source clones, that is, to small ~150 kb pieces of the genome, and to chromosomes, whose sequences were constructed by the merging of overlapping source clone sequences (Figure 11.1).

Systems integration pertains to the overall coordination of the mapping, sequencing, assembling and finishing of the genome, to the organization of the component steps for each of these processes, and to the various quality controls, validation procedures, and feedback mechanisms that ensured accuracy in the final product. In the initial stages of the project (1990–1996), systems integration focussed primarily on procedures for sequencing source clones, with an effort to maximize efficiency and throughput. Here, the "pipeline" analogy prevailed. The idea is that fluid flows smoothly from point A to point B when things are working properly and leaks, blockages, backflows or diversions disrupt the flow and, therefore, must be anticipated, attended to or prevented. In the later stages (2000–2003), the focus shifted to finishing a representative sequence for each human chromosome. The systems integration analogy at work in building chromosome sequences out of source clone sequences was more like conducting a symphony orchestra. Immense coordination and cooperation among the sequence centers was required to turn cacophony into glorious music.

11.2.2. Strategy Used for Sequencing Source Clones

Figure 11.2 summarizes the four major processes through which samples pertaining to source clones "flowed" in the genome sequencing "pipeline." The samples are of three types: biological/chemical materials (clones, DNA templates, sequencing reactions), images (fingerprint patterns, dye peaks) and words (strings of A,G,C,T representing DNA sequences). These three data types challenged prevailing procedures for acquiring, storing, accessing, analyzing and sharing data throughout the course of the human genome project. Because of the scale involved, integrating the various processes of data acquisition put significant pressure on the laboratory information management systems in place at each genome center.

11.2.2.1. Mapping (Source Clone Acquisition) In terms of source material for sequencing, the starting point for the publicly funded genome centers was not the genome-as-a-whole, as with the Celera effort, but rather pieces of representative genomes embedded in large-insert cloning vectors that enable propagation of the cloned insert in the bacterium *Escherichia coli*. Use of cloning vectors was required for the physical isolation of pieces of chromosomal DNA in a quantity sufficient for sequencing. Prior to 1995, cosmid vectors, which hold an average insert size of about 40 kilobases (kb), were typically used. After 1995, cosmids were replaced by PACs (P1 artificial chromosomes), which hold inserts around 80 kb [24], and then BACs (bacterial artificial chromosomes), which hold inserts up to about 250 kb [36, 43]. Given a human genome size of 3 gigabases, it would require 20,000 BAC clones with an average insert size of 150 kb to cover the entire genome once if they were laid end-to-end. Since genomic inserts are generated by semi-random processes, ensuring adequate clone coverage of the entire genome required the construction of clone libraries containing several "genome equivalents" of DNA inserts [8]. These clone libraries consisted of several hundred 384-well plates of frozen bacterial cultures from which any

Genome Sequencing Pipeline

Mapping
- Genetic marker identification
- Source clone library screening
- Fingerprinting and/or End sequencing
- Fluorescence in situ hybridization (used for validation)

↓ Sequence-ready large-insert source clone

Sequencing
- DNA extraction from source clone
- Shotgun library construction
- DNA template preparation
- Sequencing reactions
- Machine loading and detection
- Base-calling

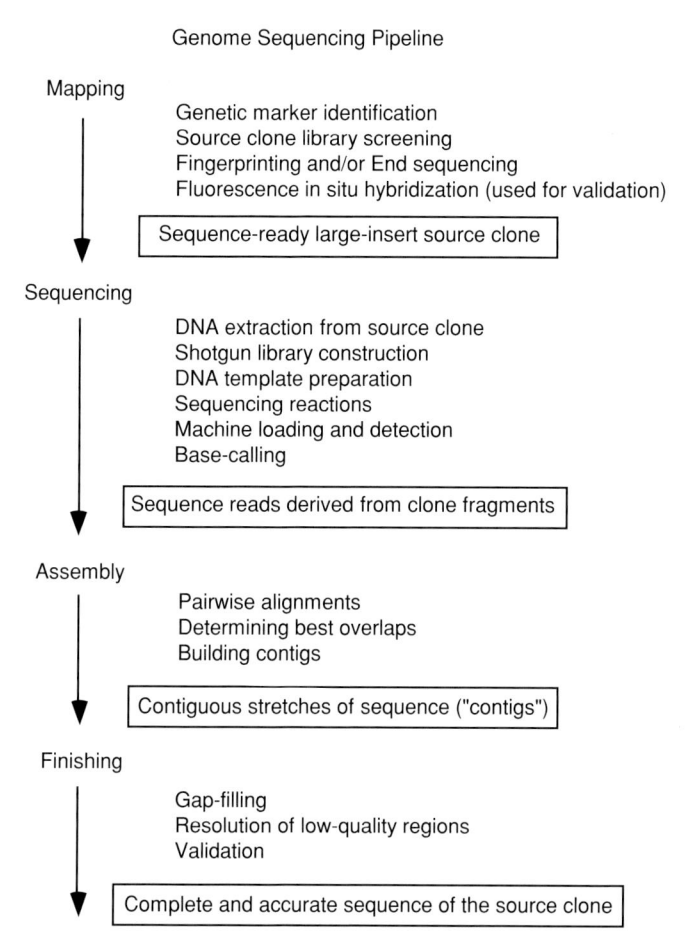

↓ Sequence reads derived from clone fragments

Assembly
- Pairwise alignments
- Determining best overlaps
- Building contigs

↓ Contiguous stretches of sequence ("contigs")

Finishing
- Gap-filling
- Resolution of low-quality regions
- Validation

↓ Complete and accurate sequence of the source clone

FIGURE 11.2. Processing of samples through the genome sequencing pipeline. Each box represents the output of an overall process (arrow) that entailed several steps (see text).

individual large-insert clone could be propagated, once its clone ID and plate address were determined.

Retrieving sequence-ready clones from a clone library was easy: one could buy a copy of the library for several thousand dollars or order individual clones from a distributor. Figuring out which clones to retrieve, however, was non-trivial. During the mid-90s, genome centers claimed entire chromosomes or portions of chromosomes for their sequencing targets. Regional sequencing required centers to identify large-insert clones containing genomic DNA from their chosen territory and not some other center's real estate: thus a need for physical mapping. In physical mapping, information is obtained from the DNA insert of a source clone that allows inferences to be made about chromosomal location.

The major strategies employed for mapping source clones prior to sequencing include:

- Library screening: Genetic markers, which are short stretches of sequence known to map to a specific chromosomal position, are used to make probes for finding matching sequences among the human genomic inserts in the clone library. This is done by hybridizing the probe to filters onto which a tiny amount of DNA from each of the clones in the library has been spotted in a known location. Matches between the probe and the filter are detected as small black circles on a film. From the position of the "positive" clones, the library addresses are "read" from the film using a location schema provided by the library manufacturer. Clone candidates are then retrieved from the library and retested with the probe to ensure that the genetic marker of interest is present. The overall screening process starting from probe design and ending with clone validation generally required at least a couple of weeks. Screens could be multiplexed (i.e., several probes combined into one hybridization) and the positive clones sorted into clusters at the validation stage.

- Restriction digest fingerprinting: Restriction enzymes (enzymes that cut DNA whenever they encounter a specific short recognition sequence such as AAGCTT) are used to generate "fingerprints." Fingerprints are distinctive patterns of DNA fragment sizes that reflect the frequency of the enzyme's recognition sequence in the region of interest. After restriction enzyme digestion of the DNA of a source clone, the fragments are separated according to size using agarose gel electrophoresis. Overlap between clones is inferred from a subset of shared fragment sizes. While this approach in and of itself does not point to chromosome location, it is used to determine the order of clones in a cluster—if any one clone in an ordered cluster is positioned on a chromosome, then the chromosomal position of the whole cluster is known.
- Clone insert end sequencing: Obtaining short sequence reads (∼500 bases) from each of the two vector-insert joints of a clone, and looking for sequence matches with already-sequenced clones or known markers. End sequence matches allow more precise positioning of overlapping clones, so long as one of the clones has already been sequenced.
- Fluorescence *in situ* hybridization (FISH): Labeling a large-insert clone with dye and hybridizing it to a metaphase chromosome spread and seeing which chromosomal band lights up under a microscope. This method was used primarily for validating a chromosomal location inferred from other mapping procedures [2]. The procedure is slow and requires a skilled technician to interpret the results.

Most genome centers or their mapping collaborators initially employed some variation of the following mapping strategy (Figure 11.3). After performing a round of library screening and identifying a cluster of clones containing a genetic marker of interest, one of the clones, called a "seed," would be sequenced. From unique sequence at the ends of the seed clone insert, new probes for screening were designed for the purpose of procuring a new batch of clones that would overlap the seed clone. Because the average distance between mapped genetic markers was greater than the average length of the clone insert, multiple probes were necessary for obtaining contiguous clone coverage of a megabase-sized region. Long stretches of overlapping clones and clone sequences were thereby obtained through an iterative screen-sequence-screen-sequence approach. Local fingerprinting and end sequencing were used to make the ordering of the clones in a cluster or region precise. In order to generate a steady supply of mapped clones for sequencing, multiple library screenings had to be done in parallel.

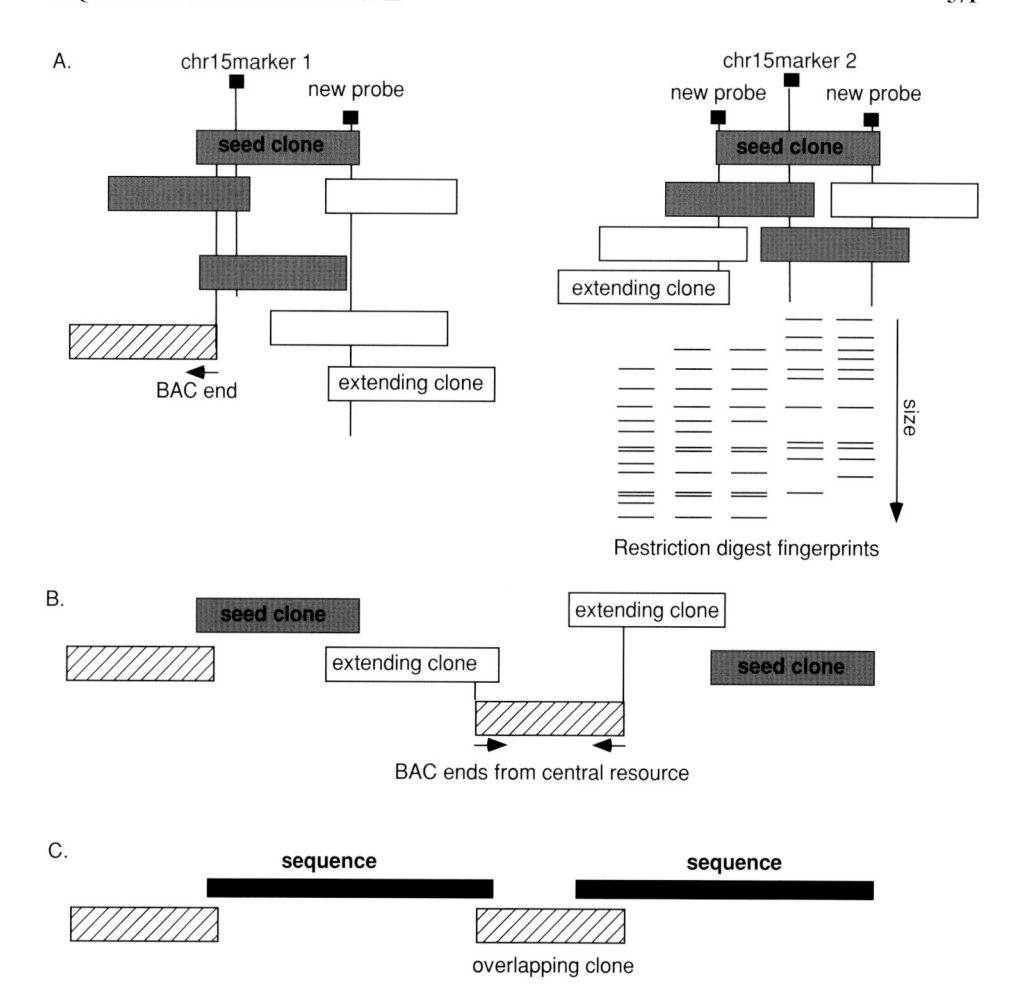

FIGURE 11.3. Construction of tiling paths. A. Clusters of clones (grey boxes) are identified from a screen of a BAC library using markers from human chromosome 15. One of the clones in each cluster is then sequenced (seed clone). From the sequences of the seed clones, new probes for screening are designed (small black boxes) and new clusters of overlapping BACs identified (blank boxes), one of which is sequenced (extending clone). The extent of overlaps are estimated using either BAC end sequencing (arrows, hatched boxes) or restriction digest fingerprinting. B. Regional tiling paths are generated from minimally overlapping clones. C. The sequences of the overlapping seed and extending clones are merged. An even longer sequence would be produced after determining the sequence of the clone joining the two clusters.

As the sequencing phase of the genome project scaled up in the mid to late '90s, it became patently clear that the slow and laborious library screening approach could not supply enough mapped clones to feed the machines, and that large-scale and centralized resources for mapping were required [38, 47]. Between 1997 and 1999, the University of Washington and The Institute for Genomic Research generated BAC end sequences from several thousand clones in two BAC libraries–RPCI11 and Caltech D [26, 51]. With the BAC end sequence resource in hand, clones with minimal overlaps to sequenced BACs could be identified by searching the genome sequence sampling database in GenBank for unique matches to BAC ends. In an independent effort, Washington University at St. Louis built a mapping resource by fingerprinting thousands of BACs from the same two libraries [22, 27]. When large numbers of fingerprinted BACs became available in 1999–2000, clones could be clustered by fragment patterns, and their approximate order within the cluster inferred. Because the method is imprecise, a highly redundant supply of fingerprinted clones was required for distinguishing true from spurious overlaps.

Use of a combination of mapping procedures enabled genome centers to construct "tiling paths," that is, ordered arrays of clones containing inserts from overlapping portions of the genome (Figure 11.3). From these tiling paths, long stretches of chromosomal sequence were reconstructed by merging the sequences of overlapping clones in the tiling path.

Even though mapping procedures were slow, labor-intensive, and tedious, they usually worked. Some problems did on occasion occur:

- The genetic marker used for a library screen turned out to map to the wrong chromosome or to more than one chromosome. This problem could be detected by genome centers fortunate enough to have in-house FISH capacity.
- Library screens, or searches of the centralized mapping resources, yielded no positive clones, thereby leaving gaps in the chromosomal tiling path.
- Mapping data based on fingerprints and end sequences gave conflicting results, meaning that the region of interest in the genome was duplicated, or was significantly different among individuals due to polymorphic variations.

Resolution, or attempts at resolution, of these problems generally occurred late in the game for the genome project, i.e., after year 2000.

11.2.2.2. Sequencing (Accumulation of Sequence Reads)   As will be discussed later in this review, there were heated debates in the early '90s over sequencing strategies, yet by about 1995, the approach called "shotgun sequencing" [10] had become widely accepted. As the size of clonable genomic inserts increased (e.g., from 35 kb cosmids to 150 kb BACs), the ratio between obtainable sequence read length (only about 400 to 1000 bases) and clone insert length decreased, meaning that large numbers of overlapping sequence reads were required to reconstruct a contiguous and accurate sequence of a source clone. In shotgun sequencing, a source clone is fragmented such that positional information of the fragments is lost, and only regained after assembly of the sequence reads generated from the subcloned fragments (Figure 11.4). The "shotgun" analogy is that prior to assembly, sequence reads splatter across a virtual consensus sequence of the source clone. Randomly generated fragments sufficient to cover the source clone many times over must be sequenced in order to ensure adequate coverage from the overlapping reads (redundancy).

Most genome centers used variations of the following generic procedures for generating the shotgun sequence reads from a source clone [39].

- Source clone DNA preparation: Preparing DNA from mapped cosmid, PAC, or BAC clones with a minimal amount of contaminating *E. coli* chromosomal DNA.
- Fragmentation: Randomly shearing the source clone DNA into short fragments using sonication, nebulization, or mechanical shearing by passage through a needle at high pressure.

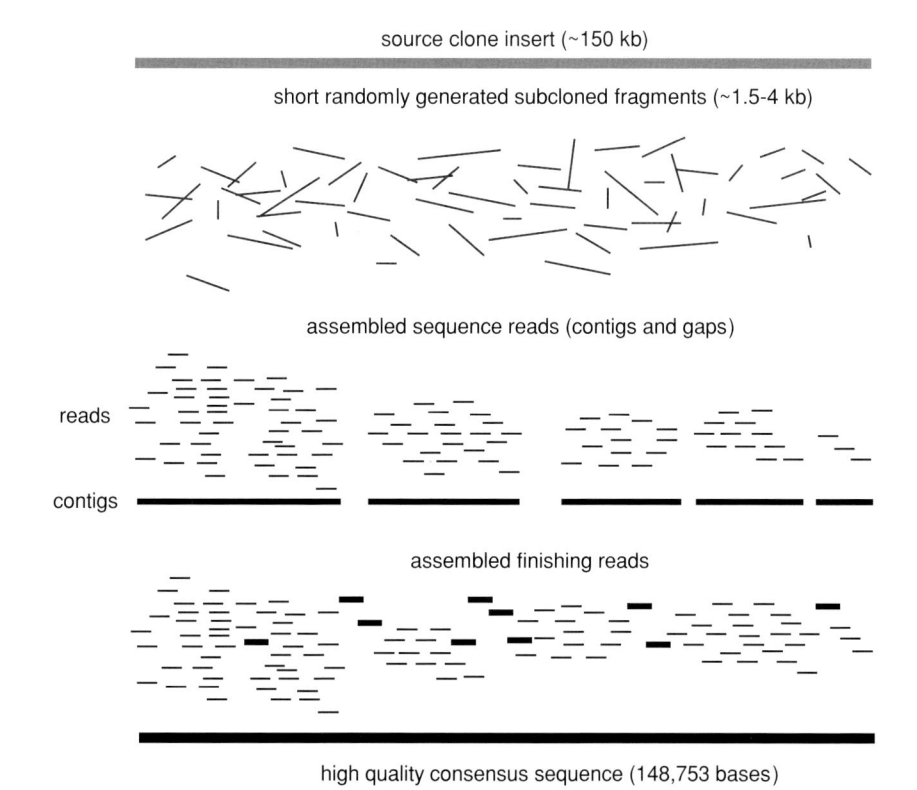

FIGURE 11.4. Strategy for shotgun sequencing. A source clone is fragmented, and the fragments of an optimal size range are subcloned into a phage or plasmid vector. After preparation of DNA from the subclones, ~500 base sequence reads are generated from one or both ends of the insert, and assembled using pairwise alignments to generate contigs. Gaps between contigs and low quality regions are resolved by obtaining additional sequence (finishing reads), after which a high quality consensus sequence for the source clone is determined from the best set of reads.

- Size selection: Purifying fragments of an optimal size range (usually 1.5–4 kb) suitable for subcloning and sequencing.
- Subcloning: Ligating fragments into viral (phage M13) vector or a plasmid (typically pUC18) vector, and transforming the ligation mixture into E coli to generate single recombinant plaques or colonies, each harboring a subclone containing a fragment of the source clone.
- Template DNA preparation: Isolating the recombinant plasmid or viral DNA from a single plaque or colony culture of E. coli.
- Sequencing: Performing sequencing reactions on the purified template DNA using premixes of primer (required for DNA replication), deoxynucleoside triphosphate (dNTP) substrates (DNA building blocks), dideoxynucleoside triphosphate substrates (ddNTPs) to terminate DNA replication at random locations, buffers, and a suitable DNA polymerase [42]. Primers are designed to be complementary to a portion of the cloning vector sequence several bases short of the vector-insert joint, so that the same

universal primer can capture the unique sequences of the inserts. Fluorescent dyes used to detect the order of nucleotides (bases) are attached either to the universal primer ("dye primer sequencing") or to the dideoxynucleoside triphosphates ("dye terminator sequencing"). In the sequencing reactions, a mixture of molecular copies of the DNA template is generated in which subsets of the mixture have terminated at each nucleotide position in the original template. Once a ddNTP is used as a building block instead of a dNTP the polymerase cannot incorporate additional bases onto that molecular chain of DNA (hence "termination"). As a result of the random incorporation of ddNTPs, the completed sequencing reaction mixture contains a nested set of molecules with lengths corresponding to the distance between the universal primer and each base position in the insert sequence. After 1990, sequencing reactions were usually performed in automated thermocycler machines in a 96-well format.

- Detection: Using electrophoresis to separate the mixture of molecules based on their size, and lasers to detect the incorporated fluorescent dye as the molecules of a given size pass by the detector. Gel electrophoresis, used in the early days of the project (1990–1998), was supplanted by capillary electrophoresis in the later stages (1998–2003). Electrophoresis and detection were performed with automated fluorescent DNA sequencers. The most commonly used commercial machines were the Applied Biosystems 373A (1990–1996), 377 (1995–1999), 3700 (1998–2002) and 3730 (2002-present), and the Molecular Dynamics/Amersham MegaBACE 1000 and 4000 capillary sequencers (1998–present).
- Base-calling: Using software to translate the dye peaks in gel or capillary images into the corresponding bases, relying upon signal-to-noise ratios and peak spacing to make the proper call. When gel electrophoresis was used, adjacent lanes of dye peaks had to be properly delineated prior to running the base-calling program. Because of crossover mistakes, the automated lane-tracking output usually needed to be tweaked by a technician using a manual override function of the software. Retracking a 96-lane gel image could take up to an hour.
- Data curation: Copying files of base-called reads to a suitable project directory pertaining to the source clone being sequenced.

Typically, genome centers used automated procedures with varying degrees of sophistication for subclone DNA template preparation, sequencing reaction set-up, sequencing, detection, and base-calling. However, the initial steps of source clone DNA preparation and shotgun library construction (fragmentation, size selection, and subcloning) were hands-on, fussy, and frequently caused difficulties due to irreproducible or non-robust protocols. The following intermittent problems were the bane of a shotgun library construction manager's existence:

- Non-random fragmentation of the source clone;
- Low yield in the ligation/transformation step;
- Large amounts of contaminating *E. coli* sequence;
- A recombinant "clone from hell" contaminating a ligation or transformation reagent such that a high percentage of the clones in the shotgun library would have the same sequence;

Average Daily Quality

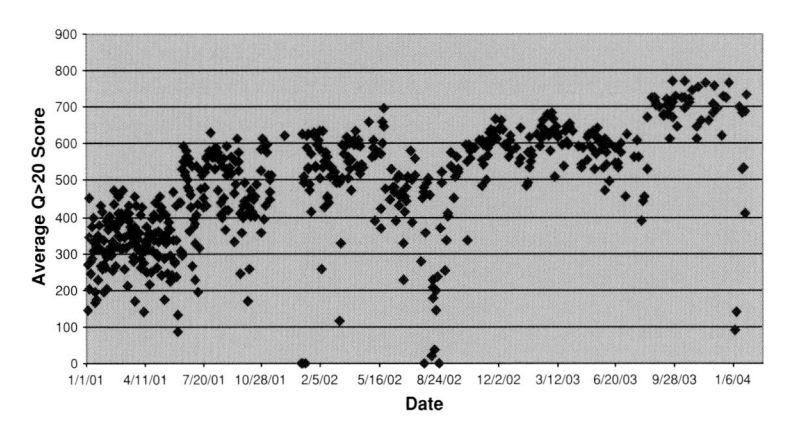

FIGURE 11.5. Fluctuations in data quality. Each dot represents an average of Phred quality scores for all reads on all machines at the Multimegabase Sequencing Center (courtesy of Scott Bloom). Quality improved with a change in sequencing reaction precipitation procedure (June, 2001), and a switch from the ABI 3700 to the 3730 sequencer (September, 2003).

- Libraries made from a "wrong" clone due to erroneous mapping information or a mistake in source clone retrieval from the clone library or a deletion of part of the insert DNA during growth of the source clone.

To detect these potential problems, a small test set of 96 reads would usually be obtained before committing to sequence a given shotgun library at high redundancy. The test set of sequence reads was also useful for verifying the map position of the source clone and for staking a claim to a chromosomal region, as will be discussed below.

Although the sequencing and detection steps were straightforward and amenable to high-throughput, multiple things could go wrong such as impure DNA templates, reagents going "bad", low or noisy signal in the detection of the fluorescent dyes, streaky gels, blank lanes, machine breakdown, sample-tracking errors, computer disk failure and the like. Thus, consistently generating high quality sequence data was difficult due to the plethora of variables in the system (Figure 11.5).

Moreover, managing the data, equipment and personnel was a perpetual juggling act. Centers were constantly recruiting and training technicians because employees would tire of boring, repetitive work and quit or transfer to a different part of the project. Some problems simply could not be anticipated or controlled. When Rick Wilson, the highly successful director of sequencing at the Washington University of St Louis genome center was queried as to his biggest challenge for data generation, he immediately responded with "romances in the lab." At Whitehead, one of the biggest problems was alleged to be handling the trash. At the Multimegabase Sequencing Center, the laboratory changed institutions twice (Caltech to University of Washington to Institute for Systems Biology) and moved five times over the course of 10 years.

11.2.2.3. Assembly In shotgun sequencing, a "consensus" sequence is constructed from the sequences of overlapping reads derived from randomly generated fragments of the source clone. In rare cases, there are sufficient data among the shotgun reads to reconstruct a completely contiguous and accurate sequence of the source clone. More typically, there are i) gaps, meaning that no overlapping reads exist for a portion of the source clone or there are ii) regions of low quality sequence insufficient to determine a clear consensus or there are iii) misassembled reads, that is, reads that appear to overlap but in actuality do not, because they derive from non-adjacent portions of the source clone. Resolution of these problems is left to "finishing" (see below).

Prior to sequencing and assembly, a determination is made regarding the number of reads to collect for a given source clone. The number might take into account the size of the insert, the desired redundancy of coverage, the average read length, the expected failure frequency, and the constraints of sample processing. For example, a genome center with a 10% read failure frequency, a target effective redundancy of 8, and an average read length of 650 bases would need to generate 13.5 reads per kilobase of source clone. A 150 kb BAC would require 2025 reads, or approximately five 384-well plates or twenty-one 96-well plates of sequencing reactions. To make it simple, such a center might assume that the average BAC size is 170 kb and sequence six 384-well plates for all the source clones that go through the system rather than attempt a tailored read number determination for each clone.

In the earlier days of the genome project, assembly was typically done starting with a random sequence read to which another random read was compared. If a sequence overlap was detected within the predetermined parameters of the assembly program (e.g., X percent identity over Y length of alignment) then the reads were combined to form a contig. If no overlap was detected, then each of the two reads formed a separate contig. The process of adding reads to pre-existing contigs or forming new contigs was repeated until all of the reads had been added. Because the sequence matches between overlapping reads were often not perfect, consensus was determined by majority rule or, if there was no majority, then the ambiguity was recorded in the consensus by a letter other than A, C, G, or T (e.g., according to the IUPAC nomenclature convention, M meant either A or C, R meant either A or G, etc.). In a phenomenon known as "consensus rot," errors and ambiguities hindered the ability to detect and assemble truly overlapping reads, with the result that a large number of relatively small contigs would result from the set of assembled reads, even at fairly high redundancy of coverage. This problem was partially ameliorated by truncating the length of sequencing reads so as to include only the "good" portions of the read.

In 1995, Phil Green at the University of Washington engineered a more effective strategy for assembly called "Phrap" (Phil's Revised Assembly Program)[2]. The first step involved calling the bases in such a way that quality scores could be associated with each base [15]. That is, the chromatogram generated by the software associated with the DNA sequencer was re-processed for the purpose of determining dye peak quality, i.e., relative spacing of neighboring peaks and signal-to-noise ratio. Each base was assigned a quality value indicative of the probability of the base having been called correctly. This base-calling program, called "Phred" (Phil's Revised Editor), would enable an assembler to give more weight to high quality bases and ignore bases that fell below an acceptable quality threshold. Using the Phred base calls and quality scores, the Phrap assembly engine performed all of the possible pairwise alignments between the input set of reads, and from the alignment

---

[2] The definitive publication on Phrap has not yet been written. See http://www.phrap.org/phrap.docs/phrap.html.

SEQUENCING THE HUMAN GENOME 377

```
                                    misassembly
        contig 13        gap       contig 47    \  /
    ─────────────    ─────    ───────────────────X──────
        →       ←              →        ←       /  \    ←
    subclone 36A2           subclone 2C4      subclone 2C4
                  →      ←
               subclone 14H9          →   →
                                  subclone 17G1
```

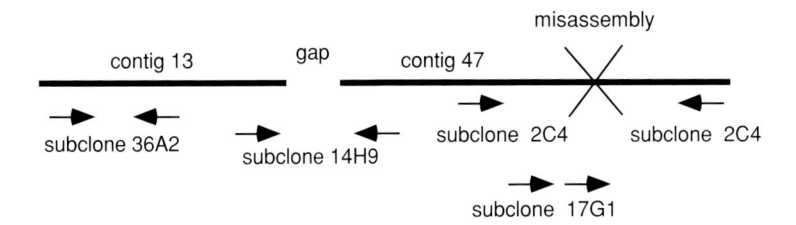

FIGURE 11.6. Plasmid end pairs. Paired ends of sublone 14H9 indicate that contig 47 is adjacent to contig 13. In subclone 2C4, the paired ends are too far apart, and in subclone 17G1, the paired ends are in the same orientation, both of which problems indicate a misassembly in contig 47.

scores determined the best matches among them. After assembly, each base was given a "Phrap score," reflecting the extent to which the base was confirmed by bases in other reads (e.g., confirmation from the opposite strand was given more weight than confirmation from the same strand.) Rather than determining a consensus based on majority rule, a "consensus" sequence for the source clone was determined by choosing the base with the highest Phred/Phrap score for each position. As a result, the Phrap assembler generated longer contigs with more accurate consensus sequences because it used the quality scores to select the best data.

Both before and since Phrap, there have been numerous sequence assemblers built for the purpose of aligning shotgun reads and constructing contigs (e.g., [1, 4, 5, 20, 29, 31, 48]). Using their assembly engine of choice, the software typically performed the following steps:

- Read processing: Assign quality scores to bases or trim reads.
- Assembly: Perform pairwise alignments between reads.
- Construct contigs: Represent the overlapping reads as a consensus sequence supported by an array of reads, with the starting and ending position of each read oriented relative to the consensus sequence.

In addition to these steps, some assemblers were able to order and orient contigs based on information derived from paired clone end sequences. That is, if both ends (i.e., vector-insert joints) of a 1–4 kb plasmid insert were sequenced, then the beginnings of those paired reads needed to be 1–4 kb apart in the assembled sequence with the sequences pointing towards each other (Figure 11.6). Using a contig editor, that is, an editor designed to display the results of the assembly, the order and orientation of contigs based on paired ends could also be determined manually.

Assuming highly redundant read coverage, the major pitfall of assembly was misassembled reads due to repetitive content in the insert of the source clone. If the sequence similarity between repeats averaged ~98% or less over a kb, Phrap could generally produce the correct answer. With lower coverage or repeats of higher sequence similarity, reads from the repeats would "pile up" in a misassembled contig and have to be sorted out in the finishing phase.

11.2.2.4. Finishing The sequence of a source clone was considered finished if a contiguous consensus sequence was accurate at each base position and the underlying assembly was correct. The accepted standard of accuracy was 99.99%, or 1 error on average in 10,000 bases. In order for a finished clone to be part of a chromosomal tiling path, further validation was required as to the integrity of the source clone DNA (i.e., no deletions or rearrangements) and the proper chromosomal assignment based on FISH, genetic marker content, or convincing overlaps with other validated clones.

Turning a set of assembled contigs into a finished sequence typically involved successive iterations of the following steps until a contiguous and accurate consensus sequence was obtained:

- Additional sequencing: Filling gaps and improving poor quality portions of the consensus due to noisy read data or sparse coverage in the assembly required the gathering of additional sequence data either by obtaining more shotgun reads, or resequencing selected clones using a different chemistry, or extending the length of a read with a custom oligonucleotide primer, or using the polymerase chain reaction (PCR) to generate sequence-ready DNA from a plasmid subclone, the source clone itself, or even human genomic DNA.
- Reassembly: Incorporating additional sequence data into a set of already assembled reads or repeating the entire assembly *de novo* upon the addition of new sequence reads.
- Editing: Determining the order and orientation of contigs, performing base-calling judgments in regions of ambiguity, breaking apart misassembled contigs, and joining overlapping contigs. This was typically done by an experienced "finisher" using a sequence contig display editor with features allowing for base-call overrides, complementation of contigs, removal of selected reads from contigs, the capacity to split contigs apart, and the creation of "fake" reads to glue contigs together.

Once the sequence of a source clone was provisionally finished, it needed to be validated to ensure that the assembly was correct. Validation procedures included satisfaction of the distance constraints imposed by paired end sequences (Figure 11.6), agreement with sequences of overlapping clones, and comparison of restriction digest fingerprint patterns predicted by the sequence to the experimentally determined patterns of the source clone.

Like mapping, finishing was a slow and laborious process requiring skilled personnel. To speed it up, several centers instituted variants of a strategy called "autofinish" [17]. Using a specified set of rules (for example, "Additional sequence data must be generated for any region in the consensus supported by two or fewer reads.") the autofinish software was designed to create a list of reads and suggested strategies to be used for generating additional sequence data. Skilled finishers stepped in after the assembly project had been subjected to several rounds of autofinish. While the finishing of most source clone sequences was straightforward, some clones posed problems, most typically gaps for which no sequence could be obtained using a variety of strategies, or misassemblies due to sequence repeats resulting from gene duplications or low complexity DNA[3]. Difficult clones often took months and sometimes years to finish!

In the last two years of the genome project, finishing was a serious problem due to a backlog of assembled source clones and political pressure to get the genome done by the deadline of April 2003. In addition to autofinish, genome centers dealt with the problem by

---

[3] Examples of low complexity DNA are homopolymer runs or long stretches of a short repeating unit of DNA, e.g., three kilobases of a consecutive 40 base repeat with minor variations.

Acc.No.	Source clone	TPF contig	Genome center	Status	Bases of overlap	Identity
AC084773.5	RP11–137H12	15CTG4_26A	WIBR	fin	33658	100.00%
AC087878.7	RP11–62C7	15CTG4_26A	WIBR	fin	53158	99.99%
AC113146.2	RP11–265N7	15CTG4_26A	WIBR	fin	53500	99.99%
AC013652.8	RP11–462P6	15CTG4_26A	WIBR	fin	48759	99.77%
AC109630.2	RP11–624L4	15CTG4_26A	WIBR	fin	43182	100.00%
AC022196.7	RP11–27M9	15CTG4_26A	WIBR	fin	1999	100.00%
AC037198.7	CTD–2033D15	15CTG4_26A	UWMSC	fin	34227	99.81%
AC023908.6	RP11–37C7	15CTG4_26A	UWMSC	fin	94495	99.98%
AC012377.5	RP11–325N19	15CTG4_26A	UWMSC	fin	91621	99.88%
AC025168.7	RP11–43D14	15CTG4_26A	UWMSC	fin	58002	99.99%
AC021755.9	RP11–521C20	15CTG4_26A	UWMSC	fin	62547	99.92%
AC025429.6	CTD–2006D8	15CTG4_26A	UWMSC	fin	62532	99.98%
AC020658.6	RP11–133K1	15CTG4_26A	UWMSC	fin	6357	99.96%
AC013356.8	RP11–64K12	15CTG4_26A	UWMSC	fin	70385	99.90%
AC091045.3	RP11–111A22	15CTG4_26A	UWMSC	fin	2747	99.81%
AC022405.5	CTD–2339L15	15CTG4_26A	UWMSC	fin	14379	99.86%
AC012476.8	RP11–532F12	15CTG4_26A	UWMSC	fin	72513	99.91%
AC025166.7	CTD–2129N1	15CTG4_26A	UWMSC	fin	43275	99.86%
AC020661.8	RP11–540011	15CTG4_26A	UWMSC	fin	63430	99.93%
AC021753.7	RP11–129I12	15CTG4_26A	UWMSC	fin	100401	99.91%

FIGURE 11.7. Portion of the chromosome 15 tiling path. This example, demarcating one of the boundaries of territory between Whitehead Institute (WIBR) and the University of Washington Multimegabase Sequencing Center (UWMSC) is taken from a National Center for Biotechnology Information web site in which the Tiling Path Files for each chromosome were recorded and updated. The color of the filled circle indicated confidence values for the position of clones in the tiling path. Because the source clones could be from different haplotypes, 100% agreement between overlapping sequences was not required.

building robots to retrieve subclones for resequencing, upping the redundancy of coverage so that few regions of the consensus would need additional work, and relaxing standards such as requiring that all regions of the consensus be covered by data from both strands and that all gaps be filled to completion.

11.2.3. Construction of the Chromosome Tiling Paths

Long stretches of chromosomal sequence were assembled by conjoining validated finished sequences from overlapping source clones (Figure 11.7). Sequence constraints determined by pairs of BAC end sequences, chromosomal marker positions, and percent identity of bases aligned in overlapping sequences were used to gauge the correctness of the tiling paths. In theory one might expect there to be 24 master sequences, one for each chromosome, when the human genome is finally finished. But in practice this is not possible due to apparently unclonable genomic DNA or to DNA that cannot be sequenced and assembled accurately such as the centromeres. Nonetheless, long stretches of contiguous sequence have been reconstructed. Chromosome 14, for example, is represented in a sequence that is 87,410,661 bases in length [19].

11.2.4. Data Sharing

Before the late-90s, the sequence of a clone was submitted to one of the official public genome databases only after it was finished. These databases are: GenBank or GSDB (the

Genome Sequence Data Base, now extinct) in the US, EMBL (European Molecular Biology Laboratory) in Europe or DDBJ (DNA Database of Japan). Data submitted to any one of these databases were also displayed by the others. When the genome project began to scale up, the International Consortium committed itself to releasing sequences prior to finishing, in a form called "working draft" [9]. A new division of the databases, called HTGS (high throughput genome sequence) was established into which assembled contigs from a source clone were to be deposited within a day of their generation. To indicate the extent of completion, annotation tags for three "phases" were defined:

- Phase 1 indicates a set of unordered assembled contigs;
- Phase 2 indicates a set of ordered and oriented contigs;
- Phase 3 indicates a finished sequence.

Annotation tags were also included for the source clone and library ID, the submitting genome center, and the chromosomal location of the sequence. Even though the assembly and/or the chromosomal map position of the working draft sequences were wrong in some instances, the genome community's decision to release the sequence data prior to finishing provided enormous benefit to researchers searching for disease genes.

With the increased use of the World Wide Web in the mid-90s, data became increasingly accessible to anyone interested. In addition to the central public databases, genome centers typically established their own web sites for sharing data and protocols. Over time, the National Center for Biotechnology Information (NCBI) and the University of California, Santa Cruz in the US, the European Bioinformatics Institute (EBI) in Great Britain, and several other organizations and companies have gathered and integrated a rich assortment of resources to facilitate understanding and utilization of the genome sequence data[4]. Along with the sequences themselves, these web sites include maps, gene identifications, gene expression results, cross-species comparisons, gene function annotations, and the like.

11.3. CHALLENGES FOR SYSTEMS INTEGRATION

At the inception of the human genome project in the late '80s, there was significant controversy over whether sequencing the genome was worth doing and if it could actually be done. At the time, the longest contiguous stretch of DNA that had been sequenced was well under 100 kb [6]. Automated fluorescent DNA sequencing had recently been invented and commercialized, representing a vast improvement over radioactive sequencing. Nonetheless, sequencing 3 billion bases would more than challenge the then-current methodologies and would likely incur significant cost. But with great optimism, it was assumed that cheap and effective new strategies for sequencing would emerge as they were needed. In the original design of the genome project, projected to last from 1990–2005, physical maps of the chromosomes would be constructed during the first 5–10 years. In parallel with this effort, revolutionary sequencing strategies leading to orders of magnitude increases in throughput

---

[4] See http://www.ncbi.nlm.nih.gov/ for NCBI; http://genome.ucsc.edu for University of California Santa Cruz; and http://www.ebi.ac.uk/ for EBI.

would be invented and implemented[5]. With the maps and the methods in place, the genome would then be sequenced during the last 5 years in high-throughput factory-style operations.

This is not exactly how things turned out. Sequence-ready physical maps were not constructed at the outset and revolutionary methods for sequencing did not appear. Nonetheless, the genome was sequenced two years ahead of schedule[1]. Along the way, the sequencing centers faced numerous challenges for systems integration due to the pioneering nature, complexity, and scale of the human genome project. These challenges fall into two broad categories: developing and applying the methodologies for sequencing source clones, and achieving overall project coordination for finishing a master sequence for each chromosome.

Between 1990 and 1996, significant progress was made regarding the choice of an overall strategy and refinement of specific sets of procedures for sequencing individual source clones. Once a mature set of procedures for shotgun sequencing was developed, the scale-up of the genome project could begin in 1997 and accelerate rapidly in 1999. While numerous challenges attended the day-to-day implementation of sequencing methods as well as the knotty issues of managing and tracking the data, from a historical perspective, the more interesting debates pertained to the initial acceptance and refinement of shotgun sequencing procedures.

Project coordination challenges came into play most noteworthily after year 2000 when the focus shifted to figuring out how to get the entire human genome sequenced, assembled, and validated. Turning multitudinous source clone sequences into polished chromosome sequences required the development of centralized resources and significant cooperation among the several genome centers doing the finishing. These challenges will be discussed in a subsequent section.

11.3.1. Methodological Challenges for Sequencing Source Clones: 1990–1997

Returning to the pipeline analogy, each of the genome centers needed to design an approach to sequencing that was capable of processing large numbers of samples through successive series of steps. The best approaches would possess the following virtues:

- Scalability: Procedures that worked on a small scale needed to carry over to a large scale if they were to be useful. Reducing the number of steps, automating as many steps as possible, increasing the sample-processing capacity for each step, instituting fail-safe sample-tracking procedures, and improving the robustness of each step all conduced to more scalable procedures.
- Cost-effectiveness: Cost-effective procedures were those with high success rates using minimal amounts of expensive reagents, equipment, labor and time.
- Good data quality: Because several steps were required to get from a mapped source clone to finished sequence, producing high quality data at each step increased the efficiency of the overall process due to greater yields and less need for backtracking.
- Elimination of bottlenecks: No matter how speedy or high-throughput any individual step was be made to be, the rate of production of the overall process could be no faster

---

[5] For the text of the various 5 year plans for the human genome project, see: http://www.ornl.gov/sci/techresources/Human_Genome/hg5yp/index.shtml

than the slowest steps. Long "cycle times" for sequencing source clones increased the managerial complexity of the overall operation.
- Adaptability: Changing technologies and policies mandated that the sequencing pipeline be somewhat flexible.

In an ideal world, optimization of all these virtues would converge on the same set of procedures, and that happened with shotgun sequencing. Along the way to developing a mature strategy, though, there were various disagreements and failed approaches, some of which will be presented herein.

11.3.1.1. Why Revolutionary Sequencing Technologies Never Got off the Ground
During the early '90s, several novel strategies for sequencing were proposed and attempted for the purpose of replacing electrophoresis-based automated fluorescent sequencing methods of detection with higher throughput approaches. At the time, the Applied Biosystems 373A sequencer produced reads of about 450 good bases in 14–16 hour runs on gels loaded with 36 or 48 lanes of sequencing reactions. At 15 reads/kb, one needed either many expensive sequencers or many days to produce enough sequence reads to assemble a 40 kb cosmid with this level of throughput. Sequencing by hybridization [13], sequencing by mass spectrometry [32], and multiplex Maxam-Gilbert sequencing [7] were among the strategies tried. These methods and others suffered from one or more of the following problems: overly high error rates; short read lengths; inability to automate sample processing; only model templates worked; or no easy way to do base-calling. Most of the revolutionary methods were not capable of producing even a tiny amount of human genome sequence data, let alone able to scale.

Electrophoresis-based detection technologies, in the meantime, improved incrementally. The number of lanes loaded onto a sequencer went from 24 to 36 to 48 to 72 to 96. The Applied Biosystems 377 sequencer introduced in the mid-90s reduced the gel run time to 7–9 hours, thus allowing for 2–3 runs a day. Improvements in the polymerases [46] and dyes [41] produced longer reads of higher quality. By 1997, one ABI 377 produced 192 reads averaging about 650 bases (124,800 bases) in a day, about an 8-fold improvement over 1992 technology (16,200 bases). In 1998, capillary electrophoresis (ABI 3700, MagaBace 1000) reduced the machine run time to a couple of hours and obviated the need for manual lane-tracking prior to base-calling. By 2003, the Applied Biosystems 3730 capillary sequencer produced about 450,000 bases a day, close to a 30-fold improvement over 1992 technology. Even though the changes were incremental as they occurred, the overall effect of improved procedures for automated fluorescent sequencing were dramatic in terms of throughput, data quality, and cost-savings.

As important as cranking out bases of high quality sequence, use of the automated sequencers integrated well with the use of robotic approaches to the upstream steps of the sequencing pipeline. By the mid-90s, the incentive to supplant electrophoresis-based detection technology with revolutionary alternative approaches was lost, and it made increasing sense to get on with the sequencing of the genome using a technology that worked [35]. Thus, the scale-up began in 1997, four years ahead of schedule. (Interestingly, in recent years there has been a resurgent interest in developing new sequencing technologies primarily aimed at resequencing portions of the genome for detecting sequence variations. A discussion of these is beyond the scope of this review.)

11.3.1.2. Why Shotgun Sequencing Became the Dominant Methodology "Shotgun sequencing," said geneticist Maynard Olson to the author of this review circa 1993, "is like buying 8 copies of a prefabricated house and constructing one house from the parts— it's inelegant and inefficient." Especially in light of the seemingly low-throughput of the sequencers and the significant expense of the reagents, having to generate subclones and sequencing reads sufficient to cover a source clone about eight times over seemed like a wasteful approach in the early '90s. Although hard to believe in hindsight, the resistance to high-redundancy shotgun sequencing was fairly vociferous in the early days of the genome project. To reduce the redundancy, more directed approaches were proposed. One of these involved using transposons to map plasmid subclone inserts, the idea being that sequencing an array of ordered inserts would require the acquisition of less sequence data (even though an additional mapping step was introduced) [45]. Other strategies advocated extensive use of oligonucleotide-directed sequencing to extend the length of the reads obtainable from phage or plasmid subclones.

Although intuitively appealing, alternatives to high-redundancy shotgun sequencing were gradually abandoned for several reasons. First, shotgun sequencing reduced the burden of finishing by vastly improving the quality of the input data. With shotgun sequencing there were fewer gaps to fill and more options for resolving discrepancies among reads, thus reducing the additional work required for resequencing. Moreover, assembly errors due to gene duplications and other repeats in the source clone could be detected and often resolved by a rearrangement of the sequencing reads. Second, the steps of shotgun sequencing were more automatable because samples could be processed in a 96-well or 384-well format using robots. Introduction of mapping or directed sequencing steps required handling and tracking individual samples. Third, shotgun sequencing was more tolerant of failure. If 96-well plates of templates or sequencing reactions were dropped on the floor or flipped into the opposite orientation or otherwise lost or mislabeled this had little effect on the downstream steps. These problems, as well as failed sequencing reactions or gel runs, could be overcome by obtaining more random reads from the shotgun library. Most genome centers built a standard failure rate into their calculations of numbers of reads/kb to sequence. In contrast to shotgunning, any procedure that relied on particular individual samples surviving through all of the steps made failures more difficult to deal with and recover from. Fourth, shotgun sequencing was fast. Once a shotgun library was made, the subsequent steps of DNA template preparation, sequencing, and assembly could happen in a few days thanks to process automation. Thus in the end, shotgun sequencing turned out to be more scaleable, cheaper, and easier to do than any of the alternative strategies.

11.3.1.3. How Shotgun Sequencing Involved Trade-Offs Within the basic framework of shotgun sequencing, genome centers had to make tactical decisions based on cost, efficiency, and data quality.

- Level of redundancy: Beyond a certain point, additional shotgun reads would fail to improve an assembly and the project would need to go into finishing. Short of that point (about 10-fold effective redundancy with 500 base reads) accumulation of more reads would help for assembly and finishing, but sequence reads were costly in terms of reagents and equipment usage. Therefore, it was tempting to accumulate shotgun reads only up to about 5-fold redundancy. Centers that shortchanged on shotgun reads tended to have "armies" of human finishers who required training in the requisite skills of bench and computer problem-solving. They also had huge backlogs of unfinished clones. Thus, saving money and time on the shotgun phase was considered by some centers to be short-sighted because the subsequent steps of finishing became more costly and slower. Other centers considered it more important to use their machine capacity to generate data for a larger number of source clones, albeit at less than optimal coverage. In the later stages of the genome project, when time-pressure was critical and reagents cheaper, most centers increased their redundancy. It was not uncommon for clones to be sequenced at 20–30 reads per kb.

- Read length: Before the Applied Biosystems 377 sequencer came on line, genome centers had to decide whether it was better to run an ABI 373A sequencer once a day and get longer reads or run the sequencer two or three times a day with shorter reads. On the one hand, shorter machine runs meant a faster accumulation of data. Also, it made no sense to obtain longer reads unless the data were of high enough quality to be useful. (Beyond a certain length, base-calling error increases dramatically). On the other hand, longer reads meant that fewer reads per kb were required to achieve a target effective redundancy, thus reducing the use of costly reagents and consumables. Longer reads had the added benefit of facilitating the assembly by giving more alignable sequence. And finally, fewer technicians were required if the sequencers were run once a day. This dispute disappeared in the late '90s with the advent of capillary sequencers capable of generating 96 750 base reads in a couple of hours.
- Template: In the early '90s single-stranded M13 phage templates were preferred over double-stranded plasmid templates for subcloning inserts because the data quality and read length were about 20% better. Plasmids had other advantages, however. Two sequencing reads could be obtained from one template, thereby reducing template preparation costs. Addition of paired plasmid reads to a shotgun assembly enabled the ordering and orienting of most of the contigs, which benefitted mapping and gene structure determination. Some genomic inserts were more stable in plasmids. And finally, because inserts up to about 4 kb could be propagated in plasmids, as opposed to the 1–2 kb inserts in M13, read lengths of individual subclones could be extended by directed sequencing with custom primers, which provided a benefit for finishing. But switching to plasmids meant not only a sacrifice in the data quality of the input reads but also a commitment to redesign the methods for DNA template preparation and, thus, an investment in time and equipment. When the genome project scaled up in 1999, some centers continued with M13, some made a wholesale switch to plasmids, and some used a mixture of the two strategies. Improvements in laboratory procedures leading to enhanced data quality have now made plasmids the template of choice.
- Sequencing reactions: Before 1991, inserts subcloned into M13 were typically sequenced using fluorescently labeled dye primers and T7 DNA polymerase ("sequenase"). Four separate sequencing reactions had to be set up for each template in a laborious multistep process involving annealing of the dye-labeled primer to the template followed by addition of the dNTP/ddNTP building blocks. The four reactions, each containing a different fluorescent dye tag, were combined prior to gel-loading. Then "cycle sequencing" with the thermophilic DNA polymerase used for PCR came on line, with the advantages that lower amounts of DNA template and reagents were required and the sequencing reactions were performed in automated

thermocyclers. Because sequenase and Taq polymerase produced different types of systematic errors, employing a combination of both methods produced a more accurate set of shotgun reads [25]. This benefit was compromised, though, by the increased labor and expense of the sequenase approach. Eventually, sequenase was abandoned. Along the same line, the development of thermophilic polymerases with a higher affinity for dye labeled dideoxy chain terminators enabled sequencing to be done in one reaction rather than four, a decided advantage over dye primers for thermocycler utilization and plasticware consumption. Yet again, because the systematic errors of the methods were different, there was benefit to using a mixture of both strategies. Some centers used one strategy for the shotgun phase and the other strategy for finishing reads, while other divided up shotgun reads between the two strategies.

- Robots: Some genome centers hired in-house engineers to build robots for DNA template preparation and sequencing reaction set-up whereas others used commercially available machines. Either way, there was continual machine obsolescence due to changing technologies and procedures. For this reason, most centers used a modular approach wherein machines could be swapped out without having to perform major modifications on the upstream or downstream procedures. Nonetheless, the decision to exchange one set of expensive machines for another was typically fraught with angst. One had to ask if the new machine was so much better in terms of throughput and/or data quality and/or cost savings that it would pay for itself before also becoming obsolete.

11.3.1.4. Why Centers did not Immediately Jump on the Phrap Bandwagon The Phred base-calling and Phrap assembly program (see above) were developed in 1995 but were not immediately adopted by all of the genome centers, although most had signed on by the time of the accelerated sequencing scale-up beginning in the summer of 1999. There was little doubt that Phrap worked better than the other prevailing assemblers. However there were two sources of resistance. One was an attachment to the way things had been done before. This was true especially for the centers that had already made a large investment in developing assembly algorithms and sequence editing software. The other source of resistance was that Phrap initially lacked an editor that finishers found useful. Phil Green's purist philosophy dictated that no editing should be done but rather that all problems with finishing should be solved by acquiring and assembling more sequence data. As a result, Consed, the sequence editor developed at the University of Washington for viewing Phrap assemblies, was initially devoid of most of the editing tools that finishers were accustomed to using such as the ability to remove reads from contigs or force join contigs. Thus, although Consed did a beautiful job of displaying base quality and the mosaic of bases used to construct the consensus sequence from shotgun reads, finishers from the other genome centers did not want to use it. Instead, several centers wrote scripts for importing Phrap assembly results into their in-house sequence editors, with the unfortunate consequence that much of the quality information provided by Phrap was not utilized effectively by the finishers. To prevent this from happening, David Gordon, the lead Consed developer at the University of Washington, gradually incorporated all of the various tools and features demanded by finishers, including a robust and sophisticated version of autofinish software [16, 17].

By year 1999, Phrap had become so popular that the International Consortium adopted a policy requiring the inclusion of Phrap base quality scores in Phase 1 or 2 draft sequence submissions in order to help users assess the likelihood of error for a given stretch of sequence. Phrap scores provided an objective measure of data quality by which all centers could be evaluated and held to a standard.

To summarize, in terms of systems integration, procedural choices had to be made with an eye to the overall goal, namely producing finished sequence in a timely, efficient, and cost-effective manner. In the early days of the genome project, there were serious disagreements over how best to do this. A mature (that is, widely utilized and agreed upon) set of methods now exists for sequencing source clones, due primarily to improvements in technology and a much better understanding of the cost, quality and efficiency trade-offs implicit in designing a system that optimizes the criteria stated above.

11.3.2. Challenges for Sequencing the Entire Human Genome: 1998–2003

As the discussion above might indicate, there was significant attention paid to developing and refining techniques for increasing the throughput and quality of sequence data. For-profit companies marketing their latest and greatest machines and reagents had an extensive interest in the genome community. Although the equipment and consumables were expensive (DNA sequencers and sample-processing robots could cost up to $350,000), increased capacity, sensitivity, and automation drove down the cost of sequencing reads to the point that a scale-up of sequencing became feasible. On the other hand, mapping source clones and finishing assembled sequences remained slow, low-tech, and labor-intensive. As of the mid-90s, genome centers were beginning to gain momentum, but there was concern about the overall pace of the project. Thus, the ground was ripe for Celera Genomics to come along in 1998 and claim that a whole genome shotgun approach would obviate source clone mapping and, furthermore, that they could sequence the genome at lesser cost and higher efficiency because a) they would have hundreds of wonderful new high-throughput Applied Biosystems capillary sequencers and b) the effort would be conducted in a single company rather than disseminated across numerous academic genome centers.

Rather than give up, as Celera suggested it should do, the International Consortium decided to speed up. Engineering first a massive scale-up and then a concerted effort to finish the genome taxed the resources of the participant genome centers and created numerous challenges for overall project coordination, some of which will now be discussed.

11.3.2.1. Before Celera: 1996–1998 Close to the end of 1997, about 60 megabases (2%) of the human genome had been finished [38]. Except for a handful of sequences longer than 500 kb, the majority of the data derived from individual cosmids or BACs containing genes of biological interest. All of the chromosomes had some sequence but no chromosome had a significant amount of sequence. The established genome centers in the US and Europe had begun to build up their sequencing operations in light of the successes of the shotgun sequencing methods described above. Meanwhile new genome centers were being formed in the interest of distributing the effort internationally and increasing the overall throughput. At around this time, the International Consortium faced three significant challenges related to project coordination:

- Establishing standards for data quality and public release.

- Distributing multimegabase-sized regions of the genome to centers based on their sequencing capacity and track record for completion, resolving territorial disputes, and ensuring that all regions of the genome would get sequenced.
- Procuring an adequate supply of mapped source clones suitable for sequencing, i.e., constructing regional physical maps.

The first two challenges were successfully addressed by HUGO, the Human Genome Organization, whose mission is to promote international discussion and collaboration on scientific, medical, legal, ethical, and social issues pertaining to the acquisition and use of human genomic data [28]. In 1996 and 1997 leaders of the larger genome centers met in Bermuda to formulate guidelines regarding data submission, data sharing, data quality standards, and the regulation of genome centers' claim rights to genomic real estate[6]. Thereafter, any genome center belonging to the International Human Genome Sequencing Consortium had to sign on to the Bermuda Principles and adapt to rule changes as the genome project proceeded. Fortunately, the software developers at GenBank, EMBL, and DDBJ cheerfully produced and updated the data submission tools required for complying with the guidelines.

As for procuring mapped clones, the methodological difficulties were described above: physical mapping is slow and labor-intensive. The mapping problem was compounded by two developments during 1996. First, in the interest of maintaining public enthusiasm and celebrating noteworthy accomplishments, the goal of mapping the genome was declared met as a result of the dense genetic marker linkage maps produced by the Whitehead Institute [21], Genethon [12], and other groups. While the construction of linkage maps was extraordinarily useful, linkage maps are not physical maps. That is, they do not provide source material suitable for sequencing the genome. Nonetheless, with mapping having been declared finished ahead of schedule and under budget, physical mapping was passed on to the sequencing centers as a largely unfunded mandate. Second, in the interest of protecting the identity of the individuals whose genomes were being sequenced, the National Institutes of Health and the Department of Energy in the United States ruled that the clone libraries must derive from anonymous donors. This ruling meant that the sequence-ready cosmid, PAC and BAC maps that had already been constructed for portions of the genome were not supposed to be used beyond 1997, by which time new approved BAC libraries would be available.

Thus, as of the end of 1997 there was a problematic disconnect in terms of systems integration:

- Less than 5% of the genome had been sequenced.
- Sequencing technologies were now sufficiently mature that it was time to scale up the genome project.
- Victory was declared for the mapping phase of the genome project without the follow-through of providing physical maps to the sequencing centers.
- The limited number of pre-existing physical maps that did exist were ruled ineligible for providing source clones for sequencing.

Centers had to solve the clone acquisition problem by i) enlisting the aid of collaborators who had longstanding interests in particular chromosomes or disease genes, ii) establishing

---

[6] For the text of the Bermuda Principles, see: http://www.hugo-international.org/hugo/meetingreports.html

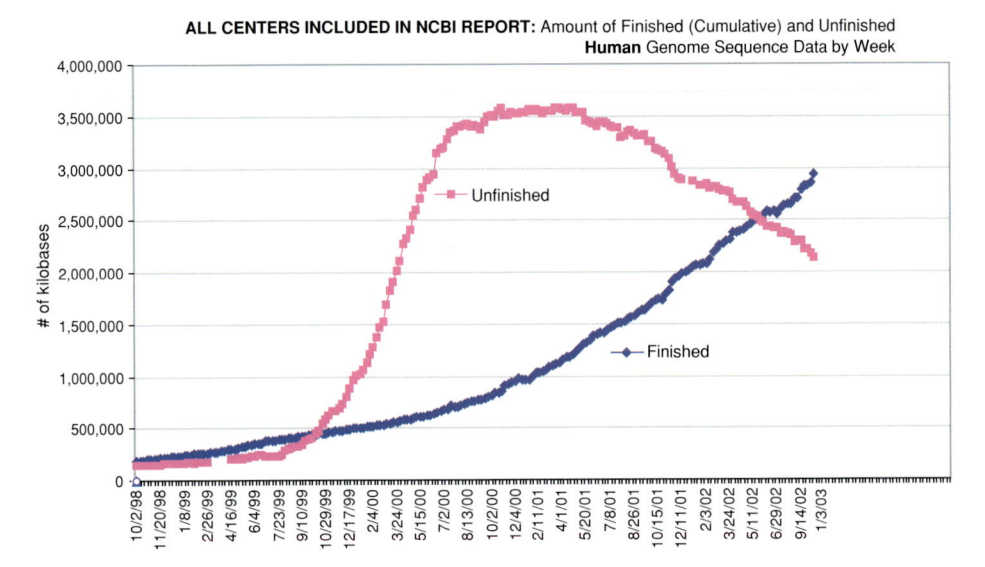

FIGURE 11.8. Scale-up of the public genome project. This figure was provided upon request from the National Human Genome Research Institute. The plot is of data deposited in GenBank between October, 1998 and January 2003.

in-house mapping groups at the various genome centers, and iii) utilizing the centralized BAC end sequence and fingerprint databases as they became available in a "map-as-you-go" approach.

With the major focus on sequencing source clones, little was being done to address the issue of long-range contiguity, that is, building stretches of finished sequence on the order of megabases. To meet their production goals, centers sequenced the clones they could get hold of by whatever mapping approaches they were set up to employ.

11.3.2.2. During Celera: 1998–2000 In May of 1998, Celera Genomics announced its bold intention to sequence the human genome using a whole genome shotgun approach [50]. That is, genomic DNA would be subcloned directly into small (2 kb and 10 kb) inserts and the regional mapping of large-insert clones would be bypassed. The International Consortium's response to the shock and awe inspired by the Celera initiative was a decision to accelerate the pace of the sequencing scale-up. Giving up on the human genome project was not a viable option because two things were important to the Consortium: One, getting the genome finished. There was skepticism that a whole genome shotgun approach would work for human DNA because of the large percentage of interspersed repeats and segmental duplications. Two, and perhaps more important, it was crucial to put the human genome sequence into the public domain for all to use without having to deal with patents and intellectual property issues.

In order to get more of the genome sequenced quickly, an intermediate goal of producing a "working draft" of at least 90% of the genome by Spring of 2000 was accepted (enthusiastically or grudgingly) by the Consortium [9]. For most centers, draft sequence consisted of multiple unordered contigs sequenced and assembled at 3.5–4.5-fold redundancy.

Figure 11.8 portrays the time course for the production of draft and finished sequence from the end of 1998 to the beginning of 2003. Between October 1999 and the end of June 2000 when the draft genome was declared "done", about 3 gigabases of unfinished sequence was produced in a heady, frenetic burst of activity that few participants in their right mind would ever want to repeat. Most of the draft (i.e., unfinished) sequence consisted of heaps of unordered contigs. Assembling these contigs into properly ordered strings of sequence representing each chromosome, and determining how much of the genome was likely to have been covered by the draft sequence, required the work of a talented group of computational biologists called the "Genome Analysis Group" [23]. How they accomplished this feat is beyond the scope of this review.

Producing the working draft of the human genome under such extreme time pressure engendered three tough challenges for project coordination:

- Determining which genome centers were likely to be successful at rapidly scaling up sequencing.
- Redistributing chromosomal territory in light of funding winners and losers and sequencing overachievers and underachievers.
- Avoiding the sequencing of redundant source clones both within and among genome centers.

In consultation with the largest genome centers, the heads of the funding agencies, especially at the National Human Genome Research Institute (NHGRI) in the US and the Wellcome Trust in Great Britain, were forced to make potentially unpopular decisions that the rest of the Consortium centers were required to live with.

11.3.2.3. How Some Sequencing Centers got Derailed and Others got Fast-Tracked Prior to the advent of Celera, the International Consortium had planned a measured pace of sequencing scale-up such that a human genome sequence would be finished by year 2005. In the US, a request for proposals (RFP) entitled "Research network for large-scale sequencing of the human genome" was posted by NHGRI in January 1998 with a due date of October 1998. To qualify, a center needed to have deposited at least 7.5 megabases of finished sequence into GenBank. In September, 1998, the RFP was modified with a due date of December 1998, an eligibility requirement of 15 megabases of finished sequence for "large" centers and 7.5 megabases for "intermediate" centers, and a commitment to complete a working draft sequence of the genome by 2001, with finishing to follow by 2003[7]. A two-tiered review process was planned, with the large centers' proposals to be reviewed in February 1999 and the intermediate centers' proposals in April 1999. After the large center grants were awarded but shortly before the review of the intermediate center proposals, the goal for completing the working draft sequence was moved ahead to the end of February 2000. This change meant that several of the proposals from the intermediate centers were reviewed unfavorably because they lacked a realistic scale-up plan for producing a significant amount of draft sequence in less than a year. As a result, some of the regional physical maps produced by these centers were never used, and personnel who had spent years already on the genome project moved on to other areas. The chromosomal territory

---

[7] See http://grants1.nih.gov/grants/guide/rfa-files/RFA-HG-98-002.html for the original proposal and http://www.genome.gov/10001023#UPDATE2 for the revised proposal.

claimed by the unfunded centers was reassigned to the survivors. The genome centers who remained in the game faced enormous logistical challenges associated with engineering a rapid scale-up: finding space, hiring and training personnel, purchasing equipment, instituting sample-tracking procedures, mapping enough clones to meet the throughput demands, and establishing adequate information technology support.

Shifting policy and goals in the middle of the funding cycle made it difficult (and some would argue impossible) for participants to write a coherent grant proposal, yet the needs were considered sufficiently pressing that hard choices were made to reduce the number of centers in the US. Five genome centers produced most of the public human genome sequence after 1998: one in England (Sanger Institute) and four in the United States (Whitehead Institute for Biomedical Research, Washington University at St Louis, the Department of Energy Joint Genome Institute, and Baylor University College of Medicine). An additional 15 centers (one in France, three in Germany, two in Japan, one in China, and eight in the US) contributed smaller amounts to the draft sequence published in 2001 [23][8].

11.3.2.4. Why Redundant Source Clones were Sequenced Even though chromosomal territory had been parceled out to the sequencing centers in a reasonably clear manner during 1999 and 2000, there was a significant amount of crossover for several reasons:

- Some of the chromosomal map information was erroneous (e.g., a marker mapped to chromosome 3 might actually be on chromosome 15 or two markers thought to be 5 megabases apart were actually 10 megabases apart).
- There was insufficient time to do proper map validation prior to sequencing.
- Some high-throughput centers sequenced random clones in order to keep their pipelines flowing.
- Tracking errors resulted in sequenced clones not actually being what they were supposed to be.
- Data that would help a center determine that a region was being sequenced elsewhere were not available or the databases that would inform a center that a region was being sequenced elsewhere were not checked or the results were ignored.

Expanding on this last point, in an effort to coordinate the sequencing, the National Center for Biotechnology Information (NCBI) established a Clone Registry database (http://www.ncbi.nlm.nih.gov/genome/clone/clonesubmit.html) to keep track of which clones the various genome centers were sequencing or intending to sequence. Since only two approved clone libraries existed, and most of the sequencing was being done from one of these libraries (RPCI11), this approach in theory made sense. The idea was that for each clone in their system, the genome centers would submit its library address (i.e., clone ID), its chromosomal location and marker content (if known) and its sequence accession number (if available) to the Clone Registry along with a status (e.g., reserved, committed, redundant, abandoned, accessioned). As an additional tracking tool, in 2000, the author of this review convinced NCBI to institute a "Phase 0" submission protocol which centers could use for obtaining an accession number for a set of unassembled sequence reads (e.g., a test plate of 96 reads) for source clones they were planning to sequence. In this way a claim on a region

---

[8] Two additional centers, one at The Institute for Genomic Research and the other housed at Applied Biosystems, also contributed to the public human genome project during the mid-90s prior to their association with Celera.

SEQUENCING THE HUMAN GENOME

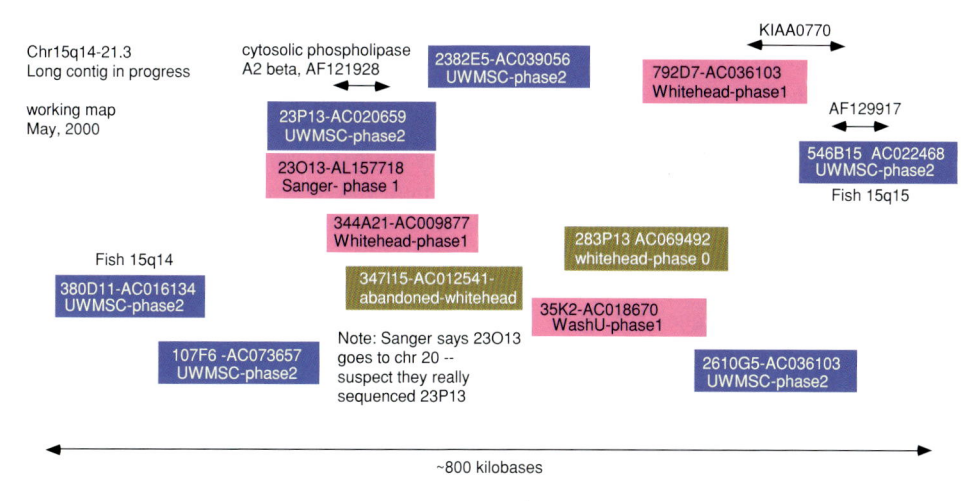

FIGURE 11.9. Portion of the chromosome 15 tiling path from the 20 megabase region assigned to the University of Washington Multimegabase Sequencing Center (UWMSC). Clones (boxes) were identified and ordered using sequence matches. Genes are indicated by double arrows.

could be established well in advance of the time that the working draft sequence of the region was completed. This was important because the cycle time for sequencing a clone, that is, the time elapsed between committing to sequence a particular clone and the assembly of its working draft sequence generally took several weeks, even in the high-throughput centers. Phase 0 sequence had the added advantage of providing map information, since even low-pass coverage could reveal genetic markers or exons of known genes. Finally, use of a sequence search tool such as BLAST would indicate overlaps between Phase 0 reads and other Phase 0, 1, or 2 draft sequences in GenBank so that centers would be aware of probable redundancies.

In practice, the scale-up happened so quickly that some sequencing centers were unable or unwilling to use the Clone Registry and GenBank databases for avoiding redundant coverage of the genome. Moreover, even if redundancies were detected, it was difficult to remove a clone from the system once it had entered the sequencing pipeline. Thus, there was significant overlap among the centers across regions of the genome sequenced as part of the working draft (see Figure 11.9 above).

However, even though it was difficult for centers to stay within their own chromosomal territory, the redundancy was useful in contributing to the overall coverage because the assembly of the draft genome was performed using information from all of the Phase 1 contigs found among overlapping BAC clones [23]. Using various metrics to assess the extent of non-redundant coverage of the genome, it was estimated that the assembled draft sequence produced by the end of June 2000 covered at least 88% of the genome, with 50% of the nucleotides in the genome being found in sequence contigs longer than 826 kb and in scaffolds (i.e., sets of ordered contigs) longer than 2.2 megabases [23].

In mid-February 2001, the International Consortium and Celera Genomics published analyses of their respective working drafts of the human genome in Nature [23] and Science [48]. Interestingly, Celera chose to supplement their in-house whole genome shotgun data with shredded public draft data prior to performing their whole genome assembly, calling into question the success of a pure whole genome shotgun approach. [33, 49][9]. It appears in retrospect, though, that Celera's strategic contribution was significant. The mouse and rat genomes are being sequenced using hybrid strategies that combine the best of clone-based and whole genome shotgun approaches. But if only draft (i.e., unfinished) sequence of a genome is desired, most centers are now using a pure whole genome shotgun approach because the assembly inaccuracies are considered to be minor, and the maps of related genomes can be used to organize the assembled scaffolds. For gene identification and comparative genomics, draft sequence is extremely useful and can be produced at considerably less cost than finished sequence.

11.3.2.5. After Celera: 2000–2003 When completion of the Consortium's working draft sequence was announced in June 2000, about 700 gigabases, or 25% of the eukaryotic portion of the genome was represented in finished high quality sequence (Figure 11.8). Chromosomes 21 [18] and 22 [14], the two smallest chromosomes, had already been finished. The other chromosomes, however, were going to require serious work to meet the finished human genome deadline of April 2003, the 50th anniversary of Watson and Crick's elucidation of the structure of DNA. Once again, the public consortium found itself in the position of facing serious challenges for systems integration:

- The working draft consisted of source clones sequenced to 4–5-fold redundancy, yet for finishing to be efficient, the redundancy needed to be 8–10-fold. Thus, additional shotgunning would need to be done.
- Some centers' drafted source clones were scattered all over the genome, while other centers' clones were fairly well regionally confined. In either case, there were significant overlaps among and between the drafted clones.
- The Bermuda Principles required that genome centers finish all the clones that they drafted.
- Centers needed to scale up their finishing operations and find ways to make finishing faster and easier.
- Territory needed to be shifted around based on each genome center's capacity for finishing.
- Uncovered portions of the genome (gaps) needed to be identified and filled using a variety of labor-intensive procedures such as screening existing libraries for source clones, making new clone libraries, subcloning PCR products, and altering sequencing strategies.
- Significant validation work was required to ensure quality and accuracy in the finished sequence for each chromosome.

In light of the daunting finishing challenges faced by the genome centers, the fact that a high quality product was completed by April 2003 is both amazing and commendable. Two more chromosomes—20 [11] and 14 [19] were published and chromosome Y followed shortly thereafter [44]. A paper detailing statistics about the nature

---

[9] Contigs retrieved from GenBank were chopped into 550 base fragments with an overlapping offset prior to the Celera assembly [48]. In retrospect, this was considered to have been a mistake, according to one Celera researcher who told the author that the Celera data alone actually gave better results.

and quality of the finished human genome sequence is planned for early 2004. Although some intractable gaps remain, 50% of the nucleotides resided in contiguous stretches of sequence (contigs) longer than 26.2 megabases as of April 2003 (and 29.1 megabases as of July 2003)[10]. This represents a 35-fold increase in contig length over the working draft sequence. As cloning, mapping, and assembly techniques continue to improve and new strategies emerge, it is anticipated that some of the currently intractable gaps will be closed. As things stand now, though, these gaps represent only a small portion of the genome.

Completion of the human genome sequence could not have been accomplished without significant cooperation between the genome centers and unrelenting encouragement from the funding agencies. In the initial stages of the genome project there was friendly competition between the centers, which turned less friendly as the pressure to produce data increased and the open range of genomic territory decreased. But then, at the end, there was extreme collegiality among the centers as it became clear that the success of all was dependent upon the success of each. Although no other genome will ever be sequenced the way the human genome was, the human genome project laid a strategic foundation for the successful completion of other cooperative sequencing efforts such as the working drafts of the mouse [30] and the rat (http://public.bcm.tmc.edu/pa/rgsc-genome.htm).

11.3.2.6. How a Sensible Strategy for Finishing the Genome Emerged Approximate chromosomal tiling paths for BAC clones whose Phase 0, 1, 2 or 3 sequences were in GenBank could be determined by sequence matches to genetic markers, BAC end sequences, and other drafted BACs. For each clone, GenBank entries provided information regarding the contributing center, the clone name, and the presumed chromosomal location for the draft sequence entry. Because of the redundancy of the sequencing during the working draft scale-up and various tracking errors in clone name or chromosomal location, the tiling paths contained a patchwork quilt of overlapping BAC clones from various centers sequenced at various levels of redundancy (Figure 11.9).

When the author presented a poster depicting all of the drafted clones mapping to a 20 megabase region of chromosome 15 at the May 2000 International Consortium strategy meeting in Cold Spring Harbor, the group lamented that maps of the entire genome were likely to look the same as the small portion portrayed in Figure 11.9. After a lively discussion, the Consortium leaders agreed to do the following:

- Abandon the rule that each center should finish all of its drafted clones.
- Assign each center specific chromosomal regions to finish.
- Encourage centers to give raw sequence reads derived from BACs outside of their own chromosomal regions to the center now responsible for the region.

Importantly, these decisions enabled centers to devise finishing strategies that were chromosome-centric rather than clone-centric, with several positive outcomes. First, filling regional gaps would now be each center's responsibility and the problems would be spread across all centers. Second, sequence reads from overlapping clones could be combined into larger assemblies with higher redundancy of coverage. This reduced the work of finishing,

---

[10] See http://genome.cse.ucsc.edu/goldenPath/stats.html at the Santa Cruz site for completion of the human genome summary statistics.

especially when data from other centers used a different sequencing chemistry that corrected errors or resolved ambiguities. Third, each center now had fewer clones to finish than it would otherwise have had in that redundant clones could simply be abandoned. Fourth, given that the 4-fold coverage of the draft sequence was insufficient for finishing, centers could start over with new clones, thereby avoiding the need to retrieve data and materials that were several months or years old. Fifth, the apportionment of the chromosomal regions and subsequent exchanges of data in some cases fostered a cooperative spirit among subsets of centers. (Interestingly, not all centers took advantage of the opportunity to use data from other centers because of the information technology effort involved with determining what data to ask for, changing file names to cohere with the center's internal data-tracking conventions, mixing datasets, and the like.)

11.3.2.7. How the Chromosome Tiling Paths and Resultant Master Sequences were Constructed and Managed Although a detailed discussion of the overarching coordination involved in the reconstruction of the chromosome sequences is beyond the scope of this review, a few points can be made. Each chromosome was assigned a "chromosome coordinator" who was responsible for receiving map and sequence information from all the contributing centers and maintaining a periodically updated file of the clones in the finishing tiling path. In parallel, Greg Schuler and his team at NCBI constructed and kept track of the so-called "NT_contigs," which were non-redundant sequences pasted together from the sequences of overlapping source clones (Figure 11.7). Gaps were identified and annotated as to type. For example, a gap could be between ordered contigs in an unfinished source clone, or between source clones (with BAC end sequence data to suggest which clones might exist to span the gap), or between source clones with no information as to gap size. Additional resources such as fosmid libraries and cosmids from telomere regions [37] have been helpful for filling gaps and resolving other finishing problems. Input has also been solicited and received from collaborators interested in genomic features such as segmental duplications [37] and disease-related genes. Periodic assemblies of the genome were constructed by Jim Kent, David Haussler and their team at University of California at Santa Cruz, at NCBI, and at the European Bioinformatics Institute.

Assemblies of the finished human genome are available from:

- UCSC : http://genome.ucsc.edu
- NCBI: http://www.ncbi.nih.gov/mapview/map_search.cgi
- Ensembl: http://www.ensembl.org

In sum, coordination of the human genome project was challenging for several reasons. First, a score of centers from several countries needed to come to agreement on goals, policies, strategies and targets. Second, the centers needed to cooperate with each other and with the centralized resource providers such as NCBI and the Santa Cruz computational team. In essence, the genome project was "open source." Third, quality assurance metrics had to be established, adhered to, and enforced. Fourth, the task was daunting in its complexity. And finally, fifth, unanticipated external factors altered the course of the project at various junctures, thereby imposing readjustments of strategies and resources.

11.4. ARE THERE LESSONS TO BE LEARNED FROM THE HUMAN GENOME PROJECT?

One could argue that the unique nature of the human genome project, the length of time it took to be done, and the various unanticipated developments that happened along the way are such that didactic generalizations of any significant import cannot be made. While this may be true, some concluding comments are nonetheless in order. First, there are some things that the leaders of the genome project did right:

- Open data release policies and sharing of protocols.
- Definition of standards for product quality and completeness.
- Establishment of centralized resources.
- Capture of the public imagination.

The decision to release unpublished sequence to the public so that others could use the data for their research has established a precedent for data sharing and an open source mentality that investigators in other areas might be encouraged to adopt. Early data release does pose two difficulties that need to be addressed. First, because career advancement in science is tied to publications, there must be a mechanism for investigators involved with large data gathering or technology development projects to receive proper credit for their contributions [40]. Second, the user groups need to be educated in the pitfalls associated with the dissemination of potentially erroneous data. For example, although the sequences of genomes being drafted now via a whole genome shotgun approach will be mostly correct, assembly errors will occur, especially in highly duplicated gene families. The sequences and analyses found in the centralized databases and posted on various web sites cannot be taken as gospel truth.

For the human genome project, the fact that there were quality standards and indications of quality in the database entries made it easier for users to evaluate the data critically. The Consortium even went so far as to conduct "quality control exercises" which involved the exchange of data among centers, with follow-up reports of detected errors. This enabled the community to develop reality-based standards and hold each other to account.

Central resources such as the genetic linkage maps, BAC end sequence database, the fingerprint contig database, assemblies of the draft genome, collections of transcribed (cDNA) sequences, and the like helped enormously with the difficult task of constructing the tiling paths for each of the chromosomes. Moreover, granting that the influence and importance of the world wide web is all-pervasive, it must be noted that the human genome project benefitted enormously from the ease of immediate access that the web has provided.

Finally, the architects of the human genome project were able to marshall public support. At key junctures, announcements of progress were made such that even if most "people on the street" did not know what the human genome project was, they had at least heard of it. A challenge for biology now is the articulation and implementation of a new vision with equal panache. Whether current contenders such as systems biology, nanotechnology, or predictive, preventive and personalized medicine will make the grade remains to be seen.

Additional potentially generalizable lessons from the human genome project include the following:

- For a procedure to be useful it must be usable.
- A process can only be as fast as its slowest step.
- Procedures must be developed and integrated in light of a goal.
- Reducing the number of players will make for a more efficient endeavor.
- External developments will enable/force changes in strategy.

At the time the genome project began it was not obvious which procedures would be best. Indeed, technology improvements in any one area would have ramifications for others. For example, with the compute power and assembly programs available in 1991, shotgun assembly of a 150 kb BAC would have been deemed impossible. Yet by 1997 it was easy, and by year 2000, 150 megabase-sized genomes were being assembled using a whole genome shotgun approach [34]. Another example: Because companies such as Applied Biosystems kept pushing the envelope in their electrophoresis-based sequencing protocols by developing better enzymes, dyes, polymers, and detection machines, a sequencing technology that some visionaries thought would be transient turned out to be dominant. Competing sequencing technologies were not useful because they were not adequately reduced to practice. The lesson is that persons developing technologies must ask themselves: What will this be used for? Does it integrate well with the overall process? Will it scale? Will ordinary technicians be able to implement the technology? During the genome project, significant amounts of money were spent developing tools and resources that were never used, primarily because they did not meet the needs of the genome centers.

Slow steps in a process can be addressed by finding ways to speed them up or by eliminating them altogether. One of the major appeals of a whole genome shotgun approach is a bypass of the up-front mapping of source clones, a step that proved to be painfully slow during the human genome project. Mapping is now done after the fact by tapping the data in centralized source clone resources such as BAC end sequences and fingerprint clusters. A whole genome shotgun approach also has the advantage of eliminating the need to prepare thousands of shotgun libraries from source clones. However, with whole genome shotgun, obtaining an assembly that is faithful to the genome of interest becomes the most difficult and potentially slowest step. This is true especially for genomes with little map information or with significant allelic variation among copies of the chromosomes.

When it comes to integrating the steps of a process in light of an overall objective, compartmentalization is a danger especially for a large scale effort. During the genome project, most centers, especially the large ones, had groups of people focussed only on one major activity. For example, there would be an informatics group, a sequence production group, a machine development group, a finishing group, a mapping group, etc. What was often missing was a group of people who understood every step of the process from a hands-on perspective and how the steps fit together. Because of compartmentalization, groups focussed on their own goals and needs, which may or may not have served the overall goal well. For example, a production group concerned only about generating sequence reads might not have paid adequate attention to data quality, because they never actually looked at or used the data. For database design reasons, a software engineer might have thought that no information such as tags for sequencing chemistry should be embedded in the file name of a sequencing read yet finishers typically wanted such information to be viewable in their sequence editor. Mappers often needed sequence data right away so they could make probes but production teams were wedded to an inflexible pipeline and would ignore the mappers' requests. Recalling the symphony analogy, genome centers needed conductors else the orchestra played inharmonious tunes.

The number of centers involved in the sequencing of the human genome was sufficiently large that overall project coordination was difficult. Genomes are now being assigned to only one or a few sequencing centers, which conduces to greater efficiency. In the early stages of the human genome project, the "let a thousand flowers bloom" approach to strategy development made sense because the best answers were unclear. Innovation from several perspectives spread across numerous research laboratories was encouraged. Once the strategy matured and was used to produce large volumes of useful data, efficiency was valued over innovation, and novel strategies could not effectively compete. Moreover, had commitments to preexisting genome centers and a genuine desire for international participation in the sequencing of the genome not been an issue, it would have made sense to have done the project in a centralized rather than distributed manner once the scale-up was deemed feasible.

As for external curve balls, the disconnect between available physical maps and increasing sequencing capacity that existed circa 1997 made the intrusion of Celera almost inevitable. Even if Celera had not entered the scene, there was demand from biologists for the genome project to speed up. Given the choice between having half of the genome all finished and all of the genome half finished, biologists wanted the latter. In other words, an error-prone working draft came to be an acceptable goal because of the strong desire for access to genes. The draft turned out to be extremely useful, and it was produced at less cost than finished sequence. If this change in scope could have been anticipated, centralized resources such as the BAC end sequences and fingerprint databases might have been established earlier. But the moral here might be that comprehensive foresight is impossible. Perhaps the best that can be hoped for is the ability and willingness of individuals and organizations to adapt to changing circumstances, which the International Consortium did rather well.

A reference sequence for the human genome is now essentially finished. Although most of the participants in the genome project will fade into obscurity, their achievement will last. A foundation has been laid for further genetic studies that will improve our understanding of what makes us human. Moreover the genome project has shown that large-scale biology and technology endeavors can be done in the context of a cooperative network of laboratories and organizations that are willing to share data and resources. Although coordination of such undertakings is difficult, in the end the benefits to productivity are potentially great. This may be the most enduring lesson of the human genome project.

ACKNOWLEDGEMENTS

The author would like to thank Chad Nusbaum, Stephen Lasky, Gane Wong, Monica Dors, and Pat Ehrman for their comments on this review, Adam Felsenfeld for helpful discussion, and her many friends in the International Consortium and Celera for sharing ideas, data, and fun times during the sequencing of the human genome.

REFERENCES

[1] S. Aparicio, J. Chapman, E. Stupka, N. Putnam et al., *Science*, 297:1301–1310, 2002.
[2] BAC Resource Consortium, *Nature*, 409:953–958, 2001.
[3] J.A. Bailey, A.M. Yavor, L. Viggiano, D. Misceo, J.E. Horvath, N. Archidiacono, S. Schwartz, M. Rocchi, and E.E. Eichler. *Am. J. Hum. Genet.*, 70:83–100, 2001.
[4] S. Batzoglou, D.B. Jaffe, K. Stanley, J. Butler, S. Gnerre, E. Mauceli, B. Berger, J.P. Mesirov, and E.S. Lander. *Genome Res.*, 12:177–189, 2002.
[5] J.K. Bonfield, K. Smith, and R. Staden. *Nucleic Acids Res.*, 23:4992–4999, 1995.
[6] E.Y. Chen, Y.C.Liao, D.H. Smith, H.A. Barrera-Saldana, R.E. Gelinas, and P.H. Seeburg. *Genomics*, 4:479–497, 1989.
[7] G.M. Church and S. Kieffer-Higgins. *Science*, 240:185–188, 1988.
[8] L. Clarke and J. Carbon. *Cell*, 9:91–99, 1976.
[9] F.S. Collins, A. Patrinos, E. Jordan, A. Chakravarti, R. Gesteland, and L. Walters. *Science*, 282:682–689, 1998.
[10] P.L. Deininger. *Anal. Biochem.*, 129:216–223, 1983.
[11] P. Deloukas, L.H. Matthews, J. Ashurst, J. Burton et al., *Nature*, 414:865–871, 2001.
[12] C. Dib, S. Faure, C. Fizames, D. Samson et al., *Nature*, 380:152–154, 1996.
[13] R. Drmanac, S. Drmanac, Z. Strezoska, T. Paunesku et al., *Science*, 260:1649–1652, 1993.
[14] I. Dunham, N. Shimizu, B.A. Roe, S. Chissoe et al., *Nature*, 402:489–495, 1999.
[15] B. Ewing and P. Green. *Genome Res.*, 8:186–194, 1998.
[16] D. Gordon, C. Abajian, and P. Green. *Genome Res.*, 8:195–202, 1998.
[17] D. Gordon, C. Desmarais, and P. Green. *Genome Res.*, 11:614–625, 2001.
[18] N. Hattori, A. Fujiyama, T.D. Taylor, H. Watanabe et al., *Nature*, 405:311–319, 2000.
[19] R. Heilig, R. Eckenberg, J.L. Petit, N. Fonknechten et al., *Nature*, 421:601–607, 2003.
[20] X. Huang and A. Madan. *Genome Res.*, 9:868–877, 1999.
[21] T.J. Hudson, L.D. Stein, S.S. Gerety, J. Ma et al., *Science*, 270:1945–1954, 1995.
[22] International Human Genome Mapping Consortium, *Nature*, 409:934–941, 2001.
[23] International Human Genome Sequencing Consortium, *Nature*, 409:860–921, 2001.
[24] P.A. Ioannou, C.T. Amemiya, J. Garnes, P.M. Kroisel, H. Shizuya, C. Chen, M.A. Batzer, and P.J. de Jong. *Nat. Genet.*, 6:84–89, 1994.
[25] B.F. Koop, L. Rowan, W.Q. Chen, P. Deshpande, H. Lee, and L. Hood. *Biotechniques*, 14: 442–447, 1993.
[26] G.G. Mahairas, J.C. Wallace, K. Smith, S. Swartzell, T. Holzman, A. Keller, R. Shaker, J. Furlong, J. Young, S. Zhao, M.D. Adams, and L. Hood. *Proc. Natl. Acad. Sci. U.S.A.*, 96:9739–9744, 1999.
[27] M.A. Marra, T.A. Kucaba, N.L. Dietrich, E.D. Green, B. Brownstein, R.K. Wilson, K.M. McDonald, L.W. Hillier, J.D. McPherson, and R.H. Waterston. *Genome Res.*, 7:1072–1084, 1997.
[28] V.A. McKusick. *Genomics*, 5:385–387, 1989.
[29] M.J. Miller and J.L. Powell. *J. Comput. Biol.*, 1:257–269, 1994.
[30] Mouse Genome Sequencing Consortium, *Nature*, 420:520–562, 2002.
[31] J.C. Mullikin and Z. Ning. *Genome Res.*, 13:81–90, 2003.
[32] K.K. Murray. *J. Mass Spectrom.*, 31:1203–1215, 1996.
[33] E.W. Myers, G.G. Sutton, H.O. Smith, M.D. Adams, and J.C. Venter. *Proc. Natl. Acad. Sci.*, 99:4145–4146, 2002.
[34] E.W. Myers, G.G, Sutton, A.L.Dulcher, I.M. Dew et al., *Science*, 287:2196–2204, 2000.
[35] M.V. Olson. *Science*, 270:394–396, 1995.
[36] K. Osoegawa, A.G. Mammoser, C. Wu, E. Frengen, C. Zeng, J.J. Catanese, and P.J. deJong. *Genome Res.*, 11:483–496, 2001.
[37] H. Riethman, A. Ambrosini, C. Castaneda, J. Finklestein, X.L. Hu, U. Mudunuri, S. Paul, and J. Wei. *Genome Res.*, 14:18–28, 2004.
[38] L. Rowen, G. Mahairas, and L. Hood. *Science*, 278:605–607, 1997.
[39] L. Rowen, S.R. Lasky, and L. Hood. *Methods in Microbiology*, 28:155–192, 1999.
[40] L. Rowen, G.K. Wong, R.P. Lane, and L. Hood. *Science*, 289:1881, 2000.
[41] B.B. Rosenblum, L.G. Lee, S.L. Spurgeon, S.H. Khan, S.M. Menchen, C.R. Heiner, and S.-M. Chen. *Nucleic Acids Res.* 25: 4500–4504, 1997.

[42] F. Sanger, S. Nicklen, and A.R. Coulson. *Proc. Natl. Acad. Sci. U.S.A.*, 74:5463–5467, 1977.
[43] H. Shizuya, B. Birren, U.J. Kim, V. Mancino, T. Slepak, Y. Tachiiri, and M. Simon. *Proc. Natl. Acad. Sci. U.S.A.*, 89:8794–8797, 1992.
[44] H. Skaletsky, T. Kuroda-Kawaguchi, P.J. Minx, H.S. Cordum *et al*. *Nature*, 423:825–837, 2003.
[45] M. Strathmann, B.A. Hamilton, C.A. Mayeda, M.I. Simon, E.M. Meyerowitz, and M.J. Palazzolo. *Proc. Natl. Acad. Sci. U.S.A.*, 88:1247–1250, 1991.
[46] S. Tabor and C.C. Richardson. *Proc. Natl. Adac. Sci. U.S.A.*, 92:6339–6343, 1995.
[47] J.C. Venter, H.O. Smith, and L. Hood. *Nature*, 381:364–366, 1996.
[48] J.C. Venter, M.D. Adams, E.W. Myers, P.W. Li *et al*., *Science*, 291:1304–1351, 2001.
[49] R.H. Waterston, E.S. Lander, and J.E. Sulston. *Proc. Natl. Acad. Sci.*, 99:3712–3716, 2002.
[50] J.L. Weber and E.W. Myers. *Genome Res.*, 7:401–409, 1997.
[51] S. Zhao J. Malek, G. Mahairas, L. Fu, W. Nierman, J.C. Venter, and M.D. Adams. *Genomics*, 63:321–332, 2000.

III

Nanoprobes for Imaging, Sensing and Therapy

12

Hairpin Nanoprobes for Gene Detection

Philip Santangelo, Nitin Nitin, Leslie LaConte and Gang Bao
Department of Biomedical Engineering, Georgia Institute of Technology and Emory University, Atlanta, GA 30332

12.1. INTRODUCTION

First hypothesized by Crick in 1958 [1], the central of dogma of biology states that DNA begets messenger RNA, which is then translated into protein. The ability to monitor this gene expression process and measure quantitatively the expression levels of mRNA can provide tremendous insight into normal and diseased states of living cells, tissues and animals, and clues to maintaining health and curing diseases. The first demonstration of mRNA being complementary to a gene and responsible for protein translation [2] was made by Hall and Spiegelman using a nucleic acid hybridization assay in which the reconstitution of the double-stranded DNA structure occurs only between perfect, or near-perfect complementary DNA strands, leading to a method of detecting complementary nucleotide sequences. In many ways this method is the precursor to both polymerase chain reaction (PCR) and other modern hybridization assays.

The development of technologies to measure gene expression levels has been playing an essential role in biology and medicine ever since the discovery of the double helical structure of DNA [3]. These technologies including oligonucleotide synthesis [4], PCR [5], Northern hybridization (or Northern blotting) [6], expressed sequence tag (EST) [7], serial analysis of gene expression (SAGE) [8], differential display [9], and DNA microarrays [10]. These technologies, combined with the rapidly increasing availability of genomic data for numerous biological entities, present exciting possibilities for understanding human health and disease. For example, pathogenic and carcinogenic sequences are increasingly being used as clinical markers for diseased states. However, the detection and identification of foreign or mutated nucleic acids is often difficult in a clinical setting due to the low abundance of diseased cells in blood, sputum, and stool samples. Consequently, the target sequence of interest is typically amplified by PCR or nucleic acid sequence-based amplification (NASBA). These assays are usually complex and prone to false-positives that hinder their clinical applications. For example, improper primer design can result in the amplification of unintended sequences or the primers could hybridize to each other and form "primer-dimer" amplicons [11]. Therefore, homogeneous assays that utilize fluorescent intercalating agents such as SybrGreen could generate a signal in the presence and absence of the pathogenic marker [12]. Alternatively, detection techniques that require the opening of the PCR tube for analysis (i.e., sequencing, electrophoresis, etc.) could lead to sample contamination. Clearly there is a need for a molecular probe that can detect nucleic acids with high specificity and generate a measurable signal upon target recognition to allow analysis in a sealed reaction tube.

Over the last decade or so, there is increasing evidence to suggest that RNA molecules have a wide range of functions in living cells, from physically conveying and interpreting genetic information, to essential catalytic roles, to providing structural support for molecular machines, and to gene silencing. These functions are realized through control of both the expression level of specific RNAs, and possibly through their spatial distribution. *In vitro* methods that use purified DNA or RNA obtained from cell lysate can provide a relative (mostly semi-quantitative) measure of mRNA expression level within a cell population; however they cannot reveal the spatial and temporal variation of mRNA within a single cell. Methods for gene detection in intact cells such as *in situ* hybridization [13] is capable of providing information of RNA expression level and localization in single cells; however, it relies on removal of the excess probes to achieve specificity, and therefore cannot be used with living cells. The ability to image specific mRNAs in living cells in real time can provide essential information on mRNA synthesis, processing, transport, and localization, and on the dynamics of mRNA expression and localization in response to external stimuli; it will offer unprecedented opportunities for advancement in molecular biology, disease pathophysiology, drug discovery, and medical diagnostics.

One approach to tagging and tracking endogenous mRNA transcripts in living cells is to use fluorescently labeled oligonucleotide probes that recognize specific RNA targets via Watson-Crick base pairing. In order for these probes to truly reflect the mRNA expression *in vivo*, they must satisfy a number of criteria: they need to be able to distinguish signal from background, convert target recognition *directly* into a measurable signal, and differentiate between true and false-positive signals. Further, these probes are required to have high sensitivity for quantifying low gene expression levels and fast kinetics for tracking alterations in gene expression in real time. In addition, they must be amenable to methods that enhance the internalization of the probes into cells with high efficiency.

In this chapter we will archive the design aspects, biological issues and challenges in developing nanostructured oligonucleotide probes for living cell gene detection. Although it is not possible to review all the relevant work in this chapter, we intend to provide extensive background information and detailed discussions on major issues, aiming to facilitate the quantitative and real-time mRNA measurements, especially in live cells and tissues.

HAIRPIN NANOPROBES FOR GENE DETECTION

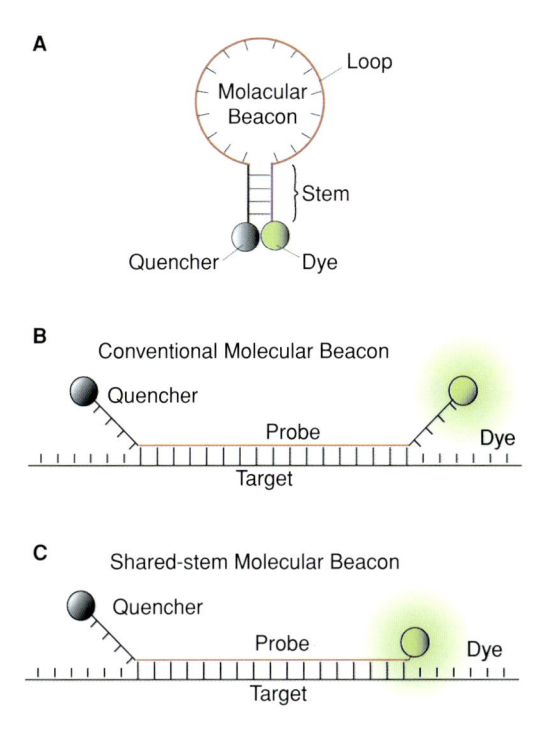

FIGURE 12.1. Illustrations of molecular beacons. (a) Molecular beacons are stem-loop hairpin oligonucleotide probes labeled with a reporter fluorophore at one end and a quencher molecule at the other end. (b) Conventional molecular beacons are designed such that the short complementary arms of the stem are independent of the target sequence. (c) Shared-stem molecular beacons are designed such that one arm of the stem participates in both stem formation and target hybridization.

12.2. NANOPROBE DESIGN ISSUES FOR HOMOGENEOUS ASSAYS

The detection and quantification of specific mRNAs require probes to have high sensitivity and specificity, especially for low abundance genes and with a small number of diseased cells in clinical samples. Further, for detecting genetic alterations such as mutations and deletions, the ability to recognize single nucleotide polymorphisms (SNPs) is essential. When designed properly, hairpin nucleic acid probes have the potential to be highly sensitive and specific. As shown in Figure 12.1, one class of such probes is known as molecular beacons, which are dual-labeled oligonucleotide probes with a fluorophore at one end and a quencher at the other end [14]. They are designed to form a stem-loop structure in the absence of a complementary target so that fluorescence of the fluorophore is quenched. Hybridization with target nucleic acid in solution or in a living cell opens the hairpin and physically separates the fluorophore from quencher, allowing a fluorescence signal to be emitted upon excitation. Thus, molecular beacons enable a homogenous assay format where background is low. The design of the hairpin structure provides an independently adjustable energy penalty for hairpin opening which improves probe specificity [15, 16]. The ability to transduce target recognition *directly* into a fluorescence signal with high signal-to-background ratio, coupled with an improved specificity, has allowed molecular beacons to enjoy a wide range of biological and biomedical applications.

A conventional molecular beacon has four essential components: loop, stem, fluorophore, and quencher, as illustrated in Figure 12.1a. The loop usually consists of 15–25 nucleotides and is selected based on target sequence and melting temperature. The stem, formed by two complementary short-arm sequences, is typically 4–6 bases long and is usually chosen to be independent of the target sequence (Figure 12.1B). Molecular beacons, however, can also be designed such that one arm of the stem participates in both stem formation and target hybridization (shared-stem molecular beacons), as illustrated schematically in Figure 12.1C. Although a molecular beacon can be labeled with any desired reporter-quencher pair, proper selection of the reporter and quencher could improve the signal-to-background ratio and multiplexing capabilities.

There are two major design issues of nanostructured molecular probes for gene detection: specificity and melting temperature. First, to ensure specificity, for each target gene, one can use NCBI BLAST [17] or similar software to select 16–20 base target sequences that are unique for the target mRNA. Secondly, since the melting temperature of molecular beacons affects both the signal-to-background ratio and detection specificity, especially for mutation detection, one has to systematically adjust the G-C content of the target sequence, the loop and stem lengths and the stem sequence of the molecular beacon to realize the optimal melting temperature. In particular, it is necessary to understand the effect of molecular beacon design on melting temperature so that, at 37°C, single-base mismatches in target mRNAs can be differentiated. In the case of homogeneous assays, secondary structure of mRNA is not an issue due to the ability to denature the mRNA structure via temperature.

The loop, stem lengths and sequences are the critical design parameters for molecular beacons, since at any given temperature they largely control the fraction of molecular beacons in each of three different conformational states: bound-to-target, stem-loop, and random-coil [16]. In many applications, the choices of the probe sequence are limited by target-specific considerations, such as the sequence surrounding a single nucleotide polymorphism (SNP) of interest. However, the probe and stem lengths, and stem sequence, can be adjusted to optimize the performance (i.e., specificity, hybridization rate and signal-to-background ratio) of a molecular beacon for a specific application [15, 18].

In general, it has been found that molecular beacons with longer stem lengths have an improved ability to discriminate between wild-type and mutant targets in solution over a broader range of temperatures. This can be attributed to the enhanced stability of the molecular beacon stem-loop structure and the resulting smaller free energy difference between closed (unbound) molecular beacons and molecular beacon-target duplexes, which generates a condition where a single-base mismatch reduces the energetic preference of probe-target binding. The competition between the two stable conformations of a molecular beacon (i.e., closed and bound to target) also explains why it has an enhanced specificity compared with linear probes. Longer stem lengths, however, are accompanied by a decreased probe-target hybridization kinetic rate. Similarly, molecular beacons with short stems have faster hybridization kinetics but suffer from lower signal-to-background ratios compared with molecular beacons with longer stems. It is interesting to note, however, that stem-less molecular beacons, which lack the short complementary arms and rely solely on the random-coiled nature and interactions between the dye and quencher to maintain a dark state, are still able to differentiate between bound and unbound states [19].

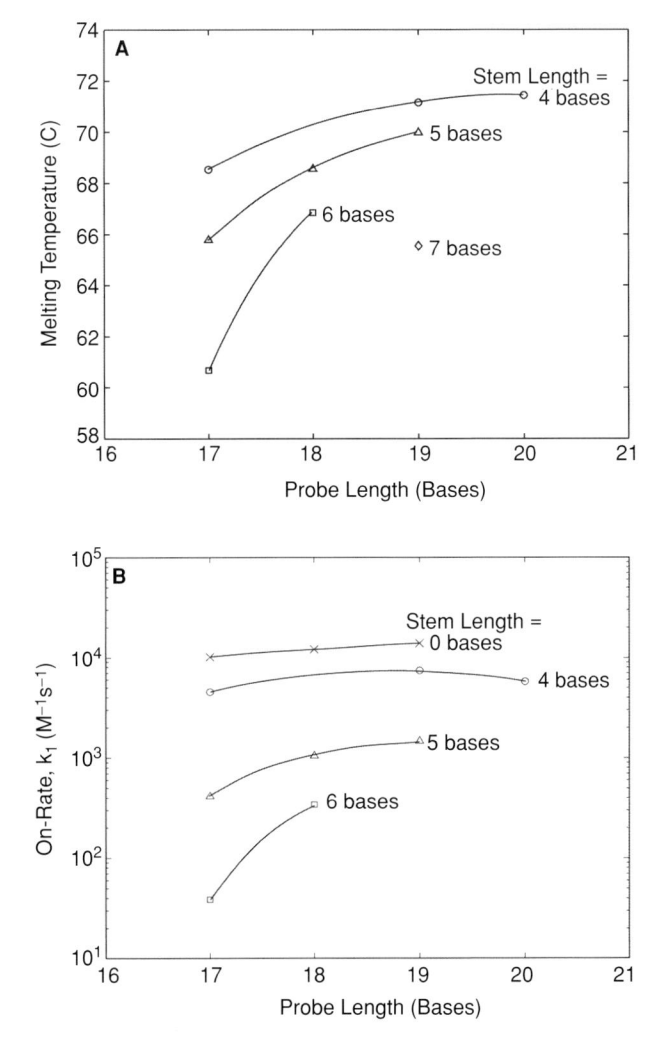

FIGURE 12.2. Structure-function relations of molecular beacons. (a) Melting temperatures for molecular beacons with different structures in the presence of target. (b) The rate constant of hybridization k_1 (on-rate constant) for molecular beacons with various probe and stem lengths hybridized to their complementary targets (Adapted from Ref. 15).

Increasing the probe length of molecular beacons results in improved target affinity and increased kinetic rates, but leads to a reduced specificity. The effect of probe length on the behavior of molecular beacons is typically less dramatic compared with that of stem length and can be used to fine-tune functionality [15]. The structure-function relationship of molecular beacons is illustrated in Figure 12.2 in which the melting temperature and kinetic on-rate constant are displayed as a function of probe length and stem length. A detailed description of the thermodynamic parameters of probe-target hybridization as determined by the structure of molecular beacons can be found in [15].

Selecting a fluorophore label for a molecular beacon as the reporter is usually not as critical as the hairpin probe design since many conventional dyes can yield satisfactory results. However, proper selection could yield additional benefits such as an improved signal-to-background ratio and multiplexing capabilities. Since each molecular beacon utilizes only one fluorophore it is possible to use multiple molecular beacons in the same assay, assuming that the fluorophores are chosen with minimal emission overlap [20]. Molecular beacons can even be labeled simultaneously with two fluorophores, i.e., "wavelength shifting" reporter dyes, allowing multiple reporter dye sets to be excited by the same monochromatic light source yet fluorescing in a variety of colors [21]. Specifically, a "harvester" fluorophore absorbs the excitation light and transfers the energy to an "emitter" fluorophore which emits fluorescence. The same "harvester" fluorophore is used with various emitter fluorophores to generate multiple colors. Another possibility is to use quantum dots (QDs) with different emission wavelengths as the reporter dye. Similar to wavelength-shifting dyes, many QDs can be excited with a single UV lamp light source [22]. However, it remains to be seen if QDs can be effectively quenched and if their functional size can be reduced to be useful when conjugated to a MB.

Clearly, multicolor fluorescence detection of different beacon/target duplexes can become a powerful tool for the simultaneous detection of multiple genes. For example, almost all cancers are caused by multiple genetic alterations in cells and the detection of cancer cells in a clinical sample would require the use of multiple tumor markers. Thus, the use of multiple molecular beacons is potentially a powerful tool in the early detection and diagnosis of cancer.

Similar to fluorophore selection, choosing the optimal quencher can also improve the signal-to-background ratio of molecular beacons. Organic quencher molecules such as dabcyl, BHQ-II^{TM} (blackhole quencher) (Biosearch Tech), BHQ-III (Biosearch Tech) and Iowa Black (IDT) can all effectively quench a wide range of fluorophores by both fluorescence resonance energy transfer (FRET) and the formation of an exciton complex between the fluorophore and the quencher [23]. In addition to organic quenchers, gold nanoparticles can also be used as quenchers [24]. It should be noted, however, that the interaction between the gold particle and the fluorophore could significantly affect the performance of molecular beacons.

12.3. IN VITRO GENE DETECTION

12.3.1. Pathogen Detection

The detection and identification of pathogens is often painstaking and fruitless due to the low abundance of diseased cells in clinical samples. The genomic sequences of the pathogen can be amplified through methods such as PCR and nucleic acid sequence-based amplification (NASBA), but the nucleic acid targets are often lost amidst other unintended products of amplification. The unique properties of molecular beacons have led to their application in a large number of homogeneous assays involving the sensitive detection of diseased states. Since most pathogens can be identified by their genomic sequences, molecular beacon-PCR assays provide a rapid and accurate method for pathogen detection and identification by simply adding molecular beacons to a typical PCR reaction tube. During the annealing stage, the molecular beacons bind to the desired target and a fluorescence signal is generated. Excess molecular beacons remain in the stem-loop structure and thus do not emit fluorescence. Therefore, the intensity of the signal is proportional to the target copy number. Further, when the temperature is increased during the extension stage, the molecular beacon melts away from the target and thus does not interfere with polymerization. It is important that the target-binding domain of the molecular beacon be designed with a melting temperature slightly above the annealing temperature for optimal results. The melting temperature of the stem should be about $10°C$ above the annealing temperature to ensure that unbound molecular beacons remain in the stem-loop conformation and, for the most part, are not open due to thermal fluctuations. Since molecular beacons can detect complementary targets during PCR amplification, subsequent handling is not necessary, allowing the use of sealed tubes and reducing the risk of carry-over contamination. Further, since molecular beacons only fluoresce in the presence of complementary targets, unintended amplification products such as "primer-dimers" and false amplicons are not detected.

It is not hard to imagine the development of simple and rapid molecular beacon-based assays for the sensitive detection of nucleic acids in a clinic to help identify specific pathogens. Molecular beacons can easily be used to differentiate between fungal pathogens such as *Candida dubliniensis* and *Candida albicans*, which possess similar phenotypic and genotypic characteristics [25, 26]. In a controlled study, the correct pathogen was identified 100% of the time following PCR amplification. Such accurate determination of relevant disease states could provide improved strategies for proper disease management. In another example, NASBA utilized with molecular beacons was able to correctly identify West Nile virus, St. Louis encephalitis, and Hepatitis B viruses with high sensitivity and specificity [27, 28]. This type of assay can easily be exploited for clinical use as a common blood screening diagnostics test.

Molecular beacons have also been found to be useful as a fast and reliable tool for the timely detection of food and water-borne pathogens, such as *Salmonella* [29]. Molecular beacons successfully differentiated between *Salmonella* and similar pathogens such as *Escherichia coli* and *Citrobacter freundii*. Therefore, detection of such pathogens can prove to be helpful in preventing bacterial disease outbreaks. Molecular beacons or other nanostructured molecular probes also have the potential to become a powerful tool in biodefense.

Using molecular beacons for the accurate detection of nucleic acid targets is not, of course, limited to the identification of pathogens but has also been extended to other assays such as the determination of the sex of embryos [30], and the differential expression of specific genes under varying environmental conditions [31]. The potential applications of molecular beacons in pathogen detection seem limitless.

12.3.2. Mutation Detection and Allele Discrimination

One of the major benefits of molecular beacons is their ability to discriminate between targets with just a single base mismatch. The stem-loop structure of molecular beacons increases the specificity of beacon/target hybridization compared with linear probes and thus may offer advantages over other reporter probes such as Taqman (Applied Biosystems) [32]. Assays requiring the detection of single nucleotide polymorphisms can be performed with either one or two molecular beacons [33]. For single molecular beacon assays, the molecular beacon must be carefully designed such that during the annealing stage, it only

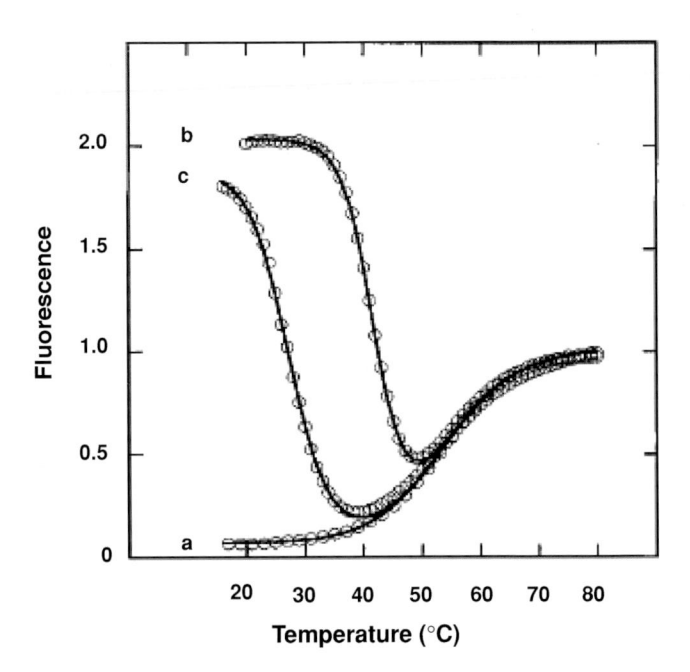

FIGURE 12.3. Thermal denaturation profiles of solutions containing molecular beacons: curve a, in the absence of targets; curve b, in the presence of a 6-fold excess of perfectly complementary targets; and curve c, in the presence of a 6-fold excess of single-base mismatched targets (Adapted from Ref. 16).

binds to the complementary target. This often requires the generation of thermal denaturation profiles to determine the "window of discrimination", as illustrated in Figure 12.3 where the best assay temperature is at 37°C. Specifically, an annealing temperature is determined that allows molecular beacons to hybridize to perfectly complementary targets but not targets with single-base mismatches. For assays utilizing two molecular beacons, one molecular beacon is designed to hybridize to the wild-type target, while the second molecular beacon is designed to hybridize to the mutant target. Each molecular beacon is labeled with a unique fluorophore with non-overlapping emission curves. Typically only the reporter fluorescence corresponding to the amplified target (wild-type or mutant) is detected. If both types of target are present and amplified, fluorescence from both probes will be detected. The competitive nature of this assay allows for high specificity and sensitivity. It has been found that even just 10 copies of a rare target can be detected in the presence of 100,000 copies of abundant target after PCR amplification [34]. A variant level of about 1% was detected following NASBA [35]. These levels of sensitivity are far superior to sequencing, which can only detect variant levels of about 10–20%.

The exact fraction of mutant alleles in a clinical sample/tissue can be determined by performing what has been dubbed "digital PCR" [36]. In this assay, the extracted DNA templates are diluted into multi-well plates such that there is only one template molecule per two wells, on average. After PCR, the fluorescent signal indicates whether the template was wild-type or mutant. The fluorescent signals from a microplate can therefore provide a digital readout of the fraction of mutant alleles. Recently, this assay has been successfully used to demonstrate the presence of allelic imbalance in colorectal tumors [37]. An alternative method to quantify heteroplasmy levels involves comparing the fluorescence intensities of the differently labeled molecular beacons during the cycle of maximum amplification [12]. This information can be obtained by taking the first derivative of the amplification curve and integrating over several of the cycles surrounding the cycle of maximum amplification. It was found that as little as 5% heteroplasmy could be measured reliably using this method.

The high sensitivity and single base specificity of molecular beacon-based assays has extended their use to many clinical and epidemiological applications. In one example, a clinical assay that detects point mutations was designed to determine whether or not *Plasmodium falciparum* samples, a parasite that causes malaria, contained an antifolate resistance-associated mutation [38]. Identification of such a mutation is extremely important for proper drug treatment. Assays have also been developed to rapidly screen blood samples for mutations in specific genes such as methylenetetrahydrofolate reductase (MTHFR) [39]. A cytosine to thymine mutation in MTHFR has been related to increased risk of cardiovascular disease and neural tube defects. The identification of such a mutation could lead to improved patient awareness. More information concerning the design and applications of molecular beacons can be found in www.molecular-beacons.org.

12.4. INTRACELLULAR RNA TARGETS

One of the most exciting and promising applications of nanostructured molecular probes such as molecular beacons is their potential use for the real-time visualization of RNA expression in living cells and tissues. For example, the ability to monitor the level of mRNA expression in living cells will provide important information concerning the temporal and spatial processing, localization, and transport of specific mRNA under various conditions. Further, detecting pathogenic markers will provide a means of locating and identifying diseased cells, allowing rapid diagnosis and prognosis of a disease.

To sensitively detect and quantify mRNA levels in live cells, it is extremely important to understand the form, distribution, and dynamics of target mRNA in live cells in order to optimize the probe design and measurement. Further, it is important to understand the impact of the probe on the cell, including its delivery, chemistry, possible toxicity, and non-specific interactions. In the following sections, we describe the basic features of mRNA in living cells, aiming to set a stage for more discussions in Section 5 on intracellular mRNA detection and quantification.

12.4.1. Cytoplasmic and Nuclear RNA

In a living cell, the functional forms of pre-mRNA/mRNA exist as ribonucleoprotein complexes (RNPs) in which numerous heterogeneous nuclear ribonucleoproteins (hnRNPs) bind to the transcript [40]. The association of these proteins begins during transcription, and there is evidence to show that some of these proteins remain bound to the mRNA all the way to the ribosome. The distribution of these proteins along the transcript, as well as the dynamic nature of the protein/RNA association, has a significant impact on the accessibility of pre-mRNA/mRNA to hairpin probes (see Figure 12.4).

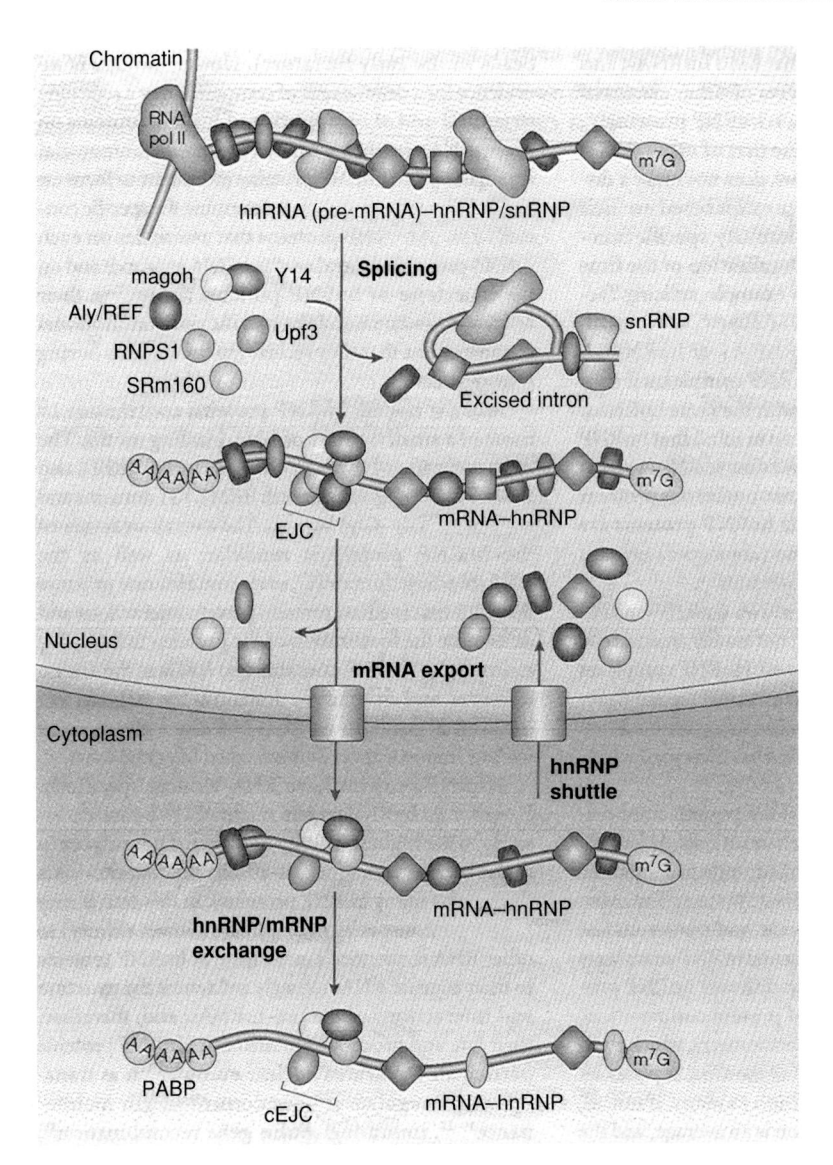

FIGURE 12.4. Illustration of HnRNP proteins and mRNP proteins along the pathway of mRNA biogenesis, together with an overview of the interaction and dynamics of the myriad of RNA-binding proteins with pre-mRNA and mRNA. EJC, exon-exon junction complex; NMD, nonsense-mediated mRNA decay; RNA pol II, RNA polymerase II; hnRNP, heterogeneous nuclear ribonucleoprotein; snRNP, small nuclear ribonucleoprotein; mRNP, mRNA-protein complex; PABP, poly(A)-binding protein; m^{7}G, methylguanosine cap (Adapted from Ref 40).

12.4.1.1. Transcription and Polyadenylation mRNA is transcribed by a protein transcription complex containing RNA polymerase II which, especially the carboxy-terminal domain, couples transcription with mRNA processing [41]. All pre-mRNA processing is a co-transcriptional event, including addition of the poly(A) tail and splicing. Even as the nascent transcript (20–25 nucleotides in length) is emerging from the transcription complex, a capping enzyme binds the 5' end, attaching the 5'-7-methylguanosine cap (reviewed in [42]. This cap is then methylated in order to stabilize the transcript against 5' exonucleolytic attack [43]. The methylated end is also a signal for nuclear cap binding proteins that facilitate the interaction between the 5' splice site of a pre-mRNA containing introns and a protein complex called the spliceosome. Depending upon a functional poly(A) signal and the terminal splice acceptor site of the terminal intron [44], the termination of transcription occurs far downstream of the poly(A) site, either before or at the same time as poly(A) is cleaved due to cleavage factors [45]. The addition of the 3' poly(A) tail of a pre-mRNA requires more than a dozen polypeptides to be present [42]. The pre-mRNA first undergoes endonucleolytic cleavage by an endonuclease at the poly(A) synthesis initiation site, followed by processive poly(A) synthesis by a poly(A) polymerase. The poly(A) tail (ranging in length from 20 to 250 nucleotides) is then bound by poly(A) binding proteins [40].

12.4.1.2. Splicing   During pre-mRNA processing, introns are removed from pre-mRNA by a ribonucleoprotein (RNP) machine (spliceosome) that contains at least 50 proteins and 5 small nuclear RNAs [46]. This spliceosome is assembled at each intron, excising the intron and then releasing it in a branched "lariat" form along with the spliced pre-mRNA. Many of the splicing proteins remain bound to mRNA after splicing [47, 48], generating a specific nucleoprotein complex that facilitates mRNA export [49]. Specifically, some of the bound proteins are members of the SR (serine-arginine) protein family of splicing factors [49], a number of which have been shown to shuttle between the nucleus and cytoplasm [47]. Proteins within this complex (such as Y14 and Mago) can target an mRNA for nonsense-mediated mRNA decay or recruit proteins such as TAP/p15 to assist in nuclear export.

12.4.1.3. Nuclear Export and Localization   Trans-acting factors (proteins that bind to *cis* elements or "zipcode" sequences within RNA) for mRNA localization are often present in granules (ribonuclear protein–RNP–complexes) that contain all components necessary for RNA processing, transport, localization, anchoring, and translation (reviewed in [50]. In other words, mRNA localization is a process initiated in the nucleus, based on the proteins that are bound during processing and accompany the mature mRNA from the nucleus to the cytoplasm. The hnRNP (heterogeneous nuclear ribonuclear protein) A/B family is one group of proteins that play a significant role in mRNA localization. Both hnRNP A1 and A2 leave the nucleus with mRNAs and then dissociate from the mRNA in the cytoplasm. Specifically, hnRNP A2 has been shown to help localize myelin basic protein (MBP) mRNA in oligodendrocytes [51–54]. In addition, it has roles in splicing, nuclear export, translational regulation, and RNA stabilization [50]. hnRNP A2 binds to a 21-nucleotide element in the 3' untranslated region (UTR) of mRNA [53, 54]. A second example of a nuclear-bound trans-acting factor is zipcode-binding protein 2 (ZBP2), which binds to the 3' UTR of β-actin mRNA [55].

The Tap protein is another protein that plays a significant role in nuclear RNA export [56–58]. This protein contains an RNA binding domain, a nuclear export signal, and a binding domain for nucleoporins (proteins that make up the nuclear pore) (reviewed in [59]. The recruitment of this protein to spliced mRNA has been proposed to be the final step required before nuclear pore binding [59].

TABLE 12.1. Summary of proteins bound to pre-mRNA.

Type of Protein	Description and function
RNA polymerase II	
Capping enzyme	
Cap binding proteins	May stay attached all the way through export from nucleus [42]
Polyadenylation machinery	At least 12 proteins including an as-of-yet unidentified endonuclease and Poly(A) polymerase, along with many other required protein factors.
Poly(A) binding proteins	
Spliceosome	A ribonucleoprotein (RNP) machine that contains at least 50 proteins and 5 small nuclear RNAs. Some of these proteins remain bound after splicing and are required for nuclear export [49].
Exon-exon junction complex	A multi-protein complex 20–24 nucleotides upstream of an exon-exon junction [40, 60]. This complex binds to as little as eight nucleotides.
hnRNP proteins	An assortment of at least 20 proteins that associate with nascent pre-mRNA; has roles in localization, processing and nuclear transport. The binding sites of some of these proteins have been identified as being in the 3' UTR of mRNAs. Others (of the hnRNP C family) associate preferentially with introns.

12.4.1.4. Distribution of RNA Between Nucleus and Cytoplasm An important issue pertaining to the measurement of mRNA levels using hairpin probes is the relative levels of RNA in the nucleus and cytoplasm. It has been reported that a large fraction of RNA (>95%) synthesized by RNA polymerase II never leaves the nucleus as mature mRNA [61]. Further, over 1/3 of the RNAs never reaches the cytoplasm due to mRNA processing events (removal of introns, transcription termination). There is also a population of primary transcripts that are not polyadenylated or transported from the nucleus (termed "nonproductive hnRNAs"), and therefore never destined to produce mRNA [61]. It is therefore possible for probes to bind to mRNAs in the nucleus that may never be translated.

12.4.1.5. Transport and Localization of mRNP Following export from the nucleus, a mRNP is often transported to specific regions within cells. It is believed that the specific localization of mRNPs has a key role in the compartmentalization of protein synthesis in the cytoplasm [62–64]. Key questions remain open in this area include: (1) Are mRNPs localized in cells? If so, is there a general cell structure where most of the mRNPs are localized? (2) What is the intracellular system along which they are being transported to their destination? (3) What factors decide specific localization of mRNPs? (4) What is the biological significance of mRNP localization in cells?

Most mRNPs are believed to be associated with the cytoskeleton, which may be used to transport and localize the mRNP to specific sites within a cell. Although the details of this process remain elusive, there is growing evidence suggesting the importance of the cytoskeleton in mRNA localization. The key evidence is that there exists a close association of polyribosomes (also referred to as polysomes) with the cytoskeleton [65–68]. In addition, studies have shown that microtubules are involved in the assembly of membrane-bound polyribosomes [69, 70]. Using drug treatment to depolymerize microtubules, membrane-bound ribosomes were prevented from initiating protein synthesis and showed a decreased level of total poly(A) mRNP and fibronectin mRNP. Although these studies did not identify the nature of the interaction between membrane (ER) bound polysomes and microtubules, they did show a close relationship between the cytoskeleton, ER bound polysomes, mRNP transport and protein synthesis. Further, using both *in vitro* reconstitution and biochemical fractionation, Hamill et al [71] provided EM evidence that a fraction of polyribosomes/mRNPs was bound to microtubules.

12.4.1.6. Localization of mRNPs At present, there is no clear consensus as to what fraction of mRNPs are localized with the cytoskeleton. Biochemical evidence provided by [72] suggests that 70–80% of mRNPs in a cell are co-localized with the cytoskeleton. Due to the lack of fluorescent or chemical tags, the verification of the biochemical results in intact cells has been limited to one or two specific mRNPs. For example, using *in situ* hybridization and biochemical fractionation, Wiseman and Hesketh [73, 74] and Russell and Dix [75] have demonstrated co-localization of mRNPs of the myosin heavy chain with cytoskeletal elements. [65, 66, 76] have shown the association of actin and poly(A) mRNPs with cytoskeletal elements, especially microtubules. Using a similar *in situ* approach, [77] have implicated cytoskeletal localization of metallothionein I. There is a clear need to better understand the localization of mRNPs, their association with specific elements in a cell and the biological significance of co-localization.

It has been shown [77, 78] that the 3' UTR plays a significant role in specific localization of certain mRNPs in cells. Specifically, using genetic engineering to perturb mRNA in the 3' UTR region, it has been shown that the 3' UTR regions of heavy chain myosin, actin, c-myc, metallothionein-I are critical for the localization of these specific mRNPs. Although the localization signal in the mRNA sequence has been identified, the structural basis of the 3' UTR, the associated binding proteins and, perhaps more importantly, the processes involved in specific localization and transport are still poorly understood. To date, proteins responsible for the specific localization of mRNA have been identified only for oskar and nanos mRNA in Drosophila oocytes [79, 80] and for actin mRNA in chicken fibroblasts [81]. In both cases, cytoskeletal elements have been shown to be involved in transport and localization of the mRNPs. See Figure 12.5a and 12.5b for proposed roles of cytoskeleton in transport and localization of mRNP.

As mentioned earlier, it is mRNP rather than mRNA alone that provides the functional unit for localized protein synthesis. In order to understand the intracellular transport and localization of mRNA, extensive studies need to be performed to track the movement of mRNP complexes. Possible approaches for doing so include labeling proteins with GFP [82], generating transgenes with high affinity binding sites for a GFP tagged proteins [80], site-specific protein labeling using FLAsH [83], and targeting mRNAs using molecular beacons [84]. These approaches in combination with high-resolution microscopy [85] or EM may reveal the detailed structural organization of mRNP transport and localization in living cells.

12.4.1.7. Degradation of mRNA One of major pathways of regulating posttranscriptional gene expression is the degradation of mRNA. Considerable focus has been placed on understanding the transcriptional controls of mRNA synthesis, but there is very limited understanding of the mechanisms responsible for controlling the rate of mRNA degradation, which affects the intracellular mRNA level. It has been suggested that a variety of physiological signals such as hormones [86, 87], iron binding proteins and cell cycle regulators [88] may have significant effects on the decay rates of specific mRNAs, and the half-life of

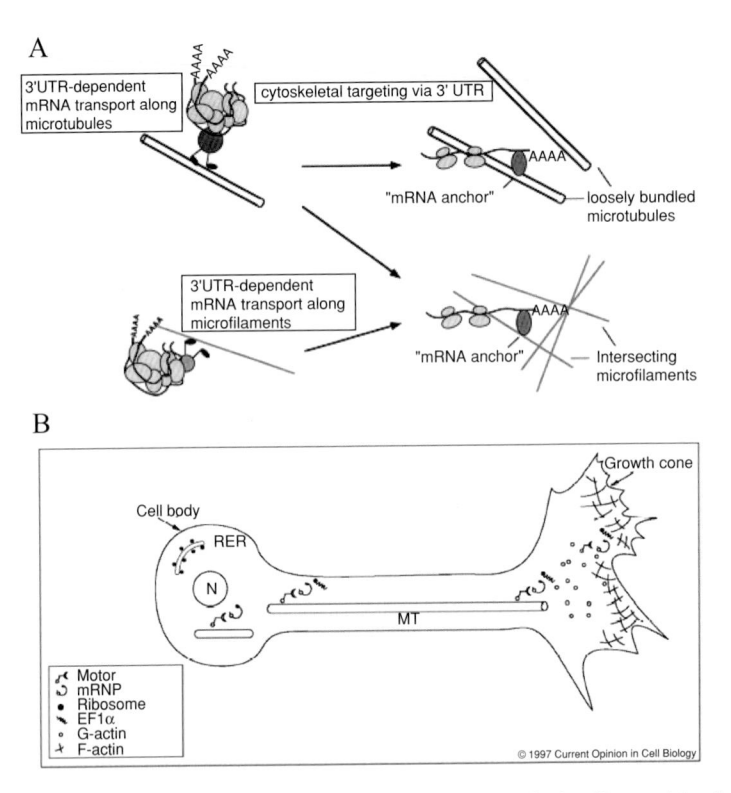

FIGURE 12.5. A) Representation of cytoskeletal elements (microtubule and microfilaments) involved in localization and transport of mRNA (Adopted from Jansen, FASEB J., 1999) and B) transport of specific m RNA along microtubule in an axon of a neuron (Adopted from Ref 66).

different mRNAs is closely related to its biological role. It has been found that specific cis acting elements rich in AU sequences, designated as the AU rich element (ARE) in the 3' UTR determine the stability of mRNA. Sequence analysis of various mRNAs with short half-lives has confirmed the presence of consensus ARE sequences. Further comparisons across species (*C. elegans* and humans) have indicated the evolutionary conservation of ARE domains [89]. ARE sequences are homologous across different mRNAs but their copy number and their clustering pattern (groups of pentamers etc.) is different. This may account for differences in mRNA lifetimes.

Association of ARE binding proteins with ARE rich domains have been proposed as the key regulatory step in controlling the turnover of RNA. Currently, we do not have a rigorous understanding of this class of proteins. Do regulatory protein-complexes such as exosomes directly bind to ARE regions? What are the structural requirements for such recognition and what is the key recognition domain? Is it structural such as hairpin loops or a combination of structural and sequence dependence? Answering these questions requires a fundamental understanding of the structural features of protein-RNA complexes, as well as the dynamics of protein-RNA interactions in the cytoplasm.

The hypothesis that ARE binding proteins mediate regulation is based on remodeling of local RNA structure, which leads to recruitment of additional *trans* acting elements such

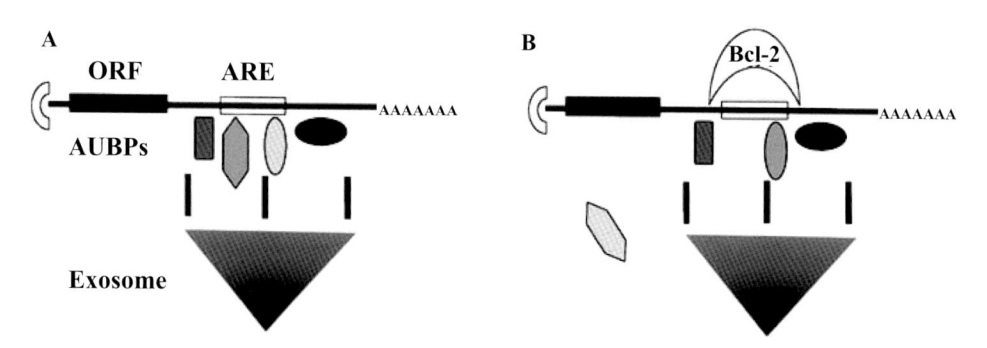

FIGURE 12.6. A model depicting the regulation of the *bcl-2* RNA decay by the Bcl-2 protein. The AUBPs and exosome associated to the ARE motif (A) are activated by the Bcl-2 protein, resulting in the displacement of an inhibitory factor (B) (Adapted from Ref 89).

as ribonucleases or helicases to degrade and unfold RNP complexes. mRNA molecules may be stabilized by the binding of signaling proteins or competing proteins, which may disrupt or interfere with the proteins involved in mRNA degradation. This model, as illustrated in Figure 12.6, has been proposed for the self-stabilization of Bcl-2 mRNA since the Bcl-2 protein binds to the ARE rich domain of its own mRNA [89].

In addition to the above ARE model of mRNA degradation, several theories have been proposed to explain the decay pathways of mRNA using a yeast-based model system. For example, [90] have proposed a relationship between mRNA stability and the length of mRNA. However, studies by [91] using microarray analysis of mRNA stability in yeast, have failed to correlate stability with the length of mRNA. Their data suggest that mRNA decay is closely related to specific functions of a given mRNA and is controlled by interactions of mRNA with specific signaling proteins.

Much remains to be elucidated regarding the processes of mRNA production, transport, localization and degradation. It is obvious, however, that at each step, from transcription to nuclear export of a mature mRNA, to mRNA decay, *proteins are always associated with pre-mRNA and mRNA*. These associations are dynamic in nature with regard to both protein type and binding sites. An mRNA molecule that gets exported to the cytoplasm is more precisely an mRNP (ribonucleoprotein complex) [40, 49, 92, 93]. Although we are still in the early stages of identifying proteins that are associated with mRNPs, some specific protein–RNA associations have been uncovered using both biochemical and genetic methods (e.g. [48, 94, 95]. Specifically, proteins have been shown to be associated at exon-exon junctions, the 3'UTR region, and the 5' cap regions of mRNA. A critical issue in mRNA detection is that nucleic acid probes may not hybridize to the target regions occupied by RNA-binding proteins. It is necessary to identify the mechanisms for mRNA transport and degradation as controlled by the RNA/protein interactions. As the steps of mRNA production, transport and localization are further clarified, it will become easier to design nanostructured hairpin probes to target regions of mRNA that are least likely to be occluded by bound proteins. Probe design should therefore take into account what is already known about proteins bound to mRNA; specifically, probes may not be designed to target the 3' UTR, the exon-exon junction, or introns due to the proven presence of protein complexes in these regions.

12.4.2. RNA Secondary Structure

RNA secondary and tertiary structures play an important role in the accessibility of nanostructured probes in living cell gene detection. This structural aspect provides a significant challenge in designing nanostructured probes as secondary structure of RNA can inhibit binding of the hairpin oligonucleotide probe to complementary sequences on the target. However at present, no existing model or software can give an accurate prediction of mRNA secondary or tertiary structure in living cells mainly due to the limited understanding of RNA-protein interactions and RNA intra-sequence interactions. Thus, most investigators have relied on empirical approaches where multiple targeting sequences are selected by "walking" along the length of the target RNA and the optimal binding sites are predicted based on the RNase H assay or gel shift assays.

A number of computational programs have been developed to predict the secondary structure of mRNA based on the sequence of mRNA and salt concentration but without considering RNA-protein interactions. The existing algorithms can be classified into two groups: (1) predictions based on minimizing free energy (e.g., Mfold, RNAsoft, Vienna RNA package); (2) probabilistic or stochastic prediction based on the alignment and phylogenetic trees of the sequences and the evolutionary conservation of critical structural components of mRNA (e.g., Pfold). For example, the algorithms initially developed by [96] based on minimizing free energy and later modified by McCaskill [97] to include base-pairing probabilities (Vienna RNA Secondary Structure Prediction) have been widely used as a structural prediction tool in the nucleic acids research community. One of the major limitations of the minimum energy-based models (e.g., Mfold) is the inability to predict pseudoknots in the secondary structure of mRNA. Although significant progress has been made in predicting the secondary structure of a naked mRNA, it is still far from making accurate predictions of mRNA structure taking into account the interactions between proteins and target mRNA in an intracellular environment. This is very problematic in the design of nanostructured oligonucleotide probes for living cell mRNA detection.

Understanding the structural changes in target mRNA upon binding of probes [98] is also important, since alterations in native structure of a target mRNA due to the binding of short oligonucleotides may change its functionality. It has been suggested that the binding of DNA probes may prevent translation by either blocking the process or by RNase H activation. To study probe/RNA interactions, tools such as OligoWalk [99] and PairFold [100] been developed. For example, OligoWalk predicts the interaction of oligonucleotide probes (both DNA/RNA) with target RNA structure based on equilibrium affinity at $37°C$, and allows the user to study the effects of probe length, concentration, and to some extent backbone chemistry (RNA or DNA) on the local and global structure of target RNA.

12.5. LIVING CELL RNA DETECTION

Various technologies and methodologies have been developed to study intracellular RNA biology by creating tagged full length RNAs or using RNA targeting probes. In most cases these tags are fluorescent, chemical (such as digoxigenin), or radioactive. For example, tagged full-length RNA has been introduced into living cells using microinjection [101–103] to monitor the localization of a specific mRNA or nuclear RNA.

Labeled linear oligonucleotide (ODN) probes of 20–50 bases have been used to study intracellular mRNA (e.g. [68, 76, 105]) via *in situ* hybridization (ISH) in which cells are fixed and permeabilized to increase the probe delivery efficiency and unbound probes are removed by washing, therefore reducing background and achieving specificity. To enhance the signal level, multiple probes targeting the same mRNA can be used (e.g. [76]). However, fixation agents and other supporting chemicals can have considerable effect on signal [106] and possibly on the integrity of certain organelles such as mitochondria. Therefore, fixation of cells, by either cross-linking or denaturing agents, combined with the use of proteases in ISH may not provide an accurate description of intracellular mRNA localization. Labeled linear ODN probes are not very useful in detecting mRNA in live cells since it is impossible to remove unbound probes by washing and thus difficult to distinguish between signal and background.

In addition to oligonucleotide probes, tagged RNA-binding proteins such as those with GFP tags have been used to detect mRNA in live cells [83]. One limitation is that it requires the identification of a unique protein, which only binds to the specific mRNA of interest. To address this issue, recently a transgene with a binding site for the phage MS2 protein was synthesized [81]. Generation of a GFP tagged phage MS2 protein in Drosophila eggs allowed the specific targeting of the nanos mRNA in a living egg system. However, there is still a significant challenge in generating transgenes with the same functionality as endogenous mRNA.

In the following sections, we discuss in more detail critical issues in living cell mRNA detection using nanostructured hairpin oligonucleotide probes, including cellular delivery of probes, intracellular probe stability and dynamics, and mRNA detection sensitivity, specificity and signal-to-background. Emphasis is placed on the design and application of molecular beacons, although the issues are common for other hairpin probes.

12.5.1. Cellular Delivery of Probes

One of the most critical aspects of measuring the intracellular level of mRNA using synthetic probes is the ability to deliver these probes into cells through the plasma membrane. In what follows we discuss the existing methods for delivering hairpin probes into live cells and the possible effects of delivery on intracellular distribution, dynamics, binding and lifetime of the probes.

The plasma membrane is quite lipophilic and restricts the transport of large and charged molecules. Therefore, it is a very robust barrier to polyanionic molecules such as hairpin oligonucleotides. Further, even if the probes enter the cells successfully, the efficiency of this process must be defined not by how many probes can enter the cell or how many cells have probes internalized, but how many probes remain functional inside cells and hybridize to their targets with high specificity. This is very different from both antisense and gene delivery applications where the reduction in level of protein expression is the final metric used to define efficiency or success. It is also important to examine how the delivery technique might impact the fate of the probes inside a cell. Since a significant amount of RNA molecules (including mRNA and rRNA) are in the cytoplasm, any delivery method aimed at measuring intracellular RNA should result in a large amount of probes in the cytoplasm.

Existing cellular delivery techniques can be divided into two categories: endocytic and non-endocytic methods. Endocytic delivery typically employs cationic and polycationic molecules such as liposomes and dendrimers, while non-endocytic methods include microinjection, and the use of cell-penetrating peptides (CPP) or steptolysin O (SLO). Here we focus on the applications of these methods to delivering oligonucleotide (ODN) probes into live cells and how delivery would impact the functionality of the probes.

12.5.1.1. Endocytic Methods ODN probes can be transported into live cells by simply incubating them in solution with cells for $\sim$2 h. The delivery is mediated by the endocytic pathway is by specific membrane surface receptors and studies have shown the existence of at least 5 major cell surface receptors that bind specifically to ODNs [107]. Fluorescently labeled ODN probes usually exhibit a punctate fluorescence pattern indicating endocytosis. Electrostatic complexes of anionic ODNs with cationic liposomes or polymeric dendrimers (such as polyamidoamine or polyethyleneimine) have been shown to enhance ODN uptake; they are designed to provide some level of nuclease protection and have in many cases incorporated pH sensitive molecules that enhance their ability to escape from endosomes. However, even with all these, the efficiency of cytoplasmic delivery of functional probes was estimated to be only 0.01–10 percent [107] due to the fact that once internalized via endocytosis, the ODN probes are predominately trapped inside endosomes and often lysosomes, and being degraded there.

During endocytosis, the receptors on the cell surface are activated by the ODN probes, which are then being enclosed in a small portion of the plasma membrane that pinches off to form an endocytic vesicle called early endosomes. The environment within early endosomes is fairly acidic, with a pH of approximately 6. As they proceed to become late endosomes, the pH continues to decrease and the newly synthesized acid hydrolases are accumulated within them. Unless escaped from the endosomes, the ODN probes will be trapped in lysosomes, which are membrane-enclosed vesicles filled with hydrolytic enzymes for intracellular digestion. Most of the active cytoplasmic nucleases are in the lysosomes, not in the cytosol itself [108]. Clearly, inside a lysosome the environment is extremely bad for ODN probes and therefore it is imperative that the ODN probes are released from the early endosomes as soon as possible. For example, viruses, both enveloped and non-enveloped, have specific mechanisms to enable their genetic material to escape from the endosomes/lysosomes and be efficiently delivered to the cytosol.

To overcome this difficulty, polyamidoamine dendrimers were designed to escape from endosomes by polymer swelling and osmotic-induced swelling, as theorized by [109]. At neutral pH, electrostatic repulsion between protonated primary amines causes the fractured dendrimer to be fully extended which, after binding electrostatically with DNA, collapse into a more compact form. Once the pH decreases within the endosome, tertiary amines become protonated and the excess polymer is released from the complex, resulting in endosomal swelling and subsequent rupture. A large excess of dendrimer is usually required to promote high delivery efficiency, possibly because the excess non-complexed dendrimers swell and help burst the endosomes.

Even with this mechanism for endosomal exit of probes, it is still unclear how nucleic acids are released from cationic molecules. For liposomes it has been suggested, that once inside a vesicle, anionic lipids from the cytoplasmic-facing monolayer diffuse into the complex, fusing with the cationic liposome and therefore releasing the oligonucleotide [110]. Dissociation of nucleic acids from cationic molecules may occur in endosomes, cytosol or the nucleus, but in most cases the mechanisms are unknown.

Many issues remain to be addressed in endocytic delivery of ODN probes, for example, how to protect ODN probes from nucleases in the endosomes, and why endocytic delivery often results in internalization of nucleic acids into the nucleus. Various studies have shown that, when internalized through the endocytic pathway, a large portion of ODN probes are trapped within intracellular vesicles such as endosomes and lysosomes, degraded by nucleases, and prevented from reaching their target [107]. Indeed, when molecular beacons are delivered into cells using polyamidoamine dendrimers and liposomes, the images are riddled with bright fluorescent punctation, indicating degradation of molecular beacons in the endosomes and questioning the protective capability of the complexed molecules. While it is advantageous that endocytic delivery tends to internalize DNA plasmid in the cell nucleus, it is problematic in delivering hairpin ODN probes for measuring cytoplasmic mRNA in that if most of the probes end up in the nucleus, cytoplasmic RNA cannot be measured.

The use of cationic molecules such as liposomes or dendrimers is not very effective in delivering ODN probes into primary cells (as opposed to cell lines). We have attempted to deliver DNA and RNA molecular beacons using both Oligofectamine (Invitrogen) and Superfect (Qiagen) into normal human dermal fibroblast (NHDF) cells and found that it was extremely unsuccessful. Both DNA and RNA beacons were trapped in intracellular vesicles and very little were delivered to the cytoplasm. In many cases they ended up in the nucleus producing a "glowing" nucleus with little localization.

12.5.1.2. Microinjection To avoid the issues with endocytic delivery, non-endocytic methods have been developed and employed. For example, oligonucleotide probes (including molecular beacons) have been delivered into cells via microinjection (e.g. [111–117]). In most of the cases the ODNs exhibited a fast accumulation in the cell nucleus, possibly due to diffusion through nuclear pores. Depletion of intracellular ATP or lowering the temperature from $37°C$ to $4°C$ did not have a significant effect on ODN nuclear accumulation, ruling out active, motor-protein driven transport. It is possible that the interactions between positively charged nuclear proteins with negatively charged ODNs are important. However, it is unclear if the rapid transport of ODN to nucleus is due to electrostatic interaction, or driven by microinjection-induced flow. There is no fundamental biological reason why ODN probes accumulate in the cell nucleus. For example, when molecular beacons targeting β-actin and vav (a protooncogene) were microinjected into living cells, it was found that more fluorescence signal was in the cytoplasm than in the nucleus [113].

It has also been observed that after internalization the fluorescence of unmodified ODNs decreases steadily and disappears within 6–10 hours [111]. However, when phosphorothioate ODNs were used fluorescence signal could be observed for up to 20 hours due was attributed to increased nuclease resistance. It is unclear why the fluorescence of the fluorophores decreases after the ODNs are being degraded. Possible reasons include quenching due to non-specific binding between the free fluorophores and proteins or metabolic degradation of fluorophores [112].

To prevent nuclear accumulation, streptavidin (60 kDa) molecules were conjugated to linear ODN probes via biotin [115]. After microinjected into cells, dual FRET linear probes could hybridize to the same mRNA target in the cytoplasm, resulting in a FRET signal. In a comparative study, linear and molecular beacon probes with DNA or 2'-*O*-methyl backbones targeting poly (A) mRNA, ribosomal RNA and small nuclear RNA were delivered into living cells using microinjection [114]. The results were compared that of in-situ hybridization with fixed cells. It was found that 2'-O-methyl molecular beacons opened quickly once inside the nucleus and that linear 2'-O-methyl probes were most promising for the detection of nuclear RNA. It is likely that 2'-O-methyl molecular beacons are more readily recognized by hairpin binding proteins; however, generally speaking, linear ODN probes have a much higher background than molecular beacons.

12.5.1.3. Cell Membrane Permeablilization Another non-endocytic method is toxin-based cell membrane permeablilization. For example, streptolysin O (SLO) is a pore-forming bacterial toxin that has been used as a simple and rapid means of introducing oligonucleotides into eukaryotic cells [118–121]. Streptolysin O belongs to the homologous group of thiol-activated toxins that are secreted by various gram-positive bacteria. SLO binds as a monomer to cholesterol and then oligomerizes into ring-shaped structures estimated to contain 50–80 subunits which surround pores of approximately 25–30 nm in diameter. This size by far exceeds the size of pores formed by other toxins, therefore allowing the influx of both ions and macromolecules. Since cholesterol composition varies between cell types, the sensitivity to SLO may vary as well. Therefore, the permeabilization protocol has to be optimized for each cell type by varying temperature, incubation time, cell number and SLO concentration. An essential feature of this technique is that the toxin-based permeabilization is reversible. This has been achieved by introducing oligonucleotides with SLO under serum-free conditions and then removing the mixture and adding normal media with serum [119, 122].

The initial work of SLO-based delivery was conducted with mouse kidney cells using 0.2 U/ml of SLO and various concentrations (0.1 to 100 μM) of antisense oligonucleotide [123]. It was found that by delivering the ODNs directly into the cytoplasm in serum-free media and bypassing endocytosis, nuclease activity is substantially lower [123]. Later, it was found that there is little cell death due to SLO when fluorescently labeled linear ODNs were delivered, and it takes 180 minutes to reach the fluorescence signal level similar to that obtained using fixed cells and conventional in-situ hybridization [124].

The SLO-based method was compared with electroporation and liposome-based method in delivering fluorescently labeled antisense ODNs into myeloid leukemia cells [121] by performing fluorescence microscopy and flow cytometry. The effect of ODN uptake on mRNA expression level was examined by Northern blotting. It was found that SLO-based permeabilization achieved an intracellular concentration of ODNs of approximately 10 times that of electroporation and liposomal-based delivery. It was further demonstrated that ODNs delivered with SLO and electroporation were in both the cytoplasm and nucleus using 20 μM of ODN concentrations, while for liposome-based delivery the ODNs were in vesicles localized near cell membrane. Delivery of ODNs with different backbone chemistries (including phosphorothioate, phosophodiester, 5-methylcytosine phosphorothioate, phosphorothioate C5-propyne pyrimidine structures and chimeric molecules composed of a central Rnase-H activating region and methylphosphonate termini) and concentrations revealed that the efficiency of SLO-based method is not sensitive to ODN backbone chemistry, which was not the case with delivery methods using cationic lipids or polymers [120, 125]. To date most SLO delivery studies are for the introduction of antisense ODNs to down-regulate the gene expression. Recently, SLO was used to deliver dual-FRET molecular beacons into normal human dermal fibroblast and MiaPaca-2 cells to measure survivin and K-ras mRNA [84]. Using a low concentration of SLO (0.2 U/ml) and low concentration of molecular beacons (~2 μM in the extracellular fluid), we found that the endocytic pathway was avoided and a localized (not diffuse) signal was observed only in the cytoplasm [84]. It seems that the combination of low permeability and low concentration of probes prevented the molecular beacons from getting into the nucleus. This is in contract to most SLO delivery studies in which much higher concentrations of ODNs (at least 10 times) as well as SLO (10–100 times) were used. We believe that the low concentrations of SLO and probes may have limited the driving force for transporting ODNs to the nucleus.

12.5.1.4. Peptide-Based Delivery As mentioned above, intracellular delivery can be very challenging as the plasma membrane forms a formidable barrier for many biomolecules. The discovery of cell penetrating peptides (CPP) has made it possible to transduce a broad range of agents, including nucleic acids, into living cells. Among the family of peptides with membrane translocating activity are antennapedia, HSV-1 VP22, and the HIV-1 Tat peptide (Tat$_{48-60}$). To date the most widely used peptides are HIV-1 Tat peptide and its derivatives due to their small size and high delivery efficiency. The Tat protein from HIV-1 virus is an 86 amino acid transactivation protein that is involved in promoting transcription of the HIV-1 virus. The Tat peptide is rich in cationic amino acids especially arginines which is very common in many of the cell penetrating peptides. However, there seems to be little homology among the CPPs. The exact mechanism for membrane translocation is currently unknown, although many hypotheses have been proposed.

It has been generally accepted that the membrane translocation mechanisms were primarily passive, involving diffusion or a membrane destabilization process that did not require receptor binding. In addition, experiments performed at 4°C seem to indicate an endocytic-independent process. Derossi et al. [126] proposed an "inverted micelle model", wherein the peptide recruits negatively charged phospholipids from the cell membrane which induces the formation of a hyrdophilic cavity. This cavity then translocates from the exterior leaflet to the interior leaflet of the plasma membrane depositing the cargo into the cytosol. It has yet to be determined whether this model is thermodynamically possible and if it would apply to all of the translocating peptides.

Schwarze et al. [127] proposed a different model for Tat peptide transduction: it involves direct penetration of the lipid bilayer as a result of the localized positive charge of the peptide and the momentum of the peptide-cargo complex drives the covalently attached cargo into the cytoplasm. After transduction, the membrane energetics would favor the reformation of the intact plasma membrane. This model takes into account the ability of the Tat peptide to transduce 40-nm particles and proteins as large as 120 kDa. However, experimental data is needed to validate this model.

Recent work has examined whether heparan sulfate (HS) proteoglycans act as receptors for internalization of the Tat peptide-cargo complex. HS is ubiquitously expressed on eukaryotic cell membranes and may be involved in the stabilization of cell membrane molecules as well as cell adhesion molecules. HS is polyanionic and its negative charges could be involved in initial peptide attraction and binding which may mediate Tat translocation. It is controversial as to the roles of HS in the mechanism of CPP. For example, Silhol et al. [128] and Violini et al. [129] provided evidence that HS is not involved in Tat uptake, while Belting [130] indicated that treatment with anti-HS antibodies or HS-degrading enzymes diminished peptide internalization, suggesting that HS proteogycans must play a role in internalization. However, Violini et al. [129] showed that fluorescently-labeled Tat peptide could not enter MDCK epithelial cells or CaCo-2 colonic carcinoma cells. Both cell types do express HS proteoglycans, suggesting that HS may not be a major player in peptide-based delivery.

A wide variety of cargo has been delivered to living cells both in cell culture and in tissue using cell penetrating peptides. For example, Allinquant et al. [131] linked Antennapedia peptide to the 5' end of DNA oligonucleotides (with biotin on the 3' end) and incubated both peptide-linked ODNs and ODNs alone with cells. By detecting biotin using streptavidin-alkaline phosphatase amplification, it was found that the peptide-linked ODNs were internalized very efficiently into all cell compartments compared with control ODNs. No indication of endocytosis was found. Similar results were obtained by Troy et al. [132] with a 100-fold increase in antisense delivery efficiency when ODNs were linked to antennapedia peptides. Astriab-Fisher et al. [133] delivered 2'-O-methyl-modified linear oligonucleotides by conjugating both Tat and Antennapedia peptides to the ODNs via a disulfide bond. The results were compared with liposome-based delivery using Lipofectin. It was found that, while the use of peptides for antisense delivery had very low toxicity and allowed for rapid distribution in cells, the intracellular accumulation of peptide-linked ODNs is still slower compared with fluorescently labeled peptides alone. It was speculated that, with the addition of the ODN cargo, the mechanism of translocation has become endocytic.

Recently, Tat peptides were conjugated to molecular beacons using three different linkages; the resulting peptide-linked molecular beacons were delivered into living to target GAPDH and survivin mRNAs [134]. It was demonstrated that, at relatively low concentrations, cells using thethe molecular beacon constructs were internalized into living cells within 30 min with nearly 100% efficiency. Further, peptide-based delivery did not interfere with either specific targeting by or hybridization-induced florescence of the probes, and the peptide-linked molecular beacons could have self-delivery, targeting and reporting functions. In contrast, liposome- (Oligofectamine) or dendrimer-based (Superfect) delivery of molecular beacons required 3–4 hours and resulted in a punctate fluorescent signal in the cytoplamic vesicles and a high background in both cytoplasm and nucleus of cells (Figure 12.7). It was clearly demonstrated that cellular delivery of molecular beacons using the peptide-based approach has far better performance compared with conventional transfection methods.

12.5.2. Intracellular Probe Stability

Although oligonucleotide hairpin probes such as molecular beacons have the potential to detect mRNA in living cells, these probes can be degraded by endonucleases or opened by hairpin binding proteins, resulting in a large amount of false-positive signals and thus significantly limiting the detection sensitivity. But what nucleases are likely to pose threats to the hairpin probes, either in the cytoplasm or nucleus? It is well known that cell organelles such as endosomes and lysosomes contain about 60 hydrolytic enzymes, most of which are active at acidic pH. These acid hydrolyases include the nucleases deoxyribonuclease II (DNase II) and ribonuclease II (RNase II) [108]. Since different cytoplasmic compartments of a cell have distinct environments with different concentration of acid hydrolyases, it is essential to deliver probes into the right place to avoid degradation. For example, if the

HAIRPIN NANOPROBES FOR GENE DETECTION

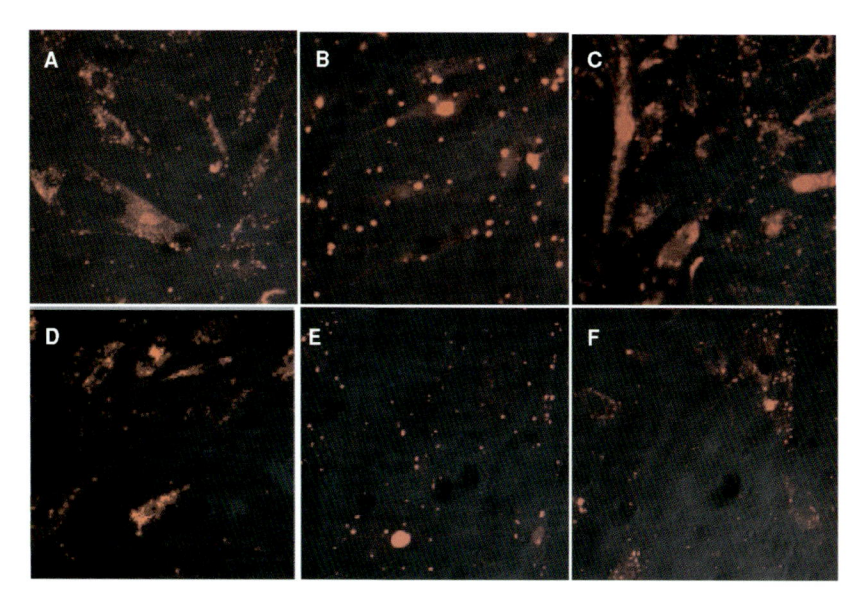

FIGURE 12.7. Cellular delivery of hairpin ODN probes using conventional transfection methods. (A–C) Fluorescence signal in HDF cells after 3.5 h transfection of unmodified GAPDH-targeting molecular beacons with (A) Superfect, (B) Oligofectamine and (C) Effectene. Note the concentrated 'bright spots' in both cytoplasm and nucleus. (D–F) Similar fluorescenc signal level was observed after 3.5 h delivery of random-sequence molecular beacons with (D) Superfect, (E) Oligofectamine and (F) Effectene. The resulting 'bright spots' in HDF cells indicate that the fluorescence signals in (A–F) were largely due to molecular beacon degradation (Adopted from Ref 134).

hairpin ODN probes are delivered into cells through the endocytotic pathway, the probes are likely to be degraded in endosomes and lysosomes where there is high concentration of nucleases. Similarly, if the hairpin ODN probes are delivered into the nucleus, it will encounter a number of nucleases involved in DNA and RNA processing and DNA repair. It is therefore important to understand the activity, specificity, and localization of nucleases in a cell.

12.5.2.1. DNases DNase II is an endonuclease that is active against both double-stranded and single-stranded DNA, with an optimal pH of 5.5 [119]. It is present in lysosomes to degrade nucleic acids that enter the cell via the endocytic pathway. It has been implicated as an apoptotic enzyme capable of chromatin degradation in the nucleus where it is translocated upon induction of apoptosis [135]. Nuc70 is also an endonuclease that has been shown to localize to the endoplasmic reticulum (ER), Golgi apparatus, and possibly secretory vesicles [136]. Upon caspase-dependent activation during apoptosis, it is translocated to the nucleus [135] for chromatin degradation, suggesting that it is inactive in the cytoplasm.

Chromatin degradation by nucleases is a hallmark of apoptosis (reviewed in [136]). Nucleases can be classified as either cation-dependent or cation-independent. For example, some nucleases are Ca^{2+}/Mg^{2+}-dependent, such as DNase I. Others are Mg^{2+}-dependent such as the caspase-activated DNases (CAD). A representative of the cation-independent class of nucleases is DNase II, which is an acid nuclease. The activity of all these nucleases is tightly regulated. CAD, for example, becomes active only after caspase-mediated cleavage of an inhibitory protein. DNase I is activated upon proteolytic cleavage from an inactive precursor. DNase II, a lysosomal acid hydrolase, is most active at low pH values.

Endo-exonucleases are Ca^{2+}/Mg^{2+}-dependent and proteolytically-activated; they may also play a role in apoptosis [137]. Endo-exonucleases primarily degrade single-stranded DNA, and exist in an inactive form in both the ER and cytosol. The active form was found primarily in the nuclei. Two precursor pools (ER and cytosolic) were proposed for the nuclease. The cytosolic pool is assumed to be involved in normal DNA repair and possibly in the nicking of supercoiled chromatin DNA. The ER-membrane bound nucleases are proposed to be a storage reserve for the enzyme to be mobilized in apoptosis. The common features of all DNAses are that their activation is usually tightly regulated by divalent ion concentration and pH and, once activated, they often translocate to the nucleus. Therefore, if the hairpin probes can be delivered through a non-endocytic pathway, DNase activation and thus probe degradation in the cytoplasm may be avoided.

12.5.2.2. RNases A wide variety of RNases have been identified that hydrolyze single-stranded RNA, double-stranded RNA, and RNA-DNA hybrid. RNases are typically concentrated in a few subcellular compartments and play different roles in RNA processing and metabolism. A specific family of RNases is a complex of 10 to 11 proteins termed "exosome" which has 3' to 5' exonuclease activity in both the nucleus and cytoplasm [138–140]. Quite a few, if not all, of the exosome proteins are exonucleases. There does not appear to be a substantial pool of free exosomal subunits; almost all the associated exonucleases in the cell are components of exosomes, even though they are all active exonucleases when isolated individually and assayed in vitro [140–142]. In the nucleus, the exosome has been shown to process 3'-extended ribosomal RNA. Exosomes may also play a role in nuclear mRNA degradation [143], and the degradation of pre-rRNA transcripts that are unable to be properly processed [144]. In the cytoplasm, the exosome is required for the 3' to 5' degradation of some poly(A) mRNAs, which may contribute to antiviral defense [145, 146]. Moore [139] proposed that this assembly of multiple exonucleases allows for both coordinated regulation and delivery of a variety of exonucleases. However, it is still unclear what controls the recognition of a specific substrate, although it has been suggested that some members of the RNA helicase family may play an important role in substrate recognition.

Acid ribonucleases (reviewed in [147, 148]) from the RNase T2 family degrade single-stranded RNA and are primarily localized in lysosomes, although members of the family have also been identified in the nucleus. The RNase A family of ribonucleases functions as antibacterial, anti-parasitic, and anti-viral agents [149], degrading single-stranded RNA. They are secretive proteins thus posing little, if any, threat to hairpin probes inside a cell.

RNase H is an enzyme that endonucleolytically degrades the RNA component of an RNA-DNA heteroduplex. Although the exact cellular function of RNase H enzymes is not fully understood, it has been suggested that Type 1 RNase H proteins are involved in processing RNA/DNA hybrids, whereas Type 2 RNase H enzymes are responsible for the removal of Okazaki primers during lagging strand DNA replication [150, 151]. It has also been demonstrated that Type 2 RNase H enzymes may play a role in the repair of misincorporated ribonucleotides [152]. The RNase H enzymes play a key role in the mechanism of inhibition of gene expression by antisense oligonucleotides. The catalytic activity of RNase

H2 in particular is associated with antisense-mediated RNA degradation [153]. It was revealed that the expression levels of both RNase H1 and RNase H2 have large variations in different human cell lines [153]. Examination of the localization of green fluorescent protein (GFP)-tagged RNase H1 and RNase H2 indicated that in most cell types, RNase H1 was distributed throughout the whole cell whereas RNase H2 was only detectable in the nucleus.

As with DNases, RNases are tightly regulated, with specific substrates and conditions necessary for activation. However, RNases are not completely characterized, so their effect on hairpin probes is difficult to predict. It is believed that within the cytosol, outside the endocytotic pathway, few, if any, RNases will pose a threat to hairpin probes. However, the nuclear environment is much more complex, and nucleases in the cell nucleus are likely to impede the use of hairpin probes in mRNA detection.

12.5.2.3. Backbone Modifications The backbone chemistry of a hairpin probe has profound implications for the behavior of the probe in the intracellular environment. For example, chemically modified oligonucleotide backbones affect probe affinity, melting temperatures, and nuclease resistance. Extensive studies of backbone modification have been performed to enhance nuclease resistance. The most common modification of the antisense probes is to replace the phosphodiester bond with a phosphorothioate bond in the oligonucleotide. In phosphorothioate oligonucleotides (PS-ODNs), a sulfur atom replaces one of the non-bridging oxygen atoms in the phosphate backbone. This modification has been shown to have greater resistance to nuclease degradation than phosphodiesters while maintaining RNase H-mediated mRNA degradation [reviewed in 154]. The melting temperature of the PS-ODN-target RNA duplex is slightly lower than the corresponding phosphodiester oligonucleotide duplex and PS-ODNs have lower affinities than the unmodified DNA oligonucleotides for the RNA target [155]. A major drawback of this backbone modification is its polyanionic backbone, which can lead to elevated protein binding and immunoresponse [154]. Excess protein binding would certainly be detrimental to the ability of a hairpin probe to hybridize to its target mRNA. It may also activate RNase H activity, which is desirable for antisense therapy, but not desirable when the probes are used for imaging mRNA expression in living cells.

Modification of the sugar moiety of an oligonucleotide forms the basis for another class of modified oligonucleotides. A number of modifications have been made at the 2' position of the sugar ring [reviewed in 156]. Adding a 2'-*O*-methyl group at this position significantly increases RNA binding affinity and enhances nuclease resistance. 2'-*O*-methyl backbone modification results in an increase in melting temperature on the order of 1°C per modified oligonucleotide [157]. Although this modification is not ideal for antisense therapy because it is RNase H-resistant (the probe-target hybrid is more an RNA:RNA duplex than a DNA:RNA duplex that RNase H recognizes), it is beneficial for a hairpin probe in that it helps avoid RNase H-mediated RNA degradation. 2'-*O*-methyl modified molecular beacons, however, may be more prone to opening by hairpin binding proteins. It may also trigger unwanted RNA interference.

Another nucleic acid analog in which the sugar ring is modified has been described as "locked nucleic acid" [LNA, reviewed in 158]. In this backbone modification, the 2'-O and 4'-C are linked by a methylene linker. This "locks" the sugar into the conformation

(C-3'-endo sugar) that is supposedly ideal for recognition of RNA. This modification results in significant increases in melting temperatures (ranging from 1°C to 8°C against DNA) and has higher specificity than unmodified DNA oligonucleotides. LNAs have higher affinity for targets primarily due to slower dissociation, as shown by stopped-flow kinetics experiments [159]. LNAs have also been shown to be stable in the presence of a number of nucleases [160].

Peptide nucleic acids (PNAs) are nucleotide analogs that also offer advantages for the design of hairpin probes. PNA is a DNA mimic where the phosphodiester-linked backbone is replaced with an N-(2-aminoethyl)glycine backbone [161] and behaves according to traditional Watson-Crick base-pairing rules. PNAs have very high binding affinity for single-stranded DNA and RNA; the melting temperature is increased by 1.45°C per monomeric PNA unit [155]. The rate of hybridization is at least as fast as DNA:DNA duplex formation. PNA:RNA duplexes are also more stable at lower salt concentrations due to the neutral character of the PNA backbone. PNA oligonucleotides exhibit both high affinity and high specificity [162] when binding to DNA. PNA oligonucleotides are completely nuclease-resistant. They also have the unique ability to "invade" a DNA duplex, although this strand displacement occurs only at salt concentrations much lower than the physiologically relevant value [reviewed in 155]. A careful study of the ability of PNA probes to bind to structured targets (i.e., those with hairpin structural motifs) showed that the higher affinity conferred by the PNA backbone enhances probe binding over a natural DNA probe [163], suggesting that PNA hairpin probes may have a distinct advantage when targeting an mRNA sequence 'buried' in the secondary and tertiary structure.

Optimization of hairpin probe design and performance requires the examination of different backbone modifications, including their combinations. Antisense research has shown that partial backbone modifications (i.e., modifying only a few oligonucleotide bases) often confer desired properties. A combination of different modifications may lead to the ideal probe backbone for a given target. One target sequence, for example, may exhibit significant secondary structure, while another may be occluded by a number of bound proteins. Hybridization to a specific target sequence, therefore, may require a specific backbone modification.

12.5.3. Intracellular mRNA Detection

One of the major challenges in measuring endogenous gene expression in living cells and tissue is the design of reporter probes with high sensitivity, specificity and signal-to-background ratio. In addition to challenges in cellular delivery and maintaining integrity of probe in cell cytoplasm, it is critical to have good target accessibility, which is largely controlled by mRNA secondary/tertiary structures and RNA-binding proteins. Specifically, an mRNA molecule in a cell usually has a folded conformation with double stranded segments. Further, mRNAs almost always have proteins bound to it, which may alter mRNA structure and prevent probe binding. Therefore, in selecting the probe sequences, it is important to avoid targeting sequences that are 'buried' inside the tertiary structure or where double stranded RNA is formed. Although predictions of mRNA secondary structure can be made using existing software, they may not be very accurate due to limitations of the biophysical models used. Further, we only have very limited knowledge of the sequences occupied dynamically by RNA-binding proteins. Therefore, for each gene to target, it may be necessary to select multiple unique sequences along the target RNA, and

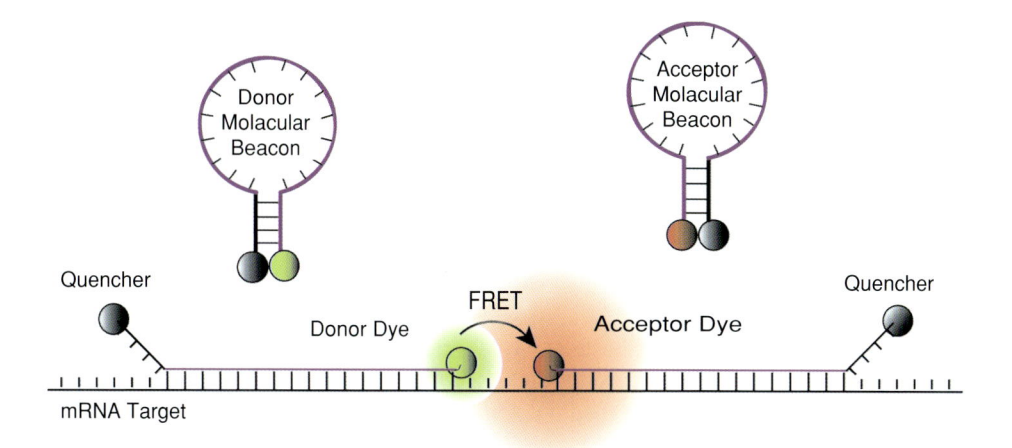

FIGURE 12.8. A schematic illustration showing the concept of dual FRET molecular beacons. Hybridization of donor and acceptor molecular beacons to adjacent regions on the same mRNA target results in FRET between donor and acceptor fluorophores upon donor excitation. By detecting FRET signal, fluorescence signals due to probe/target binding can be readily distinguished from that due to molecular beacon degradation and non-specific interactions (Adopted from Ref 84).

have corresponding molecular beacons designed, synthesized and tested in cells to achieve high signal-to-background ratio. Clearly, a better understanding of mRNA structure and RNA-protein interactions will facilitate significantly the design of hairpin probes for gene detection.

To drastically reduce the false-positive signals, two different approaches have been taken. The first is a dual molecular beacons approach which measures the fluorescent signal due to fluorescence resonance energy transfer (FRET) or luminescence resonance energy transfer (LRET) as a result of the direct interaction between two molecular beacons [18, 84, 117]. The second approach is to modify the backbone of molecular beacons to make them more resistant to endogenous nucleases, as discussed earlier. For certain applications such as long time monitoring of mRNA expression in living cells, it may require the combination of both approaches in order to achieve high signal-to-background ratio.

As shown in Figure 12.8, the dual FRET/LRET molecular beacons approach utilizes a pair of molecular beacons, one with a donor fluorophore and a second with an acceptor fluorophore. Probe sequences are chosen such that the molecular beacons hybridize adjacent to each other on a single nucleic acid target, bringing the respective fluorophores into close proximity and promoting FRET [18, 84]. The sensitized emission from the acceptor fluorophore then serves as a positive signal in the FRET based detection assay. When the donor and acceptor fluorophores are properly chosen (e.g., with minimal spectral overlap), a strong fluorescence signal will emit from the acceptor only when both molecular beacons are hybridized to the same target and FRET occurs. The fluorescence emitted from molecular beacons that are degraded or opened by protein interactions will be substantially lower than the signal elicited by the donor/acceptor FRET interaction. Thus, a true positive signal owing to probe/target binding events can be readily distinguished from false-positive signals.

To demonstrate the potential of dual FRET molecular beacons approach, an in-solution spectroscopy study was carried out [18]. Specifically, a series of molecular beacons were designed and synthesized which are in antisense orientation with respect to exon 6 and exon 7 of the human GAPDH gene, and a dabcyl quencher was attached to the 5'-end and a 6-Fam fluorophore to the 3'-end of donor molecular beacons; a dabcyl quencher was attached to the 3'-end and either a Cyanine 3 (Cy3), 6-carboxyrhodamine (ROX), or Texas Red fluorophore was attached to the 5'-end of acceptor molecular beacons. The stem sequence was designed to participate in both hairpin formation and target hybridization, as shown in Figure 12.1c. This was adopted to help fix the relative distance between the donor and acceptor fluorophores and improve energy transfer efficiency. Both the donor and acceptor moleculars were designed with a probe length of 18 bases and a stem length of 5 bases. The synthetic wild-type GAPDH target has a 4-base gap between the donor dye and the acceptor dye. Gap spacing was adjusted to 3, 5, and 6 bases by either removing a guanine residue or adding 1 or 2 thymine residues [18].

FRET measurements were carried out using a Safire microplate fluorometer (Tecan) to excite the donor beacons and detect resulting acceptor beacon fluorescence emission (500 nm to 650 nm) of a sample with equal amount (200 nM) of target, donor and acceptor molecular beacons. As shown in Fig. 12.9a, when 6-FAM (peak excitation at 494 nm)

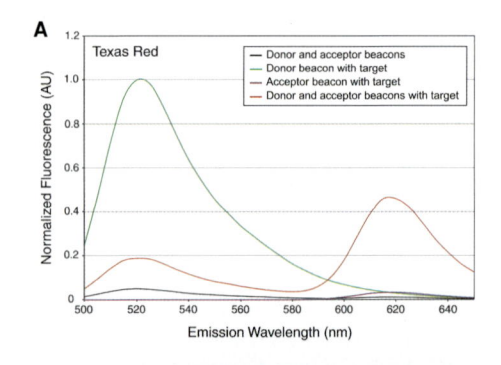

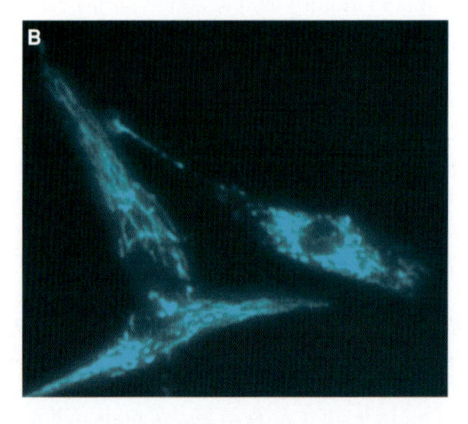

FIGURE 12.9. (A) Emission spectra for dual FRET molecular beacons with a Fam-Texas Red FRET pair. The samples were excited at a wavelength of 475 nm (Adopted from Ref 18). (B) Localization of K-ras mRNA in stimulated human dermal fibroblast (HDF) cells imaged using dual FRET molecular beacons. Note the intriguing filamentous pattern of mRNA localization.

was used as the donor dye and Texas Red (perk emission at 620 nm) as the acceptor dye, a signal-to-background ratio of about 50 was achieved. Clearly, the FRET signal should allow differentiation between the signal emitted upon target detection and false-positive signals.

When the distance between the donor and acceptor molecular beacons was increased from 3 to 6 bases, there was a slight increase in the FRET signal intensity (data not shown). This trend was found to be the same for all the acceptor fluorophores studied. It seems that a gap size of 4 or 5 bases is desirable whereas a 3-base gap is unfavorable due to possible interference between the donor and acceptor dyes.

To demonstrate the ability of molecular beacons in sensitive detection of mRNA in living cells, dual-FRET molecular beacons targeting wild-type K-ras mRNA were designed and synthesized; they were delivered into normally-growing and stimulated human dermal fibroblasts (HDF cells) using SLO. A control molecular beacon (random beacon) was also designed whose sequence has no perfect match in the mammalian genome. Both the K-ras-targeting dual FRET molecular beacon pair, and control molecular beacons have a 16-base target sequence and 5-base stem [84]. For both molecular beacon pairs, Cy3 and Cy5 fluorophores were used as the donor and acceptor, respectively. One hour after the delivery of molecular beacons using SLO, the HDF cells were excited at 545 nm (Cy3 maximum excitation), and the resulting fluorescence signal was observed at 665 nm (Cy5 maximum emission). It was found that the dual-FRET molecular beacons approach could provide fascinating images of mRNA localization [84]. As an example, Fig. 12.9b shows a fluorescence image of K-ras mRNA localization in a few stimulated HDF cells. This direct visualization of mRNA in single living cells revealed intriguing filamentous localization pattern. Evidently, K-ras mRNA molecules are not randomly distributed in the cytoplasm, but localized, possibly to a cytoskeletal component. It is very important to understand why K-ras mRNAs localize in such a way, and what are the biological implications. It is also essential to study the localization of mRNAs corresponding to different classes of proteins. It is likely that the molecular beacons approach will provide essential information on mRNA synthesis, processing, transport, localization, and dynamics in living cells.

12.6. OPPORTUNITIES AND CHALLENGES

Nanostructured molecular probes such as molecular beacons have the potential to enjoy a wide range of applications that require rapid and sensitive detection of genomic sequences. However, to date molecular beacons are used mostly as a tool for the detection of single stranded nucleic acids in homogeneous in vitro assays. For example, molecular beacons have been modified for solid phase studies [164, 165]. Surface immobilized molecular beacons used in microarray assays allow for the high throughput parallel detection of nucleic acid targets while avoiding the difficulties associated with PCR-based labeling [164, 166]. Another novel application of molecular beacons is the detection of double-stranded DNA targets using PNA "openers" that form triplexes with the DNA strands [19]. Further, proteins can be detected by synthesizing "aptamer molecular beacon" [167, 168] which, upon binding to a protein, undergoes a conformational change that results in the restoration of fluorescence.

Perhaps the most exciting application of hairpin probes including molecular beacons is living cell gene detection. As demonstrated, the dual FRET molecular beacons technique can detect endogenous mRNA in living cells rapidly with high specificity, sensitivity, and signal-to-background ratio, thus providing a powerful tool for laboratory and clinical studies of gene expression in vivo. For example, in drug discovery, this method can be used in high-throughput assays to quantify and monitor the dose-dependent changes of specific mRNA expression in response to different candidate drug molecules. In basic biological studies, this method will allow real-time visualization of the dynamics and localization of specific RNAs. The ability to detect and quantify the expression of specific genes in living cells in real-time will offer tremendous opportunities for biological and disease studies, provide another leap forward in our understanding of cell and developmental biology, disease pathophysiology, and significantly impact medical diagnostics.

There are a number of challenges in detecting and quantifying RNA expression in living cells. In addition to probe design issues and target accessibility, quantification of mRNA expression in single cells poses a significant challenge. For example, it is necessary to distinguish true and background signals, determine the fraction of mRNA molecules hybridized with probes, and quantify the possible self-quenching effect of the reporter, especially when mRNA is highly localized. Since the fluorescence intensity of the reporter may be altered by the intracellular environment, it is also necessary to create an internal control by, for example, injecting fluorescently labeled oligonucleotides with known quantity into the same cells and obtaining the corresponding fluorescence intensity. Further, unlike in RT-PCR studies where the mRNA expression is averaged over a large number of cells (usually over one million), in optical imaging of mRNA expression in living cells, only a relatively small number of cells (typically less than one thousand) are observed. Therefore, the average copy number per cell may change with the total number of cells observed due to the (often large) cell-to-cell variation of mRNA expression.

Binding between proteins and hairpin probes may cause non-specific opening of the hairpin, thus increase the background signal. There are many proteins that could bind nucleic acids in the cell cytoplasm, especially hairpin-binding proteins. Although certain modifications of the oligonucleotide backbone chemistry may help, other modifications, such as 2'-O-methyl chemistry, may increase the likelihood of probes being recognized by hairpin-binding proteins.

Another issue in living cell gene detection using hairpin ODN probes is the possible effect of probes on normal cell function, including protein expression. As revealed in the antisense therapy research, complementary pairing of a short segment of an exogenous oligonucleotide to mRNA can have a profound impact on protein expression levels and even cell fate. For example, tight binding of the probe to the translation start site can block mRNA translation. Binding of a DNA probe to mRNA can also trigger RNase H-mediated mRNA degradation. However, the probability of eliciting antisense effects with hairpin probes may be very low when low concentrations of probes (<200 nM) are used for mRNA detection, in contrast to the high concentrations (typically 20 µM; [120]) employed in antisense experiments. Further, it generally takes 4 hours before any noticeable antisense effect occurs, whereas visualization of mRNA with hairpin probes requires less than 2 hours after delivery. However, it is important to carry out a systematic study of the possible antisense effects. When 2'-O-metheyl hairpin probes are used, hybridization between the RNA-like probe and mRNA target results in double-stranded RNA, which may trigger RNA interference. This possibility should also be investigated.

ACKNOWLEGEMENT

This work was supported in part by NSF (BES-0222211), and by DARPA/AFOSR (F49620-03-1-0320).

REFERENCES

[1] F.H. Crick. *Symp. Soc. Exp. Biol.*, 12:138–163, 1958.
[2] B.D. Hall and S. Spiegelman. *Proc. Natl. Acad. Sci. U.S.A.*, 47:137–163, 1961.
[3] J.D. Watson and F.H. Crick. *Nature*, 171:737–738, 1953.
[4] J. Hachmann and H.G. Khorana. *J. Am. Chem. Soc.*, 91:2749–2757, 1969.
[5] R.K. Saiki, S. Scharf, F. Faloona, K.B. Mullis, and G.T. Horn et al. *Science*, 230:1350–1354, 1985.
[6] J.C. Alwine, D.J. Kemp, B.A. Parker, J. Reiser, and J. Renart et al. *Methods Enzymol.*, 68:220–242, 1979.
[7] M.D. Adams, M. Dubnick, A.R. Kerlavage, R. Moreno, J.M. Kelley et al. *Nature*, 355:632–634, 1992.
[8] V.E. Velculescu, L. Zhang, B. Vogelstein, and K.W. Kinzler. *Science*, 270:484–487, 1995.
[9] P. Liang and A.B. Pardee. *Science*, 257:967–971, 1992.
[10] M. Schena, D. Shalon, R.W. Davis, and P.O. Brown. *Science*, 270:467–470, 1995.
[11] H.A. Erlich, D. Gelfand, and J.J. Sninsky. *Science*, 252:1643–1651, 1991.
[12] K. Szuhai, E. Sandhaus, S.M. Kolkman-Uljee, M. Lemaitre, and J.C. Truffert et al. *Am. J. Pathol.*, 159:1651–1660, 2001.
[13] M. Buongiorno-Nardelli and F. Amaldi. *Nature*, 225:946–948, 1970.
[14] S. Tyagi and F.R. Kramer. *Nat. Biotechnol.*, 14:303–308, 1996.
[15] A. Tsourkas, M.A. Behlke, S.D. Rose, and G. Bao. *Nucleic Acids Res.*, 31:1319–1330, 2003.
[16] G. Bonnet, S. Tyagi, A. Libchaber, and F.R. Kramer. *Proc. Natl. Acad. Sci. U.S.A.*, 96:6171–6176, 1999.
[17] D.J. States, W. Gish, and S.F. Altschul. *Methods*, 3:66–70, 1991.
[18] A. Tsourkas, M.A. Behlke, Y. Xu, and G. Bao. *Anal. Chem.*, 75:3697–3703, 2003.
[19] H. Kuhn, V.V. Demidov, J.M. Coull, M.J. Fiandaca, B.D. Gildea, and M.D. Frank-Kamenetskii. *J. Am. Chem. Soc.*, 124:1097–1103, 2002.
[20] S. Tyagi, D.P. Bratu, and F.R. Kramer. *Nat. Biotechnol.*, 16:49–53, 1998.
[21] S. Tyagi, S.A. Marras, and F.R. Kramer. *Nat. Biotechnol.*, 18:1191–1196, 2000.
[22] W.C. Chan, and S. Nie. 1998 *Science*, 281:2016–2018.
[23] S.A. Marras, F.R. Kramer, and S. Tyagi. *Nucleic Acids Res.*, 30:e122, 2002.
[24] B. Dubertret, M. Calame, and A.J. Libchaber. *Nat. Biotechnol.*, 19:365–370, 2001.
[25] A. Sebti, T.E. Kiehn, D. Perlin, V. Chaturvedi, and M. Wong et al. *Clin. Infect. Dis.*, 32:1034–1038, 2001.
[26] S. Park, M. Wong, S.A. Marras, E.W. Cross, T.E. Kiehn et al. *J. Clin. Microbiol.*, 38:2829–2836, 2000.
[27] S. Yates, M. Penning, J. Goudsmit, I. Frantzen, B. van de Weijer et al. *J. Clin. Microbiol.*, 39:3656–3665, 2001.
[28] R.S. Lanciotti and A.J. Kerst. *J. Clin. Microbiol.*, 39:4506–4513, 2001.
[29] W. Chen, G. Martinez, and A. Mulchandani. *Anal. Biochem.*, 280:166–172, 2000.
[30] K.E. Pierce, J.E. Rice, J.A. Sanchez, C. Brenner, and L.J. Wangh. *Mol. Hum. Reprod.*, 6:1155–1164, 2000.
[31] R. Manganelli, E. Dubnau, S. Tyagi, F.R. Kramer, and I. Smith. *Mol. Microbiol.*, 31:715–724, 1999.
[32] I. Tapp, L. Malmberg, E. Rennel, M. Wik, and A.C. Syvanen. *Biotechniques*, 28:732–738, 2000.
[33] M.M. Mhlanga and L. Malmberg. *Methods*, 25:463–471, 2001.
[34] J.A. Vet, A.R. Majithia, S.A. Marras, S. Tyagi, and S. Dube et al. 1999. *Proc. Natl. Acad. Sci. U.S.A.*, 96:6394–6399.
[35] A. de Ronde, M. van Dooren, L. van Der Hoek, D. Bouwhuis, and E. de Rooij et al. *J. Virol.*, 75:595–602, 2001.
[36] B. Vogelstein and K.W. Kinzler. *Proc. Natl. Acad. Sci. U.S.A.*, 96:9236–9241, 1999.
[37] I.M. Shih, W. Zhou, S.N. Goodman, C. Lengauer, K.W. Kinzler, and B. Vogelstein. *Cancer Res.*, 61:818–822, 2001.

[38] R. Durand, J. Eslahpazire, S. Jafari, J.F. Delabre, and A. Marmorat-Khuong et al. *Antimicrob. Agents Chemother.*, 44:3461–3464, 2000.
[39] B.A. Giesendorf, J.A. Vet, S. Tyagi, E.J. Mensink, F.J. Trijbels, and H.J. Blom. *Clin. Chem.*, 44:482–486, 1998.
[40] G. Dreyfuss, V.N. Kim, and N. Kataoka. *Nat. Rev. Mol. Cell. Biol.*, 3:195–205, 2002.
[41] S. McCracken, N. Fong, K. Yankulov, S. Ballantyne, and G. Pan et al. *Nature*, 385:357–361, 1997.
[42] A.J. Shatkin and J.L. Manley. *Nat. Struct. Biol.*, 7:838–842, 2000.
[43] Y. Furuichi and A.J. Shatkin. *Adv. Virus. Res.*, 55:135–184, 2000.
[44] M.J. Dye and N.J. Proudfoot. *Mol. Cell.*, 3:371–378, 1999.
[45] C.E. Birse, L. Minvielle-Sebastia, B.A. Lee, W. Keller, and N.J. Proudfoot. *Science*, 280:298–301, 1998.
[46] C.A. Collins and C. Guthrie. *Nat. Struct. Biol.*, 7:850–854, 2000.
[47] J.F. Caceres, G.R. Screaton, and A.R. Krainer. *Genes. Dev.*, 12:55–66, 1998.
[48] H. Le Hir, M.J. Moore, and L.E. Maquat. *Genes. Dev.*, 14:1098–1108, 2000.
[49] M.J. Luo, and R. Reed. *Proc. Natl. Acad. Sci. U.S.A.*, 96:14937–14942, 1999.
[50] K.L. Farina and R.H. Singer. *Trends Cell. Biol.*, 12:466–472, 2002.
[51] K. Kristensson, N.K. Zeller, M.E. Dubois-Dalcq, and R.A. Lazzarini. *J. Histochem. Cytochem.*, 34:467–473, 1986.
[52] A.N. Verity and A.T. Campagnoni. *J. Neurosci. Res.*, 21:238–248, 1988.
[53] K. Ainger, D. Avossa, A.S. Diana, C. Barry, E. Barbarese, and J.H. Carson. *J. Cell. Biol.*, 138:1077–1087, 1997.
[54] K.S. Hoek, G.J. Kidd, J.H. Carson, and R. Smith. *Biochemistry*, 37:7021–7029, 1998.
[55] W. Gu, F. Pan, H. Zhang, G.J. Bassell, and R.H. Singer. *J. Cell. Biol.*, 156:41–51, 2002.
[56] J. Katahira, K. Strasser, A. Podtelejnikov, M. Mann, J.U. Jung, and E. Hurt. *Embo. J.*, 18:2593–2609, 1999.
[57] A.E. Pasquinelli, R.K. Ernst, E. Lund, C. Grimm, M.L. Zapp et al. *Embo. J.*, 16:7500–7510, 1997.
[58] C. Saavedra, B. Felber, and E. Izaurralde. *Curr. Biol.*, 7:619–628, 1997.
[59] B.R. Cullen. *Proc. Natl. Acad. Sci. U.S.A.*, 97:4–6, 2000.
[60] I.M. Palacios. *Curr. Biol.*, 12:R50–R52, 2002.
[61] D.A. Jackson, A. Pombo, and F. Iborra. *Faseb J.*, 14:242–254, 2000.
[62] G. Aakalu, W.B. Smith, N. Nguyen, C. Jiang, and E.M. Schuman. *Neuron*, 30:489–502, 2001.
[63] E.A. Shestakova, R.H. Singer, and J. Condeelis. *Proc. Natl. Acad. Sci. U.S.A.*, 98:7045–7050, 2001.
[64] J.E. Hesketh. *Exp. Cell. Res.*, 225:219–236, 1996.
[65] G.J. Bassell, R.H. Singer, and K.S. Kosik. *Neuron*, 12:571–582, 1994.
[66] G. Bassell and R.H. Singer. *Curr. Opin. Cell. Biol.*, 9:109–915, 1997.
[67] J. Hesketh, D. Jodar, A. Johannessen, K. Partridge, I. Pryme, and A. Tauler. *Biochem. Soc. Trans.*, 24:187S, 1996.
[68] P. Mahon, K. Partridge, J.H. Beattie, L.A. Glover, and J.E. Hesketh. *Biochim. Biophys. Acta.*, 1358:153–162, 1997.
[69] P.R. Walker and J.F. Whitfield. *J. Biol. Chem.*, 260:765–770, 1985.
[70] B. Zhou and M. Rabinovitch. *Circ. Res.*, 83:481–489, 1998.
[71] D. Hamill, J. Davis, J. Drawbridge, and K.A. Suprenant. *J. Cell. Biol.*, 127:973–984, 1994.
[72] S.K. Pramanik, R.W. Walsh, and J. Bag. *Eur. J. Biochem.*, 160:221–230, 1986.
[73] J.W. Wiseman, L.A. Glover, and J.E. Hesketh. *Biochem. Soc. Trans.*, 24:188S, 1996.
[74] J.W. Wiseman, L.A. Glover, and J.E. Hesketh. *Cell. Biol. Int.*, 21:243–248, 1997.
[75] B. Russell and D.J. Dix. *Am. J. Physiol.*, 262:C1–C8, 1992.
[76] G.J. Bassell, C.M. Powers, K.L. Taneja, and R.H. Singer. *J. Cell. Biol.*, 126:863–876, 1994.
[77] P. Mahon, J. Beattie, L.A. Glover, and J. Hesketh. *FEBS Lett.*, 373:76–80, 1995.
[78] E.H. Kislauskis, Z. Li, R.H. Singer, and K.L. Taneja. *J. Cell. Biol.*, 123:165–172, 1993.
[79] D. Ferrandon, L. Elphick, C. Nusslein-Volhard, and D. St Johnston. *Cell*, 79:1221–1232, 1994.
[80] K.M. Forrest and E.R. Gavis. *Curr. Biol.*, 13:1159–1168, 2003.
[81] V.M. Latham Jr., E.H. Kislauskis, R.H. Singer, and A.F. Ross. *J. Cell. Biol.*, 126:1211–1219, 1994.
[82] A.S. Brodsky and P.A. Silver. *Methods*, 26:151–155, 2002.
[83] B.A. Griffin, S.R. Adams, J. Jones, and R.Y. Tsien. *Methods Enzymol.*, 327:565–578, 2000.
[84] P.J. Santangelo, B. Nix, A. Tsourkas, and G. Bao. Dual FRET molecular beacons for mRNA detection in living cells. Submitted to *Nucleic Acids Res.*, 2004.
[85] S.W. Hell. *Nat. Biotechnol.*, 21:1347–1355, 2003.

[86] J. Ross. *Microbiol. Rev.*, 59:423–450, 1995.
[87] J. Ross. *Trends Genet*, 12:171–175, 1996.
[88] N. Heintz, H.L. Sive, and R.G. Roeder. *Mol. Cell. Biol.*, 3:539–550, 1983.
[89] A. Bevilacqua, M.C. Ceriani, S. Capaccioli, and A. Nicolin. *J. Cell. Physiol.*, 195:356–372, 2003.
[90] T.C. Santiago, I.J. Purvis, A.J. Bettany, and A.J. Brown. *Nucleic Acids Res.*, 14:8347–8360, 1986.
[91] Y. Wang, C.L. Liu, J.D. Storey, R.J. Tibshirani, D. Herschlag, and P.O. Brown. *Proc. Natl. Acad. Sci. U.S.A.*, 99:5860–5865, 2002.
[92] S. Pinol–Roma and G. Dreyfuss. *Nature*, 355:730–732, 1992.
[93] I.E. Gallouzi and J.A. Steitz. *Science*, 294:1895–1901, 2001.
[94] C.G. Burd and G. Dreyfuss. *Science*, 265:615–621, 1994.
[95] F. Stutz, A. Bachi, T. Doerks, I.C. Braun, and B. Seraphin et al. *Rna*, 6:638–650, 2000.
[96] M. Zuker and P. Stiegler. *Nucleic Acids Res.*, 9:133–148, 1981.
[97] J.S. McCaskill. *Biopolymers*, 29:1105–1119, 1990.
[98] D.J. Ecker, T.A. Vickers, T.W. Bruice, S.M. Freier, and R.D. Jenison et al. *Science*, 257:958–961, 1992.
[99] D.H. Mathews, M.E. Burkard, S.M. Freier, J.R. Wyatt, and D.H. Turner. *Rna*, 5:1458–1469, 1999.
[100] M. Andronescu, R. Aguirre-Hernandez, A. Condon, and H.H. Hoos. *Nucleic Acids Res.*, 31:3416–3422, 2003.
[101] Q. Huang and T. Pederson. *Nucleic Acids Res.*, 27:1025–1031, 1999.
[102] J.B. Glotzer, R. Saffrich, M. Glotzer, A. Ephrussi. *Curr. Biol.*, 7:326–337, 1997.
[103] M.R. Jacobson and T. Pederson. *Proc. Natl. Acad. Sci. U.S.A.*, 95:7981–7986, 1998.
[104] M. Somasundaran, M.L. Zapp, L.K. Beattie, L. Pang, and K.S. Byron et al. *J. Cell. Biol.*, 126:1353–1360, 1994.
[105] J.L. Veyrune, G.P. Campbell, J. Wiseman, J.M. Blanchard, and J.E. Hesketh. *J. Cell. Sci.*, 109 (Pt 6):1185–1194, 1996.
[106] S. Behrens, B.M. Fuchs, F. Mueller, and R. Amann. *Appl. Environ. Microbiol.*, 69:4935–4941, 2003.
[107] S. Dokka and Y. Rojanasakul. *Adv. Drug Deliv. Rev.*, 44:35–49, 2000.
[108] N.C. Price and L. Stevens. *Fundamentals of Enzymology: The Cell and Molecular Biology of Catalytic Proteins*. Oxford University Press. New York, 1999.
[109] M.X. Tang, C.T. Redemann, F.C. Szoka Jr. *Bioconjug. Chem.*, 7:703–714, 1996.
[110] O. Zelphati, F.C. Szoka Jr. *Proc. Natl. Acad. Sci. U.S.A.*, 93:11493–11498, 1996.
[111] J.P. Leonetti, N. Mechti, G. Degols, C. Gagnor, B. Lebleu. *Proc. Natl. Acad. Sci. U.S.A.*, 88:2702–2706, 1991.
[112] A.M. Koch, F. Reynolds, M.F. Kircher, H.P. Merkle, R. Weissleder, and L. Josephson. *Bioconjug. Chem.*, 14:1115–1121, 2003.
[113] D.L. Sokol, X. Zhang, P. Lu, and A.M. Gewirtz. *Proc. Natl. Acad. Sci. U.S.A.*, 95:11538–11543, 1998.
[114] C. Molenaar, S.A. Marras, J.C. Slats, J.C. Truffert, and M. Lemaitre et al. *Nucleic Acids Res.*, 29:E89–E89, 2001.
[115] A. Tsuji, H. Koshimoto, Y. Sato, M. Hirano, and Y. Sei-Iida et al. *Biophys. J.*, 78:3260–3274, 2000.
[116] J. Perlette and W. Tan. *Anal. Chem.*, 73:5544–5550, 2001.
[117] D.P. Bratu, B.J. Cha, M.M. Mhlanga, F.R. Kramer, and S. Tyagi. *Proc. Natl. Acad. Sci. U.S.A.*, 100:13308–13313, 2003.
[118] R.V. Giles, C.J. Ruddell, D.G. Spiller, J.A. Green, and D.M. Tidd. *Nucleic Acids Res.*, 23:954–961, 1995.
[119] M.A. Barry and A. Eastman. *Arch. Biochem. Biophys.*, 300:440–450, 1993.
[120] R.V. Giles, D.G. Spiller, J. Grzybowski, R.E. Clark, P. Nicklin, and D.M. Tidd. *Nucleic Acids Res.*, 26:1567–1575, 1998.
[121] D.G. Spiller, R.V. Giles, J. Grzybowski, D.M. Tidd, and R.E. Clark. *Blood*, 91:4738–4746, 1998.
[122] I. Walev, S.C. Bhakdi, F. Hofmann, N. Djonder, and A. Valeva et al. *Proc. Natl. Acad. Sci. U.S.A.*, 98:3185–3190, 2001.
[123] E.L. Barry, F.A. Gesek, and P.A. Friedman. *Biotechniques*, 15:1016–1018, 10120, 1993.
[124] S. Paillasson, M. Van De Corput, R.W. Dirks, H.J. Tanke, M. Robert-Nicoud, and X. Ronot. *Exp. Cell. Res.*, 231:226–233, 1997.
[125] B.H. Lloyd, R.V. Giles, D.G. Spiller, J. Grzybowski, D.M. Tidd, and D.R. Sibson. *Nucleic Acids Res.*, 29:3664–3673, 2001.
[126] D. Derossi, G. Chassaing, and A. Prochiantz. *Trends Cell. Biol.*, 8:84–87, 1998.
[127] S.R. Schwarze, K.A. Hruska, and S.F. Dowdy. *Trends Cell. Biol.*, 10:290–295, 2000.

[128] M. Silhol, M. Tyagi, M. Giacca, B. Lebleu, and E. Vives. *Eur. J. Biochem.*, 269:494–501, 2002.
[129] S. Violini, V. Sharma, J.L. Prior, M. Dyszlewski, and D. Piwnica-Worms. *Biochemistry*, 41:12652–12661, 2002.
[130] M. Belting. *Trends Biochem. Sci.*, 28:145–151, 2003.
[131] B. Allinquant, P. Hantraye, P. Mailleux, K. Moya, C. Bouillot, and A. Prochiantz. *J. Cell. Biol.*, 128:919–927, 1995.
[132] C.M. Troy, D. Derossi, A. Prochiantz, L.A. Greene, and M.L. Shelanski. *J. Neurosci.*, 16:253–261, 1996.
[133] A. Astriab-Fisher, D. Sergueev, M. Fisher, B.R. Shaw, and R.L. Juliano. *Pharm. Res.*, 19:744–754, 2002.
[134] N. Nitin, P.J. Santangelo, G. Kim, S. Nie, and G. Bao. Peptide-linked molecular beacons for efficient delivery and rapid mRNA detection in living cells, submitted to *Nucleic Acids Res.*, 2004.
[135] Y. Nakagami, M. Ito, T. Hara, T. Inoue, and S. Matsubara. *Acta. Oncol.*, 42:227–236, 2003.
[136] A. Urbano, R. McCaffrey, and F. Foss. *J. Biol. Chem.*, 273:34820–34827, 1998.
[137] M.J. Fraser, S.J. Tynan, A. Papaioannou, C.M. Ireland, and S.M. Pittman. *J. Cell. Sci.*, 109 (Pt 9):2343–2460, 1996.
[138] P. Mitchell and D. Tollervey. *Nat. Struct. Biol.*, 7:843–846, 2000.
[139] M.J. Moore. *Cell*, 108:431–434, 2002.
[140] A. van Hoof and R. Parker. *Cell*, 99:347–350, 1999.
[141] C. Allmang, E. Petfalski, A. Podtelejnikov, M. Mann, D. Tollervey, and P. Mitchell. *Genes. Dev.*, 13:2148–2158, 1999.
[142] P. Mitchell, E. Petfalski, A. Shevchenko, M. Mann, and D. Tollervey. *Cell*, 91:457–466, 1997.
[143] T. Kadowaki, M. Hitomi, S. Chen, and A.M. Tartakoff. *Mol. Biol. Cell.*, 5:1253–1263, 1994.
[144] N.I. Zanchin and D.S. Goldfarb. *Mol. Cell. Biol.*, 19:1518–1525, 1999.
[145] J.S. Jacobs, A.R. Anderson, and R.P. Parker. *Embo. J.*, 17:1497–1506, 1998.
[146] Y. Matsumoto, R. Fishel, and R.B. Wickner. *Proc. Natl. Acad. Sci. U.S.A.*, 87:7628–7632, 1990.
[147] R.A. Deshpande and V. Shankar. *Crit. Rev. Microbiol.*, 28:79–122, 2002.
[148] M. Irie. *Pharmacol. Ther.* 81:77–89, 1999.
[149] A. Egesten, K.D. Dyer, D. Batten, J.B. Domachowske, and H.F. Rosenberg. *Biochim. Biophys. Acta.*, 1358:255–260, 1997.
[150] B.R. Chapados, Q. Chai, D.J. Hosfield, J. Qiu, B. Shen, and J.A. Tainer. *J. Mol. Biol.*, 307:541–556, 2001.
[151] P.S. Eder, R.Y. Walder, and J.A. Walder. *Biochimie*, 75:123–126, 1993.
[152] B. Rydberg and J. Game. *Proc. Natl. Acad. Sci. U.S.A.*, 99:16654–16659, 2002.
[153] A.L. ten Asbroek, M. van Groenigen, M. Nooij, and F. Baas. *Eur. J. Biochem.*, 269:583–592, 2002.
[154] S. Agrawal. *Biochim. Biophys. Acta.*, 1489:53–68, 1999.
[155] A. De Mesmaeker, K.H. Altmann, A. Waldner, and S. Wendeborn. *Curr. Opin. Struct. Biol.*, 5:343–355, 1995.
[156] M. Manoharan. *Biochim. Biophys. Acta.*, 1489:117–130, 1999.
[157] E.A. Lesnik, C.J. Guinosso, A.M. Kawasaki, H. Sasmor, and M. Zounes et al. *Biochemistry*, 32:7832–7838, 1993.
[158] M. Petersen and J. Wengel. *Trends Biotechnol.*, 21:74–81, 2003.
[159] U. Christensen, N. Jacobsen, V.K. Rajwanshi, J. Wengel, and T. Koch. *Biochem. J.*, 354:481–484, 2001.
[160] M. Frieden, H.F. Hansen, and T. Koch. *Nucleosides Nucleotides Nucleic Acids*, 22:1041–1043, 2003.
[161] P.E. Nielsen, M. Egholm, R.H. Berg, and O. Buchardt. *Science*, 254:1497–1500, 1991.
[162] T. Ratilainen, A. Holmen, E. Tuite, P.E. Nielsen, and B. Norden. *Biochemistry*, 39:7781–7791, 2000.
[163] S.A. Kushon, J.P. Jordan, J.L. Seifert, H. Nielsen, P.E. Nielsen, and B.A. Armitage. *J. Am. Chem. Soc.*, 123:10805–10813, 2001.
[164] X. Liu and W. Tan. *Anal. Chem.*, 71:5054–5059, 1999.
[165] D. Kambhampati, P.E. Nielsen, and W. Knoll. *Biosens. Bioelectron.*, 16:1109–1118, 2001.
[166] F.J. Steemers, J.A. Ferguson, and D.R. Walt. *Nat. Biotechnol.*, 18:91–94, 2000.
[167] N. Hamaguchi, A. Ellington, and M. Stanton. *Anal. Biochem.*, 294:126–131, 2001.
[168] R. Yamamoto, T. Baba, and P.K. Kumar. *Genes Cells*, 5:389–396, 2000.

13

Fluorescent Lanthanide Labels with Time-Resolved Fluorometry in DNA Analysis

Takuya Nishioka, Jingli Yuan, and Kazuko Matsumoto
Department of Chemistry and Advanced Research Institute for Science and Engineering, Waseda University, 3-4-1 Okubo, Shinjuku-ku, Tokyo 169-8555, Japan

13.1. INTRODUCTION

Some lanthanide (Eu^{3+}, Tb^{3+}, Sm^{3+}, and Dy^{3+}) complexes are known to be luminescent. Compared with organic fluorescent compounds, the fluorescence of lanthanide chelates has several special different properties; (1) The lifetime of the lanthanide chelates is very long; that of europium(III) and terbium(III) chelates usually ranges from several hundred microseconds to more than one millisecond, and that of samarium(III) and dysprosium(III), 10 to 100 microseconds. (2) Stokes shifts are very large: the chelates are excited by UV light (310–350 nm) and emit fluorescence in the visible region. (3) The emission profiles are sharp, having a full width at half maximum (FWHM) of only ~10 nm.

In the last 20 years, lanthanide chelates have been successfully developed as fluorescence labels for highly sensitive detection of various biological molecules in time-resolved fluorometry. Lanthanide fluorescence labels have been used in time-resolved fluorometry of immunoassay (TR-FIA), DNA hybridization assay, high performance liquid chromatography (HPLC), fluorescence imaging microscopy, and other bioassays, and shown great improvement of the sensitivity compared to the conventional fluorometry using organic fluorescent labels. The high sensitivity or detectability is due to easy distinction of the specific fluorescence signal of the lanthanide labels from background signals present in most biological samples. Time-resolved fluorometry also obviates the problems associated with light scattering of the optical components. These problems cannot be solved easily by the general fluorescence method. The principles and applications of time-resolved fluorescence measurements for diagnostics and biotechnology have been reviewed [1–11].

13.2. LANTHANIDE FLUORESCENT COMPLEXES AND LABELS

The fluorescence of aquo complexes of lanthanide ions is very weak because of their very low extinction coefficients and fluorescence quantum yields, but the fluorescence can be dramatically enhanced when they form complexes with appropriate organic ligands. The fluorescence properties of these complexes strongly depend on the structure of the ligands, and the properties specific to lanthanide chelates are caused by the fluorescence mechanism of the lanthanide complexes shown in Fig. 13.1; the ligand is excited to the S_1 state at the first stage, and then to the triplet state (T_1) via intersystem crossing. Successively the energy is transferred from T_1 to the metal ion center, and finally transition from the excited state of metal ion to the ground state gives visible light emission. The strongest fluorescence of europium(III) complexes with β-diketonate ligands at 615 nm corresponds to the transition from $^5D_0$ to $^7F_2$. Among several types of ligands, β-diketone and aromatic amine ligands are known for a long time as the favorable ligands for lanthanide fluorescence. In order to use the lanthanide complexes for DNA detection and other biological assays, the complex must have a binding group to biomolecules, and isothiocyanate, sulfonyl chloride, and carboxylate of N-hydroxysuccinimide are usually used to couple with DNA, proteins or other biological molecules. In many cases, introduction of an active binding group to a fluorescent lanthanide complex is not so easy, and sometimes causes decrease of the fluorescence considerably. Therefore design and synthesis of favorable fluorescent lanthanide labels for bioassay are difficult.

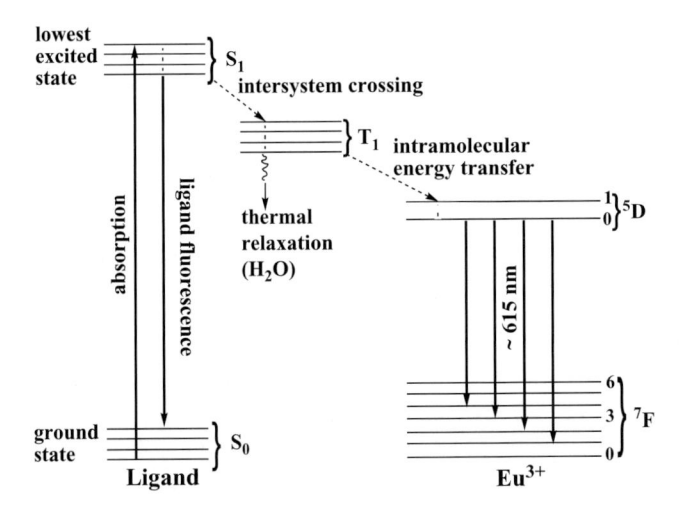

FIGURE 13.1. Mechanism of the fluorescence of Eu^{3+} chelates. Modified from *CRC Crit. Rev. Anal. Chem.*, 18, 105–154 (1987).

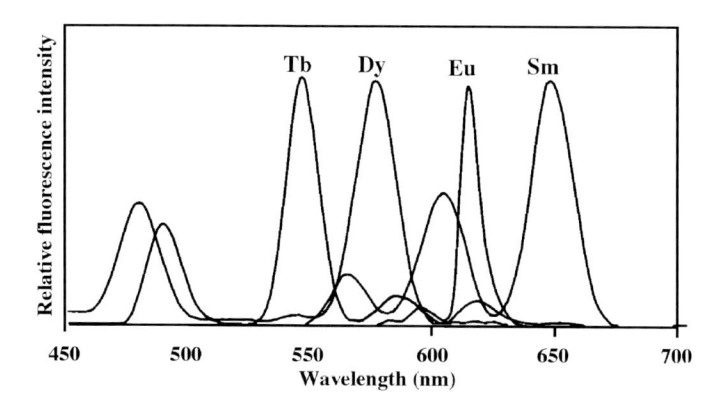

FIGURE 13.2. Emission spectra of the complexes of Eu^{3+}, Sm^{3+}, Tb^{3+}, and Dy^{3+} with PTA in the presence of 1,10-phenanthroline, Triton X-100, and Y^{3+}. Modified from *Anal. Chim. Acta*, 256, 9–16 (1992).

Since Weissman discovered in 1942 that Eu^{3+} complexes with β-diketone-type ligands emit fluorescence when excited with UV light [12], this class of complexes has been investigated intensively and is used for lanthanide analysis [13] and for laser materials [14]. In these early works, the researchers found that 2-naphthoyltrifluoroacetone (β-NTA), 2-thenoyltrifluoroacetone (TTA) and pivaloyltrifluoroacetone (PTA) are the best ligands for fluorescent complexes of Eu^{3+}, Sm^{3+}, Tb^{3+} and Dy^{3+}; among these β-NTA and TTA are effective only for Eu^{3+} and Sm^{3+}, whereas PTA is effective for all four ions at room temperature. Figure 13.2 shows the emission spectra of the Eu^{3+}, Sm^{3+}, Tb^{3+}, and Dy^{3+} complexes with PTA in the presence of 1,10-phenanthroline, Triton X-100, and Y^{3+} [15]. However, the bidentate lanthanide-β-diketonate complexes in Figure 13.2 cannot be used as labels, since there is no active binding group on these β-diketone ligands and these complexes are not very stable with the stability constants only in the order of 10^3 to 10^6 [16, 17], therefore the complexes dissociate in highly diluted solutions and the fluorescence decreases.

Recently, three chlorosulfonylated tetradentate β-diketone-type ligands were synthesized (Figure 13.3) [18–20]. They differ from other β-diketones, because the fluorescence intensities of their Eu^{3+} complexes are not weakened on attaching a sulfonyl chloride group to the ligand. Compared with the bidentate β-diketone ligands, the tetradentate structures in these ligands increase the stabilities of the Eu^{3+} complexes and also the fluorescence intensities because the ligands decrease the number of the coordinated water on the lanthanide ion and avoid the fluorescence quenching by water.

Among the three ligands, 4,4'-bis(1",1",1",2",2",3",3"-heptafluoro-4",6"-hexanedion-6"-yl)-chlorosulfo-*o*-terphenyl (BHHCT) is the most suitable label for time-resolved fluorometry. Compared with other two, BHHCT has the advantage that (i) it has only one sulfonyl chloride group, therefore cross labeling among several protein molecules does not occur, (ii) its Eu^{3+} complex maintains long fluorescence lifetime (400 to 700 μs) in various buffers, (iii) the relative fluorescence intensity ($\varepsilon\phi$) of the Eu^{3+} complex is considerably larger than those of other europium labels, and (iv) the Eu^{3+} complex (bound to bovine serum albumin (BSA)) has a relatively large stability constant (about 10^{10} M^{-1}),

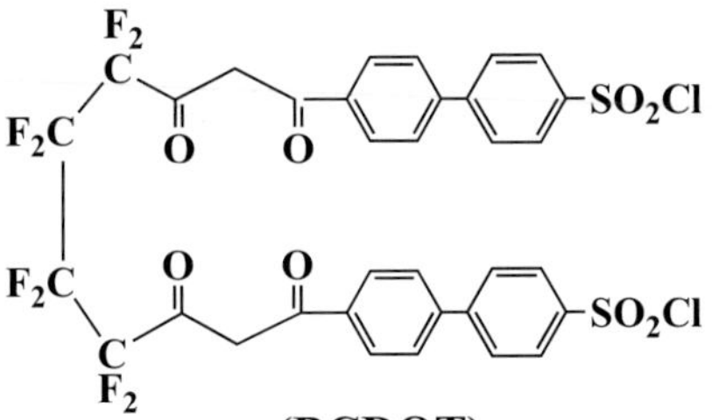

FIGURE 13.3. Structures of three chlorosulfonylated tetradentate β-diketone labels. BCDOT = 1,10-bis(4"-chlorosulfo-1',1"-diphenyl-4'-yl)-4,4,5,5,6,6,7,7-octafluorodecane-1,3,8,10-tetraone, BCOT = 1,10-bis(8'-chlorosulfodibenzothiophene-2'-yl)-4,4,5,5,6,6,7,7-octafluorodecane-1,3,8,10-tetraone, BHHCT = 4,4'-bis(1",1",1",2",2",3",3"-heptafluoro-4",6"-hexanedion-6"-yl)chlorosulfo-o-terphenyl.

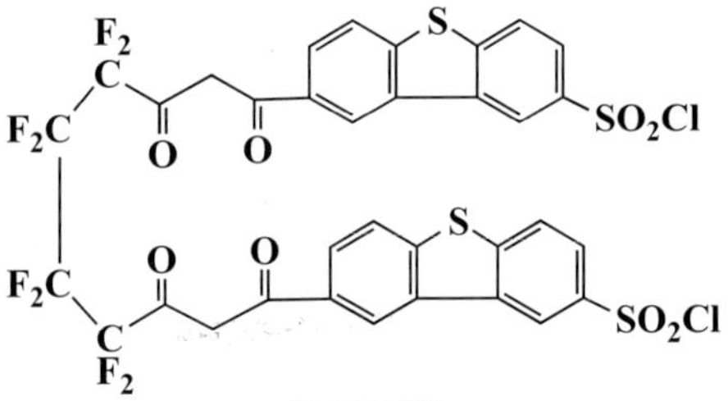

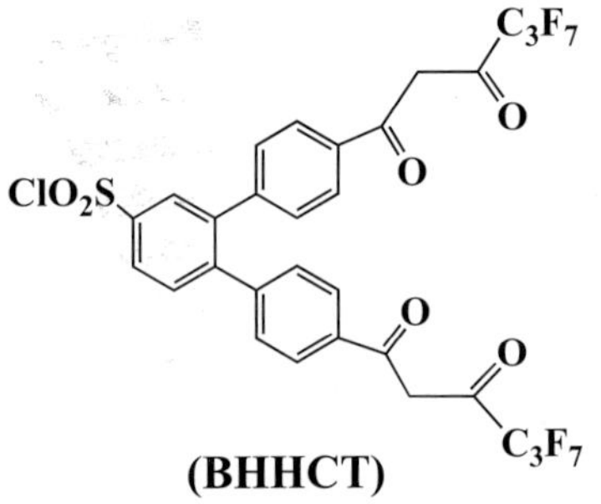

and so the complex is very stable in solutions of complex multi-component ones as used in bioassays. The ligand BHHCT is easy to conjugate to proteins through sulfonamide formation (protein-NH-SO$_2$-label). The main drawback of BHHCT is its low solubility in water-based buffers, which makes it unsuitable for direct labeling of small molecules in aqueous solution. However, BHHCT-labeled BSA, the hapten-BSA conjugate, streptavidin (SA), antibodies, and other proteins and nucleic acids are soluble in water-based buffers.

Aromatic amine derivative-type ligands mostly consist of derivatives of pyridine, 2,2'-bipyridine, 2,2',2"-terpyridine, and 1,10-phenanthroline [21–24]. Figure 13.4 shows the structures of four Eu^{3+} chelates with aromatic amine derived ligands, that can be covalently

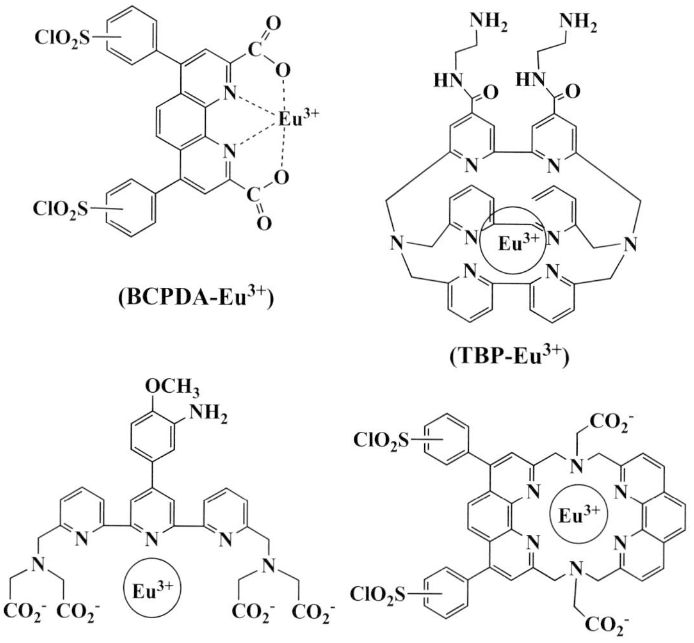

FIGURE 13.4. Structures of four fluorescent Eu^{3+} chelates with aromatic amine ligands that can be covalently bound to proteins.

bound to proteins, and among which 4,7-bis(chlorosulfophenyl)-1,10-phenanthroline-2,9-dicarboxylic acid (BCPDA)-Eu^{3+} and trisbipyridine cryptate (TBP)-Eu^{3+} are commercially used for europium fluorescence labels in TR-FIA.

13.3. TIME-RESOLVED FLUOROMETRY OF LANTHANIDE COMPLEXES

The main problem of the conventional fluorescence bioassay is the strong background signal including the fluorescence from the coexisting biological materials, the scattering light associated with Tyndall, Rayleigh, and Raman scattering, and the background luminescence from the optical components such as the cuvettes, filters and lenses. The elimination of the background signals is essential for highly sensitive detection. For this purpose, the time-resolved fluorometry using long-lived fluorescent lanthanide chelates is the most favorable method, since the background noise is short-lived with a lifetime of a nanosecond to a few microseconds, and easily removed by time-resolved measurement. The principle of the time-resolved fluorometric measurement is illustrated in Figure 13.5. After a sample is excited by a flash lamp, the fluorescence of all the molecules begins to decay exponentially. Since the short-lived background signal rapidly decreases, it can be effectively eliminated during the delay time. This permits to measure only the long-lived lanthanide fluorescence during the counting time with high sensitivity. Furthermore, the measurement is usually reiterated many times (usually one second per cuvette) to accumulate the signal and improve the signal to noise ratio.

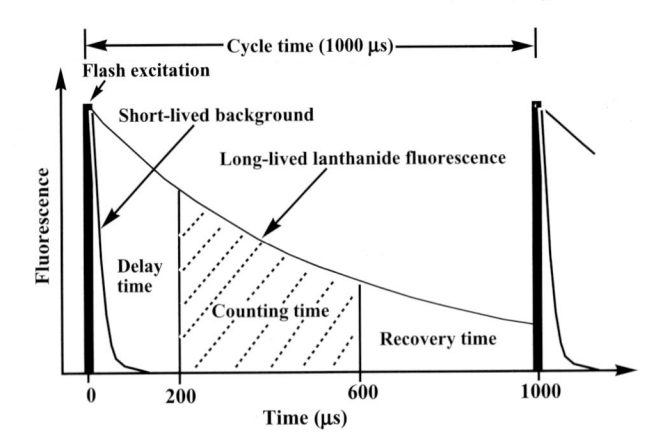

FIGURE 13.5. Principle of the time-resolved fluorometric measurement with a delay time of 200 μs, a counting time of 400 μs, and a cycle time of 1000 μs.

13.4. DNA HYBRIDIZATION ASSAY

DNA hybridization assay is one of the most widely used tools for diagnosis of infections, genetics, neoplasmic diseases, and microbial taxonomy [25–27]. A seven-color time-resolved fluorescence DNA hybridization assay was developed for the detection of amplified products of the polymerase chain reaction (PCR) of the seven human papilloma virus (HPV) types 16, 18, 31, 33, 35, 39, and 45, associated with cervix cancer [28]. In the method, seven combinations of the non-fluorescent lanthanide labels, Eu^{3+}, Tb^{3+}, Sm^{3+}, Eu^{3+}-Tb^{3+}, Eu^{3+}-Sm^{3+}, Tb^{3+}-Sm^{3+}, and Eu^{3+}-Tb^{3+}-Sm^{3+} were used to label seven HPV type-specific oligonucleotide probes. After hybridization of the labeled probes with the immobilized target strains, the Eu^{3+} and Sm^{3+} fluorescences were measured after addition of the DELFIA fluorescence enhancement solution containing β-diketone, TOPO, and Triton X-100. Then, a solution containing 4-(2',4',6'-trimethoxyphenyl)pyridine-2,6-dicarboxylic acid and cetyltrimethylammonium bromide was added to make the Tb^{3+} complex fluorescent, and the Tb^{3+} fluorescence was measured. After the measurement, the PCR products from each of the seven different viral strains were correctly assigned by monitoring the contribution from each of the three lanthanide ions to the total fluorescence. Heinonen et al. developed a triple-label time-resolved fluorescence DNA hybridization assay method for detection of PCR amplification products of seven cystic fibrosis mutations in human blood [29]. In the method, 14 kinds of allele-specific oligonucleotides labeled with Eu^{3+}, Sm^{3+}, or Tb^{3+} complexes were used as wild-type specific and mutant-specific probes. After the biotinylated PCR products collected in the SA coated microtiter wells were denatured and hybridized with the 14 probes, the Eu^{3+}, Sm^{3+}, and Tb^{3+} fluorescences were measured with the same method as described above.

In addition, the enzyme amplified time-resolved fluorometric measurement system has been used for another DNA hybridization assay. Bortolin and co-workers reported the quantitative measurement of DNA PCR products by using a digoxigenin-labeled PCR product (the specific capture probe was immobilized onto the microtiter well) or a digoxigenin-tailed specific probe (the PCR product was captured onto the microtiter well) for hybridization [30]. After the hybrids were reacted with ALP-labeled anti-digoxigenin antibody, 5'-fluorosalicylphosphate and Tb^{3+}-EDTA were added for the coloration reaction and the time-resolved fluorescence measurement. Recently, this method was further improved for higher sensitivity by using a two-round enzyme amplification method [31]. In this improved method, digoxigenin-labeled target DNA was immobilized onto the anti-digoxigenin antibody-coated microtiter wells and then hybridized with a biotinylated DNA probe. After the hybrid was reacted with horseradish peroxidase-labeled SA, a solution containing biotinylated tyramine and H_2O_2 was added to react with the horseradish peroxidase. This reaction results in the attachment of multiple biotin moieties to the solid phase. Then ALP-labeled SA was added to react with the immobilized biotin molecules. The coloration reaction and time-resolved fluorescence measurement were performed as described above. Although this method needs an additional step, it gives a 10-times enhanced signal-to-noise ratio compared with the previous method. The BHHCT-Eu^{3+}-labeled SA-BSA was also employed for the DNA hybridization assay [32].

The procedures of the conventional DNA hybridization assay are tedious and time-consuming, requiring immobilization of the target DNA on a solid support, prehybridization, hybridization, washing, and detection. In addition, the hybridization reaction in the solid-solution phase proceeds rather slow, which also makes the assay time-consuming. To circumvent this problem, the homogeneous DNA hybridization assay based on fluorescence resonance energy transfer (FRET) has been developed [33]. In this assay, two DNA probes, one labeled with biotin on the 3'-terminus and the other with Cy5 on the 5'-terminus were used. After hybridization, the BHHCT-Eu^{3+}-labeled SA was added to react with biotin. When the target DNA is present, the BHHCT-Eu^{3+} and Cy5 come close to each other, and the energy transfer occurs from the Eu^{3+} complex to Cy5. Thus the target DNA can be detected by measuring the sensitized emission of Cy5 at 669 nm with the UV-light excitation and the normal fluorescence measurement mode or the time-resolved mode. This method overcomes the interference of the background emission, and gives a high sensitivity with the detection limit of 200 pM.

Another homogeneous DNA hybridization assay method has been developed, which is based on DNA-mediated formation of a ternary complex of EDTA-Eu^{3+}-β-diketonate, to improve the detection limit [34]. The principle of the method is illustrated in Figure 13.7, in which two DNA probes are used; one labeled with the EDTA-Eu^{3+} chelate on the 5'-terminus and the other the β-diketone, 5-(4"-chlorosulfo-1',1"-diphenyl-4'-yl)-1,1,1,2,2-pentafluoro-3,5-pentanedione (CDPP), on the 3'-terminus. The two probes are complementary to the contiguous regions of the target DNA. After hybridization, two labels come close to each other to form the strongly fluorescent ternary complex of EDTA-Eu^{3+}-β-diketonate, and thus the target DNA can be detected with the detection limit of 6 pM (0.6 fmol per assay) by time-resolved fluorescence measurement. Compared with the formation of the ternary complex in the absence of the target DNA, the formation of the complex between EDTA-Eu^{3+} and β-diketonate in the presence of the target DNA is strongly enhanced because the complex is fixed firmly in the contiguous region of the DNA template after the hybridization.

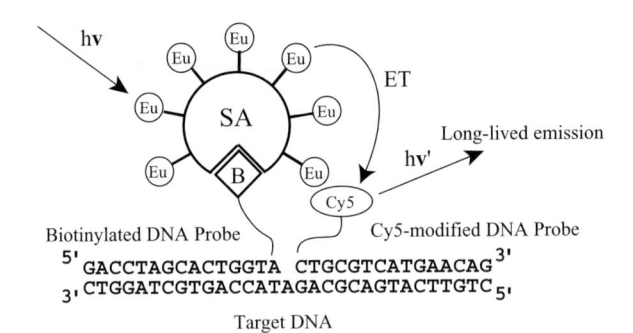

FIGURE 13.6. Principle of the homogeneous DNA hybridization assay based on fluorescence resonance energy transfer (FRET).

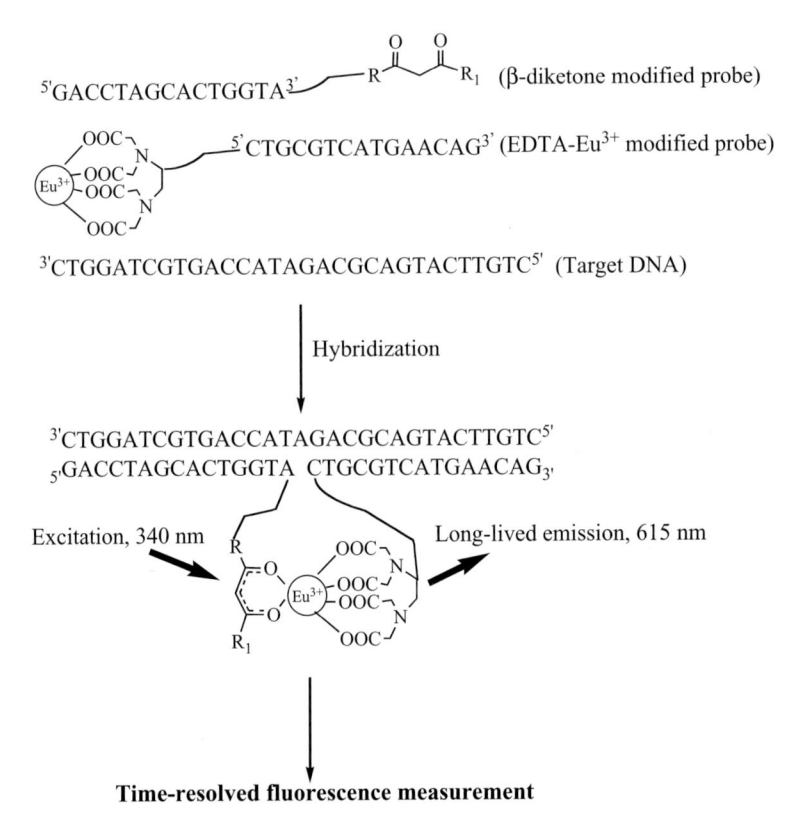

FIGURE 13.7. Principle of the homogeneous DNA hybridization assay based on DNA-mediated formation of ternary complex of EDTA-Eu^{3+}-β-diketonate.

CONCLUSION

Time-resolved fluorometry by using lanthanide fluorescence labels is growing rapidly for applications in diagnostics and biotechnology. The combination of a lanthanide label and time-resolved fluorometric detection has been proved to efficiently remove undesired background fluorescence and detect even the very weak fluorescence which could not be detected with the conventional normal fluorometry using organic dyes. Such high detectability will innovate the performance of high through-put bio-chip technology and fluorescence microscopy. In DNA chip technique, amplification of DNA by PCR is unavoidable but should be kept as little as possible in order to avoid the risk of error in DNA duplication. In fluorescence microscopy, time-resolved measurement is expected to reduce the autofluorescence of the biomaterials and would give much better contrast. In this way, the lanthanide labels are expected to innovate the diagnostic and biotechnology world. Of course, the labels still need to be improved; they must have higher quantum yields, higher molar extinction coefficients, and longer fluorescence lifetimes, in order to have higher sensitivity and simpler analysis format. In addition, development of Eu^{3+}, Sm^{3+}, Tb^{3+}, and Dy^{3+} four-color fluorescence labels is desired, since they could be applied in multiple-color time-resolved fluorescence imaging, four-color DNA sequencing, and DNA and protein microarrays. Currently Sm^{3+} and Dy^{3+} fluorescent complexes are only weakly fluorescent, and improvement of the fluorescence efficiency is highly desired. Synthesis of new fluorescent lanthanide complexes is therefore still an area that needs both more intensive and extensive effort of research.

REFERENCES

[1] E. Soini and I. Hemmilä. *Clin. Chem.*, 25:353–361, 1979.
[2] I. Hemmilä. *Clin. Chem.*, 31:359–370, 1985.
[3] E. Soini and T. Lövgren. *CRC Crit. Rev. Anal. Chem.*, 18:105–154, 1987.
[4] E.P. Diamandis. *Clin. Biochem.*, 21:139–150, 1988.
[5] E.P. Diamandis and T.K. Christopoulos. *Anal. Chem.*, 62:1149A–1157A, 1990.
[6] I. Hemmilä. *Appl. Fluoresc. Technol.*, 1:1–8, 1988.
[7] I. Hemmilä. *Scand. J. Clin. Lab. Invest.*, 48:389–400, 1988.
[8] E.F.G. Dickson, A. Pollak, and E.P. Diamandis. *Pharmac. Ther.*, 66:207–235, 1995.
[9] I. Hemmilä and S. Webb. *DDT*, 2:373–381, 1997.
[10] J. Yuan and K. Matsumoto. *Bunseki Kagaku*, 48:1077–1083, 1999.
[11] K. Matsumoto and J. Yuan. *Metal Ions in Biological Systems*. A. Sigel and H. Sigel (eds.), Marcel Dekker, New York, Basel, Chapter 6, Vol. 40, pp.191–232, 2003.
[12] S.I. Weissman. *J. Chem. Phys.*, 10:214–216, 1942.
[13] H.G. Huang, K. Hiraki, and Y. Nishikawa. *Nippon Kagaku Kaishi*, 66–70, 1981.
[14] R. Reisfeld and C.K. Jørgensen. *Lasers and Excited States of Rare Earths*, Springer, Berlin, 1977.
[15] Y.-Y. Xu and I. Hemmilä. *Anal. Chim. Acta*, 256:9–16, 1992.
[16] W.D. Horrocks and D.R. Sudnick. *J. Am. Chem. Soc.*, 101:334–340, 1979.
[17] A.G. Goryushko and N.K. Davidenko. *Zh. Neorg. Khim.*, 25:2666–2668, 1980.
[18] J. Yuan and K. Matsumoto. *Anal. Sci.*, 12:695–699, 1996.
[19] J. Yuan and K. Matsumoto. *J. Pharm. Biomed. Anal.*, 15:1397–1403, 1997.
[20] J. Yuan, K. Matsumoto, and H. Kimura. *Anal. Chem.*, 70:596–601, 1998.
[21] R.A. Evangelista, A. Pollak, B. Allore, E.F. Templeton, R.C. Morton, and E.P. Diamandis. *Clin. Biochem.*, 21:173–177, 1988.
[22] G. Mathis. *Clin Chem.*, 39:1953–1959, 1993.

[23] A.K. Saha, K. Kross, E.D. Kloszewski, D.A. Upson, J.L. Toner, R.A. Snow, C.D.V. Black, and V.C. Desai. *J. Am. Chem. Soc.*, 115:11032–11033, 1993.
[24] D. Horiguchi, K. Sasamoto, H. Terasawa, H. Mochizuki, and Y. Ohkura. *Chem. Pharm. Bull.*, 42:972–975, 1994.
[25] S. Inoue and R. Honda. *J. Clin. Microbiol.*, 28:1469–1472, 1990.
[26] T. Sekiya, M. Fushimi, H. Hori, S. Hirohashi, S. Nishimura, and T. Sugimura. *Proc. Natl. Acad. Sci. U.S.A.*, 81:4771–4775, 1984.
[27] I.C. Hsu, R.A. Metcalf, T. Sun, J.A. Welsh, N.J. Wang, and C.C. Harris. *Nature*, 350:427–429, 1991.
[28] M. Samiotaki, M. Kwiatkowski, N. Ylitalo, and U. Landegren. *Anal. Biochem.*, 253:156–161, 1997.
[29] P. Heinonen, A. Iitiä, T. Torresani, and T. Lövgren. *Clin. Chem.*, 43:1142–1150, 1997.
[30] S. Bortolin, T.K. Christopoulos, and M. Verhaegen. *Anal. Chem.*, 68:834–840, 1996.
[31] P.C. Ioannou and T.K. Christopoulos. *Anal. Chem.*, 70:698–702, 1998.
[32] K. Yoshikawa, J. Yuan, K. Matsumoto, and H. Kimura. *Anal. Sci.*, 15:121–124, 1999.
[33] S. Sueda, J. Yuan, and K. Matsumoto. *Bioconjugate Chem.*, 11:827–831, 2000.
[34] G. Wang, J. Yuan, and K. Matsumoto. *Anal. Biochem.*, 299:169–172, 2001.

14

Role of SNPs and Haplotypes in Human Disease and Drug Development

Barkur S. Shastry
Department of Biological Sciences, Oakland University, Rochester, MI 48309

Keywords: genome, sequence, disease, gene, pharmacogenetics, association

14.1. INTRODUCTION

In recent years it has become clear that genetic factors play a major role in every human disease except trauma. In many Mendelian or monogenic disorders, disease genes are identified by linkage and positional cloning methods using pedigrees and simple tandem repeat markers (di-, tri- or tetra-nucleotides). However, linkage based methods have limited power and are not readily applicable for complex disorders such as cardiovascular, osteoporosis, neuropsychiatric disorders, diabetes, asthma and cancer. This is because multiple loci are involved in these disorders and each locus contributes a small effect to disease etiology. The above complex disorders unfortunately, also occur in higher frequency and are a major social burden.

Now the whole genome sequence is a reality. The challenge faced by many investigators from around the world is how to use this massive information for practical purposes such as improving quality of life. Since many diseases are transmitted from parents to offspring, one immediate goal is to find out which gene (s) predisposes people to various diseases and how the sequence variations in a gene affect functions of its product. The second relevant question is how one could use this genetic information to predict a particular drug response or susceptibility to toxic side effects. This is because it is known that only a sub-population of patients experience side effects for certain medications while others do not. This inter-patient variability could be likely due in part to polymorphism in genes encoding drug metabolizing enzymes, drug transporters and/or receptors [3, 32]. Identification of gene defects and the provision of genetic determinants of drug efficacy and toxicity will ultimately enable physicians to optimize the use of medication effectively.

14.2. SNP DISCOVERY

It is known that between any two individuals about 99.9% of the DNA sequence are similar to one another. The remaining 0.1% DNA differs and accounts for diversity among population, disease susceptibility and drug response [71, 126]. The DNA variability is due to mutation, genetic drift, migration and selection. A recent survey suggests that some variants are common to all populations while others are distributed in a more restricted manner [103]. The most common variant is called polymorphism. These are nothing but variants at specific nucleotide positions in the genome among different people (Fig. 14.1) and are designated as single nucleotide polymorphisms (SNPs). They occur once in every thousand base pairs or less throughout the genome [16, 25]. It is estimated that the human genome may contain as many as 10 million such SNPs [104]. These variations can occur either in coding regions (cSNPs, also called non-synonymous SNPs) or outside the coding regions. According to one report, about 50% of SNPs are in the non-coding region, 25% lead to missense mutations (cSNPs) and the remaining 25% are silent mutations [19, 40]. Neutral substitutions in exons are found to occur at a much higher rate (30–60%) compared to that in non-coding regions. This could be due to the presence of relatively rich GC sequences in exons [114].

Although the frequency of SNPs across the genome is greater than any other type of polymorphism, they vary among different genomic regions, genes and different human populations [19]. As mentioned above, it is also not necessary that all SNPs will change the phenotype. However, they could influence gene expression and regulation. For instance, those SNPs that occur in protein coding sequences (cSNPs) cause an amino acid change and possibly contribute to disease susceptibility and drug metabolism. There are approximately 200,000 cSNPs in the human genome [22]. Because of their important role in disease susceptibility and drug metabolism, one of the goals of the human genome project is cataloging these variations [23] and generating an ordered high-density SNP map of the human genome. This map is made available for the public through the web-site http://snp.cshl.org. The database of SNP is also maintained by the National Center for Biotechnology Information (NCBI) and can be found in the web-site http://www.ncbi.nlm.nih.gov/

SNP

———— GATCAGC|G|ACT ———— individual A
———— GATCAGC|A|ACT ———— individual B

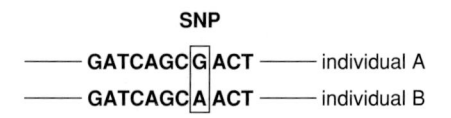

FIGURE 14.1. A Schematic illustration of single nucleotide polymorphism (SNP). Figure shows strings of nucleotides at which individuals A and B differ by just one base.

14.3. DETECTION OF GENETIC VARIATION

A wealth of technologies is available for detecting genomic variations. This short article is not meant to be a review of all methods or their uses that are available to date for SNP analysis. Readers are directed to consult several other reviews on this topic that are already published [39, 50, 52, 53]. Briefly, for small scale studies, assays used for SNP detection include: DNA sequencing, single strand conformational polymorphism [87], restriction fragment length polymorphism (RFLP), denaturing gradient gel electrophoresis (DGGE—uses a chemical denaturing gradient [34, 88]), temperature gradient gel electrophoresis (TGGE—uses a temperature gradient), constant denaturant gel electrophoresis (CDGE—employs a constant amount of denaturants), oligonucleotide ligation assay [123], allelic specific oligonucleotide hybridization [102] and nuclease mutation detection method [38, 131]. These techniques are simple and non-radioactive methods. They can detect as little as a one base difference with a scanning region of 100–1000 base pair. However, the detection limit is approximately 80% and it is likely that mutations at the ends of the fragment as well as in GC rich regions could be missed. Further improvements in these techniques such as addition of a GC rich sequence to the end of the primers (GC clamp), two dimensional DGGE and dideoxy fingerprinting coupled with single strand conformational polymorphism (SSCP), reported to increase the efficiency of detection to nearly 100% [61, 109].

In order to apply any SNP identification technique to a large-scale study, it is necessary that it should be accurate, sensitive, flexible and most importantly cost effective. For this purpose, a variety of reliable high-throughput methods have been developed. These include: structure specific nuclease invader technique [35, 73, 80], primer extension [9, 116], molecular becons—probes that fluoresce on hybridization [75, 119], TaqMan assay [28, 62], denaturing HPLC, genomic mismatch scanning, gene chip [58], nano technology, melting curve single nucleotide polymorphism [14, 70], fluorescent competitive allele specific polymerase chain reaction, constant denaturant capillary electrophoresis [11], high-density oligonucleotide arrays [29] and ligation rolling-circle amplification [79, 94, 130]. In addition, several academic and private institutions are developing a robotic version of the SNP-IT (offered by Orchid Biosciences) and MassARRAY system (offered by Sequenom). However, as expected for any technique, many of these methods have their own limitations and advantages. For instance, the gene chip method may offer the advantage of minimum sample handling and higher marker throughput but it requires a custom array for each marker set. Similarly, denaturing HPLC technology has flexibility but requires specialized equipment. Moreover, although, Beadedarray [86] and MassARRAY [99], are available for rapid genotyping, further improvements are needed to develop an inexpensive method to screen large numbers of SNPs from hundreds and thousands of patients and controls.

After the identification and mapping of a particular SNP, it is necessary to calculate the allele frequency. This can be accomplished by calculating the percentage of individuals containing the polymorphism within a population. If the frequency is greater than 5–10% then they are considered useful SNPs. A recently reported new assay which makes use of the 3'-to 5'-exonuclease proofreading activity is a useful method for typing, and allele frequency data can be obtained directly from genomic DNA samples [17]. The allele frequency database is available from the Kidd lab home page http://info.med.yale.edu/genetics/kkidd. From these data it appears that SNP allele frequencies vary considerably across ethnic groups and populations. This suggests that different SNP panels will be required for different studies depending on the origin of the population. Methods (Bayesian and likelihood) have also been developed for estimating human population growth, migration, population divergence and recombination rates for SNPs [82, 83]. The widely variable SNP frequency between genes also indicates the difference in the recombination rate across the genome [18, 81, 97].

14.4. DISEASE GENE MAPPING

By identifying and developing a high-density map of human DNA sequence polymorphism it may be possible to identify disease related genes more effectively by employing the association strategy [54, 55, 68, 98]. This is because such studies do not need a large family. In this approach, a comparison between affected and unaffected individuals will be made for a set of genetic markers. If the marker shows more prevalence in affected individuals than in unaffected, then it is considered as evidence of an association between the marker and the phenotype (reviewed in ref. [107, 108]). For this purpose, an analytical approach has been developed which also incorporates the contribution of environmental factors [132]. This method however, still needs a large sample size [56] and there are other pitfalls such as the generation of millions of genotypes. Despite this limitation, there has been some success recently in identifying the association between the APOE 4 allele and late-onset Alzheimer disease [66, 100], factor V Leiden mutation and venous thrombosis [122] and promoter polymorphism in the insulin gene and type I diabetes [10]. A partial list of diseases associated with SNPs is presented in Table 14.1. Although there are several high-throughput methods that are available for genotyping thousands of samples [45], this type of whole genome approach to mapping is still expensive. A suggested alternative approach in some cases is to use pooled DNA to quantitate the result [13] that will reduce the number of samples to be analyzed. However, such approaches are not recommended for all studies. In addition, this method gives only a rough estimate of allele frequencies and it is possible that some rare alleles (less than 5%) could be easily missed.

14.5. EVOLUTION

Genetic variants are not only considered to be responsible for inter-individual differences and in disease risk but also for molecular evolution. It is estimated that more than 99% of SNPs occur elsewhere in the genome and they do not change amino acid or regulatory sites. Hence they may not have any phenotypic effect. According to the neutral theory of evolution [48], these SNPs are not subjected to natural selection [16, 105]. Therefore, they are considered to be more stable or less mutable. However, those SNPs that do change amino acids may have phenotypic effects and might be subject to natural selection. If these SNPs are retained over time then they must have some selection advantages for individuals. The retention of variants by natural selection is an important step in evolution [60]. Hence raw evolutionary data can be generated by comparing the ratio of non-synonymous to synonymous change in several proteins in different species. This may allow us to trace the branching point of an evolutionary tree. This is an important step in evolution because

TABLE 14.1. A partial list of diseases associated with SNPs

Disease	Gene	Ref	Disease	Gene	Ref
Idiopathic PD and FTD	Tau	[65, 111]	Urinary bladder cancer	Cyclin-D1	[124]
Knee and hip Osteoarthritis	Coll.	[44]	Dyslipidemia	Lipase	[128]
Graft patency of femoropopliteal bypass	PAFAH	[121]	Oxalate stone disease	E-Cad.	[120]
Atopy and asthma	Chemokine	[36]	Lung cancer	MMP-1 p53	Zho et al., 2001; [15]
Myocardial infarction	Thrombomodulin	[89]	Systemic sclerosis	Fibrillin-1	[118]
	Prostacyclin synthase	[76]	POAG	TIGR	[24]
Severe Sepsis	TNF—α	[85]	Migraine	Insulin Receptor	[69]
Obesity	PAI—1	[41]	Arrhythmia	KCNQ1	[59]
Bipolar affective disorder	HTR 3A	[84]	Eating disorder	Melanocortin	[51]
Rheumatoid arthritis	MIF	[8]	SLE Blood Pressure	Prolactin CPB2	[113, 49]
Hyperbilirubinemia	UGT1A1	[115]	Juvenile idiopathic arthritis	MIF	[30]
Hyperandrogenism and Ovarian dysfunction	SHBG	[42]	Primary biliary cirrhosis	MBL	[67]

MIF = macrophage migration inhibitory factor; UGT1A1 = UDP glucuronosyl transferase; PAF-AH = plasma platelet-activating factor—acetylhydrolase; TNF = tumor necrosis factor; PAI = plasminogen activator inhibitor; HTR 3A = serotonin receptor gene; PD = Parkinson disease; FTD = frontotemporal dementia; SLE = systemic lupus erythematosus; CPB2 = encodes the thrombin-activable fibrinolysis inhibitor (TAFI); MBL = mannose binding lectin; SHBG = sex hormone binding globulin; MMP-1 = matrix metalloproteinase –1; POAG = primary open-angle glaucoma; TIGR = trabecular meshwork inducible glucocorticoid respose; KCNQ1 = potassium channel; Cad. = cadherin at that branch point variance has advantages for the species and fixed in the gene pool. This can be of immense value in understanding the evolution of the human genome in the future.

14.6. HAPLOTYPES

Identification of millions of SNPs across the genome is an expensive and technologically demanding task. For this reason, a much simpler method called haplotyping and understanding the haplotype structure (HapMap) has been developed [37]. A haplotype is a group of linked SNPs in the same chromosome that are inherited together—a process known as linkage disequilibrium—which is a non-random association of alleles in a chromosomal segment. Thus, a particular SNP will be in linkage disequilibrium with many other SNPs [96, 117]. In a simple sense, a haplotype (blocks of information) is a genotype of a single chromosome and it is likely to be responsible for human genetic diversity. Since a haplotype contains stretches of SNPs in a particular chromosome, they are thought to yield more accurate and predictive information on the complex biology of the genotype—phenotype relationship than a single SNP. This is because polymorphic bases can be distributed differently on the maternal and paternal chromosomes and such differences may affect gene expression and drug response by two patients. In addition, haplotype analysis may significantly reduce the amount of data to be analyzed [47] for gene identification or drug development because only a small number of SNPs are required to map the disease gene.

The haplotype structure often determines the phenotypic consequences and they can be used as genetic markers [6, 26, 27, 46]. In order to determine directly the haplotype structure of genomic DNA, a simple robust method involving a long–range polymerase chain reaction and intra-molecular ligation has been developed [72]. Recently, using gene chip sequencing arrays developed by Perlegen Sciences, it has become possible to define the haplotype structure of chromosome 21 [90]. However, linkage disequilibrium may vary between various populations and across the genome and this requires additional methods [96]. Furthermore, two companies: Parkin Elmer and Sequenom Inc. (San Diego) have launched web sites (www. snpscoring.com and www. realsnp.com) which provide information on population frequencies of SNPs in various populations, how the technology works, how to run the assay and what are the benefits. When the HapMap is completed, a simple comparison of HapMap between normal and patient DNA will provide the region of the genome that is most likely responsible for the disease. Similarly, comparison of HapMap of individuals responding differently to medication will pinpoint the genomic region valuable in personalized treatment. Thus, haplotyping can provide a better understanding of the genetic correlation to clinical response than a single SNP genotype. This is because the specific combination of multiple linked SNPs can have an additive or subtractive effect on a particular trait and this is very useful for designing a drug [5, 108]. Additionally, haplotype based SNP analysis can be applied to crop genetics [95].

14.7. DRUG DEVELOPMENT

Human genetic variation also plays a key role in determining drug toxicity and efficacy. For instance, some weight-loss drugs were found to cause cardiac valve damage and pulmonary hypertension. Similarly, non-sedating anti-histamine drugs and antibiotics produced cardiac arrhythmias [91, 106]. In fact, the incidence of serious and fatal adverse drug reactions is very high. According to one study, properly prescribed medications caused 100,000 deaths in the United States alone [57]. It has also been reported that the HLA class II haplotype determines the response to cytokine therapy in a subset of patients with renal cell carcinoma [31]. In this regard, rapidly developing pharmacogenomic research and a toxicogenomic assay may shed some light on the response of an entire genome as well as the influence of genetic variation to experimental drugs or class of drugs. Identification of the genotype of patients and correlating that with drug toxicity (Fig. 14.2) may enable clinicians to prescribe most effective drugs with diminished side effects. In short, understanding the relationship between genotypic variation and drug metabolism at the molecular level may lead to customized medications in the future. It will fit each patient's needs so that clinicians need not have to rely on a guessing game with drugs [12, 20, 59, 63, 92, 112].

The reason for this is that, it is well known that concentration of certain drugs in a patient is determined by drug metabolizing enzymes such as cytochrome P450, N-acetlytransferase and other key enzymes [33, 74]. For instance, a variation in cytochrome CYP 2D6 activity has been linked to liver cancer [2] and genetic polymorphism has been reported in this enzyme [78, 107, 108]. As mentioned earlier, an SNP map of the human chromosome suggests that some SNPs are in the exons [4, 77]. Therefore, identification of non-synonymous SNPs (those that cause amino acid change) and other SNPs in the promoter region are very useful in understanding drug response because they have a functional effect [92, 112]. Additionally, it is known that some patients respond poorly to a given medication (Fig. 14.2) while the other patients either experience incompatibility or no response at all [21, 101]. By using a high-density SNP map and comparing the pattern of those individuals who respond poorly, adversely and positively, it is possible to identify responsible polymorphism in a gene and to develop safer and more effective drugs. In summary, SNPs and haplotypes are valuable markers to identify certain individuals susceptible to certain diseases such as Alzheimer, depression and diabetes. They are also helpful to understand individual variability in drug response.

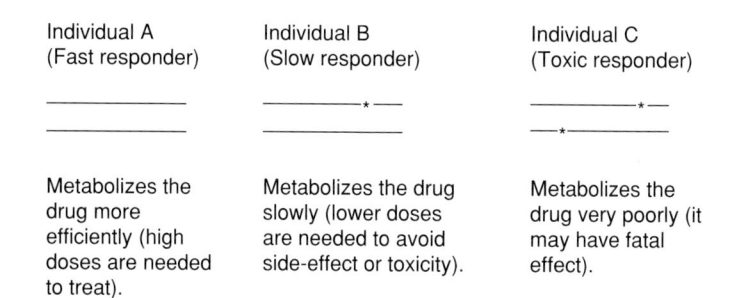

Individual A (Fast responder)	Individual B (Slow responder)	Individual C (Toxic responder)
───────────	─────*───	─────*───
───────────	───────────	──*────────
Metabolizes the drug more efficiently (high doses are needed to treat).	Metabolizes the drug slowly (lower doses are needed to avoid side-effect or toxicity).	Metabolizes the drug very poorly (it may have fatal effect).

FIGURE 14.2. Highly hypothetical representation of the relationship between the genotype and drug response. Two horizontal lines denote a pair of homologous genes, *mark indicates polymorphism in the gene which is responsible for different types of drug response.

CONCLUDING REMARKS

The human genome project is a big step forward for the identification of genes responsible for complex diseases. One major challenge for the future is to understand how genetic factors contribute to more common and complex conditions such as cancer, psychiatric disorders, diabetes and asthma. By identifying and understanding the genome—wide polymorphisms, a more complete medicinal response (adverse effect and efficacy) profile can be developed for each drug. This will ultimately enable clinicians to provide an effective therapy using the individual patient's genetic background. It is hoped that these genetic approaches will significantly reduce long-term hospitalization and care in the near future.

The use of SNP in genetic testing and development of individualized medicine is truly exciting. However, to meet this challenge, several underlying problems should be considered. For instance (a) only a small fraction (15%) of SNPs are characterized to date to carry out meaningful experiments, (b) a huge number of SNPs are needed to make a connection between disease and SNP, (c) they are not randomly distributed in the genome, (d) it is possible that some of them may not be allelic variants but cismorphism [7]. In addition, a large number of studies are required to identify haplotypes that are in linkage disequilibrium. Identification of a haplotype associated with type II diabetes mellitus [43] is a good example of this problem. It is clear that some SNPs do change the amino acid sequence or regulatory site but it is not clear how useful they are in diagnostic testing. This is because of the complexity of human disease and the involvement of environmental factors that determine the phenotype. It is also unknown how genotype alone can predict disease susceptibility. For instance, although identical twins have identical genotypes, they exhibit 50% concordance for more common diseases [64, 93, 125, 127]. Moreover, we should also consider the influence of developmental programs such as DNA methylation, X-inactivation and environmental conditions during postnatal development in causing disease susceptibility and phenotype.

Furthermore, individualized drug development also needs more challenging approaches. This is because, drug-metabolizing enzymes such as cytochrome P450 family have been well characterized but the drug receptors (or membrane transporter) have not been well studied. For instance, estrogen β and cyclooxygenase receptors and their therapeutic values are largely unknown. Additionally, these receptors are highly variable or genetically diverse in a given population which adds another limitation. Moreover, several drugs have multiple genetic effects that make the study design even more complex. Although it will be a long time before personalized medicine becomes a reality [129], pharmacogenomics plays a major role in reducing or preventing drug related deaths and hospitalization costs. Meanwhile, future research will uncover methods to make SNP markers as useful tags for medical testing.

REFERENCES

[1] R.A.H. Adan and T. Vink. *Eur. Neuropsychopharmacol.*, 11:483, 2001.
[2] J.A. Agundez, M.C. Ledesma, J. Benitez, J.M. Ladero, A. Rodriguez-Lescure, E. Diaz-Rubio, and M. Diaz-Rubio. *Lancet*, 345:830, 1995.
[3] R.B. Altmann, and T.E. Klein. *Ann. Rev. Toxicol. Pharmacol.*, 42:113, 2002.

[4] D. Altshuler, V.J. Pollara, C.R. Cowles, W.J. Van Etten, J. Baldwin, L. Linton, and E.S. Lander. *Nature*, 407:513, 2000.
[5] M.C. Athanasion, A.K. Malhotra, C.B. Xu, and J.C. Stephens. *Psychiatr. Genet.*, 12:89, 2002.
[6] J.S. Bader. *Pharmacogenomics*, 2:11, 2001.
[7] J.A. Bailey, Z.P. Gu, R.A. Clark, K. Reinert, R.V. Samonte, S. Schwartz, M.D. Adams, E.W. Myers, P.W. Li, and E.E. Eichler. *Science*, 297:1003, 2002.
[8] J.A. Baugh, S. Chitnis, S.C. Donnelly, J. Monteiro, X. Lin, B.J. Plant, F. Wolfe, P.K. Gregersen, and R. Bucala. *Genes Immunity*, 3:170, 2002.
[9] P.A. Bell, S. Chaturvedi, C.A. Gelfand, C.Y. Huang, M. Kochersperger, R. Kopla, F. Modica, M. Pohl, S. Varde, R. Zhao, X. Zhao, and M.T. Boyce-Jacino. *Biotechniques*, 32:S70, 2002.
[10] S.T. Bennet, A.M. Lucassen, S.C.L. Gough, E.E. Powell, D.E. Undlien, L.E. Pritchard, M.E. Merriman, Y. Kawaguchi, M.J. Dronsfield, F. Pociot, J. Nerup, N. Bouzekri, A. Cambon-Thomsen, K.S. Ronningen, A.H. Barnett, S.C. Bain, and J.A. Todd. *Nat. Genet.*, 9:284, 1995.
[11] J. Bjorheim, T.W. Abrahamsen, A.T. Kristensen, G. Gaudernack, and P.O. Ekstrom. *Mutat. Res. Fundamentals Mol. Mech. Mutagenesis*, 526:75, 2003.
[12] J.H. Bream, A. Ping, X. Zhang, C. Winkler, and H.A. Young. *Genes Immunity*, 3:165, 2002.
[13] G. Breen, D. Harold, S. Ralston, D. Shaw, and D. St Clair. *Biotechniques*, 28:464, 2000.
[14] G. Breen. *Psychiatr. Genet.*, 12:83, 2002.
[15] E. Boros, I. Kalina, I. Biros, A. Kohut, E. Bogyiova, J. Salagovic, and J. Stubna. *Neoplasma*, 48:407, 2001.
[16] A.J. Brookes. *Gene*, 234:177, 1999.
[17] P. Cahill, M. Bakis, J. Hurley, V. Kamath, W. Nielsen, D. Weymouth, J. Dupuis, L. Doucette-Stamm, and D. R. Smith. *Genome Res.*, 13:925, 2003.
[18] F. Cambien, O. Poirier, V. Nicaud, S.M. Herrmann, C. Mallet, S. Ricard, I. Behague, V. Hallet, H. Blanc, V. Loukaci, J. Thillet, A. Evans, J.B. Ruidavets, D. Arveiler, G. Luc, and L. Tiret. *Am. J. Hum. Genet.*, 65:183, 1999.
[19] M. Cargill, D. Altshuler, J. Ireland, P. Skalr, K. Ardile, N. Patil, C.R. Lane, E.P. Lim, N. Kalyanaraman, J. Nemesh, L. Ziaugra, L. Friedland, A. Rolfe, J. Warrington, R. Lipshutz, G.Q. Daley, and E.S. Lander. *Nat. Genet.*, 22:231, 1999.
[20] S. Chanock. *Dis. Markers*, 17:89, 2001.
[21] L.J. Cohen and C.L. DeVane. *Ann. Pharmacotherapy*, 3:1471, 1996.
[22] F.S. Collins, M.S. Guyer, and A. Chakravarti. *Science*, 278:1580, 1997.
[23] F.S. Collins, A. Patrinos, E. Jordan, A. Chakravarti, R. Gesteland, L. Walters, E. Fearon, L. Hartwelt, C.H. Langley, R.A. Mathies, M. Olson, A.J. Pawson, T. Pollard, A. Williamson, B. Wold, K. Buetow, E. Branscomb, M. Capecchi, G. Church, H. Garner, R.A. Gibbs, T. Hawkins, K. Hodgson, M. Knotek, M. Meisler, G.M. Rubin, L.M. Smith, R.F. Smith, M. Westerfield, E.W. Clayton, N.L. Fisher, C.E. Lerman, J.D. McInerney, W. Nebo, N. Press, and D. Valle. *Science*, 282:682, 1998.
[24] E. Colomb, T.D. Nguyen, A. Bechetoille, J.C. Dascotte, F. Valtot, A.P. Brezin, M. Berkani, B. Copin, L. Gomez, J.R. Polansky, and H.J. Garchon. *Clin. Genet.*, 60:220, 2001.
[25] D.N. Cooper, B.A. Smith, H.J. Cooke, S. Niemann, and J. Schmidtke. *Hum. Genet.*, 69:201, 1985.
[26] J. Couzin. *Science*, 296:1391, 2002.
[27] M.J. Daly, J.D. Rioux, S.F. Schaffner, T.J. Hudson, and E.S. Lander. *Nat. Genet.*, 29:229, 2001.
[28] F.M. De La Vega, D. Dailey, J. Ziegle, J. Williams, D. Madden, and D.A. Gilbert. *Biotechniques*, 32:S48, 2002.
[29] S.L. Dong, E. Wang, L. Hsie, Y.X. Cao, X.G. Chen, and T.R. Gingeras. *Genome Res.*, 11:1418, 2001.
[30] R.P. Donn, E. Shelley, W.E.R. Ollier, and W. Thomson. *Arthritis and Rheumatism.*, 44:1782, 2001.
[31] J.A. Ellerhorst, W.H. Hildebrand, J.W. Cavett, M.A. Fernandez-Vina, S. Hodges, N. Poindexter, H. Fisher, and E.A. Grimm. *J. Urology*, 169:2084, 2003.
[32] W.E. Evans and J.A. Johnson. *Ann. Rev. Genomics Hum. Genet.*, 2:9, 2001.
[33] W.E. Evans and M.V. Relling. *Science*, 286:487, 1999.
[34] R. Fodde and M. Losekoot. *Hum. Mutat.*, 3:83, 1994.
[35] L. Fors, K.W. Lieder, S.H. Vavra, and R.W. Kwiatkowski. *Parmacogenomics*, 1:219, 2000.
[36] A.A. Fryer, M.A. Spiteri, A. Bianco, M. Hepple, P.W. Jones, R.C. Strange, R. Makki, G. Tavernier, F.I. Smilie, A. Custovic, A.A. Woodcock, W.E.R. Ollier, and A.H. Hajeer. *Genes and Immunity*, 1:509, 2000.
[37] S.B. Gabriel, S.F. Schaffner, H. Nguyen, J.M. Morre, J. Roy, B. Blumenstiel, J. Higgins, M. DeFelice, A. Lochner, M. Faggart, S.N. Liu-Cordero, C. Rotimi, A. Adeyemo, R. Cooper, R. Ward, E.S. Lander, M.J. Daly, and D. Altshuler. *Science*, 296:2225, 2002.

[38] M.M. Goldrick. *Hum. Mutat.*, 18:190, 2001.
[39] I. Gut. *Hum. Mutat.*, 17:475, 2001.
[40] M.K. Halushka, J.B. Fan, K. Bentley, L. Hsie, N.P. Shen, A. Weder, R. Cooper, R. Lipshutz, and A. Chakravarti. *Nat. Genet.* 22:239, 1999.
[41] J. Hoffstedt, L.L. Andersson, L. Persson, B. Isaksson, and P. Arner. *Diabetologia*, 45:584, 2002.
[42] K.N. Hogeveen, P. Cousin, M. Pugeat, D. Dewailly, B. Soudan, and G.L. Hammond. *J. Clin. Invest.*, 109:973, 2002.
[43] Y. Horikawa, N. Oda, N.J. Cox, L. Xiangquan, M. Orho-Melander, M. Hara, Y. Hinokio, and T.H. Lindner. *Nat. Genet.*, 26:163, 2000.
[44] T. Ikeda, A. Mabuchi, A. Fukuda, A. Kawakami, R. Yamada, S. Yamamoto, K. Miyoshi, N. Haga, H. Hiraoka, Y. Takatori, H. Kawaguchi, K. Nakamura, and S. Ikegawa. *J. Bone Mineral Res.*, 17:1290, 2002.
[45] S. Jenkins and N. Gibson. *Comp. Funct. Genomics*, 3:57, 2002.
[46] G.C.L. Johnson, L. Esposito, B.J. Barratt, A.N. Smith, J. Heward, G. DiGenova, H. Ueda, H.J. Cordell, I.A. Eaves, F. Dudbridge, R.C.J. Twells, F. Payne, W. Hughes, S. Nutland, H. Stevens, P. Carr, E. Tuomilehto-Wolf, J. Tuomilehto, S.C.L. Gough, D.G. Clayton, and J.A. Todd. *Nat. Genet.*, 29:233, 2001.
[47] B. Jordan, A. Charest, J.F. Dowd, J.P. Blumenstiel, R-F. Yeh, A. Osman, D.E. Housman, and E.S. Lander. *Proc. Natl. Acad. Sci. U.S.A.*, 99:2942, 2002.
[48] M. Kimura. *The Neutral Theory of Molecular Evolution*. Cambridge, Cambridge University Press, 1983.
[49] M.L. Koschinsky, M.B. Boffa, M.E. Nesheim, B. Zinman, A.J.G. Hanley, S.B. Harris, H. Cao, and R.A. Hegele. *Clin. Genet.*, 60:345, 2001.
[50] V.N. Kristensen, D. Kelefiotis, T. Kristensen, and A.L. Borresen-Dale. *Biotechniques*, 30:318, 2001.
[51] T. Kubota, M. Horie, M. Takano, H. Yoshida, K. Takenaka, E. Watanabe, T. Tsuchiya, H. Otani, and S. Sasayama. *J. Cardiovasc. Electrophysiol.*, 12:1223, 2001.
[52] P.-Y. Kwok. *Pharmacogenomics*, 1:95, 2000.
[53] P.-Y. Kwok. *Ann. Rev. Genomics Hum. Genet.*, 2:235, 2001.
[54] E. Lai. *Genome Res.*, 11:927, 2001.
[55] E. Lai, J. Riley, I. Purvis, and A. Roses. *Genomics*, 54:31, 1998.
[56] E.S. Lander and N.J. Schork. *Science*, 265:2037, 1994.
[57] J. Lazarou, B.H. Pomeranz, and P.N. Corey. *JAMA*, 279:1200, 1998.
[58] B. Lemieux. *Curr. Genomics*, 1:301, 2000.
[59] T.D. Levan, J.W. Bloom, T.J. Bailey, C.L. Karp, M. Halonen, F.D. Martinez, and D. Vercelli. *J. Immunol.*, 167:5838, 2001.
[60] D.A. Liberles. *Genome Biol.*, 2:1, 2001.
[61] Q. Liu and S.S. Sommer. *PCR Methods Appl.*, 4:97, 1994.
[62] K.J. Livak, J. Marmaro, and J.A. Todd. *Nat. Genet.*, 9:341, 1995.
[63] H.D. Lohrer and U. Tangen. *Pathobiol.*, 68:283, 2000.
[64] N. Martin, D. Boomsma, and G. Machin. *Nat. Genet.*, 17:387, 1999.
[65] E.R. Martin, W.K. Scott, M.A. Nance, R.L. Watts, J.P. Hubble, W.C. Koller, K. Lyons, R. Pahwa, M.B. Stern, A. Colcher, B.C. Hiner, J. Jankovic, W.G. Ondo, F.H. Allen, C.G. Goetz, G.W. Small, D. Masterman, F. Mastaglia, N.G. Laing, J.M. Stajich, R.C. Ribble, M.W. Booze, A. Rogala, M.A. Hauser, F.Y. Zhang, R.A. Gibson, L.T. Middleton, A.D. Roses, J.L. Haines, B.L. Scott, M.A. Pericak-Vance, and J.M. Vance. *JAMA*, 286:2245, 2001.
[66] E.R. Martin, J.R. Gilbert, E.H. Lai, J. Riley, A.R. Rogala, B.D. Slotterbeck, C.A. Sipe, J.M. Grubber, L.L. Warren, P.M. Conneally, A.M. Saunders, D.E. Schmechel, I. Purvis, M.A. Pericak-Vance, A.D. Roses, and J.M. Vance. *Genomics*, 63:7, 2000.
[67] M. Matsushita, H. Miyakawa, A. Tanaka, M. Hijikata, K. Kikuchi, H. Fujikawa, J. Arai, S. Sainokami, K. Hino, I. Terai, S. Mishiro, and M.E. Gershwin. *J. Autoimmunity*, 17:251, 2001.
[68] J.J. McCarthy and R. Hilfiker. *Nat. Biotechnol.*, 18:505, 2000.
[69] L.C. McCarthy, D.A. Hosford, J.H. Riley, M.I. Bird, N.J. White, D.R. Hewett, S.J. Perontka, L.R. Griffiths, P.R. Boyd, R.A. Lea, S.M. Bhatti, L.K. Hosking, C.M. Hood, K.W. Jones, A.R. Handley, R. Rallan, K.F. Lewis, A.J.M. Yeo, P.M. Williams, R.C. Priest, P. Khan, C. Donnelly, S.M. Lumsden, J. O'Su'livan, C.G. See, D.H. Smart, S. Shaw-Hawkins, J. Patel, T.C. Langrish, W. Feniuk, R.G. Knowles, M. Thomas, V. Libri, D.S. Montgomery, P.K. Monaco, C.F. Xu, C. Dykes, P.P.A. Humphrey, A.D. Roses, and J.J. Purvis. *Genomics*, 78:135, 2001.
[70] J.L. McClay, K. Sugden, H.G. Koch, S. Higuchi, and I.W. Craig. *Anal. Biochem.*, 301:200, 2002.

[71] D.C. McClean, I. Spruill, S. Gevao, E.Y.S. Morrison, O.S. Bernard, G. Argyropoulos, and W.T. Garvey. *Hum. Biol.*, 75:147, 2003.
[72] O.G. McDonald, E.Y. Krynetski, and W.E. Evans. *Pharmacogenet.*, 12:93, 2002.
[73] C.A. Mein, B.J. Barratt, M.G. Dunn, T. Siegmund, A.N. Smith, L. Esposito, S. Nutland, H.E. Stevens, A.J. Wilson, M.S. Phillips, N. Jarvis, S. Law, M. de Arruda, and J.A. Todd. *Genome Res.*, 10:330, 2000.
[74] U.A. Meyer and U.M. Zanger. *Ann. Rev. Pharmacol. Toxicol.*, 37:269, 1997.
[75] M.M. Mhlanga and L. Malmberg. *Methods*, 25:463, 2001.
[76] T. Nakayama, M. Soma, S. Saito, J. Honye, J. Yajima, D. Rahmutula, Y. Kaneko, M. Sato, J. Uwabo, N. Aoin, K. Kosuge, M. Kunimoto, K. Kanmatsuse, and S. Kokubun. *Am. Heart J.*, 143:797, 2002.
[77] J.C. Mullikin, S.E. Hunt, C.G. Cole, B.J. Moltimore, C.M. Rice, J. Burton, L.H. Matthews, R. Pavitt, R.W. Plumb, S.K. Sims, R.M.R. Ainscough, J. Attwood, J.M. Bailey, K. Barlow, R.M.M. Bruskiewich, P.N. Butcher, N.P. Carter, Y. Chen, C.M. Clee, P.C. Coggill, J. Davies, R.M. Davies, E. Dawson, M.D. Francis, A.A. Joy, R.G. Lambel, C.F. Langford, J. MaCarthy, V. Mall, A. Moreland, E.K. Overton-Larty, M.T. Ross, L.C. Smith, C.A. Steward, J.E. Sulston, E.J. Tinsley, K.J. Turney, D.L. Willey, G.D. Wilson, A.A. McMurray, I. Dunham, J. Rogers, and D.R. Bentley. *Nature*, 407:516, 2000.
[78] M. Nakajima, Y. Fujiki, K. Noda, H. Ohtsuka, H. Ohkuni, S. Kyo, M. Inoue, Y. Kuroiwa, and T. Yokoi. *Drug Metabol. Dispositt.*, 31:687, 2003.
[79] J.R. Nelson, Y.C. Cai, T.L. Giesler, J.W. Farchaus, S.T. Sundaram, M. Ortiz-Rivera, L.P. Hosta, P.L. Hewitt, J.A. Mamone, C. Palaniappan, and C.W. Fuller. *Biotechniques*, 32:S44, 2002.
[80] M. Neville, R. Selzer, B. Aizenstein, M. Maguire, K. Hogan, R. Walton, K. Welsh, B. Neri, and M. de Arruda. *Biotechniques*, 32:S34, 2002.
[81] D.A. Nickerson, S.L. Taylor, K.M. Weiss, A.G. Clark, R.G. Hutchinson, J. Stengard, V. Salomaa, E. Vartiainen, E. Baerwinkel, and C.F. Sing. *Nat. Genet.*, 19:233, 1998.
[82] R. Nielsen. *Genet.*, 154:931, 2000.
[83] R. Nielsen and M. Slatkin. *Evolution*, 54:44, 2000.
[84] B. Niesler, T. Flohr, M.M. Nothen, C. Fischer, M. Rietschel, E. Franzek, M. Albus, P. Propping, and G.A. Rappold. *Pharmacogenet.*, 11:471, 2001.
[85] G.E. O'Keefe, D.L. Hybki, and R.S. Munford. *J. Trauma-injury Infect. Crit. Care*, 52:817, 2002.
[86] A. Oliphant, D.L. Barker, J.R. Stuelpnagel, and M.S. Chee. *Biotechniques*, 32:S56, 2002.
[87] M. Orita, H. Iwahana, H. Kanazawa, K. Hayashi, and T. Sekiya. *Proc. Natl. Acad. Sci. U.S.A.*, 86:2766, 1989a.
[88] M. Orita, Y. Suzuki, T. Sekiya, and K. Hayashi. *Genomics*, 5:874, 1989b.
[89] H.Y. Park, T. Nabika, Y. Jang, H.M. Kwon, S.Y. Cho, and J. Masuda. *Hyperten. Res.*, 25:389, 2002.
[90] N. Patil, A.J. Berno, D.A. Hinds, W.A. Barrett, J.M. Doshi, C.R. Hacker, C.R. Kautzer, D.H. Lee, C. Marjoribanks, D.P. McDonough, B.T.N. Nguyen, M.C. Norris, J.B. Sheehan, N. P. Shen, D. Stern, R.P. Stokowski, D.J. Thomas, M.O. Trulson, K.R. Vyas, K.A. Frazed, S.P.A. Fodor, and D.R. Cox. *Science*, 294:1719, 2001.
[91] M. Pirmohamed and B.K. Park. *Trends Pharmacol. Sci.*, 22:298, 2001.
[92] M. Pitarque, O. Von Richter, B. Oke, H. Berkkan, M. Oscarson, and M. Ingelman-Sundberg. *Biochem. Biophys. Res. Comm.*, 284:455, 2001.
[93] R. Plomin, M. U. J. Owen, and P. McGuffin, Science 264, 1733, 1994.
[94] X.Q. Qi, S. Bakht, K.M. Devos, M.D. Gale, and A. Osbourn. *Nucl. Acids Res.*, 29:U68, 2001.
[95] A. Rafalski. *Curr. Opinion Plant. Biol.*, 5:94, 2002.
[96] D.E. Reich, M. Cargil, S. Bolk, J. Ireland, P.C. Sabeti, D.J. Richter, T. Lavery, R. Kouyoumjian, S.F. Farhadian, R. Ward, and E.S. Lander. *Nature*, 411:199, 2001.
[97] M.J. Rieder, S.L. Taylor, A.G. Clark, and D.A. Nickerson. *Nat. Genet.*, 22:59, 1999.
[98] J.H. Riley, C.J. Allan, E. Lai, and A. Roses. *Pharmacogenomics*, 1:39, 2000.
[99] C.P. Rodi, B. Darnhofer-Patel, P. Stanssens, M. Zabeau, and D. Van den Boom. *Biotechniques*, 32:S62, 2002.
[100] A.D. Roses. *Ann. Rev. Med.*, 47:387, 1996.
[101] M.V. Rudorfer, E.A. Lane, W.H. Chang, M. Zhang, and W.Z. Potter. *Br. J. Clin. Pharmacol.*, 17:433, 1984.
[102] R.K. Saiki, P.S. Walsh, C.H. Levenson, and H.A. Erlich. *Proc. Natl. Acad. Sci. U.S.A.*, 86:6230, 1989.
[103] B.A. Salisbury, M. Pungliya, J.Y. Choi, R.H. Jiang, X.J. Sun, and J.C. Stephens. *Mutat. Res. Fundamentals Mol. Mech. Mutagenesis*, 526:53, 2003.
[104] R.S. Schifreen, D.R. Storts, and A.M. Buller. *Biotechniques*, 32:S14, 2002.

[105] N.J. Schorx, D. Fallin, and S. Lanchbury. *Clin. Genet.*, 58:250, 2000.
[106] F. Sesti, G.W. Abbott, J. Wei, K.T. Murray, S. Saksena, P.J. Schwartz, S.G. Priori, D.M. Roden, A.L. George, and S.A.N. Goldstein. *Proc. Natl. Acad. Sci. U.S.A.*, 97:10613, 2000.
[107] B.S. Shastry. *J. Hum. Genet.*, 47:561, 2002.
[108] B.S. Shastry. *Int. J. Mol. Med.*, 11:379, 2003.
[109] V.C. Sheffield. *Proc. Natl. Acad. Sci. U.S.A.*, 86:232, 1989.
[110] M.M. Shi. *Clin. Chem.*, 47:164, 2001.
[111] M.J. Sobrido, B.L. Miller, N. Havlioglu, V. Zhukareva, Z.H. Jiang, Z.S. Nasreddine, V.M.Y. Lee, T.W. Chow, K.C. Wilhelmsen, J.L. Cummings, J.Y. Wu, and D.H. Geschwind. *Arch. Neurol.*, 60:698, 2003.
[112] M. Spiecker, H. Darius, T. Hankeln, M. Soufi, H. Friedl, J.R. Schaefer, E.R. Schmidt, D.C. Zeldin, and J.K. Liao. *Circulation*, 104(supl):314, 2001.
[113] A. Stevens, D. Ray, A. Alansari, A. Hajeer, W. Thomson, R. Donn, W.E.R. Ollier, J. Worthington, and J.R.E. Davis. *Arthritis Rheumatism*, 44:2358, 2001.
[114] S. Subramanian and S. Kumar. *Genome Res.*, 13:838, 2003.
[115] J. Sugatani, K. Yamakawa, K. Yoshinari, T. Machida, H. Takagi, M. Mori, S. Kakizaki, T. Sueyashi, M. Negishi, and M. Miwa. *Biochem. Biophys. Res. Comm.*, 292:492, 2002.
[116] A.C. Syvanen. *Hum. Mutat.*, 13:1, 1999.
[117] P. Taillon-Miller, I. Bauer-Sardina, N.L. Saccone, J. Putzel, M. Laitnen, A. Cao, J. Kere, and G. Pilia. *Nat. Genet.*, 25:324, 2000.
[118] F.K. Tan, N. Wang, M. Kuwana, R. Chakraborty, C.A. Bona, D.M. Milewicz, and F.C. Arnett. *Arthritis Rheumatism*, 44:893, 2001.
[119] S. Taygi, D.P. Bratu, and F.R. Kramer. *Nat. Biotechnol.*, 16:49, 1998.
[120] F.J. Tsai, H.C. Wu, H.Y. Che, H.F. Lu, C.D. Hsu, and W.C. Chen. *Urologia Int.*, 70:278, 2003.
[121] N. Unno, T. Nakamura, H. Mitsuoka, T. Saito, K. Miki, K. Ishimaru, J. Sugatani, M. Miwa, and S. Nakamura. *Surgery*, 132:66, 2002.
[122] J. Voorberg, J. Roelse, R. Koopman, H. Buller, F. Berends, J.W. Tencate, K. Mertens, and J.A. Vanmourik. *Lancet*, 343:1535, 1994.
[123] U. Wandegren, R. Kaiser, J. Sanders, and L. Hood. *Science*, 241:1077, 1988.
[124] L.Z. Wang, T. Habuchi, T. Takahashi, K. Mitsumori, T. Kamoto, Y. Kakehi, H. Kakinuma, K. Sato, A. Nakamura, O. Ogawa, and T. Kato. *Carcinogenesis*, 23:257, 2002.
[125] R. Waterland and C. Garza. *Am. J. Clin. Nutr.*, 69:179, 1999.
[126] M.K. Weabman, C.C. Huang, J. DeYoung, E.J. Carlson, T.R. Taylor, M. de la Cruz, S.J. Johns, D. Stryke, M. Kawamoto, T.J. Urban, D.L. Kroetz, T.E. Ferrin, A.G. Clark, N. Risch, I. Herskowitz, and K.M. Giacomini. *Proc. Natl. Acad. Sci. U.S.A.*, 100:5896, 2003.
[127] K.M. Weiss and J.D. Terwilliger. *Nat. Genet.*, 26:151, 2000.
[128] X.Y. Wen, R.A. Hegele, J. Wang, D.Y. Wang, J. Cheung, M. Wilson, M. Yahyapour, Y. Bai, L.H. Zhuang, J. Skaug, T.K. Young, P.W. Connelly, B.F. Koop, L.C. Tsui, and A.K. Stewart. *Hum. Mol. Genet.*, 12:1131, 2003.
[129] S.J. Wiezorek and G.J. Tsongalis. *Clinica Chemica Acta*, 308:1, 2001.
[130] T. Yoshino, H. Takeyama, and T. Matsunaga. *Electrochem.*, 69:1008, 2001.
[131] R. Youil, B.W. Kemper, and R.G.H. Cotton. *Proc. Natl. Acad. Sci. U.S.A.*, 92:87, 1995.
[132] L.P. Zhavo, S.Y.S. Li, and N. Khalid. *Am. J. Hum. Genet.*, 72:1231, 2003.
[133] Y. Zhu, M.R. Spitz, L. Lei, G.B. Mills, and X.F. Wu. *Cancer Res.*, 61:7825, 2001.

15

Control of Biomolecular Activity by Nanoparticle Antennas

Kimberly Hamad-Schifferli

Department of Mechanical Engineering and the Division of Biological Engineering, Massachusetts Institute of Technology, 77 Massachusetts Avenue 56-341C, Cambridge, MA 02139

15.1. BACKGROUND AND MOTIVATION

Biological molecules are now being increasingly viewed as machines due to the remarkable complexity, accuracy, and efficiency of their functions. Consequently, there is a developing effort to harness the engineering of Nature by going beyond characterizing biological systems and investigating methods for direct manipulation. This effort has been furthered by the enormous progress in understanding the complex mechanisms and structures of biological systems. Largely fueled by the advances in biological and biochemical tools and also characterization techniques such as x-ray crystallography and NMR, this knowledge has reached a point where one can describe the structure and mechanisms of large and complex biological molecules in molecular detail. Consequently, direct manipulation of biomolecular activity is an attractive route for the development of new types of hybrid systems that utilize the engineering of Nature.

In order to give a sense of how others have approached this problem, four examples of harnessing biological systems are described below. It is important to note that these examples do not comprise an exhaustive list as this field is changing rapidly, and numerous new techniques are being discovered continually. Instead, it is intended to give a sense of the diverse approaches to manipulating biological machines.

15.1.1. ATP Synthase as a Molecular Motor

ATP Synthase (also called proton-translocating ATP synthase or proton pumping-ATPase) is a membrane protein which is the most abundant enzyme [65]. Its function is

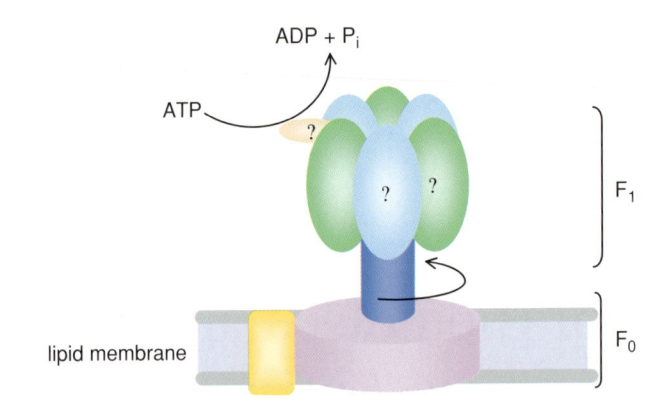

FIGURE 15.1. Schematic of ATP Synthase, consisting of the F_1 and F_0 subunits.

to create proton gradient across a membrane using the energy gotten from converting ATP to ADP. Alternatively, it can have the opposite function and use the energy from a proton gradient to synthesize ATP from ADP. ATP Synthase outperforms artificial motors, which operate at an efficiency of 35% to 96% in electrical to mechanical energy conversion. It is also an important enzyme as ATP is universally used as fuel in living organisms. It is a large protein (500kD) and has a complex structure. It is composed of the subunits F_0, which spans the membrane, and the F_1 subunit which sits above the membrane (Figure 15.1). The F_1 subunit is comprised of the α and β subunits which are arranged with three fold symmetry. In catalyzing these reactions, the entire F_1 subunit undergoes a rotational motion. The protein acts as a molecular stepper motor with ATP as fuel, generating torques of ~100pN·nm. Although there are many motor proteins that have been well studied, rotation is an unusual motion in proteins, and outside of bacterial flagellum examples of molecular rotary motors are rare. Through x-ray crystallography of the F_1 subunit [1], a molecular level description of the catalysis and associated motion has been determined. Because the efficiency of this motion is remarkable with respect to artificial motors, and sophisticated knowledge of the mechanism has been studied, it is desirable to harness the conformational change which can be simply induced by introduction of ATP. In order to harness this rotational motion and exploit the enzyme as a molecular motor, the F_1 subunit was engineered with histidines in key positions which act as attachment points to a surface. In addition, one of these sites serve as an attachment point for streptavidin to attach a Ni bar (150 nm by 1400 nm) so the rotational motion can be visualized by microscopy [54]. The engineered enzyme is immobilized on a surface at specific attachment sites put in by fabrication. ATP is introduced by flushing a solution over the chip, and in the presence of ATP the Ni propeller arm spins, as viewed by microscopy. This motion is performed with calculated efficiencies of 50%–80% depending on the ATP concentration, and with torques of 19–20pN·nm. However, utilizing it in its natural environment is desirable, and in the form of a fabricated device it is difficult to envision implementation in real biological systems. However, this example illustrates that knowledge of the structure and mechanism of even sophisticated enzymes coupled with rational design can result in harnessing biological machines in novel ways. Ultimately, one could create devices that exploit single biomolecules for mechanical applications.

15.1.2. Biological Self Assembly of Complex Hybrid Structures

Biology exploits principles of self-assembly to construct complex structures out of both organic and inorganic materials with great precision. On the other hand, we have constructed artificial systems largely employing top down approaches, such as microfabrication. However, these approaches are limited in constructing complex devices and machines of nanometer size dimensions. Also, they are much less efficient than how Nature achieves complex structures, and often require severe reaction conditions or harsh or toxic reagents. Consequently, there has been a great effort towards discovering how to utilize biological systems to construct inorganic/biological systems or at least mimic the key processes. Of particular interest is self-assembly on nanoscale dimensions, which would have important ramifications for device applications.

One compelling example of biological self assembly is biomineralization, or the crystallization of inorganic structures. This is seen exemplified in natural materials such as hard shells, bone, and teeth, and also in organisms such as a diatoms and magnetotactic bacteria. Biological systems have determined routes for making novel structures that are composites of inorganic and organic material, which often exceed the mechanical properties of pure inorganic materials. If created artificially, these structures often require high temperatures and pressures, in stark contrast to their biological counterparts which are synthesized in atmospheric conditions. This unique synthesis is achieved by utilization of proteins that recognize specific crystal faces and thus control the crystallization morphology [2, 11, 12, 20]. Biological strategies for assembling materials is much more complex than the simple crystallization we can do artificially. As a result, many have studied the proteins that can recognize crystal surfaces [53]. In order to exploit the properties of these proteins, bacteriophages have been genetically engineered so that they display these peptides on their coats. This allows use of the bacteriophage to crystallize nanostructures of unusual composition such as semiconductors (ZnS, GaAs) [60] and magnetic materials (CoPt, FePt) [44]. Also, the protein coats of viruses have been recognized as unique encapsulations for nanoparticle mineralization, as they are well defined in shape and size. A natural occurring example of this is the iron storage protein ferritin, which is a spherical protein made of 24 subunits. It stores iron in the form of a 6nm ferric oxohydroxide particle in its 7nm cavity. As an artificial analogue of ferritin, the cowpea chlorotic mottle virus (CCMV) has been manipulated to mineralize nanoparticles of novel materials [21]. The shell of the CCMV consists of 180 monomer proteins assembled into an icosahedron with a cavity ~18nm in diameter, making it a suitable vessel for mineralization of nanoparticles [40, 60]. By changing chemical conditions of the interior of the encapsulation, 15nm nanoparticles of paratungstate ($H_2W_{12}O_{42}^{10-}$) could be formed. The resulting self-assembled nanoparticles were highly crystalline, as evidenced by crystalline lattice fringes in TEM, and had narrow size distributions.

Many have recognized that DNA has great potential for organizing biomolecules and inorganic/organic structures on the nanometer length scale [52]. DNA has intrinsic length scales that make it a suitable candidate for this purpose—it is Ångstroms in width, and can be as long as millimeters in length [38, 46, 47]. More importantly, the chemistry of base pairing in DNA imparts specificity of hybridization, and DNA strands can find their complements in solution amongst thousands of other candidates. Advances in chemical modification of DNA have resulted in numerous methods for linking it to small molecules, surfaces, and nanoparticles [43]. Thus, many have looked to use DNA as a means of spatially organizing nanostructures. Gold nanoparticles of two different sizes (5 nm and 10 nm) are covalently attached to DNA strands of different sequences. By dictating the sequence of the complement added to the solution, different spatial arrangements of nanoparticles can be achieved. For example, both homo- and hetero-dimers can be created, as well as different arrangements of trimers. Through rational sequence design, spatially extended structures of DNA [38, 61] as well as unusual structures such as loops, hairpins, cubes, and truncated octahedra can be self-assembled [19]. Furthermore, these structures have been decorated with gold nanoparticles, and structures were ascertained through atomic force microscopy (AFM) [62] (Figure 15.2b).

While structures with complexity approaching fabricated transistors have not been created yet, this is an important step towards using biology to template and spatially organize molecules and objects for applications in devices.

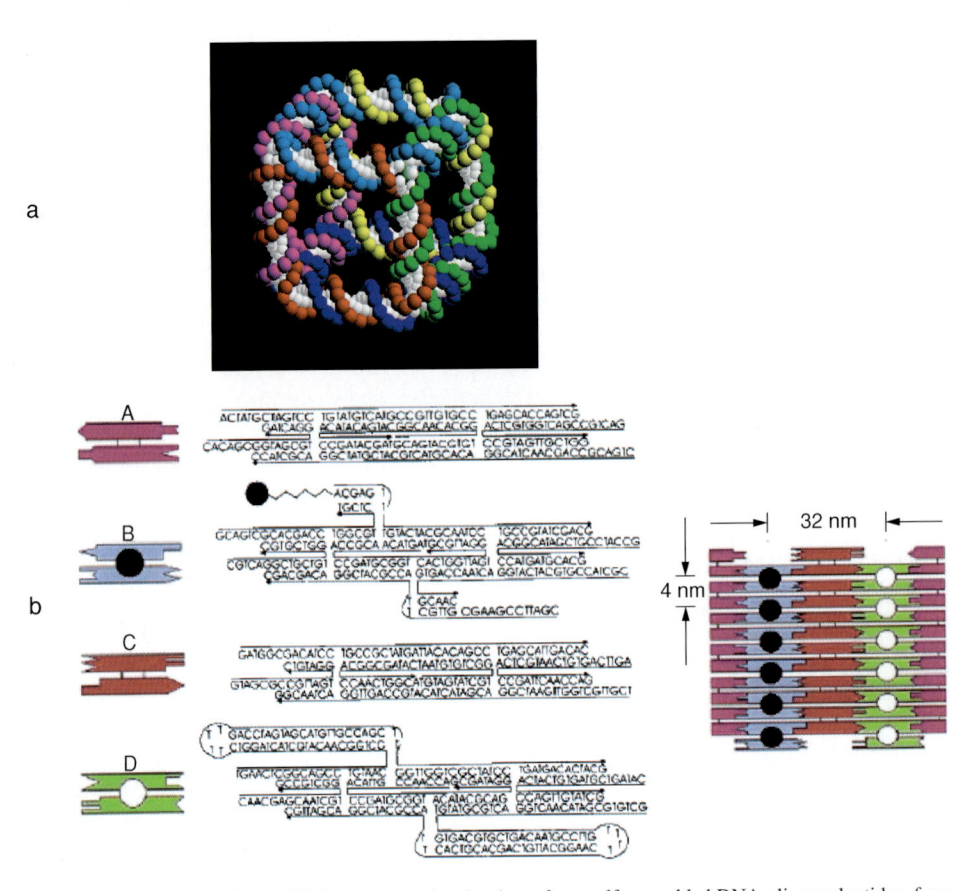

FIGURE 15.2. DNA self assembled structures, a) cube shape from self-assembled DNA oligonucleotides, from [19], b) DNA tiling decorated with nanoparticles, from [62].

15.1.3. DNA as a Medium For Computation

Biological systems are attractive as computational media. All organisms can be viewed as computers, as even simple organisms such as bacteria can accept inputs (i.e., toxins located nearby) and respond according to the information encoded in their DNA (i.e., the bacterium moves away). Biological systems all have superior computational efficiencies, processing power, and information storage densities over traditional computers, in addition to the ability to fully self-replicate, which so far has not been reproduced artificially. Consequently, exploiting biological for computational purposes has been pursued.

The first experimental realization of a biological system as a computer was by Adelman [3] in which the process of DNA hybridization in solution was used to solve a model mathematical problem. The traveling salesman problem ascertains the best route between multiple cities without repeating cities, and is normally intractable for traditional computers where the number of cities is large. Here, the possible mathematical solutions were represented by DNA strands of different sequences. A connection between two cities was represented by formation of a hybrid pair between those two points. All possible combinations were formed in solution, but one could filter out combinations that do not follow the constraints of the problem. If the problem was solved successfully, DNA of a certain length and sequence would remain. Adleman showed that using DNA to solve the traveling salesman problem was successful for a simplified version, but in theory it can be utilized on a larger version for which the solution is unattainable if approached by traditional computers. This approach offers new capabilities to computation, namely massively parallel computing. Because everything is in solution, large numbers of different answers can be sampled simultaneously. As a result, other formats for computation have been realized in which mathematical problems are represented by DNA tiles that form extended two dimensional structures according to whether they are a solution or not [61]. In addition, analogs of binary logic functions have been created out of DNA enzymes [56], where "1" and "0" are represented by fluorescence states of a product of the DNA enzyme. Also, simple circuit elements such as ring oscillators and bistable switches have been constructed using transcriptional machinery [23, 24].

15.1.4. Light Powered Nanomechanical Devices

One novel way to manipulate biomolecules is by decorating them with organic molecules which isomerize upon optical excitation, allowing light to control the biomolecule. Perhaps the most well studied molecule with this property is azobenzene (Figure 15.3a) which rotates the NNC bond angle by 120° in isomerization. It goes from the trans form to cis under optical excitation at $\lambda = 365$ nm, and in the reverse direction with $\lambda = 420$ nm or thermal excitation. Many have viewed this molecule as a molecular bistable switch powered by light. In addition, azobenzene can be linked to specific sites on a biomolecule (R groups), enabling the use of light to induce a drastic conformational change in the biomolecule.

Along these lines, azobenzene has been used to photoactivate DNA dehybridization [5]. It can be chemically attached to a phosphate backbone and substituted as a base in DNA during solid phase synthesis. In the trans form it does not perturb formation of a hybrid between its parent strand and its complement. However, in the cis form, it perturbs

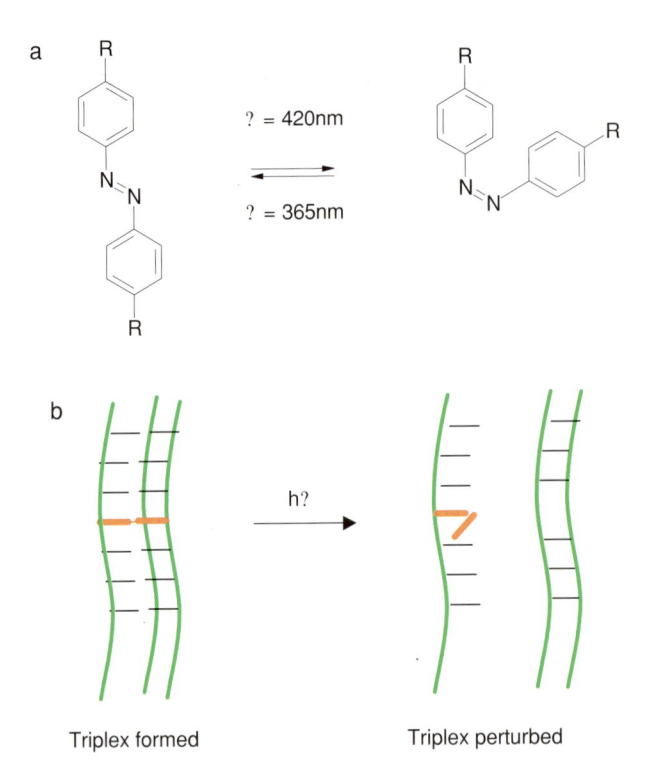

FIGURE 15.3. Light powered isomerization of a molecule. a) isomerization of azobenzene moiety upon light. b) utilization of isomerization of azobenzene to perturb DNA triplex formation.

the local structure in the DNA and formation of the hybrid pair is impeded. This has been used to manipulate DNA triplex formation by light (Figure 15.3b). Azobenzene is incorporated into one site on a DNA 13mer that can form a triplex structure with two other oligos. Upon UV illumination, the azobenzene converts from trans to cis, which in its non-planar form perturbs the local structure in the DNA in such a way that base pairing between the 13mer and its complements is weakened. As a result, the triplex is destabilized, as measured by melting curves of the complex with and without illumination [5]. In addition, perturbation of hybrid formation by azobenzene has been used to control the length of DNA transcription in vitro. The azobenzene is incorporated into a 12mer blocking strand that hybridizes to a specific site on a template downstream from the primer. With the blocking strand present and the azobenzene in the trans form, DNA polymerase lacking strand displacement and also 5' to 3' exonuclease activity is physically blocked from reading the message, so a truncated message is transcribed. Upon UV illumination, the azobenzene isomerizes into the cis form, causing the blocking strand to dehybridize from the template and allowing the polymerase to transcribe a full-length copy [63].

Azobenzene has also been used to control the conformation of proteins [32, 36, 42]. Amino acids containing the molecule have been incorporated into proteins at specific sites during synthesis or through site directed mutagenesis, allowing phototriggering of structural changes in proteins or photo-deactivitation of protein activity. One example has used azobenzene to control the activity of Ribonuclease S [32, 42]. RNase S mutants containing the azobenzene amino acid analog at key sites in the S-peptide portion of the protein were generated by chemical synthesis. By UV illumination, the catalysis rate (k_{cat}) of the mutant containing the azobenzene analog can be reduced by approximately $1/2$ as it switches the azobenzene from the trans to cis form. This is a working example by which light can control the activity of a protein.

Azobenzene has also been exploited as an optically powered molecular machine and being explored for use as a nanomechanical device [31]. In this case, multiple azobenzene units were incorporated into a polymer chain which was attached to an atomic force microscope cantilever. Under illumination, the azobenzene changes conformation, pulling the cantilever. The resulting average length change per molecule is 2.8nm. They then use this polymer to convert light into mechanical work in cycles, by repeatedly pulling the cantilever. The overall optical to mechanical efficiency, η, of 0.1 is obtained.

All of these examples show that knowledge of biological molecules has now reached a point where detailed, molecular-level knowledge of sophisticated and complex systems and mechanisms are attainable. This has been possible through recent advances in x-ray crystallographic techniques and various spectroscopies such as NMR, in combination with extensive mutational analysis and chemical design of inhibitors and binders. As a result, we are at a point where we can use this information and engineer Nature's systems so that they can be manipulated to perform functions that either do not occur naturally or are much more efficient than their artificial counterpart.

15.2. NANOPARTICLES AS ANTENNAS FOR CONTROLLING BIOMOLECULES

In designing a way to control biomolecules, the technique for manipulation must be compatible in vitro and in vivo for real utility. Thus, it should be applicable in the complex and crowded three-dimensional environments in cells. In addition, this means of control should be direct, reversible, selective, and not specific to one type of biomolecule but universally applicable.

A new technique using nanoparticles as antennas for controlling the activity of biological molecules addresses many of these issues [26, 27]. It utilizes an adapation of induction heating, a technique which is employed industrially to heat metals in a rapid and non-contact manner [48, 67]. Basically, a metal part is placed in a coil to which a current is applied (Figure 15.4, left). This generates a magnetic field in the part. In order to oppose this magnetic field (Lenz's Law), eddy currents are induced in the metal part. If the applied current is alternating, the magnetic field is changing direction, and the induced eddy currents oscillate in direction rapidly. This rapidly heats the metal part by joule heating.

In order to apply this to biological molecules for control, the metal species to be heated are nanoparticles. These nanoparticles are attached to a biological molecule in such a way that the normal functioning of the biomolecule is not affected (Figure 15.4, right). An external alternating magnetic field induced alternating eddy currents in the metal particle. The particle transfers the heat the biomolecule, inducing denaturation and thus shutting off the activity. Once the field is turned off, the biomolecule dissipates the heat into solution and refolds, resuming activity. Thus, the nanoparticle acts like a localized heat source which induces denaturation. Since the alternating magnetic field only heats the metal or magnetic

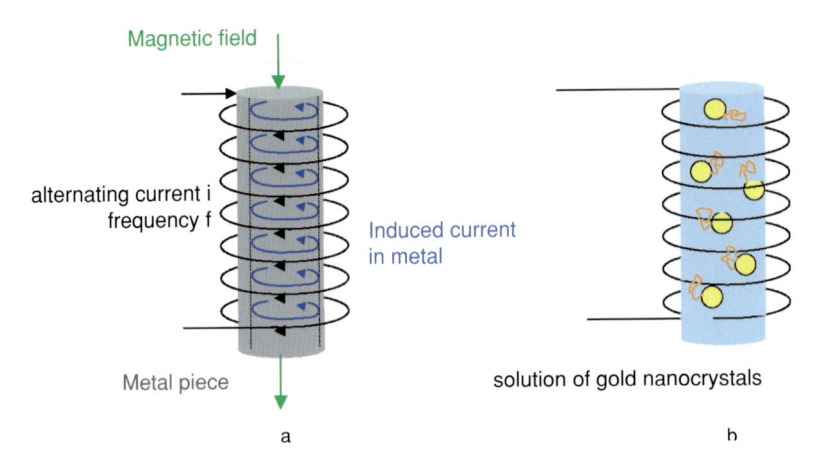

FIGURE 15.4. a) General schematic for induction heating of metal parts, b) induction heating of nanoparticles attached to biomolecules in solution.

particle directly and not the biomolecule or solution, it is a way to directly control the attached biomolecule. Heat dissipation of proteins into solution is fairly rapid [41], and for limited powers control of biomolecules reversibility is possible.

Nanoparticles are ideal systems for antennas for many reasons. First, the size of nanoparticles is about the size of the biomolecules to control, proteins and DNA (nanometers). Secondly, nanoparticles are soluble due to the solution phase synthesis used to synthesize particles. As a result, the nanoparticle-biomolecule hybrid is soluble, so the protein or DNA molecule can carry its nanoparticle antenna inside of a cell, which is like solution (albeit much more crowded/higher concentration [31]). This has important ramifications for in vivo feasibility. Thirdly, the chemistry to link nanoparticles to biomolecules has been explored. A covalent link between the two is required, not simply electrostatic adsorption because the protein may become detached when nanoparticle is heated. Fortunately, there are many options for achieving this chemical link, and there is extensive previous work in which nanoparticles have been successfully attached to proteins and DNA without perturbing the function, both in vitro and in vivo. Finally, nanoparticles have been studied for their interesting size and material dependent properties [4, 7, 28, 55], and some of these properties may be exploited in developing antennas.

This technique is ideal for controlling biological function because nearly all biomolecules denature with heat, giving this the potential to be universally applicable. In addition, magnetic fields can be used in tissue. Typically tissue creates a problem for optical techniques as it blocks out wavelengths shorter than 800 nm, the majority of visible spectrum where most dyes absorb and emit. However, some have been able to circumvent this obstacle by using chromophores that that absorb and emit in the infrared (IR). Core shell nanostructures of gold shells on and silica nanoparticles are excitable in the IR due to their novel structures [30]. Also, nanocrystals composed of semiconductors with small band gaps (CdTe, GaAs) absorb and emit in the IR. These IR fluorescent nanocrystals are biocompatible and have been used successfully for imaging through tissue.

CONTROL OF BIOMOLECULAR ACTIVITY BY NANOPARTICLE ANTENNAS

Nanoparticles have already been incorporated into biological systems for many other purposes. Typically they are employed as sensors, where they can give an optical readout of the state of hybridization of DNA or indicate whether or not a species has bound to its receptor. Metal nanoparticles have a surface plasmon resonance that is sensitive to its immediate surroundings, so a change such an antibody binding to its target or DNA hybridizing to its complement [22, 57] can shift this resonance. In some formats, nanoparticles provide a highly sensitive means of detection, prohibiting the need for PCR, which is not an immediate readout and may introduce errors. Semiconductor nanocrystals have been used as fluorescent tags that are more robust than organic dyes [39, 51]. Other forms of using nanoparticles to control processes in biological systems have been in hyperthermia, where magneticl particles are collected in tissue and heated similarly by an alternating magnetic field to burn tumors. In addition, magnetic particles have been utilized as "magnetic tweezers" to pull on cell surfaces to determine mechanics of the membranes. In addition, magnetic particles have served as sensors of DNA hybridization [34, 49].

There are three general schemes of using RFMF heating of nanoparticles to control the activity of biological molecules (Figure 15.5). First, heat from the nanoparticles can be used to directly change the conformation of a protein, turning off activity (Figure 15.5a).

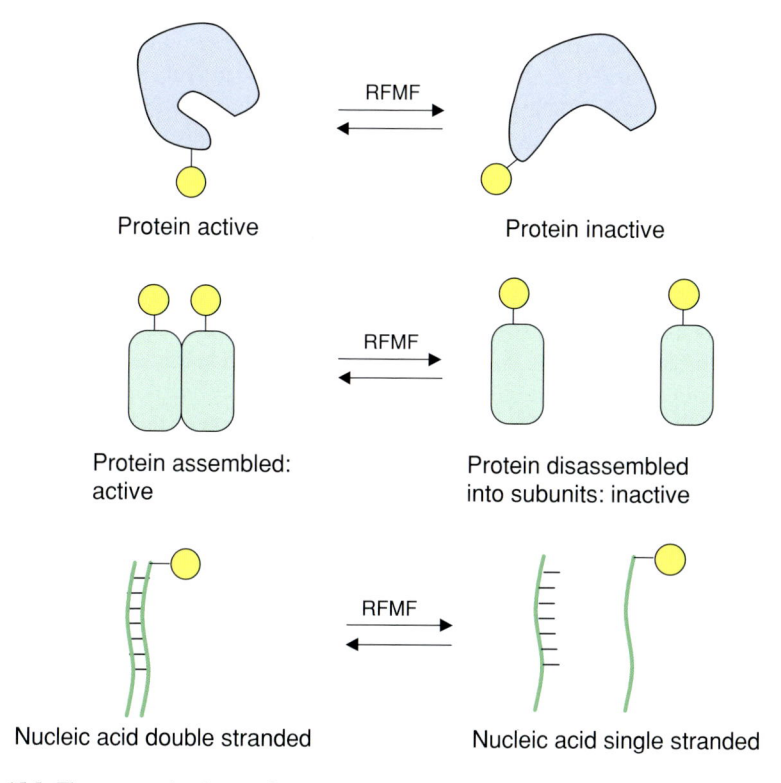

FIGURE 15.5. Three general schemes for controlling the activity of biomolecules. a) directly controlling the conformation of a biomolecule, b) using nanoparticle heating to break up non covalent bonds between subunits of the protein, altering activity, c) using nanoparticle heating to denature nucleic acids, altering activity.

Alternatively, it can be used to control proteins made of subunits which are held together by noncovalent bonds (Figure 15.5b). Upon heating by the RFMF, the nanoparticle breaks these noncovalent bonds, separating the subunits and turning the protein off (or on). For control of proteins, the nanoparticle must be labeled in a specific spot on the protein. Thirdly, the nanoparticles can be used to break the hydrogen bonds in nucleic acids and denature the pair (Figure 15.5c). This last example will be described in detail below.

15.2.1. Technical Approach

Nanoparticles of high crystalline quality, soluble, and well-defined sizes are required for their use as antennas. There are many routes for solution phase synthesis [10, 13, 14, 33, 45, 64], which typically yields particles that are soluble in organic solvents. However, water solubility is key for utility in biology, so methods of ligand exchange and surface functionalization have been developed [33, 45].

The linkage between nanoparticle antenna and the biomolecule to be controlled must be a covalent link. Although electrostatic adsorption has been used successfully to link proteins to inorganic nanoparticles, it may not be sufficient here as heating the particle may cause detachment from the biomolecule. In addition, precise placement of nanoparticles is desirable for control of proteins. Fortunately, multiple chemistries for attaching gold nanoparticles to biomolecules have been developed. The most commonly used chemistry is through a thiol attached to the DNA or protein which can react directly with gold to form a gold-thiol bond, which has been used successfully in many instances [22, 66]. Secondly, one can change the ligand on the surface of the nanoparticle to a moiety that can react with a group on a biomolecule to form covalent bond [29]. For example, the reaction between a primary amine with a sulpho N-hydroxysuccinimide (NHS) ester can form a peptide bond [25, 50]. Thus, the DNA can be appended with the primary amine on the 5' or 3' end, and the particle would have a ligand that contains the NHS ester. Alternatively, the DNA can be functionalized with a thiol group, which can react with a maleimide ligand on the nanoparticle to form a thioether bond. In general, reaction with a ligand on the surface is a method that is not specific to gold nanoparticles, and can be utilized with particles of different materials. Replacement of ligands on nanoparticle surfaces can be done by standard ligand exchange [45]. In addition, advances in DNA synthesis have facilitated these chemistries, as reactive groups can conveniently be incorporated during solid-phase synthesis. Furthermore, nanoparticles with reactive ligands are commercially available (Nanoprobes, Inc.).

Purification of the nanoparticle-bioconjugate species from unlabeled DNA/protein and free nanoparticles can be achieved by agarose gel electrophoresis. Gel electrophoresis has been established as a technique which can separate strands of DNA according to size with a resolution of a few base pairs. Consequently, it is an appropriate technique to separate DNA- and protein-nanoparticle conjugates [59, 66]. In order to obtain the purified conjugates in solution, bands were cut out of the gel matrix and the DNA-gold conjugates were extracted by centrifugation of the isolated bands through spin filters. A simpler method to purify the bioconjugate from free nanoparticles is by ethanol precipitation. If the nanoparticles have suitable surface ligands, only the DNA will precipitate from solution, permitting separation. In this case the nanoparticles must be in large excess of the DNA, as unlabeled DNA will also precipitate.

Here the system for induction heating of the particles in solution is similar to those used for industrial induction heaters, which typically consists of a generator that sends alternating currents through a coil, creating alternating magnetic fields [49, 67]. The current was applied to coils that had multiple turns (10^1 to 10^2), which were constructed of wire wrapped around a plastic cuvette/tube holder. The holder had an open structure to maximize light passage through the sample for optical experiments, and had a cross section of $\sim 1 cm^2$ so that microfuge tubes could fit inside. An RF signal generator (Hewlett Packard 8648C) generated currents with frequencies from 30kHz to 1GHz with an output power of 1 mW, which was then amplified to result in an output power from 0.4 W–4 W. This is an estimate for the ultimate output power as losses can occur from setup architecture. In order to eliminate effects from heating of the sample by the coil, the entire coil was placed in a large water bath at room temperature (T = 22 °C, volume $\sim$1 L). Samples were either in PCR tubes or quartz cuvettes if spectroscopic measurements were being performed, and sample volumes were typically in the range of 180–200 μL.

15.2.2. Dehybridization of a DNA Oligonucleotide Reversibly by RFMF Heating of Nanoparticles

The first molecule to be controlled was a DNA hairpin dehybridization, to demonstrate that the nanoparticle can be heated and denature an attached biomolecule. The DNA loop hairpin (also known as a molecular beacon [58]), which is a single stranded nucleic acid that is self complementary on the ends (Figure 15.6a). The loop/hairpin is utilized because the self-complementary structure facilitates rehybridization on rapid timescales [9], so a test of reversibility is feasible. The 1.4 nm gold nanoparticle was attached to the DNA via a functional group that was appended to one of the bases in the loop region. This group was an amine, which can react with an NHS ester on the surface of the nanoparticles. Here one needs a way to monitor the hybridization state of the molecule and determined if it is in single- or double-stranded form. In order to do so, DNA hyperchromicity can be exploited, which is the property of DNA by which the optical absorption of the bases increases when going from double stranded to single stranded form [16]. This occurs at 260nm, the wavelength characteristic of base absorption. Thus, the sample is put in a coil in an optical absorption spectrometer and the optical absorption at 260 nm is monitored while turning on and off the radio-frequency magnetic field at 15 second intervals.

The resulting optical absorption values for a $\sim$0.1μM solution of the nanoparticle-labeled molecular beacon are shown in Figure 15.6b, upper curve. When the RFMF is turned on, the optical absorption increases, and when it is turned off, the optical absorption decreases to its original value. This indicates that the DNA is dehybridizing with application of the field, and rehybridizing in the absence of the field. As a result, this shows that the control of the state of hybridization of the DNA is fully reversible. Furthermore, control experiments in which the same oligo without a nanoparticle is exposed to the field show no response (Figure 15.6b, lower curve), demonstrating that neither the solution nor the DNA molecule itself is responsive to the RFMF.

15.2.3. Determination of Effective Temperature by RFMF Heating of Nanoparticles

Since the nanoparticle produces some sort of heating, it is essential to get an estimate on what temperature it creates in its local environment. It should be an appropriate temperature

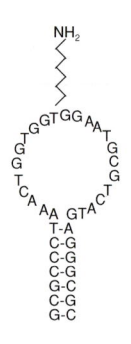

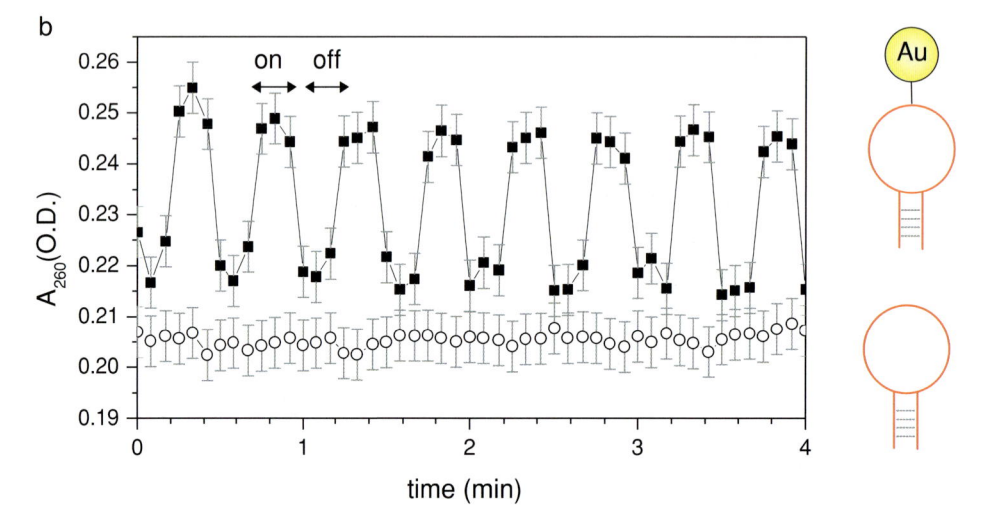

FIGURE 15.6. Using RFMF heating of nanoparticles to control denaturation of a molecular beacon. a) schematic of the molecular beacon and its attachment point for the nanoparticle (amine), b) optical absorption of nanoparticle-molecular beacon in the RFMF (upper curve) and only the molecular beacon in the RFMF (lower curve).

for denaturation of DNA or proteins. Based on the above results, the nanocrystals evidently create a higher temperature which may be localized to the DNA oligo. In order to determine this effective temperature increase, samples that were exposed to the alternating magnetic field were compared to those exposed to fixed global temperatures. In this case, a two-phase system was utilized in order to avoid temperature dependent effects on optical properties (Figure 15.7a). Nanocrystals were linked to a 12mer of DNA which had the 5' end functionalized with a fluorophore, which served as a means to count the oligo. It was hybridized to a solid support. This was achieved by using a complement that had a biotin on one end, which could be captured onto streptavidin coated agarose bead. Because the beads were large (diameter ~100μm), they settled to the bottom of the tube, comprising the solid phase. The supernatant above it acted as the solution phase. If the oligo was denatured by either heat or the RFMF, it could diffuse into the supernatant, where it could be removed from the solid phase and quantified by fluorescence spectroscopy.

One sample of the two phase system was prepared and aliquoted into separate tubes. Each tube was exposed to a fixed global temperature using a 1L water bath. One tube

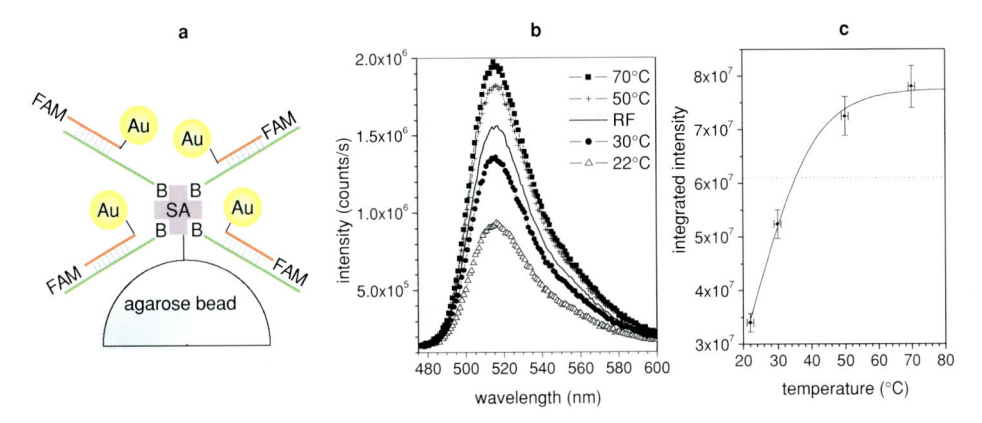

FIGURE 15.7. Determining the effective temperature of RFMF heating of a nanoparticle. a) schematic of the two phase system utilized. The species to be denatured is the DNA 12mer with an nanoparticle on the 3' end and a fluorophore (FAM) on the 5' end. b) fluorescence spectra of the supernatant c) integrated areas of the fluorescence spectra of the thermally heated samples. The RFMF peak (dotted line) extrapolates to a temperature of ~35°C.

was placed in the coil with the RFMF at room temperature. Following heat or RFMF, the samples were spun down and the supernatant was removed. Each supernatant was then measured by fluorescence spectroscopy, which yielded a relative concentration of DNA present in the sample. Figure 15.7b shows the spectra from the supernatant of each aliquot. The fluorophore, FAM, has an emission maximum at 515nm. For samples exposed to higher temperatures, the fluorescence intensity increased as more DNA was denatured. The sample that was exposed to the RFMF (thick black line) had a fluorescence intensity between that of the 30°C and 50°C samples. In order to quantify the effective temperature of the RFMF sample, the fluorescence intensity peaks were integrated and plotted as a function of incubation temperature (dots), shown in Figure 15.7c. The curve is fit to a sigmoidal (line) characteristic of DNA melting curves [17]. The intensity for the RFMF sample was extrapolated (dotted line) to a temperature of 35°C, or +13°C above ambient temperature. This is the effective temperature increase the DNA oligo experiences from the RFMF. It is important to note that this temperature jump is sufficient to partially denature many proteins in biology. However, using this technique to dehybridize longer strands of DNA which have elevated melting temperatures may be difficult using the parameters utilized in the experiments described here. One way to achieve denaturation of longer oligos would be to increase the delivered power to the coil, which would increase the heating of the nanoparticle. Also, using multiple nanocrystals on the DNA can increase heating and permit control of longer oligos. However, functionalization becomes difficult as the only available points in the middle of the chain are off the bases themselves and nanoparticles attached to these sites may interfere with formation of the hybrid pair.

15.2.4. Selective Dehybridization of DNA Oligos by RFMF Heating of Nanoparticles

This technique will be useful only if can be used selectively, i.e., control the molecule with the nanoparticle antenna and while not affecting surrounding molecules that have no

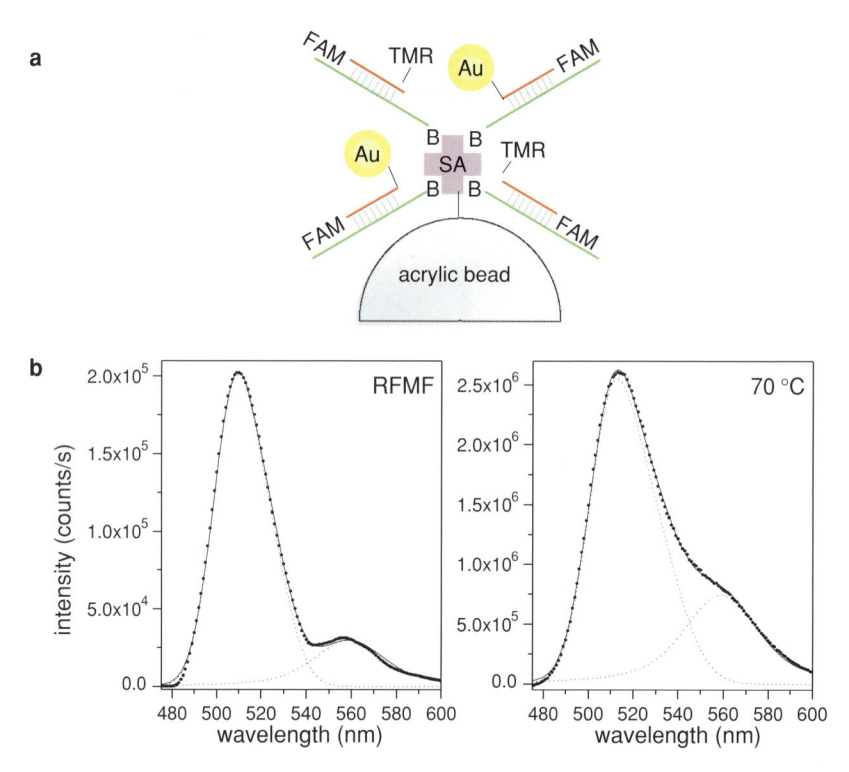

FIGURE 15.8. Determining selective denaturation of DNA by RFMF heating of nanoparticles. a) schematic of the two phase experiment utilized. The two species to be denatured is a nanoparticle labeled 12mer (identical to that in Figure 15.7) and another labeled with tetramethylrhodamine on the 3' end. b) Fluorescence spectra (points) and integrated areas (dotted lines) of the sample thermally heated to 70°C (right) and the one heated by the RFMF (left).

nanoparticles attached. In order to do so, we used a similar twophase system described above, but used a mixture of oligos with and without nanoparticles attached (Figure 15.8a). These oligos are identical in sequence to the one with the nanoparticles (identical 12mer, same melting temperature) but instead of a nanoparticle on the 3' end, there is another dye that is spectroscopically distinct from FAM. The oligo to be denatured was mixed with another oligo that was identical in length and sequence and melting temperature but had a fluorophore instead of a nanoparticle attached to the 3' end. The ratio of oligos with nanoparticles to those without was roughly 1:1. The fluorophore for this second oligo was tetramethylrhodamine (TMR), which is spectroscopically distinct from FAM ($\lambda_{absorption} = 555$ nm, $\lambda_{emission} = 563$ nm) and thus can be distinguished in fluorescence spectroscopy (Figure 15.8). One sample was immersed in a 70°C water bath, and the other was exposed to the RFMF at room temperature. Fluorescence spectra are shown in Figure 15.8b where the ratio of the fluorescence intensity at 515nm and 563nm are compared. The globally heated sample had a higher ratio of TMR: FAM, while the sample that was exposed to the RFMF had a lower ratio. Integration of the peaks to quantify the ratios showed that the sample exposed to the global heating has approximately 50:50 gold vs. TMR labeled oligos, while the RFMF sample has a ratio of 80:20. This indicates that denaturation of the gold labeled oligo was enhanced in the RFMF. However, its denaturation is not completely exclusive. One possible explanation is that is that streptavidin on the bead is tetrameric, and thus has the capacity to bind four biotins and thus four complementary oligos. Consequently, each streptavidin on average would possess a mixture of gold-labeled and TMR-labeled oligos. As a result, the TMR-labeled oligos were estimated to be on the order of 10nm or less away from a gold-labeled oligo. At these distances it is expected that through-solution heating will occur, resulting in non-specific denaturation.

The experiment shows that selectivity is limited by the degree of heat localization around the nanoparticle. This presents a challenge for implementation in cells which have extremely crowded environment, with protein concentrations on the order of 300mg/mL. As a result, the space between proteins is on the order of nanometers [37], with not many solvent layers between them. Therefore, heat localization will have to be on order of nanometers for it not to affect other proteins in solution. The study of heat localization around nanoparticles in solution is currently a work in progress. Classical heat transfer equations indicate that for a point source embedded in a medium, heat by conduction is fairly localized [8], but these descriptions are based on continuum approaches, and may not necessarily apply to nanoscale systems. Experimentally it has been determined that heat conductivity coefficients for solutions of metal nanoparticles are much higher than bulk systems, and cannot simply be described in terms of volume fractions. [15, 18, 35] As a result, this problem is currently of interest to the heat transfer field as these solutions of nanoparticles could be used as heat management fluids that remove heat from electronic devices.

CONCLUSIONS AND FUTURE WORK

Control of biomolecular activity by heating of nanoparticles by RFMF has been shown to be reversible and selective for the simple case of DNA denaturation. Nanoparticle antennas are promising as a tool for studying and utilizing biological systems as they can be universally applied to biological systems and have foreseeable compatibility in vivo. Future work includes control of proteins that perform complex processes. This will be a challenge as specific placement of the nanoparticle on the protein is required. This is easily achievable when labeling DNA since end functionalization or incorporation of unique bases is possible during solid phase synthesis for short oligos. However, proteins are much more complex than DNA, and knowledge of the three-dimensional structure is crucial for application of this technique. For proper control, the nanoparticle must be close enough to the active site so that heat from it changes its structure, but on the other hand it must not perturb the protein's function when no field is applied. Furthermore, control of the chemistry so that the nanoparticle is at one amino acid site and not at all of them is essential. Other issues like protein denaturation and sticking to the surface of the nanoparticle must also be addressed, most likely through manipulation of the nanoparticle surface chemistry. Clearly, this technique is much more powerful if applicable in cells or organisms. While this has not yet been tested, introduction of nanoparticles into cells and organisms has been successful in various instances. Nanoparticles have been ingested by cells, injected into tissues, and have had no effect on the functioning of the organism. Consequently, this is a promising new tool that has potential to control a broad range of biological functions in relevant biological environments. It could have important ramifications for diagnostic tools and pharmaceutical applications, as it may enable reversible and specific control of disease related species remotely. It represents but one step towards fully utilizing the richness and complexity of the machines of Nature.Hope

REFERENCES

[1] P. Abrahams, A.G.W. Leslie, R. Lutter, and J.E. Walker. Structure at 2.8 Å resolution of F_1-ATPase from bovine heart mitochondria, *Nature*, 370:621–628, 1994.

[2] L. Addadi and S. Weiner. Interactions between acidic proteins and crystals: Stereochemical requirements in biomineralization. *Proceedings of the National Academy of Sciences*, Vol. 82, pp. 4110–4114, 1985.

[3] L.M. Adleman. Molecular Computation of Solutions to Combinatorial Problems. *Science*, 266:1021–10–24, 1994.

[4] A.P. Alivisatos. Semiconductor clusters, nanocrystals, and quantum dots. *Science*, 271, 933–937, 1996.

[5] H. Asanuma, X. Liang, T. Yoshida, A. Yamazawa, and M. Komiyama. Photocontrol of triple-helix formation by using azobenzene-bearing oligo(thymidine), *Angewandte Chemie-International Edition In English*, 39:1316–1318, 2000.

[6] H. Asanuma, T. Yoshida, T. Ito, and M. Komiyama. Photo-responsive oligonucleotides carrying azobenzene at the 2'-position of uridine. *Tetrahed. Lett.*, 40:7995–7998, 1999.

[7] M.G. Bawendi, M.L. Steigerwald, and L.E. Brus. The quantum mechanics of larger semiconductor clusters (Quantum Dots). *Ann. Rev. Phys. Chem.*, 41:477–496.

[8] A. Bejan. *Heat Transfer*, New York, John Wiley & Sons, Inc., 1993.

[9] G. Bonnet, O. Krichevsky, and A. Libchaber. Kinetics of conformational fluctuations in DNA hairpin-loops. *Proceedings of the National Academy of Sciences*, Vol. 95, pp. 8602–8606, 1998.

[10] L.O. Brown and J.E. Hutchison. Controlled growth of gold nanoparticles during ligand exchange. *JACS*, 121:882–883, 1999.

[11] S. Brown. Protein-mediated particle assembly. *Nano Lett.*, 1:391–394, 2001.

[12] S. Brown, M. Sarikaya, and E. Johnson. A Genetic Analysis of Crystal Growth. *J. Mol. Biol.*, 299:725–735, 2000.

[13] M. Brust, J. Fink, D. Bethell, D.J. Schiffrin, and C. Kiely. Synthesis and reactions of functionalized gold nanoparticles. *J. Chem. Soc., Chem. Commun.*, 16:1655, 1995.

[14] M. Brust, M. Walker, D. Bethell, D.J. Schiffrin, and R. Whyman. Synthesis of thiol-derivatized gold nanoparticles in a two-phase liquid-liquid system. *J. Chem. Soc., Chem. Commun.*, 1994:801, 1994.

[15] D.G. Cahill, W.K. Ford, K.E. Goodson, G.D. Mahan, A. Majumdar, H.J. Maris, R. Merlin, and S.R. Philpot. Nanoscale thermal transport. *J. Appl. Phys.*, 93:793–818, 2003.

[16] C.R. Cantor and P. Schimmel. *Biophysical Chemistry Part I: The conformation of biological macromolecules*. San Francisco, W.H. Freeman, 1980a.

[17] C.R. Cantor and P. Schimmel. *Biophysical Chemistry Part II: Techniques for the study of biologial structure and function*. San Francisco, W. H. Freeman, 1980b.

[18] G. Chen. Particularities of heat conduction in nanostructures. *J. Nanopart. Res.*, 2:199–204, 2000.

[19] J. Chen and N.C. Seeman. The synthesis from DNA of a molecule with the connectivity of a cube. *Nature*, 350:631–633, 1991.

[20] J.H. Collier and P.B. Messersmith. Phospholipid strategies in biomineralization and biomaterials research. *Ann. Rev. Mat. Res.*, 31:237–263, 2001.

[21] T. Douglas and M. Young. Host–guest encapsulation of materials by assembled virus protein cages. *Nature*, 393:152–155, 1998.

[22] R. Elghanian, J.J. Storhoff, R.C. Mucic, R.L. Letsinger, and C.A. Mirkin. Selective colorimetric detection of polynucleotides based on the distance-dependent optical properties of gold nanoparticles. *Science*, 277:1078–1081, 1997.

[23] M.B. Elowitz and S. Leibler. A synthetic oscillatory network of transcriptional regulators. *Nature*, 403:335–338, 2000.

[24] T.S. Gardner, C.R. Cantor, and J.J. Collins. Construction of a genetic toggle switch in *Escherichia coli*. *Nature*, 403:339–342, 2000.

[25] J.F. Hainfield and R.D. Powell. New frontiers in gold labeling. *J. Histochem. Cytochem.*, 48:471–480, 2000.

[26] K. Hamad-Schifferli. DNA Hybridization: Electronic control of. In J.A. Schwarz, C. Contescu, and K. Putyera (eds.), Encyclopedia of Nanoscience and Nanotechnology. New York, Marcel Dekker, 2003.

[27] K. Hamad-Schifferli, J.J. Schwartz, A. Santos, S. Zhang, and J.M. Jacobson. Remote electronic control of DNA hybridization through inductive heating of an attached metal nanocrystal. *Nature*, 415:152–155, 2002.

[28] J.R. Heath and J.J. Shiang. Covalency in semiconductor quantum dots. *Chem. Soc. Rev.*, 27:65–71, 1998.

[29] G.T. Hermanson. *Bioconjugate Techniques*, San Diego, Academic Press, 1996.

[30] L.R. Hirsch, R.J. Stafford, J.A. Bankson, S.R. Sershen, B. Rivera, R.E. Price, J.D. Hazle, N.J. Halas, and J.L. West. Nanoshell-Mediated Near-Infrared Thermal Therapy of Tumors Under Magnetic Resonance Guidance. *Proceedings of the National Academy of Sciences*. Vol. 100, pp. 13549–13554, 2003.

[31] T. Hugel, N.B. Holland, A. Cattani, L. Moroder, M. Seitz, and H.E. Gaub. Single-molecule optomechanical cycle. *Science*, 296:1103–1106, 2002.

[32] D.A. James, D.C. Burns, and G.A. Woolley. Kinetic characterization of ribonuclease S mutants containing photoisomerizable phenylazophenylalanine residues. *Protein Eng.*, 14:983–991, 2001.

[33] N.R. Jana and X. Peng. Single-phase and gram-scale routes toward nearly monodisperse Au and other noble metal nanocrystals. *JACS*, 125:14280–14281, 2003.

[34] L. Josephson, J.M. Perez, and R. Weissleder. Magnetic nanosensors for the detection of oligonucleotide sequences. *Angewandte Chemie-International Edition In English*, 40:3204–3206, 2001.

[35] P. Keblinski, S.R. Phillpot, S.U.S. Choi, and J.A. Eastman. Mechanisms of heat flow in suspensions of nano-sized particles (nanofluids). *Internat. J. Heat Trans.*, 45:855–863, 2002.

[36] J.R. Kumita, O.S. Smart, and G.A. Woolley. Photo-Control of Helix Content in a Short Peptide. *Proceedings of the National Academy of Sciences*, 97:3803–3808, 2000.

[37] H. Kuthan. Self-organisation and orderly processes by individual protein complexes in the bacterial cell. *Prog. Biophys. Mol. Biol.*, 75:1–17, 2001.

[38] T.H. LaBean, H. Yan, J. Kopatsch, F. Liu, E. Winfree, J.H. Reif, and N.C. Seeman. Construction, analysis, ligation, and self-assembly of DNA triple crossover complexes. *JACS*, 122:1848–1860, 2000.

[39] D.T. Larson, W.R. Zipfel, R.M. Williams, S.W. Clark, M.P. Bruchez, F.W. Wise, and W.W. Webb. Water-soluble quantum dots for multiphoton fluorescence imaging in vivo. *Science*, 300:1434–1436, 2003.

[40] S.-W. Lee, C. Mao, C.E. Flynn, and A.M. Belcher. Ordering of quantum dots using genetically engineered viruses. *Science*, 296:892–895, 2002.

[41] T. Lian, B. Locke, Y. Kholodenko, and R.M. Hochstrasser. Energy flow solute to solvent probed by femtosecond IR spectroscopy: malachite green and heme protein solutions. *JPC*, 98:11648–11656, 1994.

[42] D. Liu, J. Karanicolas, C. Yu, Z. Zhang, and G.A. Woolley. Site-specific incorporation of photoisomerizable azobenzene groups into Ribonuclease S. *Bioorg. Medi. Chem. Lett.*, 7:2677–2680, 1997.

[43] C.J. Loweth, W.B. Caldwell, X. Peng, A.P. Alivisatos, and P.G. Schultz. DNA-Based assembly of gold nanocrystals. *Angewandte Chemie-International Edition In English*, 38:1808–1812, 1999.

[44] C. Mao, D.J. Solis, B.D. Reiss, S.T. Kottmann, R.Y. Sweeney, A. Hayhurst, G. Georgiou, B. Iverson, and A.M. Belcher. Virus-based toolkit for the directed synthesis of magnetic and semiconducting nanowires. *Science*, 303:213–217, 2004.

[45] C.B. Murray, D.J. Norris, and M.G. Bawendi. Synthesis and characterization of nearly monodisperse CdE (E = S, Se, Te) semiconductor nanocrystallites. *JACS*, 115:8706–8715, 1993.

[46] C.M. Niemeyer. Progress in "engineering up" nanotechnology devices utilizing DNA as a construction material. *Appl. Phys. A.*, 68:119–124, 1999.

[47] C.M. Niemeyer. Nanoparticles, proteins, and nucleic acids: biotechnology meets materials science. *Angewandte Chemie-International Edition In English*, 40:4128–4158, 2001.

[48] M. Orfeuil. *Electric Process Heating: Technologies/ Equipment/ Applications*. Columbus, Ohio, Battelle Press, 1987.

[49] J.M. Perez, L. Josephson, T. O'Loughlin, D. Högemann, and R. Weissleder. Magnetic relaxation switches capable of sensing molecular interactions. *Nat. Biotechnol.*, 20:816–820, 2002.

[50] J.E. Reardon and P.A. Frey. Synthesis of undecagold cluster molecules as biochemical labeling reagents. 1. Monoacyl and mono[*N*-succinimidooxy)succinyl] undecagold clusters. *Biochem.*, 23:3849–3856, 1984.

[51] S.J. Rosenthal, I. Tomlinson, E.M. Adkins, S. Schroeter, S. Adams, L. Swafford, J. McBride, Y. Wang, L.J. DeFelice, and R.D. Blakely. Targeting cell surface receptors with ligand-conjugated nanocrystals. *JACS*, 124:4586–4594, 2002.

[52] N.C. Seeman and A.M. Belcher. Emulating Biology: Building Nanostructures from the Bottom Up. *Proceedings of the National Academy of Sciences*, 10:1073, 2002.

[53] W.J. Shaw, J.R. Long, J.L. Dindot, A.A. Campbell, P.S. Stayton, and G.P. Drobny. Determination of statherin N-terminal peptide conformation on hydroxyapatite crystals. *JACS*, 122:1709–1716, 2000.

[54] R.K. Soong, G.D. Bachand, H.P. Neves, A.G. Olkhovets, H.G. Craighead, and C.D. Montemagno. Powering an inorganic nanodevices with a biomolecular motor. *Science*, 290:1555–1558, 2000.

[55] M.L. Steigerwald and L.E. Brus. Semiconductor crystallites: a class of large molecules. *Acc. Chem. Res.*, 23:183–188, 1990.

[56] M.N. Stojanovic, T.E. Mitchell, and D. Stefanovic. Deoxyribozyme-based logic gates. *JACS*, 124:3555–3561, 2002.

[57] A.T. Taton, C.A. Mirkin, and R.L. Letsinger. Scanometric DNA array detection with nanoparticle probes. *Science*, 289:1757–1760, 2000.

[58] S. Tyagi, and F.R. Krame. Molecular beacons: probes that fluoresce upon hybridization. *Nat. Biotechnol.*, 14:303–308, 1996.

[59] S. Wang, N. Mamedova, N.A. Kotov, W. Chen, and J. Studer. Antigen/Antibody immunocomplex from CdTe nanoparticle bioconjugates. *Nano Lett.*, 2:817–822, 2002.

[60] S.R. Whaley, D.S. English, E.L. Hu, P.F. Barbara, and A.M. Belcher. Selection of peptides with semiconductor binding specificity for directed nanocrystal assembly. *Nature*, 405:665–668, 2000.

[61] E. Winfree, F. Liu, L. Wenzler, and N. Seeman. Design and self-assembly of two-dimensional DNA crystals. *Nature*, 394:539–544, 1998.

[62] S. Xiao, F. Liu, A.E. Rosen, J.F. Hainfeld, N.C. Seeman, K. Musier-Forsyth, and R.A. Kiehl. Selfassembly of metallic nanoparticle arrays by DNA scaffolding. *J. Nanopart. Res.*, 4:313–317, 2002.

[63] A. Yamazawa, X. Liang, H. Asanuma, and M. Komiyama. Photoregulation of the DNA polymerase reaction by oligonucleotides bearing an azobenzene. *Angewandte Chemie-International Edition In English*, 39:2356–2357, 2000.

[64] H. Yang, J.E. Reardon, and P.A. Frey. Synthesis of undecagold cluster molecules as biochemical labeling reagents. 2. Bromoacetyl and maleimido undecagold clusters. *Biochemistry*, 23:3857–3862, 1984.

[65] M. Yoshida, E. Muneyuki, and T. Hisabori. ATP synthase—a marvellous rotary engine of the cell. *Nat. Rev.: Mol. Cell Biol.*, 2:669–677, 2001.

[66] D. Zanchet, C.M. Micheel, W.J. Parak, D. Gerion, and A.P. Alivisatos. Electrophoretic isolation of discrete Au nanocrystal/DNA conjugates. *Nano Lett.*, 1:32–35, 2001.

[67] S. Zinn and S.L. Semiatin. *Elements of Induction Heating, Design Control, and Applications*. Metals Park, A.S.M. International, 1988.

16

Sequence Matters: The Influence of Basepair Sequence on DNA-protein Interactions

Yan Mei Wang, Shirley S. Chan and Robert H. Austin
Dept. of Physics, Princeton University, Princeton, NJ 08544, USA

16.1. INTRODUCTION

The sequencing of the human genome, along with the 200-odd other genomes that have been sequenced, does not represent the solution to a puzzle but rather the necessary introduction to a bigger puzzle. That puzzle is how all the 30,000-odd some genes in the human genome are expressed and controlled in a proper sequence for a cell to function. We can hardly address this enormous problem in this brief review, but instead wish to concentrate on one very small but very important aspect of this problem: physical aspects to how proteins are able to achieve base-pair specific recognition. By "physical aspects" we mean that the proteins distort (strain) the DNA helix when they bind, and if this strain is a function of the sequence of the distorted region then the basepair dependent free energy associated with the strain provides a way to discriminate amongst the basepairs.

Our own group has been wrestling with getting DNA containing transcription factors into nanochannels in an attempt to read the proteins coding for genes on a single molecule basis [6, 7, 55, 56]. A major issue that will occur as we shrink structures to the nanometer scale [11] is the deformation of the DNA double helix due to the presence of the transcription factors, and the influence of sequence on the protein binding specificity. We will not address directly the issues of nanotechnology here that are needed to make structures small enough to care about these DNA deformations, but instead hope to lay down some of the basic physical concepts needed for analyzing deformed molecules at the nanoscale.

There are three levels of protein-DNA specificity. At a coarse level is the formation of chromatin by the binding of proteins to DNA. While not exquisitely sequence dependent, chromosomes do form very ordered structures and there certainly is a basic sequence dependence to how proteins are guided to form chromosomes during mitosis. At the next level of specificity and complexity are proteins such as restriction enzymes which cut DNA at certain sites. These proteins do not have the on-off control that transcription factors have but they do certainly cut at precision sites with great biological importance. Finally we have the transcription factors, which show the highest specificity of all: binding constants to certain sites approach 10^{-14} molar, and are 10^7 times higher than non-specific binding constants to dsDNA.

Consider for example how the λ phage exists within a bacteria. The λ phage is an extremely well known bacteriophage which can operate as a genetic switch: the phage can quietly exist within the genome of its host *E. coli* (this mode is called the lysogenic phase). However, upon sensing of damage to the genome of the bacterium the phage can switch to a "lytic" phase by exising itself from the host genome and turning off normal control of the phage expression. This removal of the normal control of phage expression results in exponential growth of the phage within the host bacteria causing finally lysis of the bacteria. In order for this switch in expression to occur, there must be a change in the gene expression of phage. In the lysogenic phase a protein called cI binds to a four-fold sequence called $O_{L,n}$ and $O_{R,n}$ where L or R stands for the left or right side respectively of the cI repressor gene, and $n = 1 - 4$ for the 4 possible binding sites. Under conditions of lysogeny the expression of cI results in a negative feed-back loop: the cI binds to the O_R and O_L control sequences. Binding of cI to the O_R sequence prevents expression of a protein called cro to the right and allows for expression of cI to the left side of the operator sequence. Binding of cI to the O_L sequences prevents expression to the left side of O_L called N. If something happens to the cI repressor protein, the O_R and O_L repressor sites are no longer occupied with three results: RNA polymerase cannot initiate cI transcription without cI binding to O_R, and the proteins cro and N can now be expressed as the RNA polymerase moves to the right and left respectively of the operator sites. An excellent review of this process can be found in the book "Genetic Switch" by Mark Ptashne [46].

Perhaps the above brief description can give the reader some idea of the amazing amount of highly specific control transcription factors must exert on gene expression in order for an organism to successfully adapt to changing conditions. How does this happen physically, that is, what is the 3-D structure of the protein that allows this to happen? An excellent introduction to this subject can be found in a review article by Luscombe et al. [34]. Generally, transcription factors can be grouped into 4 basic classes, although there are many exceptions to the following list: (1) helix-turn-helix, (2) zinc finger, (3) leucine, (4) helix-loop-helix. For all 4 cases the basic structure consists of two DNA binding sections which are separated by a linker region. The linker region does not contact the DNA helix. Why is this particular structure so predominantly chosen? We can guess that perhaps part of the recognition process involves some sort of straining of the helix in the non-contacted region. Figure 16.1 shows some examples of how proteins can deform the double helix upon binding.

This bending of the helix which would make some sense as a means of enhancing specific if in fact the elastic properties of DNA are a function of the basepair composition, for then the strain energy would stored in the strained helix would vary as a function basepair sequence in the strained sequence. In order to test this supposition three subjects must be

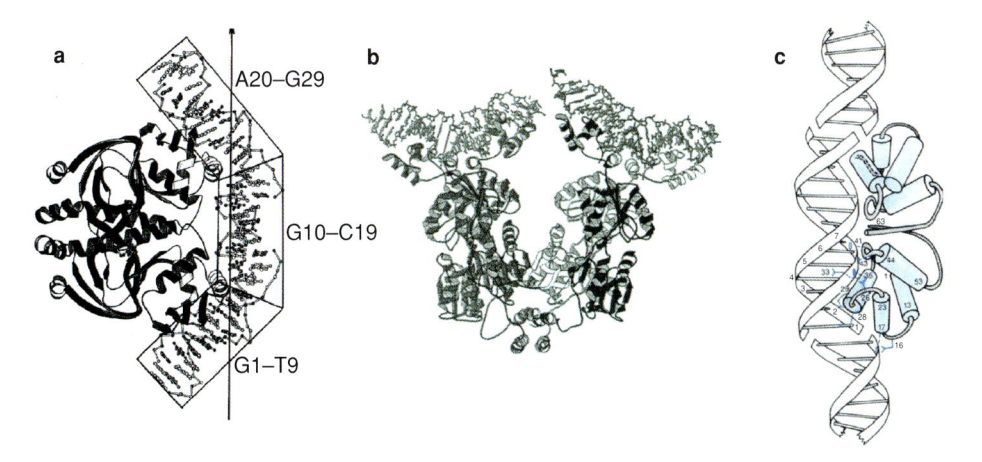

FIGURE 16.1. Dimer binding of (a) CAP, (b) LacR protein and (c) phage 434 repressor protein to DNA. The bending angles are 90° in CAP, 42° in LacR protein and 25° in phage 434 repressor protein. LacR protein in nature binds as tetramers to two isolated DNA. LacR protein and phage 434 repressor protein bind to DNA with helix-turn-helix motif. Adapted from Refs. [17] [31] and [1].

addressed: (1) we have to have a mathematical formalism for the strain energy stored in a deformed helix; (2) we have to have measurements of the elastic constants of the helix as a function of basepair sequence; (3) we need to bring the previous parts together and show that experiments support the supposition that sequence-dependent elastic properties of DNA and strain can explain a significant part of the free energy of sequence dependent specificity in transcription factors.

The structure of Deoxyribose Nucleic Acid (DNA) was first proposed by Watson and Crick in 1953 [57]. In their model, DNA is a double helix of anti-parallel sugar-phosphate chains held together by a stack of complementary base-pairs (bp). Base-pairs contain four base: adenine (A) and guanine (G) (purine), cytosine (C) and thymine(T) (pyrimidine). Purine bases always pair with pyrimidine bases; i.e., A pairs with T (A·T) and G with C (G·C). Purine and pyrimidine in a base pair are connected by hydrogen bonds—two for A·T and three for G·C. This DNA is called a classical Watson-Crick B-form DNA. In real life, the Watson-Crick DNA of perfect helix form with all its bases, sugars and phosphates at their assigned locations, rarely exists. What exists is a double helix with local (of order base-pair separation distance) structural variations brought about by generic differences of the stacking four bases. Since each base has its own unique composition and thus electronic distribution, it is conceivable that for different base sequences, equilibrium separation and rotation between bases would be different. For example, it would not be hard to imagine that DNA homo-polymer such as polyd(G·C) where the long purine bases reside on one chain having different conformation from alternating DNA polymer such as polyd(GC).

It is now generally accepted that properties of DNA (structural, mechanical etc.) depend on base-pair sequence. While sugar phosphate backbone ensures the continuous double helical form of DNA, it is mainly the sequence of bases that determines DNA local geometrical properties. In order to discuss the sequence's effects on DNA, it is necessary to first define all relevant base movements in DNA. Translational and rotational parameters for relative displacements of base to base or base-pair to base-pair were established and standardized in 1989 [16, 18]. Figure 16.2 shows all coordinate parameters alone which translations

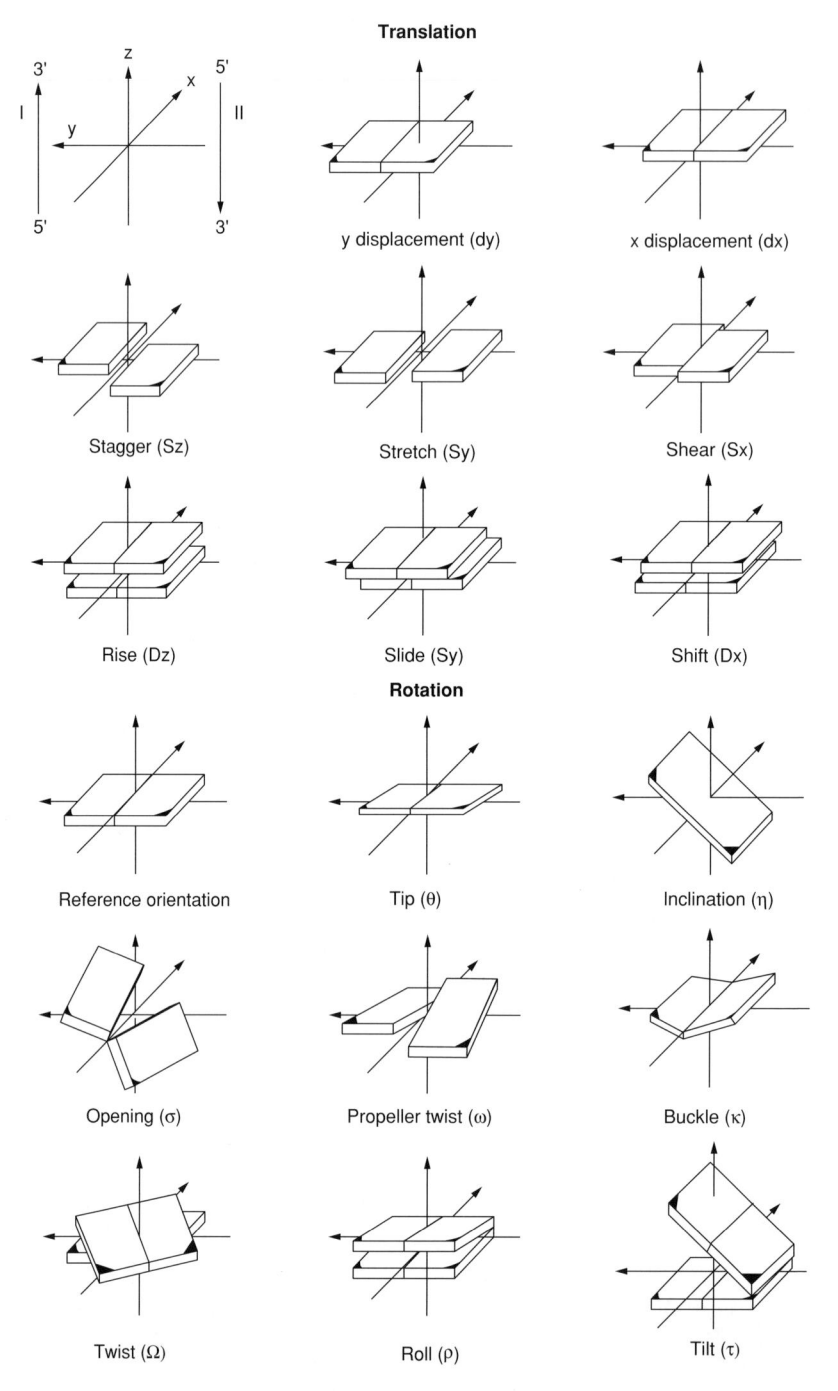

FIGURE 16.2. Translational and rotational movements of bases and base-pairs. Shaded edges are minor groove sides of DNA and black corners are sites where sugar attaches to bases. Strands on the left run from 5′ to 3′ in the +z direction and strands on the right in the −z direction. Propeller twist, helical twist, roll and tilt shown here are all positive rotations. Adapted from Ref. [3].

and rotations of bases and base-pairs are performed. Among these translational and rotational parameters, twist (propeller twist and helical twist), roll, tilt and slide are sufficient in describing sequence dependent DNA properties and DNA-protein interactions in this article.

16.2. GENERALIZED DEFORMATIONS OF OBJECTS

We relied heavily in this section on Landau and Lifshitz Vol. 7 (Elasticity) [32] and an informative article by Goldstein, Powers and Wiggins [22]. We first note that it is possible to write the displacement $\delta \vec{r}$ of vector $\vec{r}$ when it is rotated through an angle $\delta\phi$ as a cross-product of the vector $\delta\vec{\phi}$ and $\vec{r}$:

$$\delta\vec{r} = \delta\vec{\phi} \times \vec{r} \qquad (16.1)$$

which you can convince yourself is true by taking the vector at some finite angle ϕ to the x axis and rotating a bit through the angle $\delta\phi$. Now, imagine that an orthonormal coordinate system $\vec{e}_i$ is attached to the filament at some point. Let $\vec{e}_1$ and $\vec{e}_2$ be the x and y axis respectively in the cross-sectional plane of the filament and let $\vec{e}_3$ be the z axis parallel to the local tangent, ie, $\vec{e}_3$ is the tangent vector. The *deformation* of the filament is the local rotation $d\vec{\phi}$ of this coordinate system as we move an arc length element ds along the filament. It is common to use the vector $\vec{\Omega}$ to represent this deformation:

$$\vec{\Omega} = \frac{d\vec{\phi}}{ds} \qquad (16.2)$$

The strains of the object are then proportional to the rate at which the coordinate system rotates:

$$\frac{\partial \vec{e}_i}{\partial s} = \vec{\Omega} \times \vec{e}_i \qquad (16.3)$$

Those elements of $\vec{\Omega}$ which are parallel to the $\vec{e}_3$ axis are defined as *torsional* deformations, while those elements along $\vec{e}_1, \vec{e}_1$ are bending deformations.

We can make this more explicit. Note that $d\vec{e}_3/ds$ is the curvature of the filament, and is equal to $1/R$, where R is the radius of curvature of the filament. Hence:

$$\frac{d\vec{e}_3}{ds} = \vec{\Omega} \times \vec{e}_3 = \frac{\vec{n}}{R} \qquad (16.4)$$

where $\vec{n}$ is a unit vector in the plane of curvature and perpendicular to the tangent $\vec{e}_3$:

$$\vec{n} = R\frac{d\vec{e}_3}{ds} \qquad (16.5)$$

If we cross the tangent vector $\vec{e}_3$ with both sides of Eq. 16.4 and use the vector identity:

$$\vec{A} \times (\vec{B} \times \vec{C}) = \vec{B}(\vec{A} \bullet \vec{C}) - \vec{C}(\vec{A} \bullet \vec{B}) \qquad (16.6)$$

we obtain an expression for the strain vector $\vec{\Omega}$ in terms of the bending and twisting components:

$$\vec{\Omega} = \frac{1}{R}(\vec{e}_3 \times \vec{n}) + \vec{e}_3(\vec{e}_3 \bullet \vec{\Omega}) \tag{16.7}$$

Although many of the strains seen in biology at the molecular level are large enough that nonlinear terms are significant it is still useful to consider the small strain limit where the stress is proportional to the strain. In a linear system the stress elastic energy should go as the square of the strain vector derived in Eq. 16.7. We would expect in the small strain limit that we can write the elastic energy F of a strained filament as:

$$F = \int \left[\frac{1}{2}\frac{EI_x}{R_x^2} + \frac{1}{2}\frac{EI_y}{R_y^2} + \frac{1}{2}C\Omega_3^2 \right] ds \tag{16.8}$$

where E is the Young's Modulus and C is the twisting rigidity and we have explicitly considered that the bending surface moment of inertia I can be different in the two orthogonal directions:

$$I_x = \int x^2 dx dy$$
$$I_y = \int y^2 dx dy \tag{16.9}$$

Note that in Eq. 16.8 we have quite explct expressions for the bending energy in terms of the Young's modulus and the surface moment of inertia, but not for the shear rigidity in terms of a shear modulus G and a factor dependent on the structure of the filament.

We now relate the torsional rigidity C to the shear modulus G. When you twist a filament, the strain is due to the rotation rate $d\phi/ds$ of the local coordinate system, and the linear assumption is that the stress (force/area) is due to this strain. In the case of twist distortions, what matters is the restoring torque that acts along the $\vec{e}_3$ direction. Consider a cylinder of radius R and length L which is fixed at one end. If we apply a shearing torque $\vec{\tau}$ at the free end of the cylinder the cylinder responds by twisting (shearing) in such a way that $\frac{d\phi}{ds}$ is a constant. Suppose we are at the free end of the rod and consider a small ring of width dr, a distance r from the axis of the rod. The ring is twisted through an angle α relative to the fixed end of the rod, and the tangential movement δs of the ring due to the shear is $\delta s = r\alpha$. Thus, the strain $\epsilon(r)$ and stress $\sigma(r)$ at this position are:

$$\epsilon(r) = \frac{r\alpha}{L}$$
$$\sigma(r) = \frac{Gr\alpha}{L} \tag{16.10}$$

where G is the shear modulus. The magnitude of the restoring torque $d\tau$ is:

$$d\tau = r \times \sigma(r) dA = \frac{2\pi G\alpha}{L} r^3 dr \tag{16.11}$$

Integration over the entire cross-section yields the total restoring torque τ:

$$\tau = \frac{2\pi G \alpha}{L} \int r^3 dr = \frac{G\alpha}{L} I_P \qquad (16.12)$$

where I_P is defined as the *polar* moment of inertia:

$$I_P = 2\pi \int r^3 dr \qquad (16.13)$$

Note that I_P varies as the fourth power of the radius of the rod, as does the surface moment of inertia I_A. We now see that since α/L is the rotational rate Ω_3 of the coordinate system and that the twisting rigidity C can be written in general as:

$$C = \frac{GI_p}{L} \qquad (16.14)$$

A final note. There is a connection between the thermally induced bending and twisting dynamics of the DNA molecule and the elastic modulii. Thermal energy stores $\frac{k_B T}{2}$ of energy per degree of freedom in a system, where k_B is Boltzmann's constant and T is the temperature in Kelvin. This thermal energy results in a bending and twisting persistence length p_B and p_T respectively [32]. These lengths are basically the average radius of curvature due to thermal induced bending and the average distance over which the helix twists through an RMS angle $<\phi^2>^{1/2} \sim 2\pi$. The bending persistence length is:

$$p_B = \frac{EI_a}{k_B T} \qquad (16.15)$$

while the twisting persistence length is:

$$p_T = \frac{GI_p}{k_B T} \qquad (16.16)$$

The bending persistence length is quite easy to observe since it results in the deformation of the backbone of the dsDNA molecule giving rise to a random walk aspect to the contour of the molecule which can be directly measured as the radius of gyration R_g of the polymer. R_g is basically the radius of the glowing blob that a genomic length dsDNA molecule appears as in a microscope. For a non-self avoiding random walk R_g is given for a polymer of contour length L and bending persistence length p_B by [19]:

$$R_g = \left[\frac{Lp_B}{6}\right]^{1/2} \qquad (16.17)$$

TABLE 16.1. Elastic properties of defined length DNA sequences

Sequence	E (dyne cm^{-2})	p_B (bp)
poly(dG)·poly(dC)	2.3×10^9	400
poly(dA-dC)·poly(dT-dG)	1.4×10^9	250
poly(dA)·poly(dT)	8.2×10^8	150

16.3. SEQUENCE DEPENDENT ASPECTS TO THE DOUBLE HELIX ELASTIC CONSTANTS

The next question is if there is a DNA sequence dependence to E and G. This sequence dependence could come from at least two interactions: "on diagonal" terms due to the differences in the hydrogen bonding patterns of G· C and A · T basepairing (denoted by the symbol ·), and "off-diagonal" terms due to differences between the basepair stacking interactions of nearest neighbor bases, such as G-C, denoted by the hyphen. The effect of sequence on DNA elastic properties (torsional and bending stiffness) was investigated in our group with a few hundred bp long DNA segments using triplet anisotropy decay techniques [25]. In this work DNA was treated as a stack of base-pairs connected by springs. Table 16. compares E for different three sequences: poly(dG)·poly(dC), poly(dA-dC)·poly(dT-dG), poly(dA)·poly(dT) and the predicted bending persistence length p_B as calculated from Eq. 16.15. These measurements showed that AT rich sequences where the most flexible and GC rich sequences where the least flexible.

Other experimental techniques have been used in studying DNA flexibility: they are scanning force microscopy [49], nuclear magnetic resonance [38], fluorescence polarization anisotropy (FPA) measurements [21] and electron paramagnetic resonance [40]. There is by no means an agreement in the literature that the values we cite in Table 16.1 are gospel truth. Hagerman in his 1988 review article [23] dismissed the results shown above as "probably artifactual". But, we will see that this story has continued in spite of his scepticism. There are probably strong basepair dependences to DNA physical properties, as a function of sequence [5], that is no doubt more complex than the simple picture presented above, but to deny the existence of any substantial basepair dependence to the elastic properties of DNA is surely wrong.

That there are substantial sequence dependences is clear from other areas of research. There is a thermodynamic literature which connects melting points of DNA oligonucleotides and basepair composition, a more indirect approach than the physical approaches discussed above but indicative of the substantial dependence of basepair composition on the physical properties of DNA. Thermodynamic measurements which measure the heat exchange ΔQ when molecules interact basically yield the values for the enthalpy changes ΔH. Enthalpy is not the same as the potential energy ΔU. Enthalpy is one of the possible "free energy" potential measurements of a system. The free energy of a system is the *extactable* energy you can remove from a system subject not only to the constraints of the conservation of energy E, but also subject to the constraint that a physical system will always want to be in the macrostate of maximum entropy S, it wants to be as disordered as it can be. Thus, enthalpy changes ΔH are more complex quantities than the simpler elastic potential energy

TABLE 16.2. Nearest Neighbor Thermodynamics

Interaction	$\Delta H°$ cal/M	$\Delta S°$ cal/K-M	$\Delta G°$ cal/M
AA/TT	9.1	24.0	1.9
AT/TA	8.6	23.9	1.5
TA/TA	6.0	16.9	0.9
CA/GT	5.8	12.9	1.9
GT/CA	6.5	17.3	1.3
CT/GA	7.8	20.8	1.6
GA/CT	5.6	13.5	1.6
CG/GC	11.9	27.8	3.6
GC/CG	11.1	26.7	3.1
GG/CC	11.0	26.6	3.1

changes we discussed earlier but do contain within them some (unknown) elements of the total thermodynamic differences between basepairs in DNA.

Breslauer's group at Rutgers University has been the leader in the field of measuring basepair dependent thermodynamic properties of DNA, predominantly in basepair specific specific heat changes [4, 54]. By analyzing the basepair dependent melting temperatures of oligonucleotides it has been possible to construct a table of basepair-basepair stacking interactions which include as we have mentioned both the elastic potential energy effects we discussed above and entropic issues which we have ignored. Table 16.2 is extracted from [4] and gives in tabular form the thermodynamic variances in stacking interactions. As the authors state in [4], "base sequence and not base composition determines stability". Note also that the sequence AA/TT, what we will call an "A-tract" in the next section, has an anomalous stability. The subtle consequences of A-tracts will be clarified next, but the point here is that the Gibbs free energy $\Delta G°$ varies by factors of 3 with basepair sequence, hardly a small effect.

16.4. SEQUENCE DEPENDENT BENDING OF THE DOUBLE HELIX AND THE STRUCTURE ATLAS OF DNA

There is more to the picture of the conformational complexity of DNA than just the variation of the elastic constants with basepair sequence as is clear from the thermodynamic results. DNA is held together with hydrogen bonds, and these bonds can connect atoms not only directly across the basepairs but also off-diagonally to neighboring groups. In order to maximize the energy of these bonds, the basepairs typically twist out of the plane and can form what is called a bifurcated hydrogen bond. Not all bases can do this, the AA-TT sequence has the least amount of steric clash and so can do this most readily if the As are all on one side and the Ts are all on the other: this is called a homo-A tract. The designation of a homo A-tract is poly (dA)·poly (dT). So, strangely, although we would expect that a random sequence of A·T basepairs to be the most flexible, this rule breaks down for A tracts because of the cross-coupling of the bifurcated hydrogen bond which can enhance the rigidity of the helix, at least at temperatures below 30 C. The formation of the bifurcated hydrogen bond results in a propellor twist to the bairpairs as they rotate out of the plane of the helix. A review of the basic idea of the bifurcated hydrogen bond can be found in [15]. As we will discuss later, this birfurcated hydrogen bond can be broken ("melt") long before the double helix loses its integrity, as has been shown by a combination of circular dichroism, scanning calorimetry and Raman spectroscopy [9, 10].

The structural implications of these off diagonal interactions, so reminiscent of the Ising model in condensed matter physics, can be quite significant. In our case, we are concerned with the phenomena of bent DNA, which occurs when "A-tracts" consisting of 4 or more consecutive runs of A bases are phased with the 10 bp repeating pattern of the B-type double helix. A phased A-tract bends the helix of dsDNA into a circle of 50 Å radius at temperatures below 30 C. The bend occurs because there is a cross-linking hydrogen bond (the bifurcated hydrogen bond) that forms between neighboring A_n and T_{n+1} bases, giving rise to a propellor twist to the plane of the basepairs. This twist gives rise to a coherent bend of the helix if the A tract is phased to the helix repeat [14].

There are two main groups which have greatly extended the theoretical work we have described here. The Olson group at Rutgers University has for one of its goals understanding the influence of chemical architecture on the conformation, properties, and interactions of nucleic acids. They have made great progress in clarify the role of local structure induced by the primary base sequence and the binding of proteins on the overall folding of DNA and RNA [41, 42]. The work in this group in primarily theoretical and has combined a variety of computational approaches (Monte Carlo and molecular dynamics simulations, potential energy calculations, developments and applications of polymer chain statistics, finite element analysis, systematic molecular modeling) with new developments in polymer theory. Following up on this work, there is a group in Denmark, the Center for Biological Sequence Analysis at the Technical University of Denmark, which has turned the analysis of basepair sequence dependence of DNA structure and elasticity to a very formal level [29, 44]. They have used the rules that we have outlined in this review paper to analyze DNA sequences based upon the structural, positioning and dynamic information we have outlined in this paper. Genomic length sequences of DNA are presented in terms of the intrinsic curvature of DNA (phased A-tracts), stacking energies (as determined thermodynamically), and a set of parameters related to the variation in the elastic constants of the basepairs as discussed above. Figure 16.3 shows in graphic form how these parameters vary with sequence in the E. coli chromosome.

16.5. SOME EXPERIMENTAL CONSEQUENCES OF SEQUENCE DEPENDENT ELASTICITY

16.5.1. Phage 434 Binding Specificity and DNase I Cutting Rates

The realization that there is in fact a basepair dependence to many physical properties of DNA outlined above naturally leads to the question if biology is aware of and uses this phenomena? Probably the most direct indication of sequence dependence to the elasticity of DNA comes from relatively simple studies of basepair dependence of the binding constants of dimeric transcription factors which have a "non-contacted" region between the dimers. This non-contacted region is over distorted by either shearing (twisting) or bending strain. The binding of phage 434 repressor protein had been a subject of interest since the binding

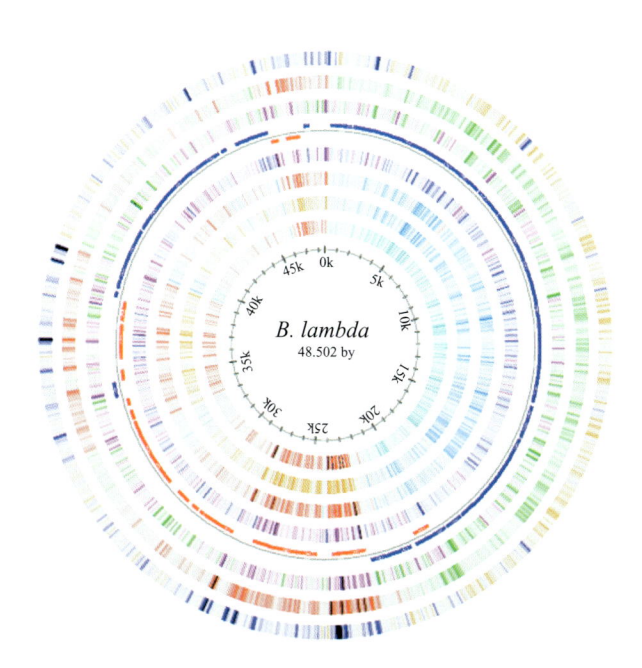

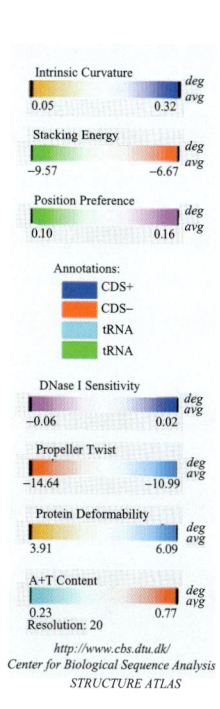

FIGURE 16.3. This "Structural atlas" provides predicted DNA structural features in the E. coli chromosome. The outermost wheel shows predicted DNA curvature based upon phased A-tracts position. The numbers on the inside of the innermost wheel is the position relative to zero minutes measured in millions of base pairs (Mbp). Taken from [44].

efficiency was observed to depend on operator sequence. Phage 434 repressor binds to 14-bp operator sites on DNA (ACAATATATATTGT) as a dimer at two ends bridged by central four base-pairs that are not in contact with the protein as we showed in (Figure 16.1 [1]). The bound phage 434 repressor bends and overtwists the center non-contacting four base-pairs. It was observed that a sequence variation in the bridging four base-pairs could change the binding affinity of the 434 repressor up to 50-fold (and 17-fold for phage 434 Cro protein) [45]. This significant 50-fold decrease in binding affinity is induced by changing the center 7·8 base-pair from T·A to C·G or G·C.

In general, with a detailed X-ray structure of the noncontacted region it would be possible to compute the strain that each basepair has compared to the expected structure of the undistorted Watson-Crick structure, and from this strain one could compute using the expression for the strain energy dU. The simplest physical way to understand how strain in the non-contacted regions could influence the binding affinity of a transcription factor is to break up the free energy change ΔG when a transcription factor binds to DNA into a (complicated) part due to all the electrostatic and specific ligand binding aspects, which may be by far the most important aspect to base-pair specific DNA recognition by transcription factors, and the elastic deformation terms. At a very simple level, the strained non-contacted DNA segment can be viewed as a series of N springs of varying spring constant k_i and varying amounts of generalized deformation [26]. The value of the (local) spring constant is set by values for the basepair dependent modulii E and G. In practice that is an impossibly difficult task at present given our uncertainly with how the modulii do vary with basepair composition. A simple way to start is to assume that the basepair springs add serially that there are no long range interactions amongst the basepairs, which is assuredly wrong. In the case of no long range interactions, the rigidity of the total spring simply becomes the weighted average of the basepair composition, f_{xy}, where f_{aa} corresponds to the fraction of basepairs where 2 A's are next to each other, etc. The bending rigidity B and the torsional rigidity C of a length of duplex DNA of length L becomes:

$$\frac{1}{C} = \frac{L}{I_P}\left(\frac{f_{aa}}{G_{aa}} + \frac{f_{ag}}{G_{ag}} + \frac{f_{gg}}{G_{gg}}\right)$$
$$\frac{1}{B} = \frac{L}{I_A}\left(\frac{f_{aa}}{E_{aa}} + \frac{f_{ag}}{E_{ag}} + \frac{f_{gg}}{E_{gg}}\right) \quad (16.18)$$

where I_A is the surface moment of inertia of the object, I_P the polar moment of inertia and E_{xy} and G_{xy} are the sequence specific Young's and shear elastic constants discussed above.

We live on a rather cold planet, and it is surprising how little strain one needs to impose on a section of DNA before the elastic energy terms become similar in magnitude to $k_B T = 1/40$ eV for $T = 300$ K. We can assume that if the 434 repressor has a fixed conformation which demands that a certain amount of strain energy U be invested in the bound complex. If the sequence denoted by n changes that binding energy by an amount ΔU then the binding coefficient K_n will be decreased by the Boltzmann factor relative to the no strain case K_0:

$$K_n = K_0 exp[-\Delta U/k_B T] \quad (16.19)$$

In the case of Ptashne's work, the relative binding coefficients of different basepair sequences was measured by varying the non-contacted region sequences, so we instead calculate the ratio of binding coefficients for different sequences 1 and 2:

$$log\left[\frac{K_0}{K_n}\right] = \frac{U_0 - U_n}{k_B T} \quad (16.20)$$

Since, as we remarked above, we live on a cold planet it doesn't take much strain for these factors to become significant and it is easiest to simply find the twist angle ϕ and bending radius R which can fit the observed binding coefficients and ask if they are reasonable. The following table is the analysis of the 434 repressor non-contacted sequences studied by Ptashne and his colleagues [45] in this elastic stain picture.

We have listed the sequences as enumerated by Ptashne et al., not in a logical sequence of effective binding constants. At the simplest level one can check to see if the strain model explains the data by simply looking for a linear correlation between the changes in the bending or the twisting elastic coefficients $B_n - B_0$ or $C_n - C_0$ with the log of the ratio of the measured binding constants $log(K_n/K_1)$. We plot in Figure 16.4 the results plotting ΔC vs $log(K_0/K_n)$.

TABLE 16.3. Repressor Binding Coefficients

Name	Sequence	f_{AA}	f_{AG}	f_{GG}	C	B	(K_0/K_n)	$\log(K_0/K_n)$
0	ATAT	1	0	0	0.37	1.15	1	0
1	TTAA	1	0	0	0.37	1.15	1.5	0.4
2	CTAG	1/3	2/3	0	0.57	1.8	5	1.6
3	GTAC	1/3	2/3	0	0.57	1.8	5	1.6
4	GTAT	2/3	1/3	0	0.45	1.4	2.5	0.9
5	AATT	1	0	0	0.37	1.15	1	0
6	AAAT	1	0	0	0.37	1.15	0.3	−1.2
7	ACGT	0	2/3	1/3	0.9	2.8	50	3.9
8	AGCT	0	2/3	1/3	0.9	2.8	50	3.9
9	AGAT	1/3	2/3	0	0.57	1.8	7	1.9

It is also possible to curve fit the experimental binding coefficents to the predicted ones by assuming either a bending radius R in the strained repressor complex or a twist angle ϕ. The same linear fit as observed in Figure 16.4 can be made by asssuming a pure twist angle of 32° along the 4 basepair uncontacted region or a bend radius of 33 Å. Of course, not all agree with this kind of analysis. Torsional measurements using time-resolved fluorescence polarization anisotropy techniques have lead to conclusions, on the other hand, that the torsional constant variation of the operators is not the cause of the phage 434 repressor protein binding affinity change [21], although it is hard to imagine that the agreement of the binding constants with the predicted values shown in Figure 16.4 is just by chance. This issue has never been resolved unfortunately as the various authors have gone their various scientific ways.

A further test of the influence of the sequence dependence of DNA elasticity can be found by studying the influence of induced bends in DNA by monomeric DNA binding proteins. This field of studies assumes that in order to bind functionally to the DNA certain proteins must induce a bend in the DNA. This field of studies was started by the Austin

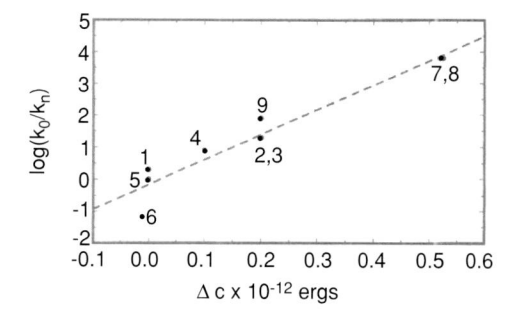

FIGURE 16.4. Relationship (calculation and experiments) between changes in the torsional constant (ΔC) and protein binding constant K_0/K_n. Numbers in the plot represent different four base sequences in the center of operator: 0-ATAT, 1-TTAA, 2-CTAG, 3-GTAC, 4-GTAT, 5-AATT, 6-AAAT, 7-ACGT, 8-AGCT and 9-AGAT. Adapted from [26].

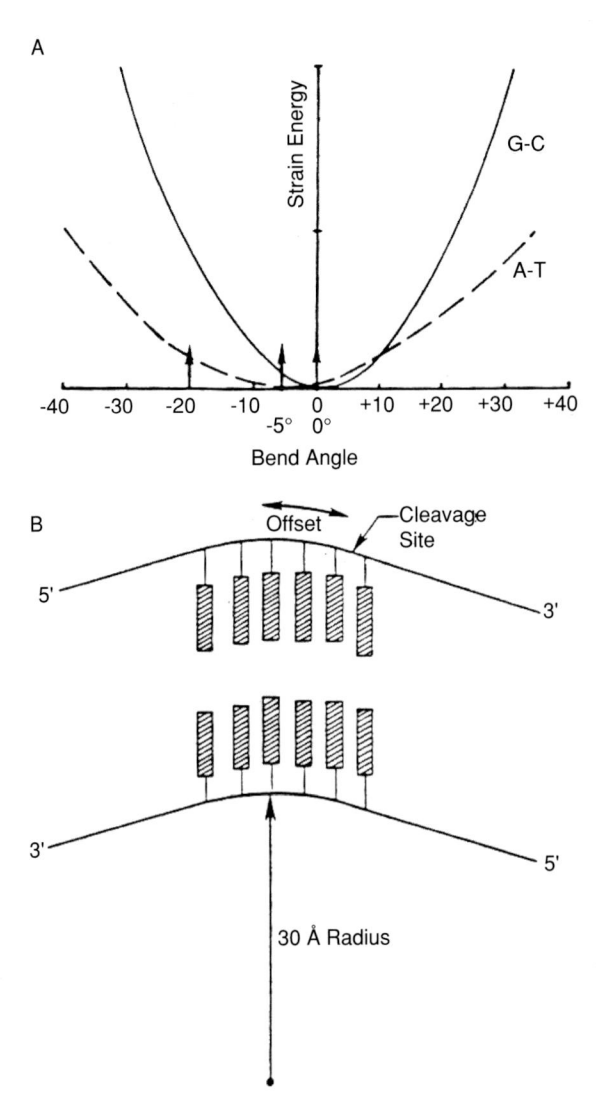

FIGURE 16.5. (A). Potential energy diagram for a bend sequence of DNA as a function of bend angle for a sequence which is completely random GC in composition or random AT. (B). The expected bend angle of 30° put into the double helix and the projected cutting site of DNase I. Taken from [27].

lab when they analyzed the cutting rates of the protein DNase I as a function of basepair sequence [27]. DNase I has cutting rates which vary over a range of 500 depending on the basepair sequence, and is known to bend DNA through an angle of about 30°. The natural inclination is to again assume that variations in the elastic energy of deformation gives rise to changes in the binding constant and hence changes in the cutting rates due to simple aspects of residence times. Fig. 16.5 taken from [27] shows how this combination of bending of the DNA by the DNase I and the strain energy caused by this bending of the helix can with good accuracy fit the known DNase I cutting rates in Fig. 16.6.

SEQUENCE MATTERS: THE INFLUENCE OF BASEPAIR SEQUENCE 491

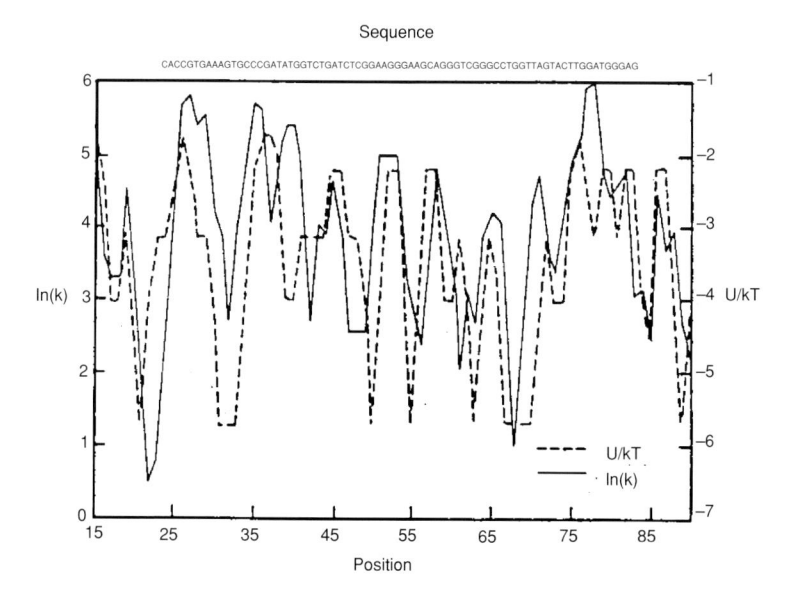

FIGURE 16.6. The observed (solid line) and predicted (dashed line) of DNase I cutting rates for the sequence shown at the top of the Figure. Taken from [27].

16.5.2. Nucleosome Formation: Sequence and Temperature Dependence

The basic flexibility rules we have discussed above break down for A tracts because of the cross-coupling of the bifurcated hydrogen bond unique to A-tracts. These cross-bonds can enhance the rigidity of the helix, at least at temperatures cold enough for the bonds to form. In fact, it has long been believed that poly(dA)poly(dT) duplex will not form nucleosome core particles [51] [48] [30]. However, it is now clear that poly(dA)·poly(dT) at temperatures below approximately 30°C exists in a conformational isomer that is different from normal B-type helix [24] [8] [9]. Most nucleosome reconstitution experiments are done at temperatures well below 30°C. It is logical to infer that if nucleosome reconstitution experiments are carried out at temperatures above the structural phase transition temperature of poly(dA)·poly(dT) that the homopolymer could in fact successfully reconstitute into nucleosomes.

There is evidence in the literature from three different groups of investigators using sequences containing large A_n tracts that it is possible to reconstitute nucleosomes at 37°C using an exchange method from native donor nucleosome core particles [33] [47] [20]. It has also been found that poly(dA)·poly(dT) and kinetoplast DNA fragments with phased oligo-dA tracks have a unique temperature-dependent structure below the global melting point of the duplex [8]. The broad temperature range over which the structure changes can be monitored by circular dichroism spectroscopy [24] [8], gel mobility change [8], and transient optical techniques [8]. Chan et al. [9] measured the thermodynamic heat capacity change associated with this pre-global melting transition using a synthetic 45 base pair DNA with 4 phased tracts of $(dA)_5$ showing that a true structural phase change occurs.

It seems plausible from the results discussed above that the unusual low temperature structure formed in A_n tracts could be responsible for the inability of the homopolymer to form nucleosomes, and that the temperature dependent structure of A_n tracts will be reflected in a similar temperature dependent nucleosome reconstitution profile.

16.5.2.1. Experimental Results of Nucleosome Reconstitution   The fractionated homopolymer d(A)·d(T) and and alternating copolymer d(AT)·d(TA) fractions were prepared by a sonication process in order to make sufficiently short fragments (200 bp) to ensure mononucleosome formation. The sonication approach gives inherently a distribution of length fragments rather than a sharp monodispersed band like restriction DNA fragments. We have chosen three length fragments for the reconstitution experiments: 180 ± 20 bp, 200 ± 20 bp, and 250 ± 30 bp. The reconstitution protocol is a modifcation of the procedure described in Ref. [37]. Poly(dA)·poly(dT) was purchased from Boehringer Mannheim, suspended in 10 X TE buffer (TE buffer is 10 mM Tris/HCl, 1 mM EDTA, pH 7.6), and incubated at 20°C below the melting temperature of 70°C to optimize duplex hybridization. The DNA was then sonicated for 15 min at 4°C in order to break the polymers into shorter fragments. The resulting material was deproteinized with phenol, ethanol precipitated, resuspended in TE buffer, then length fractionated by a 1-meter Phamacia Sephacryl S500 column eluting with 50 mM Tris/HCl, 1 mM EDTA, pH 7.6. Aliqouts were taken to provide fractions, the mean length and width of each fraction was determined by electrophoresis on a 1.8% agarose gel. The fractionated samples were stored in 10 mM Tris, pH 7.5, 10 μM EDTA, at concentrations in the range of 0.25–0.45 mg/ml. 146 bp DNA was isolated from chicken erythrocyte core particles which had been digested by micrococcal nuclease to give DNA fragments with natural random sequences [35]. The 146 bp samples contained over 90% 146 bp fragments and the balance shorter fragments. The 146 bp sample was dissolved in 50 mM Tris, pH 8, 1 mM EDTA, at a concentration of 0.35 mg/ml. Histone octamers were isolated from H1-depleted chicken erythrocyte chromatin by hydroxyapatite dissociation method [2] and kept in 50 mM Tris, pH 8 and 2M NaCl at a concentration of 0.62 mg/ml.

The mono-nucleosomes were reconstituted by first mixing the octamer with DNA (calculated with 146 bp molar equivalent in concentration) at about 1.0:1.1 ratio and 1 M NaCl at 40, 24 and 4°C. An aliquot of the same volume of buffer (50 mM Tris, pH 8, 1 mM EDTA) was added to the sample in 4 steps every 1–2 Hr, to dilute the NaCl concentration down to 0.6 M. Then, 10–15 μl of sterilized mineral oil was added only to the top layer of the 40°C sample to prevent vaporization within the tube. Each sample was kept at 0.6 M NaCl and its corresponding temperature over night. The last volume of buffer which diluted NaCl to 0.1 M was pre-equilibrated at each temperature before adding to the sample. The final volume of each sample was about 50 to 70 μl. After at least one hour of incubation time, the samples and controls were analyzed by 5% polyacrylamide gels (1.5 mm thick, 14 cm long, 2–3 Hr of run time at 120–140 Volt DC) in TBE buffer, pre-equilibrated at 40, 24 and/or 4°C. The temperature of the gels during the runs could heat up by about 2 or 3°C. The gels were soaked in 0.2 μg/ml ethidium bromide solutions for 30 min after electrophoresis and the UV excited fluorescence was recorded on Polaroid instant films and negatives.

The results of these reconstitution experiments support the idea that at physiological temperatures poly(dA)·poly(dT) does indeed change its flexibility. The results are shown in Fig. 16.7. We summarize the results here. Experiments were done at three different

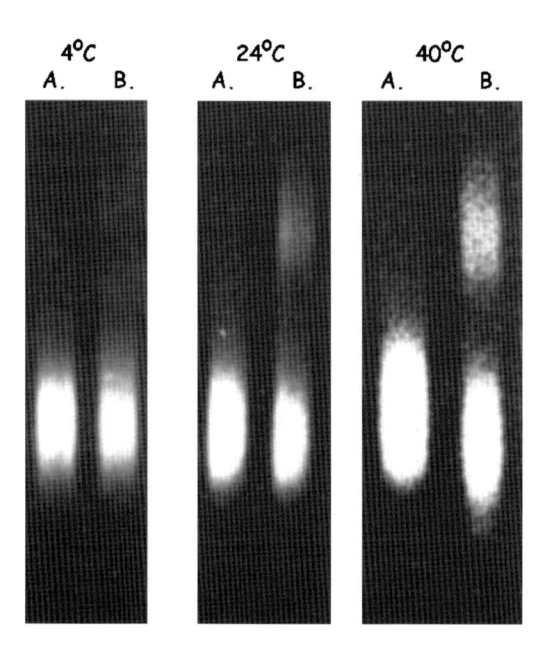

FIGURE 16.7. Gel analysis of nucleosome reconstitution experiments. Lane A for each temperature is the 200 bp poly(dA)·poly(dT) sample in the absence of histones, and lane B is the 200 bp poly(dA)·poly(dT), under the incubation conditions outlined in the text.

temperatures: 4°C, 24°C and 40°C. Although the disproportionation of duplex into triplex is a possibility at the high salt concentrations used in the reconstitution procedure, the maximum temperature of the reconstitution procedure (40°C) is well below the melting point of either duplex (dA)·(dT) or triplex (dT)·(dA)·(dT) [43] and so the duplex will be stable during the reconstitution procedure.

The gel results shown clearly show that poly(dA)·poly(dT) does not reconstitute to form nucleosomes at 4°C, but an increasing fraction of reconstituted nucleosomes is seen at 24°C and 40°C. Control experiments with chicken erythrocyte fragments show approximately constant reconstitution at all three temperatures (data not shown). Interestingly, we also found that the observed reconstitution efficiency depended strongly on the temperature of gel used to assay for reconstitution: the 40°C incubated poly(dA)·poly(dT) reconstitution experiment analyzed by a 4°C gel run showed substantially less reconstitution than the 40°C run gel, indicating that the low temperature DNA phase can reform even in the nucleosome and result in unfolding of previously reconstituted nucleosomes.

There are several ramifications to this work over and above the fact that poly(dA)·poly(dT) can in fact reconstitute into nucleosomes if reconstitution is done at temperatures above the structural phase transition temperature of the homopolymer. Most importantly, we note that the structural transition temperature of poly(dA)·poly(dT) is approximately 35°C, and this implies that A_n tract structural changes can have an important, variable impact on chromatin structural integrity at physiological conditions. The fact that reconstituted (dA)·(dT) nucleosomes unfold when cooled emphasizes this point.

CONCLUSIONS

The field of sequence-depend properties of DNA elasticity has been slow to develop. In spite of some fairly convincing early experiments indicating that sequence matters not only in the Watson-Crick basepair sense but also in terms of modulating the structure and flexibility of the double helix, doubts cast on the subject seemed to stop work in this area. However, as more and more structures of protein-DNA complexes are solved, it has become clear that very the double helix is physically distorted by binding of proteins, giving rise to a significant elastic energy term to the free energy of DNA-protein interactions. As a consequence of this, there has arisen a significant theoretical effort to compute in very general terms how these base-pair sequence effects, at many different levels of neighboring interactions and phased nature of position in the helix, can influence the interaction between proteins that distort the double helix and their binding constants. Further, it became clear that sequence matters not only in elastic properties but also in static conformations as long as the basepair sequences were phased to the helix repeat.

Many questions still remain unanswered; for example, it would be fascinating to observe DNA bending (protein induced or intrinsic) directly using single molecule fluorescent techniques, such as fluorescence resonance energy transfer technique [52]. Now with the availability of optical 1.5-nm localization ability [50], it should be interesting to examine bent protein-DNA complex using fluorescent-fusion-protein and DNA dye. As the ability to confine the DNA in nanostructures, we will be able to add an additional strain to the DNA molecule which will compete with the induced strain by binding of proteins, and we will be able to understand at a deeper level how the basepair sequence dependent rigidity of the double helix influences the way that proteins interact with the helix.

This work was supported by grants from DARPA (MDA972-00-1-0031), NIH (HG01506), NSF Nanobiology Technology Center (BSCECS9876771, the State of New Jersey (NJCST 99-100-082-2042-007) and US Genomics.

REFERENCES

[1] J.E. Anderson, M. Ptashne, and S.C. Harrison. Structure of the repressor-operator complex of bacteriophage 434. *Nature*, 326:846, 1987.
[2] K.S. Bloom and J.N. Anderson. Fractionation and characterization of chromosomal proteins by the hydroxyapatite dissociation method. *J. Biol. Chem.*, 253:4446, 1978.
[3] V.A. Bloomfield, D.M. Crothers, and I. Tinoco. *Nucleic Acids: Structure, Properties, and Functions*: J. Stiefel (ed.), University Science Books, Sausalito, CA, 2000.
[4] K.J. Breslauer, R. Frank, H. Blocker, and L.A. Marky. Predicting DNA Duplex Stability from the Base Sequence. *Proceedings of the National Academy of Sciences of the United States of America*. Vol. 83, p. 3746, 1986.
[5] C.R. Calladine and H.R. Drew. Principles of Sequence-Dependent Flexure of DNA. *J. Mol. Biol.*, 192:907, 1986.
[6] Cao Han, Yua Zhaoning, Wang Jian, Chen Erli, Wua Wei, O. Jonas Tegenfeldt, H. Robert Austin, and Y. Stephen Chou. Fabrication of 10 nm enclosed nanofluidic channels. *Appl. Phys. Lett.*, 81:174, 2002.
[7] Cao Han, O. Jonas Tegenfeldt, H. Robert Austin, and Y. Stephen Chou. Gradient nanostructures for interfacing microfluidics and nanofluidics. *Appl. Phys. Lett.*, 81:3058, 2002.
[8] S.S. Chan, K.J. Breslauer, M.E. Hogan, D.J. Kessler, R.H. Austin, J. Ojemann, J.M. Passner, and N.C. Wiles. Physical studies of DNA premelting equilibria in duplexes with and without homo dA·dT Tracts: correlations with DNA binding. *Biochemistry*, 29:6161, 1990.

[9] S.S. Chan, K.J. Breslauer, R.H. Austin, and M.E. Hogan. Thermodynamics and premelting conformational changes of phased (dA)$_5$ tracts. *Biochemistry*, 32:11776, 1993.
[10] S.S. Chan, R.H. Austin, I. Mukerji, and T.G. Spiro. Temperature-dependent ultraviolet resonance Raman spectroscopy of the premelting state of dA center dot dT DNA. *Biophys. J.*, 72:1512, 1997.
[11] S.Y. Chou and P.R. Krauss. Imprint lithography with sub-10 nm feature size and high throughput. *Microelect. Eng.*, 35:237, 1997.
[12] *CRC Handbook of Chemistry and Physics*, 84th Edn., CRC Press, Boca Raton, FL, USA.
[13] N.R. Cozarelli, T. Boles, and J. White. *Topology and its Biological Effects*. Cold Spring Harbor, 1990.
[14] D.M. Crothers, T.E. Haran, and J.G. Nadeau. Intrinsically Bent DNA. *J. Biol. Chem.*, 265:7093, 1990.
[15] R.E. Dickerson and T.K. Chiu. Helix bending as a factor in protein/DNA recognition. *Biopolymers*, 44:361, 1997.
[16] R.E. Dickerson. Definitions and nomenclature of nucleic acid structure parameters. *Euro. Mol. Biol. J.*, 8:1, 1989.
[17] R.E. Dickerson. DNA bending: the prevalence of kinkiness and the virtues of normality. *Nuc. Acids Res.*, 265:1906, 1998.
[18] S. Diekmann. Definitions and nomenclature of nucleic acid structure parameters. *J. Mol. Biol.*, 205:787, 1989.
[19] M. Doi and S.F. Edwards. Theory of Polymer Dynamics, Academic Press, NY, 1977.
[20] K.R. Fox. Wrapping of genomic polydA·polydT tracts around nucleosome core particles. *Nuc. Acids Res.*, 20:1235, 1992.
[21] B.S. Fujimoto and J.M. Schurr. Dependence of the torsional rigidity of DNA on base composition. *Nature*, 344:175, 1990.
[22] R.E. Goldstein, T.R. Powers, and C.H. Wiggins. Viscous nonlinear dynamics of twist and writhe. *Phys. Rev. Lett.*, 80:5232, 1998.
[23] P.J. Hagerman. Flexibility of DNA. *Ann. Rev. Biophys. Biophys. Chem.*, 17:265, 1988.
[24] J.E. Herrera and J.B. Chaires. A premelting conformational transition in Poly(dA)-Poly(dT) coupled to daunomycin binding. *Biochemistry*, 26:1993, 1989.
[25] H. Hogan, J. LeGrange, and B. Austin. Dependence of DNA helix flexibility on base composition. *Nature*, 304:752, 1983.
[26] M.E. Hogan and R.H. Austin. Importance of DNA stiffness in protein-DNA binding specificity. *Nature*, 329:263, 1987.
[27] M.E. Hogan, M.W. Roberson and R.H. Austin. DNA flexibility variation may dominate DNase I cleavage. *Proc. Natl. Acad. Sci. U.S.A.*, 86:9273, 1989.
[28] F. Jacob and J. Monod. Genetic regulatory mechanisms in synthesis of proteins. *J. Mol. Biol.*, 3:318, 1961.
[29] L.J. Jensen, C. Friis, and D.W. Ussery. Three views of microbial genomes. *Res. Microbiol.*, 150:773, 1999.
[30] G.R. Kunkel and H.G. Martinson. Nucleosomes will not form on double-stranded RNA or over Poly(dA)·poly(dT) tracts in recombinant DNA. *Nucleic Acids Res.*, 6:6869, 1981.
[31] M. Lewis et al. Crystal structure of the lactose operon repressor and its complexes with DNA and inducer. *Science*, 271:1247, 1996.
[32] E.M. Lifshitz, L.D.L. *Theory of Elasticity*, Pergamon Press, New York NY, 1981.
[33] R. Losa, S. Omari, and F. Thoma. Poly(dA)·poly(dT) rich sequences are not sufficient to exclude nucleosome formation in a constitutive yeast promoter. *Nucleic Acids Res.*, 18:3495, 1990.
[34] N. Luscombe, S.E. Austin, H.M. Berman, and J.M. Thornton. An overview of the structures of protein-DNA complexes. *Gen. Biol.* 1, 1 (2000).
[35] L.C. Lutter. Kinetic analysis of deoxyribonuclease I cleavages in the nucleosome core: evidence for a DNA superhelix. *J. Mol. Biol.*, 124: 391, 1978.
[36] N.L. Marky and G.S. Manning. *Biopolymers*, 31:1557, 1991.
[37] K.W. Marvin, P. Yau, and E.M. Bradbury. Isolation and characterization of acetylated histones H3 and H4 and their assembly into nucleosomes. *J. Biol. Chem.*, 265:19839, 1990.
[38] K. McAteer and M.A. Kennedy. NMR evidence for base dynamics at all TpA steps in DNA. *J. Biomol. Struct. Dyn.*, 17:1001, 2000.
[39] K. Nadassy, I. Tomás-Oliveira, I. Alberts, J. Janin, and S.J. Wodak. Standard atomic volumes in double-stranded DNA and packing in protein–DNA interfaces. *Nucleic Acids Res.*, 29:3362–3376, 2001.
[40] T.M. Okonogi et al. Sequence-dependent dynamics of duplex DNA: the applicability of a dinucleotide model. *Biophys. J.*, 83:3446, 2002.

[41] K. Wilma Olson, A. Andrey Gorin, Lu, Xiang-Jun, M. Lynette Hock, and B. Victor Zhurkin. DNA Sequence-dependent deformability deduced from protein-DNA crystal complexes. *Proc. Natl. Acad. Sci., U.S.A.*, 95:11163, 1998.

[42] K. Wilma Olson and B. Victor Zhurkin. Modeling DNA deformations. *Curr. Opin. Struct. Biol.*, 10:286, 2000.

[43] Y.W. Park and K.J. Breslauer, Drug binding to higher ordered DNA structures: netropsin complexation with a nuclei acid triple helix. *Proc. Natl. Acad. Sci. U.S.A.*, 89:6653, 1992.

[44] G.A. Pedersen, L.J. Jensen, K.E. Nelson, S. Brunak, and D.W. Usssery. A DNA structural atlas for escherichia coli. *J. Molec. Biol.*, 299:706–709, 2000.

[45] G.B. Koudelka, S.C. Harrison, and M. Ptashne. Effect of noncontacted bases on the affinity of 434 operator for 434 repressor and Cro. *Nature*, 326:886, 1987.

[46] M. Ptashne. *Genetic Switch: Phage Lambda and Higher Organisms*. Blackwell Science, Malden, MA USA, 1992.

[47] H.L. Puhl, S.R. Gudibande, and M.J. Behe. Poly[d(A·T)] and other syntheitc polydeoxynucleotides containing oligoadenosine tracts form nucleosomes easily. *J. Mol. Biol.*, 222:1149, 1991.

[48] D. Rhodes. Nucleosome cores reconstituted from poly(dA-dT) and the octamer of histones. *Nucleic Acids Res.*, 6:1805, 1979.

[49] A. Scipioni, C. Anselmi, G. Zuccheri, B. Samori, and P.D. Santis. Sequence-dependent DNA curvature and flexibility from scanning force microscopy images. *Biophys. J.*, 83:2408, 2002.

[50] A. Yildiz et al. Myosin V Walks hand-over-hand: Single fluorophore imaging with 1.5-nm localization. *Science*, 300:2061, 2003.

[51] R.T. Simpson and P. Kunzler. Chromatin and core particles formed from the inner histones and synthetic polydeoxyribonucleotides of defined sequence. *Nucleic Acids Res.*, 6:1387, 1979.

[52] K. Steinmetzer, J. Behlke, S. Brantl, and M. Lorenz. CopR binds and bends its target DNA: a footprinting and fluorescence resonance energy transfer study. *Nucleic Acids Res.*, 30:2052, 2002.

[53] D. Strahs and M. Brenowitz. DNA conformational-changes associated with the cooperative binding of CI-repressor of bacteriophage-lambda to O-R. *J. Mol. Biol.*, 244:494, 1994.

[54] D. Szwajkajzer and K. Breslauer. The influence of sequence on the thermodynamics of DNA melting and ligand-binding properties for a family of octameric duplexes. *Biophys. J.*, 64:A281, 1993.

[55] R.H. Austin, J. Tegenfeldt, H. Cao, S. Chu, and E.C. Cox. Scanning the controls: genomics and nanotechnology. *IEEE Transac. Nanotechnol.*, 1:12, 2002.

[56] O. Tegenfeldt Jonas, Prinz Christelle, Cao Han, Chou Steven, W. Reisner Walter, Riehn Robert, Mei Wang Yan, C. Cox Edward. C. Sturm James, Silberzan Pascal, and H. Austin Robert. The dynamics of genomic-length DNA molecules in 100-nm channels. *Proc. Natl. Acad. Sc. U.S.A.*, 101:10979, 2004.

[57] J.D. Watson and F.H.C. Crick. Molecular structure of nucleic acids: a structure for deoxyribose nucleid acid. *Nature*, 171:737, 1953.

17

Engineered Ribozymes: Efficient Tools for Molecular Gene Therapy and Gene Discovery

Maki Shiota[1], Makoto Miyagishi[1,2], and Kazunari Taira[1,2]

[1]Department of Chemistry and Biotechnology, School of Engineering, The University of Tokyo, 7-3-1 Hongo, Tokyo 113-8656, Japan
[2]Gene Function Research Center, National Institute of Advanced Industrial Science and Technology (AIST), Central 4, 1-1-1 Higashi, Tsukuba Science City 305-8562, Japan

17.1. INTRODUCTION

Ribozymes are catalytic RNA molecules that cleave RNAs with high specificity. Hammerhead ribozymes are small and particularly versatile catalytic RNA molecules that cleave RNAs at specific sites (Figure 17.1A, left). The rapidly developing field of RNA catalysis is of particular current interest not only because of the intrinsic catalytic properties of ribozymes but also because of the potential utility of ribozymes as therapeutic agents and specific regulators of gene expression [16, 17, 45, 88, 89, 90, 105]. However, despite extensive efforts, the efficiency of ribozyme *in vivo* has generally been too low to achieve the desired biological effects. Unlike *in vitro*, conditions *in vivo* are very complex and many parameters must be taken into account, in particular conditions, the interactions of a ribozyme or its gene with intracellular proteins, which seem to be significant. Many modifications of and improvements in ribozymes, as well as methods for the introduction of ribozymes into cells, have been developed in attempts to exploit ribozymes *in vivo*.

Our group developed an efficient ribozyme-expression system that allows the efficient inactivation by a ribozyme of a specific gene *in vivo*, and we have applied this system to efforts at gene therapy and the functional analysis of genes. In this review, we describe our efforts to generate two novel ribozymes, namely, an allosterically controllable ribozyme and

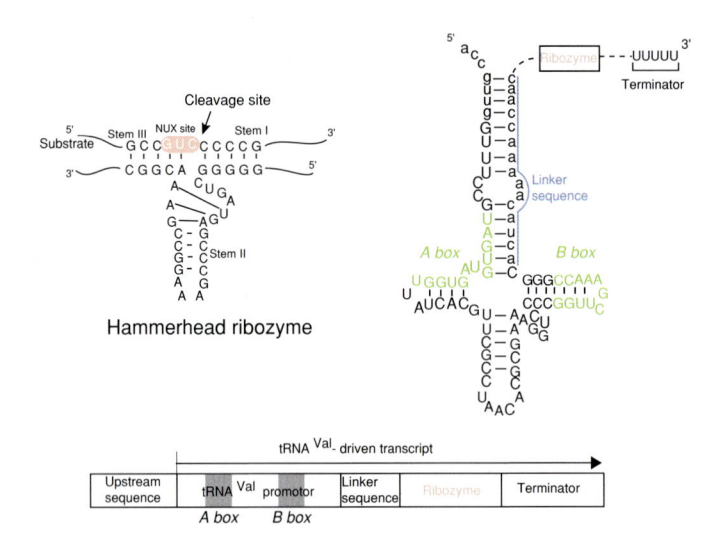

FIGURE 17.1. Secondary structure of the hammerhead ribozyme (on the left). The hammerhead ribozyme consists of a substrate-binding region (stems I and III) and a catalytic core with a stem-loop II region. Using a construct in which the substrate-binding region of the ribozyme is complementary to the target RNA, we can create molecular scissors that cleave RNA in a site-specific manner. When the catalytic core captures the catalytically indispensable Mg^{2+} ions, cleavage occurs only at a NUX triplet (N can be any base; X can be A, C or U). tRNA^{Val}-driven transcript (on the right). RNA polymerase III recognizes a promoter that is located within the gene for the tRNA sequence being transcribed. The ribozyme is linked downstream of the partially modified human tRNA^{Val} through a linker, as shown below the secondary structures.

an RNA-protein hybrid ribozyme. We chose the name "maxizyme" for the allosterically controllable ribozyme, which has sensor arms that recognize sequences in its target mRNA. In the presence of such target sequences exclusively, it forms a cavity that can capture the catalytically indispensable Mg^{2+} ions that allow it to cleave the target mRNA [49–51, 101, 102]. In the RNA-protein hybrid ribozyme, the cleavage activity of a hammerhead ribozyme is coupled with the unwinding activity of an RNA helicase. This hybrid ribozyme can cleave its target mRNA extremely efficiently, regardless of the secondary structure of the target RNA [107, 113].

17.2. METHODS FOR THE INTRODUCTION OF RIBOZYMES INTO CELLS

If synthetic ribozymes are to work inside cells in the human body, they must be delivered to individual cells and access their targets without degradation. However, ribozymes are rapidly destroyed in the gastrointestinal tract and in the blood. In addition, ribozymes are nucleic acids with high molecular weights, and it is unlikely that ribozymes themselves can be taken up directly by cells.

For the administration of ribozymes in a therapeutic setting *in vivo*, ribozymes must be carried by vectors, which can efficiently infiltrate the phospholipid bilayer of cell membranes. Many virus-based vectors have been shown to have potential as vehicles for gene delivery *in vivo*, and these viruses include adenovirus, herpesvirus, retrovirus, adeno-associated virus, and lentivirus [24, 43, 57, 70, 74, 87, 116]. Rapid advances in viral vector technology have led not only to improved efficiency of introduction of ribozymes into cells but also to more precise control of transduction, such as tissue-specific transduction. The various viral vectors are derived from detoxified viruses, but their exploitation has been fraught with problems related to production, immunogenicity and safety [27, 40, 117]. In particular, it is difficult to prepare and purify many viral vectors.

Artificial non-viral vectors tha are currently being developed involve naked nucleic acids, either alone or in a variety of molecular conjugates with, for example, liposomes, polymers, and polypeptides [1, 10, 72, 86]. Compared to the use of viral vectors, the use of non-viral vectors is cost-effective and their synthesis is straightforward and safe. However, the efficiency of introduction of these vectors is sometimes relative low. It is unclear whether viral vectors or non-viral vectors are the best carriers of ribozymes into cells and different goals might be better met with one system or the other.

Available methods for the administration of ribozymes *in vivo* include direct administration of chemically synthesized ribozymes and the introduction of ribozymes via plasmids that carry a gene for the ribozymes, which is transcribed inside cells. In the former method, it is necessary to stabilize ribozymes against nucleolytic degradation by chemical modification, such as thio modification. However, such modifications can increase both toxicity and cost, and large amounts of ribozyme are needed to ensure that sufficient numbers of ribozymes reach the target cells. Thus, methods using plasmids have received the most attention because a stable form of DNA can be administered and the ribozyme can be synthesized constitutively inside cells.

Once a ribozyme has been introduced ito cells, the environment within the cells must be taken into account to ensure that the ribozyme will function effectively.

17.3. RIBOZYME EXPRESSION SYSTEMS

17.3.1. The pol III System

The efficacy of a ribozyme after it has been introduced via a plasmid into cells, depends on five factors: (1) the amount of ribozyme that is transcribed; (2) the stability of the ribozyme after transcription; (3) the subcellular localization of the ribozyme; (4) the activity of the ribozyme itself; and (5) the accessibility to the ribozyme of the target mRNA. Among these factors, the first three are critically controlled by the expression system.

When the gene for a ribozyme is introduced into a cell, it can be transcribed by the transcriptional machinery to produce active ribozymes. Initially, the RNA polymerase II system (pol II) was usually used for the expression of ribozymes. In this system, transcripts are automatically modified, and a cap and a poly(A) tail are added at the 5' end and the 3' end, respectively. Thus, the stability of transcripts is guaranteed and, moreover, the transcripts are exported from the nucleus to the cytoplasm as mature mRNAs. However, while this pol II expression system is suitable for the transcription of long RNAs (several hundred to several thousand bases), it is not particularly suitable for the transcription of short RNAs, such as ribozymes. Moreover, for accurate transcription by the pol II system, extra sequences must be added to an otherwise compact ribozyme. These extra sequences might decrease the ribozyme's activity by forcing it to assume a higher-order structure or to bind to cellular proteins. Thus, we have focused our attention on the RNA polymerase III (pol III) system, which is involved mainly in the transcription of short RNAs, such as tRNA [20]. The rate of transcription by pol III is two to three orders of magnitude higher than that by pol II [12]. In addition, shorter extra sequences are needed. Therefore, the pol III system is ideal for the expression of ribozymes. Among available pol III systems, we chose to focus on the tRNA transcription system. In the pol III system, the promoter is located within the tRNA sequence that is transcribed and, thus, it is inevitable that a portion of the tRNA becomes incorporated into the ribozyme. Since tRNAs are exported to the cytoplasm from the nucleus as mature tRNAs with trimmed 5' and 3' ends, it was postulated that tRNA-attached ribozymes would not be exported to the cytoplasm. However, as noted below, tRNA-attached ribozymes are exported efficiently to the cytoplasm in mammalian cells. Moreover, the additional tRNA sequence does not cause any loss of ribozyme activity but, rather, appears to have a positive effect on ribozyme activity [34, 35, 42].

17.3.2. Relationship Between the Higher-Order Structure of Ribozymes and their Activity

In our ribozyme-expression system, the ribozyme is linked downstream of a partially modified human tRNA^{Val} via a linker (Figure 17.2). The higher order structure of ribozymes obviously affects their stability and ribozymes with fewer exposed regions in a single-strand structure are more stable than others against intracellular nucleolytic degradation. In our system, we include a small stem-loop structure at the 3' end to promote stability. By contrast, if a double-stranded region is too long, it might be recognized by other nucleases. To prevent such problems, the double-stranded region corresponding to the linker is designed with a bulge.

In our ribozyme-expression system, the higher-order structure of the ribozyme is strongly affected by the length of the linker. To determine how the higher-order structure might affect the activity and stability of our ribozyme [42], we constructed three different types of expression system, in which ribozymes targeted to the same sequence (a sequence that is relatively strongly conserved in HIV-1) were linked to the promoter of the gene for the tRNA via linkers with different sequences (Figure 17.2A). When we predicted secondary structures by Zucker's method [120], we found that the ribozymes had secondary structures that were similar to those of tRNAs but that they had different structures in the substrate-binding region.

In ribozymes (Rz) 1 through 3 in Figure 17.2A, the degree of freedom of the substrate-binding region increases in that order (a single strand has a higher degree of freedom than a double strand and, thus, binds more efficiently to the substrate). For ribozymes, which are RNA enzymes, ease of binding to the substrate is an important determinant of activity, and we can predict that a higher degree of freedom of the substrate-binding region should result in higher activity. In experiments designed to test this prediction, the differents in activity *in vitro* against a short substrate were confirmed. We also examined the stability of the ribozymes in cells. Plasmids encoding each ribozyme were introduced into cells, and levels of ribozymes in cells were monitored by Northern hybridization. The most stable ribozyme, Rz2, was 26-fold more abundant than the most unstable ribozyme, Rz1, and also 5-fold more abundant than Rz3 (Figure 17.3A). It is unclear why these structures, which are so similar overall, have such very different stabilities. The differences might be due to differences in the extend to which each ribozyme in the cell is recognized by a degradative nuclease.

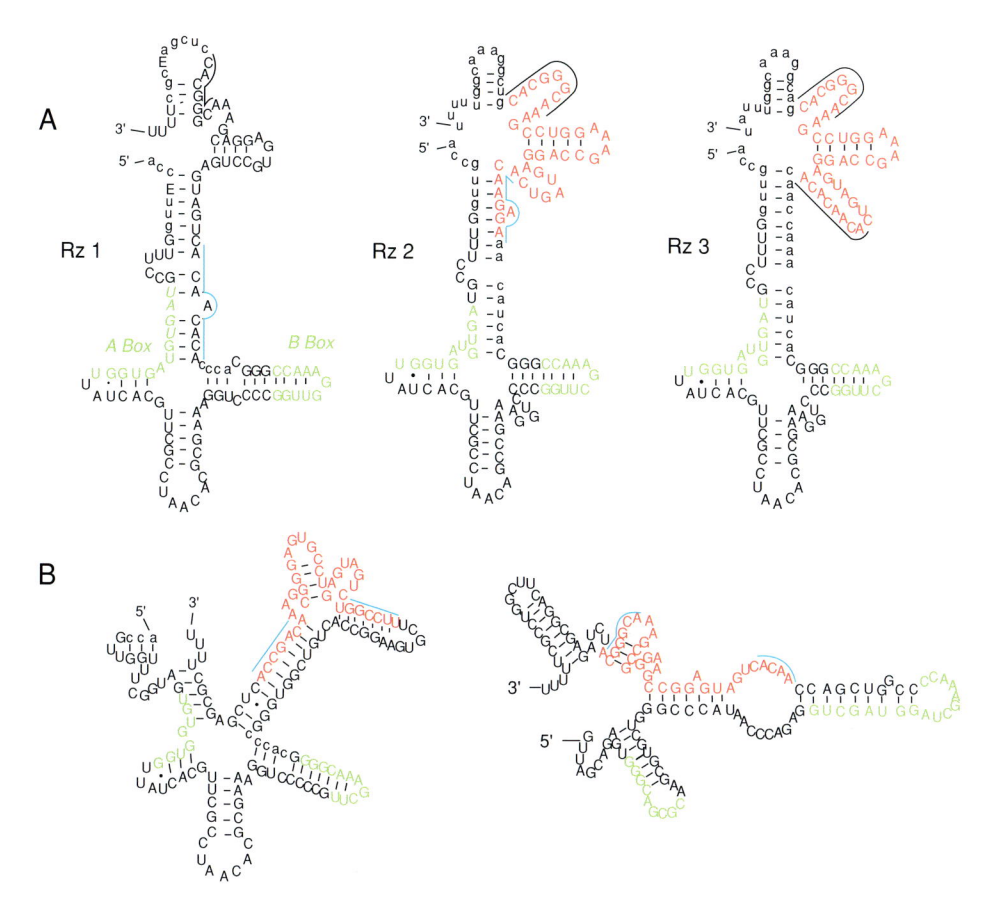

FIGURE 17.2. (A) Secondary structures of tRNA-Rz that were efficiently exported to the cytoplasm. Sequences in linker regions are indicated by lower-case letters. Ribozyme sequences are shown in red and substrate-binding sites are underlined in blue. (B) Secondary structures of tRNA-Rz that accumulated in the nucleus.

We identified the ribozyme that was the most active in the cell using a gene for luciferase as a reporter gene. Luciferase catalyzes a reaction that yields a chemiluminescent product [41]. We designed an assay in which the ribozyme was targeted to the 5'-untranslated region of the gene for luciferase, in cultured cells. But results suggested a correlation between the activity and the intracellular stability of the ribozyme. We also studied the activity of ribozymes against HIV in cultured cells. We infected cells that carried a ribozyme expression system with HIV-1 and then we monitored viral protein (p24) synthesis as an index of viral replication (Figure 17.3B). The results were similar to those obtained in the luciferase assay, confirming that intracellular stability is an important determinant of the efficacy of ribozymes in cells.

17.3.3. Subcellular Localization and Efficacy of Ribozymes

The subcellular localization of a ribozyme after transcription is another important factor that affects the ribozyme's activity [6, 33, 97]. The mRNA that is the ribozyme's target is

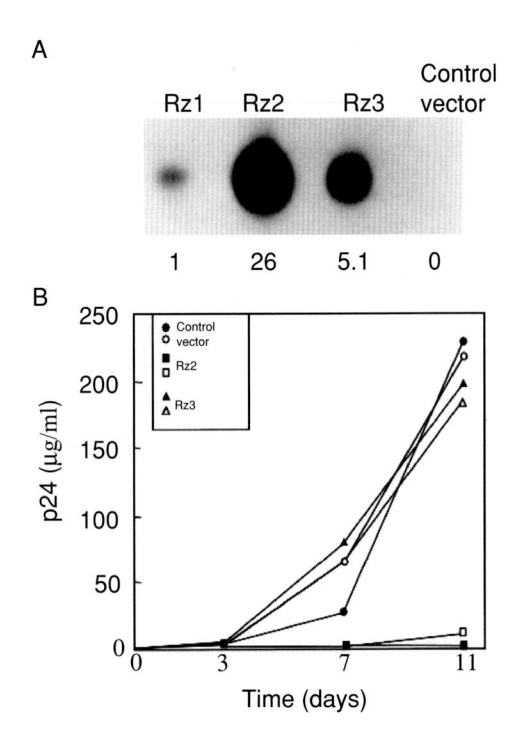

FIGURE 17.3. Stability of tRNA^{Val}-ribozymes in cultured cells. (A) Steady-state levels of expression of tRNA^{Val}-ribozymes. The photograph shows the result of Northern blotting analysis with the probes specific for the ribozyme (B) Inhibitory effects in cultured cells of tRNA^{Val}-driven ribozymes on the expression of p 24.

transcribed in the nucleus and, after splicing, it is transported into the cytoplasm, where it is translated into protein. Nuclear precursors to mature mRNAs (pre-mRNA) might be less accessible to ribozymes than mature cytoplasmic mRNAs because pre-mRNAs form complexes with heterogeneous nuclear proteins and small nuclear ribonuclear proteins and they interact with various RNA-binding proteins, for example, proteins involved in splicing and in the export of processed mRNAs. It is also likely that higher-order structures of mRNAs are disrupted more effectively in the cytoplasm than in the nucleus by various RNA helicases [113]. Thus, ribozymes and their target mRNAs should be colocalized in the cytoplasm if ribozymes are to be effective.

We found initially that tRNA^{Val}-driven ribozymes with high levels of activity were exported efficiently to the cytoplasm, while similarly expressed tRNA-ribozymes with low levels of activity accumulated in the nucleus [74]. Then we attempted systematically to identify the cellular compartment in which a ribozyme acts most effectively [33]. We designed several types of functional RNA targeted to the junction site of the chimeric *BCR-ABL* mRNA that causes chronic myelogenous leukemia (CML). CML occurs as a consequence of reciprocal chromosomal translocations that result in the formation of a fused *BCR-ABL* gene [71, 77]. To examine the correlations between nuclear localization and/or the transport of functional RNAs and their activity *in vivo*, we used two kinds of promoter, the promoter of the gene for tRNA^{Val} described above and a U6 promoter. Transcripts expressed under

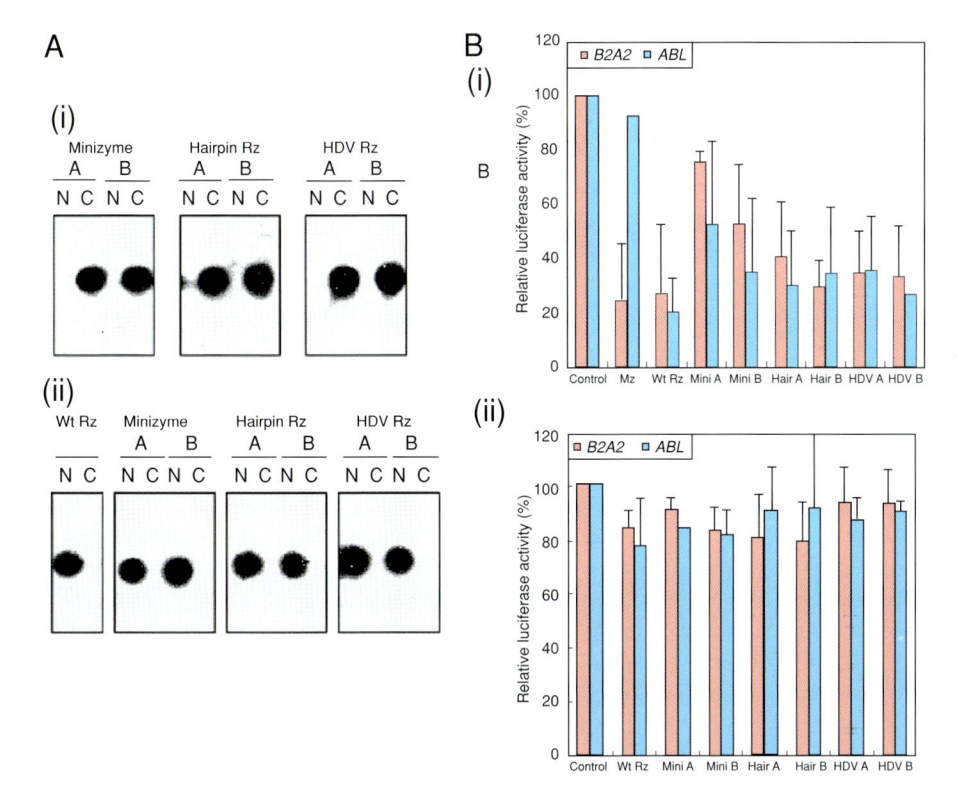

ENGINEERED RIBOZYMES: EFFICIENT TOOLS FOR MOLECULAR GENE THERAPY 503

FIGURE 17.4. (A) Nuclear localization of functional RNAs. The steady state levels of tRNA^{Val}-driven ribozymes (i) and U6-driven ribozymes (ii) and their localization are shown. Approximately the same levels of expression of functional RNAs from both promoters were observed. N, nuclear fraction; C, cytoplasm fraction. (B) Inhibitory effects in cultured cells of tRNA^{Val}-driven ribozymes (i) and U6-driven ribozymes (ii) on the expression of chimeric genes for *BCR-ABL*-luciferase and *ABL*-luciferase. B2A2 is consisting of exon 2 of *BCR* and exon 2 of *ABL*.

the control of these promoters are located in the cytoplasm and the nucleus, respectively [14, 78].

Figure 17.4A shows the steady-state levels of tRNA^{Val}-driven ribozymes and U6-driven ribozymes and their localization. We detected approximately the same level of expression of each functional RNA from each promoter and, without exception, tRNA^{Val}-driven ribozymes were localized in the cytoplasm and U6-driven ribozymes were localized in the nucleus. We then estimated the activities of the various functional RNAs in cultured cells. Figure 17.4B shows the inhibitory effects of tRNA^{Val}-driven ribozymes and U6-driven ribozymes on the expression, in cultured cells, of chimeric genes for *BCR-ABL*-luciferase and *ABL*-luciferase. As noted above, a decrease in luciferase activity indicated the cleavage of transcripts by ribozymes. Without exception, the tRNA^{Val}-driven ribozymes, which had been exported to the cytoplasm, had inhibitory effects, whereas U6-driven ribozymes, which had remained in the nucleus, were completely ineffective, despite the fact that both types of ribozyme were targeted to the identical site and both had similar activity *in vitro*. Thus, the cytoplasmic localization of tRNA-attached ribozymes is clearly critical for high-level intracellular activity.

17.3.4. Mechanism of the Export of tRNA-Ribozymes from the Nucleus to the Cytoplasm

In our hands, all tRNA^{Val}-ribozymes transcribed by our pol III expression system were located in the cytoplasm. However, there are reports that ribozymes transcribed by a similar system that also exploits a tRNA promoter accumulate in the nucleus [6, 23]. A ribozyme that was designed to target the same HIV sequence as ours ([6]; Figure 17.2B, right) was reported to be localized in the nucleus. In addition, the extend of inhibition of expression of the target mRNA by the ribozyme was substantially lower than we observed. We have also been able to design tRNA-type ribozymes (for a different application) that accumulate in the nucleus without being transported into the cytoplasm; the secondary structures of such a ribozyme is shown in Figure 17.2B (left).

When we compare the secondary structures of the ribozymes that are transported to the cytoplasm with those of ribozymes that accumulate in the nucleus, we can easily see that the structures are different (Figures 17.2A and 2B). In particular, the ribozyme that is transported to the cytoplasm (Figure 17.2A) assumes a cloverleaf structure that is similar but not identical to that of a tRNA, while the ribozyme that accumulate in the nucleus (Figure 17.2B) has a very different structure. These results suggest that a tRNA^{Val}-ribozyme can be transported to the cytoplasm only when the structure of the tRNA^{Val}-ribozyme resembles that of a tRNA.

Rapid progress has been made in efforts to understand the mechanism involved in the export of tRNAs to the cytoplasm [3, 4, 25, 46, 63, 65]. The transport of tRNAs requires a tRNA-binding protein called exportin-t (Xpo-t) and the Ran GTPase, and rapid transport requires the hydrolysis of GTP. It appears that only mature tRNAs, with accurately trimmed 5' and 3' ends and an attached 3' CCA end are recognized by Xpo-t [3, 4, 46, 63, 65]. Aminoacylation of each tRNA also appears to be critical for the export of tRNAs from the nucleus to the cytoplasm in *Xenopus* oocytes [65] and in yeast [25]. In *Xenopus* oocytes, immature tRNAs with several extra nucleotides at the 3' end are not recognized by Xpo-t and, as a result, they are not exported to the cytoplasm. This phenomenon suggests the existence of a proofreading mechanism in cells whereby only tRNAs that are usable in the cytoplasm, with mature 5' and 3' ends, can be exported to the cytoplasm. However, in our studies of tRNA^{Val}-ribozymes, which might be considered equivalent to a kind of immature tRNA because of the extra sequences at their 3' ends, we found that such ribozymes were efficiently exported to the cytoplasm in mammalian cells [49–54, 42, 33, 35, 111].

We investigated the discrepancy between the reported observations that led to the proposal of the existence of a proofreading mechanism and our own observations of the efficient export to the cytoplasm of tRNA^{Val}-ribozymes [55]. We first considered the possible existence of an alternative pathway for the export of tRNAs. The existence of an additional pathway for tRNA export had already been suggested in yeast and we can assume that tRNA export is essential for cell survival. However, contrary to our expectations, we found evidence to suggest that Xpo-t was probably involved in the transport of tRNA^{Val}-ribozymes in somatic cells and, moreover, that a mechanism similar to that for the recognition of the tertiary structure of tRNAs was involved in the interaction of Xpo-t with tRNA^{Val}-ribozymes [54]. Nevertheless, in *Xenopus* oocytes, tRNA-attached ribozymes were not exported to the cytoplasm, as we might have predicted from the proofreading hypothesis. Further investigations revealed that the Xpo-t/RanGTP complex did not interact with tRNA-attached ribozymes in oocytes even though we detected such an interaction *in vitro* and, more importantly, in several lines of somatic cells. These findings hinted at the presence of some kind of inhibitor in *Xenopus* oocytes rather than suggesting the involvement of an alternative pathway in somatic cells. We found subsequently that a nuclear extract prepared from *Xenopus* oocytes did, indeed, strongly inhibit the export of tRNA-attached ribozymes in somatic cells, and this observation suggests the existence of a strong inhibitor(s) in *Xenopus* oocytes specifically.

It seems likely that the export of tRNAs in *Xenopus* oocytes is subject to a special kind of regulation. Moreover, the proofreading mechanism that is operative in *Xenopus* oocytes seems to involve a specific inhibitor(s) that appears specifically to recognize immature tRNAs and, thus, structures such as tRNA-attached ribozyme. In somatic cells, when the choice of linker and ribozyme sequence is made appropriately, a tRNA-attached ribozyme seems to be recognized as a mature tRNA by Xpo-t and to be exported by Xpo-t to the cytoplasm.

17.4. RNA-PROTEIN HYBRID RIBOZYMES

17.4.1. *Accessibility to Ribozymes of their Target mRNAs*

If ribozymes are to function effectively inside cells, high-level expression, intracellular stability, and efficient export to the cytoplasm are essential. However, even ribozymes that have been improved in each of these respects are sometimes ineffective, probably because they are unable to reach their target. It is likely that the rate-limiting step *in vivo* for the cleavage by a ribozyme of a phosphodiester bond is the association and annealing of the ribozyme with its target site [33].

To overcome the problem of accessibility, computer-generated predictions of secondary structure are typically used to identify targets that are most likely to have an open conformation [120]. However, these predictions are often inaccurate because of unpredictable RNA-protein interactions that change the structures of RNAs in cells. To circumvent this problem in a similar system, some researchers have applied an unwieldy systematic approach that involves huge numbers of candidate molecules [68, 84]. Such an approach tends to be both expensive and laborious. To avoid dependence on either of these approaches, we attempted to develop a ribozyme that would be able to access any chosen target site regardless of local secondary structure.

17.4.2. *Hybrid Ribozymes that Efficiently Cleave their Target mRNAs, Regardless of Secondary Structure*

We postulated that it might be useful to design a ribozyme that could recruit a protein that could, in turn, eliminate any interfering secondary structure in the target mRNA, thereby making any site in the target mRNA accessible to the ribozyme. To create such a ribozyme, we tried to link a ribozyme to an RNA helicase, a member of a class of proteins with nonspecific RNA-binding, sliding, and unwinding activities [13, 30, 59, 108]. We introduced an RNA motif, the constitutive transport element (CTE). The CTE appears to interact with RNA helicases both *in vitro* and *in vivo* [8, 26, 28, 32, 60, 92, 103, 104, 115] and was discovered as a cytoplasmic transport signal for D-type retroviral RNA. We postulated that

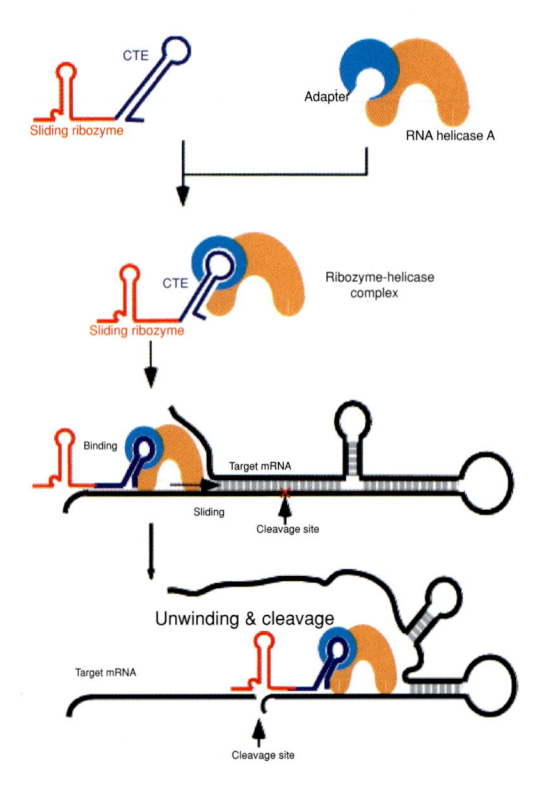

FIGURE 17.5. Schematic representation of the way in which a CTE-Rz coupled to an RNA helicase might cleave a hidden target site upon the unwinding of local secondary structure.

an RNA helicase coupled to a ribozyme via the CTE might efficiently guide the ribozyme to its target site by resolving any inhibitory structure in the target mRNA, with the resultant efficient cleavage of the mRNA.

Figure 17.5 shows a schematic representation of the way in which a CTE-ribozyme sequence coupled to an RNA helicase might cleave a sequestered target site after the unwinding of local secondary structure. We designed a tRNA^{Val}-ribozyme, with a pol III-mediated expression system, and attached the CTE sequence to the 3' end. We chose the TAR region of the long terminal repeat (LTR) of HIV-1 as the target of the ribozyme [41, 51]. The TAR region does not mutate because it is essential for the replication of HIV-1. Thus, the TAR region might be an effective target in any potential gene therapy against HIV-1. However, the stem structure of the TAR region prevents access by functional RNAs, such as antisense RNAs and ribozymes. If our hybrid CTE-ribozyme could disrupt the stem structure of the TAR region, locate its target and cleave the mRNA, this ribozyme might be an effective drug against HIV-1 irrespective of any mutations in the virus.

We prepared CTE-connected and unconnected ribozymes that were targeted to the TAR region, as shown in Figure 17.6A. The target gene, which consisted of the LTR of HIV-1 and a gene for luciferase, was stably expressed in HeLa cells. Figure 17.6B shows the suppression of the activity of the LTR-driven luciferase by the CTE-ribozyme. TAR Rz4 and TAR Rz5 were designed to target sites that we predicted would be inaccessible within the

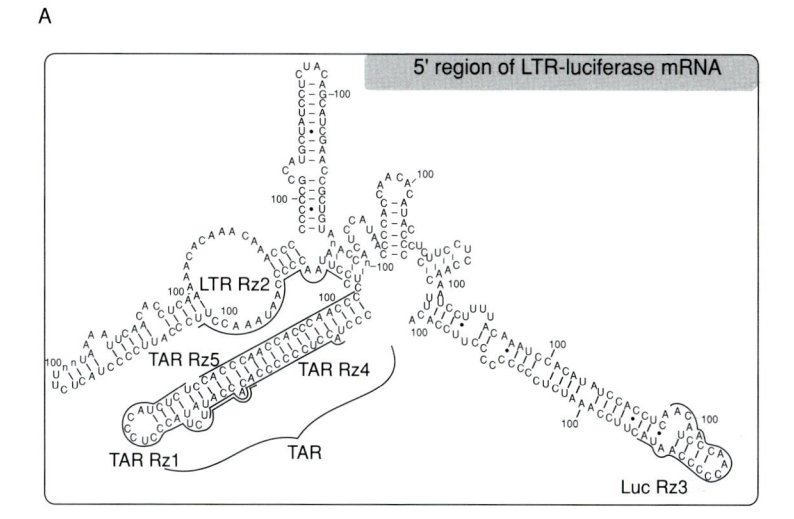

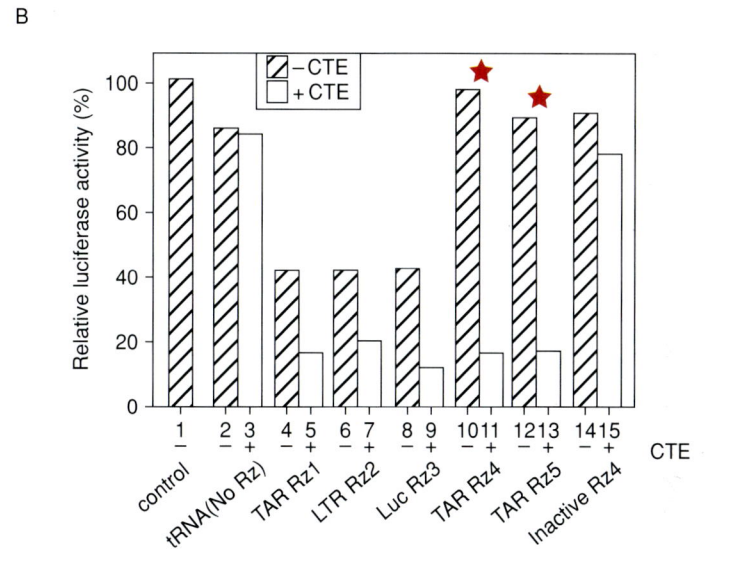

FIGURE 17.6. (A) The secondary structure (predicted by the MulFold program; 35) of the 5' region of a long terminal repeat-luciferase target mRNA. (B) Suppression of LTR-driven luciferase activity by CTE-Rz. Stars indicate results obtained with ribozymes targeted to relatively inaccessible sites in the target. LTR-Luc HeLa cells were transiently transfected with Tat alone (lane 1) or with Tat and the indicated tRNA-based Rz construct. Luciferase activity, as an indicator of Rz activity, is reported as a percentage of the Tat-only control. The values are the means of at least three independent assays. Because the assays involved transient transfections, there was some variability. The standard errors were within 10% when data were recorded on the same day and within 10–25% when they were recorded on different days. However, in all cases, the CTE-Rz always had significantly greater activity than the non-CTE-Rz.

well-documented stable stem structure of the TAR region. These ribozymes without the CTE (non-CTE ribozymes) had no effects on the levels of activity of the reporter luciferase. When the CTE was attached, these ribozymes strongly inhibited the luciferase activity (Figure 17.6B, lanes 11 and 13), with an 80% reduction in reporter activity. Furthermore, these CTE-coupled ribozymes had higher activity than non-CTE ribozymes that were designed to target "open" sites (TAR Rz1, LTR Rz2, Luc Rz3). Attachment of the CTE to other ribozymes also enhanced their activity. In particular TAR CTE-Rz4 and CTE-Rz5 had suppressive activities similar to those of TAR CTE-Rz1, LTR CTE-Rz2, and Luc CTE-Rz3. These results suggest that addition of the CTE sequence might allow all ribozymes to attack their targets efficiently.

We tested the general applicability of the CTE-Rz by targeting several endogenous targets, such as the mRNA for mouse procaspase-3 (CPP 3), and we obtained similar results [113]. It appears that CTE-ribozymes have specificity, strong activity, and general utility. However, the most important observation was that CTE-ribozyme were able to cleave their target mRNAs at any site, regardless of the predicted secondary or tertiary structure. All of our CTE-ribozymes has strong activity in cultured cells, and, in many cases, they were very active when the respective parental ribozymes, without the CTE, were inactive. Therefore, the hybrid CTE-ribozyme should have broad applicability because it is extremely easy to design and use. By contrast, previously developed ribozyme technologies require specialized skills. Furthermore, the improved efficacy of the CTE-ribozymes makes them even more suitable than earlier ribozymes for a wide range of applications, as described below. Having improved the efficacy of our ribozyme by eliminating constraints related to selection of target sites, we have produced a powerful tool both for basic research and for therapeutic interventions. We have also demonstrated that a poly(A) tail can be used instead of a CTE with similar results (Kawasaki, 2002).

17.5. MAXIZYMES: ALLOSTERICALLY CONTROLLABLE RIBOZYMES

Several years ago, we developed a completely new type of ribozyme as a result of efforts to shorten the hammerhead ribozyme. We succeeded in creating an allosteric ribozyme, the maxizyme, which functions as a dimer with significant specificity and activity both *in vitro* and *in vivo* [49–51, 101, 102]. This dimeric allosteric ribozyme allowed the first successful demonstration of an artificially created enzyme with potential utility as a biosensor not only *in vitro* but also *in vivo* (see below). As discussed below, we developed novel dimeric RNA motifs as a result of studies of "minizymes" that were aimed at shortening ribozymes. The designation "minizyme" is applied to shortened (minimized) ribozymes, but this designation also had a negative connotation as a ribozyme with extremely low activity (minimum). By contrast our novel ribozymes have extremely high activity in cells [49–51]. Therefore, we chose the name "maxizyme" for these new highly active minimized dimeric ribozymes [minimized, active, x-shaped (functions as a dimer), and intelligent (allosterically controllable) ribozyme].

17.5.1. Shortened Hammerhead Ribozymes that Function as Dimers

The cleavage of RNA by ribozymes has a general requirements for the presence of a divalent metal ion such as a magnesium ion [7, 11, 15, 18, 21, 62, 64, 93, 109–112, 119].

The catalytic domain of a hammerhead ribozyme captures these catalytically indispensable Mg^{2+} ions. The key to the creation of various allosteric ribozymes is the ability to control the capture of Mg^{2+} ions via a conformational change in the catalytic core of the ribozyme. For example, it is possible to replace RNA by the more stable DNA in the substrate-binding region or to shorten the stem-loop II region. However, for the potential application of ribozymes in a clinical setting, ease of design and economics dictate that smaller size is preferable. A smaller version of the hammerhead ribozyme, namely, a minizyme, has been made by replacing the stem-loop II region by a short linker [2, 19, 66, 106]. Unfortunately, such minizymes have quite low activity, as compared to the parental ribozymes. However, we found that a minizyme that lacked the entire stem-loop II region was essentially as active as the "wild-type" parent [2]. Kinetic and NMR analyses revealed that the shortened ribozyme was essentially inactive as a monomer but had extremely high catalytic activity as a dimer (Figure 17.7A) [47, 49]. This ribozyme, which is a "dimeric minizyme", was renamed "maxizyme" [49–51, 101, 102].

We extended our studies of dimeric minizymes by designing a heterodimeric system composed of two different monomers, maxizyme left (MzL) and maxizyme right (MzR) [47, 50, 51]. In this system, shown in Figure 17.7B, the substrate is cleaved only when MzL and MzR form a dimer. Since such maxizymes have two substrate-binding regions, we were able to convert our heterodimeric maxizyme into an allosteric version that was able to function as a sensor.

17.5.2. Design of an Allosterically Controllable Maxizyme

Ribozymes can target essentially any RNA at specific sites but there is a minimum required cleavable sequence: cleavage occurs only after the sequence NUX (where N is any base and X is A, C, or U) [94]. In some cases, such a cleavable triplet sequence is not available at a suitable position in the target RNA. We have focused on such a case that involves a chimeric mRNA of clinical importance. Chimeric mRNAs are generated by chromosomal translocations and they are fusion mRNAs in which the first part and the second part are derived from two different genes. They are often involved in the pathogenesis of disease. A well known example of a pathogenic chromosomal translocation is the Philadelphia chromosome, which causes chronic myelogenous leukemia (CML) [71, 77]. The Philadelphia chromosome is the result of a reciprocal translocation that involves the *BCR* and *ABL* genes, and the product of translocation is a fusion mRNA. Fusion mRNAs of this type are tumor-specific and pathogenetically important and, thus, they are obvious targets for nucleic acid therapeutics [95]. However, because of the absence of a NUX sequence near the site of fusion in the target mRNA, conventional ribozymes cannot distinguish the chimeric mRNA from the normal parental mRNAs [48, 50, 112]. Both the *BCR* gene and the *ABL* gene are important for cell survival. Thus, when we design ribozymes that might cleave the chimeric mRNA, we must be sure to avoid cleavage of normal mRNAs, which share sequences with the abnormal fusion mRNA.

There have been many attempts to cleave chimeric mRNAs specifically using ribozymes, but it is extremely difficult to cleave only the chimeric mRNA without affecting the normal parental mRNAs [29, 82]. As mentioned above, maxizymes can bind to two different target sites. We took advantage of the two substrate-binding regions to design a maxizyme with one substrate-binding region that corresponds to the abnormal junction

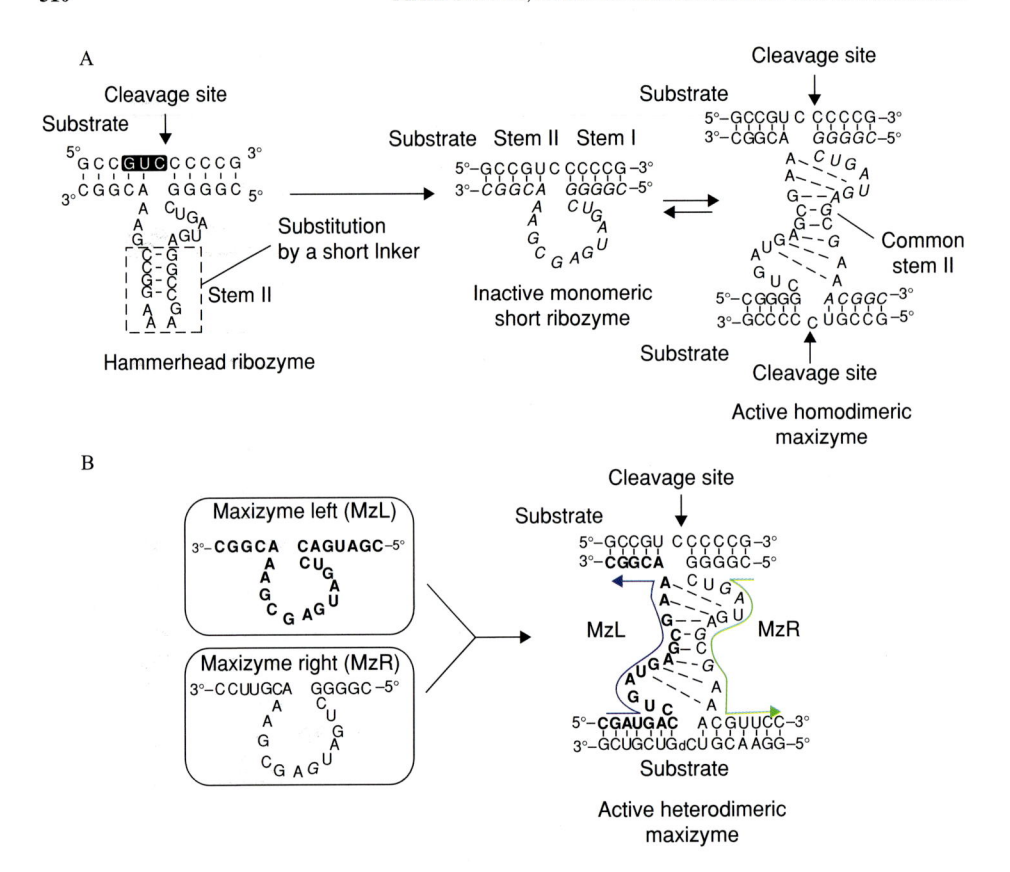

FIGURE 17.7. (A) Development of an allosterically controllable maxizyme. Secondary structures of the parental hammerhead ribozyme and a conventional minizyme. The homo dimeric maxizyme is active. The monomeric minizyme, namely, a hammerhead ribozyme with a deleted stem-loop II region, is inactive. (B) The heterodimeric (MzL and MzR) maxizyme has two different substrate-binding sites: one is complementary to the sequence of interest (activator or inhibitor) and the other is complementary to a cleavable sequence.

sequence and a second binding region that corresponds to an efficient cleavage site even though the site was at some distance from the junction. One substrate-binding site functioned as the "eye" or sensor that was able to distinguish the chimeric mRNA from the normal mRNA, while the other served as the "scissors" that actually cleaved the target (Figure 17.8B, left).

In this system, dimers that act as the switch for the cleavage activity form only when there is binding to the junction sequence and to the cleavable sequence that contains the NUX triplet. Moreover, the base paring in the stem II region influences the stability of the dimeric structure. If the base pairing is very stable, the dimer will form even in the absence of substrate, and the cleavage sequence is cleaved regardless of whether or not the junction sequence is present. By contrast, if the base paring is unstable, the dimer will not form and cleavage activity will be minimal. Control of base-pair stability and the construction of base sequences that allows formation of an active ribozyme only in the presence of the chimeric mRNA are the keys to the success of this system.

ENGINEERED RIBOZYMES: EFFICIENT TOOLS FOR MOLECULAR GENE THERAPY

FIGURE 17.8. (A) Formation of an active or an inactive maxizyme via dimerizations that are regulated allosterically by specific effector sequences. The heterodimer (MzL and MzR) can have two different binding sites: one is complementary to the sequence of interest (activator or inhibitor) and the other is complementary to a cleavable sequence. In order to achieve high substrate-specificity, the maxizyme should be in an active conformation only in the presence of the abnormal *BCR-ABL* junction (on the left), while the conformation should remain inactive in the presence of normal *ABL* mRNA (on the right) and in the absence of the *BCR-ABL* junction (on the right).

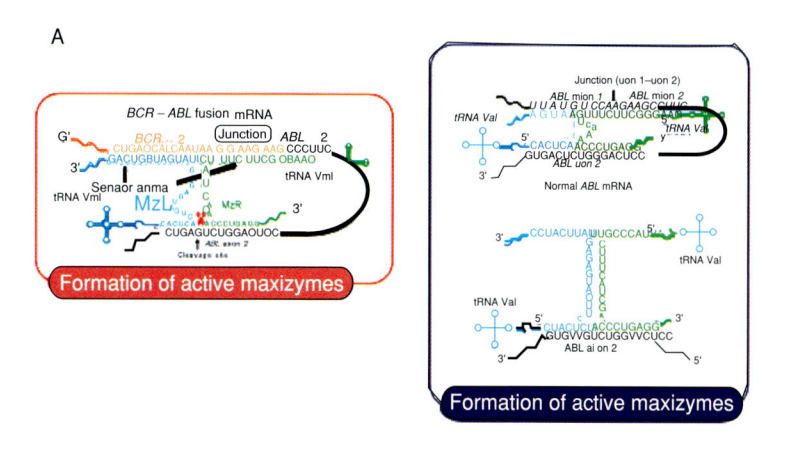

Figure 17.8A shows the secondary structure of the ribozyme that we designed to attack the fusion mRNA. The active dimer is formed only when the two substrate-binding regions of the ribozyme bind correctly to the two positions on the chimeric mRNA (Figure 17.8B). In the presence of the normal *ABL* mRNA, which should not be cleaved, the structure of the central active region changes, so that only the inactive structure is formed. In addition, the ribozyme is designed such that, if only a monomer binds to the cleavage site of the substrate-binding sequence, only the inactive type of ribozyme is generated (Figure 17.8A). This inactive structure does not support the capture of the catalytically essential magnesium ion, and cleavage does not occur. When we synthesized the MzL and MzR ribozymes and evaluated their substrate specificity *in vitro*, we found that only the chimeric mRNA was cleaved with an extremely high degree of specificity. Thus, our artificial allosteric enzyme

cleaved the target RNA as the result of a structural change that occurred only in the presence of the oncogenic chimeric mRNA with the junction sequence [50].

We evaluated our maxizyme (MzL plus MzR) in cultured mammalian cells. To ensure the efficient formation of dimers in cells, we used the promoter of the gene for a tRNA that is recognized by RNA polymerase III, as described above. We introduced a conventional ribozyme, each individual monomer of the maxizyme, and both monomers together into cells derived from patients with CML. We demonstrated that the active maxizyme was a heterodimer and that cleavage by the heterodimeric maxizyme, rather than an antisense effect, was responsible for the specific suppression of expression of the *BCR-ABL* mRNA. Moreover, this maxizyme had significantly higher activity than the "wild-type" ribozyme from which it was derived.

17.5.3. Inactivation of an Oncogene in a Mouse Model

There have been many attempts, using various approaches, to construct artificial allosteric enzymes, but success *in vivo* has been minimal. The maxizyme was, to our knowledge, the first artifical allosteric enzyme to function in cells in culture.

We next examined the anti-tumor effect of our maxizyme in animals [101]. We used a retroviral system for the expression of the maxizyme in leukemic cells. We subcloned two tRNA^{Val}-driven expression cassettes, which corresponded to each component of the heterodimer, in tandem in the retroviral vector. A line of CML cells (BV173) was transduced either with a control vector, in which the maxizyme sequence had been deleted, or with the maxizyme-encoding vector. We then injected a bolus of each line of transduced BV173 cells into the tail veins of mice.

The differences between the two groups of mice were reflected in their mortality rates (Figure 17.9). All of the mice injected with control BV173 cells died of diffuse leukemia, confirmed at necropsy, after 6 to 13 weeks (median survival time, 9 weeks), whereas mice injected with maxizyme-transduced BV173 cells remained healthy. Our results indicate that each subunit of the maxizyme, introduced by the retroviral vector, was produced at the appropriate concentration to support dimerization *in vivo*. Also, the maxizyme apparently functioned successfully in animals, cleaving *BCR-ABL* mRNA with exceptional efficiency.

At present, use of kinase inhibitors [22] and allogeneic transplantation are the only effective therapies for CML, with only half of all patients, on average, being eligible for the latter treatment because of the limited availability of donors and age restrictions. Our results raise the possibility that our maxizyme might be useful for purging bone marrow cells in cases of CML treated by autologous transplantation, when it would presumably reduce the incidence of relapse by decreasing the tumorigenicity of contaminating CML cells in the transplant.

17.5.4. Generality of the Maxizyme Technology

The maxizyme is of considerable interest because it imparts a sensor function to short ribozymes that act as a dimer. Using this sensor function, we can specifically cleave abnormal chimeric mRNA exclusively, without affecting the normal mRNA, which should not be cleaved. Maxizyme technology is not limited to the disruption of the abnormal chimeric gene

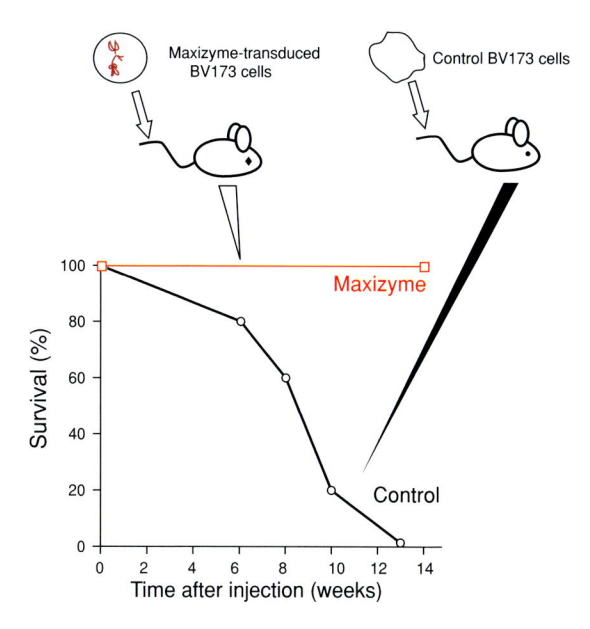

FIGURE 17.9. The antitumor effects of the maxizyme in a murine model of chronic myelogenous leukaemia (CML). The survival of animals was monitored daily for more than 20 weeks after inoculation; all control mice died within 13 weeks, whereas maxizyme-treated mice remained disease-free for the entire period of the investigation.

in CML. Abnormal chimeric genes generated from reciprocal chromosomal translocations are frequently found in several types of leukemia. Maxizymes have been used successfully to cleave only abnormal target mRNAs, which lack a NUX cleavage site at the junction, in the case of acute lymphoblastic leukemia (ALL) and acute promyelocytic leukemia (APL), without any damage to the products of normal genes [102]. Abnormal chimeric mRNAs are also generated by errors in splicing. A maxizyme can be designed to act against each possible transcript and should be able to distinguish its target specifically from other mRNAs. We have already constructed various maxizymes that target different chimeric genes, and each of them is very active and highly specific [50, 102]. Thus, maxizymes appear to be powerful gene-inactivating agents with allosteric functions that allow them to cleave any type of chimeric mRNA specifically. To our knowledge, the maxizyme is the first artificial, allosteric enzyme whose activity has been demonstrated at the animal level, highlighting its potential utility in a clinical setting [31, 102]. It should be noted, however, that the transcripts of chimeric genes of the type discussed here can also be destroyed by small interfering RNAs (siRNAs; [81, 91]).

17.6. IDENTIFICATION OF GENES USING HYBRID RIBOZYMES

As noted in section 4.2, we developed a hybrid ribozyme that coupled the cleavage activity of hammerhead ribozymes with the unwinding activity of RNA helicase and was able to cleave its target mRNA extremely efficiently, regardless of the secondary structure

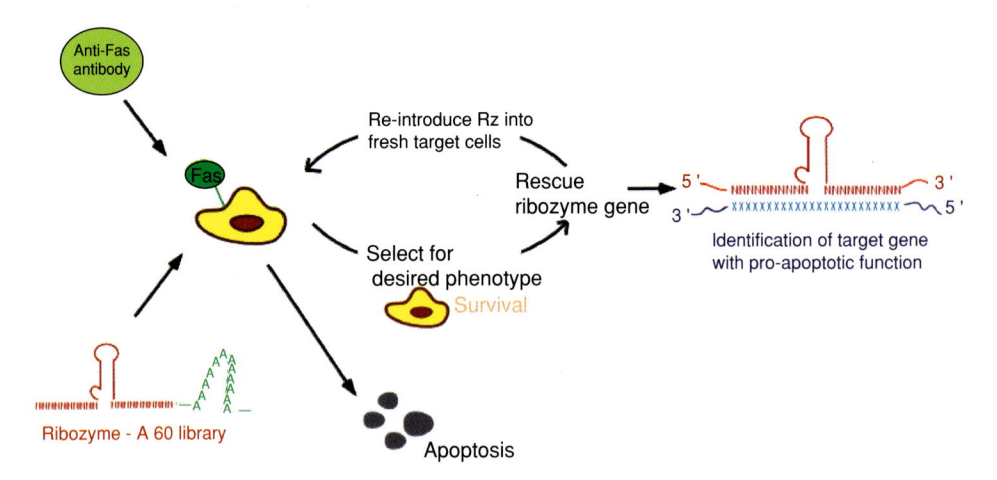

FIGURE 17.10. Schematic diagram of the application of the gene discovery system to the Fas-induced apoptosis using libraries of poly(A)-connected hybrid-Rz. Hela-Fas cells that expressed the randomized Rz-60 libraries were treated with the Fas-specific (Anti-Fas) antibodies.

of the RNA. We attempted to use this novel ribozyme not only to cleave specific known target mRNAs but also to identify genes associated with specific phenotypes in cells. This can be accomplished using ribozymes with randomized binding arms, as has been done with hairpin ribozyme [5, 44, 61, 114]. The sequence of the human genome has become available, and it will be extremely valuable to have methods for the rapid identification of important genes. We conducted this function analysis by using hybrid ribozymes, which coupled cleavage activity with the unwinding activity of an endogenous RNA helicase [100]. We demonstrate that ribozyme of this type are able to cleave the target mRNA at a chosen site, regardless of the putative secondary or tertiary structure in the vicinity of the target site, and thus they can be used for rapid identification of functional genes in the post-genome era.

We attached a poly(A) 60 sequence to the 3' end of a tRNA^{Val}-driven ribozyme (Rz-A60) instead of the CTE sequence [75, 113]. This poly(A) sequence interacts with endogenous RNA helicase eIF4AI via interactions with poly(A)-binding protein (PABP) and PABP-interacting protein-1 (PAIP). We demonstrated that this complex was able to unwind an RNA duplex substrate effectively, and cleaved otherwise inaccessible target sites. We used these ribozymes to discover the functions of unknown genes. Since our hybrid ribozymes can attack structured sites, they can attack mRNAs with high-level efficiency. If libraries of hybrid ribozymes with randomized binding arms are introduced into cells, the genes associated with any changes in phenotype can be readily identified by sequencing the specific ribozyme clone [100]. Figure 17.10 shows a schematic representation of our gene-discovery system.

We established a novel system for screening functional genes in the signaling pathway of Fas-induced apoptosis in HeLa-Fas cells using randomized Rz-A60-expression libraries. In this system, we randomized 10 nt in each substrate-binding arm of Rz-A60, and then retroviral vectors that carried the randomized Rz-A60-expression libraries were introduced into HeLa-Fas cells. After treatment of these HeLa-Fas cells with Fas-specific antibodies, which normally induces apoptosis, cells that survived were collected and the genomic DNA was isolated from each clone. Sequencing of the randomized region of Rz-A60 in each genomic DNA enabled us rapidly to identify genes involved in the Fas-induced apoptotic pathway.

We identified a variety of genes with pro-apoptotic functions, such as genes for FADD, caspase 8, caspase 9 and caspase 3. In the absence of the poly(A) tail, we would not have identified genes for FADD and caspase 8 in our first screening with the randomized Rz libraries, since only poly(A)-connected ribozymes targeted to FADD mRNA or caspase 8 mRNA affected the expression of target genes. Our study demonstrated the successful application of a hybrid Rz to gene discovery. Using this gene discovery system, we have also identified many factors [76] involved in other apoptotic pathways [36–38], metastasis [98, 99], Alzheimer's disease [79, 80], and the roles of microRNAs [56].

17.7. SUMMARY AND PROSPECTS

As discussed above, it is now possible to achieve high-level activity of ribozymes *in vivo* and to cleave any specific target RNA in the cell using either our maxizyme or a hybrid ribozyme. These ribozymes also efficient tools for the analysis of gene function.

Our efficient ribozyme-expression systems are also applicable to the expression of small interfering RNAs (siRNAs), which induce the sequence-dependent degradation of a cognate mRNA via RNA interference (RNAi). The application of RNAi in mammals has the potential to allow the systematic analysis of gene expression and also the therapeutic silencing of gene expression [67].

Since siRNAs are 21- to 23-nt RNA duplexes with 2- or 3-nt overhanging 3' ends, the pol III system, which is suitable for efficient transcription of small RNAs, is also useful for the expression of siRNAs. Another advantage of the pol III system is that transcription terminates at four or more T residues, leaving 1-4 U residues at the 3' terminus of the nascent RNA. These properties allow use of DNA templates to synthesize small RNAs with structural features close to those of siRNAs that are active *in vivo*.

A number of groups have developed plasmid-based siRNA-expression systems using pol III promoters [9, 59, 69, 83, 85, 96, 118]. Two pol III promoters have been used predominantly, the U6 promoter and the H1 promoter. As mentioned above, tRNA^{Val}-driven ribozymes are transcribed at high levels, and the transcripts are transported to the cytoplasm without exception when an appropriate linker is used. Our work indicates that RNAi in mammalian cells occurs in the cytoplasm [39]. Therefore, the tRNA^{Val} expression system should also be ideal for the expression of siRNA. In fact, tRNA-dsRNAs, in which a short hairpin structure is attached to a tRNA-like ribozyme, is efficiently transported to the cytoplasm, and effectively induces RNAi-mediated gene silencing (Kuwabara, 2003). tRNA-dsRNAs should be powerful tools for studies of the functions of genes in mammalian cells and they might also be useful as therapeutic agents as ribozymes.

Now we can employ two versatile gene knock-down tools; ribozymes and siRNAs, as usage for any purpose.

REFERENCES

[1] S.F. Alino, J. Crespo, M. Bobadila, M. Lejarreta, C. Blaya, and A. Crespo. *Biochem. Biophys. Res. Comm.*, 204:1023, 1994.
[2] S.V. Amontov and K. Taira. *J. Am. Chem. Soc.*, 118:1624, 1996.
[3] G.-J. Arts, M. Fornerod, and I.W. Mattaj. *Curr. Biol.*, 6:305, 1998a.
[4] G.-J. Arts, S. Kuersten, P. Romby, B. Ehresmann, and I.A. Mattaj. *EMBO J.17*, 7430, 1998b.
[5] C. Beger, L.N. Pierce, M. Kruger, E.G. Marcusson, J.M. Robbins, P. Welsh, P.J. Welch, K. Welte, M.C. King, J.R. Barber, and F. Wong-Staal. *Proc. Natl. Acad. Sci. U.S.A.*, 98:130, 2001.
[6] E. Bertrand, D. Castanotto, C. Zhou, C. Carnonnelle, G.P. Lee, S. Chatterjee, T. Grange, R. Picket, D. Kohn, D. Engelke, and J.J. Rossi, *RNA*, 3:75, 1997.
[7] K.R. Birikh, P.A. Heaton, and F. Eckstein. *Eur. J. Biochem.*, 245:1, 1997.
[8] I.C. Braun, E. Rohrbach, C. Schmitt, and E. Izaurralde. *EMBO J.*, 18:1953, 1999.
[9] T.R. Brummelkamp, R. Bernards, and R. Agami. *Science*, 296:550, 2002.
[10] N.J. Caplen, E.W. Alton, P.G. Middleton, J.R. Dorin, B.J. Stevenson, X. Gao, S.R. Durham, P.K. Jeffery, M.E. Hodson, and C. Coutelle *et al. Nature Med.*, 1:39, 1995.
[11] C. Carola and F. Eckstein. *Proc. Natl. Acad. Sci. U.S.A.*, 3, 274 (1999).
[12] M. Cotton and M.L. Birnstiel. *EMBO J. 8*, 3881, 1989.
[13] J. de la Crutz, D. Kressler, and P. Linder. *Trends Biochem. Sci.*, 24:192, 1999.
[14] D. Das, D. Henning, Wright, and R. Reddy. *EMBO J. 7*, 503, 1998.
[15] J.A. Doudna. *Curr. Biol.*, 8:495, 1998.
[16] F. Eckstein, and D.M.J. Lilley, *Nucleic Acids and Moleculad Biology: Catalytic RNA.* (Springer-Verlag, Berlin, 1996), vol. 10.
[17] R.P. Erickson and J. Izant. *Gene Regulation: Biology of Antisense RNA and DNA.* Raven Press, New York, NY, 1992.
[18] Famulok. *Curr. Opin. Struc. Biol.*, 9:324, 1999.
[19] D.J. Fu, F. Benseler, and L.W. McLaughlin. *J. Am. Chem. Soc.*, 116:4591, 1994.
[20] E.P. Geiduschek and G.P. Tocchini-Valentini. *Annu. Rev. Biochem.*, 57:873, 1988.
[21] R.F. Gesteland, T.R. Cech, and J.F. Atkins. *The RNA World.* Spring Harbor Laboratory Press, NY, 1999.
[22] J.M. Goldman and J.V. Melo. *New Engl. J. Med.*, 349:1451, 2003.
[23] P.D. Good, A.J. Krikos, X.L. Li, N.S. Lee, L. Giver, A. Ellington, J.A. Zaia, J.J. Rossi, and D.R. Engelke. *Gene Ther.*, 4:45, 1997.
[24] E.M. Gorden and W.F. Anderson. *Curr. Opin. Biotechnol.*, 5:611, 1994.
[25] Grosshand, E. Hurt and G. Simos. *Genes Dev.*, 14:830, 2000.
[26] P. Grüter, C. Tabernero, C. von Kobbe, C. Schmitt, C. Saavedra, A. Bachi, M. Wilm, B.K. Felber, and E. Izaurralde. *Mol. Cell*, 1:649, 1998.
[27] Y.J. Hernandez, J. Wang, W.G. Kearns, S. Loiler, A. Poirier, and T.R. Flotte. *J. Virol.*, 73:8549, 1990.
[28] C.A. Hodge, H.V. Colot, P. Stafford, and C.N. Cole. *EMBO J.*, 18:5778, 1999.
[29] H. James, K. Mills, and I. Gibson. *Leukemia*, 10:1054, 1996.
[30] E. Jankowsky, C.H. Gross, S. Shuman, and A.M. Pyle. *Nature*, 403:447, 2000.
[31] T. Kanamori, K. Nishimaki, S. Asoh, Y. Ishibashi, I. Takata, T. Kuwabara, K. Taira, H. Yamaguchi, S. Sugihara, T. Yamazaki, Y. Ihara, K. Nakano, S. Matsuda, and S. Ohta. *EMBO J.*, 22:2913, 2003.
[32] Y. Kang and B.R. Cullen. *Genes Dev.*, 13:1126, 1999.
[33] Y. Kato, T. Kuwabara, M. Warashina, H. Toda, and K. Taira. *J. Biol. Chem.*, 276:15378, 2001.
[34] H. Kawasaki, J. Ohkawa, N. Tanishige, K. Yoshinari, T. Murata, K.K. Yokoyama, and K. Taira. *Nucleic Acids Res.*, 24:3010, 1996.
[35] H. Kawasaki, R. Eckner, T.P. Yao, K. Taira, R. Chiu, D.M. Livingston, and K.K. Yokoyama. *Nature*, 393:284, 1998.
[36] H. Kawasaki and K. Taira. *EMBO Rep.*, 3:443, 2002a.
[37] H. Kawasaki and K. Taira. *Nucleic Acids Res.*, 30:3609, 2002b.
[38] H. Kawasaki, R. Onuki, E. Suyama, and K. Taira. *Nat. Biotechnol.*, 20:376, 2002c.
[39] H. Kawasaki and K. Taira. *Nucleic Acids Res.*, 61:700, 2003.
[40] M.A. Kay, L. Meuse, A.M. Gown, P. Linsley, D. Hollenbaugh, A. Aruffo, H.D. Ochs, and C.B. Wilson. *Proc. Natl. Acad. Sci. U.S.A.*, 94:4684, 1997.

[41] S. Koseki, J. Ohkawa, R. Yamamoto, Y. Tanabe, and K. Taira. *J. Control Rel.*, 53:159, 1998.
[42] S. Koseki, T. Takebe, K. Tani, S. Asano, T. Shioda, Y. Nagai, T. Shimada, J. Ohkawa, and K. Taira. *J. Virol.*, 73:1868, 1999.
[43] I. Kovesdi, D.E. Brough, J.T. Brude, and T.J. Wickham. *Curr. Opin. Biotechnol.*, 8:583, 1997.
[44] K. Kruger, P.J. Grabowski, A.J. Zaug, J. Sands, D.E. Gottschling, and T.R. Cech. *Cell*, 31:147, 1982.
[45] G. Krupp and R.K. Gaur. *Ribozyme: Biochemistry and Biotechnology.* Eaton Publishing, MA, 2000.
[46] U. Kutay, G. Lipowsky, E. Izaurralde, F.R. Bischoff, P. Schwarzmaier, E. Hartmann, and D. Görlich. *Cell.*, 1:359, 1998.
[47] T. Kuwabara, S.V. Amontov, M. Warashina, J. Ohkawa, and K. Taira. *Nucleic Acids Res.*, 24:2302, 1996.
[48] T. Kuwabara, M. Warashina, T. Tanabe, K. Tani, S. Asano, and K. Taira. *Nucleic Acids Res.*, 25:3074, 1997.
[49] T. Kuwabara, M. Warashina, M. Orita, S. Koseki, J. Ohkawa, and K. Taira. *Nat. Biotechnol.*, 16:961, 1998a.
[50] T. Kuwabara, M. Warashina, T. Tanabe, K. Tani, S. Asano, and K. Taira. *Mol. Cell*, 2:617, 1998b.
[51] T. Kuwabara, M. Warashina, A. Nakayama, J. Ohkawa, and K. Taira. *Proc. Natl. Acad. Sci. U.S.A.*, 96:1886, 1999.
[52] T. Kuwabara, M. Warashina, and K. Taira. *Curr. Opin. Chem. Biol.*, 4:669, 2000a.
[53] T. Kuwabara, M. Warashina, and K. Taira. *Trends Biotechnol.*, 18:462, 2000b.
[54] T. Kuwabara, M. Warashina, S. Koseki, M. Sano, J. Ohkawa, A. Nakayama, and K. Taira. *Nucleic Acids Res.*, 29:2780, 2001a.
[55] T. Kuwabara, M. Warashina, M. Sano, H. Tang, F. Wong-Staal, E. Munekata, and K. Taira. *Biomacromol.*, 2:1229, 2001b.
[56] T. Kuwabara et al. *Cell*, submitted for publication.
[57] D.S. Latchman. *Mol. Biotechnol.*, 2:175, 1994.
[58] C.-G. Lee, P.D. Zamore, M.R. Green, and J. Hurwitz. *J. Biol. Chem.*, 268:16822, 1993.
[59] N.S. Lee, T. Dohjima, G. Bauer, H. Li, M.J. Li, A. Ehsani, P. Salvaterra, and J. Rossi. *Nat. Biotechnol.*, 20:500, 2002.
[60] J. Li, H. Tang, T.M. Mullen, C. Westberg, T.R. Reddy, D.W. Rose, and F. Wong-Staal. *Proc. Natl. Acad. Sci. U.S.A.*, 96:709, 1999.
[61] Q.X. Li, J.M. Robbins, P.J. Welch, F. Wong-Staal, and J.R. Barber. *Nucleic Acids Res.*, 28:2605, 2000.
[62] D.M.J. Lilley. *Curr. Opin. Struct. Biol.*, 9:330, 1999.
[63] G. Lipowsky, F.R. Bischoff, E. Izaurralde, U. Kutay, S. Scharfer, H.J. Gross, H. Beier, and D. Görlich, *D. RNA*, 5:539, 1999.
[64] D.M. Long and O.C. Uhlenbeck. *Proc. Natl. Acad. Sci. U.S.A.*, 91:6977, 1994.
[65] E. Lund and J.E. Dahlberg. *Science*, 282:2082, 1998.
[66] M.J. McCall, P. Hendry, and P.A. Jennings. *Proc. Natl. Acad. Sci. U.S.A.*, 89:5710, 1992.
[67] M.T. McManus and P.A. Sharp. *Nat. Rev. Genet.*, 3:737, 2002.
[68] N. Milner, K.U. Mir, and E.M. Southern. *Nat. Biotechnol.*, 15:537, 1997.
[69] M. Miyagishi and K. Taira. *Nat. Biotechnol.*, 20:497, 2002.
[70] A. Mountain. *A. Trends Biotechnol.*, 18:119, 2000.
[71] A.J. Muller, J.C. Young, A.M. Pendergast, M. Pondel, N.R. Landau, D.R. Littman, and O.N. Witte. *Cell. Biol.*, 11:1785, 1991.
[72] R.J. Mumper, J.G. Duguid, K. Anwer, M.K. Barren, H. Nitta, and A.P. Rolland. *Pharm. Res.*, 13:701, 1996.
[73] J.A.H. Murray. *Antisense RNA and DNA.* Wiley-Liss Inc., New York, NY, 1992.
[74] L. Naldini. *Curr. Opin. Biotechnol.*, 9:457, 1998.
[75] B. Nawrot, S. Antoszczyk, M. Maszewska, T. Kuwabara, M. Warashina, K. Taira, and W.J. Stec. *Eur. J. Biochem.*, 270:3962, 2003.
[76] D.L. Nelson, E. Suyama, H. Kawasaki, and K. Taira. *TARGETS*, 2:191, 2003.
[77] P.C. Nowell and D.A. Humgerford. *Science*, 132:1497, 1960.
[78] J. Ohkawa and K. Taira. *Human Gene Ther.*, 11:577, 2000.
[79] R. Onuki, A. Nagasaki, H. Kawasaki, T. Baba, T.Q.P. Ueda, and K. Taira. *Proc. Natl. Acad. Sci. U.S.A.*, 99:14716, 2002.
[80] R. Onuki, Y. Bando, E. Suyama, T. Katayama, H. Kawasaki, T. Baba, M. Tohyama, and K. Taira. *EMBO J.*, submitted for publication, 2003.
[81] K. Oshima, H. Kawasaki, Y. Soda, K. Tani, S. Asano, and K. Taira. *Cancer Res.*, 63:6809, 2003.
[82] C.J. Pachuk, K. Yoon, K. Moelling, and L.R. Coney. *Nucleic Acids Res.*, 22:301, 1994.
[83] P.J. Paddison, A.A. Caudy, and G.J. Hannon. *Proc. Natl. Acad. Sci. U.S.A.*, 99:1443, 2002.

[84] V. Patzel and G. Sczakiel. *Nat. Biotechnol.*, 16:64, 1998.
[85] C.P. Paul, P.D. Good, I. Winer, and D.R. Engelke. *Nat. Biotechnol.*, 20:505, 2002.
[86] J.C. Perales, T. Ferkol, M. Molas, and R.W. Hanson. *Eur. J. Biochem.*, 226:255, 1994.
[87] J.E. Rabinowits and J. Samulski. *Curr. Opin. Biotechnol.*, 9:470, 1998.
[88] J.J. Rossi and N. Sarver. *Trends Biotechnol.*, 8:179, 1990.
[89] J.J. Rossi. *Trends Biotechnol.*, 13:301, 1995.
[90] N. Sarver, E.M. Cantin, P.S. Chang, J.A. Zaida, P.A. Ladne, D.A. Stephens, and J.J. Rossi. *Science*, 247:1222, 1990.
[91] M. Scherr, K. Battmer, T. Winkler, O. Heidenreich, A. Ganser, and M. Eder. *Blood*, 101:1566, 2003.
[92] C. Schmitt, C. von Kobbe, A. Bachi, N. Pante, J.P. Rodrigues, C. Boscheron, G. Rigaut, M. Wilm, B. Seraphin, M. Carmo-Fonseca, and E. Izaurralde. *EMBO J.*, 18:4332, 1999.
[93] W.G. Scott. *Curr. Opin. Chem. Biol.*, 3:703, 1999.
[94] T. Shimayama, S. Nishikawa, and K. Taira. *Biochemistry*, 34:3649, 1995.
[95] E. Shtivelman, B. Lifschitz, R.P. Gale, B.A. Roe, and J. Canaani. *Cell*, 47:277, 1986.
[96] G. Sui, C. Soohoo, B. Affarel, F. Gay, Y. Shi, W.C. Forrester, and Y. Shi. *Proc. Natl. Acad. Sci. U.S.A.*, 99:5515, 2002.
[97] B.A. Sullenger and T.R. Cech. *Science*, 262:1566, 1993.
[98] E. Suyama, H. Kawasaki, T. Kasaoka, and K. Taira. *Cancer Res.*, 63:119, 2003a.
[99] E. Suyama, H. Kawasaki, M. Nakajima, and K. Taira. *Proc. Natl. Acad. Sci. U.S.A.*, 100:5616, 2003b.
[100] K. Taira, M. Warashina, T. Kuwabara, and H. Kawasaki. *Functional hybrid molecules with sliding ability.* Japanese Patent Application H11-316133, 1999.
[101] T. Tanabe, T. Kuwabara, M. Warashina, K. Tani, K. Taira, and S. Asano. *Nature*, 406:473, 2000a.
[102] T. Tanabe, I. Takata, T. Kuwabara, M. Warashina, H. Kawasaki, K. Tani, S. Ohta, S. Asano, and K. Taira. *Biomacromol.*, 1:108, 2000b.
[103] H. Tang, G.M. Gaietta, W.H. Fischer, M.H. Ellisman, and F. Wong-Staal. *Science*, 276:1412, 1997.
[104] H. Tang and F. Wong-Staal. *J. Biol. Chem.*, 275:32694, 2000.
[105] P.C. Turner. *Methods in Molecular Biology: Ribozyme Protocols.* Humana Press, Totowa, NJ, 74, 1997.
[106] T. Tuschl and F. Eckstein. *Proc. Natl. Acad. Sci. U.S.A.*, 90:6991, 1993.
[107] R. Wadhwa, H. Ando, H. Kawasaki, K. Taira, and S.C. Kaul. *EMBO Rep.*, 4:595, 2003.
[108] J.D.O. Wagner, E. Jankowsky, M. Company, A.M. Pyle, and J.N. Abelson. *EMBO J.*, 17:2926, 1998.
[109] N.G. Walter and J.M. Burke. *Curr. Opin. Chem. Biol.*, 2:24, 1998.
[110] M. Warashina, D.M. Zhou, T. Kuwabara, and K. Taira. *Ribozyme structure and function: Comprehensive Natural Products Chemistry.* Elsevier Science Ltd., Oxford, vol. 6, p. 235, 1999.
[111] M. Warashina, T. Kuwabara, and K. Taira. *Structure*, 8:207, 2000a.
[112] M. Warashina, Y. Takagi, W.J. Stec, and K. Taira. *Curr. Opin. Biotechnol.*, 11:354, 2000b.
[113] M. Warashina, T. Kuwabara, Y. Kato, M. Sano, and K. Taira. *Proc. Natl Acad. Sci. U.S.A.*, 98:5572, 2001.
[114] P.J. Welch, E.G. Marcusson, Q.K. Li, C. Begar, M. Kruger, C. Zhou, M. Leavitt, F. Wong-Staal, and J.R. Barber. *Genomics*, 66:274, 2000.
[115] C. Westberg, J.P. Yang, H. Tang, T.R. Reddy, and F. Wong-Staal. *J. Biol. Chem.*, 275:21396, 2000.
[116] N. Wu and M.M. Ataai. *Curr. Opin. Biotechnol.*, 11:205, 2000.
[117] Y. Yang, Q. Li, H.C. Ertl, and J.M. Wilson. *J. Virol.*, 69:200, 1995.
[118] J.Y. Yu, S.L. DeRuiter, and D.L. Turner. *Proc. Natl. Acad. Sci. U.S.A.*, 99:6047, 2002.
[119] D.M. Zhou and K. Taira. *Chem. Rev.*, 98:991, 1998.
[120] M. Zucker. *Methods Enzymol.*, 180:262, 1989.

About the Editors

Professor Mauro Ferrari is a pioneer in the fields of bioMEMS and biomedical nanotechnology. As a leading academic, a dedicated entrepreneur, and a vision setter for the Nation's premier Federal programs in nanomedicine, he brings a three-fold vantage perspective to his roles as Editor-in-Chief for this work. Dr. Ferrari has authored or co-authored over 150 scientific publications, 6 books, and over 20 US and International patents. Dr. Ferrari is also Editor-in-Chief of Biomedical Microdevices and series editor of the new Springer series on Emerging Biomedical Technologies.

Several private sector companies originated from his laboratories at the Ohio State University and the University of California at Berkeley over the years. On a Federal assignment as Special Expert in Nanotechnology and Eminent Scholar, he has provided the scientific leadership for the development of the Alliance for Cancer Nanotechnology of the National Cancer Institute, the world-largest medical nanotechnology operation to date. Dr. Ferrari trained in mathematical physics in Italy, obtained his Master's and Ph.D. in Mechanical Engineering at Berkeley, attended medical school at The Ohio State University, and served in faculty positions in Materials Science and Engineering, and Civil and Environmental Engineering in Berkeley, where he was first tenured. At Ohio State he currently serves as Professor of Internal Medicine, Division of Hematology and Oncology, as Edgar Hendrickson Professor of Biomedical Engineering, and as Professor of Mechanical Engineering. He is Associate Director of the Dorothy M. Davis Heart and Lung Research Institute, and the University's Associate Vice President for Health Science, Technology and Commercialization.

Dr. Mihri Ozkan is currently an Assistant Professor in the Department of Electrical Engineering at UC-Riverside with a research focus in nanotechnology and its applications in biology and engineering. She received her Ph.D. degree in the Department of Electrical and Computer Engineering at UC-San Diego and her M.S. degree in the Department of Materials Science and Engineering at Stanford University. She has over four years of industrial experience including at Applied Materials, Analog Devices and at IBM Almaden Research Center. Her awards and honors include "Emerging Scholar Award of 2005" by the American Association of University Women, "Invited participant of Kecks Future Initiative" (2005) by the National Academy of Science, "Regents Faculty Excellence Award" (2001 and 2004), "Visionary Science Award" (2003), "Technical Ingenuity Award" (2003), "Research Leadership Award" (2003), "Selected US team member in US-Japan Nanotechnology Symposium" (2003), and "Best graduate student awards" from the Materials

Research Society, the Society of Biomedical Engineering and Jacobs School of Engineering (1999, 2000, 2001). Dr. Ozkan's research is recognized as "frontier research" by the *Virtual Journal of Nanoscale Science & Technology* (edited by Dr. David Awschalom) and featured many times in public newspapers, on the cover of journals, online news sites and newsletters. She is an active board member and treasurer in the International Society for BioMEMS and Biomedical Nanotechnology. Her editorial activities include the Journal of Sensors and Actuators B and the Journal of Biomedical Microdevices. She holds more than 25 patent disclosures and about 8 US-patents.

Professor Michael J. Heller began his position at University of California, San Diego in July 2001. He has a joint appointment between the department's of Bioengineering and Electrical and Computer Engineering (ECE). His experience (academic and industrial) includes many areas of biotechnology and biomedical instrumentation, with particular expertise in DNA synthesis, DNA microarray diagnostics and optoelectronic based biosensor technologies. Dr. Heller has been the co-founder of three high-tech companies: Nanogen, Nanotronics and Integrated DNA Technologies. Dr. Heller's most recent work involved the development of an integrated microelectronic array based system for genotyping, genetic and infectious disease diagnostics, protein analysis, cell separations and for nanofabrication applications. Dr. Heller has a respectable publication record, and has been an invited speaker to a large number of scientific conferences and meetings related to DNA microarrays, biosensors, lab-on-a-chip devices, bio-MEMS and nanotechnology. He has over 30 issued US patents related to microelectronic chips, microarrays and integrated devices for DNA hybridization, miniaturized sample to answer diagnostic devices, biosensors, genomics, proteomics, nanotechnology and nanofabrication, nano-based DNA optical storage and for fluorescent energy transfer in DNA nanostructures. Dr. Heller has been a panel member for the NAS(NAE) Review of National Nanotechnology Initiative 2001–2002; the NAS(NAE)—Engineer for the 2020 - 2001/2002; the White House (OSTP) National Nanotechnology Initiative 1999/2000; and has also been involved in a number of NSF Nanotechnology Workshops.

Index

AbetaP-mediated toxicity, modulator of, 89
Acetonitrile, 164, 170, 359
Acrylamide polymers, 61
β-actin, 32, 413, 421
Active microelectronic array hybridization technology, 141
Acute lymphoblastic leukemia (ALL), Affymetrix microarray, 40, 513
Acute myeloid leukemia (AML), 40
Advanced array technology, 360
Affymetrix GeneChip arrays, 42
Affymetrix GeneChip microarrays, 37–38, 44
   profiling of transcripts in human cancer cell lines, 44
Affymetrix microarrays, 37, 40, 43
   applications of, 40
ALP-labeled anti-digoxigenin antibody, 443
Aminopropyltriethoxysilane, 170
Aminosilane, 61, 170
Amplification
   ligation rolling-circle, 449
   nucleic acid sequence-based, 404, 408
   plots for serial dilutions of GAPDH target, 33
Amplified, biotin labeled RNA (aRNA), 38
Anomalous positive DEP effect, 113
Anti permissive molecules, 61
Antibodies, 6, 9, 14, 18, 48, 50–51, 128–130, 132, 195, 214–216, 219–220, 222–224, 227, 300–301
Antibody-modified ion channel switches, 50
Antibody production, 50
Antigen, 48
APMA gated Na^+ ion channels, 72, 83
APMA-gated channels, 73
Apoptosis, 69, 79–80, 84, 87–88, 91, 93, 118, 425–426, 514
   condensation of chromatin due to, 84, 86, 91
   inducement of, 118
   induction of, 425
   slow onset of, 87
   vs. necrosis, 84, 88

Apoptotic bundles, 80, 83–84, 86
Applied Biosystems 7900HT sequence detection system, 25
assay
   precision of, 33
   reproducibility of, 34
data analysis,
   real-Time, 31, 35
   standard curve method, 31
   threshold cycle value, 31
Aptamers, 50, 129
ARE *See* AU rich element
Array architecture, general principles of, 172
"Arrays of microarrays", 24
Arrays. *See specific arrays*
Assembly, 66
Atomic force microscopy, 163, 219–221, 297, 462
   biotin-streptavidin system with, 290
   covalent protein
      Shiff base formation, 290
   poly His tagged proteins
      poly-His-Ni $^{2+}$ specific interaction, 290
ATP Synthase, as molecular motor, 459
Attenuated total reflection–Fourier transform infrared spectroscopy, 163
AU rich element (ARE), 416
   model of mRNA degradation, 417
Autofinish, software for generating additional sequence data, 378, 385
Autoimmune disorders, protein arrays for diagnosis of, 133
Aziridine polymerization, 173, 181

Bacillus anthracis, sensitivity study for, 18
Bacterial DNA, identification of, 17
Basepair sequences, relative binding coefficients of, 488
B-cell library, 41
Beadedarray, 449
Benzoin ethyl ester, 165
Benzophenone, 165

BeWo trophoblasts, 122
Biacore, 50, 162, 221
Bio compact disk assay, 360
Bio-CD workstation, 361
Biochemical Oxygen Demand (BOD) sensor, 53
Biochip technologies, 95
Biological agents, rapid identification of, 10, 15–16
Biological toxins for biological warfare applications
  cholera toxin B (CTB), 14
  staphylococcal enterotoxin B(SEB), 14
Biological warfare agents, rapid identification of, 4, 15–16
Biological weapons, threat of, 98
Biomarker identification, 127
Biomimetic receptors, 49
Biomineralization, biological self assembly of, 461
Biomolecular activity by nanoparticle antennas, control of
  ATP synthase as molecular motor, 459
  complex hybrid structures, biological self assembly of, 461
  DNA as medium for computation, 463
  nanomechanical devices, light powered, 463
  nanoparticles as antennas for controlling biomolecules, 465
    dehybridization of DNA oligonucleotide reversibly by RFMF heating of nanoparticles, 471
    determination of effective temperature by RFMF heating of nanoparticles, 469
    selective dehybridization of DNA oligos by RFMF heating of nanoparticles, 471
    technical approach, 468
Biomolecules, diverse population of, 127
Bioparticles, 16, 63, 104, 107, 113, 118–119
  intrinsic physico-chemical properties of, 119
Bioparticles, dielectric polarizability of, 107
Bioreceptor component classification of, 48
Bioreceptors, 48, 95
Biosensing System, 66
Biosensors
  affinity, 48
  antibody based, 50
  catalytic, 48
  cell based, 52
  characteristics of, 49
  definition of, 47–48
  enzyme-based, 51–52
  fluorescence based cell biosensors, 53–54
  ion channels, 51
  IUPAC definition of, 47–48
  nucleic acid based, 49, 51
  properties of, 48
  resonant mirror-based biosensor (Lab Systems), 50
  surface plasmon resonsance biosensor (BIAcore), 50
  tissue based, 49–50
  use in
    defense applications, 49
    environmental, 49–52, 66
    medical, 49
    toxicological, 49
Biothreat defence, 104
Biotin molecules, 144, 443
Black Hole Quencher^{TM}, 28
BLAST, sequence search tool, 391, 406
Blood
  DEP spectrum of, 116–117
  screening diagnostics test, 409
Blood cells
  human, 122
  TWD theoretical model for manipulation, separation and characterization of, 119
Boc-chemistry, 188
"Bottom-up" process, 138–139, 146–147
Bristol-8 B lymphoblastoid cells, 122
Brownian force, 63
Bubble generation pumps, 330

Cancer cells, 104, 109, 118–119, 122–123, 286, 300, 408
  DEP spectrum of, 116–117
  TWD theoretical model for manipulation, separation and characterization of, 119
CapB gene, sensitivity study for, 18
Capillary electrophoresis, 312, 331, 374, 382, 449; *See also* Electrophoresis
  constant denaturant, 449
Capillary valving techniques, 340
Carbohydrate–lectin interactions, 297
Carbon nanotubes, 137–139, 146–147
Carbonate, N, N-disuccinimidyl, 173
6-Carboxyfluorescein
  2, 7-dimethoxy-4, 5-dichloro-, 26
  hexacholoro-, 26
  tetrachloro-, 26
Carboxylate, N-succinimidyl-*trans*-4-(maleimidylmethyl)cyclohexane-1–170
Carboxytetramethylrhodamine6-, 28
CD applications, 339
  automated cell lysis, 344
  enzyme-linked immunosorbant assays, 341
  capillary valving on CD, 336, 340–341
    burst frequency, 70, 332, 336–337
    critical burst condition, 336
  cellular based assays, 342
  integrated nucleic acid sample preparation, 356
  analytical measurements, 332, 359
  multiple parallel assays, 341

PCR amplification, 356
MALDI MS analysis, 358
CD centrifugal microfluidic platform, 338
CD fluid propulsion, 333
CD photo detector, 360
Cell
  dielectric properties of, 12, 56, 61, 63–64
  dielectrophoretic (DEP) force, 12
Cell activation, 116
Cell adhesion and growth, 61
Cell based biosensors, 52, 57, 96, 98
Cell bursting, 120, 122
  selective, 124
Cell culture
  neuron, 59, 67
  primary osteoblast, 68
Cell cultures, neuroblastoma, 88, 121
Cell death, 64, 80, 83, 88, 91, 114–115, 422
Cell differentiation, 68
Cell line, H19-7, from ATCC, 65
Cell manipulation on a CMOS chip, 109
Cell networks, ordered, 61
Cell patterning techniques by
  bio-microelectronic circuits, 60
  microfabrication schemes, 60
  micro-contact printing (μCP), 61
  topographical method, 60
Cell penetrating peptides (CPP), 420, 423–424
Cell physiometry tools based on dielectrophoresis, 103–123
  dielectrophoresis, 104–107
  dielectric polarizability of bioparticles, 107
  dynamics of interfacial polarization, 107–112
  surface charge effects, 113–116
  other physiometric effects, 116–118
  traveling wave dielectrophoresis (TWD), 118–120
  controlling possible DEP-induced damage to cells, 120–123
Cell separation, 12, 61, 63, 121, 145–146
  dielectrophoretic, 12, 61
Cell therapeutics, 103
Cell types, parameters for dielectrophoretic patterning, 65
Cell-based assays, 54
Cell-based biosensors, 52, 57, 96, 98
Cell-based functional genomics, 156
Cellular delivery techniques
  endocytic, 419–421, 423–425
  non-endocytic, 419–422, 426
Cellular metabolism based biosensors, 55
  cytosensor microphysiometer, 55
  in cancer research, 55
  microfabrication technology used in, 55, 65
  microfluidics, 55
Cellular microorganism based sensors, 52
  in environmental treatment processes
    biochemical oxygen on demand, 52
Cellular proteins, diversity of, 127
Cellular sensors, impedance based, 56
  electric cell-substrate impedance sensing (ECIS) technique, 56
  noninvasive assay of cultured cell adhesion, 56
Cellulose, amino derivatization of, 189
Centrifuge based fluidic platforms, 329
  CD applications, 339
    automated cell lysis on CD, 344
    CD Platform for ELISA, 223, 340
    cellular based assays on CD platform, 342
    integrated nucleic acid sample preparation and PCR amplification, 356
    modified commercial CD/DVD drives in analytical measurements, 365–359
    multiple parallel assays, 341
    sample preparation for MALDI MS Analysis, 358
    two-point calibration of an optode-based detection system, 339
  compact disc or micro-centrifuge fluidics, 333
  simple fluidic function, 334
    mixing of fluid, 334
    packed columns, 339
    valving, 318, 331
    volume definition (metering) and common distribution channels, 338
Cervical carcinoma cells, separation from blood, 145
Charged coupled device (CCD), man-made, 138–139, 146
Chemical agents, selection of, 69
Chemical agent sensing, 58, 70
  signature pattern for control experiments, 70
Chemical analytes, visualization of physiological changes due to effect of
  ethanol on neurons, 80
  ethanol on osteoblasts, 80
  hydrogen peroxide on neurons, 83
  hydrogen peroxide on osteoblasts, 84
  pyrethroid on neurons, 86
  pyrethroid on osteoblasts, 88
  EDTA on neurons, 89
  EDTA on osteoblasts, 91
Chemical microarrays, 289–292, 296, 298, 300–302
  applications, 300
    cell-binding studies, 300
    cell signaling, 230, 298, 300, 302
    diagnostic studies, 301
    drug discovery, 302
  characterization, 299
  post-translational modification, 297–299, 324
Chemical warfare agents, VX and soman (GD), 57

Chip assembly, 66–67
Chips designed for particles and or biomolecular microseparations, 5
Chromatogram DNA sequencer, 376
Chronic myelogenous leukemia, causes of, 502, 509
Clark-type probe for dissolved oxygen, 53
Clausius-Mosotti factor, 105
Clone cluster, construction of tiling path of, 371
Clone insert end sequencing, 370
Clone library, retrieving sequence-ready clones from, 369–370, 375
Clones. *See also* Source clones
CMOS, 14, 156
    400-site chip, 16
    cell manipulation on, 109
    fabrication technology, 138
    second generation developed at Nanogen, 17
CMOS array, electronic assay for fl-SEB and fl-CTB on, 14
CODIS. *See* Combined DNA index system
Coherent surfaces, stepwise synthesis on, 185
Colorectal tumors, allelic imbalance in, 411
Combined DNA index system, 10
Complementarity determining regions (CDRs), 224
Complementary metal-oxide-semiconductor field-effect transistors, *See* CMOS
Complex hybrid structures, biological self assembly of, 461
"Consensus rot", 376
"Consensus" sequence, in shotgun sequencing, 372, 375
Contact printing process, peptide arrays of contact tip deposition printing, 200
    dip-pen nanolithography, 173, 200, 202
    micro contact printing, 61, 173, 200
    pin-and-ring printing, 200, 202
Contig, sequence overlap, 366, 372
Cosmids, 368, 372, 386, 394
Counterions
    field-induced fluctuations and mobility of, 116
    relaxations, 113–114, 116
CSF1PO, STR loci, 11
CTE-ribozyme sequence coupled to RNAhelicase, 514
Cytosensor Microphysiometer®, principle of, 506

DABCYL, 28
Dark Quencher™, 28, 31
DELFIA fluorescence enhancement solution, 442
Deoxyribose nucleic acid (DNA), structure of Watson-Crick B-form, 479
DEP behavior of cell suspensions, monitoring of, 114
DEP effect, *anomalous* positive, 113–114
DEP field-flow fractionation technique, 118

DEP separation for U937 and PBMC mixture, procedure of, 13
DEP spectrum of mammalian cells, 117
DEP traps, 109
    by extruded quadrupolar traps, 109
    by high-density electrode arrays, 109
    by zipper electrodes, 109
    positive and negative, principle of generation of, 62
DEP. *See also* Dielectrophoresis
    selective separation and detection of bacteria by, 115
Detection probes
    hybridization probes, 30
    hydrolysis probes, 26
    molecular beacon probes, 30
    scorpion probes, 30
Diagnostic applications,
    DNA genotyping, 141
Diaminocyclohexane, 1, 2-plasma deposition of, 165
Dielectric dispersions, (α- and β-, 114
Dielectrophoresis, 70, 103, 145; *See also* DEP
    behavior through electrophoresis, 111
    forces, 12–14
    frequency-dependent behavior of, 107, 108, 112
Dielectrophoresis for cell patterning, 61
    basis of dielectrophoresis, 62
    dielectric properties of cells, 64
    effect of electric fields on cells, 64
    microelectrodes, dielectrophoresis, 64
Differential display, 23
Diffuse large B-cell lymphoma (DLBCL), 41
Digoxigenin-labeled PCR product, 443
Digoxigenin-tailed specific probe, 443
Diketonate, β-, 443
Dipole moment, 107
    induced, 110
Dip-Pen nanolithography/scanning probe lithography, 173
Displacement flux densities, 108
DNA
    constructs, 137
    elastic properties, effect of sequence on, 494
    filament, kinetics of deformation of, 489–491
    flexibility, experimental techniques, 492
    hybridization, 31
    hyperchromicity, 469
    intercalating dyes, ethidium bromide and SYBR Green, 25
    labeling of targets and amplification,
        by single fluorophore experimental designs, 37
        by two fluorophore experimental designs, 39
    medium for computation, 463
    nanocomponents and other nanofabrication applications using EFAD devices, 137–156
    probes, 4

thermodynamic measurements of, 484
  basepair dependent, 488
  twist of, rotational and translational parameters, 489
DNA analyses, 318
  fragment separations, 318
  sample purification, 318
  sequencing, 324
  integration of PCR, 322
DNA analysis, fluorescent lanthanide labels with time-resolved fluorometry in, 443
  DNA hybridization assay, 448–451
  forensic analysis, implementation of SNP assays in, 11
  rapid, 16
  lanthanide fluorescent complexes and labels, 438
  time-resolved fluorometry of lanthanide complexes, 441
DNA genotyping diagnostic applications, Nanogen microelectronic arrays, 141
DNA hairpin dehybridization, 469
DNA hybridization assays, 144, 442–444
DNA inserts
  "genome equivalents" of, 368
  of source clone, 374
DNA micro spot-array hybridization assays, 360
DNA microarray technologies, 3–4, 14
  for measuring gene expression, 24
DNA oligos, dehybridization by RFMF heating, 471
DNA-protein interactions, influence of basepair sequence on, 494
  generalized deformations of objects, 481
  double helix and structure atlas of DNA, sequence dependent bending of, 485
  double helix elastic constants, sequence dependent aspects to, 484
  sequence dependent elasticity, some experimental consequences, 486
    phage, 434 binding specificity and DNase I cutting rates, 486
    nucleosome formation, sequence and temperature dependence, 491–492
DNase II endonuclease in endocytic pathway, 425
cDNA arrays, 42
  comparison with oligonucleotide microarray expression profiles, 44
  microscale printing of DNA, 36
  preparation of sample, 36
  printing of PCR products, 35
cDNA microarrays, profiling of transcripts in human cancer cell lines, 44
dsDNA molecule, deformation of backbone of, 483
Drug development, 452
Drug discovery, 103
Dulbeco modified eagle medium (DMEM), 68

Dye (s)
  DNA intercalating
    ethidium bromide, 25
    SYBR Green, 25
  Hoechst, 79–82, 84, 86, 88, 90, 93, 95, 97
  fluorescent reporter dye, 26, 28
  quencher dye, 26, 28
  reporter, wavelength shifting, 408
  trypan blue, 65

"Eberwine" strategy, 39
EDTA
  on neurons, visualization of physiological changes due to effect of, 93–95
  on osteoblasts, visualization of physiological changes due to effect of, 80
EDTA sensing, 76
  single neuron sensing, 76
  single osteoblast sensing, 76
Eigen vectors, 71, 73
Elastomers, polydimethylsiloxane (PDMS), 61, 202
Electric cell-substrate impedance sensing (ECIS), 56
Electric field array devices, DNA nanocomponents and other nanofabrication applications, 137
  active microelectronic array hybridization technology, 140–141, 144–145, 147
  electric field assisted nanofabrication process, 146–153
  integration of optical tweezers for manupilation of live cells, 146
Electrical sensing, 50, 52, 65
Electrical sensing cycle, 70
ElectroCapture™ assay, 18
  PKA assay, 18–19
Electrofusion and electroporation, limitations in, 124
Electrokinesis, 109
Electrokinetic pumping, 312
Electron paramagnetic resonance (EPR), to study DNA flexibility, 484
Electronic assay for fl-SEB and fl-CTB on CMOS array, 14
Electronic hybridization, 7, 11
Electronic microarray technology, applications in genomics and proteomics 3–18
  applications, 5
    single nucleotide polymorphisms (SNPs)-based diagnostics, 10–11
    forensic detection, 3, 10–12
    gene expression profiling, 9, 12–13
    cell separation, 12, 61
    electronic immunoassays, 14
    electronic microarray technology and applications, miniaturization of, 14–18
    proteomics in, 18

Electronic microarray technology (*cont.*)
  overview of, 4
    nanochip array and nanochip workstation 5–6, 16
    nanochip electronic microarrays, capabilities of, 8
Electronic microarrays
  applications of, 8
  cell separation, 12, 61, 63, 121, 145–146
  forensic detection, , 3, 10–12
  gene expression profiling, 41–42
  molecular diagnostics, 10
Electronic multiplexing, 10
Electronic pumps
  dielectrophoretic, 312
  electrohydrodynamic, 312
  electrokinetic, 312
Electroosmotic fluid flow, 106
Electro-osmotic pumping, 312
Electrophoresis
  capillary, constant denaturant, 449
  gel
    denaturing gradient, 449
    temperature gradient, 449
Electrophoresis-based detection technologies, for sequencing, 382
ELISA. *See* Enzyme linked immunosorbant assay
Environmental chamber, 66
Enzyme linked immunosorbant assay (ELISA), 50, 340
Enzyme-based biosensors, 51
Enzymes, 48, 51–52, 96, 137–138, 140, 163, 173, 204
Epitope mapping
  B-cell, peptide arrays by, 298
  multipin technology
    anti-β endorphin, 285
    monoclonal antibodies, 128, 214–215
  polyclonal serum, 224
Epoxy-modified surfaces, chemistry of, 193
Ethanol, 71
  ethylene diamene tetra acetic acid (EDTA), 70
  hydrogen peroxide, 69
  on neurons, visualization of physiological changes due to effect of, 80–82
  on osteoblasts, visualization of physiological changes due to effect of, 81, 83–84
  pyrethroid, 69
Ethanol sensing, 71
  single neuron sensing, 71
  single osteoblast sensing, 71
Ethidium bromide and SYBR Green, DNA intercalating dyes, 24
5′-Exonuclease activity of Taq DNA polymerase, 28
Export of tRNAs in *Xenopus* oocytes, 504

Extra cellular multiple-site recording probes, 59
Extracllular potential based biosensors, 58
  microelectrode array technology, 60
  signature patterns of specific chemical agent, 60
Extruded quadrupolar traps, 109

FAM^{TM}, 26
FarWestern analysis, 133
Fast Fourier Transformation (FFT) analysis, 65, 69
Fetal cells
  DEP behavior of, 118
  in maternal blood, 145
FFT. *See* Fast Fourier Transformation analysis
Fibroblast growth factor, basic (bFGF), 68
Field effect transistors (FET), 61
Field flow fractionation, 109
Field-induced cell destruction, on set of, 124
Filter Binding assay, radioactive, 19
Fingerprinting, restriction digest 376
FISH. *See* Fluorescence *in situ* hybridization
FLAG epitope peptides, 171
FLAG-tag, 194
Fluid flow, synchronized pulses of, 109
Fluo-3, 80
Fluorescence based cell biosensors, 53
  fluorescence imaging, 54
  fluorescent reagents
    fluorescence resonance energy transfer, 28, 54
    green fluorescent protein, 54
Fluorescence biosensor, 54
Fluorescence DNA hybridization assay, 442
Fluorescence energy transfer techniques (FRET), 54
Fluorescence *in situ* hybridization (FISH), 370
Fluorescence polarization (FP) PKA assay, antibody-based, 19
Fluorescence polarization anisotropy (FPA), 484
Fluorescence resonance energy transfer (FRET), 28, 30, 54, 296, 408, 429, 443
Fluorescence spectroscopy, 54, 470–472
Fluorescence–kinetic detection methods, 25
  gene, 33
Fluorescent competitive allele specific polymerase chain reaction, 449
Fluorescent detection system, 144
Fluorescent reporter dye, 26
Fluorophore with non-overlapping emission curves, 410
Fluorophore, 37
  acceptor/donor, 30–31, 55
Fluorosilanes, 61
Forensic DNA profile
  database of, 10
  polymorphic STR loci, 10

INDEX

Forster resonant energy transfer
  mechanism, 146
  process, 139
Fourier transformation (FFT) techniques, 65
"Frequency modulation", 69
FRET. *See* Fluorescence resonance energy transfer
FUN effect, 123
Fungal pathogens in *Candida dubliniensis*, 429

GABAA activation, 76
GAPDH target, amplification plots for serial dilutions of, 33
Gene chip sequencing arrays, 452
Gene detection, *in vitro*, 403
  allele discrimination, 409
  mutation detection, 409
  pathogen detection, 408
Gene expression analysis TaqMan hydrolysis probes in, 29
Gene expression analysis, application of technologies for, 23
Gene expression of phage, 486
Gene expression profiling methods
  with cDNA arrays, 41
  with oligonucleotide arrays, 42
Gene expression profiling utilizing microarray technology and RT-PCR, 23
  amplification efficiency of target/reference genes, 32, 34
  analytical sensitivity and dynamic range, 33–34
  cDNA arrays, 35–36
  comparative threshold method, 31
  microarrays, 32
  printed oligonucleotide microarrays, 36Real-Time PCR (RT-PCR), 23–25
  qualification of gene panels using RT-PCR, 32
  real-time RT-PCR summary, 32
  reproducibility and precision, 34
  standard curve method, 31–32
  technology platforms, 35
Gene expression, posttranscriptional, regulation of, degradation of mRNA, 421
Gene fluorescent detection of, 3
Gene sequencing, shotgun, 378–389
GeneChip microarrays, 37
General purpose interface bus (GPIB) control, 67
Genes, identification of, using hybrid ribozymes, 513
Genetic marker, clones cluster of, 370
Genetic variation, detection of, 449
Genome project coordination, challenges for, 394
  after Celera, 392
  before Celera, 386
  during Celera, 388
Genome sequencing pipeline, 369
Genomic DNA, haplotype structure of, 452

Genomics and proteomics, application of electronic microarray technology in, 3–18
Genomics, nanochip technology application in, examples of, 4
Genotyping analysis, 145
Genotyping, TaqMan hydrolysis probes in, 29
Glass surface modification/activation via starbust dendrimer coating, 179
Glass surfaces, multistep functionalization of, 194
Glyceraldehydes-3-phosphate dehydrogenase (GAPDH), 32
Goniometry, 163
Gyrolab MALDI SP1 sample preparation CD, 358

H19-7 cell line from ATCC, 65
H19-7 cells, 67–68
Hagen–Poiseuille equation, 335
Hairpin nanoprobes for gene detection, 403
  in vitro gene detection, 408
    pathogen detection, 408
    mutation detection and allele discrimination, 409
  living cell RNA detection, 418
    cellular delivery of probes, 419
    intracellular probe stability, 419
    intracellular mRNA detection, 411, 428
  nanoprobe design issues for homogeneous assays, 405
  intracellular RNA targets, 411
    cytoplasmic and nuclear RNA, 411
    RNA secondary structure, 418
    opportunities and challenges, 431
Hairpin probe design, 408
Hammerhead ribozyme, catalytic domain of, 498, 508
Haplotyping, 452
HapMap, 452
Harvester fluorophore, 408
HCV diagnostic applications, TaqMan hydrolysis probes in, 29
Heterogeneous nuclear ribonucleoproteins, 411
HEX^{TM}, 26
High throughput phenotyping, 156
High throughput screening (HTS) situations, 18, 54
High-density electrode arrays, 109
High-density oligonucleotide arrays, 449
High-risk patients, identification by gene expression profiling, 41
High-throughput screening (HTS) kinase assays, 18
HIV diagnostic applications, TaqMan hydrolysis probes in, 29
HnRNPs. *See* Heterogeneous nuclear ribonucleoproteins
Hoechst dye, 79–80
Homogeneous DNA hybridization assay, 443
"Hostboards", 140
HPLC technology, denaturing, 449

HTS. *See* High-throughput screening kinase assays
Human cancer cell lines (the NCI-60 panel), 44
Human disease and drug development, SNPs role, and haplotypes in, 447
  detection of genetic variation, 449
  disease gene mapping, 450
  drug development, 452
  evolution, 450
  haplotypes, 452
  SNP discovery, 448
Human genetic diversity, 452
Human genome sequencing, approaches in
  construction of chromosome tiling paths, 379
  data sharing, 379
Human genome, research network for large-scale sequencing of, 389
Human genome, sequencing, 365
  approaches used to sequence human genome, 365
    overview, 366
      strategy used for sequencing source clones, 368
      construction of the chromosome tiling paths, 379
      data sharing, 379
    challenges for systems integration, 365
      methodological challenges for sequencing source clones:, 1990–1997, 381
      challenges for sequencing the entire human genome:, 1998–2003, 386
    are there lessons to be learned from the human genome project?, 395
Hybridization probes
  acceptor/donor fluorophore, 30
  quanti probe, 31
Hybridization probes, 30–31
Hydrogen peroxide sensing, 72
  single neuron sensing, 71
  single osteoblast sensing, 71
Hydrolysis probes, 26, 28

IL-1, expression levels of, increased, after LPS treatment, 14
Immobilization of peptides
  chemoselective, 194
  non-selective, 191
Immunoassays
  electronic, 14
  electric field driven, 14
Immunohistochemistry, 77
  in situ synthesized oligonucleotide microarrays, 37
Insulin-like growth factor receptor (IGF-IR), human type I, 68
Integrated circuit (IC) technology, 95
Interfacial Polarization, dynamics of, 107
International Prognostic Indicator, 41

Intracellular potential based biosensors, 57
  chemical warfare agents
    soman (GD), 57
    VX, 57
Ion channel switches, antibody-modified, 50
Ion channels biosensors, 50
Ion-selective optode detection, 339
Irradiation with UV-light, 193
Irradiation, 182

JOE^{TM}, 26

"Knock down" effect *in-vivo*, 70

"Lego" blocks for nanofabrication, 147
"Lymphochip", 41
"Pick & Place" fabrication process, 137
"SELEX" for systematic evolution of ligands by exponential enrichment, 129
"Top-down" process, 138
Lanthanide fluorescent, complexes\labels of, 438
Laser excitation sources, 144
LDPS. *See* light directed peptide synthesis
LED arrays, 189
Leica SP2 UV confocal microscope, 78
Leiden, 10
Leukemia classification, 40
Library screening, 286, 369
Library types, 203
  de novo approaches, 203
  protein sequence-derived libraries, 204
Ligands
  aromatic amine derivative-type, 440
  chlorosulfonylated tetradentate β-diketone-type, 439
LightCycler^{TM}, 31
Light-directed peptide synthesis (LDPS), 187–188
Limit of detection (LOD), 34
Limit of quantitation (LOQ), 34
Lipopolysaccharide-mediated differentiation, 12
Lithography, Dip-Pen nanolithography/scanning probe, 58

Magnetic tweezers, 467
MALDI. *See* Matrix-assisted laser desorption ionization sample preparation
Maleimidylhexanoate, N-succinimidyl-6-, 170
Maleimidylpropionate, N-succinimidyl-3-, 170
Mapped clones for sequencing, 370
Mapping source clones, strategies for
  clone insert end sequencing, 370
  fluorescence in situhybridization, 370
  library screening, 286, 369
  restriction digest fingerprinting, 370
Mapping, 131, 162–163

INDEX

Mapping, disease gene, 450
Mass spectrometry, 130
MassARRAY system, by Sequenom 449
Matrix-assisted laser desorption ionization sample preparation, 287–288
Maxizyme, active/inactive conformation of, 508
Maxizyme, allosterically controllable, design of 509
Maxizymes, 508
  design of, 509
  generality of, 512
  inactivation of oncogene, 510
  shortened hammerhead ribozymes, 508
Maxizyme-transduced BV173 cells, 512
Maxwell stress tensor, 107
Maxwell–Wagner interfacial polarization, 107
Mechanical pumps
  bubble generation, 312
  osmotic pressure, 312
  pneumatic pressure, 312
  syringe drive, 312
  thermal expansion, 312
MeCP2 mutations associated to Rett syndrome, 10
Medical diagnostics, 95, 103
Melting curve single nucleotide polymorphism (MCSNP), 449
Membrane capacitance, 110
Membranes,
  acrylic-acid-modified 166
  methylester-modified membrane, 166
MEMS (Micro-Electromechanical Systems), 156
MEMS devices
  high density data storage devices, 140
Mendelian disorders, identification of disease genes, 447
Methacrylate polymers, 61
MGB EclipseTM (Epoch Biosciences) probe, 31
Michael addition, 170
Micro total analysis systems, 357
Microarray analysis of mRNA stability in yeast, 417
Microarray and fluidic chip for extracellular sensing, 47
  antibody based biosensors, 50–51
  biosensing system, 51
  cell based biosensors, 52
  cell culture, 67
    neuron culture, 67–68
    primary osteoblast culture, 68
  cell patterning techniques, 60
  cellular metabolism based biosensors, 55
  cellular microorganism based sensors, 52
  chemical agents, selection of, 65
    ethanol, 69
    hydrogen peroxide, 69
    pyrethroid, 70
    ethylene diamene tetra acetic acid (EDTA), 70
  chemical agent sensing, 70
    signature pattern for control experiments, 70–71
  dielectrophoresis for cell patterning, 61
  dielectrophoresis, basis of, 62
  dielectrophoresis, microelectrodes and, 63–64
    EDTA sensing, 76
    single neuron sensing, 71
    single osteoblast sensing, 71
  electrical sensing cycle, 70
  ethanol sensing, 71
    single neuron sensing, 71
    single osteoblast sensing, 71
  environmental chamber, 66
  enzyme based biosensors, 51–52
  experimental measurement system, 67–68
  extra-cellular potential based biosensors, 59–60
  fluorescence based cell biosensors, 53–54
  hydrogen peroxide sensing, 73
    single neuron sensing, 73–74
    single osteoblast sensing, 74–75
  impedance based cellular sensors, 55–57
  intracellular potential based biosensors, 57–58
  ion channel biosensors, 51
  nucleic acid based biosensors, 51
  physiological changes due to effect of chemical analytes, visualization of, 80
    ethanol on neurons, 80–82
    ethanol on osteoblasts, 81, 83–84
    hydrogen peroxide on neurons, 83, 85–86
    hydrogen peroxide on osteoblasts, 85, 87–89
    pyrethroid on neurons, 88–91
    pyrethroid on osteoblasts, 91–93
    EDTA on neurons, 89
    EDTA on osteoblasts, 91
  pyrethroid sensing, 74
    single neuron sensing, 71
    single osteoblast sensing, 71
  immunohistochemistry, 77
  signal processing, 68
Microarray system, compact disc-based
  fiber-optic microarray biosensor, 295
  immunoassays, 14
  piezoelectric inkjet applicator, 295
Microarray technologies, application of, 1
  electronic, applications in genomics and proteomics, 3–18
Microarrays
  applications, 40
  cDNA arrays, 35
  in situ synthesized oligonucleotide, 37
  protein interaction, 224
  spotting techniques, 132
  printed oligonucleotide, 36
Micro-centrifuge fluidics, 333

Micrococcus lysodeikticus, DEP frequency response for, 114–115
Micro-contact printing (μCP), for promoting cell adhesion, 61
Microelectrodes, 63
Microelectronic arrays
　for bioresearch, 137
　for DNA clinical diagnostics, 140
Microelectronic arrays, applications of, 145
Microelectronic arrays, by Nanogen, for DNA genotyping diagnostic applications, 141
Microelectronic MEMS devices, heterogeneous integration of, 140
Microfluidic channel networks, 181
Microfluidic devices
　biomedical, 351
　clinical diagnostic applications, 351
Microfluidics propulsion techniques, 331
Microlithography, 146
Microlocations, 14
Micromanipulators, 67
Microorganisms, immobilised mixed culture of, 16, 52
Micro-patterning
　contact printing techniques, 173
　micro-mirror mediated patterning, 181
　photolithografic technologies, 179
Microspheres, 147–149
Microzoom^{TM} optical probe, 66
Miniaturized electronic microarray system, 16
Miniaturized quantitative cell-based assays, 349
Minor groove binder (MGB), 31
Molecular beacon probes
　configuration, 30
　point mutations detection by, 30
　polymorphisms detection by, 30
　probe-target duplex, 30
Molecular beacons, components of
　fluorophore, 30
　loop, 30
　stem, 30
　quencher, 412
Molecular gene therapy and gene discovery, engineered ribozymes, efficient tools for, 505
Molecular probes, nanostructured, 4
Motherboards, 140, 147
MPC. *See* mini-Pepscan cards
MRNA, expression of, inhibition by ribozyme, 512
MRNA, secondary structure, computational programs to predict, 424
Multiarray-based biochip technology, 93
Multiplex hybridization analysis, 11
Multiplex PCR-amplified DNA, 360

Multistep functionalization
　glass surface, 173
　pretreated titan surfaces, 168
Myc-tag, 195

Nano chip rmicroarray, 100-site, 4
NanoChip array
　assays format, 9
　capabilities of, 8
　cartridge of, 16
　fabrication of, 4–5
　permeation layer of, 5
　workstation of, 5
NanoChip microarray, active hybridization technology and passive hybridization technologies, 4
Nanochip technology application in genomics, examples of, 11
NanoChip^{TM} cartridge, 7, 143
　assembly, 143
NanoChip^{TM} molecular biology workstation, 143
NanoChip workstation, 7–8
Nanoelectronics
　applications for, 138
　nanoelectronic devices, 138
Nanofabrication,
　bottom-up processes for, 138–139
　DNA chips, 142
　Electric field assisted process, 146
Nanogen's 400-site NanoChiparray and cartridge, 16
Nanogen's electronic microarray technology, 5
Nanogen's microelectronic DNA chip device, 8
Nanogen's Nanochip Workstation, 8
Nanogen's portable electronic microarray detection, 10, 16
Nanomechanical devices, light powered, 463
Nanoparticles, 137
Nanoparticles, as antennas, 465
　solution phase synthesis of, 466
　with reactive ligands, 468
Nanoparticles, DEP behavior of, 116
Nanoparticles, in controlling biomolecules, 473
　dehybridization of DNA by RFMF heating of nanoparticles, 471
　determination of effective temperature by RFMF heating of nanoparticles, 469
　selective dehybridization of DNA by RFMF heating of nanoparticles, 471
　technical approach, 468
Nanoparticles, RFMF heating of, 467
Nanoprobe design issues for homogeneous assays, 405
Nanoprobes for imaging, sensing and therapy, 401
Nanospheres, 140, 147–148
　DNA sequence, 140
　production of, 140
Nanotubes, 137–138, 146–147

INDEX

Nanowires, 137
NASBA. *See* Nucleic acid sequence-based amplification
Necrosis, 84, 88, 224–225, 231
Neuroblastoma cell cultures, 88
Neuronal cultures on microelectrode array, 59
Neuron-EDTA frequency spectrum, 78
Neuron-ethanol frequency spectrum, 72
Neuron-hydrogen peroxide frequency spectrum, 74
Neuron-pyrethroid frequency spectrum, 76
Neurons and osteoblasts, parameters for positive and negative DEP for, 65
NMDA gated channels, 73–74, 83, 86
   activation of, 73–74, 86
NMDA gated channels, 83, 86
NMDA receptor gated Cl^- and Na^+ ion channels, 76
NMDA receptors
N-methyl-d-aspartate (NMDA) receptor dependent, 80
Non contact printing, peptide arrays of
   by piezoelectric device
   piezoelectric ink-jets, 204
Non-fluorescent ("dark") quencher, 28
Northern blotting, 23, 403, 422
Nuclear magnetic resonance (NMR), to study DNA flexibility, 287, 459, 465, 509
Nuclease invader technique, 449
Nuclease mutation detection method, 449
Nucleic acid based biosensors, 51
Nucleic acid fragmentation, 93
Nucleic acid probes, microscale printing of, 36
Nucleic acids, 48, 139, 167, 404, 409, 418, 420–421, 423, 425
Nucleosome formation, 491
Nucleosome reconstitution, 491–492
NUX cleavage site, 513
NUX triplet, 510

OBOC (one-bead one-compound)
   peptide libraries, 284
   encoded, OBOC small molecule combinatorial libraries, 287–288
   combinatorial library methods and chemical microarray techniques, 283
OBOC peptide libraries, 283–284
Oligocarbamates, 188
Oligonucleotide (ODN) probes, 419–420
Oligonucleotide microarray expression profiles, 44
Oligonucleotide microarray, 36–37, 39, 43–44, 200, 228, 299
Oligonucleotide microarrays, *in situ* synthesized
   Affymetrix GeneChipmicroarrays, 37–38, 42, 44
   manufacture by light-directed method, 37
Oligonucleotide microarrays, methods for production of, 39

Oligonucleotide probes, microinjection of, 30, 36–37, 144, 404–405, 418–419
OligoWalk Probe/RNA interactions tools, 418
Oncogene, inactivation of, 512
One step grafting procedure, 167
One-bead one-compound (OBOC) combinatorial libraries to chemical microarrays, 287
   chemical microarray, application of, 297
      cell-binding studies, 300
      diagnostic studies, 301
      drug discovery and cell signaling, 300
      non-biological applications, 304
      post-translational modification, enzyme-substrate and inhibitor studies, 299
      protein binding studies, 298
   detection methods in chemical microarrays, 299
      detection methods to identify post-translational modification of proteins, 299
      identification and characterization of bound proteins, 296
   encoded OBOC small molecule combinatorial libraries, 287
   peptide and chemical microarrays, 289
   OBOC peptide libraries, 284
      CD, microfluidics, fiber optic microarray, multiplex beads, 295
      immobilization methods for pre-synthesized libraries, 289
      in situ synthesis of microarrays, 292
One-bead one-compound. *See* OBOC
Optical tweezers, integration of, for manipulation of live cells, 153
Ordered cell networks, 61
Organic quencher molecules, 408
Osmotic pressure pumps, 312
Osteoblast-EDTA frequency spectrum, 79
Osteoblast-ethanol frequency spectrum, 73
Osteoblast-hydrogen peroxide frequency spectrum, 75
Osteoblast-pyrethroid frequency spectrum, 77
Osteoblasts and neurons, parameters for positive and negative DEP for, 66
Osteoblasts, 87
Oxidized silicon wafers, 142
Oxygen electrode, Clark-type, 53
Oxygen, dissolved, Clark-type probe for, 53

PACs (P1 artificial chromosomes), 368
Paired mismatch (or MM), 37
Parasites, TWD theoretical model for manipulation, separation and characterization of, 119
Pathogen isolation, two-level-stacked microlaboratory, 16
Pathogens, rapid identification of, 16
Patterned surfaces, generalization of, 173, 181
PBFI ester, staining with, 88

PBFI, AM ester of, 88
PCR amplicon, biotinylated, 8–9, 143
PCR amplicons, 143
PCR cycle number, 28
PCR cycle number, 28
Pentafluorophenylesters, 167
Pentanedione, 5-(4"-chlorosulfo-1',
    1"-diphenyl-4'-yl)-1, 1, 1, 2, 2-pentafluoro-3,
    5-(CDPP), 443
Peptide arrays in proteomics and drug discovery, 161
  applications of, 161
    antibodies, 162
    application of peptide arrays: miscellaneous, 228
    enzyme-substrate and enzyme-inhibitor
      interactions, 226
    peptidomimetics, 231
    protein-protein interactions, 224
  assays for, 224
    read-out, 219
    screening, 215
  generation of, 128
    coherent surfaces and surface modification,
      163–178
    generation of micro-structured surfaces, 173
    peptide array preparation, 182
    techniques for array production with
      pre-synthesized peptides, 200
  library types, 200
    de novo approaches, 210
    protein sequence-derived libraries, 204
Peptide arrays, 182
  architectures
    gel pad arrays of, 172
    gold electrodes arrays of, 56–57
  assays, 215
    read-out, 219
    screening, 215
  cell binding, 300
  micro-structured surfaces, 173
  preparation, 173
    chemoselective reactions, 196
    contact printing, 61, 173, 200
    non contact printing, 129, 200
  pre-synthesized peptides, 200
  screening 215
  surface modification, 163
Peptide arrays, applications of
  4 α-helix bundle mini-protein of, 217, 229
  antibodies, 162
  antibody paratope mapping of, 224
  bibliography of, 162, 231, 265
  chaperone activity, 225
  characterization of peptide ligand, 229
  DNA binding, 228
  enzyme-inhibitor interactions, 226

enzyme–substrate interactions, 95
linear epitopes, 222
metal ion binding peptides, 229
optimization of peptidic ligands, 229
peptide ligands, 210–211, 213
polyclonal antibody epitope mapping, 224
protein interaction domain-ligand interactions,
  206
Peptide arrays, read out of,
  chemoluminescence, 162, 219
  chromogenic, 162, 219
  fluorescence, 162, 219
  label-free, 162, 219
  radioactivity, 130, 162, 219
Peptide libraries
  split-mix synthesis method, 283–284, 289
Peptide ligands, *de novo* approaches for identification
    of, 211
  combinatorial library, 231, 285
  random scan, 285
Peptide microarrays, 163, 197, 199, 221, 226–228,
    230
  immobilization methods, 289
  in situ synthesis, 185, 191, 200, 289
Peptide nucleic acids (PNAs) nucleotide analogs, 167,
    190, 228
Peptide sequence,
  combinatorial explosion of 210
Peptide synthesis on coherent surfaces, principle of,
    186
Peptide
  SPOT synthesis 162–163, 190–192
Peptide/ligand interaction, 171
Peptidomimetrics, 266
Perfect match (or PM) features, 37, 431
Periodic granular rimming flow, 350
Phage 434 binding specificity, 486
Phage, gene expression of, 478
Phage-display libraries, 128
Phil's revised assembly program (Phrap), 376,
    385
Phil's Revised Editor (Phred), 376
Phosphorylation, 18, 173, 220–221, 227
Photo-induced graft copolymerization, 165
Photolithographic masks, 178
Photolithography for patterning of surface free
    energies, 183
Photolithography techniques, 61
Photolithography, "topdown" process of, 147
Photonic crystals, 153
Photoresist coatings, 178
Photosensitizers, 165
Phrap score, 377, 386
Phrap. *See* Phil's revised assembly program
Phred. *See* Phil's Revised Editor

INDEX

Physiometric Effects, 118
PhysioNetics, 111
PKA assay, 18–19
PKA inhibitors, 19
Plastic devices, 312, 314, 316
  microfluidic devices, 311
  direct fabrication, 316
    laser ablation, 316
    mechanical machining, 316
  replication
    casting, 316
    compression molding, 316
    embossing, 316
    injection molding, 316
Plastic microfluidic devices for DNA and protein analyses, 315 detection, 316
  DNA analyses, 318
    integrating PCR and DNA fragment separations, 318
    DNA sequencing, 320
    DNA sample purification, 321
  materials, 312
  electrokinetic pumping, 312
  plastic devices, 312
    device fabrication, 316
    pumping and detection, 316
  protein analyses, 311
    enzymatic digestion for protein mapping, 324
    isoelectric focusing for studying protein interactions, 323, 326
Pneumatic pressure pumps, 312
Polarizability factor, 108, 119
Polarization, double layer, 107
Polyethylene, 162, 165, 167, 170, 284
Poly-L-Lysine, 61, 181
Poly-lysine coated glass slides, 173
Polymers, amino modified, 162
Polymers, methacrylate and acrylamide, 61
Polymorphism discrimination assays, 31
Polymorphism
  restriction fragment length, 449
  single strand conformational, 449
Polymorphism, melting curve single nucleotide, 449
Polypropylene, 162, 165–166
Polystyrene, 147, 165
Polyurethane films, 164
Power spectral density analysis, 68–69
Printed cDNA microarrays, methods for production of, 39
Printed oligonucleotide microarrays
  manufacturing of synthetic DNA, 139
  pre-synthesized DNA probe, 36
  synthetic oligonucleotide probes, 36
Probe-target binding, 406
Probe-target hybridization kinetic rate, 406

Probe-target hybridization, thermodynamic parameters of, 407
Protein analyses, 311
  enzymatic digestion, 324
  isoelectric, 323
Protein chip array, 298
Protein interactions, detection by
  ELISA-based sandwich approach, 130
  surface plasmon resonance method, 130, 162, 296, 299
  rolling circle amplification method, 130, 449
Protein microarrays, 127
  for analysis of proteins involved in recombination & DNA repair
    protein expression microarrays, 130–132
    protein interaction arrays, 132–133
  generation of, 128
    proteins, 128
    antibodies, 128–129
    surface chemistry, 129
    microarray production, 129
    detection, 130
  protein arrays, 133
Protein microarrays, classification of, 127
Protein sequence-derived libraries
  amino acid substitution, 206–207, 209
  combinatorial deletion, 208
  cyclization scans, 208
  duotope scan, 204, 206
  hybritope scan, 204
  overlapping peptides, 204
  substitutional analysis, 207–208
  trancation, 208
  types of, 210
Protein synthesis, localization of mRNPs in, 414
Protein-DNA interactions, 132
Protein-DNA specificity, levels of, 478
Protein-protein interactions, 224
Proteins, 128
Proteins, ARE binding, 422
Proteins, bound, detection of
  grating couplers, 296
  label-free optical techniques, 296
  reflectometry, 296
  surface plasmon resonance, 296
Proteins, fluorescence, 55
Proteomic analysis by protein array, concept of, 127
Proteomics and genomics, application of electronic microarray technology in, 3–18
Proteomics, applications in, 18
Prothrombin, 10
Psoralen, crosslinker agent, 152
Pyrethroid sensing, 75
  single neuron sensing, 75
  single osteoblast sensing, 76

Pyrethroid, visualization of physiological changes due to effect of
　on neurons, 88–91
　on osteoblasts, 91, 91–93
Pyroccocus furious crystal structure, 131
Pyrrole-modified peptides, electrochemical copolymerization of, 197

Qiagen quantiprobe™, 27
Qiagen, 31
QuantiProbe™, 31
Quantitative assay of specific RNA target, 29
Quantum dots, 137–140
Quantum dots, 140
Quick Spin protein desalting column, 345

RAD51B-histone interactions, 133
Reactive oxygen species, 84
Reads
　misassembled, 376
　overlapping, 376
　shotgun, 376
Real time RT-PCR, 31
Real-time PCR, mechanism of, 32
Reciprocal chromosomal translocations, 502, 513
Recombinant his-tagged proteins, 129
Reporter dyes, wavelength shifting, 408
Resonant mirror-based biosensor (Lab Systems), 50
Restriction digest fingerprinting, 370
Restriction enzymes, 370, 478
Restriction fragment length polymorphism (RELP), 449
Reverse transcription-polymerase chain reaction (RT-PCR), 23
RFLP *See* Restriction fragment length polymorphism
RFMF heating of nanoparticles, 469
RFMF heating of nanoparticles, determination of effective temperature by, 469
Ribozyme expression systems, 499, 515
　functional analysis of genes, 497
　gene therapy, 497
　pol III system, 499
Ribozyme, subcellular localization of, 501
Ribozymes into cells, methods for introduction of, 506
Ribozymes
　administration of, 498–499
　RNA-protein hybrid, 498, 505
Ribozymes, as therapeutic agents, 515
Ribozymes, engineered, efficient tools for molecular gene therapy and gene discovery, 505
　maxizymes, allosterically controllable ribozymes, 508
　　allosterically controllable maxizyme, design of, 509
　　maxizyme technology, generality of, 512

　　oncogene, inactivation in mouse model, 512
　　hybrid ribozymes, identification of genes using, 513–514
　　shortened hammerhead ribozymes as dimers, 508
　methods for introduction of ribozymes into cells, 497–499
　ribozyme expression systems, 499
　　higher-order structure, and their activity, relationship between, 149, 499
　　Pol III system, 499–500
　　subcellular localization and efficacy, 501
　　tRNA-ribozymes, export from nucleus to cytoplasm, mechanism of, 504
　RNA-protein hybrid ribozymes, 505
　　accessibility to ribozymes of their target mRNAs, 505
　　hybrid ribozymes that efficiently cleave their target mRNAs, 505
Ribozymes, secondary structures of, 500
Ribozymes, subcellular localization and efficacy of, 499, 501
　factors affecting efficacy, 507
RMS power, 69
RNA constructs, 137
RNA detection, living cell, 411
　cellular delivery of probes, 419
　intracellular probe stability, 419, 424
　intracellular RNA detection, 434
RNA polymerase T7, mediated amplification of, 12, 38
RNA processing, RNases roles in, 412–413
RNA protection, 23
RNA study
　ABI Prism 7900HT SDS, 32
　analytical sensitivity, 33
　　limit of detection (LOD), 34
　　limit of quantitation (LOQ), 34
　bias percentage 33
RNA targets, intracellular, 411
RNA viral load tests testing, TaqMan hydrolysis probes in, 29
RNA, secondary structure of, 418
RNA: DNA duplexes, detection of, 37–38
RNAm localization in nucleus, 413–414
RNAm, splicing of, 412–413
RNA-protein hybrid ribozymes, 505
RNAs, functional, in cultured cells, activities of, 511
Rolling circle amplification (RCA), 130, 449
RRNA pre-transcripts, degradation of, 426

Saccharomyces cerevisiae, proteome of, 128
SBFI, AM ester of
Scanning force microscopy, to study DNA flexibility, 484
Scorpion™ Primer/Probes, 26, 30

Screen-sequence-screen-sequence approach for clone cluster, 370
Secondary ion mass spectroscopy, 163
Sensitizers, 165
Separation buffer, 65, 68
Sequence data, hierarchical approach of
  assembly, reconstructing sequences, 373
  finishing, filling gaps of, 367, 378
  mapping, source clone, 378
  sequencing, raw data, 366–367
Sequence dependent elasticity, experimental consequences of, 486
Sequence dependent
  aspects to double helix elastic constants, 484
  bending of double helix
  structure atlas of DNA, 485
Sequence-ready clones, retrieving from clone library, 369
Sequencing by hybridization, 382
Sequencing human genome, approaches in
  construction of chromosome tiling paths, 385
  data sharing, 385
Sequencing of human genome, large-scale, research network for, 389
Sequencing source clones, methodological challenges, 381
Sequencing source clones, strategy used for, 368, 381–382, 386, 388
Sequencing technologies, revolutionary, 382
Sequencing, 382
Serial analysis of gene expression (SAGE), 23, 403
Short tandem repeat (STR) forensics analysis, 137
Short tandem repeats (STRs), 10
Short tandem repeats, 10
Shotgun sequence reads generation from source clone, generic procedures for
  base-calling, 374, 378, 382
  data curation, 374
  detection, 374
  fragmentation, 372
  sequencing, 372
  size selection, 373
  source clone DNA preparation, 372
  subcloning, 373
  template DNA preparation, 373
Shotgun sequencing of gene, 372–373, 375, 381–383, 386
Shotgun sequencing, basic framework of
  level of redundancy, 383
  read length, 372
  robots for DNA template preparation, 385
  sequencing reactions, 368, 373
  template, 383
Signal processing, 60, 68–69
Signature pattern for single osteoblast, 75

Signature pattern vector (SP), 69
Signature pattern vector (SPV) of cell, 70
Signature patterns of cell based biosensors, 52
Signature patterns of specific chemical agent, 52, 60
Signature Patterns, 47, 52, 60, 65, 98
Silicon wafers, transformation into mercapto-modified surfaces, 197
Silicon, oxidized, 170
Single fluorophore experimental design, 38
Single nucleotide polymophisms (SNPs), 10
Single nucleotide polymorphism, 405, 409, 412
Single nucleotide polymorphisms (SNPs), 23, 448
Single platinum micro-electrode pad on the NanoChip rmicroarray, 6
Single strand conformational polymorphism (SSCP), 449
SNP discovery, 448
SNP genotyping analysis, 145
SNP genotyping, 30
SNP identification technique, 449
SNP. *See* Single nucleotide polymophisms, 10
SNP-IT, robotic version of, 449
SNPs, diseases associated with, 10, 447
SODA. *See* synthesis on defined areas
Soman and VX (GD), chemical warfare agents, 57
Source clone DNA preparation, 372, 374
Source clone, gaps of, 393
Source clones acquisition, 368
Source clones
  strategies for mapping of
    clone insert end sequencing, 370
    FISH, 370, 372, 378
    library screening, 286, 369–371
    restriction digest fingerprinting, 370–371
Spearman rank correlation coefficient, 42–43
SPOT concept, application of, 189
SPOT synthesis, 167
SPOT synthesis, 162, 167, 189–190, 210, 214, 219, 222–223, 225–227, 230–231, 298
SPOT. *See* spot synthesis of peptides
SSCP. *See* Single strand conformational polymorphism
Stacked microlaboratory, fabrication of stacked structure, 15–16
Stainless steel foils, 171
Stainless steel foils, functionalization of, 171
Staphylococcus enterotoxin a (SEA) and b (SEB), 14, 17
Starbust dendrimer coating, 179
Stem cells
  DEP spectrum of, 116, 122
  DEP behavior of, 113–116
Stem sequences, 30
  fluorescent PCR primers, 30
  hairpin loop probe structure, 30
  SNP genotyping, 30, 145

STR loci, multiplex hybridization analysis of, 10–11
STR. *See* Short tandem repeats
Strand displacement amplification (SDA), 9, 17
Streptolysin O (SLO), poreforming bacterial toxin, 422
Striatial cholingeric internurons, extracellular potentials of, 58
Super ATM and Super T^{TM}, 31
Superbase^{TM}, 31
Superposition-TWD, 119
Surface activation
   polyethylene, 165
   polypropylene, 165
   polystyrene, 164
Surface charge effects, 113–116
Surface modification, 162–163, 165, 173, 196, 321
   γ-radiation, 164
   adsorption of structured α-helical peptides, 173
   corona treatment, 164
   graft copolymerization, 164–165
   hydrophilic surface modification, 165
   hydroxy-functionalized surfaces, 167
   Maleinimide-modified, 170
   non-selective immobilization, 191, 193
   silicon wafers, 142, 197
   silylation, 164
Surface plasmon resonance (SPR), 130
Surface plasmon resonsance biosensor (BIAcore), 50
Surface tensin arrays, generalization of, 185
Surface tension arrays, generation of, 179, 185
Surface-enhanced laser desorption/ionization time-of-flight mass spectrometry (SELDI-TOFMS), 162
SYBR Green assay, 25
SYBR green DNA and ethidium bromide, intercalating dyes, 25
Syringe drive pumps, 312
Systems integration, challenges for, 380
   challenges for sequencing human genome, 381
   methodological challenges for sequencing, 381

T cells, 120
   TWD velocity for, 119
T lymphocytes, membrane capacitance values of, 118
T lymphocytes, separation of, 120
TAMRA^{TM}, 28
Taq DNA polymerase, 5'-exonuclease activity of, 28
Taq DNA polymerase, exonuclease activity of, 28
TaqMan assay, 449
TaqMan expression profiles, 42
TaqMan hydrolysis probes, 29
TaqMan probes, 30
TaqMan real-time RT-PCR, 28
TaqMan real-time RT-PCR, 28–29

target amplification and labeling, 37
   single fluorophore experimental designs, 37–39
TASμ *See* Micro Total Analysis Systems
Tat peptide transduction, 423
Teichuronic acid, 114
Template DNA preparation, 373
Tetramethylrhodamine, 472
TET^{TM}, 26
TGF-β gene in PBMC, expression level of, 13
TGF-β, expression level of, 13
TGF-β, expression levels of, increased, after LPS treatment, 13
TH01, STR loci, 11
Thermal denaturation profiles of window of discrimination, 410
Thermal expansion pumps, 312
Three-dimensional coatings
   acrylamide gel pads, 173
   agarose films, 173
   gelatine pads, 173
   hydrogels, 173
   semi-wet gels, 173
Threshold cycle (CT), 28
Thrombosis associated SNPs,
   factor V (Leiden), 10
   factor II (prothrombin), 10
   methylenetetrhydrofolate reductase (MTHFR), 10
Tiling-path arrays, 189
Time-resolved fluorescence polarization anisotropy(TR-FPA) techniques for torsional measurements, 489
Time-Resolved fluorometry, 437
TMR. *See* Tetramethylrhodamine
TMR: FAM, ratio of, 472
TNF-α, expression level of, 13
TNF-α, expression levels of, increased, after LPS treatment, 13
Torsional measurements, TR–FPA techniques for, 489
TPOX, STR loci, 11
Transcription by the pol II system, 499
Transcription factors
   helix-loop-helix, 478
   helix-turn-helix, 478
   leucine, 478
   zinc finger, 478
"Transcriptome", 24
Transducers in biosensors, types of
   electrochemical, 48
   mass-sensitive, 48
   optical, 48
   thermometric, 48
Transport of tRNA-type ribozymes in cytoplasm, 504
Traveling Wave Dielectrophoresis, 118
   junction of, 119
   models of, 119

Trypan blue dye, 65
TWD junction, 119
TWD velocity for T cells, 121
TWD. *See* Traveling Wave Dielectrophoresis
Two fluorophore experimental design, 39
  two fluorophore experimental designs, 39–40
    applications, 40–41
  gene expression profiling methods, comparison of, 41
    cDNA arrays and other gene expression profiling methods, 41–42
    oligonucleotide arrays and other gene expression profiling methods, 42–44
    cDNA and oligonucleotide microarray expression profiles, 44

U937 cells expression levels of IL-1, TNF-α, and TGF-β, increased, after LPS treatment, 12–14
U937 cells, expression level of, 12–14

Vapor phase photo-grafting, 165
Variable angle spectral ellipsometry, 163
Vertical Cavity Surface Emitting Lasers (VCSEL), 153 Laguerre mode, 155
Very large scale immobilized polymer synthesis, 187
VIC®, 26

Visualization of physiological changes due to effect of chemical analytes effect of, 80
  EDTA on neurons, 89
  EDTA on osteoblasts, 91
  ethanol on neurons, 80
  ethanol on osteoblasts, 80
  hydrogen peroxide on neurons, 83
  hydrogen peroxide on osteoblasts, 84
  pyrethroid on neurons, 86
  pyrethroid on osteoblasts, 88
Vollum, sensitivity study for, 18
VX and soman (GD), chemical warfare agents, 57

Water-borne pathogens, in *Salmonella*, 409
Wavelength shifting reporter dyes, 408

X-ray photoelectron spectroscopy, 163

Yeast, 105, 107
Yeast, TWD theoretical model for manipulation, separation and characterization of, 119
Yeast, viable and unviable, DEP separation of, 65
Yersinia pestis (plague), DNA identification, 17

Zebrafish, 91
Zipper electrodes, 109

Abbreviated Table of Contents

List of Contributors	xvii
Foreword	xxi
Preface	xxiii

I. Application of Microarray Technologies 1

1. Electronic Microarray Technology and Applications in Genomics and Proteomics 3
Ying Huang, Dalibor Hodko, Daniel Smolko, and Graham Lidgard

2. Gene Expression Profiling Utilizing Microarray Technology and RT-PCR 23
Dominick Sinicropi, Maureen Cronin, and Mei-Lan Liu

3. Microarray and Fluidic Chip for Extracellular Sensing 47
Mihrimah Ozkan, Cengiz S. Ozkan, Shalini Prasad, Mo Yang, and Xuan Zhang

4. Cell Physiometry Tools based on Dielectrophoresis 103
Ronald Pethig

5. Hitting the Spot: The Promise of Protein Microarrays 127
Joanna S. Albala

6. Use of Electric Field Array Devices for Assisted Assembly of DNA Nanocomponents and Other Nanofabrication Applications 137
Michael J. Heller, Cengiz S. Ozkan, and Mihrimah Ozkan

7. Peptide Arrays in Proteomics and Drug Discovery 161
Ulrich Reineke, Jens Schneider-Mergener, and Mike Schutkowski

8. From One-Bead One-Compound Combinatorial Libraries to Chemical Microarrays 283
Kit S. Lam, Ruiwu Liu, Jan Marik, and Pappanaicken R. Kumaresan

II. Advanced Microfluidic Devices and Human Genome Project 309

9. **Plastic Microfluidic Devices for DNA and Protein Analyses** 311
   Z. Hugh Fan and Antonio J. Ricco

10. **Centrifuge Based Fluidic Platforms** 329
    Jim V. Zoval and M.J. Madou

11. **Sequencing the Human Genome: A Historical Perspective On Challenges For Systems Integration** 365
    Lee Rowen

III. Nanoprobes for Imaging, Sensing and Therapy 401

12. **Hairpin Nanoprobes for Gene Detection** 403
    Philip Santangelo, Nitin Nitin, Leslie LaConte, and Gang Bao

13. **Fluorescent Lanthanide Labels with Time-Resolved Fluorometry In DNA Analysis** 437
    Takuya Nishioka, Jingli Yuan, and Kazuko Matsumoto

14. **Role of SNPs and Haplotypes in Human Disease and Drug Development** .. 447
    Barkur S. Shastry

15. **Control of Biomolecular Activity by Nanoparticle Antennas** 459
    Kimberly Hamad-Schifferli

16. **Sequence Matters: The Influence of Basepair Sequence on DNA-protein Interactions** 477
    Yan Mei Wang, Shirley S. Chan, and Robert H. Austin

17. **Engineered Ribozymes: Efficient Tools for Molecular Gene Therapy and Gene Discovery** 497
    Maki Shiota, Makoto Miyagishi, and Kazunari Taira

About the Editors 519
Index 521